AF572701

RESERVOIR SEDIMENTOLOGY

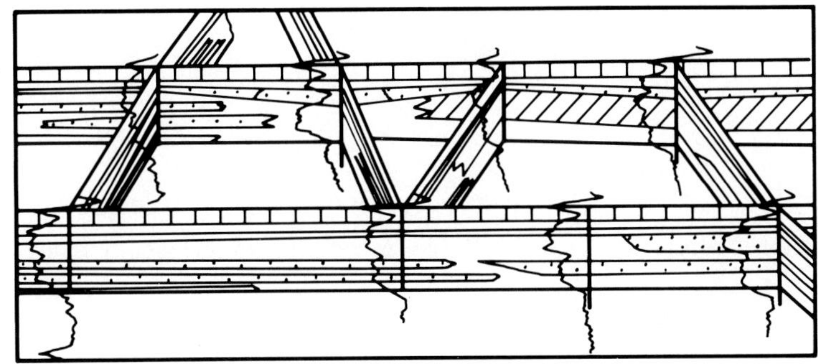

Edited by

Roderick W. Tillman
Consulting Sedimentologist, Tulsa, Oklahoma
and
K. J. Weber
Shell International Petroleum Maatsehappÿ,
The Hague, The Netherlands

SOCIETY OF ECONOMIC PALEONTOLOGISTS AND MINERALOGISTS

Barbara H. Lidz, Editor of Special Publications

Special Publication No. 40

Tulsa Oklahoma, U.S.A. ISBN: 0-918985-69-2 *AUGUST 1987*

A Publication of
The Society of Economic Paleontologists and Mineralogists

ISBN 0-918985-69-2

P.O. Box 4756
Tulsa, OK 74159-0756

Printed in the United States of America

PREFACE

This volume is a collection of papers which focus on the sedimentology of siliclastic sandstone and carbonate reservoirs. The papers were selected to show how detailed sedimentologic descriptions, when combined with engineering or other subsurface geologic techniques, yield reservoir models which may be used for reservoir management during field development and during secondary or tertiary enhanced oil recovery.

In all the papers the framework for the field descriptions relies heavily on full-diameter cores. In addition to conventional 4-inch-diameter cores, frozen and rubber-sleeve cores were utilized in one or more of the studies.

In addition to cores, at least one other geologic or engineering technique is integrated into each study presented in this volume. This integration of sedimentologic descriptions with other techniques gives rise to synergism. Techniques for study of fractures in reefs contribute to the understanding of Norman Wells field in Canada. Oriented cores are incorporated into studies of shelf sandstones, deltas and fluvial channels. Intepretation of dipmeter results, using "sedimentologic intuition," contributes to the understanding of deep-water deposits in the Woodbine Sandstone. Comparison of large-scale bedding features in outcrops in Arizona allows modeling of aeolian sandstone in the North Sea. Mining techniques, including the statistical technique of Kriging, allow modeling of tidal and estuarine sandstones in oil sands of Canada.

The papers in this volume were prepared and, in some cases, rewritten several times so that they could be easily understood by sedimentologists, subsurface geologists and petroleum engineers. Many of the techniques and types of descriptions presented in this volume are used by geologists in major petroleum company laboratories; however, these techniques have not yet routinely been made a requirement of field studies by subsurface geologists and engineers. It is our hope that the field studies presented here may serve as examples for future field studies.

This volume is divided into four sections. The sections are: Introduction; Marine Sandstone and Carbonate Reservoirs; Shoreline, Deltaic and Fluvial Reservoirs; and Aeolian Reservoirs. The introductory paper by Ebanks stresses the principals involved in reservoir description and management of Enhanced Oil Recovery (E.O.R.). The problems of reservoir heterogeneity and the timeliness of geologic input into field studies are stressed. The need for periodic communication, during the course of the field study, among sedimentologists, subsurface geologists and reservoir engineers is emphasized. The *scale* of heterogeneity features in reservoirs is a theme of all the papers in this volume. Heterogeneities described range from as large as lithologic and facies changes to as small as shale lenses and secondary pore-filling clays and cements.

Precise quantitative geologic input into computer simulation modeling has, to date, not been a subject of great concern in published papers; however, its importance is now beginning to be recognized. Obtaining geologic input data for quantitative modeling requires that a significant change in "geologic mentality" take place. The nature of geologic data has traditionally been qualitative; however, to relate to engineering problems and models, geologic data must be presented in a quantitative format.

MARINE SANDSTONE AND CARBONATE RESERVOIRS

In this section, the papers are equally divided among sandstones and carbonates. The sandstone papers involve shelf and deep-water reservoirs, whereas the carbonate papers stress the whole range of carbonate reservoirs and the influence of fractures on reef reservoirs.

Tillman and Martinsen describe a Shannon Sandstone shelf-ridge complex from Hartzog Draw field which produces from Upper Cretaceous sandstones in Wyoming. The shelf ridge is subdivided into a variety of productive and nonproductive facies. Production in the field is characterized according to whether it is in *Central Ridge* or *High-* or *Low-Energy Ridge-Margin Sandstone Facies*. The production characteristics and position of these facies within the field are emphasized as are the criteria for defining each of these and other nonproductive facies. Oriented cores allow documentation of flow directions (southerly) for the bottom currents that formed the shelf ridge. A model is produced in which the distribution of the three producing facies is related to formative processes (obliquely-impinging, southerly-flowing currents).

The paper by Phillips utilizes cores and dipmeters to define the internal geometry of turbidite channel sandstones in the Permian Cherry Canyon sandstones in New Mexico. Dipmeter patterns, indicating a variety of features, are documented by comparison of dipmeter and core data. By combining dipmeter data and "sedimentologic intuition," several geologic aspects can be recognized including unconformities, slump faults, contorted beds, "massive" sandstones, and stacked channels. Drape over channels, recognized on dipmeters, is used to predict channel orientations. The presence and distribution of turbidites exhibiting Bouma sequences are utilized in describing the reservoir. The integration of dipmeters and core data yield a very useful deep-water turbidite channel model.

Jardine and Wilshart document, using eight field examples, a wide range of carbonate reservoir types. According to these authors carbonates form about 20 percent of the world's sedimentary rocks and contain 40 percent of the world's oil. Their documentation of six fields in moderate detail and two fields in greater detail produce useful models for carbonate reservoirs ranging in age from Devonian to Triassic. Nearshore cyclic carbonates are encountered in Boundary Lake and Steelman fields. A fractured open platform reservoir is described at Quirk field, and a platform-margin reservoir is described at Clarke Lake field. Four fields described in the paper, Golden Spike, Judy Creek, Norman Wells and Redwater fields, are reef bioherms. The geologic data used in developing this series of carbonate models focus on depositional environment, porosity characteristics and diagenetic character. These types of data were then integrated with engineering data. Models for each of these reservoir types were used in computer simulation studies to forecast reservoir behavior under various depletion methods and to assist in ongoing reservoir surveillance studies.

Irish and Kempthorne discuss the Devonian age Norman Wells field reef complex from northern Canada and the effect of abundant fractures on the reservoir. A synergistic geological and engineering study for the reservoir is presented. The reasons for entrapment of oil in the extreme updip end of the reef buildup are discussed. Limestone conglomerates are recognized as important facies in their reservoir descriptions, and the effects of reefal cycles on production are discussed.

Results from pressure buildups, pulse testing and interference tests were integrated into two five-layer geological-engineering models. One model was for "reef margin" and one was for the "reef interior." Layer one, in both models, represented a widespread stromatoporoid thicket or foreslope facies. Layer two, in both models, was lagoonal. Other layers vary between models and include additional foreslope facies, supratidal deposits, reef margin facies and backreef facies. Optimum well spacing, fracture occurrence and spacing, and directional permeability variations were also integrated into proposed pool development and forecasting models.

SHORELINE, DELTAIC AND FLUVIAL RESERVOIRS

Two papers discussing reservoirs, which are in part shoreline deposits, stress local variabilities (heterogeneity). One of these papers involves a regional study and the other is a detailed mining study in which closely spaced core holes form the basis for a predictive reservoir model of Athabasca oil sands in Alberta, Canada.

Rennie, in his study of Athabasca oil sands in the Sandelta project in northern Alberta, recognized 22 geologic facies; however, only three of these facies are reservoirs, tidal channels, distributary channels and fluvial/estuarine sands. These environments were grouped by Rennie under the headings of Delta Marsh and Estuarine environments. Palynology and other paleontological analyses were used to corroborate environments of deposition based on sedimentary structures. A three-dimension "computer block model" was constucted primarily by using descriptions of frozen core. The core was frozen in order to preserve the very friable sandstones. Blocks in the model were coded according to facies type, and a geostatistical technique called Kriging was used to extrapolate oil saturation and other reservoir property values. The predictability of the model was proved by the drilling of numerous additional wells following completion of the predictive model.

Wightman, Pemberton and Singh, in their paper on Lower Cretaceous upper Mannville sandstones of Alberta, seek to present evidence that *anastomosing fluvial* deposits in the Mannville, referred to in the literature by numerous authors, are commonly not formed as have been described. They propose an alternate exploration. Their evidence from trace fossils present in cores is used to corroborate results obtained from detailed descriptions of physical and lithologic features observed in numerous cores. They were able to develop two models; one in the west involves mostly eastward-flowing channels. The model developed for the eastern area involved amalgamation of channel and shoreline deposits and stacking of crevasse splays in interdistributary bays.

As an outgrowth of their study, a series of criteria for recognition of brackish water deposits is also documented. Recognition of brackish water deposits using ichnology, palynology (microplankton) and the presence of syneresis cracks, pyrite and siderite aided greatly in their environment of deposition reconstruction.

Papers by Tillman and Jordan and by Kruit document reservoir heterogeneity in pilot studies in deltaic reservoirs. Production at El Dorado field, which produces from the shallow Permian Admire sandstone in Kansas, was initiated in 1915. Secondary waterflood has also produced significant amounts of oil. In 1975 a project was initiated to determine whether a tertiary micellar flood might yield additional oil from El Dorado field. A pilot project, which was designed to test two different micellar fluids in side-by-side pilots, was initiated. The early prognosis of a relatively homogeneous reservoir in both pilot areas was contradicted by geologic and engineering techniques employed in a synergistic study of 51 acres included in the pilot study.

Division of the reservoir into sedimentary facies indicated that channel-fill sandstones form most of the reservoir intervals within the pilot area; however, it was determined that the channel deposits were *not* uniformly distributed in plan view and varied significantly in thickness and continuity from area to area. Results from oriented cores, pulse testing and transmissability testing, when integrated with environment of deposition maps, cross sections and fence diagrams, yield a heterogenous model, which indicates that most of the remaining oil-in-place is in *High- and Low-Energy Distributary Channel Sandstones* and that crevasse-splay sandstones which formed in interdistributary bays were of only minor importance.

Kruit, in a study of Miocene sandstones in East Tia Juana field in Venezuela, recognized fluvial-deltaic reservoirs overlain by shallow marine deposits. Descriptions of rubber-sleeve cores, along with log shapes calibrated using cores, were used to construct models for steam injection enhanced oil recovery techniques. Documentation of the variation in production among the various producing facies, which included active-channel fill, abandoned-channel fill, natural levee and flood basin deposits, allowed construction of a predictive model. Geologic criteria for recognition of channel facies and inundation planes are given, and various geologic-engineering maps are utilized to predict results of "steam drive" and "steam soak" recovery techniques.

Jordan and Tillman, in describing the Pennsylvanian age Bartlesville Sandstone in southern Kansas, dispute earlier barrier island depositional environmental interpretations. Environments of deposition recognized in five oriented cores strongly indicate that the Bartlesville Sandstone reservoir in Seeley-Wicke field is a series of stacked *Distributary Channel Sandstones*. Associated crevasse-splay, overbank and delta plain deposits are also recognized in cores and on logs.

Criteria are given for recognition of stacked sand-on-sand channel contacts. The concept of stacked channels is supported by vertical variability of permeability. Dip-angle and dip-azimuth for high-angle trough crossbeds were recorded in five oriented cores. Substantial agreement in four of the five oriented cores indicates that current flow in the reservoir was parallel to the elongation of reservoir (SSW) rather than perpendicular, as would be expected if the reservoir were a barrier island.

AEOLIAN SANDSTONE

One major problem encountered by sedimentologists who are limited to subsurface cores for environment of deposition modeling is related to the scale of the features encountered in reservoir sandstones. Crossbed sets in aeolian sandstones are typically several meters thick and may extend several ten's or hundred's of meters laterally. In order to understand better the degree to which lateral set boundaries affect aeolian reservoir heterogeneity, Weber studied outcrops in De Chelly Canyon, Arizona. He used conclusions from the outcrop studies to predict the scale of features in the Permian Rotliegendes Sandstone in Leman field in the North Sea. In both outcrop and subsurface, he recognized foreset laminae formed on the lee slope of transverse dunes which were relatively uniform geologically and in terms of porosity and permeability. Bottomset zones were found to have substantially poorer reservoir properties.

Roderick W. Tillman

ACKNOWLEDGMENTS

This volume is the result of the efforts of a dedicated group of authors and reviewers from a relatively wide range of specialties including siliclastic sedimentology, carbonate sedimentology, siliclastic and carbonate petrology, geological engineering, reservoir management, reservoir engineering and clay mineralogy.

We acknowledge, with special thanks, the technical reviews of papers in this volume by Don Bebout, Mike Boyles, Jim Ebanks, Frank Ethridge, Ralph Hunter, Susan Jackson, John Hobson, Bob Loucks, Don McCubbin, Eric Mountjoy, Karen Porter, Roger Slatt, Dar Spearing, Jim Steidman, Robert Weimer and Chuck Williamson.

CONTENTS

INTRODUCTION

PART I. MARINE SANDSTONE AND CARBONATE RESERVOIRS

PART II. SHORELINE, DELTAIC AND FLUVIAL RESERVOIRS

PART III. AEOLIAN RESERVOIRS

GEOLOGY IN ENHANCED OIL RECOVERY

W. J. EBANKS, JR.
ARCO Oil and Gas Company, Inc. PRC-G112 2300 West Plano Parkway Plano, Texas 75075

ABSTRACT: Geologists have an important role to play in the management of oil and gas reservoirs and, especially, in the implementation of enhanced oil recovery (E.O.R.) projects. Cooperative work by engineers and geologists is needed to maximize the ultimate recovery of producing fields, because many of the problems that arise in their development are related to reservoir heterogeneity, mineralogy, or petrophysics. Geologists should be involved in all aspects of an E.O.R. project, from selection of an appropriate process for a particular reservoir, through reservoir description and simulation of performance, to avoidance or solution of problems during the installation, operation, and evaluation of the project.

Among the larger scaled geological factors that affect reservoir performance are the distribution and attitude of the reservoir rocks and enclosing less permeable strata, the arrangement of facies within the reservoir interval, its layering or stratification, presence of shale "breaks" within the reservoir interval, natural fractures, and the influence of a hydrodynamic pressure gradient on the movement of fluids within the reservoir. Smaller scaled features also are of great importance. These include the texture and mineralogy of the reservoir rock, the types and arrangement of clays, and the geometry of the pore system.

Perhaps the most unique contribution a geologist can make to understanding the production performance of a reservoir or of individual injection or production wells in an E.O.R. project is his ability to recognize and predict the distribution of different "rock types" through the application of conceptual models of facies distribution and an understanding of diagenetic modification of those patterns of rock facies.

INTRODUCTION

The subdiscipline, or specialty, of production geology, sometimes referred to as exploitation geology, is the application of geological studies to problems of the development of an oil and gas field, following its discovery and confirmation. Typically, geologists who work in this capacity interact frequently with petroleum reservoir engineers and petrophysicists in performing more detailed kinds of studies of reservoir rocks than are needed for the exploration process. Increasingly, geologists are being included in task groups within companies, or as consultants, to do detailed descriptions of oil reservoirs as an important step in better reservoir management and especially in implementing enhanced oil recovery (E.O.R.) projects.

Enhanced oil recovery includes several very expensive and difficult, and so very risky, processes (Table 1), which, when applied to an oil reservoir, are expected to result in the production of more oil than would have been produced by more conventional primary and secondary methods. When applied after waterflooding or some other secondary recovery process, enhanced recovery may be referred to as tertiary recovery. In fact, however, most of these E.O.R. processes would work even better if applied earlier in the life of a field. There is actually more variety than the simple classification in Table 1 would suggest, because there are variations required for most of the processes in order that they may be tailored to specific reservoir conditions.

The purpose of this paper is to increase the awareness of the reader of the important contribution geologists can make to the success of E.O.R. projects and to encourage cooperation and communication between geologists (production geologists and sedimentologists) and engineers. This will be accomplished by describing the role geologists play in engineering-geology teams and discussing some examples of the application of geology to engineering problems and the oil recovery process.

GEOLOGY-ENGINEERING COOPERATION

Case studies and the collective experience of oil industry personnel have shown that most E.O.R. projects that have failed to perform as expected have done so for reasons that are related more to geological problems than to engineering problems. In fact, it has been said that our inability to describe reservoir rocks in sufficiently quantitative terms to predict the patterns of fluid flow in them is potentially the "Achilles heel" of E.O.R. operations (Geffen, 1976). It was the engineering community who recognized the need to include geology in the planning and execution of E.O.R. projects, and they used the term "synergism" to describe the close interaction of geologists and engineers in this area. Traditionally, geologists and engineers have been educated differently and have worked separately; as a consequence, their approaches to problems have developed differently. The geologist thinks in terms of continuously varying characteristics, such as portrayed by an isopach map of an oil reservoir; the engineer thinks in terms of discretely varying elements of a geometric array of data, such as the elements of a digital mode of a portion of the reservoir (Fig. 1). It is encouraging to see that in recent years there have been purposeful attempts to bring together these separate disciplines.

An enhanced oil recovery field test should be a well controlled experiment, analogous to any carefully designed fluid-displacement experiment performed in a laboratory. In the E.O.R. test, however, the test apparatus and even the laboratory are located in a remote subsurface environment, and the experiment can be monitored only indirectly. Unfortunately, in an E.O.R. project, millions of dollars can be spent before it is obvious that the experiment is faulty and should be terminated. For this reason, a great deal of effort should be spent in learning all that we can about the E.O.R. "test apparatus," that is, the reservoir rock, before expensive and technically difficult oil displacement experiments are attempted in the subsurface.

Since it will be a team of engineers who ultimately designs and applies the E.O.R. processes, the burden will be on the geologists to provide geologic information in a form that can be used by the engineers in their reservoir-simulation studies. On the other hand, an appreciation of the types of geologic variability that may be encountered in

TABLE 1.—CLASSIFICATION OF ENHANCED OIL RECOVERY PROCESSES ACCORDING TO TYPE OF RESERVOIR ENERGY OR FLUID INJECTED.

Thermal	Chemical	Miscible
Steam Injection	Surfactant/Polymer Flooding	CO_2 Injection
In Situ Combustion	Polymer Flooding	Hydrocarbon Injection
	Alkaline Flooding	N_2 Injection

natural reservoirs and how this may affect the movement of subsurface fluids should encourage engineers to look to geologists for assistance in specific cases. There have been barriers to open communications between the two disciplines in the past. Some engineers have been naive concerning the natural complexity of sedimentary facies and the effects of diagenesis on the rocks. Geologists commonly are reluctant to assign numbers to variable rock properties on the cell-by-cell scale necessary for mathematical portrayal of the reservoir. Further, there is a common reluctance on the part of some geologists to change an earlier interpretation or map when new data in the form of engineering tests rather than geologic observations are received. Geologists who have been involved in reservoir description with engineers understand the dynamic nature of geologic interpretations, and their counsel is worth noting: "To adequately define these (reservoir) heterogeneities requires joint geological and engineering studies" (McCaleb, 1978); "Throughout his work, the geologist requires assistance and guidance from the engineer" (Harris, 1975); ". . . there must be a constant feedback between the reservoir engineer, the petrophysical engineer, and the geologist in formulating the ultimate model of the reservoir and of course in monitoring the actual performance of the project" (Diehl, 1976); "Continued monitoring of operating performance is essential to assure that the geological model is valid" (Jardine and others, 1977).

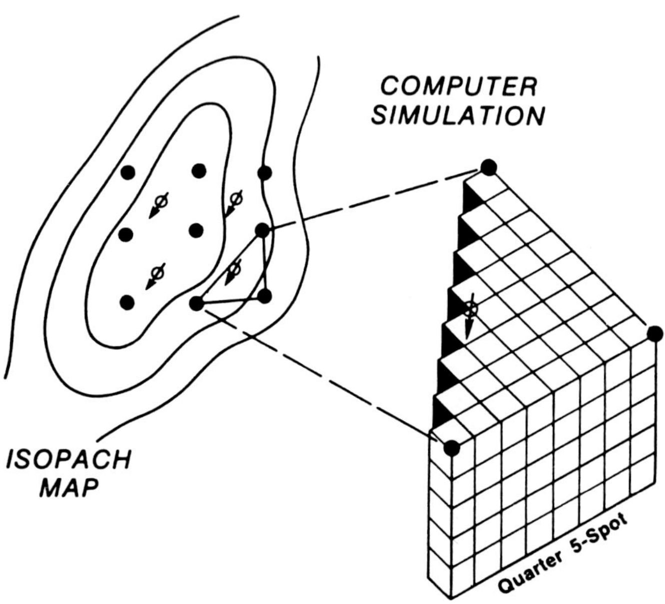

FIG. 1.—Diagram of a small enhanced recovery project symbolizing the differences in the way an oil reservoir is visualized by geologists and engineers who respectively attempt to represent it by maps and by numerical simulation.

As a result of their training, geologists are well suited to interpreting the subtle variations in microscopic and macroscopic properties of rocks. This insight has three main areas of application in enhanced oil recovery: process selection, reservoir simulation, and problem avoidance (or solution).

Process selection.—

One of the most common applications of geologic data that pertains to specific reservoirs or, in some cases, to classes of reservoirs is in the "screening" process, by which it is determined which E.O.R. process is most suited to a particular reservoir. This matching of reservoirs and E.O.R. processes should be done early in order to facilitate planning and to determine the logistical and financial feasibility of the E.O.R. project that is planned.

Basic descriptors of the rocks and of the fluids (Table 2) are used routinely in this planning (National Petroleum Council, 1976). Sometimes the screening of a reservoir is carried out in two stages, an initial stage in which basic physical and chemical properties are considered and a more detailed stage in which reservoir facies, variability of porosity and permeability, presence of fracturing, and other factors are considered. In the end, if screening criteria indicate that more than one E.O.R. process may be appropriate for a particular reservoir, the process which will probably yield the most oil is the one which is chosen unless there are overriding economic factors that affect the choice. A potential danger in this screening procedure can arise if insufficient weight is given to potential problems that would result from reservoir rock composition or facies changes within the project area. Unfortunately, in the early planning phase of operations, insufficient geologic data may be available to make judgments concerning these factors, especially if the field was first developed hurriedly and with little attention given to documenting reservoir geology.

Reservoir simulation.—

Thorough reservoir description leads to accurate reservoir simulation. This, in turn, has application in two areas of engineering practice: history matching of reservoir performance, and predictive modeling of future fluid production. History matching of previous fluid production is commonly performed with numerical simulation models of a reservoir to verify the accuracy of petrophysical parame-

TABLE 2.—FACTORS USED IN "SCREENING" OF RESERVOIRS TO DETERMINE MOST APPLICABLE PROCESS OF ENHANCED OIL RECOVERY.

Rock Descriptors	Fluid Descriptors
Lithology	Oil Gravity
Thickness	Water Salinity
Depth	Temperature
Porosity	Oil Saturation
Permeability	Oil Viscosity
Transmissibility	Gas/Oil Ratio
Environment of Deposition	Drive Mechanism
Heterogeneity	
Fracturing	

ters, pressure gradients, and other variables in the equations of fluid flow. Predictive modeling is used to forecast future reservoir performance under certain operating conditions and to develop economic forecasts by which projects can be compared and assigned priorities.

Reservoir description can never account for all of the variability of nature. Consequently, another opportunity for engineer-geologist interaction exists in the effort to model a reservoir accurately. No purely geological model, which may be as much as 90% conceptual, is as good as a combination engineering-geological model, which contains a substantial amount of factual information. That is, a geological model of reservoir quality is improved by consideration also of data from pressure tests, production history, and petrophysical analyses (Fig. 2). The feedback in flow of information between geologists and engineers in this type of situation is bound to result in a more precise, less ambiguous model of reservoir performance than would have been possible by either discipline acting singly.

In Wasson Field, which produces oil from the San Andres dolomite in West Texas, economic and technologic limitations and logistical problems caused a delay in implementation of fieldwide waterflooding until the 1960s. When waterflood performance did not attain expected rates of oil recovery, reservoir geology studies were initiated that led to a much more complicated concept of reservoir layering in which the continuity of porous intervals was much less than originally assumed (Fig. 3). Incorporation of the revised concept into the engineering design greatly improved project performance (Ghauri and others, 1974). A similar experience has been cited in management of the West Seminole San Andres Unit, West Texas (Barrett and others, 1977; Harpole, 1979).

In a geological-engineering study of the Loudon field, which produces oil from the Mississippian Weiler sandstone in central Illinois, Harris (1975) has described the dependence of the geologist on the engineer for information used to develop a geologically-based mathematical model of a reservoir. The geological facies recognized must be assigned petrophysical characteristics based on core analysis, and questions about continuity of reservoir rocks or the presence of barriers to flow of fluids can best be answered by the results of interwell pressure testing.

Problem avoidance (or solution).—

Planning for the avoidance of "problems," usually in the form of formation damage, begins for the engineer-geologist team before a well is drilled. A geologist's knowledge of rock composition and his ability to infer patterns of changing composition from the results of previous drilling ideally suit him to advising in these areas. Much of the

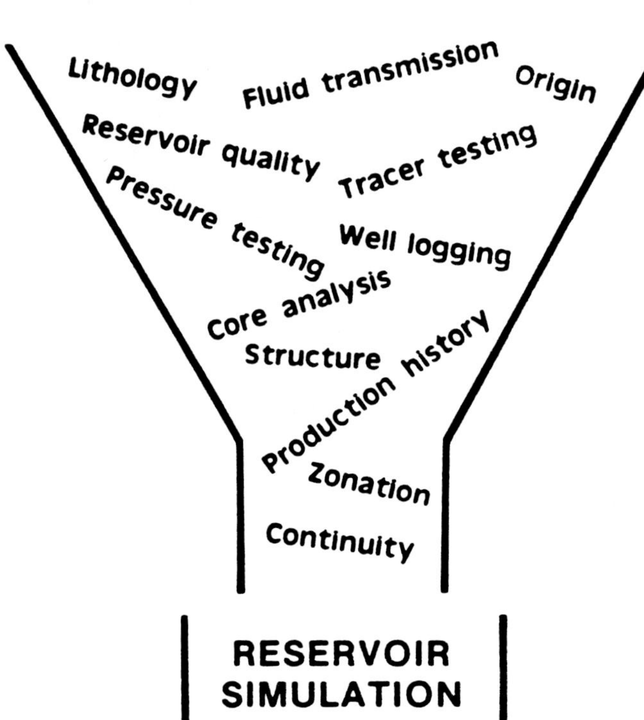

FIG. 2.—Interplay of geologic and engineering data in a reservoir simulation. Information on the geological and engineering aspects of an oil reservoir is used to produce a less ambiguous model of the reservoir's behavior than could be done using only one or the other type of data.

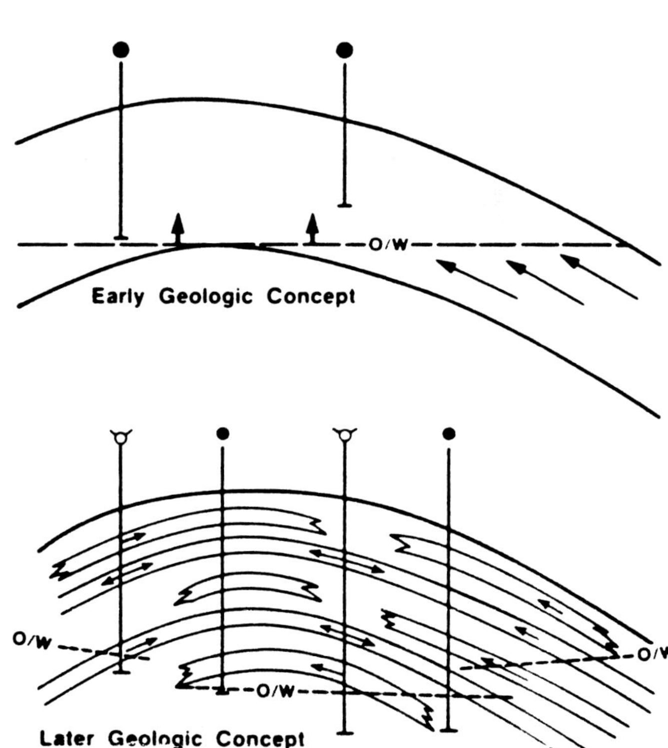

FIG. 3.—Reservoir heterogeneity that was diagnosed from experience gained from the operation of a waterflood (Ghauri and others, 1974). New insights concerning the complexity of a reservoir may, in some cases, be useful in designing a later enhanced oil recovery project.

engineering practice in the design of drilling fluids, well treatment and completion methods, and design of fluids for injection and improved recovery of oil is based on hard-won, trial-and-error experience in a particular producing formation. The rise in acceptance by industry of well treatments that are "tailor-made" for particular projects is indicative of the recognition of the variability of rock composition and its importance to the well completion process. Completion procedures that minimize formation damage are of great importance, because balancing volumes and pressures of injected fluids among the wells in a pattern in order to effect maximum areal sweep of the reservoir is a necessity for efficient operation of any E.O.R. process.

If, in drilling, a potentially productive formation is exposed to fluids that react with or disturb the minerals present, the rocks may become incapable of producing the oil and gas they contain or of accepting injection fluids. The problem of fluid sensitivity, especially among clay minerals, is well known (Hewitt, 1963; Almon and Davies, 1978, 1981). Iron-bearing minerals also are sensitive to non-native, especially acidic, fluids introduced to a formation after drilling (Smith and others, 1969). Fortunately, in laboratory experiments that closely simulate an E.O.R. fluid displacement process, the mineralogy of test cores should experience the same unfavorable reactions to injected fluids that occur in the subsurface. Measures can then be taken to avoid these reactions or to allow for their effects in evaluating the E.O.R. process.

Another task of the engineer in which there is a need for proper geologic input is the estimation of remaining recoverable oil-in-place before initiation of an E.O.R. project, which bears heavily on the economic attractiveness of the project. Commonly, these estimates are based on production history and calculations of original oil-in-place supplemented by well tests and specialized logging of wells in the project area. Failure to recognize the distribution, as well as the amount, of oil remaining may lead to unduly optimistic estimates of expected oil recovery. If facies analysis establishes that most of the remaining oil is located in a portion of the reservoir or in a type of pore system that cannot be reached by an anticipated E.O.R. process, failure of the project to perform as expected is assured (Basan and others, 1978).

GEOLOGICAL FACTORS

In order to emphasize the variety of ways in which geologists can contribute to the successful planning and execution of enhanced oil recovery projects, the remainder of this discussion consists of examples of geological factors that affect performance, that is, the production of fluids from a reservoir. Where possible, case studies are cited to provide the reader with more in-depth discussion than is given here.

Geology of oil reservoirs should be addressed at two levels, or scales. First, the distribution of the rocks that form the trap, reservoir and non-reservoir rocks, should be identified and mapped. Second, the three-dimensional variations in the properties of these rocks should be determined and portrayed numerically, as average values and as ranges of values.

Reservoir distribution.—

Attitude, or shape, and continuity of porous rock in the subsurface are important qualities of reservoir rock which we commonly quantify through the use of maps of structure and thickness of selected rock units. These "shape" qualities provide information about the limits of the reservoir and about its boundary conditions.

A structural trap, whether formed by folds or faults in the rocks, or by both, has a very different character than does a stratigraphic trap. The height of the oil column, type of depletion mechanism, the area over which any bottom-water drive is effective, the nature of the discontinuities within the reservoir, and the integrity of the seal around the trap may be different for these two types of traps (Exum and Harms, 1968); and these factors may influence plans for in-fill drilling, fluid injection patterns, and levels of pressure application in an E.O.R. project. In some cases, there are significant differences in these factors, even within a reservoir in the same field (Harms, 1966; McCaleb and Wayhan, 1969).

Maps of reservoir shape may be used in combination with analyses of cores and logs to draw other maps of reservoir capacity and transmissibility. Combinations of factors, such as the product of "net" sandstone, average porosity and hydrocarbon saturation, or the product of average permeability and thickness divided by fluid viscosity, can be contoured as a means of quantifying reservoir properties (Harris, 1975). If a geologist prepares these maps, he or she should be influenced by patterns of geometry dictated by the environment of deposition of the reservoir rock. Preferred directions of reservoir elongation and thickening often can be inferred on the basis of a knowledge of the environment(s) of deposition. These judgmental factors must be recognized as interpretive and be kept flexible, to be changed if additional information necessitates it.

Maps of reservoir geometry are essential to describe the reservoir adequately for simulation, but they are only as good as the data on which they are based, that is, widespread samples only inches in diameter separated by hundreds or thousands of feet of interwell distance. It is in this interwell area where the geologist may be able to improve on "blind" extrapolation of rock information. Use of a technique of mapping known as "facies-biased" contouring is a powerful tool for extending interpretations of reservoir continuity and quality into undrilled areas from points of nearby control (Swanson, 1979).

Facies changes.—

Some of the most significant advances which have been made in the study of sedimentary rocks in recent years have been in improving our understanding of the internal variability of types of rock bodies which were formed under different conditions (Fig. 4). Through studies of modern sand bodies, we have learned much about the general differences in texture, composition, and internal structure of, for instance, a sand deposited in an alluvial valley and one formed as a barrier island on the margin of the sea (Pryor, 1973; LeBlanc, 1977; Sneider and others, 1978). Further studies of this type, especially on ancient counterparts of

SANDSTONE
SEDIMENTARY ENVIRONMENT
CLASSIFICATION

• Continental	**• Coastal**	**• Marine**
Alluvial	Estuarine	Inner Shelf
Fluvial	Deltaic	Outer Shelf
Lacustrine	Barrier Bar-Island	Deep Marine
Eolian	Intertidal	
	Chenier	

FIG. 4.—Simplified classification of environments of clastic sediment deposition. Recognition of these sedimentary environments will allow geologic models of reservoir characteristics to be applied to rocks in the subsurface. Prediction of the behavior of fluids between wells that penetrate different parts of the reservoir may depend on predictions of rock quality based on these models.

the modern sands, suggest that natural variability, even within each genetic class of sandstone, is great. This variability, of course, lessens the value of ideal models for prediction of sand-quality distribution; however, these ideal models still allow generalizations which can assist an engineer in interpreting apparently anomalous test data, in design of a pattern of injection and production wells to maximize recovery of oil, or in other applications pertaining to E.O.R. operations.

Sneider and others (1977) have described in some detail a study of part of Elk City field, Oklahoma, which produces oil and gas from Pennsylvanian sandstones and conglomerates. The study, made in preparation for a waterflood of the field, facilitated mapping the distribution of porous and non-porous rocks and predicting response of various wells to waterflooding. By correlating the texture of the sandstones with their petrophysical properties and well log signature, vertical sequences of genetically distinct rock types were recognized and mapped. What was earlier considered to be one producing sandstone actually consists of four different types of sandstone, each with its own set of characteristics (Fig. 5). Understanding of the sequence of events leading to the formation of such a complex reservoir, with several major producing zones separated from one another by impermeable layers, is extremely valuable as a guide in the design of a supplemental recovery project, and it is a contribution that is uniquely geological.

On a smaller scale, a similar problem of understanding the uneven reaction to an E.O.R. process of various wells in a small five-spot pattern is easier if it is known that they penetrate rocks representing a complex delta plain environment, perhaps with an original distribution of subenvironments, such as shown in Figure 6 (Pusch and others, 1976; Jordan and Tillman, 1982, and this volume). It would be expected that flow along the more permeable channel complex would be favored by reservoir fluids over flow across this trend into less permeable types of sediments. Design of injection-production patterns where this type of additional information is considered will greatly improve the chances of dealing successfully with reservoir heterogeneity. Optimization of well spacing and the choices of pattern for injection and production wells are much more soundly based when patterns of areal variability of the reservoir rock are considered.

NET SAND ISOPACH

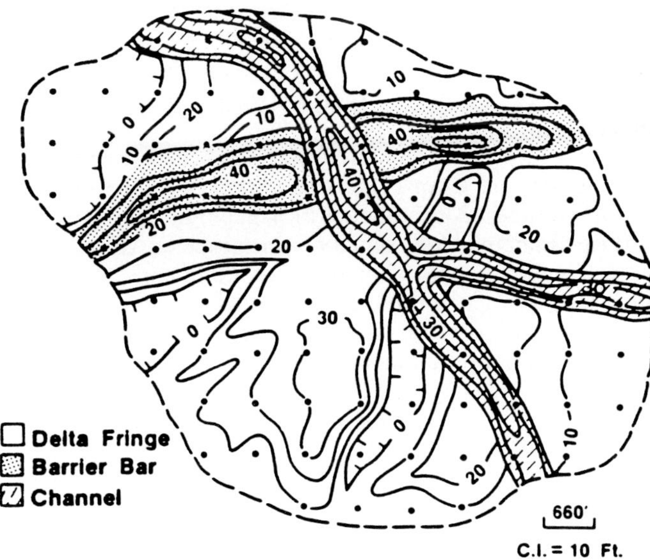

FIG. 5.—Paleogeologic map of environments of sandstone deposition, Elk City field, Oklahoma; adapted from Sneider and others (1977). The thicknesses of oil-bearing sandstones and areas of the field that produce from sandstones with different grain sizes and textures and different sedimentary structures are in part inherent to the sedimentary environment in which they were deposited.

In combined engineering-geological studies of reservoirs, it usually is the geologist's task to determine the minimum number of types of reservoir rock that must be delineated in order to describe accurately the reservoir's behavior in transmitting fluids. As in the study of the Elk

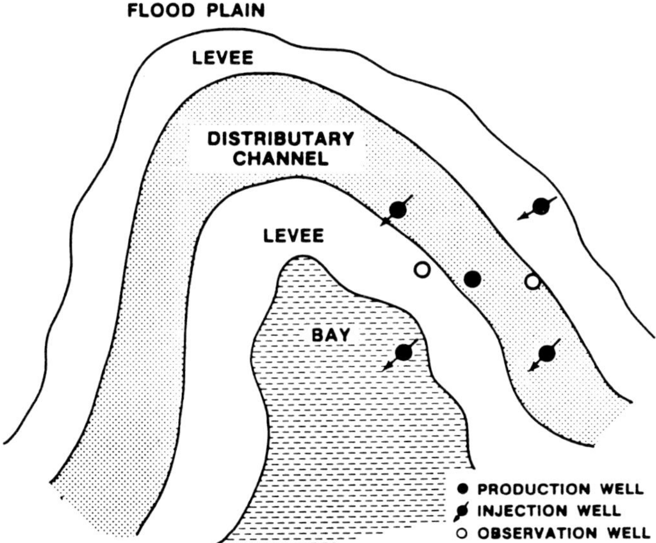

FIG. 6.—Hypothetical five-spot E.O.R. project area superimposed on facies pattern of a common reservoir type. The distributary channel sandstone facies would be expected to have good reservoir properties. The levees would have low porosity and permeabilities and the bay would include abundant shale. The effects of this facies distribution are discussed in the text.

City field cited above, the starting point for this task is usually a detailed reservoir description and facies analysis.

This task of "rock-typing" in carbonate reservoirs is usually more complicated than it is in sandstones because of the complexity and frequent variations in the fabric of carbonate rocks. It is necessary to recognize not only the rocks that had similar origins, but also the rocks in which various fluids are expected to behave similarly.

Langston and Chin (1968) have described the complex distribution of 13 lithofacies in a geologic study of subsurface Rainbow Reef oil pools in northern Alberta (Fig. 7A). Combining this knowledge of lithofacies with measurements of the capillary properties of the rocks and their relative permeability to various fluids, the authors were able to group various lithofacies such that only five reservoir facies were necessary to describe the pool quantitatively (Fig. 7B). Geologic considerations provide an accurate method of determining fluid flow, so that genetic rock units may be translated into fluid flow units for reservoir simulation.

Similar use of geologic reservoir models and an understanding of a complex diagenetic history of the carbonate reservoir rocks enabled Roehl (1967) to characterize the Ordovician Stony Mountain and Silurian Interlake supratidal to subtidal units in fields along the Cabin Creek Anticline of the Williston Basin. A combination of lithologic and petrophysical parameters is necessary to recognize significantly different types of rocks in these complexly stratified reservoirs. More recently, study of the Red River Formation, which occurs beneath the Stony Mountain in this same area, using techniques similar to those of Roehl, has led to recognition of similar sequences of cyclic carbonates in this unit (Ruzyla and Friedman, 1981). Reservoir performance during efforts at enhanced recovery in these reservoirs will require understanding of the complex interstratification of producing and non-producing intervals.

Stratification.—

Stratification of reservoirs is a problem with which engineers have learned to deal, on a statistical basis, with limited success; input of geologic insight may enhance the understanding of this characteristic of reservoirs. For instance, an idealized model of the vertical sequence of types of sediment deposited by a meandering stream affords us a means of understanding at least one kind of stratification sequence (Fig. 8). The various layers within the sequence affect the flow of fluids according to their characteristics. In a point bar sandstone reservoir, the combination of a ripple-bedded, silty, very fine-grained sandstone overlying a crossbedded, coarser grained sandstone will result in retardation of flow higher in the bed and deflection of flow in the direction of dip of the lower trough crossbeds (Fig. 9; Hewitt and Morgan, 1965; Weber, 1980).

This geometry and textural variation could complicate attempts to fireflood this reservoir by causing uneven advance of the combustion front and possibly causing the front to move into the water at the base of the reservoir, an effect that is opposite to the usual tendency of hot combustion gases to rise and propagate between wells at the top of the reservoir. In a chemical flood, the permeability contrast of this stratified reservoir could cause a loss of control of mobility of injected chemicals. Because of the tendency of fluids to avoid the zone of lower permeability, there would be very low vertical-sweep efficiency within the reservoir. Because the chemicals injected would not contact all of the residual oil in the sandstone, estimates of recoverable oil would be greatly inflated and the economic viability of the project would be in question.

The degree of interbedding of permeable and non-permeable beds in a producing formation is of great interest to engineers who attempt to predict the performance of reservoirs. Randomly distributed, discontinuous shale barriers within a reservoir sandstone may or may not affect oil recovery depending on their dimensions. Richardson and others (1977) showed with computer models that in a reservoir that undergoes vertical drainage, the presence of discontinuous horizontal shale barriers of 240-ft (73 m) width results in the recovery of 5–7% less oil than would be recovered from the same reservoir if barriers of only 60-ft (18 m) width were present, and 8–13% less oil than if no barriers to vertical flow were present. This effect is attributed to retardation of vertical drainage and capillary entrapment of the oil above the shale barriers. These authors note correctly that far more attention has been paid to the continuity of sandstones than to the shale partings and beds that occur between and within them.

Zeito (1965) has recorded the number and apparent lateral dimensions of shale "breaks" within several sandstone outcrop sections. The frequency of occurrence of shales in a vertical interval is variable within the classes of environments studied, but continuity of shales increases systematically from distributary channels, to deltaic, to marine sandstones. Weber (1980) has expanded on the work of Zeito and added observations on the small shale lenses and textural changes within distributary channel-fill sandstones. He provides a prototype model for anisotropy of horizontal and vertical permeability distribution in a festoon crossbedded sandstone.

Pryor and Fulton (1978) devised continuity indices to describe the varying discontinuity of sandstones in different parts of a Holocene deltaic complex. They noted that the number of interbed impermeable boundaries increases upward in fluvial point bar sand bodies, but it decreases upward in distributary mouth bar sands. Distributary mouth bar (fluvial-marine) sands also are more continuous laterally than fluvial point bar sands, and both of these types of sand bodies are more continuous than prodelta sands.

Shale "breaks," as barriers to vertical flow, can have an important bearing on the success of enhanced recovery processes in which oil is displaced by gas, coning of water into the bottoms of producing wells is important, or gravity-drainage is an important mechanism of oil production. Where shale "breaks" coalesce laterally, they may be important in any fluid displacement process. Enhanced oil recovery projects, with their close well spacing and with good lithologic control from logs and cores, should offer nearly ideal conditions in which to study further this problem of sandstone-shale interbedding and to improve on sedimentologic models of subsurface reservoirs.

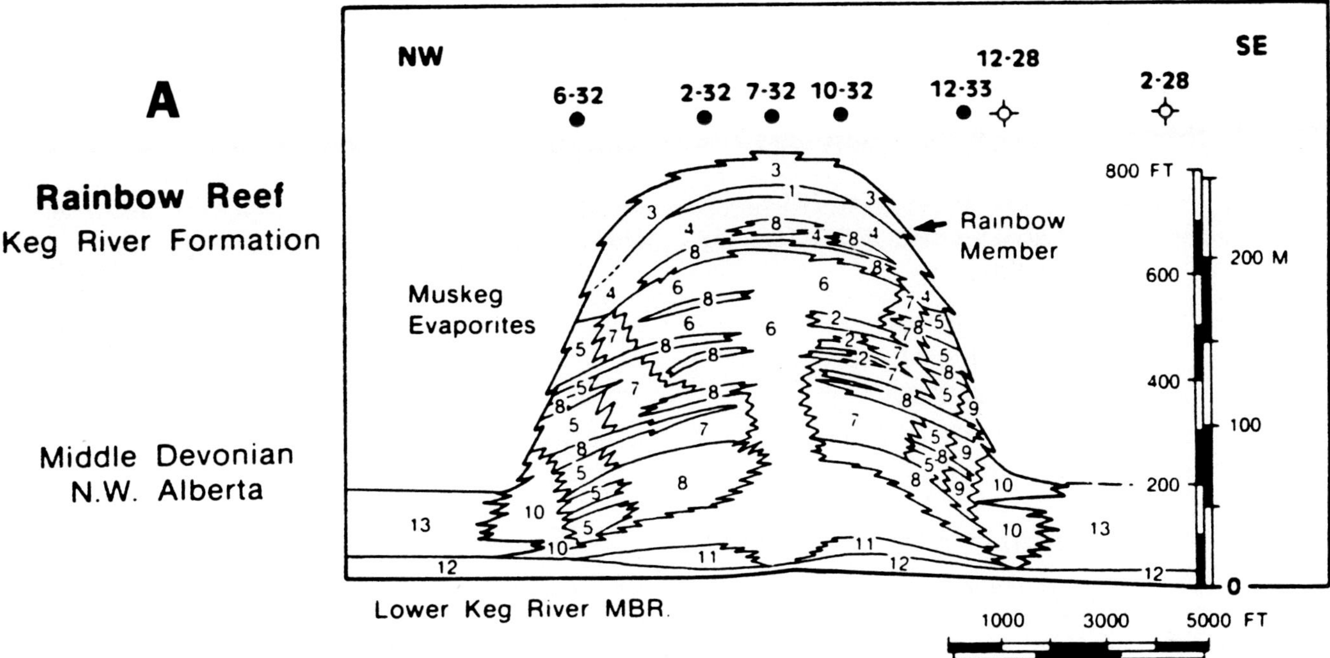

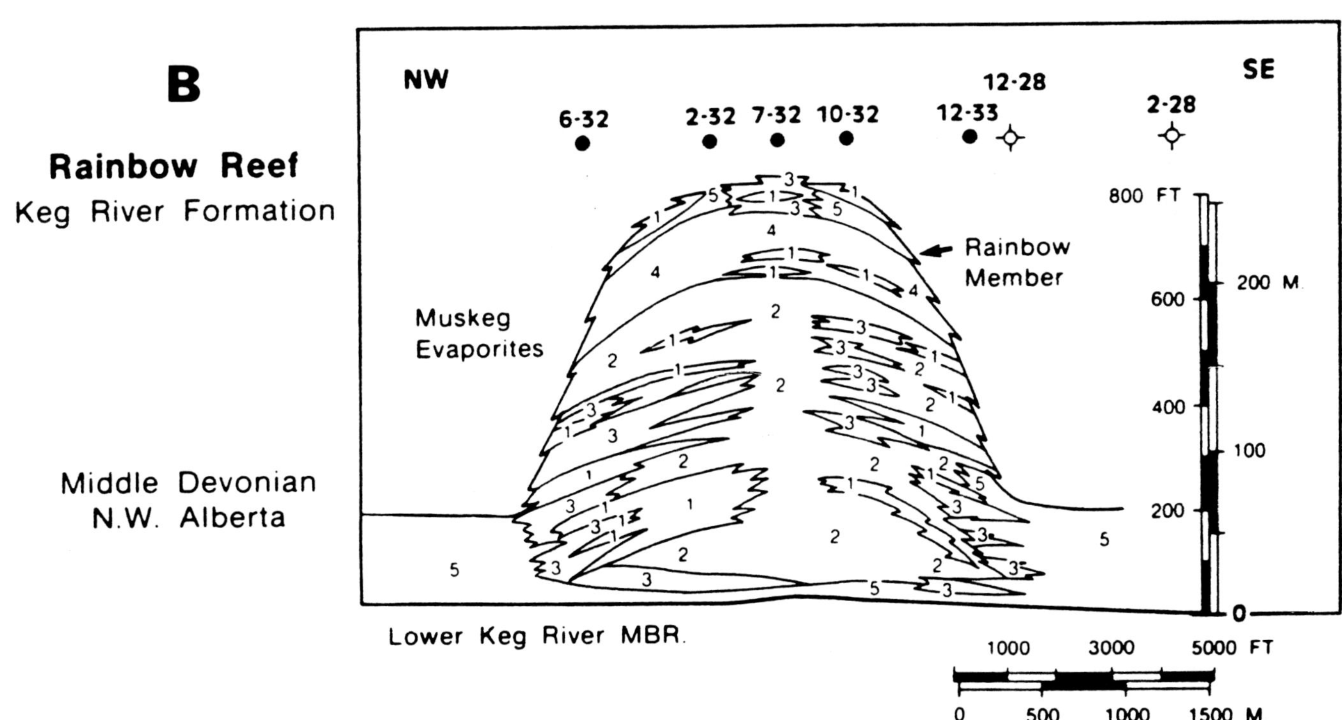

FIG. 7.—(A) Sedimentary facies of a pinnacle reef in the Rainbow Member of the Keg River Formation, northern Alberta. Thirteen facies are identified. Modified from Langston and Chin (1968). (B) Facies of the same Rainbow Reef as in (A), showing the extent of units in which the flow of fluids would be expected to be similar. Facies outlined in (A) have been grouped here on the basis of both facies and flow characteristics to reduce the complexity for numerical simulation.

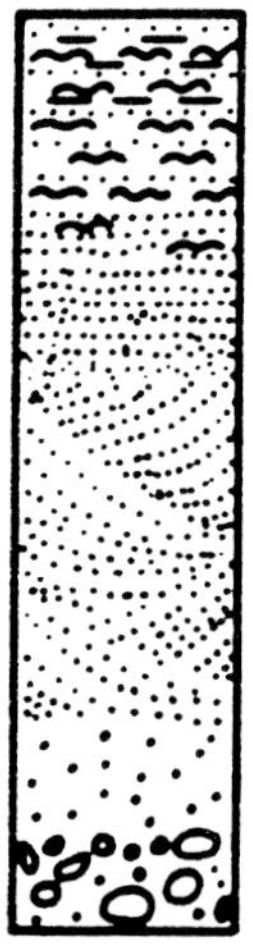

POINT BAR MODEL	ROCK	STRUCTURE	PERMEABILITY
	•siltstone, very fine grained, muddy sandstone	horizontal laminae, ripple bedded	very low
	•silty, fine grained sandstone, poorly sorted	ripple bedded, parallel bedded	low to moderate
	•fine-medium grained sandstone well sorted	cross bedded	moderate to high
	•medium-coarse grained sandstone and conglomerate, poor-moderate sorting	massive or cross bedded	low to moderate

FIG. 8.—Point bar geologic model showing the sequence of rock textures and structures in a reservoir consisting of a single point bar deposit. The influence of stratification on the horizontal permeability of the rock is also indicated; the potential effects of diagenesis are disregarded.

Hydrodynamics.—

A factor that is sometimes ignored during the planning stages of an E.O.R. project, but which should enter into the design of the fluid injection pattern, is the condition of the subsurface hydrologic regime. If a pressure gradient of sufficient magnitude is present and a condition of hydrodynamic flow of the formation water exists, then there may be a deflection of the intended flow of injected chemicals, which would drastically alter the apparent efficiency of areal

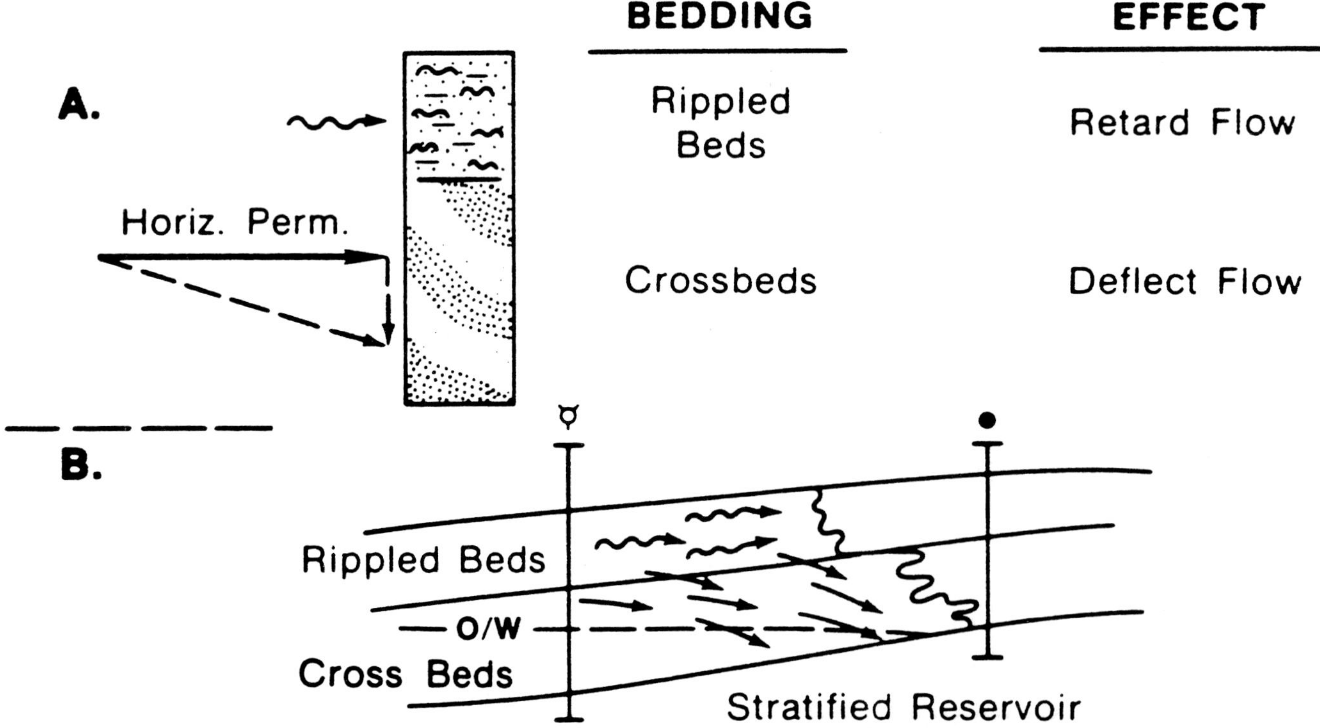

FIG. 9.—(A) Diagram showing the effects of sedimentary structures and textures on the flow of fluids in a point bar sandstone reservoir. The crossbedded unit is coarser grained and is inferred to have better reservoir properties than the overlying rippled unit. (B) Uneven advance of an injected fluid in the reservoir is shown. Permeability variations resulting from differences in reservoir quality are the major cause of flow patterns of this type.

sweep of a project (Strange and Talash, 1976; Widmyer and others, 1978). Depending on the gradient in a particular situation, it may be impossible to oppose the natural flow; where this is the case, only a small fraction of the chemicals intended to enter the interwell area between injection and production wells may actually do so (Fig. 10). A reservoir fluid drift of as little as 29.6 ft (9 m) per year (0.0304 psi per foot pressure gradient) was noted as adversely affecting areal sweep at El Dorado field, Kansas (Zetik and Tucker, 1979). There probably have been many instances where the phenomenon was not recognized, even though it may have caused unexpected, even disappointing, results in E.O.R. projects. At the very least, a project-wide initial pressure gradient should be considered as a possible explanation for anomalous interwell flow of fluids, and, preferably, this factor should be investigated before injection is begun, when perturbations of the pressure field are at a minimum.

Fracturing.—

Fractures in a reservoir rock can have a strong but unpredictable effect on the performance of an E.O.R. project unless data are available that show the directional effects of fracture orientation. In some cases, histories of well performance during waterflooding are available and may be used to determine fracture orientation in an area where an enhanced recovery project is planned. In other cases, where problems in controlling flow of fluid have developed after initiation of a project, injection of chemical tracers or pressure transient testing may be employed to deduce the patterns of fracture alignment within a project area (Fig. 11; Trantham and others, 1979).

The paths of high permeability that fractures provide reduce the ability of injected chemicals to contact some parts of the reservoir and result in early breakthrough of fluids in certain pairs of wells, greatly affecting areal sweep efficiency. Knowledge of geologic conditions in an area could lead to anticipation of fracture orientation. This would help in interpreting data from interwell testing and may even lead to avoiding problems through proper placement of patterns of injection-production wells or the use of gelled diverting agents to reduce fracture-permeability in some wells.

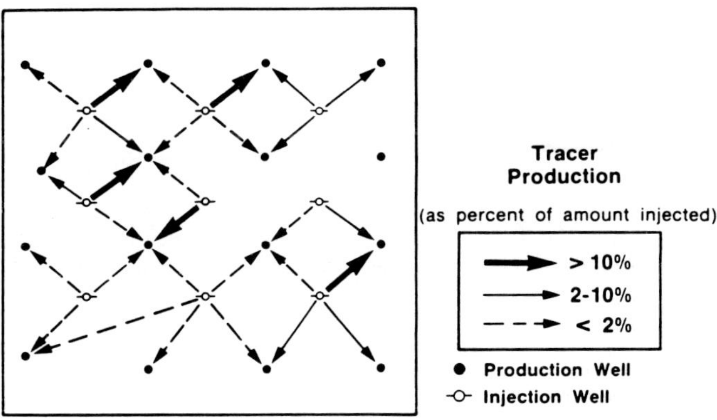

FIG. 11.—Diagram of the production of tracers among the wells in an E.O.R. project area. Preferential flow of fluids bearing tracers toward certain wells suggests alignment of fractures in the quadrant between those wells and a nearby injection well. Several different radioactive tracers are used in the various injection wells to enable this interpretation to be made. Modified from Trantham and others (1979).

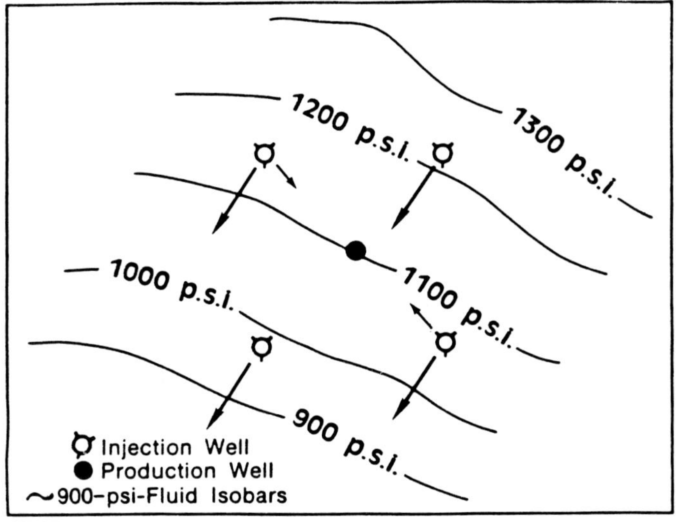

FIG. 10.—Diagram of a hypothetical pattern of injection and production wells in a reservoir in which an existing hydrodynamic gradient would be expected to cause fluid drift and uneven sweep of the interwell areas by injected fluids.

Fractures may actually be beneficial where they provide more balanced fluid injection rates in areas of a field where injectivity is low. Artificial hydraulic fracturing is sometimes used to develop more balanced fluid injection, although this is a risky procedure in an area with closely spaced wells, as in many E.O.R. projects.

Textural properties.—

In addition to the large-scale factors such as facies changes, reservoir stratification, hydrodynamic gradients, and fracturing, a number of small-scale features also affect the flow of fluids in rocks. Even more fundamental than the small-scale features are the effects which microscopic anisotropies of the rock pore structure have on fluids and gases (Table 3).

Only limited experimental data are available, but grain size of detrital sediments is known to affect parameters such as porosity and permeability (Graton and Fraser, 1935; Fraser, 1935; Krumbein and Monk, 1942; Pettijohn and others, 1973). Permeability decreases with decreasing grain

TABLE 3.—FACTOR CONTROLLING PORE GEOMETRY AND SURFACE AREA IN POROUS ROCK.

Original	Diagenetic
Grain size	Grain alteration
Grain sorting	Grain dissolution
Grain packing	Pressure solution
Grain shape	Grain overgrowths
Intergranular matrix	Cements
Clay laminae	Authigenic clays

size in artificially packed aggregates of spheres. Larger pore connections in uniformly coarser sands result in higher permeability than is found in fine sand. Porosity should remain unaffected by changes in grain size in moderately well sorted sediments, but, as a result of the tendency toward poor sorting in coarse sands, porosity is often highest in the more fine-grained sands (e.g., Sneider and other, 1977). Increase in sorting correlates with increase in porosity and increase in permeability in artificially packed aggregates. Pryor (1973) has shown that some of these relationships between texture and porosity-permeability apply to natural sand deposits and some do not. Moreover, there are differences in "style" of packing between sands deposited as river bars and those forming beaches and dunes. These differences result in corresponding differences in the relationships of texture to porosity-permeability among these classes of sand bodies.

These generalizations from studies of unconsolidated sediments may be nullified completely by the effects of diagenesis, especially cementation. In the end, careful study of the reservoir rocks of each project is necessary to establish the parameters that control rock fabric and the flow of fluids through the rock.

Pore geometry.—

Relative permeability of a reservoir rock to different fluids that pass through it during the life of an enhanced recovery project is a very important parameter for predicting fluid recoveries at various points in time and for judging the efficiency of displacement of oil from the reservoir. Morgan and Gordon (1970) have noted the importance of the pore geometry of rocks in determining their relative permeability characteristics. In situations where wettability and the history of changes in saturation of the rocks are the same, rocks with different pore geometries may be expected to have different relative permeabilities (Fig. 12). Because of the large number of variables affecting the pore geometry and surface area in rocks, it is not sufficient, in choosing samples for measurement of relative permeability, to be guided by only one characteristic, for instance absolute permeability to air. Texture, amount and type of cements and interstitial clays, angularity and packing of the grains, and size distribution of pores should also be considered as part of the "rock-typing" process and as an indication of what samples are representative of the reservoir.

Carbonate reservoirs are noted for the complexity of their pore networks. The range of shapes and sizes of carbonate particles (with some even being hollow), the variability of cementation patterns, and the susceptibility of carbonate rocks to modification early and late in their history, result in rock fabrics which are quite different from those commonly found in quartz-rich clastic rocks such as sandstones (Fig. 13). Compare, for instance, the pore structure of the sandstone consisting of rounded quartz grains and small amounts of interstitial clay with the carbonate composed of angular crystals of fossil fragments. The dolomite has a complex, bimodal pore-size distribution with pore throats that are more like slots than tubes (Wardlaw, 1976). Another example of compound pore structure in carbonates occurs in carbonate mudstones with scattered, hollow, fossil particles and throughgoing fractures (Fig. 13). If study of these types of pore systems is not undertaken, attempts at fluid injection and enhanced oil recovery in carbonates will be on "shaky foundations" and, almost surely, will depend largely on luck for success.

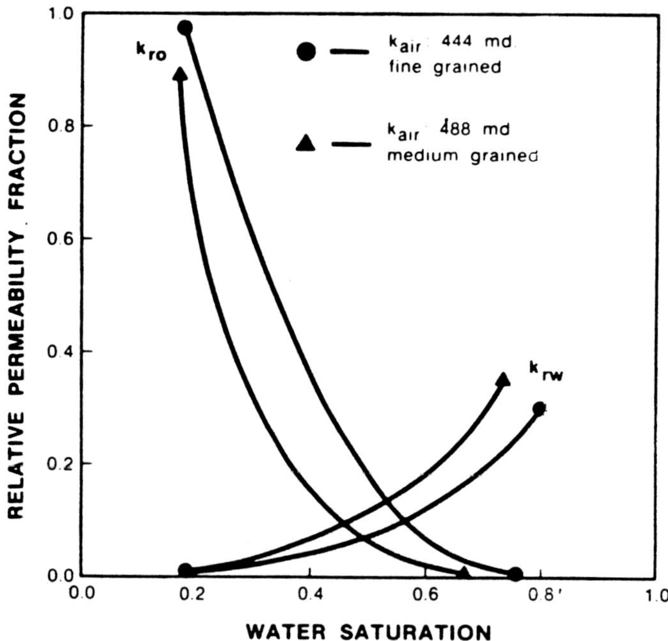

FIG. 12.—Relative permeability to oil and water of two samples of sandstone from the same reservoir; the samples are known to have nearly identical absolute permeability to air. The variability between the samples indicates that other factors such as grain size (shown here), sorting, packing, and surface area have considerable effect on the relative permeability. Modified from Morgan and Gordon (1970).

In Bindley field, western Kansas, which produces oil from a Mississippian dolomite reservoir, primary facies include porous rocks with two entirely different types of pore systems (Ebanks and others, 1977). Predictions of values of permeability from measurements of porosity (Fig. 14) are not good in this example because of the wide range of permeabilities which may correspond to very small ranges of porosity. One of the keys to understanding the distri-

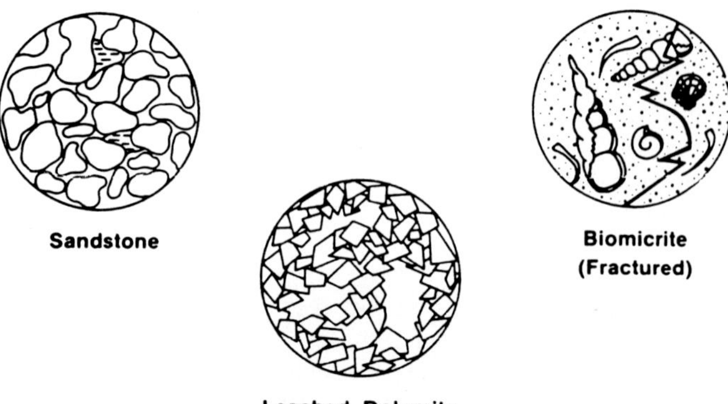

FIG. 13.—Schematic differences in microscopic pore geometry in rocks of different compositions.

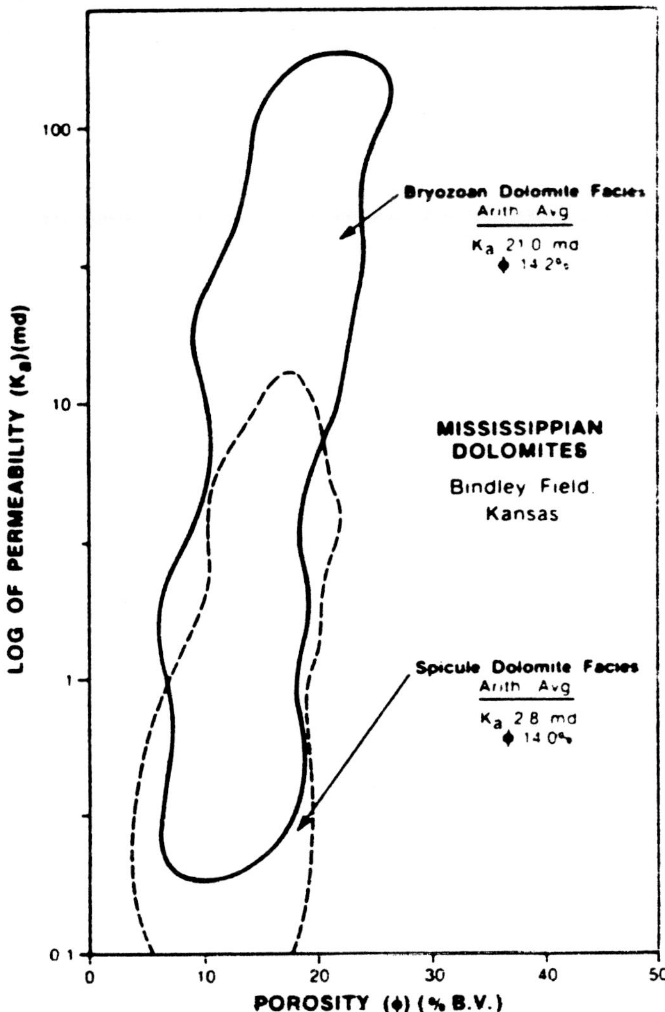

FIG. 14.—Plot of log-permeability to air *versus* porosity in Mississippian dolomites from Bindley field, Kansas. Significant differences in the range of values of these parameters occur in the two different carbonate reservoir facies.

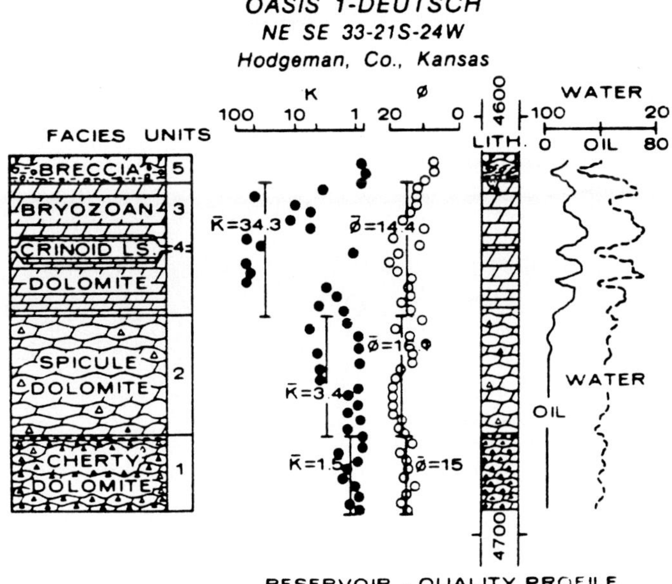

FIG. 15.—Vertical sequence of carbonate facies in Bindley field, Kansas plotted to show their different reservoir rock properties and their relation to accumulation of oil. The bryozoan dolomite has small vuggy pores in a medium-crystalline dolomite matrix, whereas the spicule dolomite has very small spicule molds in a more dense dolomite matrix (Ebanks and others, 1977).

bution of different pore systems is the recognition of the differences in the primary facies (Fig. 15). The bryozoan dolomite facies comprises relatively coarsely crystalline subhedral dolomite with a well connected network of vuggy pores. These pores result from enlargement of molds left from the dissolution of fragments of fenestrate bryozoans. In contrast, the spicule dolomite facies consists of isolated molds of sponge spicules in a matrix of finely crystalline, anhedral dolomite. Although both facies are oil-stained in cores, and they have the same average porosity, only the bryozoan dolomite is sufficiently permeable to produce oil. In an enhanced recovery project, little, if any, recoverable oil reserves should be assigned to the spicule dolomite facies.

Sandstones, which may be less susceptible to drastic modification after deposition than are carbonates, commonly show diagenetic effects such as overgrowth cementation, precipitation of chemical cement, partial dissolution of grains, and alteration of particles to clay. These diagenetic changes may seriously affect not only the amount, but also the kinds of pores in the rock (Fig. 16). Following diagenetic changes such as these, a higher percentage of the pores will be isolated from adjacent pores or connected by smaller pore throats, overall pore sizes will be reduced, and the tortuosity of the pore system will be increased. Besides these geometrical changes, rocks in which authigenic clays are formed will experience enormous increases in internal surface area and the creation of microporosity. Microporosity may in turn cause large increases in apparent water saturation (Pittman, 1979).

Large surface area is detrimental to the effectiveness of surfactant and polymer solutions in enhanced recovery processes, because the tendency of these chemicals to adsorb on rock surfaces and to be lost from circulation depends

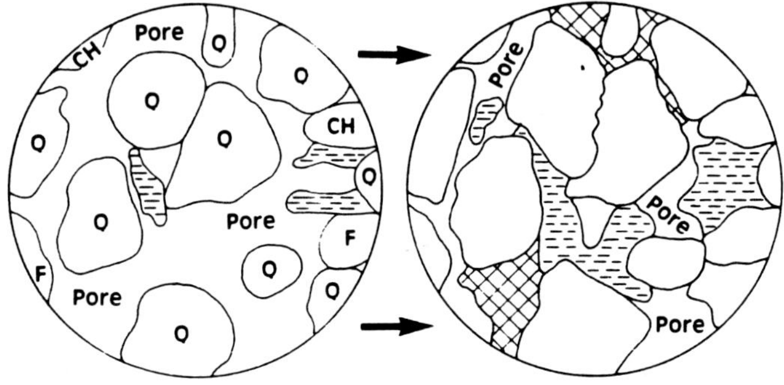

FIG. 16.—Reduction in porosity in sandstones as a result of cementation and growth of authigenic minerals in the pores. Not only the amount, but also the size and arrangement of pores are affected by diagenesis.

partly on surface area of the pore system to which they are exposed (Kalpakci and others, 1981). Microporosity, so small that it can be resolved in samples of rock only at high magnifications with a scanning electron microscope, is also important in evaluation of formations from wireline logs and from the results of formation tests because of the tendency of water in these small pores to be immobile (Kieke and Hartman, 1974). Whether in sandstone or carbonate rocks, water in water-wet micropores is held more tightly by capillary forces than is the water in larger, intergranular or vuggy pores. As a result, water saturation of the rocks, as calculated from wireline logs or analysis of cores, seems anomalously high and may not indicate the capability of the rocks to produce oil with little or no water. The types and distribution of diagenetically altered reservoir rocks cannot always be predicted ahead of drilling. In E.O.R. project areas, however, where the density of well control is usually great and where cores are usually available, geologists should be able to relate patterns of alteration of rock fabric to other, more predictable, features of the rock and to improve the ability of engineers to model the reservoir successfully.

Permeability and porosity of the Rotliegendes Sandstone in the North Sea province varies with both location and composition of the diagenetic mineral assemblage (Stalder, 1973). Correlations of these two parameters, and, consequently, the ability to predict rock permeability from log-derived values of porosity depend on whether or not one considers the dominant clay mineral present (Fig. 17). Kaolinite-cemented sandstones are more permeable than are illite-cemented sandstones with the same porosity. Whether the origin of this characteristic is related to primary facies (Robinson, 1981) or to diagenetic history (Almon, 1981) is in dispute. Nevertheless, geologists have been able to relate these parameters to aspects of the rock sequence that can be mapped and have been able to predict trends in reservoir quality that relate to reservoir performance.

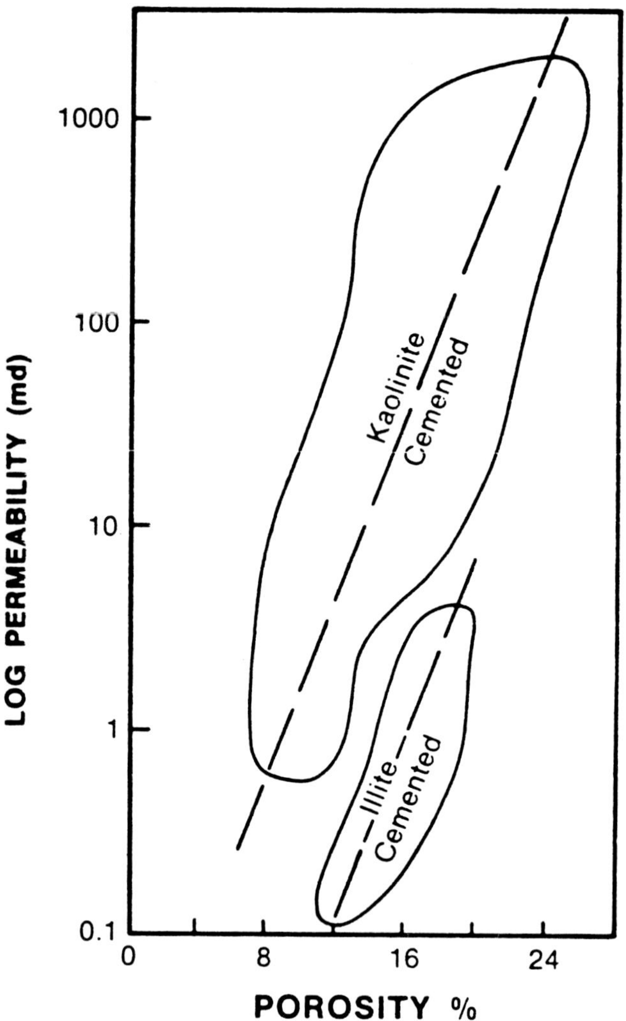

FIG. 17.—Sandstones of the Rotliegendes Formation in the North Sea in which discrete families of properties relating to the types of authigenic clay cement are evident. Adapted from Stalder (1973).

Interstitial clay.—

The importance of clays in affecting the performance of reservoir rocks is well established. Clays can affect reservoir performance adversely in at least four ways (Table 4). Since almost all sandstones contain some clay, these clay-related factors must be considered. Carbonate rocks, in contrast, rarely contain much clay, except as distinct beds or laminae of shale.

The role of clays in formation damage during or after drilling was mentioned earlier. One of the beneficial effects of clays in E.O.R. processes in sandstones is the effect they evidently have in promoting the deposition of coke to sustain combustion in fire-flooding projects. Different clays have quite different characteristics. Swelling clays, such as smectite or degraded illite, tend to reduce rock permeability when contacted by fresh water; kaolinite is notorious for becoming dislodged and moving about within rock pore systems to lodge in small passages, thereby reducing permeability; illite creates more tortuous paths through pore systems by bridging gaps between grains; and chlorite is a source of iron in pore water of formations that have been treated by acids during post-drilling stimulation. Pittman and Thomas (1978) have described the ways that clays occur in rock pore networks, Neasham (1977) has classified the effects that clays with differing morphologies have on the flow of fluids in pore systems, and Hower (1974) and Almon and Davies (1978) have described the influence of clays on various aspects of hydrocarbon production and well stimulation. It is worth emphasizing again the enormous increases in surface area that rock pore systems experience

TABLE 4.—SIMPLE CLASSIFICATION OF TYPES OF PROBLEMS CAUSED BY CLAYS IN RESERVOIRS.

Swelling or slaking clay
Movement of fines in pores
Increased surface area of pores
Reactions with fluids in rock

when they are the sites of growth of authigenic clays and the detrimental effect this can have on the efficiency of chemical floods and on log analysis.

SUMMARY

Because of the economic risk inherent in an enhanced oil recovery (E.O.R.) project, evaluation of a reservoir for suitability for E.O.R. and as a means of predicting its performance must begin with an accurate description of the reservoir. This description, if it is to address all of the factors that are important to the success of a project, should be the result of a combined geological and engineering study, with continuing feedback between the two disciplines to arrive at the least ambiguous portrayal of the reservoir as is possible. Good reservoir description has also been recognized as important to maximizing ultimate recovery of oil and gas in some of the largest fields discovered in recent years, Prudhoe Bay (Wadman and others, 1978) and Jay (Shirer and others, 1978) fields, well before the time when they will be candidates for E.O.R. processes.

Geologists should be involved in selection of a suitable E.O.R. process for a particular reservoir, in the accurate simulation of performance of the reservoir, and in avoiding problems in implementing an E.O.R. process or helping to solve problems that develop during its operation. Variations in composition and quality of reservoir rocks and the effects they will have on the distribution of fluids within them are best interpreted and predicted from geologic data.

Many geologic factors contribute to the behavior of an E.O.R. project. Large-scale features, such as the arrangement of facies, layering or stratification, the number and arrangement of shale "breaks," natural fractures, and hydrodynamic pressure gradients, can have a marked influence on the movement of fluids within a reservoir. Smaller scale features, such as grain size and sorting, types and arrangements of clays, geometry of pores, effects of diagenesis, and composition of the reservoir rock affect fluid-rock interactions and efficiency of displacement of one fluid by another on a microscopic level.

Finally, an understanding of trends in rock composition and variability enable a geologist to recognize "rock types" that differ significantly from each other and which should be the basis for subdividing a reservoir sequence for numerical simulation. This same understanding also enables a geologist to interpolate between wells the nature of changes in the rocks and, so, to guide a reservoir engineer in modeling the performance of an E.O.R. project through the choice of suitable constraints on the variables he is modeling. A geologist, in order to be effective in this role, must be willing to express his or her descriptions of the reservoir rock in numerical form that is useful to the engineer counterpart.

REFERENCES

ALMON, W. R., 1981, Depositional environment and diagenesis of Permian Rotliegendes sandstones in the Dutch sector of the southern North Sea, *in* Longstaffe, F. J., ed., Clays and the Resource Geologist: Mineralogical Association of Canada, Short Course Handbook, v. 7, May, 1981, p. 119–147.

———, AND DAVIES, D. K., 1978, Clay technology and well stimulation: Transactions of Gulf Coast Association of Geologists, v. 28, p. 1–6.

———, AND ———, 1981, Formation damage and the crystal chemistry of clays, *in* Longstaffe, F. J., ed., Clays and the Resource Geologist: Mineralogical Association of Canada, Short Course Handbook, v. 7, May, 1981, p. 81–103.

BARRETT, D. D., HARPOLE, K. J., AND ZAAZA, M. W., 1977, Reservoir data pays off: West Seminole San Andres Unit, Gaines County, Texas: Society of Petroleum Engineers, SPE Paper 6738, 52nd Annual Meeting, p. 1–5.

BASAN, P. B., MCCALEB, J. A., AND BUXTON, T. S., 1978, Important geological factors affecting the Sloss field micellar pilot project: Society of Petroleum Engineers, SPE Paper 7047, Fifth Symposium on Improved Methods for Oil Recovery, p. 111–114.

DIEHL, A. L., 1976, Is there a role for the geologist in enhanced recovery?: Proceedings, Petroleum Society of Canadian Institute of Mining, Paper No. 7625, p. 1–6.

EBANKS, W. J., Jr., EUWER, R. M., AND ZELLER, D. E. N., 1977, Mississippian combination trap, Bindley field, Hodgeman County, Kansas: American Association of Petroleum Geologists Bulletin, v. 61, p. 309–330.

EXUM, F. A., AND HARMS, J. C., 1968, Comparison of marine-bar with valley-fill stratigraphic traps, western Nebraska: American Association of Petroleum Geologists Bulletin, v. 52, p. 1851–1868.

FRASER, H. J., 1935, Experimental study of the porosity and permeability of clastic sediments: Journal of Geology, v. 43, part 1, p. 910–1010.

GEFFEN, T. M., 1976, Methods for recovering more oil from known fields, *in* Gary, J. H., and Golden, J. O., eds., Research on Petroleum, Natural Gas, and Oil Shale—the University's Role: Workshop, Colorado School of Mines, Golden, CO, p. 27–36.

GHAURI, W. K., OSBORNE, A. F., AND MAGNUSON, W. L., 1974, Changing concepts in carbonate waterflooding—West Texas Denver Unit Project—an illustrative example: Journal of Petroleum Technology, p. 595–606.

GRATON, L. C., AND FRASER, H. J., 1935, Systematic packing of spheres, with particular relation to porosity and permeability: Journal of Geology, v. 43, part 1, p. 785–909.

HARMS, J. C., 1966, Stratigraphic traps in a valley fill, western Nebraska: American Association of Petroleum Geologists Bulletin, v. 50, p. 2119–2149.

HARPOLE, K. J., 1979, Improved reservoir characterization—a key to future reservoir management for the West Seminole San Andres Unit: Society of Petroleum Engineers, SPE Paper 8274, 54th Annual meeting, p. 1–8.

HARRIS, D. G., 1975, The role of geology in reservoir simulation studies: Journal of Petroleum Technology, p. 625–632.

HEWITT, C. H., 1963, Analytical techniques for recognizing water-sensitive reservoir rocks: Journal of Petroleum Technology, p. 813–818.

———, AND MORGAN, J. T., 1965, The Fry *in situ* combustion test—reservoir characteristics: Journal of Petroleum Technology, v. 17, p. 337–353.

HOWER, W. F., 1974, Influence of clays on the production of hydrocarbons: Society of Petroleum Engineers, SPE Paper 4785, Symposium on Formation Damage Control, p. 165–172.

JARDINE, D., ANDREWS, D. P., WISHART, J. W., AND YOUNG, J. W., 1977, Distribution and continuity of carbonate reservoirs: Journal of Petroleum Technology, p. 873–885.

JORDAN, D. W., AND TILLMAN, R. W., 1982, Reservoir description, *in* Van Horn, L. E., ed., El Dorado Micellar-Polymer Demonstration Project, Seventh Annual Report, September 1980–August 1981: U.S. Department of Energy, DOE/ET/13070-79, p. 55–73 and Appendix F.

KALPAKCI, B., KLAUS, E. E., DUDA, J. L., AND NAGARAJAN, R., 1981, Permeability modification of porous media by surfactant solutions: Society of Petroleum Engineers, SPE Paper 9930, Regional Meeting, Bakersfield, CA, p. 483–491.

KIEKE, E. M., AND HARTMAN, D. J., 1974, Detecting microporosity to improve formation evaluation: Journal of Petroleum Technology, p. 1080–1086.

KRUMBEIN, W. C., AND MONK, G. D., 1942, Permeability as a function of the size parameters of sedimentary particles: American Institute of Mining and Metallurgical Engineers Technical Publication 1492, p. 153–163.

LANGSTON, J. R., AND CHIN, G. E., 1968, Rainbow Member facies and related reservoir properties, Rainbow Lake, Alberta: American Association of Petroleum Geologists Bulletin, v. 152, p. 1925–1955.

LeBlanc, R. J., Sr., 1977, Distribution and continuity of sandstone reservoirs: Journal of Petroleum Technology, p. 776–804.

McCaleb, J. A., 1978, The role of the geologist in reservoir description: contribution to notes for American Association of Petroleum Geologists Reservoir Fundamentals School, p. 6.

———, and Wayhan, D. A., 1969, Geologic reservoir analysis, Mississippi Madison Formation, Elk Basin field, Wyoming-Montana: American Association of Petroleum Geologists Bulletin, v. 53, p. 2094–2113.

Morgan, J. T., and Gordon, D. T., 1970, Influence of pore geometry on water-oil relative permeability: Journal of Petroleum Technology, p. 1199–1208.

National Petroleum Council, 1976, Enhanced Oil Recovery: National Petroleum Council, Washington, D.C., 231 p.

Neasham, J. W., 1977, The morphology of dispersed clay in sandstone reservoirs and its effect on sandstone shaliness, pore space, and fluid flow properties: Society of Petroleum Engineers, SPE Paper 6858, 52nd Annual Meeting, p. 1–3.

Pettijohn, F. J., Potter, P. E., and Siever, R., 1973, Sand and Sandstone: Springer-Verlag, New York, 618 p.

Pittman, E. D., 1979, Porosity, diagenesis, and productive capability of sandstone reservoirs, *in* Scholle, P. A., and Schluger, P. R., eds., Aspects of Diagenesis: Society of Economic Paleontologists and Mineralogists Special Publication 26, p. 159–174.

———, and Thomas, J. B., 1978, Some applications of scanning electron microscopy to the study of reservoir rock: Society of Petroleum Engineers, SPE Paper 7550, 53rd Annual Meeting, p. 1–4.

Pryor, W. A., 1973, Permeability-porosity patterns and variations in some Holocene sand bodies: American Association of Petroleum Geologists Bulletin, v. 57, p. 162–189.

———, and Fulton, K., 1978, Geometry of reservoir-type sand bodies in the Holocene Rio Grande Delta and comparison with ancient reservoir analogs: Society of Petroleum Engineers, SPE Paper 7045, Symposium on Improved Oil Recovery, p. 81–84.

Pusch, W. H., Almon, W. R., Flesch, G. A., and Kellerhals, G. E., 1976, Coring and Core Analysis, *in* Rosenwald, G. W., Miller, R. J., and Vairogs, J., eds., El Dorado Micellar-Polymer Demonstration Project, Second Annual Report, July 1975–May 1976: Energy Research and Development Administration Technical Information Center, p. I-15 to I-25 and Appendix.

Richardson, J. G., Harris, D. G., Rossen, R. H., and Van Hee, G., 1977, Synergy in reservoir studies: Society of Petroleum Engineers, SPE Paper 6700, 52nd Annual Meeting, p. 1–6.

Robinson, A. E., 1981, Facies types and reservoir quality of the Rotliegendes Sandstone, North Sea: Society of Petroleum Engineers, SPE Paper 10303, 56th Annual Meeting, p. 1–3.

Roehl, P. O., 1967, Stony Mountain (Ordovician) and Interlake (Silurian) facies analogs of Recent low-energy marine and subaerial carbonates, Bahamas: American Association of Petroleum Geologists Bulletin, v. 51, p. 1979–2032.

Ruzyla, K., and Friedman, G. M., 1981, Geological heterogeneities important to future enhanced recovery in carbonate reservoirs of Upper Ordovician Red River Formation at Cabin Creek Field, Montana: Society of Petroleum Engineers, SPE/DOE Paper 9802, Second Joint Symposium on Enhanced Oil Recovery, p. 403–407.

Shirer, J. A., Langston, E. P., and Strong, C. B., 1978, Application of fieldwide conventional coring in the Jay-Little Escambia Creek unit: Journal of Petroleum Technology, p. 1774–1780.

Smith, C. F., Crowe, C. W., and Nolan, T. J., 1969, Secondary deposition of iron compounds following acidizing treatments: Journal of Petroleum Technology, p. 1121–1129.

Sneider, R. M., Richardson, F. H., Paynter, D. D., Eddy, R. E., and Wyant, I. A., 1977, Predicting reservoir rock geometry and continuity in Pennsylvanian reservoirs, Elk City field, Oklahoma: Journal of Petroleum Technology, p. 851–866.

———, Tinker, C. N., and Meckel, L. D., 1978, Deltaic environment reservoir types and their characteristics: Journal of Petroleum Technology, p. 1538–1546.

Stalder, P. J., 1973, Influence of crystallographic habit and aggregate structure of authigenic clay minerals on sandstone permeability: Geologie en Mijnbouw, v. 52, p. 217–220.

Strange, L. K., and Talash, A. W., 1976, Analysis of Salem low-tension waterflood test: Society of Petroleum Engineers, SPE Paper 5885, Symposium on Improved Oil Recovery, p. 605–612.

Swanson, D. C., 1979, Deltaic deposits in the Pennsylvanian Upper Morrow Formation of the Anadarko Basin, *in* Hyne, N. J., ed., Pennsylvanian Sandstone of the Mid-Continent: Tulsa Geological Society, Tulsa, OK, p. 115–168.

Trantham, J. C., Threlkeld, C. B., and Peterson, H. L., 1979, Reservoir description for a surfactant/polymer pilot in a fractured, oil-wet reservoir—North Burbank Unit Tract 97; Society of Petroleum Engineers, SPE Paper 8432, 54th Annual Meeting, p. 1–5.

Wadman, D. H., Lamprecht, D. E., and Mrosovsky, I., 1978, Reservoir description through joint geologic-engineering analysis: Society of Petroleum Engineers, SPE Paper 7531, 53rd Annual Meeting, p. 1–6.

Wardlaw, N. C., 1976, Pore geometry of carbonate rocks as revealed by pore casts and capillary pressure: American Association of Petroleum Geologists Bulletin, v. 60, p. 245–257.

Weber, K. J., 1980, Influence on fluid flow of common sedimentary structures in sand bodies: Society of Petroleum Engineers, SPE Paper 9247, 55th annual meeting, p. 1–7.

Widmyer, R. H., Frazier, G. D., Strange, L. K., and Talash, A. W., 1978, Low tension waterflood at Salem Unit—post-pilot evaluation: Society of Petroleum Engineers 5th Symposium on Improved Methods of Oil Recovery, p. 459–463.

Zeito, G. A., 1965, Interbedding of shale breaks and reservoir heterogeneities: Journal of Petroleum Technology, p. 1223–1228.

Zetik, D. F., and Tucker, J. R., 1979, Performance prediction, *in* Rosenwald, G. W., ed., El Dorado Micellar-Polymer Demonstration Project, Fourth Annual Report, September 1977–August 1978: U.S. Department of Energy, BETC-1800-40, I-18 to I-36.

PART I
MARINE SANDSTONE AND CARBONATE RESERVOIRS

SEDIMENTOLOGIC MODEL AND PRODUCTION CHARACTERISTICS OF HARTZOG DRAW FIELD, WYOMING, A SHANNON SHELF-RIDGE SANDSTONE

R. W. TILLMAN[1]

Exploration and Production Laboratory, Cities Service Oil and Gas Company, Tulsa, Oklahoma 74102

AND

R. S. MARTINSEN[2]

Cities Service Oil and Gas Company, Denver, Colorado

ABSTRACT: Hartzog Draw field is a stratigraphically controlled oil reservoir which produces from the Upper Cretaceous Shannon Sandstone at depths from 9,000 to 9,600 ft (2,727 to 2,910 m). The producing interval consists of a large mid- to outer shelf sand-ridge complex deposited well below effective normal wave base more than 100 mi (160 km) from shore. The productive interval in the shelf-ridge complex has a maximum thickness of 60 ft (18 m), and averages 20 ft (6 m) in thickness. The field is 22 mi (35 km) long and is as much as 3-1/2 mi (5.8 km) wide. Since its discovery in 1975, over 175 primary production wells were completed on 160-acre spacing. Initial oil-in-place was estimated to be 350 million barrels. Secondary waterflood was initiated in 1981 and 115 additional infield wells were to be drilled by the end of 1985.

The reservoir is completely enveloped in shale, has a solution gas drive, no water table and no produced formation water. Net pay is primarily a product of porosity, permeability and thickness of the sandstone, and is directly related to sedimentary facies. Detailed studies of five cores located in the northern, eastern, and central portions of the field allow definition of nine facies. Three of the facies are primarily high-angle crossbedded sandstones; the other facies show a variety of low-energy features including ripples and abundant burrowing. The *Central Ridge Facies,* a high-angle trough crossbedded slightly glauconitic quartz sandstone, is a consistently high quality reservoir. The *High-Energy Ridge-Margin Facies,* a crossbedded highly glauconitic sandstone containing siderite and clay rip-up clasts is also a relatively high quality reservoir; the *Low-Energy Ridge-Margin Facies,* which consists of interbedded ripples and troughs, and the *Inter-Ridge Facies (Shaly),* a rippled interbedded sandstone and shale, generally are marginal quality to non-reservoirs.

The average porosity for the field is 12% and the average permeability is 12 md. Higher mean values are recorded in the producing intervals of the five cores studied. Values for the *Central Ridge Facies* are 15% and 15 md and for the *High-Energy Ridge-Margin Facies* are 14% and 19 md.

Sandstone isopach maps and cross sections perpendicular to the field elongation show that the field is asymmetrical and considerably steeper on the northeast flank. Paleocurrent flow directions inferred from oriented cores indicate a southerly flow of the currents responsible for deposition of the Hartzog Draw shelf-ridge complex.

LOCATION AND GENERAL RESERVOIR CHARACTERISTICS

Hartzog Draw field, which is located on the gently sloping east flank of the Powder River Basin, Wyoming (Fig. 1), is the largest of numerous linear, northwest- to southeast-trending, Upper Cretaceous, stratigraphically trapped oil fields in the basin (Fig. 2). The present structure conforms to a general southwest regional dip of about 1–2°, typical of that portion of the basin (Fig. 3), and has little influence on production. The paleostructural configuration of the basin, however, may have had a strong influence on deposition of the reservoir and will subsequently be discussed in more detail.

Production in Hartzog Draw field is from the Upper Cretaceous (Campanian) Shannon Sandstone Member of the Cody Shale (Fig. 4). The Shannon Sandstone is a producing reservoir in more than 26 fields in the Powder River Basin (Table 1). Several of the fields, including Pine Tree, have undergone extensive development since Table 1 was compiled. The Shannon Sandstone and the Sussex Sandstone, which occurs about 300 ft (91 m) above the Shannon and is depositionally similar, are both marine sandstones completely enveloped in marine shale (Fig. 5) and are thus separated by shale facies from the overlying and laterally equivalent continental and nearshore facies of the Mesaverde Formation. The shoreline during Shannon time was approximately 80 to 100 mi (134 to 167 km) west of Hartzog Draw field.

Present addresses:

[1]Consulting Sedimentologist, 4555 S. Harvard, Tulsa, OK 74135

[2]Consultant, 3901 Gray Gables Road, Laramie, Wyoming 82070

An isopachous map of the net sandstone thickness shows the narrow, linear character of the sandstone body (Fig. 6), which is 22 mi (35 km) long and ranges from 1 to 3.5 mi (1.7 to 5.8 km) in width. In individual wells reservoir sandstone thicknesses average 25 ft (7.6 m) but range from less than 5 to more than 65 ft (1.5 to 19.7 m) in thickness. The reservoir facies are developed within a 60- to 90-ft-thick (17.2 to 27.8 m) sequence of burrowed to bioturbated, rippled, interbedded sandstone and shale, which in turn is completely enveloped in a dark gray, commonly silty, shale facies of the Cody Shale. Net pay thickness ranges from 1 to 62 ft (0.3–19 m) and closely parallels the sandstone thickness. Whereas as much as 25% porosity and 100 md permeability have been measured in cores, the field as a whole averages only 12% porosity and 12 md permeability (Martinsen, 1981). The best porosities and permeabilities generally are associated with thick net pay sandstones and thus tend to occur in the more central and eastern portions of the field. Initial potentials flowing in excess of 3,000 barrels of oil per day (BOPD) are not uncommon for the central field wells. Wells on the northeastern side of the field tend to have better porosities and permeabilities than do wells on the western side. This difference is probably related more to depositional setting (the eastern side being the "seaward" and upcurrent side of the sand body) than to the present structural setting.

The Shannon sandstones in the Hartzog Draw area are from 9,000 to about 9,800 ft (2,740–2,987 m) deep. Initial reservoir pressure was measured at about 5,000 psi, which indicates the reservoir was slightly overpressured. During primary production, average reservoir pressure *locally* dropped to about 1,000 psi, which is below the calculated

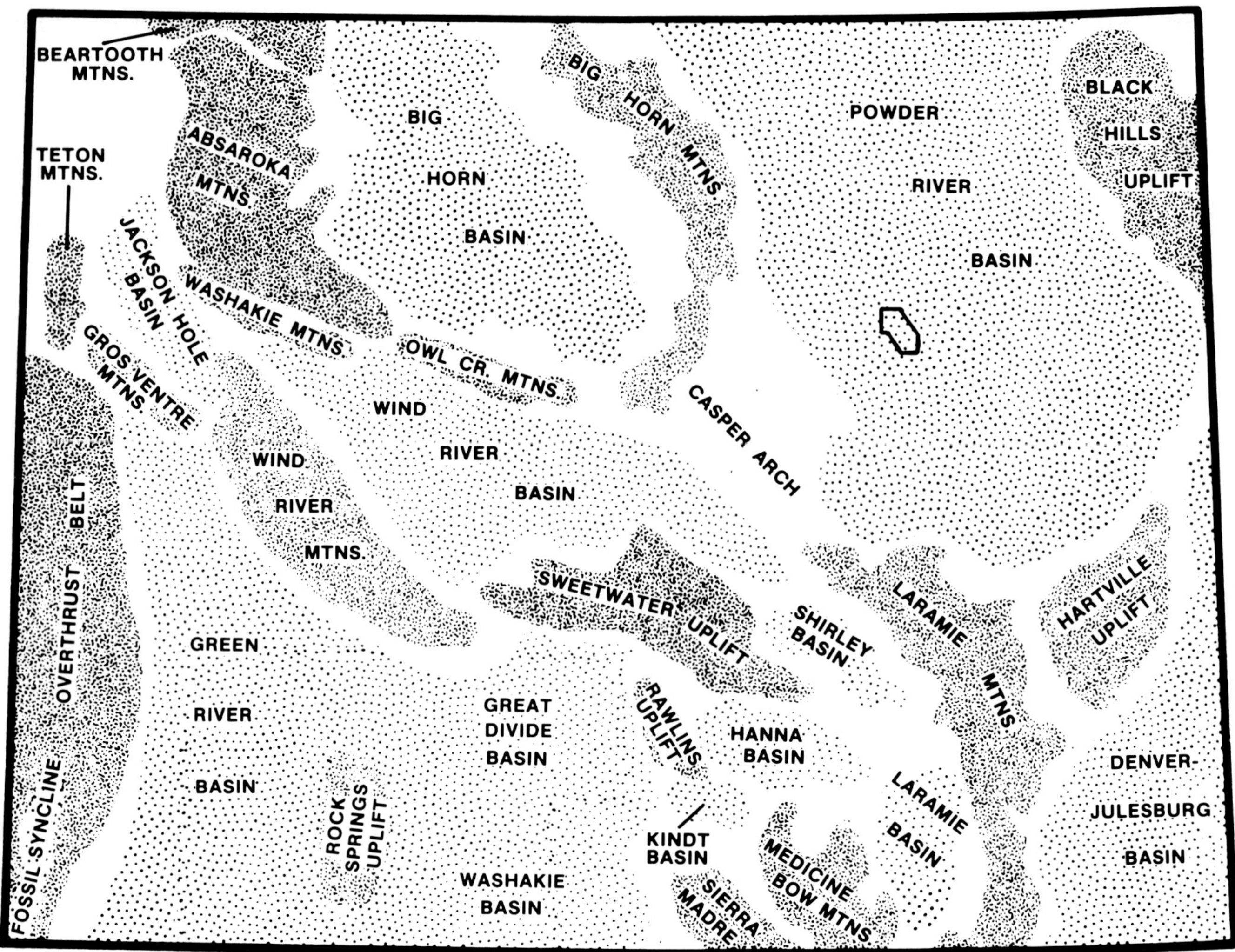

FIG. 1.—Map of Wyoming showing basins, uplifts and mountain ranges. Hartzog Draw field is in the area outlined in the Powder River Basin.

bubble point pressure of 1,550 psi. The field has no apparent water table and did not produce water during primary production, although log-calculated water saturations averaged 30–40%. The oil produced is 36° API gravity with initial solution gas-oil ratio of 292 standard cubic feet per stock tank barrel. The primary reservoir drive mechanism is solution gas, and only about 21% or 75 million barrels of the approximately 350 million barrels of oil-in-place was initially expected to be recovered during primary production. Production through 1981 was 32 million stock tank barrels. Secondary recovery was begun in 1982, as described by Hunt and Hearn (1982) and Hearn and others (1984).

FIELD DEVELOPMENT HISTORY

Hartzog Draw field was discovered in August 1975 with the completion of Southland Royalty's No. 1 Bud Christensen (Sec. 34 T46N R76W, Campbell County, Wyoming). Nearest production to the wildcat at that time was in Heldt Draw field, 2-1/2 mi (4 km) to the west (Fig. 7). Heldt Draw is a small field (approximately 4 MMBO ultimate recovery), which, prior to discovery of Hartzog Draw, was among the best known of the stratigraphically trapped Shannon oil fields. The discovery well for Hartzog Draw field, the No. 1 Bud Christensen (6341), was a deep test (12,000 ft, 3,657 m) drilled to the Lower Cretaceous Muddy Formation. At that time the potential for significant reserves was considered to be much greater for the Muddy Sandstone than for the Upper Cretaceous Shannon Sandstone. After attempts to complete the well in the Muddy failed, the well was completed from the Shannon Sandstone flowing 425 BOPD and 156 MCFGPD from about 14 ft (4.2 m) of pay at a depth of 9,200 ft (2,788 m). A confirmation well, the Cities Service Federal AB-1A (6334), drilled shortly thereafter, was completed flowing 440 BOPD and 133 MCFGPD from 5 ft (1.5 m) of pay (Tillman and Martinsen, 1985). What is now designated Hartzog Draw field was originally known as both East Heldt Draw and South Pumpkin Buttes field.

During the first year of development (through September 1976), 16 wells were drilled (Fig. 8), and initial projections of ultimate field size and oil recovery were modeled after Heldt Draw. Although net pay thicknesses encountered in

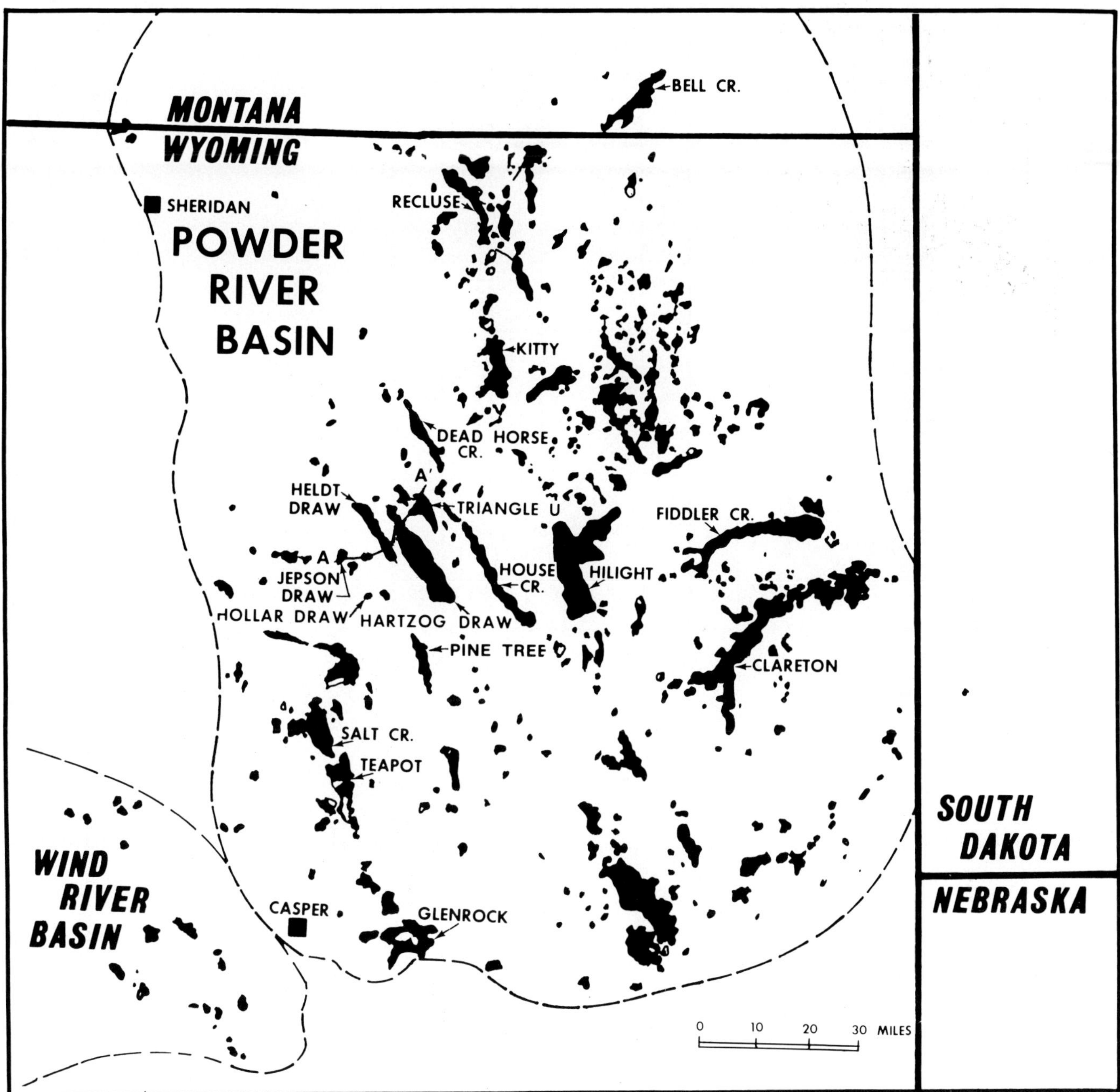

FIG. 2.—Powder River Basin Upper Cretaceous oil fields. Production from the Shannon Sandstone is from generally northwest-southeast-trending fields, including Hartzog Draw, Heldt Draw, Jepson Draw, Hollar Draw and others. Sussex Sandstone production is also from northwest-southeast-trending fields, including House Creek field.

Hartzog Draw had already exceeded those in Heldt Draw field, the field drew little notice. In December 1976 two significant wildcats (NE Sec. 14 T44N R75W, and NE Sec. 36 T44N R75W) were drilled, which, when incorporated into a shelf-ridge model, suggested the probable extension of the field 11 mi (17.6 km) to the south-southeast. This projection indicated the field would be twice as long as the nearby Heldt Draw field. These two new wells both exceeded the maximum thickness and initial potential of the wells to the north. By October 1977 (Fig. 9), Hartzog Draw extended 22 mi (37 km) in length and from 1 to 3-1/2 mi (1.7 to 5.8 km) in width. At that time, 140 successful wells had been completed on 160-acre spacing, with no dry development wells (Martinsen and Tillman, 1978, 1979). Most of the wells drilled during the time between January to October 1977 potentialed flowing for more than 1,000 BOPD, and several flowed in excess of 3,000 BOPD. The highest recorded initial potential was 3,408 BOPD. Nearly all wells were hydraulically fractured and required artificial lift shortly after being completed. Ultimately, 177 producing wells were

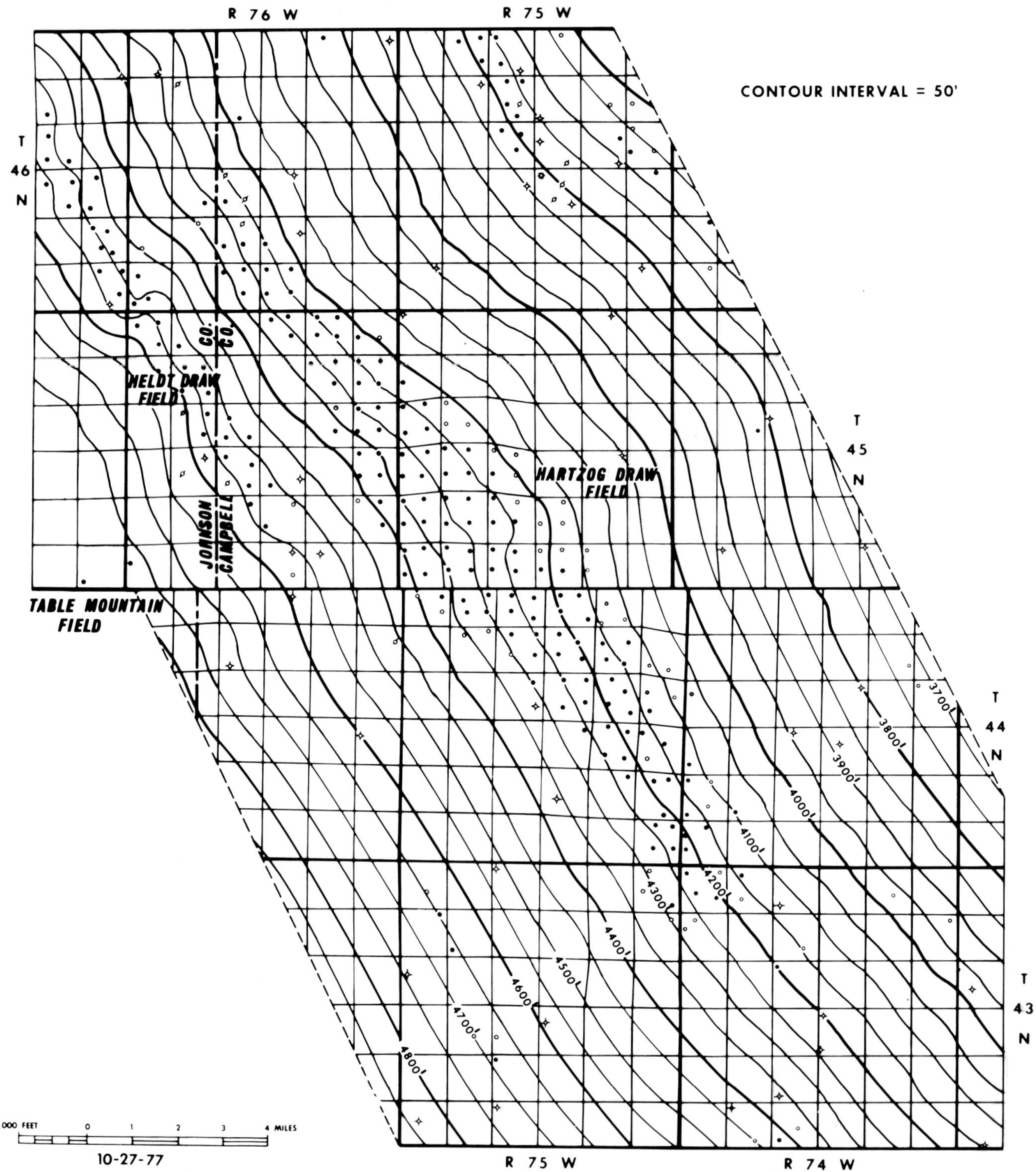

FIG. 3.—Shannon structure map. Dip in this part of the Powder River Basin is generally monoclinal toward the southwest.

completed, and approximately 32 million barrels of oil were recovered during primary field development.

Hartzog Draw field was unitized for secondary recovery in 1980 (Hunt and Hearn, 1982). Waterflood development began in 1981 and included an infill drilling pilot project in the center of the field (Hearn and others, 1984). Excellent waterflood response supported further infill drilling; by the end of 1985 approximately 115 infill wells will have been drilled covering the major portion of the field. The waterflood pattern is a 160-acre five-spot. Predicted pri-

WESTERN POWDER RIVER BASIN

MILLION YEARS BEFORE PRESENT	STAGES	LITHOLOGIC UNITS		
66	MAASTRICHTIAN	LANCE FM		
		FOX HILLS		
72.5	CAMPANIAN	LEWIS		
		MESA VERDE FM	TEAPOT SANDSTONE	
			UNNAMED SHALE MEMBER	
			PARKMAN SANDSTONE	
		CODY SHALE	STEELE SHALE	SUSSEX SANDSTONE
				SHANNON SANDSTONE
				FISHTOOTH MEMBER
84	SANTONIAN		NIOBRARA SHALE	

FIG. 4.—Upper Cretaceous stratigraphic section, western Powder River Basin. The Shannon Sandstone is a part of the Campanian age Cody Shale. Ages are from Ryer and McPhillips (1983).

mary and secondary oil recovery is 128 million stock tank barrels, or 39% of the original oil-in-place. More than 18 operators and 80 working interest owners have been involved in the development of Hartzog Draw field. No working interest owner holds more than a 16% interest in the Unit. Cities Service Oil and Gas Corporation is the Unit operator.

TABLE 1.—SHANNON SANDSTONE CUMULATIVE PRODUCTION, POWDER RIVER BASIN, WYOMING, THROUGH DECEMBER 1981. (J. J. MORRIS, PERS. COMMUN.).

Field	Discovery Date	BO	MCFG	BW
Shannon	1889	572,254	713,151	784,428
Big Muddy	1916	545,449	50,297	0
Teapot East	1929	3,198,145	5,972	678,373
Cole Creek	1940	9,099,144	224,090	1,413,572
Sussex	1948	4,749,130	1,891,210	1,408,906
Meadow Creek	1949	44,299	1,048	8,827
Meadow Creek North	1949	248,833	9,546,488	6,863
Meadow Creek East	1951	123,083	2,032	268
Sussex West	1951	16,451,602	8,333,593	13,311,776
Ash Creek	1952	7,002,758	162,887	1,809,891
Ash Creek South	1954	7,436,058	149,498	3,872,076
Heldt Draw	1973	3,673,053	2,604,809	148,543
Holler Draw	1974	1,273,912	565,787	5,789
Jepson Draw	1974	911,464	55,713	283
Indian Creek	1974	1,540,719	2,396,426	7,175
Hartzog Draw	1975	32,694,789	14,034,747	85,252
Culp Draw	1975	564,553	510,657	4,793
Flying E	1975	46,440	81,819	133
Pine Tree	1976	82,217	108,081	0
Pumpkin Buttes	1976	389,073	253,326	1,602
Sievers	1976	31,577	12,895	0
Table Mountain	1977	927,870	674,820	663,526
Nipple	1977	77,578	0	0
Collins	1978	632,966	353,210	556
Hatch	1980	8,103	7,928	33
Phillips Creek	1981	45,654	54,256	330
TOTALS:		92,370,723	42,794,740	24,212,995

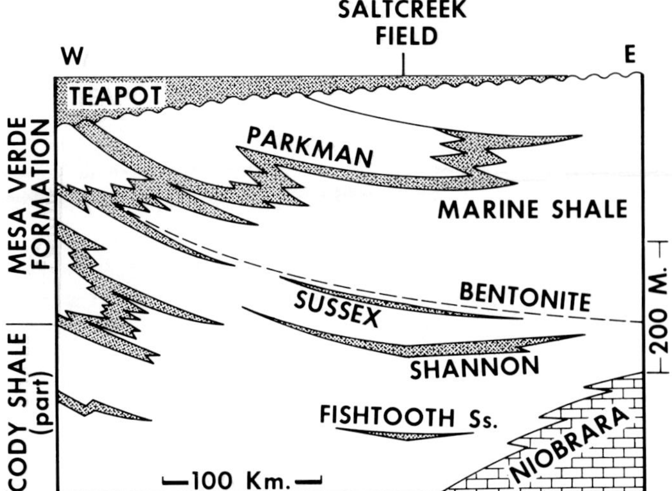

FIG. 5.—Stratigraphic reconstruction of the Upper Cretaceous in central Wyoming. Dashed line is a bentonite marker which is time synchronous. The Shannon Sandstone pinches out both shoreward and seaward and is approximately 100 mi (160 km) offshore from time equivalent shoreline sandstones. (After Gill and Cobban, 1966).

HARTZOG DRAW FACIES

Three lithologies (sandstone, siltstone, and shale) and nine facies are associated with the Shannon shelf-ridge complex at Hartzog Draw field. The reservoir sandstones' mean grain sizes range from upper fine to lower medium (175 to 250 μm) and generally occur in a coarsening-upward sequence. Earlier workers subdivided the sandstone into two facies (Table 2A), thin bedded and crossbedded (Spearing, 1975). In this study, however, subdivision into nine statistically definable facies aids greatly in describing and recognizing the variation in distribution and intensities of depositional processes which were active during the growth of the shelf-ridge complex (Fig. 10). Ranganathan and Tye (1986) have used a facies subdivision which indicates four sandstone facies types (Table 2A) for cores in Hartzog Draw. Only two of these facies are directly comparable with the terminology used herein. Others of Ranganathan's and Tye's facies occur in more than one of our facies.

Typically, the base of the sequence at Hartzog Draw is a bioturbated shaly siltstone (*Bioturbated Shelf Siltstone Facies*), which is overlain by a rippled and burrowed interbedded sandstone and shale unit (*Inter-Ridge Facies, Shaly*) (Tillman and Martinsen, 1979). These two basal units are complementary to one another in that where burrowing exceeds 50% by volume, the unit is designated as a burrowed or bioturbated (75% or more burrowed) siltstone; where burrowing is less than 50% and more than 50% of the unit is rippled interbedded sandstone and shale, it is designated as *Inter-Ridge Facies (Shaly)*. This relationship is clearly shown in the Cities Service AB-1A core (Tillman and Martinsen, 1985, Fig. 13).

The units overlying the *Inter-Ridge Facies (Shaly)* are commonly crossbedded and vary considerably from location to location. Where the change in energy level is gradual across the boundary between the *Inter-Ridge* and crossbedded facies, as is observed in the western (down-

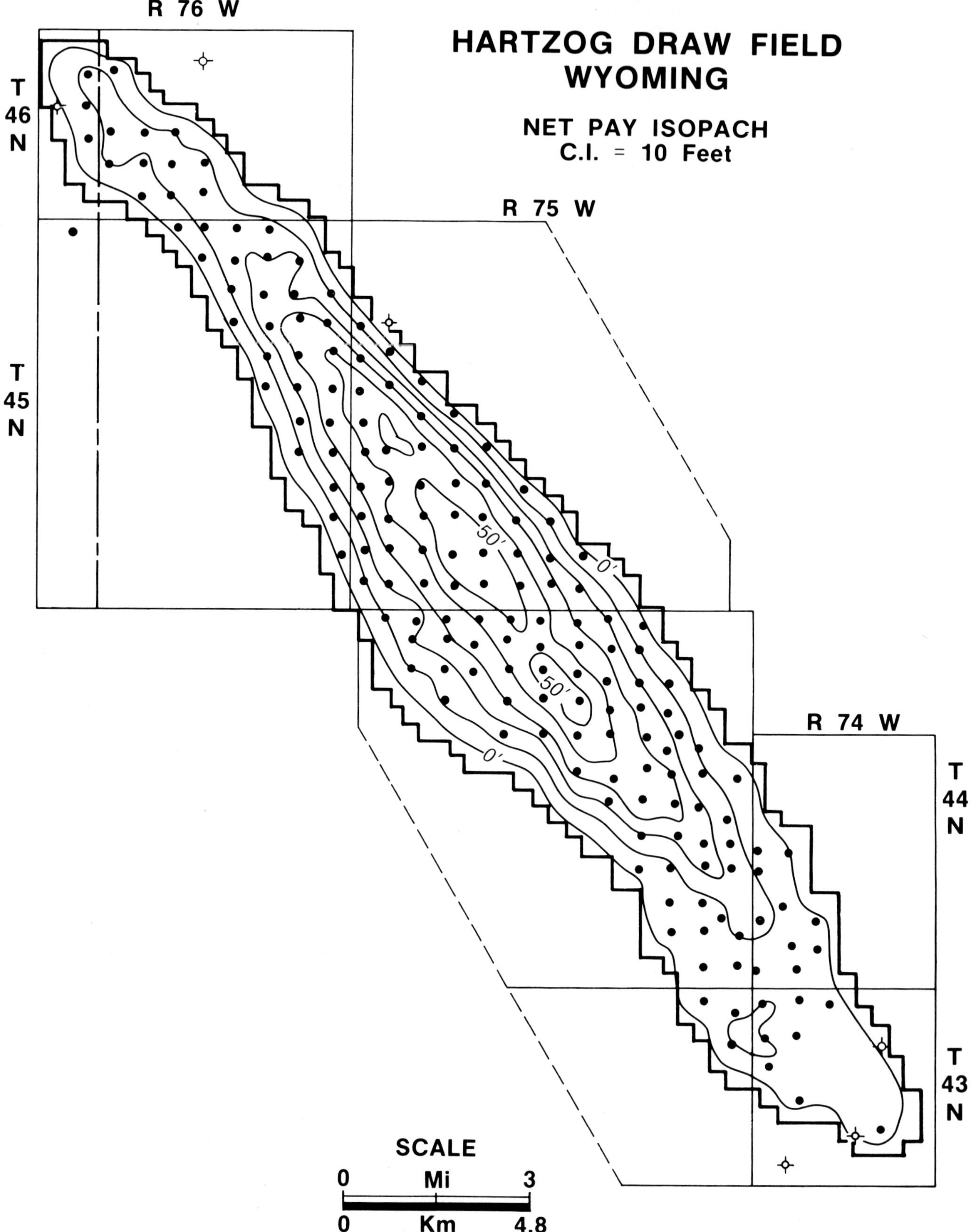

FIG. 6.—Net pay isopach map of Shannon Sandstone, Hartzog Draw field. Contour interval is 10 ft (3 m). Minimum porosity included in net pay zone was 11% using a 2.71 grain density. (Modified from Tillman and Martinsen, 1979, and Hunt and Hearn, 1982).

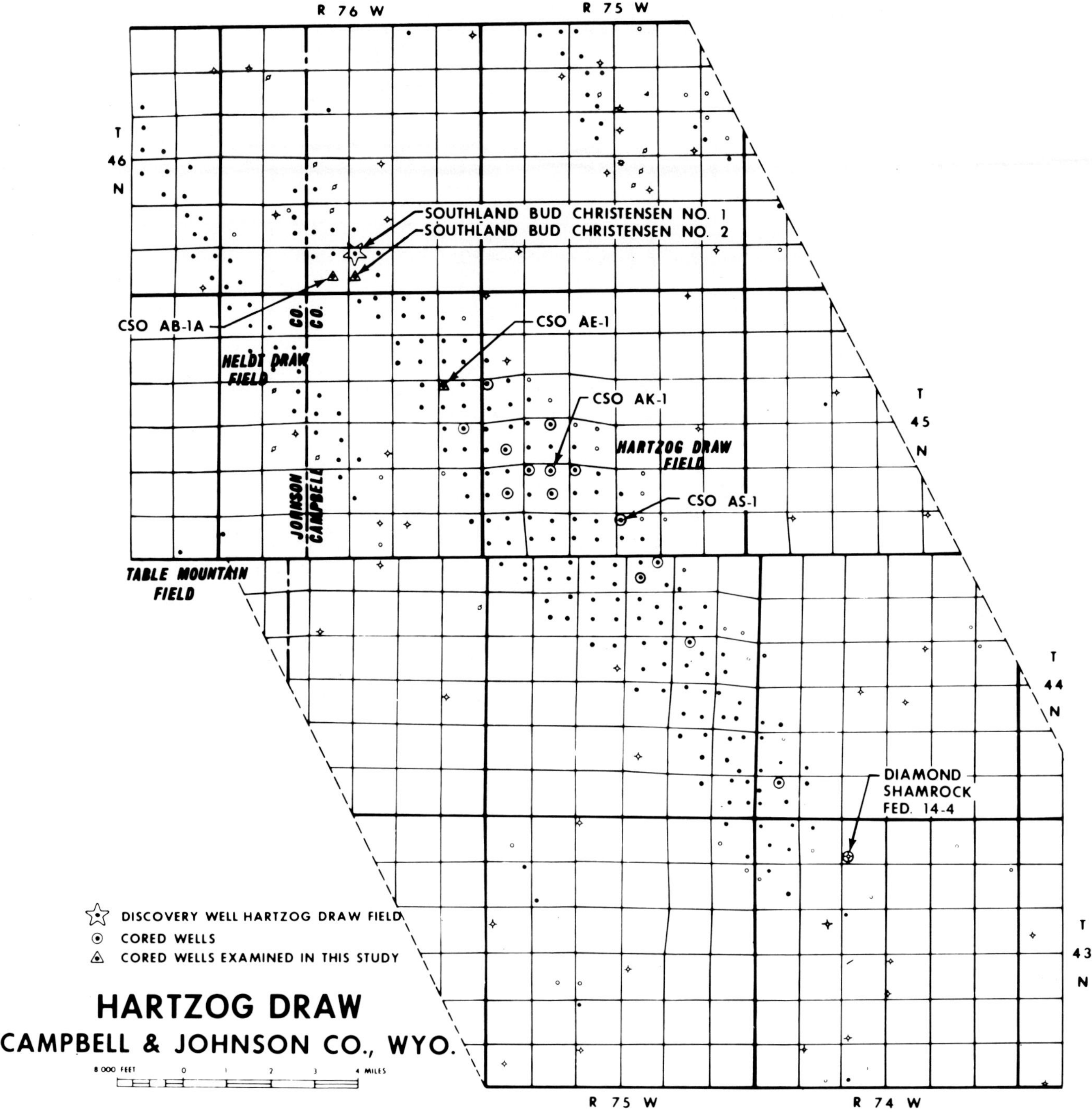

FIG. 7.—Well locations, Hartzog Draw field and Heldt Draw field, Wyoming. Hartzog Draw field discovery well is marked by a star. Wells cored during primary producing phase are indicated by circles. Locations of the cored wells discussed in the text are shown.

current) part of the shelf-ridge complex, the facies overlying the *Inter-Ridge Facies (Shaly)* is commonly a sandstone composed of interbedded trough (planar-tangential) and rippled (current) sandstone beds. Interbedding of these forms commonly occurs at least once in every succession of four beds. This facies, which exhibits interbedding of troughs and ripples, is designated as a *Low-Energy Ridge-Margin Facies*. Where a more abrupt change in the intensity of the depositing currents occurred above the *Inter-Ridge Facies (Shaly)*, there may be an abrupt change from rippled to an almost totally trough crossbedded unit containing abundant glauconite, and shale and siderite rip-up clasts. This facies is designated as a *High-Energy Ridge-Margin Facies*. In the central and eastern portions of the shelf-ridge complex, a relatively clean, highly trough (planar-tangential) crossbedded unit occurs above the *Inter-Ridge Facies (Shaly)* and is designated as the *Central Ridge Facies*. This highly crossbedded facies is characterized, in addition to

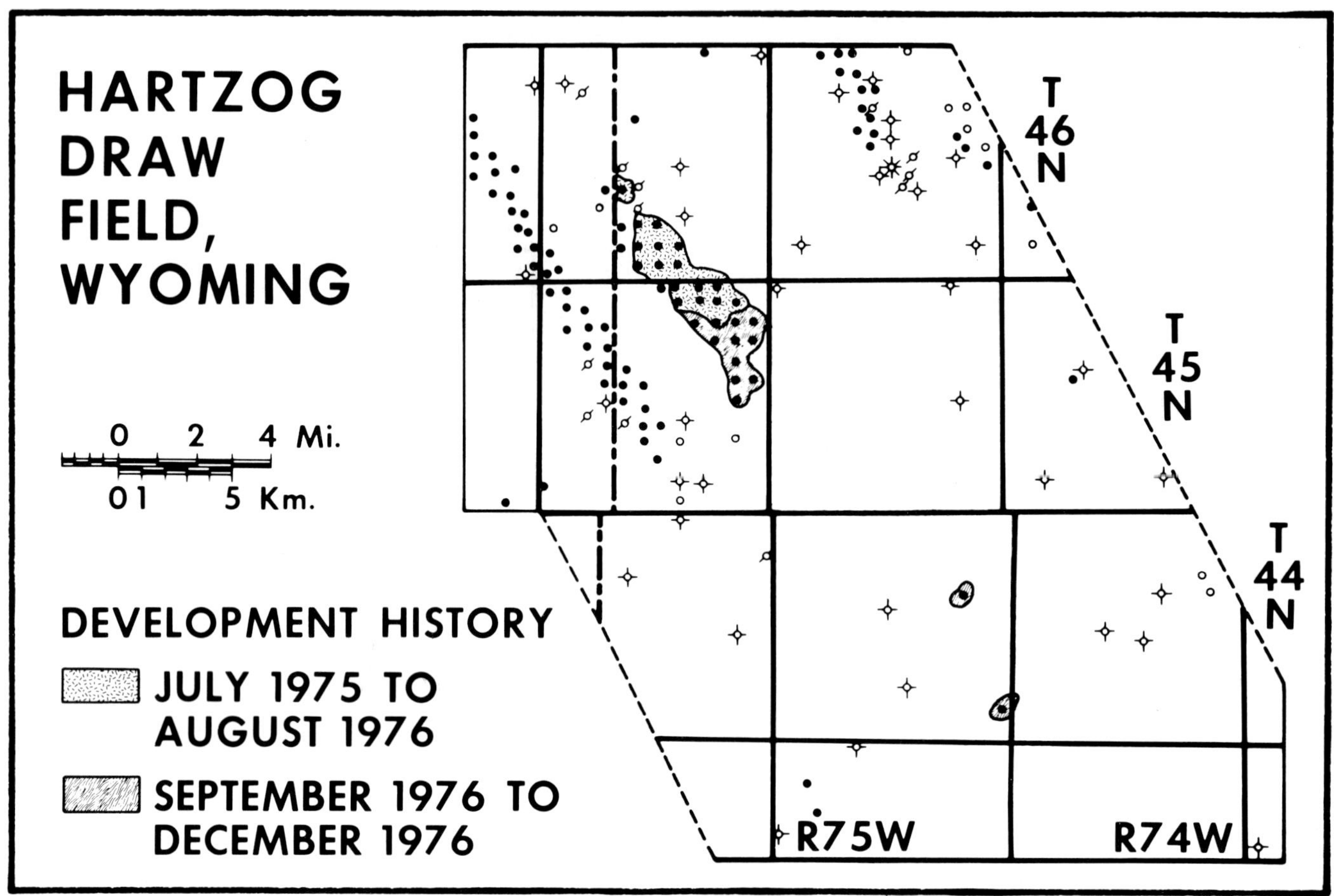

FIG. 8.—Hartzog Draw field development history from discovery in July 1975 through December 1976. Note the two wells in T44N R75W. These wells allowed some operators to predict the length of the field at an early stage of development.

its high degree of crossbedding, by the sharp, predominantly horizontal bed boundaries occurring every 0.5 to 1.5 ft (0.2–0.5 m) vertically. Similar horizontal to subhorizontal bed boundaries are also observed in the other facies comprising the shelf-ridge complex. This horizontality in bedding surfaces is in strong contrast to the crossbedded cut-and-fill current deposits normally associated with fluvial and distributary channels.

Overlying and lateral to the *Central Ridge Facies* may be any one of the previously described facies, or the upper contact may be a burrowed to bioturbated silty shale (*Burrowed Shelf Shale Facies*). The shale, overlying the ridge complex, generally shows a decrease upward, over several feet, in the amount of silt and burrowing.

Lateral facies changes within the shelf-ridge complex are controlled in large part by the variations in the intensity of depositing currents. Measurement of transport directions in oriented cores in the central and eastern part of the field indicates that currents impacted on the shelf-ridge complex obliquely and flowed toward the south (Fig. 11). The eastern margin of the shelf-ridge complex commonly contains a relatively high percentage of *High-Energy Ridge-Margin Facies*, which grades laterally (westward) to higher and higher percentages of *Central Ridge Facies*. West of the central portion of the shelf-ridge, a lateral change from *Central Ridge* to *Low-Energy Ridge-Margin Facies* is most common (Tye and others, 1986). A similar, but less predictable sequence of lateral changes occurs from north to south within Hartzog Draw field.

The crossbedded sandstones in Hartzog Draw field may be divided into three facies, each of which is characteristic of one (or two) areas of the field. Variations in the following components serve as a basis of separating the crossbedded facies.

(1) Sand/shale ratio
(2) Crossbedding, percentage
(3) Ripples, percentage
(4) Shale (and siderite) rip-up clasts, percentage
(5) Glauconite, percentage
(6) Reworked material, percentage
(7) Mean grain size
(8) Burrowing, percentage

The mean percentages of the most significant variables used in separating the high-energy crossbedded facies are given in Table 2B. Three cored wells, which illustrate all the facies observed in Hartzog Draw field, are described in detail in this study. Photographs, detailed descriptions and

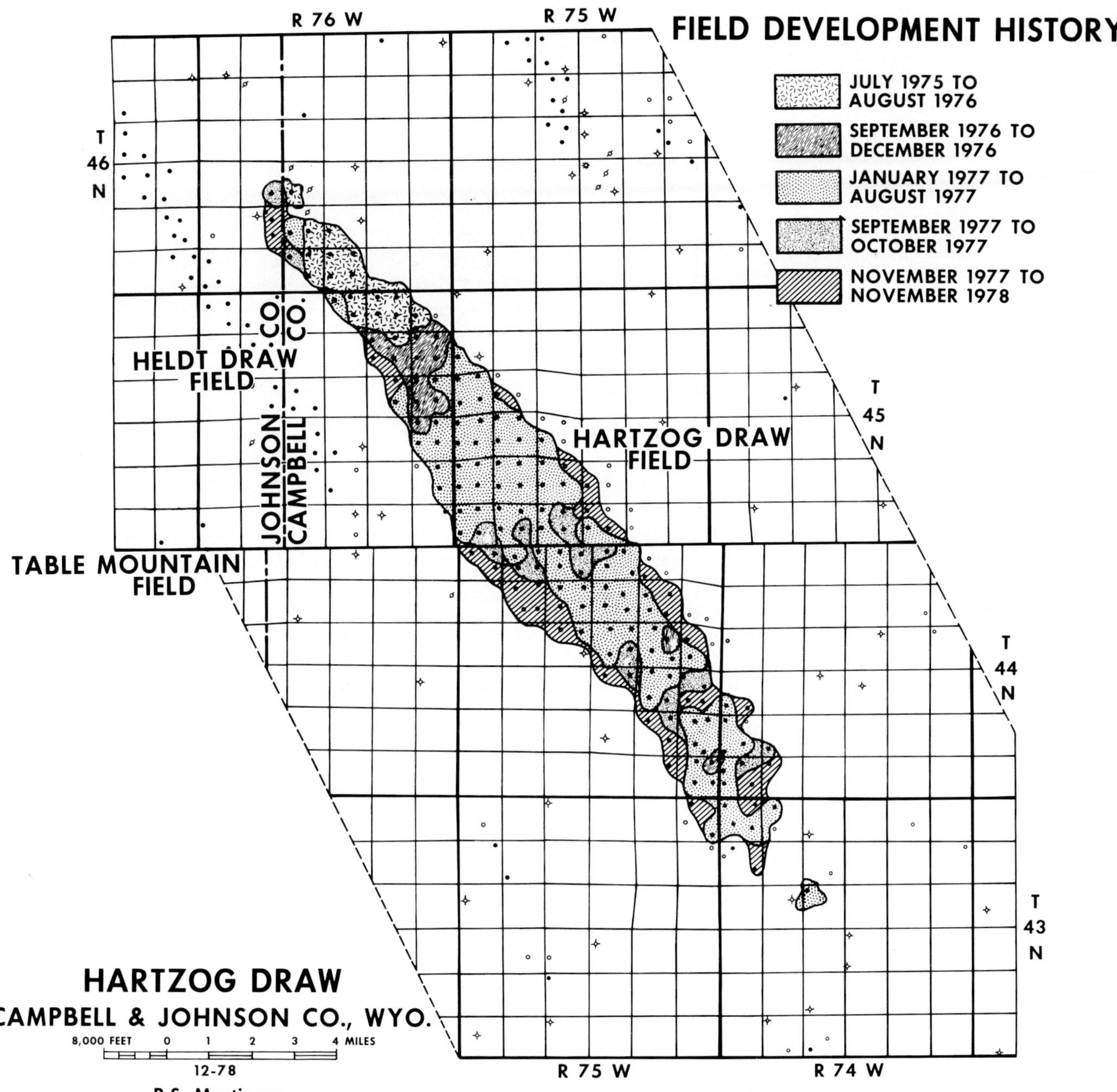

FIG. 9.—Sequential Hartzog Draw field development history. July 1975 to the end of primary production drilling in November 1978. Compare with Figure 8. Additional infield wells (not shown) were drilled during secondary production.

annotated logs for each core are included in Appendix A of this paper.

The *Central Ridge Facies* is best illustrated in the Cities Service Federal AE-1 (Figs. A1–A4). Similar detailed descriptions of this facies in two other wells are given by Tillman and Martinsen (1985). Thin, *High-Energy Ridge-Margin Facies* are well illustrated in the Southland Royalty Bud Christensen No. 2 core (Figs. A5–A8). The Cities Service Federal AS-1 has as its major reservoir facies a typical thick *High-Energy Ridge-Margin Facies* (Figs. A9–A12).

Central Ridge Facies

Description.—

The *Central Ridge Facies* is composed primarily of trough to planar-tangential high-angle, lower medium (200 μm) mean grain size sandstones (Fig. 12). Bed boundaries are nearly horizontal and beds range from 0.5 to 1.5 ft (15–45 cm) in thickness. Glauconite, shale clasts and siderite clasts occur in small amounts in this as well as other coarse-grained facies. Calcite cement varies in a somewhat random man-

TABLE 2A.—SHANNON FACIES NOMENCLATURE 1976 TO PRESENT.

FACIES NOMENCLATURE			
Present Terminology	Tillman and Martinsen, 1984	Ranganathan and Tye, 1986	Spearing, 1976
Central Ridge Facies	Central Bar Facies	Facies A (and C)	Cross bedded sandstone facies
Central Ridge (Planar-Laminated) Facies	Central Bar (Planar Laminated) Facies	Facies C	
High-Energy Ridge-Margin Facies	Bar Margin Facies (Type 1)	Facies B	
Low-Energy Ridge-Margin Facies	Bar Margin Facies (Type 2)	Facies C and A	
Inter-Ridge Facies	Interbar Facies	Facies D	Ripple bedded sandstone facies

NOTE: Facies names have been changed in this paper to emphasize the ridge (rather than bar) terminology for offshore sand deposits.

ner throughout the facies and locally forms a very white sandstone. Small amounts of quartz and siderite cement and clay also occur (Ranganathan and Tye, 1986).

The *Central Ridge Facies* is the best reservoir facies and, when compared to the other crossbedded facies, has the highest sand/shale ratio (9.2:1), the highest percentage of crossbeds (66%), the lowest percentage of glauconite (8%), the least ripples (10%), and the lowest percentage of rip-up clasts (7%), Table 2B. Rip-up clasts in this facies are commonly shale rip-up clasts, although rounded siderite clasts are also present. The percentage of ripples-on-troughs (10%) is equal to the *Low-Energy Ridge-Margin Facies* (10%) and is slightly lower than for the *High-Energy Ridge-Margin Facies* (15%). The grain size is coarser (215 μm) than the *Low-Energy Ridge-Margin Facies* and slightly finer than the mean size (225 μm) of the *High-Energy Ridge-Margin Facies*. The percentage of burrowing (4%) is usually less than any other crossbedded facies; diversity is also commonly less than in the other facies. Horizontal laminations are almost totally absent in all three crossbedded facies. A detailed summary of all the characteristics of those units designated as *Central Ridge Facies* is given in Table A1.

This unit attains a maximum thickness of about 40 ft (12 m) but more commonly it is about 20 ft (6 m) thick. This thickness exceeds that of any other facies, particularly in the central portions of the shelf-ridge complex. Commonly, beds consisting predominantly of trough crossbeds are stacked four to 10 high (or higher). This contrasts strongly with the interbedding of crossbeds and ripples in *Low-Energy Ridge-Margin Facies* and the common occurrence of beds containing abundant rip-up-clast beds in the *High-Energy Ridge-Margin Facies*. Individual bed thicknesses (0.5 to 2 ft, 0.15 to 0.6 m) are only slightly greater than for other facies.

In describing cores for this study, the minimum described unit thickness is 1 ft (0.3 m). Occasionally, thin intervals (<1 ft, 0.3 m) more characteristic of other facies occur within the *Central Ridge Facies;* however, these are not designated separately in order to prevent excessive fragmentation of units. This minimum unit thickness of 1 ft (0.3 m) was chosen, in part, because units less than 1 ft (0.3 m) thick are seldom observable on subsurface logs.

Fifty-eight percent of the *Central Ridge Facies* contacts with overlying and underlying units are with one or the other of the *Ridge-Margin Facies*. The facies overlying the *Central Ridge Facies* are *Ridge-Margin Facies* in 66% of the occurrences. *Shelf Silty Shale* also overlies the *Central Ridge Facies* in 17% of the occurrences. *High-Energy Ridge-Margin Facies* occurs in 33% of the contacts at the base of the *Central Ridge Facies*. *Central Ridge Facies, Low-Energy Ridge-Margin Facies, Inter-Ridge Facies (Shaly)* and *Burrowed Shelf Siltstones* are each observed in 16–17% of the facies underlying the *Central Ridge Facies*. Documentation of overlying and underlying facies occurrences for *Central Ridge Facies,* as well as other facies, are summarized in Tables 3A and 3B.

The contacts above and below the *Central Ridge Facies* are relatively sharp; an average of 0.4 to 0.5 ft (12 to 15 cm) of transition is observed. The range of "contact thickness" is from 0.01 to 1.0 ft (0.3 to 30 cm; Table A1).

Interpretation.—

As will be discussed in more detail later, the *Central Ridge Facies* occurs most commonly in the thickest (central) part of the field (Fig. 6). Processes responsible for forming the high-angle crossbedded megaripples which form most of this facies were periodic, and commonly the upper contact of individual beds was eroded to form a nearly horizontal bedding surface. These subhorizontal bedding surfaces readily distinguish shelf sands of this type from most high-angle channel-fill deposits. Most channel-fill deposits have significant amounts of cut-and-fill rather than the subhorizontal bedding surfaces common in shelf sandstones. The relatively uniform thickness of crossbed sets and near absence of low-energy sedimentary structures (ripples) suggest that deposition of this facies took place under uniformly high-energy current flow conditions.

The fact that very thick sequences of this facies (40 ft, 12 m) occur indicates that, once begun, the processes which formed this facies persisted (periodically?) for a relatively long period of time. Outcrop studies as well as subsurface studies indicate that this facies most commonly grades laterally into one or the other of the *Ridge-Margin Facies*.

Central Ridge (Planar Laminated) Facies

Description.—

The *Central Ridge (Planar Laminated) Facies* is characterized by a relatively high percentage of planar to low-angle laminated beds. In the one cored unit assigned to this facies (Unit 7, CSO AE-1, Fig. A3), the percentage of subhorizontal planar lamination is not so critical as is the fact

FIG. 10.—Shannon facies summary. Nine facies are summarized in terms of lithology, sedimentary structures, burrowing, reservoir potential and occurrence frequency in Hartzog Draw field.

SHANNON FACIES SUMMARY

	CENTRAL-RIDGE FACIES	CENTRAL-RIDGE (PLANAR LAMINATED) FACIES	HIGH-ENERGY RIDGE-MARGIN FACIES	LOW-ENERGY RIDGE-MARGIN FACIES
LITHOLOGY	Fine to medium grained quartzose sandstone, moderately glauconitic; rare siderite clasts and shale rip-up clasts.	Fine to medium grained quartzose sandstone.	Predominately medium grained sandstone, abundant shale and limonite rip-up clasts and lenses, commonly very glauconitic.	Fine-grained sandstone with only rare shale interbeds. Fewer clasts and lenses and less glauconitic than High-Energy Ridge-Margin Facies.
SEDIMENTARY STRUCTURES	Predominantly moderate angle trough and planar-tangential cross bedding. Trough sets commonly horizontally truncated.	Mostly sub-horizontal plane-parallel laminated sandstone, 0.5′-thick laminasets. Minor shale and sandstone ripples.	Mostly moderate angle troughs, some current ripples, shale clasts rarely show preferred orientations.	Sequences of several beds of troughs interbedded with sequences of several rippled beds.
BURROWING	Sparse	Sparse	Sparse	Sparse
RESERVOIR POTENTIAL	Excellent	Limited?	Good	Moderate to Good
SUBSURFACE OCCURENCES HARTZOG DRAW FIELD	Common	Very uncommon	Common	Common

	INTER-RIDGE FACIES (SHALEY)	INTER-RIDGE SANDSTONE FACIES	BIOTURBATED SHELF-SANDSTONE FACIES	BIOTURBATED SHELF-SILTSTONE FACIES	SHELF SILTY-SHALE FACIES
LITHOLOGY	Thinly interbedded fine to very fine-grained silty sandstone and silty shale, slightly glauconitic.	Fine-grained sandstone. Virtual absence of silty shale. Slightly glauconitic	Silty, fine-grained sandstone. Up to 15% shale, primarily associated with burrows. Slightly glauconitic.	Shaly, slightly sandy dark gray siltstone, traces to moderate amounts of glauconite.	Silty dark gray shale; rare thin (1/8″ thick) silty sandstone lenses.
SEDIMENTARY STRUCTURES	Predominantly horizontal ripple-form bedding surfaces marked by interbedded shales. Trace of wave ripples; current ripples predominate	Predominantly horizontal ripple-form bedding surfaces. Bedding commonly indistinct. Trace of wave ripples; current ripples predominate.	Few physical structures preserved. Mottled appearance. Some ripple-form horizontal beds up to 8 inches thick. Trace of distinct ripples and small troughs.	Few physical structures preserved. Scattered thin rippled sand and horizontal laminasets. Bedding commonly destroyed.	Common sub-horizontal laminae. Bedding surfaces indistinct, horizontal. Rare current ripples.
BURROWING	Moderate to locally high	Low to Moderate	Mottled to distinctly burrowed. More than 75% burrowed.	More than 75% Burrowed	Low to moderate
RESERVOIR POTENTIAL	Limited	Limited	Limited	None	None
SUBSURFACE OCCURENCES HARTZOG DRAW FIELD	Common	Uncommon	Common	Moderately Common	Common

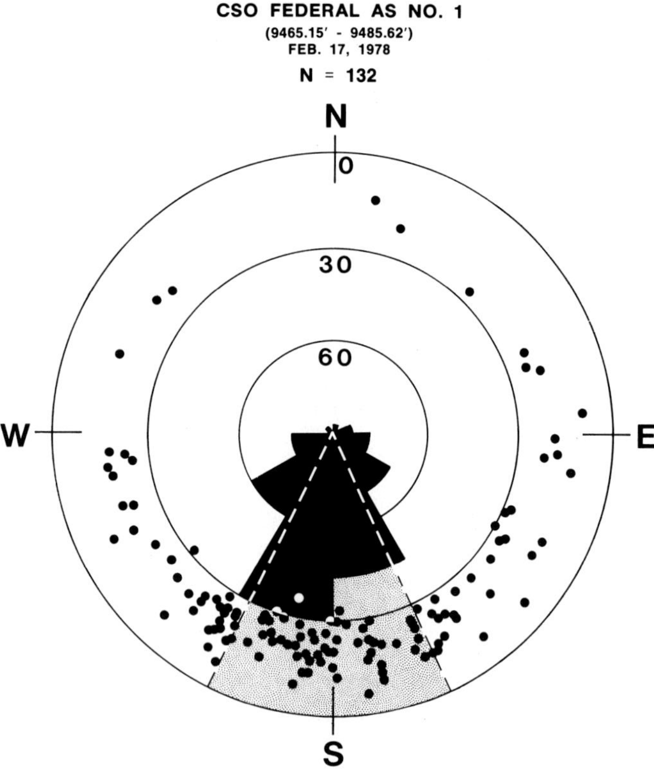

FIG. 11.—Current transport directions in conventionally oriented core from Cities Service Federal AS-1, Shannon Sandstone 9,465.1–9,485.6 ft (2885–2891 m). Total number of data points is 132. Orientation of core by Sperry-Sun, 1977. Mean transport direction is 181° (south). Directional pattern cells shown in black are 30° wide; maximum data points per cell, 40. Two-thirds (one standard deviation) of the flow directions measured on trough crossbedding are within an arc of 52° and are shown in the shaded area between dashed lines. Angle of dip of crossbed readings increases from 0° at the edge of the circle to 90° at the center. The dip angle on a large number of the trough crossbeds is about 25°; transport directions are recorded at the point on the core where the lowest point on the base of a trough, or lamina set within a trough, occurs.

that the subhorizontal laminations are recurrent through nearly a foot (0.3 m) of section (Fig. A1). Near the south end of the Shannon outcrops at Salt Creek anticline, this facies is well exposed (Tillman and Martinsen, 1986). In cores and outcrop, rippling is also associated with this facies in the form of thin interbedded sandstones and shales. Interbedded trough crossbedding is rare.

The only other facies that contains significant horizontal laminations is the *Inter-Ridge Facies (Shaly)*. In this facies 2- to 3-in.-thick (5–7.6 cm) subhorizontal to horizontal cosets are also locally present (Fig. A11), but they never approach a foot (0.3 m) in thickness. These thin planar laminated sandstones in the *Inter-Ridge Facies (Shaly)* are also certainly the result of storm deposits similar to those reported in the North Sea (Reineck and Singh, 1971; Tillman and Reineck, 1975).

The *Central Ridge (Planar Laminated) Facies* averages 4% glauconite, but where concentrations of as much as 30% glauconite occur on some laminations, portions of the sandstone have a conspicuous green color. Shale is present in amounts averaging 11%. Most of the shale is interbedded with sandstone; a trace of shale clasts is observed (Table A1).

Burrowing is about 20% by volume. Five burrow types indicate a moderate diversity. No burrows were identified by name; 10% of the burrows are white, sand-filled, 1/4- to 1/2-in. (0.6–1.3 cm) horizontal burrows; most of the other burrows are smaller in diameter and have a nearly horizontal orientation. Two percent of the unit is bioturbated (75% or more burrowed). Fifty-eight percent of the total section cored has one or more burrows.

Above the planar laminated facies is a *Burrowed Shelf Siltstone Facies,* and underlying it is a *High-Energy Ridge-Margin Facies*. The lower contact of the facies is very sharp. The upper portion of the facies is rippled and sparsely burrowed and grades into a *Burrowed Shelf Siltstone Facies*.

Interpretation.—

The processes associated with the formation of the subhorizontal planar laminations in Hartzog Draw field may have involved increased flow (high regime flow) associated with strong storm currents. They may also indicate the results of topographic upbuilding of the bar so as to put the top of the bar in a position where wave-generated currents could affect the bottom. This latter possibility seems less likely, however. In the CSO AE-1, this facies occurs at the top of the shelf-ridge sequence. Whether this facies contributes to the reservoir pay interval is undetermined.

High-Energy Ridge-Margin Facies

Description.—

The *High-Energy Ridge-Margin Facies* is distinguished from the *Central Ridge Facies* by the presence of significant amounts (18% by volume) of black to dark gray rip-up shale clasts and oblate siderite rip-up clasts. No other facies averages more than 10% rip-up clasts; the *Central Ridge Facies* averages only 3% shale clasts (Table 2B). The facies averages 75% sandstone, 22% shale, and 3% siltstone (Table A2). The shale occurs predominantly as shale clasts (mean = 15%). Seven percent laminated shale is observed. Glauconite is more conspicuous in this facies than in any other facies and ranges from 15–20% of the sediment. On some laminae, glauconite comprises nearly 75% of the grains (Fig. 13B) and, in such instances, gives the sandstones a deep green color. Rounded to oblate buff colored 1/2- to 3/4-in.-diameter (1.3–1.9 cm) siderite pebbles are commonly present in amounts ranging from a trace to 8% (by volume; Fig. 13A and B). Lenses or beds of siderite observed in outcrops (Tillman and Martinsen, 1984) are also observed in cores, particularly in this facies. Erosion of these siderite lenses produces the oblate siderite rip-up clasts which occur in all the crossbedded facies but are slightly more common in this facies, both in cores and in outcrop.

Ripples (Fig. 13D) may comprise as much as 20% of the unit, but average 11%. Burrowing does not exceed 8% by volume in this facies, and, typically, the burrowing is concentrated near the top of the unit. A very low diversity of burrow types was observed.

The association of facies with other facies can be quan-

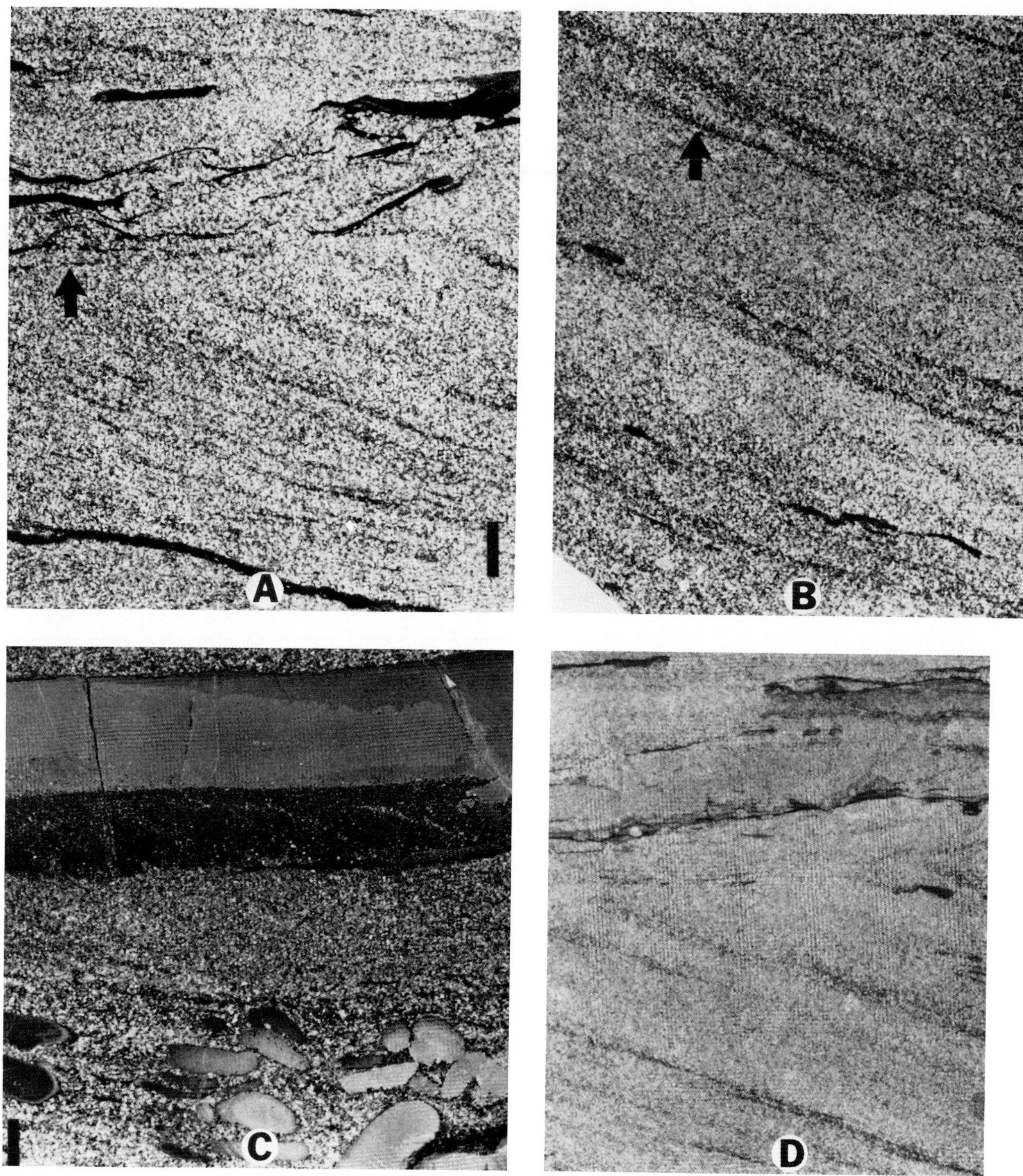

FIG. 12.—*CENTRAL RIDGE FACIES*, Shannon Sandstone. (A) Trough crossbedding. Top of trough truncated (arrow) and overlain by rippled thin trough set containing deformed shale clasts. A third trough set is at the top. *Central Ridge Facies*, CSO AB-1A, Unit 4, 9,255.0 ft (2820 m). (B) High-angle trough crossbedding. Dark streaks are glauconite. Note thinning of lamina set from left to right (above arrow); typical of trough lamination. *Central Ridge Facies*, CSO AE-1, Unit 3, 9,158.0 ft (2791 m). Scale = 1 cm. (C) Typical siderite lens (gray) overlying darker silty shale lens. Siderite lens contains some silt. Siderite pebbles (lower right) well rounded to oblate, the result of ripping up of bedded siderite followed by minor transport, rounding and redeposition. *Central Ridge Facies*, Southland Royalty Bud Christensen No. 2, Unit 2B, 9,179.9 ft (2798 m). (D) Truncated trough crossbedding in interval of 27.2 ft (8.3 m) of continuous *Central Ridge Facies* (Tillman and Martinsen, 1985). Note variation in thickness of lamina sets even over 4-in.-wide (10 cm) core; typical of trough lamination. Trace of white, sand filled, 1/8-in.-diameter (0.3 cm) burrows at lamina set contact in upper one-third of photo. Cities Service AK-1, Unit 4A, 9,363.6 ft (2854 m).

TABLE 2B.—COMPARISON OF AVERAGE PERCENTAGES OF FEATURES WHICH DIFFERENTIATE THE THREE HIGH-ENERGY CROSSBEDDED FACIES.

	Sandstone	Cross beds	Ripples	Shale Total	Shale Clasts	Glauconite	Reworked Total	Reworked Clasts
CENTRAL RIDGE FACIES	92	66	10	5	3	8	7	7
HIGH-ENERGY RIDGE-MARGIN FACIES	75	47	11	22	15	16	21	17
LOW-ENERGY RIDGE-MARGIN FACIES	64	25	44	30	6	15	14	14
INTER-RIDGE FACIES (SHALY)	58	3	61	31	tr	3	4	tr

Note: Inverse relationship between sandstone vs. shale and total crossbeds vs. ripples. Reworked clasts include both shale and siderite. For more detailed compilation, refer to Appendix Tables A1 and A2.

TABLE 3A.—PROBABILITIES OF OVERLYING FACIES OCCURRENCES.

FACIES (Below / Above)	Central Ridge Facies	Central Ridge (Planar Laminated)	High-Energy Ridge-Margin	Low-Energy Ridge-Margin	Inter-Ridge Facies (Shaly)	Bioturbated Shelf Siltstone	Shelf Silty Shale
CENTRAL RIDGE FACIES $N = 6$	17%	—	50%	16%	—	—	17%
CENTRAL RIDGE (PLANAR LAMINATED) $N = 1$	—	—	—	—	—	100%	—
HIGH-ENERGY RIDGE-MARGIN $N = 6$	33%	17%	—	—	—	16%	33%
LOW-ENERGY RIDGE-MARGIN $N = 4$	25%	—	50%	—	—	25%	—
INTER-RIDGE FACIES $N = 6$	17%	—	—	33%	—	50%	—
BIOTURBATED SHELF SILTSTONE $N = 7$	14%	—	15%	15%	28%	—	28%
SILTY SHALE $N = 3$	—	—	—	—	100%	—	—

TABLE 3B.—PROBABILITIES OF UNDERLYING FACIES OCCURRENCES.

FACIES (Above / Below)	Central Ridge Facies	Central Ridge (Planar Laminated)	High-Energy Ridge-Margin	Low-Energy Ridge-Margin	Inter-Ridge Facies (Shaly)	Bioturbated Shelf Siltstone	Shelf Silty Shale
CENTRAL RIDGE FACIES $N = 6$	17%	—	33%	16%	16%	17%	—
CENTRAL RIDGE (PLANAR LAMINATED) $N = 1$	—	—	100%	—	—	—	—
HIGH-ENERGY RIDGE-MARGIN $N = 6$	50%	—	—	33%	—	17%	—
LOW-ENERGY RIDGE-MARGIN $N = 4$	25%	—	—	—	50%	25%	—
INTER-RIDGE FACIES $N = 6$	—	—	—	—	—	33%	66%
BIOTURBATED SHELF SILTSTONE $N = 7$	—	14%	14%	14%	43%	—	14%
SILTY SHALE $N = 3$	33%	—	—	—	—	66%	—

FIG. 13.—*RIDGE MARGIN FACIES.* (A) Rounded siderite pebbles (gray) and shale rip-up clasts (black) in moderate angle-bedded trough crossbeds. The angular shale clasts contrast strongly with the rounded siderite clasts. Note faint laminations parallel to the long dimension of the clasts in both shale and siderite clasts. Dark grains forming some laminations are high concentrations (40–70%) of glauconite. *Low-Energy Ridge-Margin Facies,*

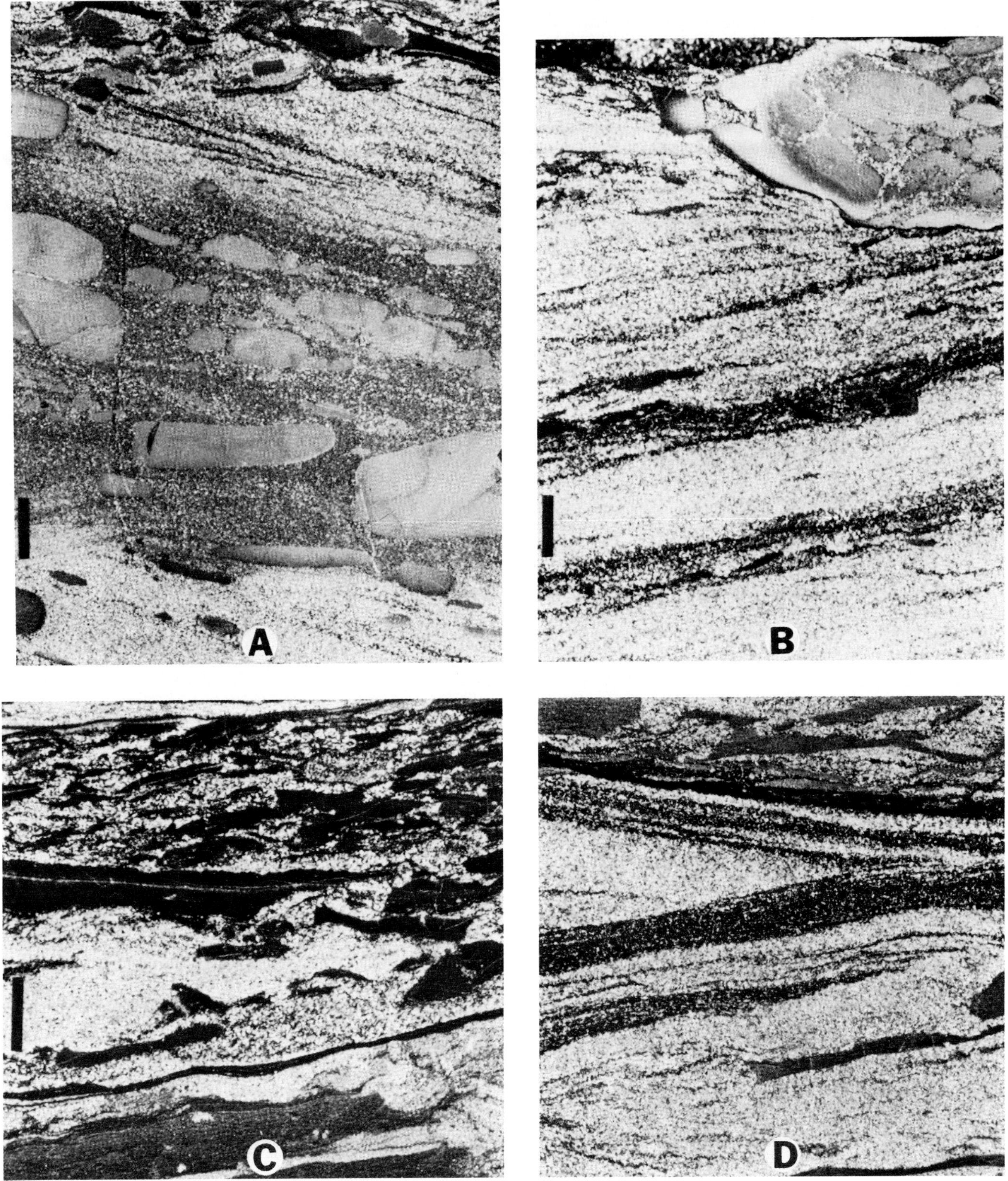

CSO AE-1, Unit 4, 9,155.7 ft (2790 m). Scale = 1 cm. (B) Composite rounded siderite clast containing rounded siderite clasts. Trough crossbedding. Dark colored laminations are high concentrations of glauconite (40–70%). *High-Energy Ridge-Margin Facies,* Southland Royalty Bud Christensen No. 2, Unit 5, 9,164.0 ft (2793 m). (C) Angular shale rip-up clasts concentrated in trough crossbedding. *High-Energy Ridge-Margin Facies.* Southland Royalty Bud Christensen No. 2, Unit 5, 9,164.0 ft (2793 m). Some shale clasts show some soft-sediment deformation. (D) Laminated shales draping rippled trough crossbedding. Note truncation of some shale laminations. Deformed rip-up shale clasts at top. *Low-Energy Ridge-Margin Facies,* CSO AB-1A, Unit 7, 9,244.4 ft (2818 m).

tified. Forty-two percent of the *High-Energy Ridge-Margin Facies* contacts (overlying plus underlying) are with *Central Ridge Facies,* 17% with *Central Ridge (Planar Laminated),* and 33% each with *Low-Energy Ridge-Margin Facies, Burrowed Shelf Siltstone Facies* and *Shelf Silty Shale.* The most common (33%) facies types to occur above this facies are the *Central Ridge Facies* and *Shelf Silty Shale* (Table 3A). Facies below this facies include *Central Ridge Facies* (50%) and *Low-Energy Ridge-Margin* (33%; Table 3B). The average transition interval for the lower and upper contacts is 0.5 and 0.4 ft (15 cm and 12 cm), respectively. The range of transition thicknesses is from 0.03 to 1.0 ft (0.9 to 30 cm).

Interpretation.—

Bedding and lithologic characteristics of the *High-Energy Ridge-Margin Facies* are similar to the *Central Ridge Facies,* but the deposition of abundant rip-up shale fragments suggests a local access to *Inter-Ridge Facies (Shaly)* or *Low-Energy Ridge-Margin Facies* shales, which were presumably deposited during periods of quieter flow. During periods of increased flow, the bedded shales were ripped up and redeposited. The abundances and size of these clasts suggest that the currents which deposited these units were very competent; as large as 3 × 2 in. (6.6 × 3.3 cm) angular shale clasts (Fig. 13C) were redeposited most commonly within the trough bedded sandstones. The presence of relatively low percentages (7%) of bedded shale within the crossbedded portion of the *High-Energy Ridge-Margin Facies* suggests a periodicity of deposition and a wide range of depositional energies. The sequence of events leading to deposition of siderite rip-up clasts is (1) erosion of the lenses which were originally deposited on the sea bottom, (2) short transport, and (3) deposition as clasts in sandy matrix.

This facies contributes significantly to production at Hartzog Draw field. This facies occurs in association with the *Central Ridge Facies* in the central and northern parts of the field and forms thick sections of reservoir sandstone along the east flank of the field. The total pay zone in the Cities Service Federal AS-1 (Figs. A9–12) located along the eastern boundary of the field (Fig. 7) is interpreted as a *High-Energy Ridge-Margin Facies.* The productive aspects of this facies are discussed in a subsequent section of the paper.

Low-Energy Ridge-Margin Facies

Description.—

This facies is characterized by sequences of beds of cross-laminated sandstones interbedded with rippled beds (Fig. 13D). The major difference between this facies and the other crossbedded facies is the presence of a significant number of rippled sandstone beds. This facies averages 44% ripples; the percentage of ripples ranges from 34–61%. Typically, the rippled beds are interbedded with high-angle crossbeds in a manner where several rippled beds or crossbeds occur before the bedding type changes. The unit is designated as a *Low-Energy Ridge-Margin Facies* because the abundant current ripples imply periods of relatively slow flowing currents and it contains a higher degree of burrowing (13%) than any other crossbedded facies. Burrowing is inferred to take place during relatively quieter periods. The burrowing is commonly associated with the rippled beds and is of low to moderate diversity. Excellent examples of this facies crop out at Salt Creek field (Tillman and Martinsen, 1984, Fig. 22).

Sandstone comprises 64% of the facies and only about 35% of the unit is crossbedded, including 10% ripples-on-troughs. The average grain size is 175 μm and the sandstone averages 15% glauconite. Interbedded shale is also more common (30%) in this facies than in the *Central Ridge* or *High-Energy Ridge-Margin Facies,* whereas shale clasts are less common (mean = 6%) than in the *High-Energy Ridge-Margin Facies* and siderite clasts are equally common (Fig. 13A). Laminated siderite drapes or lenses are slightly more common in this facies than in any other.

This facies differs from both the *Central Ridge* and *High-Energy Ridge-Margin Facies* in its finer mean grain size and higher percentage of ripples. It also differs from the *High-Energy Ridge-Margin Facies* in that it contains fewer rip-up clasts, and concentrations of glauconite which impart a green color, are not nearly as abundant.

The *Low-Energy Ridge-Margin Facies* is in contact (underlying and overlying) with each of four facies an equivalent number of times. The facies are *Central Ridge Facies, High-Energy Ridge-Margin, Inter-Ridge Facies (Shaly),* and *Bioturbated Shelf Siltstone.* The most common overlying facies is the *High-Energy Ridge-Margin Facies* (50%, Table 3A) and the most common underlying facies is the *Inter-Ridge Facies (Shaly)* (Table 3B). The average thickness of the contact transition interval for the lower and upper contacts is 0.07 and 0.3 ft (2 and 9 cm), respectively. The ranges in thicknesses are 0.01 to 0.2 ft and 0.03 to 1.0 ft (0.3 to 30 cm), respectively.

Interpretation.—

This facies is interpreted to be a transitional facies between the two higher energy facies, the *Central Ridge* and *High-Energy Ridge-Margin Facies,* and the lower energy *Inter-Ridge Facies (Shaly).* Whereas the average sand size (175 μm), high frequency of ripple bedding, higher percentage of burrows and presence of bedded shale (Table 2B) indicate times of relatively low-energy deposition, periodic increases in energy produced crossbedded sandstones. The fact that this facies occurs most commonly in the western and southern parts of Hartzog Draw indicates that it was deposited downcurrent from the sandier more typically crossbedded *Central Ridge Facies* and represents the sheltered deposits on the lee side of the ridge.

Inter-Ridge Facies (Shaly)

Description.—

The *Inter-Ridge Facies (Shaly)* is a non-reservoir facies and is so named because it is the predominant facies *between* Hartzog Draw and Heldt Draw fields. It also forms the lower "foundation" for the reservoir facies in both Hartzog and Heldt Draw fields, where it underlies all the crossbedded facies. This facies consists of interbedded rippled silty shale and rippled fine- to very fine-grained sandstone with minor to major amounts of burrowing (Fig. 14). It is distinguished from the *Inter-Ridge Sandstone Facies* by the presence of interbedded shale. Fine-grained (150 μm) sandstone is present in amounts averaging 58% and ranging from 35 to 78% sandstone (Table A2). This facies averages 13% silt and the remainder (31%) is silty shale. Fine- to medium-grained sandstone forms a few of the relatively thicker lenses and fills some of the larger burrows.

This facies contrasts strongly with the *Central Ridge* and *Ridge-Margin Facies*. No sandstone beds exceeding 4 in. (10 cm) in thickness were observed, and in the cores, interbedded shale and rippled sandstone (60% ripples) are the prominent features (Fig. 14). The sandstones are commonly 2- to 3-in.-thick (5 to 7.5 cm) lenses or beds composed almost entirely of current ripples. A few symmetrical (wave) ripples are also observed. High-angle crossbedding aver-

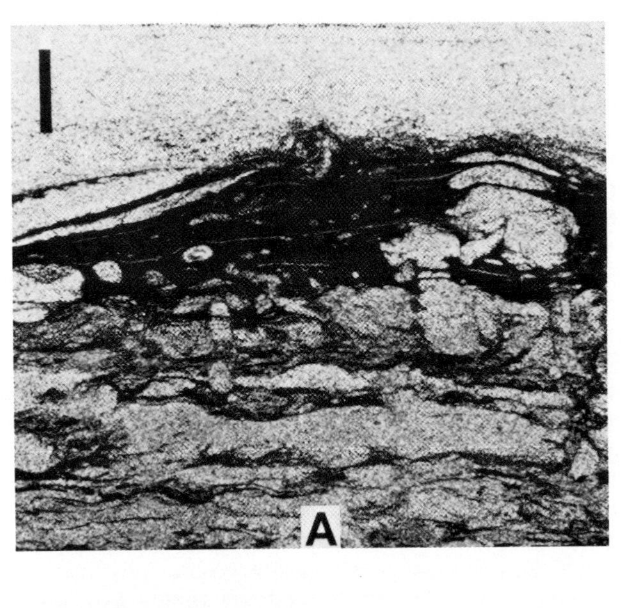

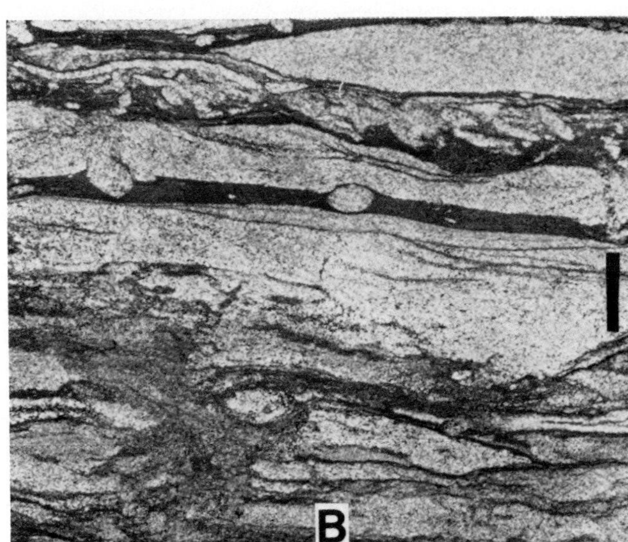

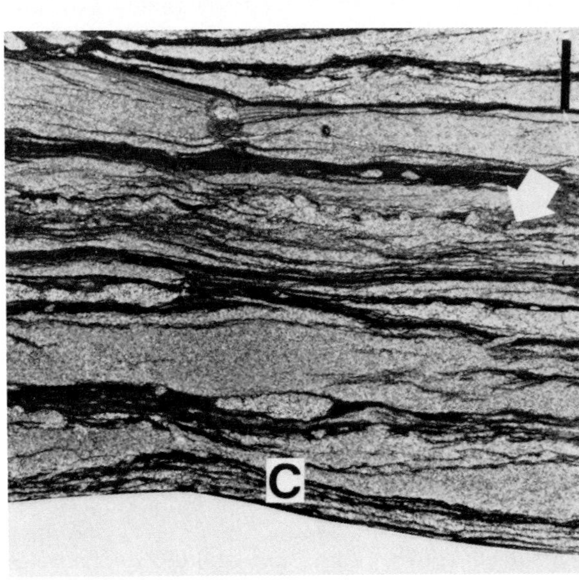

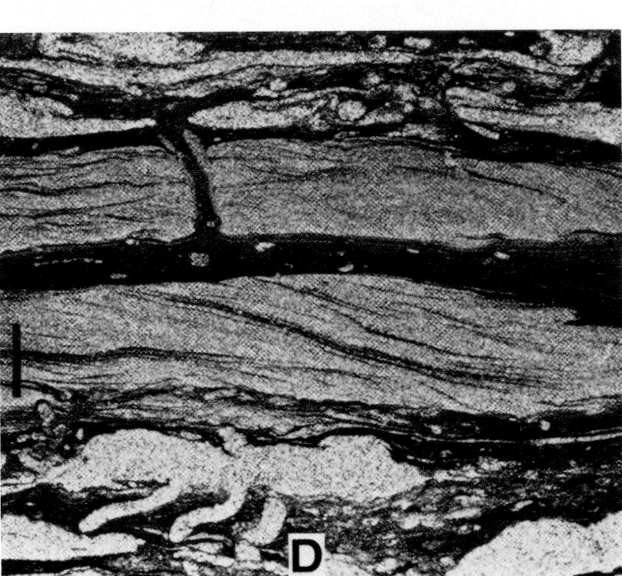

FIG. 14.—*INTER-RIDGE FACIES (SHALY)*. (A) Burrowed interbedded and interlaminated fine-grained sandstone and silty shale overlying a 3-in.-thick (7.6 cm) subhorizontally laminated sandstone (only top portion of sandstone is in photo). Basal sandstone typical of what are interpreted as storm deposited sandstones. This type of 2- to 4-in.-thick (5–10 cm) horizontally laminated sandstone occurs randomly in the *Inter-Ridge Facies (Shaly)*. Burrows are variable in size (1/16- to 1/4-in.-diameter, 0.1–0.6 cm) and orientation (oblique to horizontal). *Inter-Ridge Facies (Shaly)*, CSO AE-1, Unit 1, 9,192.4 ft (2802 m). (B) Burrowed interbedded and interlaminated fine-grained sandstone and silty shale. Rippled sandstone, silty shale mainly formed as drapes over relief of ripples. Burrows are mostly oblique, some scattered horizontal burrows. *Inter-Ridge Facies*, Southland Royalty Bud Christensen No. 2, Unit 1, 9,211.3 ft (2808 m). (C) Rippled interbedded and interlaminated fine-grained sandstone (150 μm) and shale. Horizontal (feeding?) burrows concentrated along some laminations (arrow). Typical association of *Inter-Ridge Facies (Shaly)*. CSO AB-1A, Unit 2A, 9,297.9 ft (2834 m). (D) Current-rippled fine-grained sandstone interbedded with laminated draped to slightly rippled silty shale. This association is typical of *Inter-Ridge Facies (Shaly)* but here occurring in a *Bioturbated Shelf Siltstone*. A variety of burrow types occurs in this photo, including shale-filled vertical and sand- and silt-filled horizontal burrows; bifurcating sand-filled burrows are present near the bottom. Southland Royalty Bud Christensen No. 2, Unit 2, 9,191.8 ft (2802 m). Scale = 1 cm.

ages 7% and does not exceed 10%. Glauconite is present in percentages ranging from 1 to 10% in the sandstone lenses but is much less conspicuous than in facies containing more sandstone. The glauconite in this facies is usually finer grained than that observed in the crossbedded facies. Scattered occurrences of subhorizontally laminated, nearly massive 1- to 2-in.-thick (2.5 to 5 cm) sandstone lenses or beds are observed. These horizontally to subhorizontally laminated beds are seldom, if ever, more than 0.3 ft (9 cm) thick. Planar laminated beds are always thin in the *Inter-Ridge Facies,* whereas those in the *Central Ridge (Planar Laminated) Facies* exceed 1 ft (30 cm).

Local, highly burrowed (mottled) areas are observed, and scattered burrowing is characteristic throughout. The percentage of burrowing, by volume, ranges from 20 to 40% and averages 34%. Burrow diversity is commonly high (10–12 burrow types) and most burrows are sand- or silt-filled (Figs. 14C and D). One-fourth-inch-diameter (0.6 cm) oblique and horizontal burrows are most common. Smaller (1/8 in., 0.3 cm) and larger (1/2–3/4 in., 1.3–1.9 cm) burrows also occur. *Teichichnus, Chondrites, Planolites* and *Terabellina* ("donut burrows") are locally identified (Fig. 15B).

This facies grades into *Burrowed Shelf Siltstone* when burrowing exceeds 50% and into *Bioturbated Shelf Siltstone* when burrowing exceeds 75%. Bioturbated intervals less than 1 ft (30 cm) thick are included in this facies. When bioturbated intervals exceed 1 ft (30 cm), they are designated as a separate facies.

A wide variety of facies overlie and underlie the *Inter-Ridge Facies (Shaly).* Underlying the *Inter-Ridge Facies (Shaly)* are *Shelf Silty Shale* (66%) and *Bioturbated Shelf Siltstone* (33%). Overlying this facies are *Low-Energy Ridge-Margin Facies* (33%), *Bioturbated Shelf Siltstone* (50%) and *Central Ridge-Facies* (17%). In all instances, *Shelf Silty Shale Facies* are overlain by *Inter-Ridge Facies (Shaly).*

The average thickness of transition for the lower and upper contact is 0.3 and 0.5 ft (9 and 15 cm), respectively. The range of transition intervals is from 0.03 to 1.0 ft (0.9 to 30 cm).

Interpretation.—

Cyclical deposition is the rule in this facies. The products resulting from alternations between slow currents carrying sand and very slow currents carrying silty clay are observed throughout. The fact that the major physical structures throughout are low-amplitude ripples, 53 to 65% of the volume of the rock, indicates that relatively low-energy conditions prevailed throughout. Asymmetrical current ripples far exceed symmetrical (wave?) ripples, indicating that directional currents were the main depositional mechanism. Local burrowing almost obscures the earlier formed ripples; this diminishes significantly the percentages of ripples observed.

The thin sandstone beds in the *Inter-Ridge Facies (Shaly)* are interpreted to be the result of storm deposits similar to those reported in the north Sea (Reineck and Singh, 1971; Tillman and Reineck, 1975). These thin beds commonly are unburrowed and represent very rapid deposition.

The fact that the unit as a whole is so highly burrowed suggests that it was deposited relatively slowly, slowly enough for the burrowers to have had time to rework significant portions of the facies. The high diversity of burrow types suggests that it was a non-stressed environment that was conducive to a wide variety of different bottom and subbottom dwellers.

Bioturbated (or Burrowed) Shelf Siltstone Facies

Description.—

The aspect most characteristic of the *Bioturbated Shelf Siltstone Facies* is that it is more than 75% burrowed. Where burrowing is more than 50% and less than 75%, the term "burrowed" is substituted for "bioturbated." The facies is from 10 to 25% very fine-grained sandstone (125 μm) and averages 37% siltstone and 39% silty shale. In outcrop, *Bioturbated Shelf Sandstone Facies* are observed in which the mean grain size is approximately 100 μm (Tillman and Martinsen, 1984). Most of the highly burrowed cored facies have finer mean sizes; most are finer than 37 μm. Physical structures preserved in the core average 22% and range from 10 to 33% (Table A3).

The percentage of burrowing (by volume) ranges from 67 to 90; when the amount of burrowing exceeds 75%, the term "bioturbated" is used. In the six cores studied, the percentage of the core which was bioturbated and assigned to this facies, ranges from only 10 to 75%. Note that the core may be 100% burrowed in some intervals, allowing for much less burrowing elsewhere within the facies and still having enough burrows to be designated as a bioturbated facies. The diversity of burrow types is moderate to high (6–10 types) and the orientation of the burrows is also diverse. Most burrows are horizontal to oblique; only rare vertically oriented burrows occur. Traces to 2% of *Chondrites* (Fig. 16B) and traces of *Teichichnus* and *Terabellina* ("donut burrows") are observed. A trace of possible *Zoophycos* was also observed (Figs. 15C, D). Most of the burrow types observed in this facies are not identified by name. Burrowing is pervasive throughout the facies (Fig. 16); the vertical interval in which no burrows are observed ranges from zero to only 4% or, stated another way, in 96% or more of the section at least one burrow can be observed in each one-tenth-ft (3 cm) length of core.

Physical structures remaining after burrowing are mostly ripples (5–26%) and horizontal laminations (trace to 10%). Deformation by physical processes is uncommon, but obliteration of sedimentary structures by burrowing is typical throughout significant intervals.

The contacts between this facies and those adjacent may be sharp to gradational (Table A3). Of the upper contacts observed, most are typically sharp and have an average transition interval of only 0.14 ft (4 cm). The range of upper contact transition intervals is from 0.01 to 0.2 ft (0.3 to 6 cm). The lower contact in most cases is more gradational; the average transition thickness is 0.4 ft (12 cm), and the range is from 0.01 to 0.7 ft (0.3 to 21 cm).

This facies is associated above and below with a wide variety of facies; however, the most common association (43%, 28%) is with *Inter-Ridge Facies (Shaly).* The *Inter-Ridge Facies (Shaly)* is most common both above (28%)

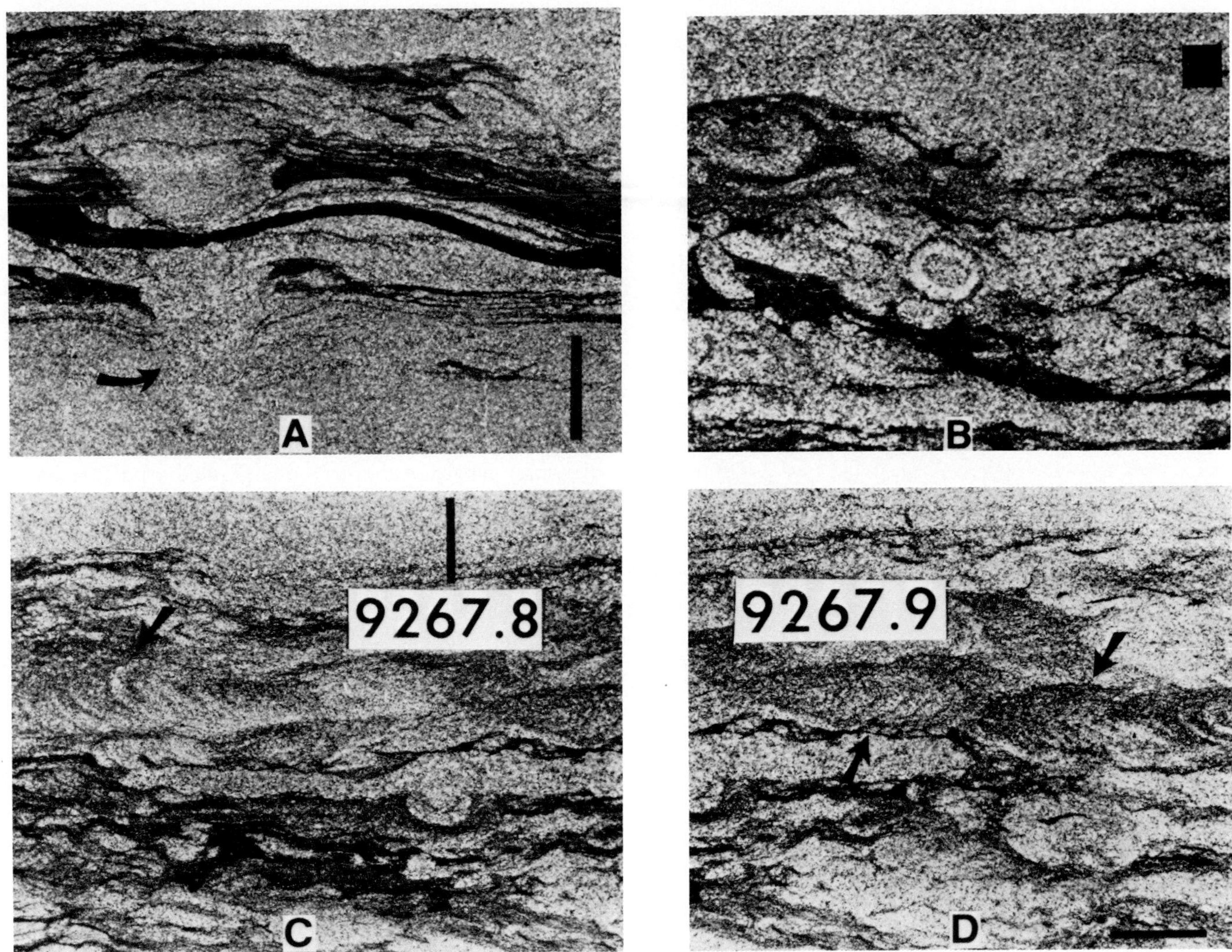

FIG. 15.—*BURROW TYPES, INTER-RIDGE FACIES AND BIOTURBATED SHELF SILTSTONE.* (A) Large vertical burrow (oblique section) in sandy shaly siltstone. Note: black area cutting *across* burrow is a break in the core. *Inter-Ridge Facies (Shaly)*, CSO AB-1A, Unit 2A, 9,296.5 ft (2834 m). (B) Silt-lined *Terabellina* ("donut burrow") in burrowed and slightly deformed siltstone. *Inter-Ridge Facies*, CSO AE-1, Unit 2, 9,181.0 ft (2798 m). (C, D) *Zoophycos*-type burrows (arrows) in siltstone. *Bioturbated Shelf Siltstone Facies*, CSO AB-1A, Unit 3B.

and below (43%) this facies (Table 3B).

The *Bioturbated Shelf Siltstones* are generally found irregularly distributed. In the six cores studied, these relatively thin, highly burrowed units are commonly found "draped" over other facies (Figs. 31, 32); as a result, individual time-equivalent lenses of this facies may have local relief of more than 10 ft (3 m).

Interpretation.—

This facies commonly is somewhat gradational with the *Inter-Ridge Facies (Shaly)* and, typically, this facies results from a thorough burrowing of the *Inter-Ridge Facies (Shaly)*. In this facies only very rarely (2%) are high-angle crossbeds, which are typical of the higher energy facies, recognized. This facies is slightly shalier (39% *vs.* 31%) and siltier (25% *vs.* 20%) and contains less sandstone (22% *vs.* 58%) than the *Inter-Ridge Facies (Shaly)* (Table A3). This suggests that burrowers are responsible for adding abundant clay-size material to the section. Commonly, burrows are silt-filled. The silt may have filtered down into open burrows or it may have been ingested by the burrowing animal and then added to the burrow fill.

It is expected that many of the burrowers in this facies captured clays from the bottom waters, and, after passing the clays and attached organic morsels through their gut, added clay to the near-surface bottom sediments. The clay was added to the sediment in the form of fecal material or as burrow linings. Extensively burrowed siltstones may be limited to the area near the submerged marine bars; however, cores lateral to the field are not available to verify this hypothesis.

It is presumed that the currents that were active in moving sand into place to form the shelf-ridge complex could also have brought in nutrients for the burrowers in the form of organic detritus. The detritus would be scattered through the finer grained laminations within the units and, as a result, commonly the burrowers are concentrated in these laminations.

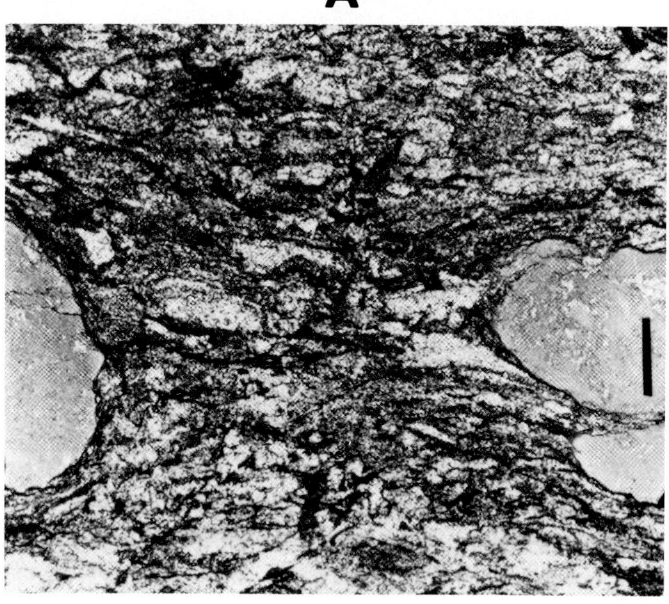

FIG. 16.—*BIOTURBATED SHELF SILTSTONE FACIES.* (A) This *Bioturbated Shelf Siltstone Facies* is more than 75% burrowed. One-half-in.-diameter (1.3 cm) horizontal sand-filled burrows as well as smaller oblique sand-filled burrows can be observed. Note that physical bedding and laminations are essentially absent; a general subhorizontal lineation of dark shale fragments suggests subhorizontal lamination, but all other laminations have been destroyed. CSO AE-1, Unit 8, 9,141.4 ft (2786 m). Scale = 1 cm. (B) *Bioturbated Shelf Siltstone Facies* exhibiting a mottled appearance. Rounded silty clasts (right and left) are siderite clasts which contain disseminated silt (sand?) grains. Siderite clasts are not common in this facies. Possible *Chondrites* burrows (black) can be observed in the lower left central portion of the photo. Physical bedding features are absent; only a faint subhorizontal lineation is visible. Burrows here are smaller diameter and show less diversity than in section at 9,141.4 ft in (A) above. CSO AE-1, Unit 8, 9,140.8 ft (2786 m).

The extensive burrowing typical of the facies is believed to occur during periods of slow deposition. Thorough bioturbation (burrowing greater than 75%) probably is the result of a series of periods of non-deposition or at least periods where physical deposition was unable to keep pace with the burrowers. High concentrations of burrows occur only locally in other facies in the shelf-ridge complex (Figs. 17A–C).

Where cores are not available and correlation is based entirely on logs, the thin shelf sandstone and siltstone (burrowed) facies between wells within the overall ridge complex are very difficult to correlate, especially where they occur intimately associated with other sandstone facies. The origin suggested here almost precludes tracing individual lenses of this facies very far laterally *within* the shelf-ridge complex. Below the ridge, however, where it is interbedded with the *Inter-Ridge Facies (Shaly),* and lateral to the shelf-ridge complex, the *Bioturbated Shelf Siltstone Facies* may be locally correlatable between wells.

The reservoir potential of this facies is very poor due to the shaly nature of the very fine-grained sandstone and its high degree of bioturbation. Nowhere is it indicated as net pay in the cores studied to date.

Shelf Silty Shale Facies

Description.—

Shelf Silty Shale Facies occurring above and below the major sandstone facies in the shelf-ridge complex are of two types: (1) rippled (flaser) and *laminated* silty shale, which is moderately burrowed, and (2) a silty shale, massive in appearance, in which the silt is *disseminated,* and small (1/8-in., 0.3-cm-diameter) shale-filled burrows are relatively abundant.

Disseminated Silty Shale Subfacies (Type 1)

The disseminated silty shale subfacies has been observed only *below* the shelf-ridge complex. This facies occurs at the base of two cores, CS AB-1A and CS AS-1 (Fig. A11). This subfacies averages 46% (by volume) burrowing and the average total vertical interval in the cores, which shows an absence of burrowing, is 20% (Table A4). Most of the burrows are less than 1/8-in. (0.3 cm) diameter, clay-filled and oblique, and give the sediment a mottled texture (Fig. 18). As much as 2% silt-lined *Terabellina* "donut burrows" are observed. The diversity of burrow types is moderate.

Physical structures are not readily visible in this subfacies; however, shaly to silty, diffuse, slightly rippled, subhorizontal laminations may be recognized in amounts averaging 53% by volume. Ripples form about 23% of the sediment.

Shelf Silty Shale Subfacies (Type 2), (Rippled, Laminated and Silt-Filled Burrowed Shale Subfacies)

The highly rippled and laminated siltstone and silt-filled burrow subfacies occurs only *above* or lateral to the shelf-ridge complex in the cores where it is observed (Figs. A3, 19).

In this subfacies, physical structures predominate, comprising 86% of the volume of the unit (Table A4). Conspicuous lenticular, silty to sandy, asymmetrical, low-amplitude, current-rippled lamina sets form 68% of the structures

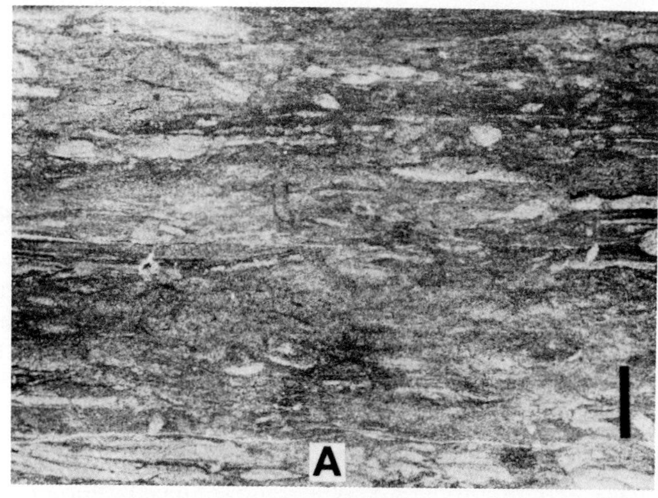

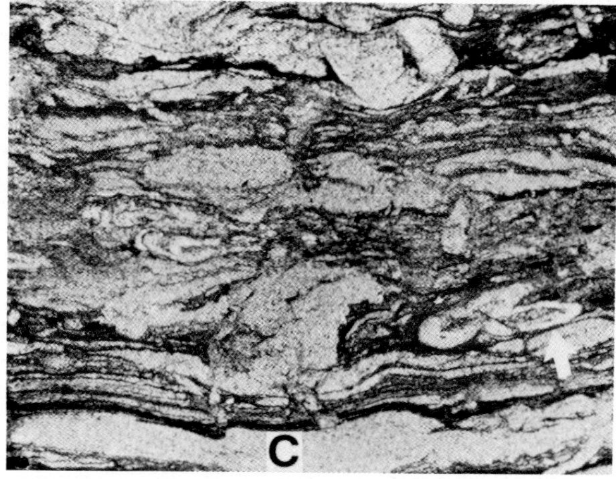

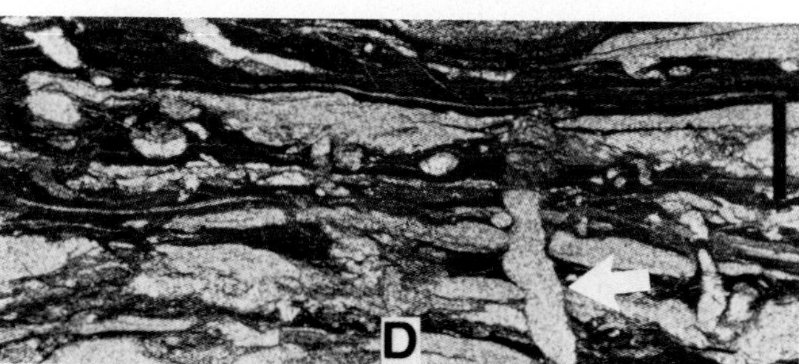

FIG. 17.—*BURROW TYPES, SHANNON SHELF RIDGE COMPLEX*. (A) Burrowed, mottled, silty shale with small diameter silt- and sand-filled burrows. Typical of parts of the *Shelf Silty Shale Facies (Type 2)* which overlies the shelf-ridge complex. CSO AB-1A, Unit 9, 9,222.5 ft (2811 m). (B) One-half-cm-diameter white horizontal (and oblique) burrows typical only of certain segments of the *Central Ridge Facies*. Burrow fillings slightly lighter in color (white) than surrounding medium- to fine-grained sandstone. CSO AE-1, Unit 3, 9,178.5 ft (2798 m). (C) Bioturbated (more than 75% burrowed) interval in interbedded silty shale and sandstone *Inter-Ridge Facies (Shaly)*. Silt-lined burrows (arrow) show compaction and flattening. At least six to eight burrow-types in photograph, Southland Royalty Bud Christensen No. 2, Unit 1, 9.208.7 ft (2807 m). (D) Interbedded rippled to deformed fine-grained sandstone and silty shale. Horizontal and vertical sand-filled burrows. Although not readily visible in photograph, vertical burrow (arrow) shows "geopetal fabric" with sand filling lower portion of burrow and shaly silt filling top portion of burrow. Typical of some of the moderately burrowed portions of the *Bioturbated Shelf Siltstone Facies*. CSO AE-1, Unit 1, 9,186.0 ft (2800 m). Length of bar scale is 1 cm.

observed. Very few of the ripples observed are symmetrical ripples (Fig. 20A). Fifteen to 25 percent of the facies is composed of subhorizontal silty laminae which are lenticular to discontinuous in nature (Figs. 20B, C).

The conspicuous silt-filled burrows (Figs. 20C, D) and distinct ripples and laminations of this subfacies distinguish it from the diffuse wispy texture observed in the silty shale subfacies below. The "shale-on-shale" contact between these two facies may be observed in the core and on the log in the Diamond Shamrock 14-4 Hartzog Federal (SWSW Sec. 4 T43N R74W) near 9,358 ft (2,890 m; Fig. 19). Above this point, the physically laminated unit grades into what might be termed a thin "shelf-ridge margin" sequence, whereas below the typical diffuse wispy texture predominates. This well is located near the east edge of the productive shelf-ridge complex along the southeast margin of the field (Fig. 7). The log characteristics of the two shales are sufficiently different so as to be discernible on SP, resistivity and gamma ray logs (Fig. 19B).

In all cores and subsurface cross sections, the *Shelf Silty Shale Facies (Type 1)* is overlain by the *Inter-Ridge Facies (Shaly)*. The upper facies contact is very sharp and the transition interval is from 0.01 to 0.04 ft (0.3 to 1.2 cm).

The *Shelf Silty Shale Facies (Type 2)*, which overlies the shelf-ridge complex, in all cases overlies a *Burrowed Shelf Siltstone Facies*. The contact between the two facies is sharp; the transition interval between facies is 0.06 to 0.2 ft (1.8 to 6 cm).

Interpretation.—

The diffuse wispy lower *Disseminated Silty Shale Subfacies (Type 1)* is assumed to be the normal shale for this part

A

B

FIG. 18.—*DISSEMINATED SHELF SILTY SHALE FACIES (TYPE 1).* (A) Pre-shelf-ridge complex subfacies. Wispy mottled silty shale with very faint laminations and tiny shaly silt-filled burrows. Bright white subhorizontal features on the photo are breaks in the core. CSO AB-1A, Unit 1, 9,306.0 ft (2837 m). (B) Non-silt laminated shale at base of Cities Service AB-1A, Unit 1, 9,305.4 ft (2836 m). Includes scattered hairlike burrows. Contrast with relatively abundant silty layers in Figure 29. Bright white subhorizontal features on photo are breaks in core. This is an X-radiograph; width of photo 4 in. (10 cm).

of the shelf and would be expected in areas of the shelf where there are no shelf-ridge buildups. The more rippled and distinctly *Laminated and Burrowed Silty Shelf Shale Subfacies (Type 2)* is assumed to be the result of conditions which contributed to building the shelf-ridge complex.

The fact that the shale lateral to, and above, the ridge facies contains as much as 75% current ripples indicates that it was deposited by bottom currents rather than directly from suspension. Presumably, the same bottom currents that deposited the trough crossbedded sandstones of the *Central Ridge* and *Ridge-Margin Facies* were responsible for deposition of the fine-grained sands, silts and clays lateral to and above the shelf-ridge complex. Why the sands were localized in the area of the shelf-ridge complex, instead of being a sheet over a broader area of the shelf, is unclear. Apparently, lateral to the sandy shelf-ridge facies, the currents were of lower velocity and moved mostly flocculated clay and silt which was deposited as the *Shelf Silty Shale Facies*.

This facies is a sealing facies above, below and lateral to the productive shelf-ridge complex. Recognition of the shelf silty shale subfacies in cores which contain little or no reservoir sandstones may aid in lateral and vertical predictability of reservoirs of the type deposited at Hartzog Draw field.

RESERVOIR CHARACTERISTICS

Reservoir characteristics in Hartzog Draw field are very variable. Controls on reservoir quality include depositional facies distribution, distribution of detrital clay, distribution of secondary clay, distribution of calcite and silica cement and frequent consequent secondary dissolution-enhanced porosity (Hearn and others, 1984). The reservoir is completely enveloped in shale, initially had a solution gas drive, no water table and no produced formation water. Even zones that calculate water saturations of over 65% from logs do not produce water.

Porosities and permeabilities calculated from the five cored wells described in this study indicate a strong hierarchical variation from one facies type to another. Regardless of position within the field, a general decrease in reservoir quality, as observed in mean values of porosity and permeability, can be seen from the highest quality facies, *Central Ridge Facies* (15%, 15 md) and *High-Energy Ridge-Margin Facies* (14%, 19 md), to *Low-Energy Ridge-Margin Facies* (11%, 11 md). *Inter-Ridge (Shaly)* (5%, 2 md) and *Burrowed Shelf Siltstone Facies* (7%, 3 md) are non-reservoir facies (Figs. 21, 22). Within some facies, a nearly equally strong variation in reservoir quality is observed *within* individual facies. Those within-facies variations are attributed primarily to diagenetic effects.

Reservoir characteristics for the field as a whole were detailed by Martinsen (1981), Hunt and Hearn (1982) and Hearn and others (1984, 1986). Average porosity within the borders of the field (Fig. 6) is 12% and the average permeability is 12 md (Martinsen, 1981). These average values are less than the average values for the *Central Ridge Facies* and *High-Energy Ridge Margin* in the cored wells

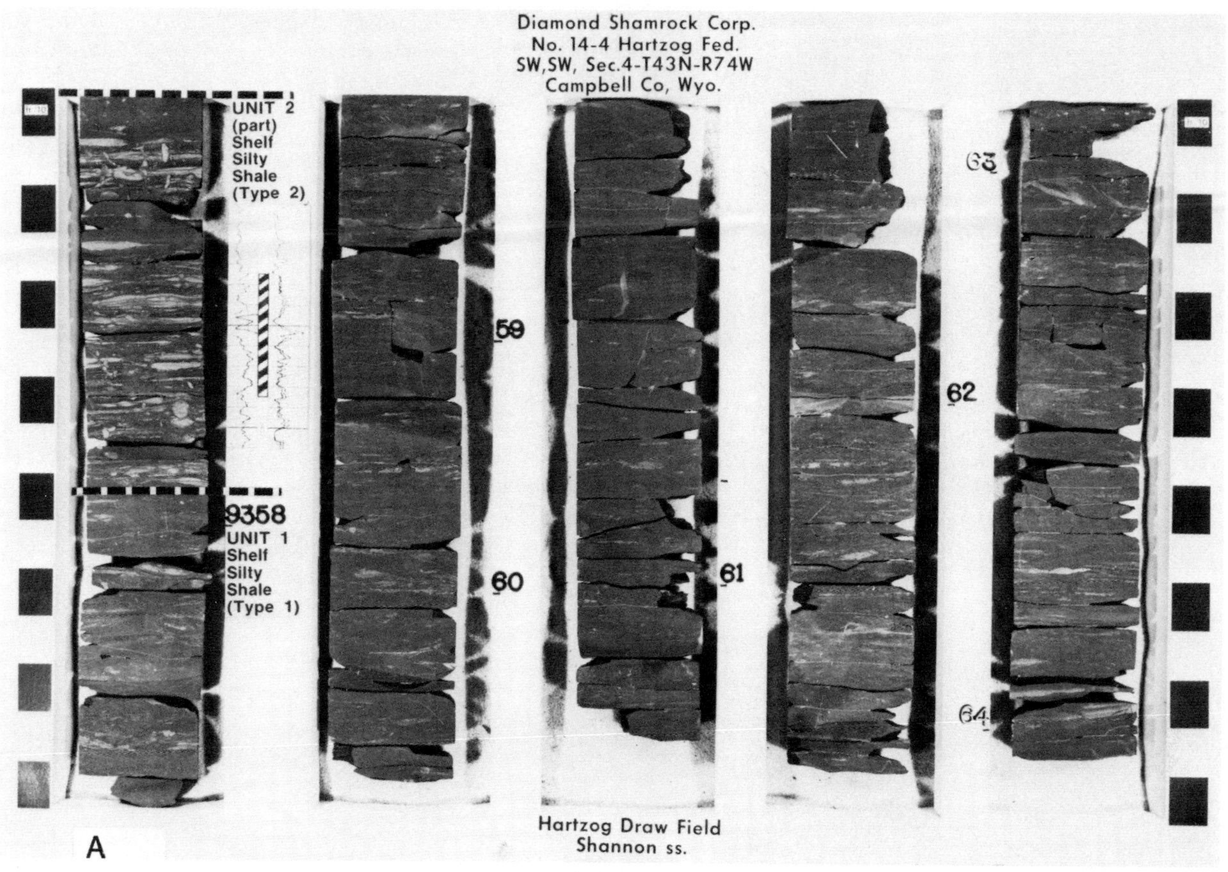

Company	DIAMOND SHAMROCK CORP.	Drillers T.D.	9467		
Well	#14-4 HARTZOG FEDERAL	Log F.R.	9458		
Field	HARTZOG DRAW	Log T.D.	9464		
County	CAMPBELL	Elevations:			
State	WYOMING	K.B. 5114	D.F.	G.L.	5098

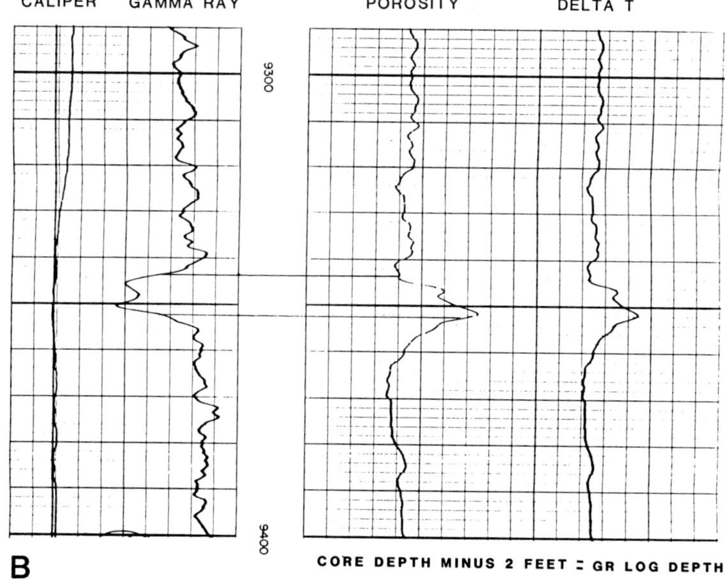

FIG. 19.—(A) Contact between *Rippled and Laminated Shelf Silty Shale Subfacies (Type 2)* and *Disseminated Shelf Silty Shale Subfacies (Type 1)*. Upper part of core typical of units above and lateral to the shelf-ridge complex; below is subfacies typical of units below the complex. The contact is at 9,358 ft (2852 m). This core, from the Diamond Shamrock Corp. No. 14-4 Hartzog Federal, SWSW Sec. 4 T43N R74W, Campbell County, Wyoming, is located east of the eastern edge and near the south end of the field (Fig. 7). This well is a dry hole immediately east of productive wells in the field. (B) Log showing variation in log character between *Type 1 (below) and Type 2 subfacies (above)*. Contact is at 9356′.

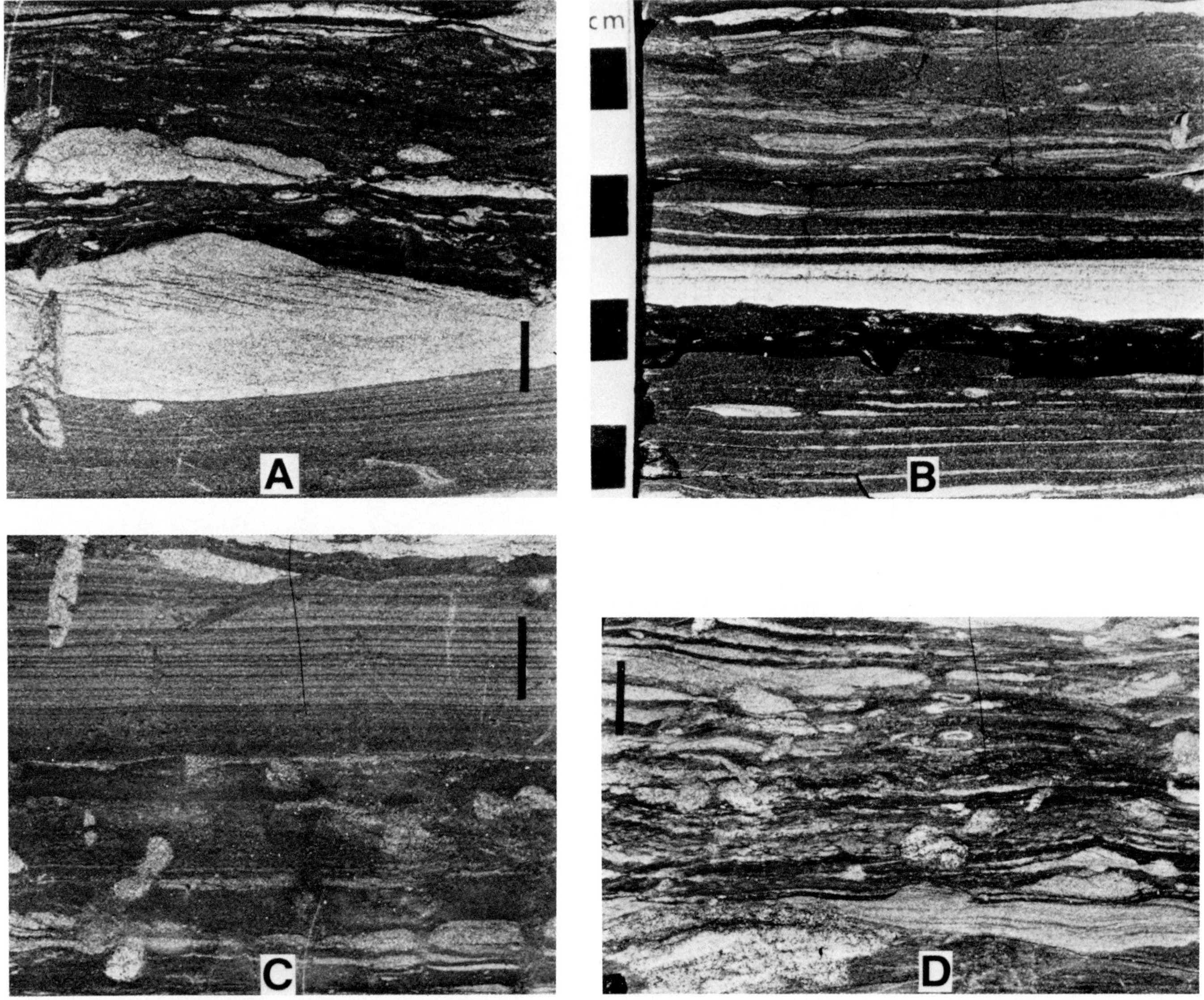

FIG. 20.—*RIPPLED AND LAMINATED SHELF SILTY SHALE FACIES (TYPE 2).* This facies is interpreted to have been deposited concurrent with and following Hartzog Draw shelf-ridge complex deposition. (A) Rippled very fine-grained sandstone displaying external symmetry and internal asymmetry (flaser- to lenticular-bedded). Interbedded with subhorizontally finely laminated siltstone below and silty shale above. A typical association in this subfacies. In area of photo, 10% (by volume) obliquely and horizontally burrowed. This and all photos in this figure are in sharp contrast with those from the pre-shelf-ridge complex subfacies (Type 1; Fig. 19). CSO Federal AE-1, Unit 9, 9,137.9 ft (2785 m). Scale = 1 cm. (B) Subhorizontally laminated very fine-grained sandstone, interlaminated with silt and silty shale. This is located 4 ft (1.2 m) above the base of the *Shelf Silty Shale Facies (Type 2)* in the CSO AB-1A at 9,235.5 ft (2815 m; Unit 9). Presumably the currents that deposited this silty shale were lateral to the stronger currents depositing the sand of the *Central Ridge Facies* to the south-southeast. (C) Horizontal interlaminated silt and silty shale overlying burrowed and rippled lenticular silty lenses. Some burrow filling and ripples near the top are very fine-grained sandstone; remainder are filled with silt-size material. Rapid deposition of horizontally laminated interval may have prevented extensive burrowing. CSO AB-1A, Unit 9, 9,236.2 ft (2815 m). Scale = 1 cm. (D) Discontinuous slightly rippled silt lenses interlaminated with silty shale. Flattened silt-lined *Terabellina* ("donut" burrows) indicate compaction. Slightly glauconitic, very fine-grained sandstone fills some of burrows in left portion of photo. CSO AB-1A, Unit 9, 9,237.0 ft (2815 m). Scale = 1 cm.

studied and are more than the average of all the other facies (Figs. 21, 22). Core and log porosity are in good agreement after core porosity is corrected for overburden pressure (Hearn and others, 1984). Mean values of porosity and permeability (uncorrected for overburden) are given on a unit by unit basis in Table A5. Tables A6, A7 and A8 show the variations *within* units in three of the cored wells (CS AE-1, Southland BC-2 and CS AS-1). Initial pressures in the field were 5,000 psig; bubble point pressure was 1,550 psig and during primary production, pressure dropped in some areas to about 1,000 psig. The drive for the field during primary production was dissolved gas, and the initial gas-oil ratio was 292 SCF/STB. Initial oil-in-place was 350 million barrels. The maximum sand thickness in the field is 60 ft (18.2 m); average thickness is 20 ft (6 m). The total shelf-ridge sequence including rippled and burrowed sand-

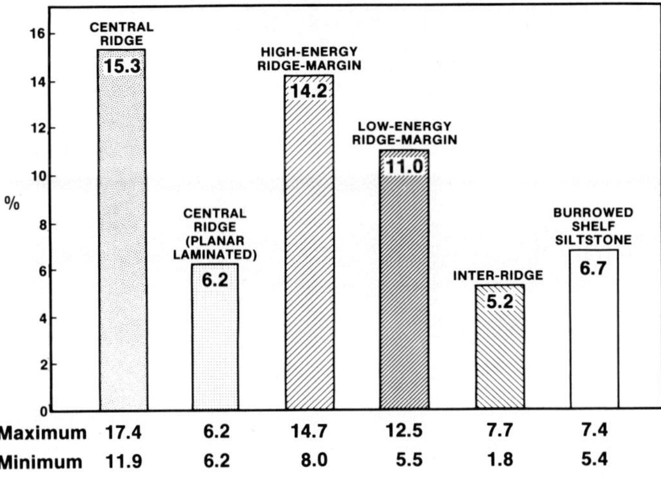

FIG. 21.—Porosity values for individual facies types, based on the five cores from Hartzog Draw field which were studied in detail; includes cores from northern, eastern, and central areas of the field. Mean porosities are plotted, maximum and minimum porosities are indicated. Figure computed from 204 measurements.

stones and silty shales ranges from 60 to 90 ft (18.2 to 27.3 m). In addition, the approximate position within the field (central, east, west, south) strongly influences the reservoir quality. The minimum net pay isopach parameter used by the field unitization committee was 9% porosity cutoff (2.67 grain density). In general, net sand and net pay maps (Fig. 6) correspond very closely.

Studies of primary oil production (Fig. 23) show a strong asymmetry in production relative to field geometry. Highest production (greater than 80,000 barrels per well-bore porosity foot) is observed (1) in a discontinuous band near the northeastern margin of the field in the middle one-third of the field length, (2) near the center of the field in the northwestern one-third of the field length, and (3) as discontinuous patches in the southeastern one-third. Lowest production is consistently along the west side of the field in the central two-thirds of the length of the field. The overall pattern then is one with a strong asymmetry of high production values to the east, moderate to excellent production (40,000–80,000 barrels per well-bore porosity-foot through 1980) extending over the northeastern two-thirds of the field, and only marginal production values over most of the western one-third of the field.

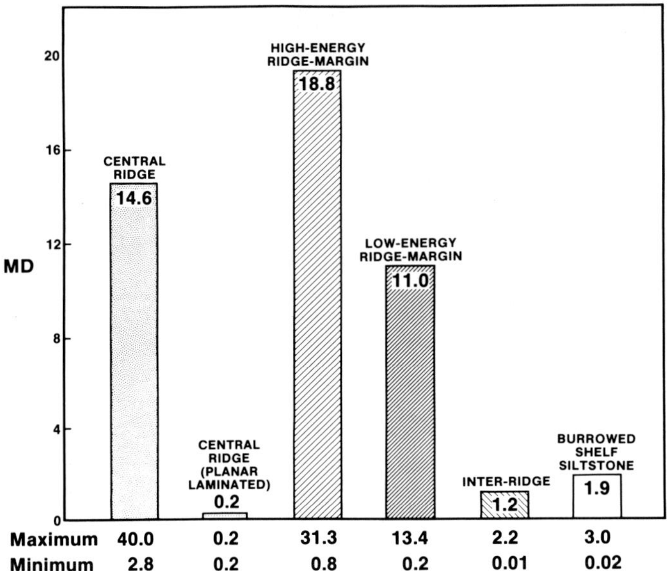

FIG. 22.—Permeability values for individual facies types, based on five cores from Hartzog Draw field; includes three cores from northern, one from eastern, and one from central area of field. Mean permeabilities are plotted; maximum and minimum permeabilities are indicated. Figure compiled from 204 measurements.

Charles Hearn (pers. commun.) agrees with general conclusions made in the discussion of Figure 23 but believes that cumulative production at a given point in time (1980) in a given well depends, in addition to reservoir quality, on (1) how long the well has been in production, (2) the efficiency of the well completion, and (3) efficiency of production equipment. In general, wells from the northern part of the field were completed earlier, accounting for some of the higher production recorded in the northern part.

Plots showing the vertical variation in facies type, core analysis porosity, permeability and oil saturation show significant variability. Included in this paper are plots showing detailed porosity and permeability distribution in the Cities Service Federal AE-1 (Fig. 24), Southland Royalty Bud Christensen No. 2 (Fig. 25) and Cities Service Federal AS-1 (Fig. 26). Similar plots for the Cities Service Federal AB-1A and AK-1 wells are included in Tillman and Martinsen (1985). In all these plots the *Central Ridge Facies,* and to a lesser degree the *High-Energy Ridge-Margin Facies,* are prime reservoir facies. The *Inter-Ridge Facies (Shaly)* in all the plots is such a low quality reservoir that it is, for all practical purposes, regarded as non-reservoir.

During April to June 1978, which was during the early stages of planning for secondary recovery, a pressure survey was taken in a number of producing wells in Hartzog Draw field. Pressure values were recorded for individual wells and then wells were grouped by the authors into "four-well blocks." This survey supports the concept of somewhat discontinuous reservoir sandstones within parts of the shelf-ridge complex. Average pressures measured varied from 500 to 5,000 psig. Strong contrasts between adjacent wells were recorded throughout the field. These contrasts were very obvious in the northwest area of the field, where virgin pressures (5,000 psig) were recorded only a mile or two from pressures as low as 500 to 1,000 psig (Fig. 27). These pressure differences, along with the facies variations observed in cores, strongly indicate that Hartzog Draw field is not a uniform reservoir, as might be deduced from the relatively gradational thickness variations indicated on the net pay isopach map (Fig. 6).

Detrital clay occurs in the field in two main types: rip-up clasts and rippled to draped laminae. Siderite occurs in rip-up clasts and discontinuous lenses. Only the rippled-to-draped shale laminae have a marked effect on controlling variations in ratios of vertical to horizontal permeabilities. The lateral extent of each of these features is readily visible

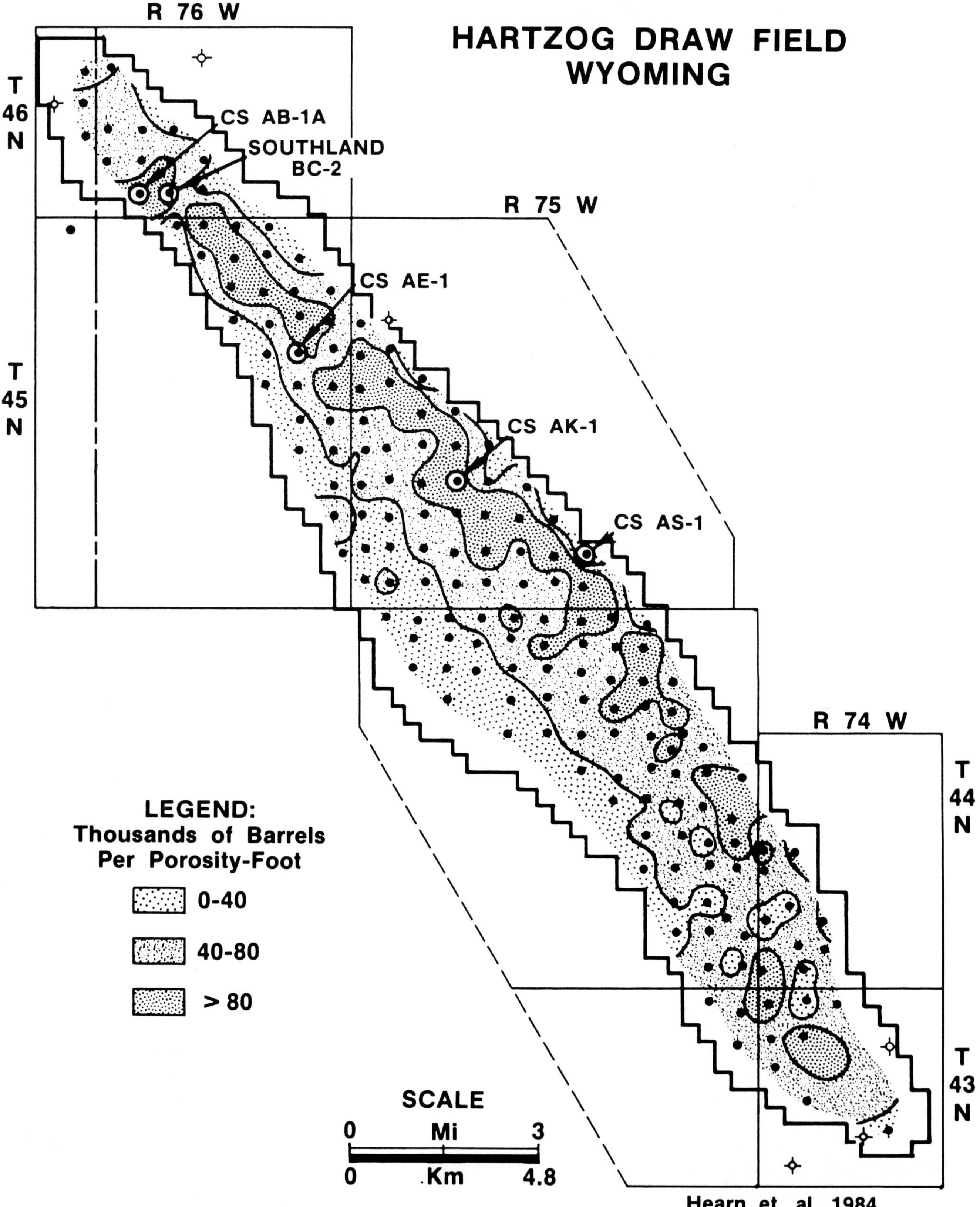

FIG. 23.—Cumulative primary oil production contour map for Hartzog Draw Unit. Cumulative oil production through 1980, per well-bore porosity-foot (from Hearn and others, 1984; Fig. 2). Note the strong asymmetry in production; better production areas are skewed strongly to the northeast. See text for discussion.

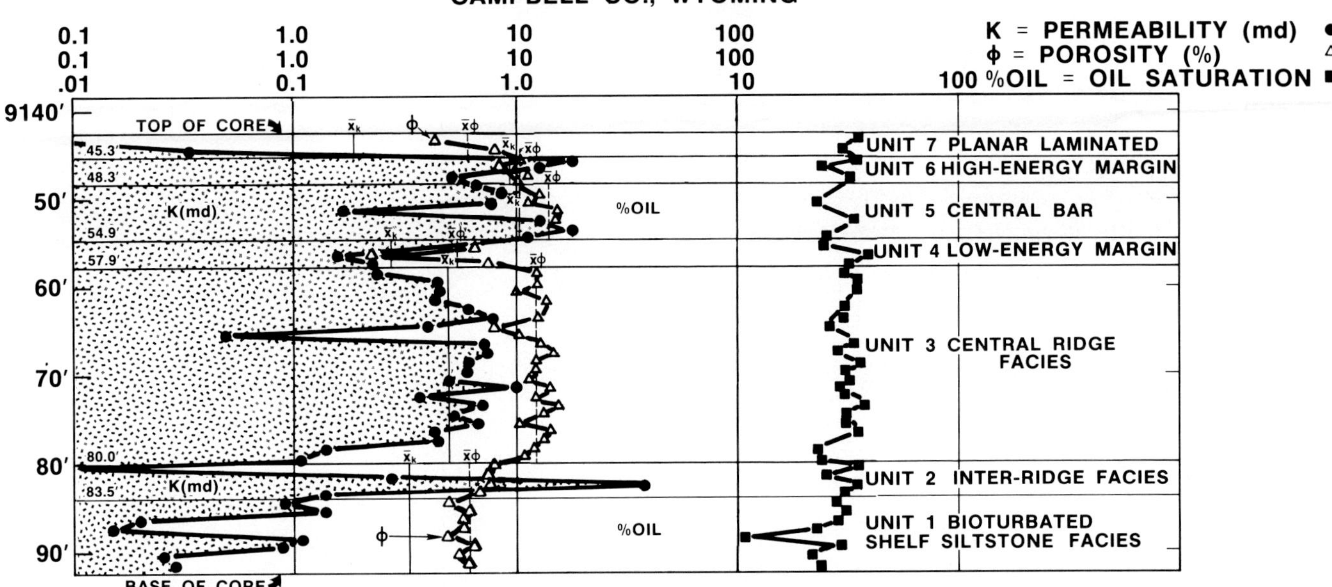

FIG. 24.—Production characteristics of individual facies recognized in slabbed core. Semi-log plot of porosity, permeability and oil saturation for cored interval in Cities Service Federal AE-1. Note that porosities (open triangles) are similar throughout the *Central Ridge* and in most of the *Ridge Margin Facies;* however, both porosity and permeability (circles) are significantly less in the *Inter-Ridge Facies (Shaly)*. Average values of porosity (x̄ϕ) and permeability (x̄k) for each unit are also indicated. Values plotted here are tabulated in Table A6; facies are the same as those described in Figures A1 to A4.

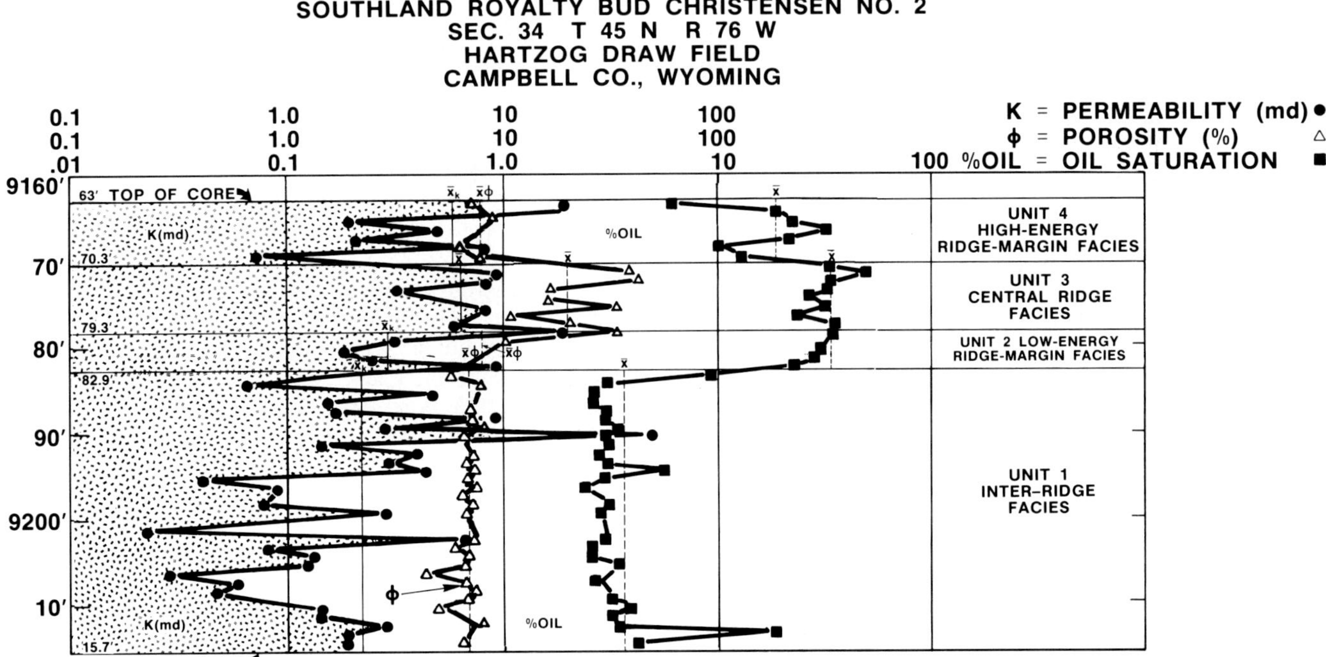

FIG. 25.—Production characteristics of individual facies recognized in slabbed cores. Semi-log plot of porosity (open triangles), permeability (circles) and oil saturation (squares) in Southland Royalty Bud Christensen No. 2. Note that porosities are significantly higher in the *Central Ridge Facies* than any of the others. Average values of porosity (x̄ϕ), permeability (x̄k) and oil saturation (x̄) are indicated for each unit. Values plotted here are tabulated in Table A7; facies are the same as those described in Figures A5 to A8.

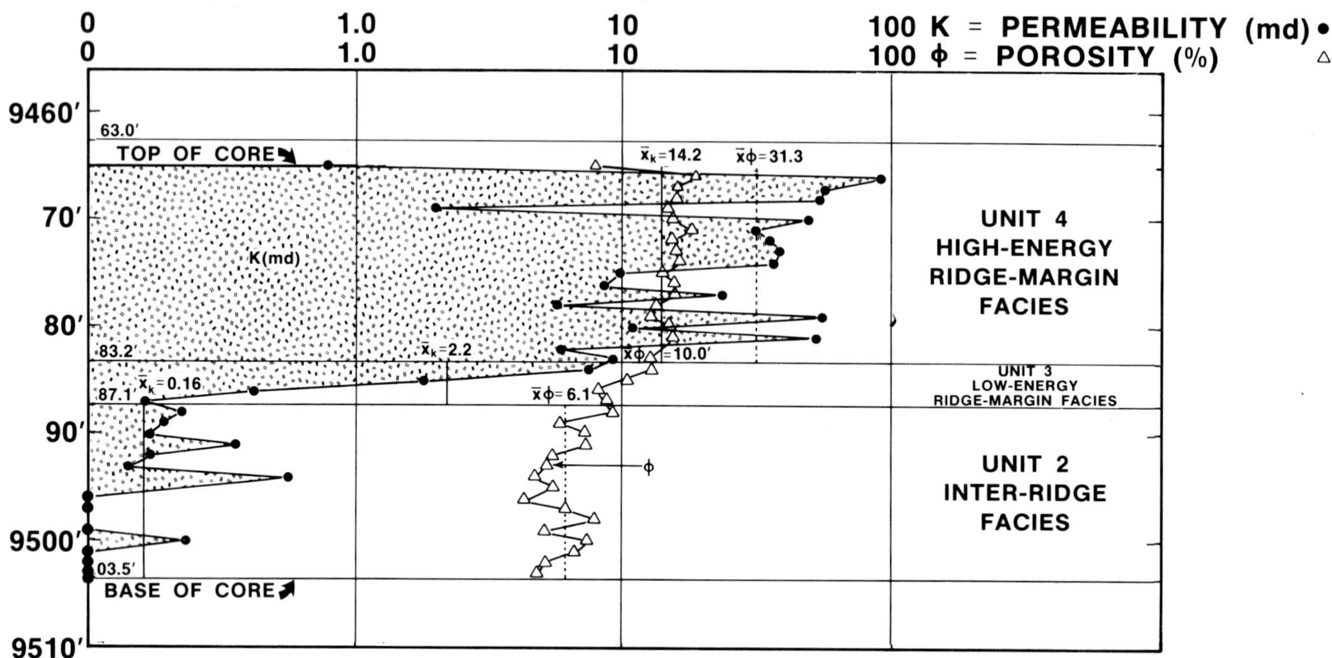

FIG. 26.—Production characteristics of individual facies recognized in slabbed core from Cities Service Federal AS-1. Semi-log plot of porosity (open triangles) and permeability (circles). Oil saturation values were not calculated. This is the only cored well studied which has as its main producing interval a *High-Energy Ridge-Margin Facies*. The absence of a *Central Ridge Facies* in this well may be responsible for the lower primary production values recorded in Figure 23. Values plotted here are tabulated in Table 8B; facies are those described in Figures A9 to A12.

in outcrop (Tillman and Martinsen, 1984). Siderite lenses (Fig. 12C) and siderite clast concentrations (Fig. 13A) have lateral extents of only a few meters and will not effect reservoir continuity between wells drilled even as close as 80-acre spacing. In contrast, shale laminae associated with rippled sandstone in *Low-Energy Ridge-Margin Facies* and *Inter-Ridge Facies (Shaly)* are more continuous and as an aggregate extend for hundreds of meters. In addition, calcite-cemented beds as much as 2 ft (0.6 m) thick can be recognized in cores. Whether cemented zones can be correlated more than a mile, as suggested by Hearn and others (1984), is doubtful, especially when correlation of similar calcite-cemented zones in nearly continuous outcrops cannot be done. More recent work on the Shannon by W. J. Ebanks Jr. (pers. commun.) also suggests that these calcite-cemented zones may not extend as far as earlier indicated.

Two families of mercury injection curves are recognized in a very limited suite of samples from two of the northern wells in the field (Fig. 28). Pore throat apertures were calculated using the formula

$$Pc = \frac{2\gamma \cos\theta}{r}$$

where

Pc = capillary pressure (atm)
γ = surface tension of mercury (480 dynes/cm)
θ = contact angle of mercury (140°)
r = radius of pore throat aperture (μm).

The six samples from the *Central Ridge Facies* all have similar curve shapes (Fig. 28). In the *Central Ridge Facies,* the percentage of the porosity which might be termed "macroporosity" (in excess of 0.5 μm) varies from about 42 to 52% of the porosity present. Microporosity (0.1 to 0.5 μm) into which mercury is injected is about 15%. The maximum percentage of void space occupied at the end of the test (155 psi) is about 55–65% of the porosity of the *Central Ridge Facies*. If entry level into the pores is assumed to be at about the 10% pore space level (E. Pittman, pers. commun.), then the entry level for this facies is quite small and ranges from 3 to 5 μm.

The samples in these wells assigned to the *High-Energy Ridge-Margin Facies* contain no true macroporosity and display a distinctly different pattern than the *Central Ridge Facies*. Entry level size ranges from 0.15 to 0.4 μm, and at the end of the test (155 psi), only 12 to 40% of the porosity is occupied by mercury.

The grain compositions of shaly (clayey) *vs.* less shaly facies in Hartzog Draw field were described by Ranganathan (*in* Hearn and others, 1984) and are indicated in Table 4. Recognition of present and former distribution of carbonate in Hartzog Draw field may be important in pre-

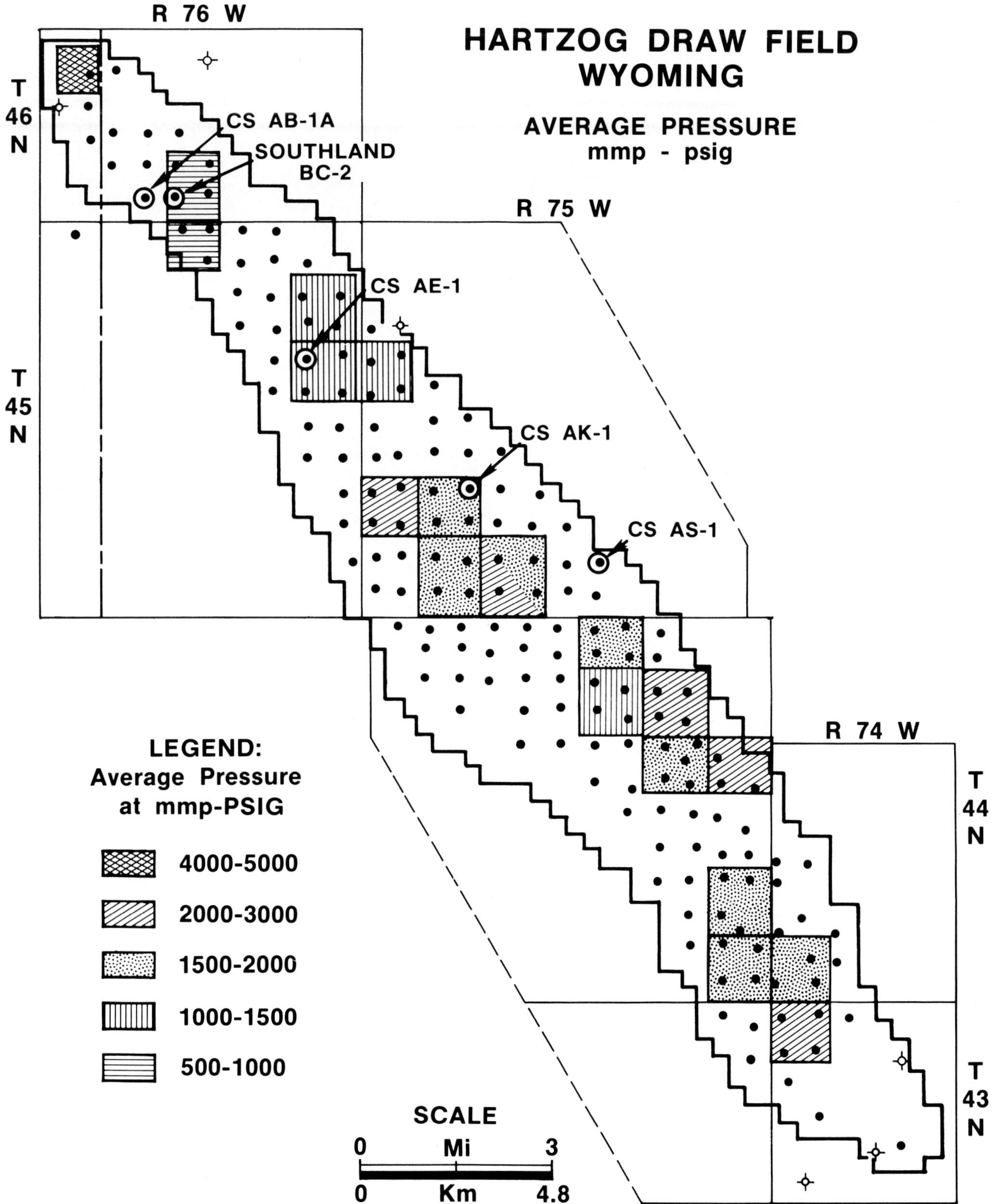

FIG. 27.—Pressures determined for individual well measurements in one or more wells per section. Data from 1978 pressure buildup survey of 21 wells. Squares (3- to 4-well blocks) show areas of similar pressures. See text for discussion.

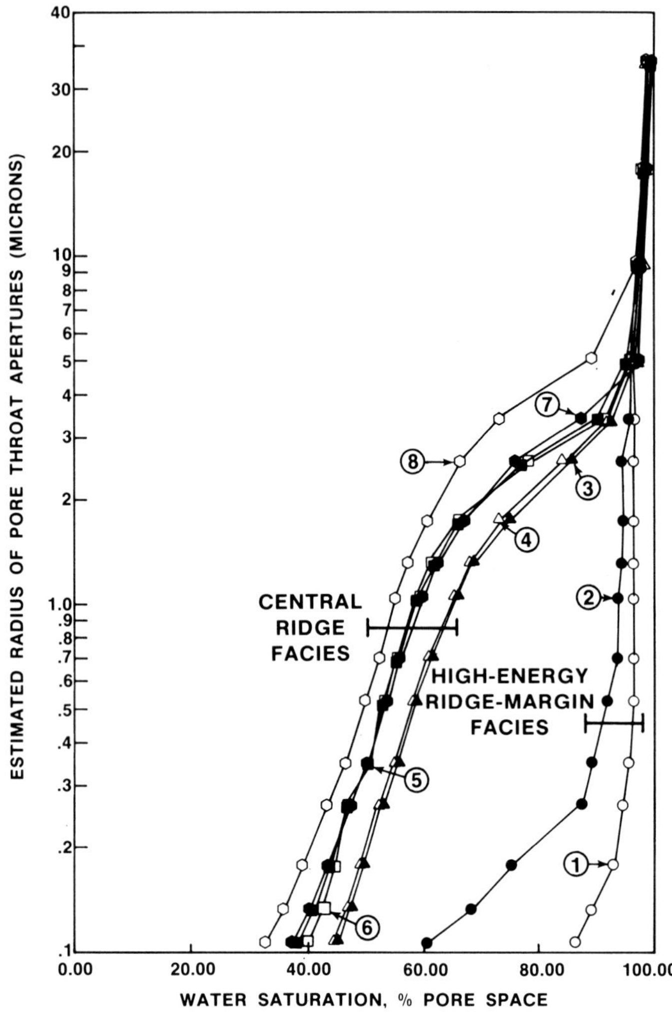

FIG. 28.—Mercury injection capillary pressure data plot of pore throat apertures *versus* percentage of pore space invaded by mercury at varying pressures ranging from 0 to 155 psi. Samples are from the Cities Service Federal AB-1A and the Southland Royalty Bud Christensen No. 2 wells. Note two families of curves. Numbers 1 and 2 are *High-Energy Ridge-Margin Facies* samples and the remainder are from *Central Ridge Facies*. Samples are as follows: *High-Energy Ridge-Margin Facies,* #1—CS AB-1A 9,258 ft (2822 m), #2—CS AB-1A 9,250 ft (2819 m); *Central Ridge Facies,* #3—CS AB-1A 9,255 ft (2821 m), #4—CS AB-1A 9,253 ft (2820 m), #5—S BC-2 9,170 ft (2795 m), #6—S BC-2 9,173 ft (2796 m), #7—S BC-2 9,176 ft (2797 m), #8—S CB2 9,179 ft (2798 m).

dicting porosity distribution, especially solution-enhanced porosity. Carbonate-cemented zones as much as 5 ft (1.5 m) thick occur, and whether they may extend for a mile or less within the crossbedded facies is a major concern. Carbonate minerals may replace some framework grains, notably plagioclase, feldspar and chert. Porosity enhancement has resulted from dissolution of part of the carbonate, especially calcite. Dissolution of porosity-occluding calcite, however, probably has only enhanced porosity somewhat, and relict primary intergranular porosity is dominant in the *Central Ridge Facies* (W. J. Ebanks Jr., pers. commun.).

TABLE 4.—PETROGRAPHIC COMPONENTS, CEMENT TYPES AND POROSITY TYPES OF FACIES LOW IN CLAY (<3%) AND HIGH IN CLAY (20%). FROM HEARN AND OTHERS (1984).

Mineral		Percentage of Clay in Rock <3%	20%
Quartz		52%	40%
Plagioclase feldspar		6%	4%
Rock fragments		10%	11%
Chert		5%	3%
Glauconite		5%	3%
Misc. Grains		5%	9%
Cement		8%	8%
Cement types:	(1) carbonate (2) silica		(1) carbonate (quartz minor) (2) pore lining clays
Porosity types:	(1) dissolution enhanced (2) moldic		(1) dissolution-enhanced (2) very low porosity

In facies which are more clay-rich, detrital clay and pore-lining diagenetic clay as well as quartz overgrowths and intergranular calcite cements all reduce porosity and permeability. Most of the detrital clay is illite. Diagenetic clays are illite, illite/smectite and chlorite.

Chlorite is more important in the higher energy crossbedded facies (Ranganathan and Tye, 1986). A light coating of chlorite commonly covers the grains and may have inhibited complete cementation in the *Central Ridge* and *High-Energy Ridge-Margin Facies* (Fig. 29A). In some facies limited quartz overgrowths accompany or precede the chlorite growth (Fig. 29B). In other cases, partial pore filling by the silica overgrowths appears to follow chlorite deposition (Figs. 29C, D). Dolomite as well as siderite also occurs as cement (Figs. 29E, F).

CROSS SECTIONS

Four cross sections are included in this paper. One section shows the regional relationships between Jepson Draw, North Holler Draw, Heldt Draw, and Hartzog Draw fields (Fig. 30). Two reflect detailed facies as correlated laterally from cored wells (Figs. 31, 32). One section, running the entire length of the field, attempts only to correlate similar lithologies as determined primarily from log character (Fig. 33). All the cross sections are hung on a bentonite marker which occurs about 30 ft (9 m) below the reservoir and

FIG. 29.—Pore distribution and pore filling, Hartzog Draw field. (A) Typical *High-Energy Ridge-Margin Facies* grain and pore distribution. A light coating of chlorite covers the center grain and scattered silica overgrowths partially fill the pores. (45 md, 16% Ø) CSO Federal AK-1, 9,373.5 ft (2857 m). Scanning electron microscope photo, magnification 200×. (B) Quartz crystal overgrowths and chlorite rims on grains in *Central Ridge (Planar Laminated) Facies*. CSO Federal AE-1, 9,145.0 ft (2787 m). SEM photo, 500×. (C) Silica overgrowths on grains in *High-Energy Ridge-Margin Facies,* right half of photo. Glauconite grain in lower left. Diagenetic effects are mild enough in this facies that porosity is good (16%). CSO Federal AK-1, 9,373.5 ft (2857 m). SEM photo 160×. (D) Euhedral silica overgrowths on earlier formed chlorite coating of detrital grain. *Low-Energy Ridge-Margin Facies,* CSO Federal AK-1, 9,388.0 ft (2861 m). SEM photo, 500×. (E) Siderite (white) in association with platy chlorite and silica overgrowths in *Central Ridge Facies*. CSO Federal AK-1, 9358.5 ft (2853 m). SEM photo, 4,000×. (F) Dolomite cement partially filling pores lined with chlorite plates and rosettes in *Central Ridge Facies*. CSO Federal AK-1, 9358.5 ft (2853 m). SEM photo, 2,000×.

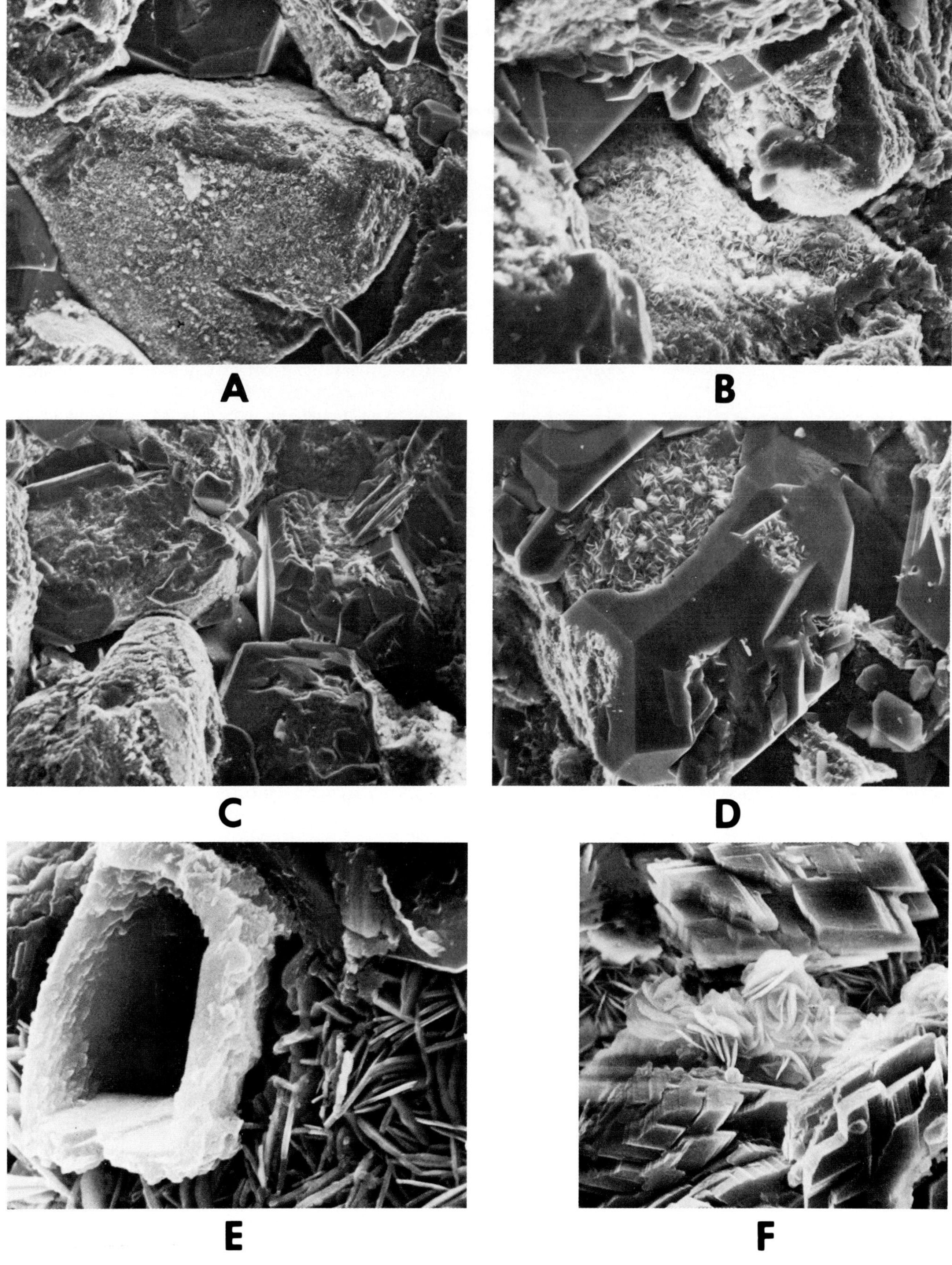
A
B
C
D
E
F

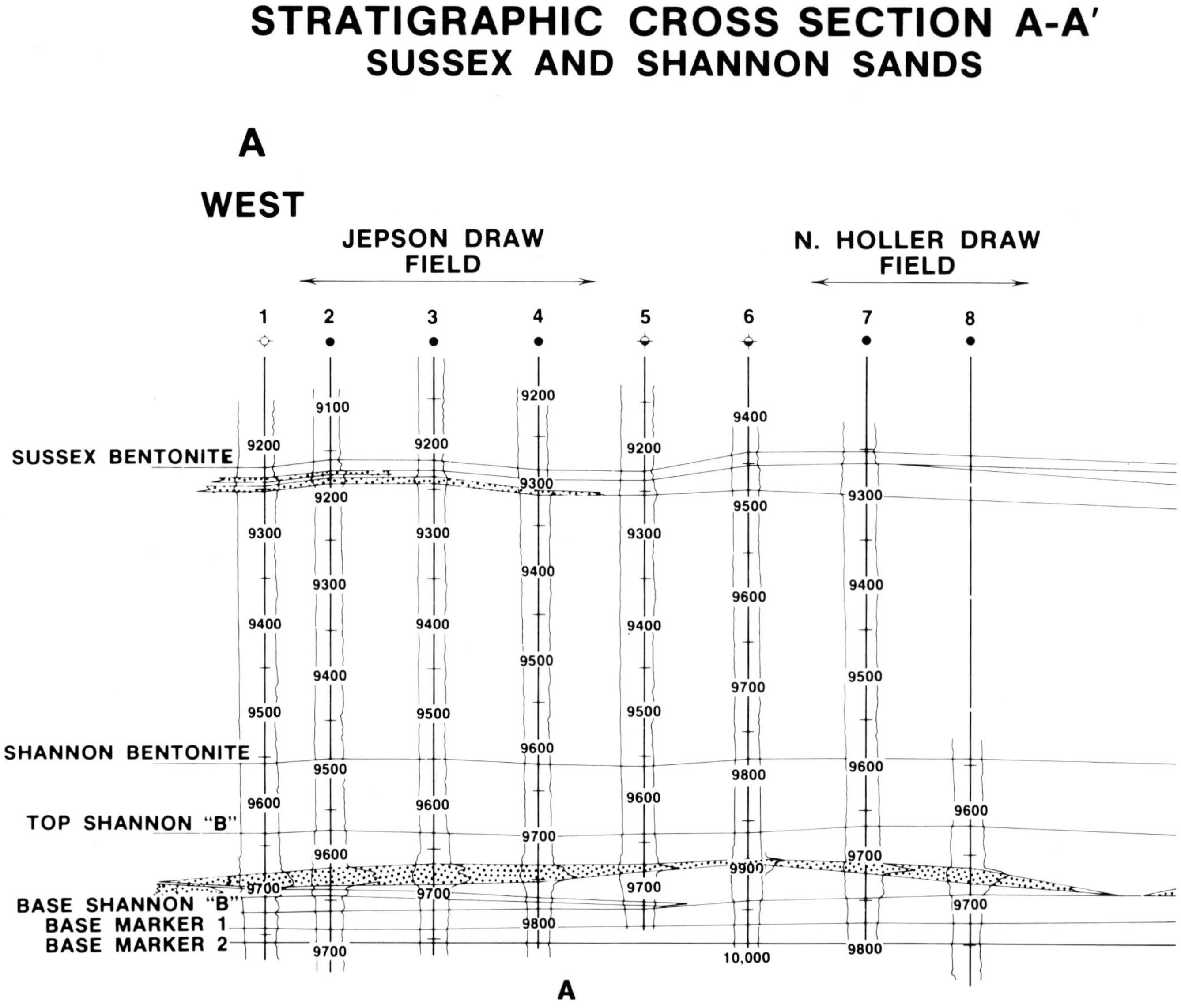

FIG. 30.—(A) Regional west-to-east cross section A-A′ which includes four oil fields which produce from the Shannon Sandstone. Contrast the horizonality of the "Base Markers" 1 and 2 and the "Shannon Bentonite" with general asymmetry of the sandstones in the producing fields. Sand pattern is "clean" porous sandstone. (B) Index map showing location of cross section A-A′.

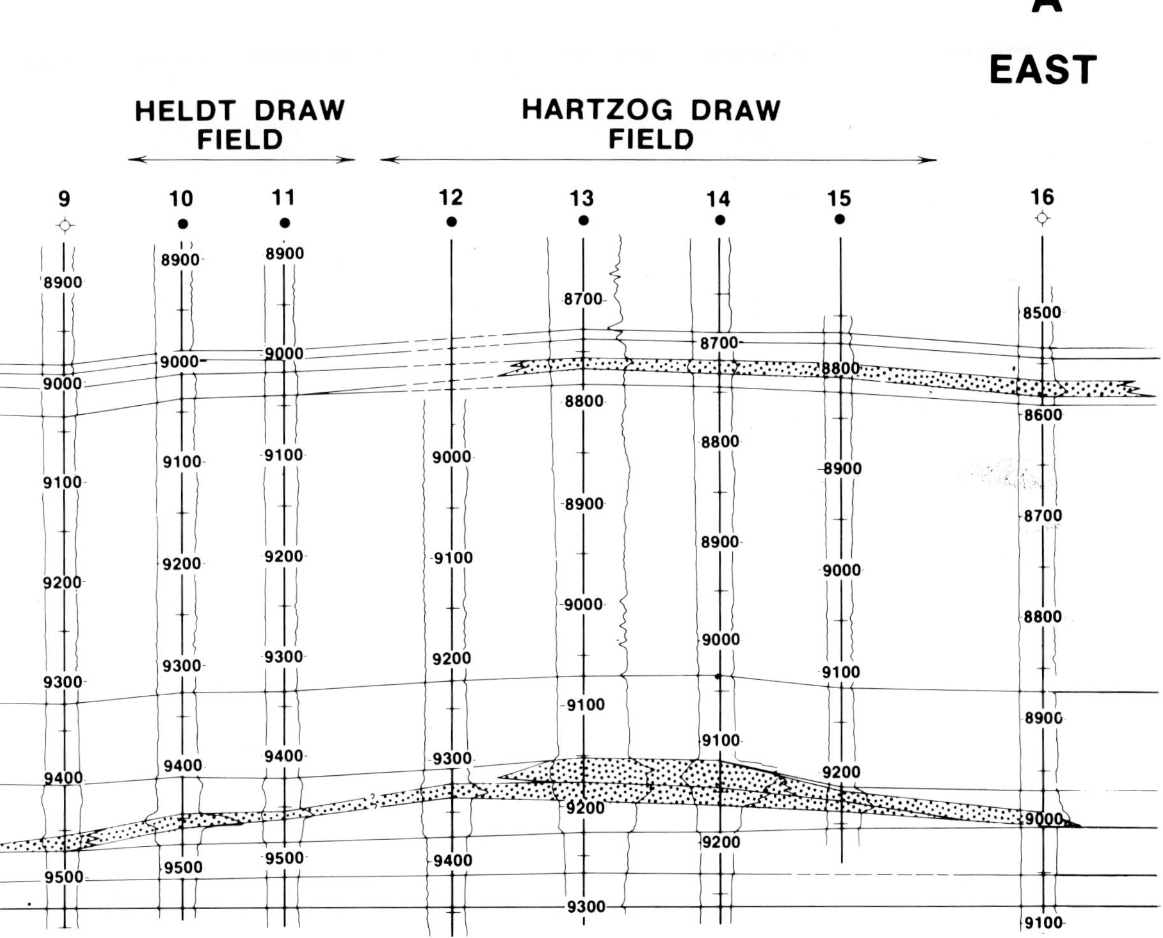
A'
EAST
HELDT DRAW
FIELD
HARTZOG DRAW
FIELD
9
10
11
12
13
14
15
16

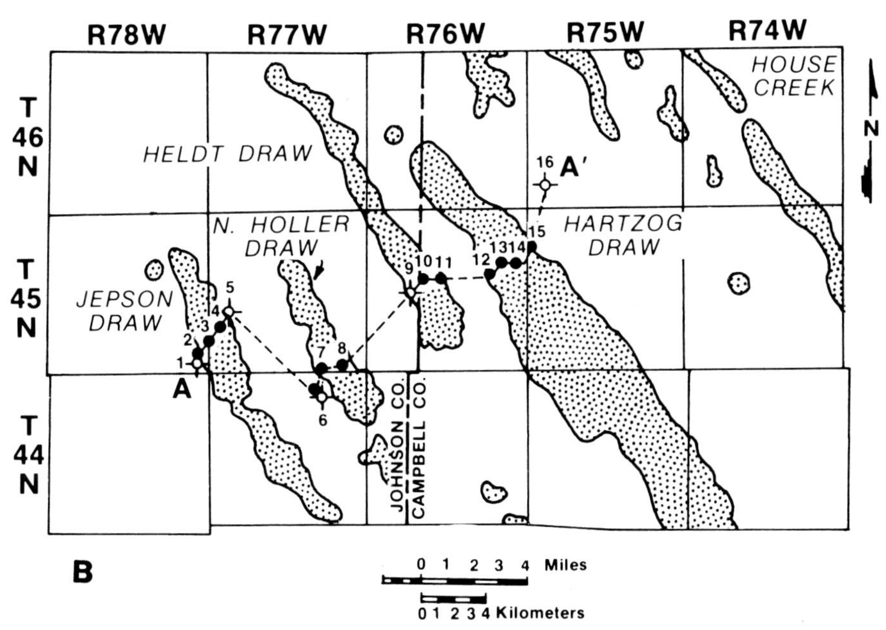
R78W
R77W
R76W
R75W
R74W
T
46
N
T
45
N
T
44
N
HOUSE
CREEK
HELDT DRAW
N. HOLLER
DRAW
HARTZOG
DRAW
JEPSON
DRAW
A
A'
JOHNSON CO.
CAMPBELL CO.
N
B
0 1 2 3 4 Miles
0 1 2 3 4 Kilometers

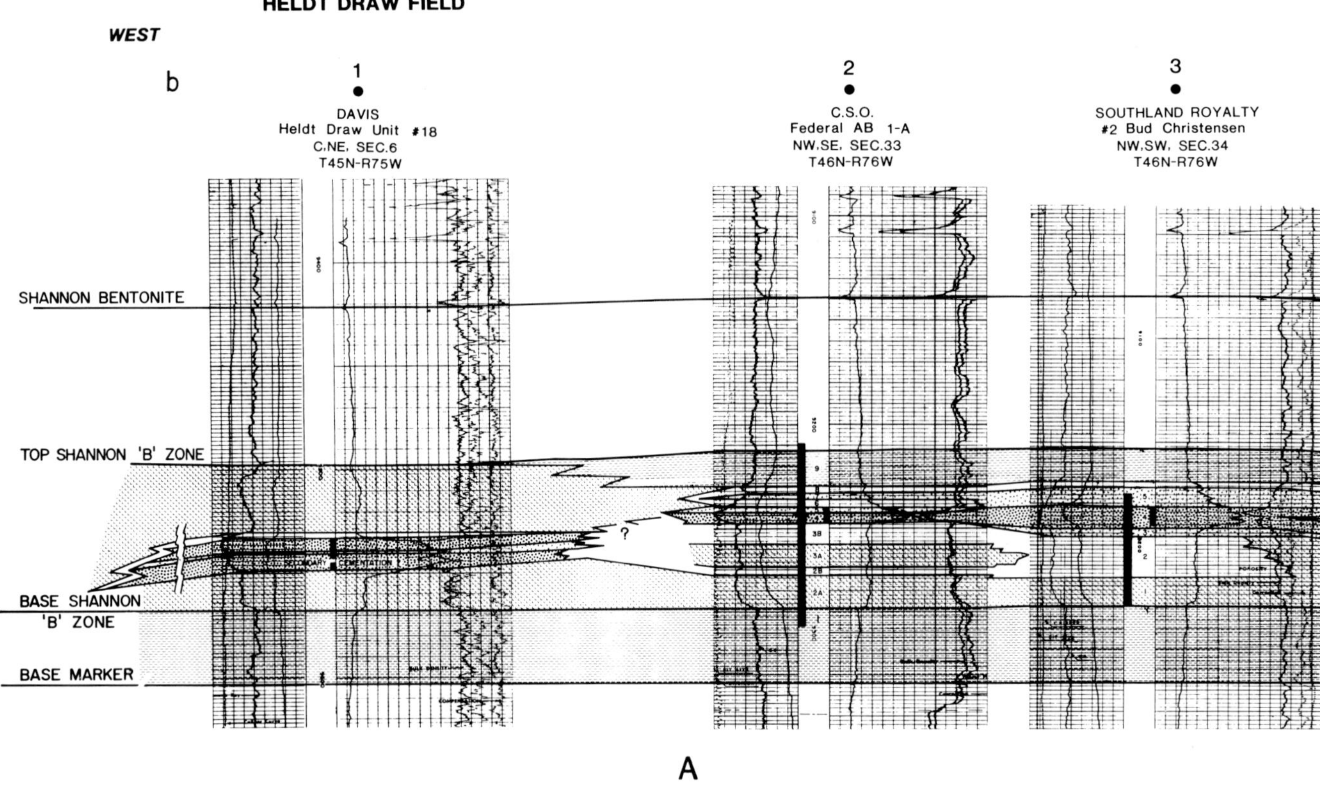

FIG. 31.—West-to-east cross section b-b′ across Heldt Draw and the northern part of Hartzog Draw fields. Discovery well (4) and cored well (Cities Service Federal AB-1A) are included in cross section. (A) Log cross section. Datum is lowest "Base Marker." (B) Summary of facies in (A). Note that Hartzog Draw field is asymmetrical, steeper on the east. Datum is bentonite marker at base. The base of the crossbedded sandstones in both fields slopes in opposite directions (non-horizontal). Location of cross section shown in Figure 34.

HARTZOG DRAW FIELD
EAST
4
SOUTHLAND ROYALTY
#1 Bud Christensen
NW,NW, SEC.34
T46N-R76W
5
SOUTHLAND ROYALTY
#4 Bud Christensen
NW,NE, SEC.34
T46N-R76W
6
C.S.O.
Federal AU #1
NE,NE, SEC.22
T46N-R76W
b'
LEGEND: FACIES
CENTRAL BAR FACIES
BAR MARGIN FACIES (TYPES I & II)
INTERBAR FACIES
BURROWED TO BIOTURBATED SHELF SILTSTONE FACIES
SHELF SILTY SHALE FACIES

HELDT DRAW FIELD
HARTZOG DRAW FIELD
1
CSO AB-1A
SO. LAND B.C. 2
4
5
6
W
E
2
3
FACIES
Central Ridge Facies
Ridge Margin Facies
Inter-Ridge Facies
Bioturbated Facies
Silty Shale Facies
DATUM; BENTONITE MARKER
RWT/RM
2 MILES
1.1 MILES
0.5 MILES

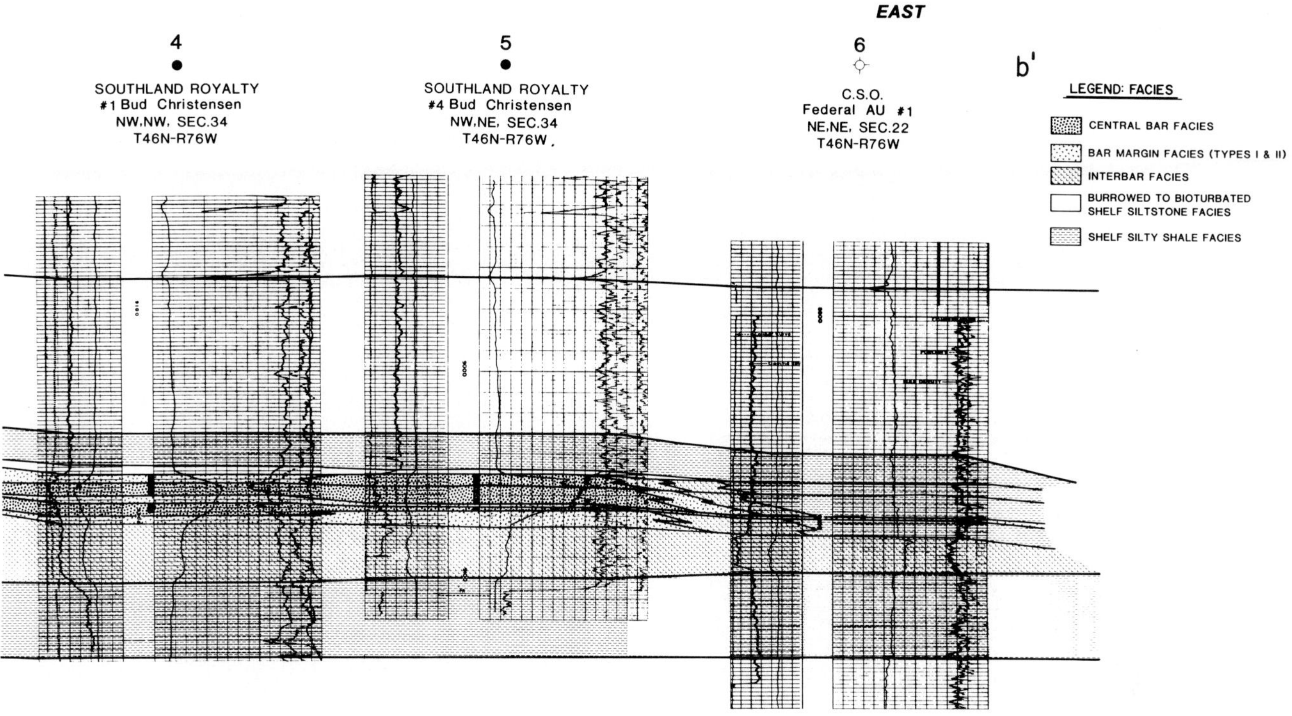

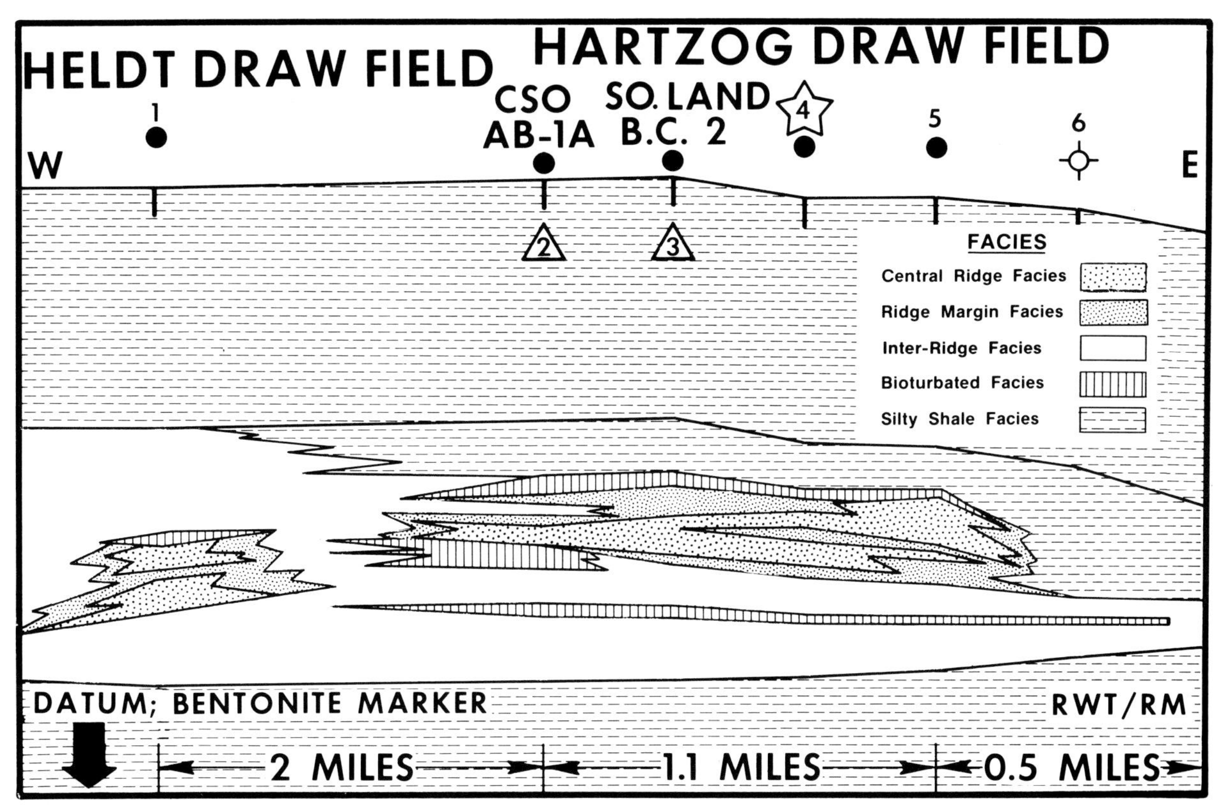

B

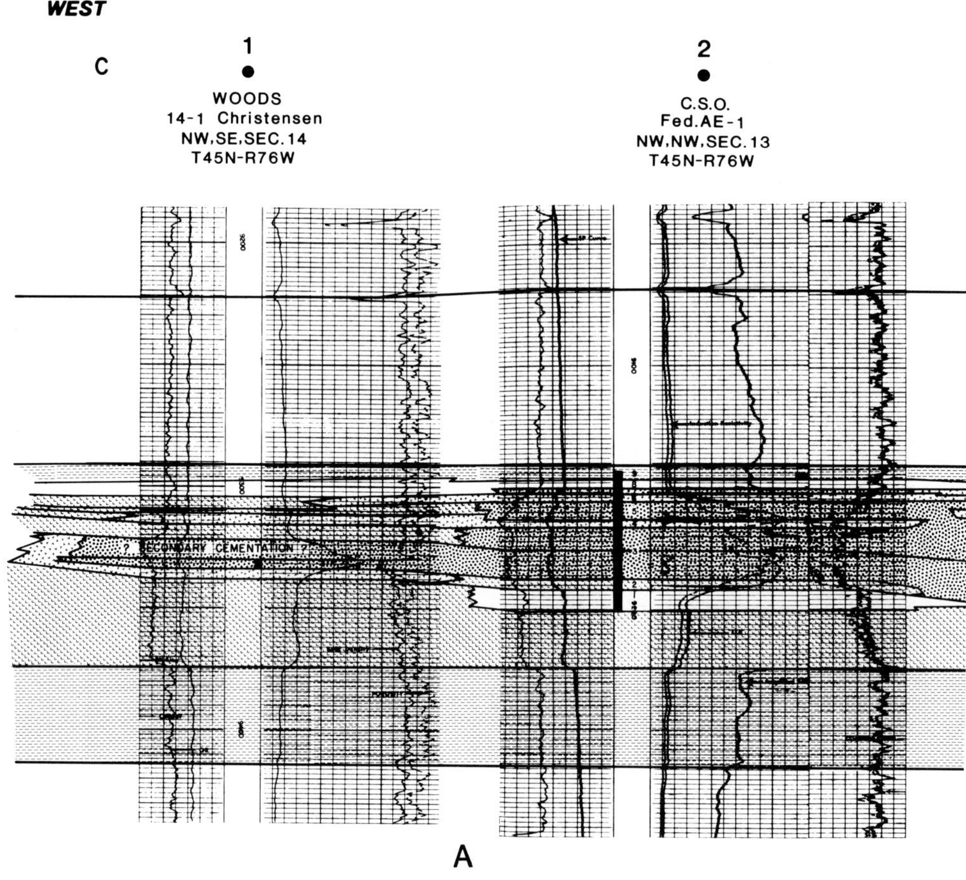

FIG. 32.—West-to-east cross section c-c′ across north-central part of Hartzog Draw field. Cored well, Cities Services Federal AE-1, is included in cross section. (A) Log cross section. (B) Summary of facies in (A). Location of cross section shown in Figure 34.

EAST
3
4
C'
TEXACO
Christensen A-1
NW,NE,SEC.13
T45N-R76W
SOUTHLAND ROYALTY
S. Pumpkin Fed. 1/7
NW,SW,SEC.7
T45N-R75W
LEGEND: FACIES
CENTRAL BAR FACIES
BAR MARGIN FACIES (TYPES I & II)
INTERBAR FACIES
BURROWED TO BIOTURBATED SHELF SILTSTONE FACIES
SHELF SILTY SHALE FACIES
SHANNON BENTONITE
TOP SHANNON "B" ZONE
BASE SHANNON "B" ZONE
BASE MARKER 1

HARTZOG DRAW FIELD
CSO AE-1
W
E
1
3
4
2
FACIES
Central Ridge Facies
Ridge Margin Facies
Inter-Ridge Facies
Bioturbated Facies
Silty Shale Facies
DATUM; BENTONITE MARKER
RWT/RM
2.2 MILES

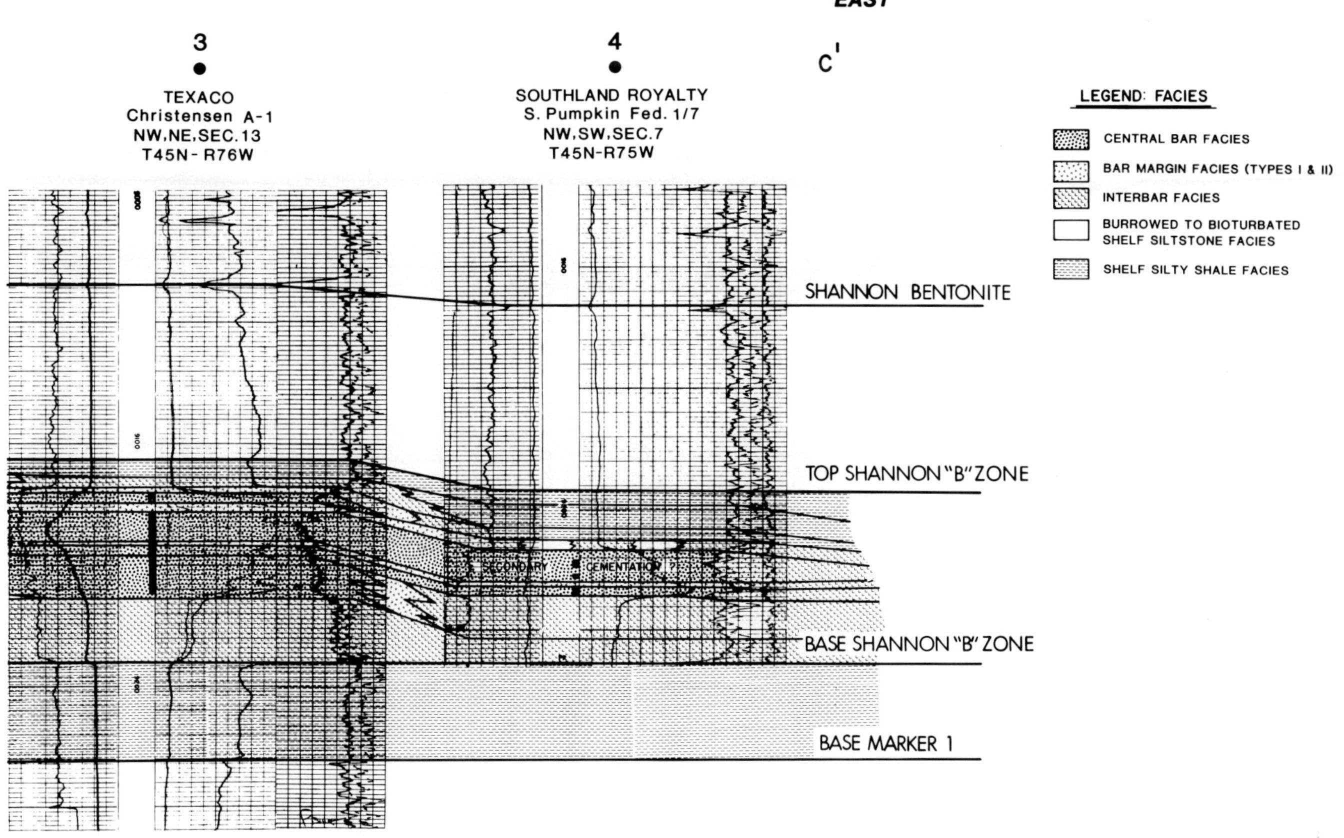

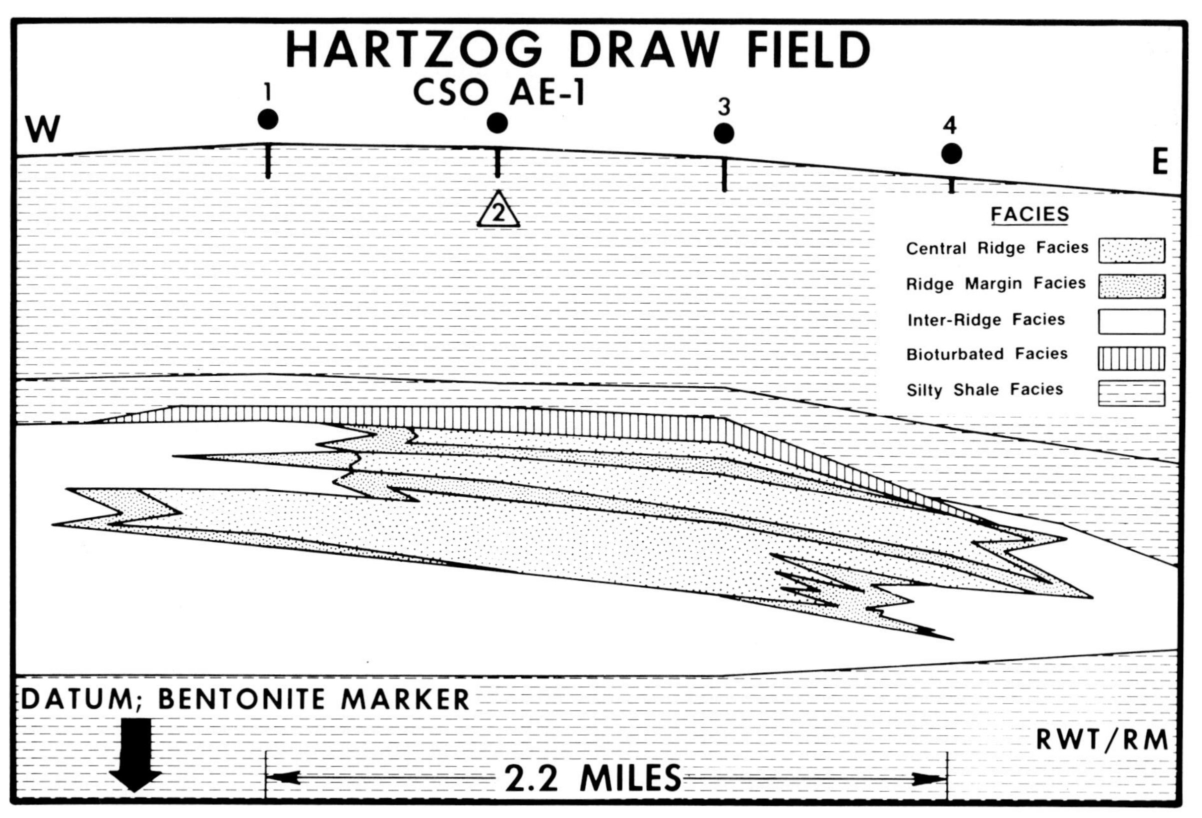

B

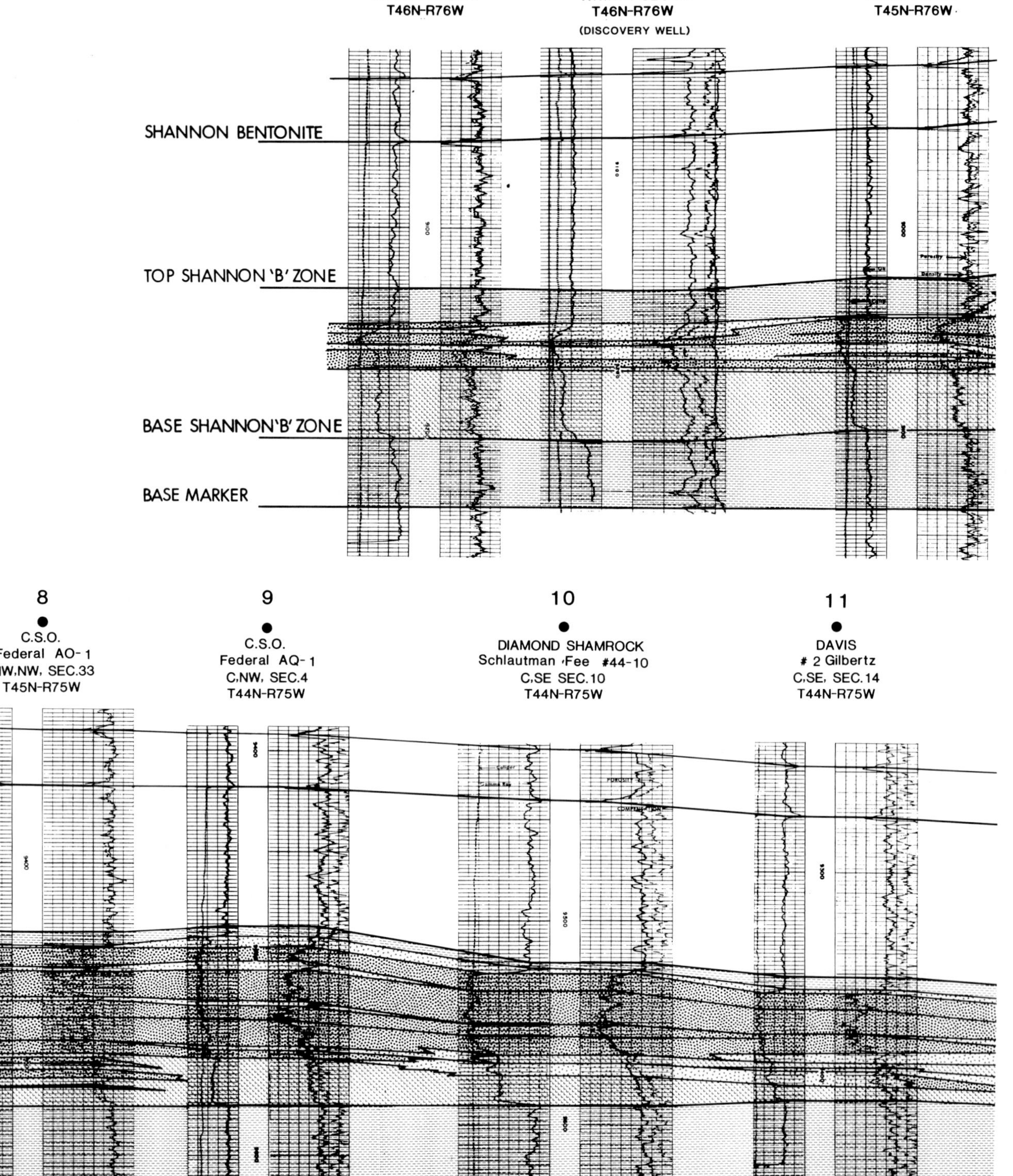

FIG. 33.—(A) North-south *lithofacies* cross section (D-D′) through Hartzog Draw field. Lithofacies more

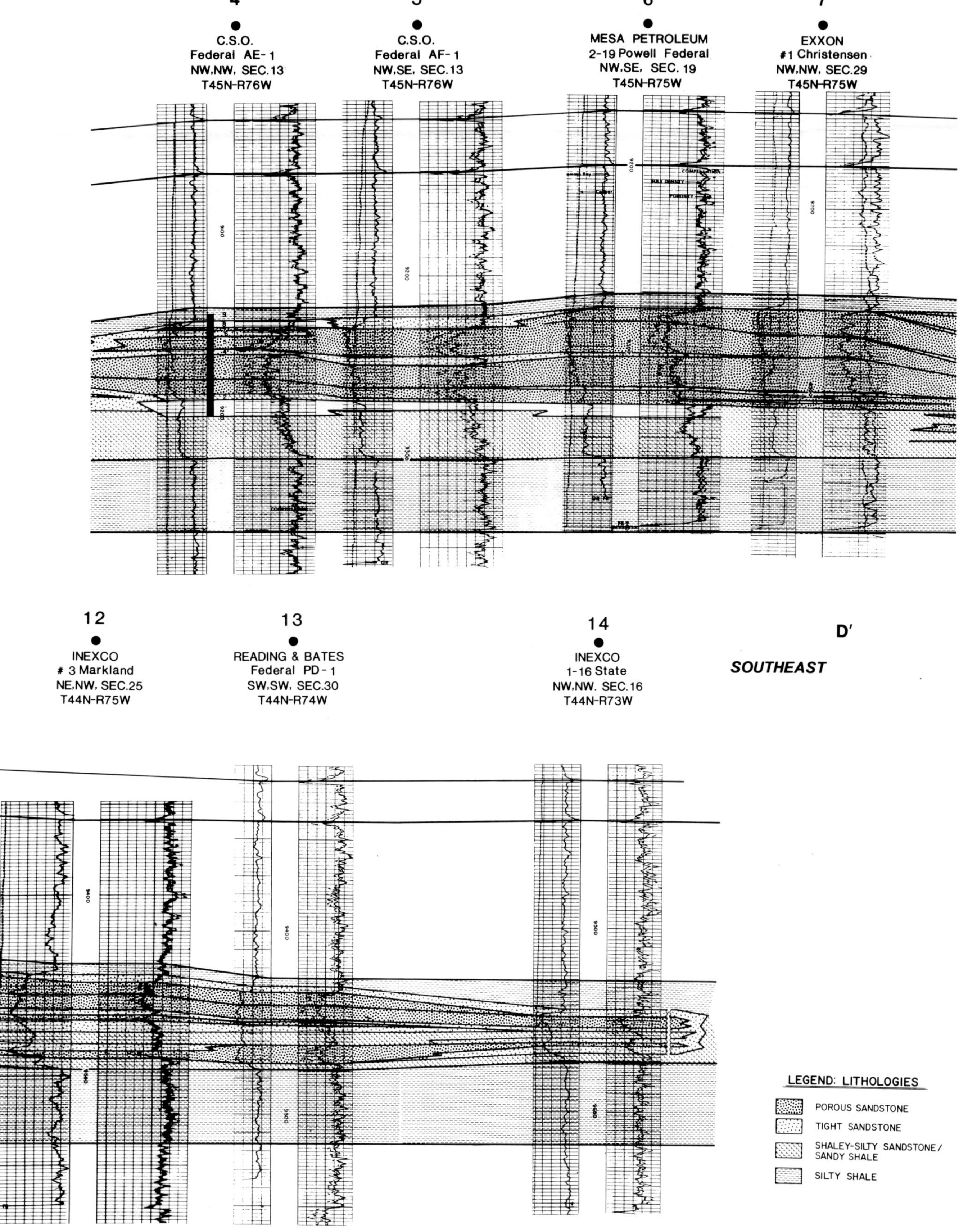

generalized here than in detailed east-west cross sections. Location of cross section is shown in Figure 34.

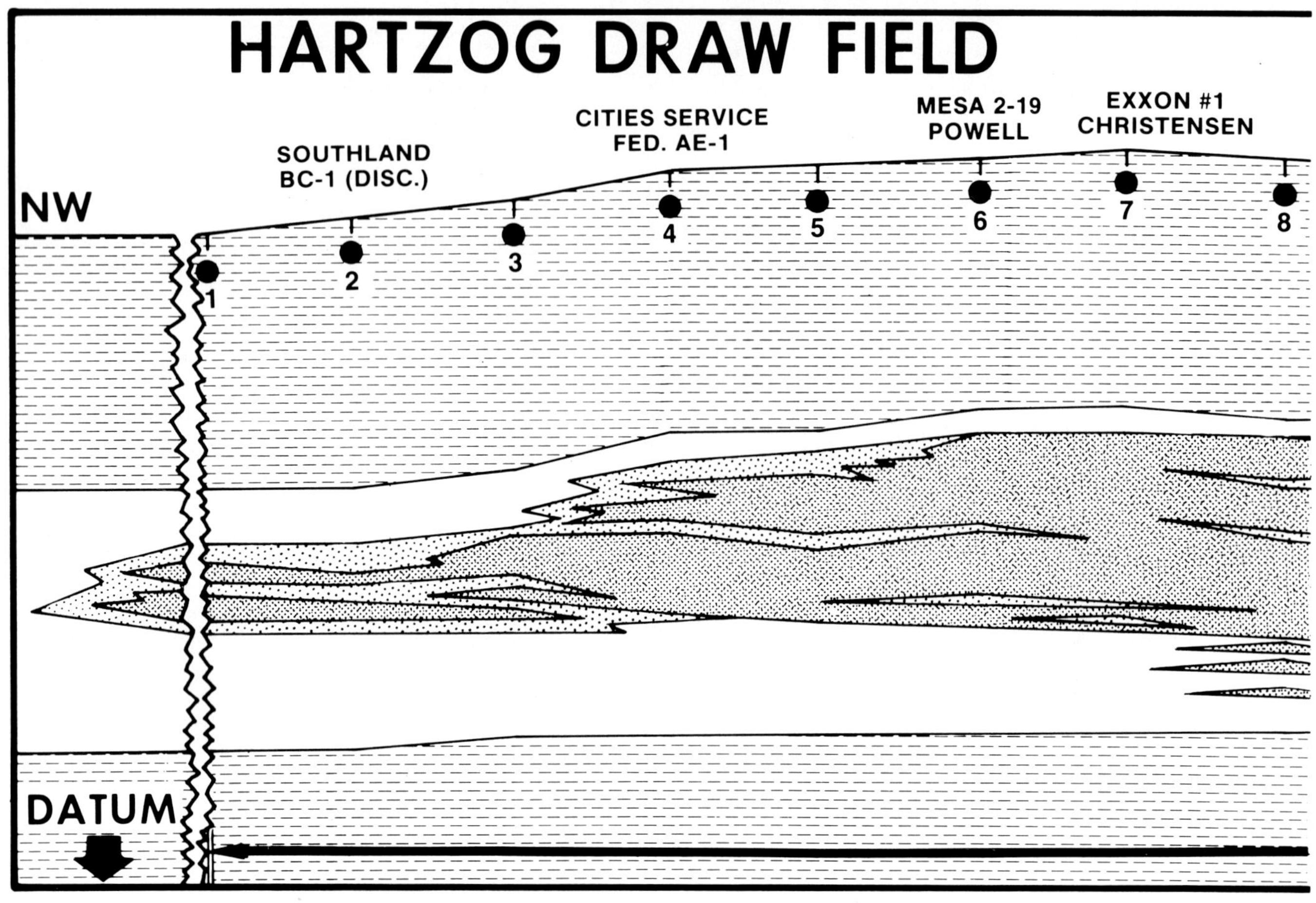

FIG. 33.—(B) Generalized north-south litho

below the base of the Shannon "B" sand zone. An upper Shannon bentonite marker occurs about 80 to 90 ft (24–27 m) above the top of the Shannon "B" sand zone.

Construction of Cross Sections

Cross Section A-A′ (Fig. 30), a subregional cross section, is constructed on an east-west line, roughly perpendicular to the strike of four Shannon producing fields in the Hartzog Draw field area, and includes both the Shannon and Sussex sandstone intervals. Only porous or "clean" sandstone and "tight" or "dirty" sandstones are indicated on the section (intervals of shale are inferred). Lithologic interpretations for this cross section are based on the amount of deflection on the spontaneous potential and resistivity curves.

Cross sections B-B′ (Fig. 31) and C-C′ (Fig. 32) run transversely west-to-east across the northern portion of the field (Fig. 34). In these two cross sections, facies observed in cores have been extrapolated laterally using similarity of log shapes and patterns to correlate facies observed in the cored wells. All correlations in these two cross sections assume an optimistic interpretation of the size and continuity of individual sandstones. Consequently, lateral continuity of individual sandstones between wells is indicated, unless otherwise suggested by core and/or log interpretations. Individual sandstones, which comprise the reservoir facies, however, may not always be laterally continuous on 160-acre spacing. Also, due to the inability of electric logs to recognize accurately measured properties of beds less than 2 ft (0.6 m) thick, sandstones which may contain vertical and horizontal barriers to flow may commonly be lumped together vertically on these cross sections. Gamma-ray and porosity logs were the primary logs used for detailed correlation. Spontaneous potential and resistivity log curves were also used.

A typical SP and gamma-ray log through the shelf-ridge complex is shown in Figure A4 (CS Fed. AE-1). The cored interval in this well does not extend down into the *Shelf Silty Shale Facies;* however, this facies was cored in other wells (Fig. 18; Tillman and Martinsen, 1985). The interval below 9,226 ft (2,795.7 m) is typical of the *Shelf Silty Shale Facies*. The interval from 9,190 to 9,226 ft (2,785.2 to 2795.8 m) is typical of both the *Inter Ridge Facies* (Shaly) and *Bioturbated Shelf Siltstone Facies*. These two facies cannot be easily differentiated on logs. The *Central Ridge Facies* commonly shows a stronger deflection to the left than any other facies. Both the *Low-* and *High-Energy Ridge Margin Facies* commonly have lower SP and higher gamma-ray values than does the *Central Ridge Facies*.

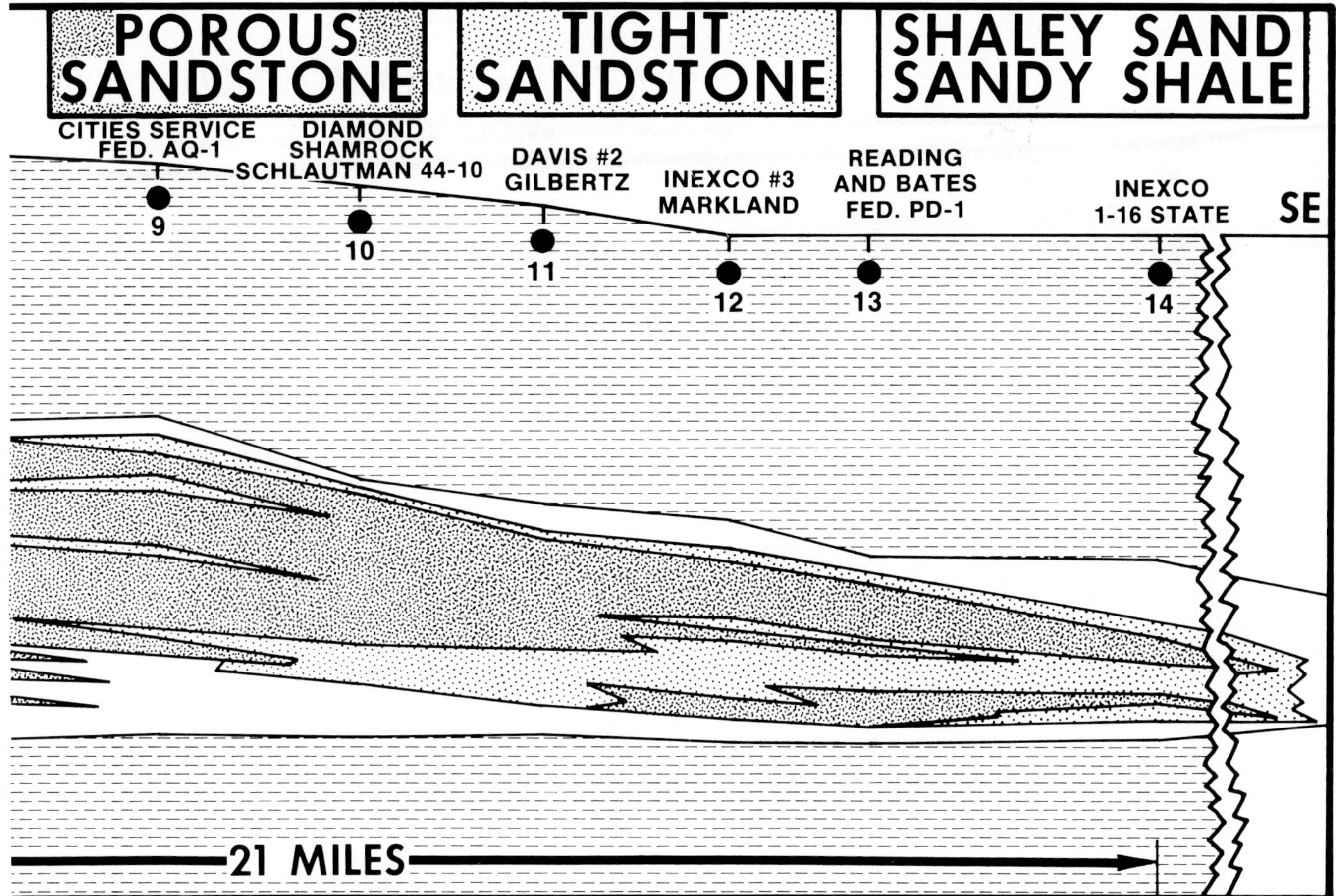

facies cross section. Summary of Figure 33A.

Cross section D-D′ (Fig. 33) runs the length of the field and roughly depicts the internal as well as overall geometry of the Shannon Sandstone. This north-south cross section does not depict individual facies because at the time this cross section was constructed, only cores from the northern portion of the field were available. An attempt to extrapolate individual facies the entire length of the field, over 22 mi (35 km), was felt to be inappropriate. Instead, it was believed that a rough approximation of the distribution of groups of related facies could be achieved by delineating lithologies interpreted from logs. As will be seen in the following discussions, the pattern of deposition presented in this north-south lithologic cross section is very similar to the patterns presented in the core-calibrated facies cross sections.

In this cross section porous sandstone is defined as having greater than 11% effective porosity (assuming a grain density of 2.71) and corresponds mostly with the *Central Ridge Facies* but may include porous *High-Energy Ridge-Margin Facies,* rarely very sandy, porous *Low-Energy Ridge-Margin Facies,* and, even more rarely, *Inter-Ridge Facies (Shaly).* Tight sandstones (as observed on logs) correspond commonly with the *Low-Energy Ridge-Margin Facies* and minor *Inter-Ridge Facies.* Tight sandstones also occur as calcite "streaks" within the *Central Ridge Facies* (W. J. Ebanks Jr., pers. commun.). Those units designated as "shaly silty sandstone/sandy shale" are primarily *Inter-Ridge Facies* and *Burrowed Shelf Siltstone Facies.*

Observations

Several important features of the field are apparent from these cross sections. The most obvious feature is that the reservoir at Hartzog Draw is not homogeneous either laterally or vertically but consists of a stacked to interfingering sequence of sandstones. In the central portions of the field, "individual" *Central Ridge* and *Ridge-Margin* sandstone units are thicker, and "sand-on-sand" contacts are more common than on the edges of the field, where the sandstone units are thinner and interfinger with the more silty, shaly and burrowed facies. As porosity and permeability are related directly to facies type and to degree and type of diagenesis, these cross sections suggest that many of the sandstones within the reservoir are in poor productive continuity with one another, especially near the edges of the field. Understanding facies distributions, therefore, is important in order to understand the potential fluid flow patterns in the reservoir.

A second feature which can be observed on the cross sections is the asymmetry of the ridge complex (Figs. 30,

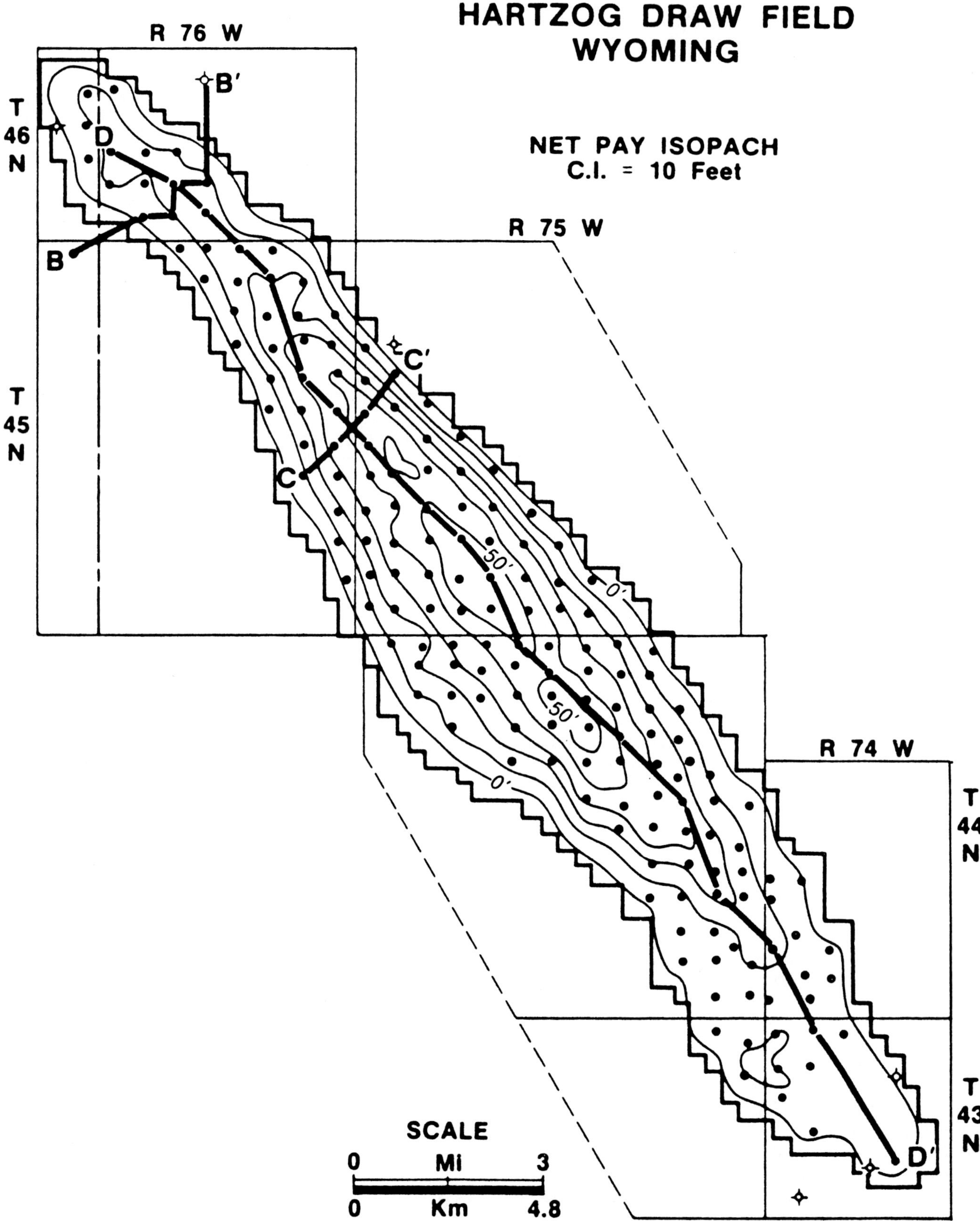

FIG. 34.—Net pay isopachous map showing the location of detailed cross sections B-B′, C-C′, D-D′ (Figs. 31, 32, 33).

31). The east flank of the field (especially the northeast flank) is much steeper than the west side. Because the sharp decrease in thickness of the pay sandstones on the east side of the field was easily mapped, very few thin pay zone wells were drilled on the east during development of the field. Those wells drilled on the east flank, however, need not encounter a thick pay section to be productive, since the quality of the reservoir facies is significantly greater along the east flank than along the west flank of the field. Where a successful hydraulic fracturing treatment in easterly wells was achieved, conduits to the rapidly thickening pay section to the west were opened. An example of anomalously high initial production is the Exxon Harper Federal No. 3 (5173, SW Sec. 17 T45N R75W) with a calculated net pay of only 1 ft and yet potentialed for 88 BOPD. A significant number of very marginally productive wells which have thin net pay intervals was drilled on the west side of the field. In the western and southwestern portion of the field the overall sandstone thickness decreases less abruptly and the better producing *Central Ridge* and *High-Energy Ridge-Margin Facies* are thin or absent.

Third, the base of the *Central Ridge/Ridge-Margin Facies* sandstones systematically changes position vertically within the Shannon "B" sand zone. The base of the "clean sandstone" rises from east to west at Hartzog Draw, and from west to east at Heldt Draw field (Figs. 30, 31). It also rises from south to north (in an upcurrent direction) in the greater Hartzog Draw area (Fig. 33), and from north to south (in a downcurrent direction) in the greater Salt Creek field area (Tillman and Martinsen, 1984). Recognition of these "rise-and-fall" relationships is very important for delineating sandstone trends. It has, in the past, been common practice in this area to employ "layer cake" stratigraphy to designate Shannon sandstones as upper or lower. This approach is incorrect because the sands were not deposited in a "layer cake" fashion, but are depositionally inclined. We believe the "rise-and-fall" of the base of the cross bedded sandstones within the Shannon interval is typical of Shannon sandstone-type deposition and probably represents deposition in response to one or more of the following: positive paleo-seafloor topography, asymmetry of the ridge geometry during formation, and/or hydrodynamic flow such that the high-energy sands are limited to certain areas which migrate as the ridge grows. Slightly different stratigraphic position of reservoir sandstone development in nearby wells, therefore, does not necessarily indicate that the two wells represent separate sandstone bed sequences and/or facies. Conversely, a similar stratigraphic position in adjacent or nearly adjacent wells doesn't necessarily indicate that a reservoir sandstone in the two wells is part of a single sandstone bed sequence and/or facies.

HARTZOG DRAW DEPOSITIONAL MODEL

Paleogeographic Setting

Belemnites, where present in the Upper Cretaceous, are probably the most common and reliable age dating criteria in the Rocky Mountains. Gill and Cobban (1966, 1969, 1973) dated the Shannon Sandstone Formation of the Cody Shale as Lower Campanian on the basis of the occurrence of "smooth" *Baculites sp*. They indicate slightly younger *Baculites sp*. (weak flank ribs) are typical of the Sussex Sandstone. Based on the occurrence (mostly in outcrop) of these two *Baculites* types, they have drawn paleogeographic maps indicating the location of the shorelines during and following Shannon Sandstone deposition (Fig. 35). As can be seen, the shoreline during Shannon time lies more than 100 km to the west of Hartzog Draw field.

Subsurface samples from the *Shelf Silty Shale Facies* in the Cities Service Federal AB-1 well (Units 1 and 8) were submitted for paleoenvironmental analysis to D. Dailey at the Cities Service Research Laboratory. He reports this facies yields a mixed agglutinated calcareous foraminiferal fauna suggestive of probable middle shelf (60–300 ft, 18.2–90.9 m) depths. The dominant faunal elements are illustrated in Figure 36 and their inferred depth distribution is shown in Figure 37. Comparison of the species identified at Hartzog Draw with those at Salt Creek (Tillman and Martinsen, 1984) indicates significant differences in the species but an overall similarity of aspect. In the Upper Cretaceous of the Rocky Mountains, specific foraminiferal morphotypes are thought to occur at water depths both shallower and deeper than the assemblages described in the Shannon. These forms are illustrated in Figure 38. Except possibly for *Ammobaculites,* morphotypes characteristic of only shallow depth (littoral and inner shelf) are absent, suggesting that Shannon foraminifera reflect depths greater than the inner shelf zone. Likewise, architectural forms typically associated with deeper shelf and slope environments, i.e., praebuliminids, valbulinerids, etc., are largely absent, implying a maximum depth limit of middle shelf for deposition of the Shannon shelf-ridge sequence at Hartzog Draw field.

Evidence of the presence of high-energy conditions dur-

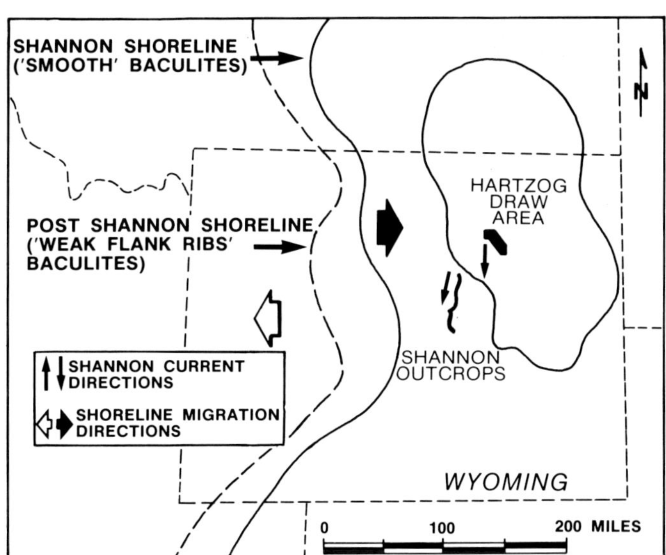

FIG. 35.—Shoreline locations during a portion of Upper Cretaceous time. Both shorelines shown are Lower Campanian. The Shannon shoreline is based on occurrences of *Basculites sp*. (smooth) and the post-Shannon (Sussex) shoreline is based on the distribution of *Baculites sp*. (weak flank ribs). Modified from Gill and Cobban (1973).

COMMON FORAMINIFERAL SPECIES SHANNON SANDSTONE, CODY FORMATION

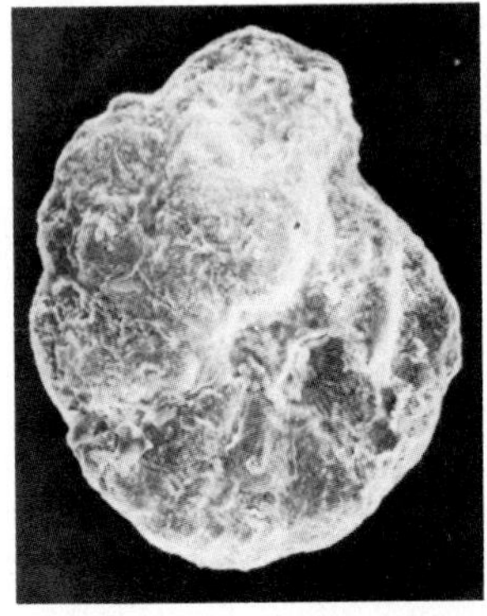
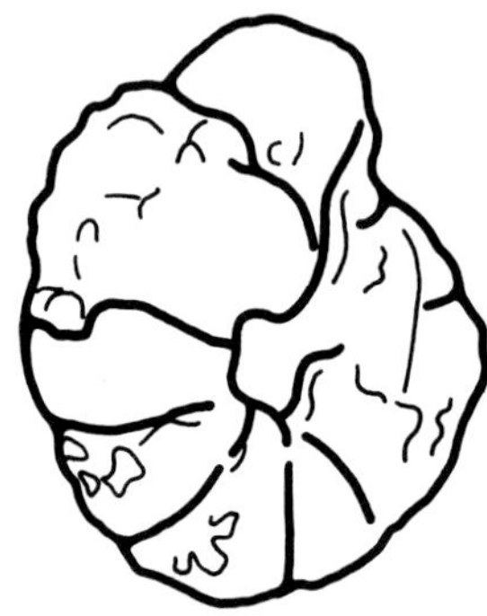

HAPLOPHRAGMOIDES SP.

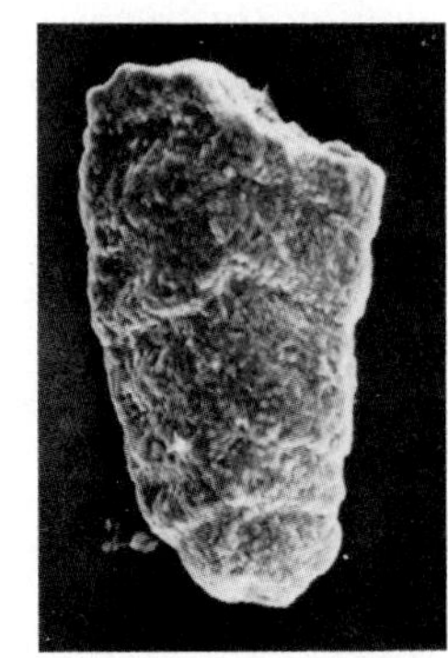
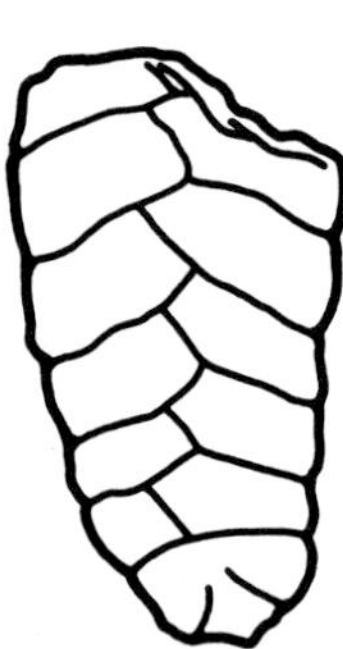

SPIROPLECTAMMINA SEMICOMPLANATA

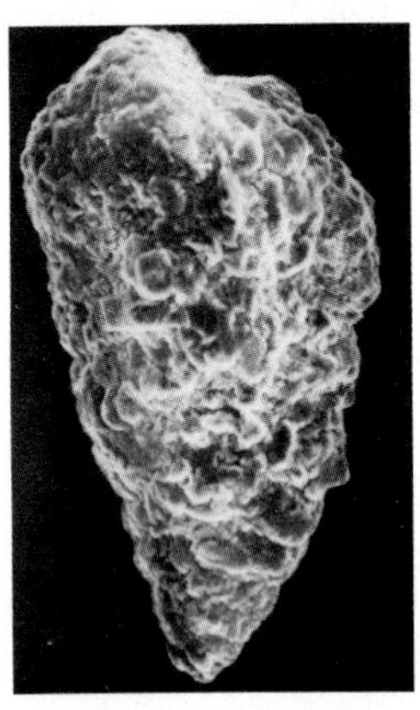

PSEUDOBOLIVINA ROLLAENSIS

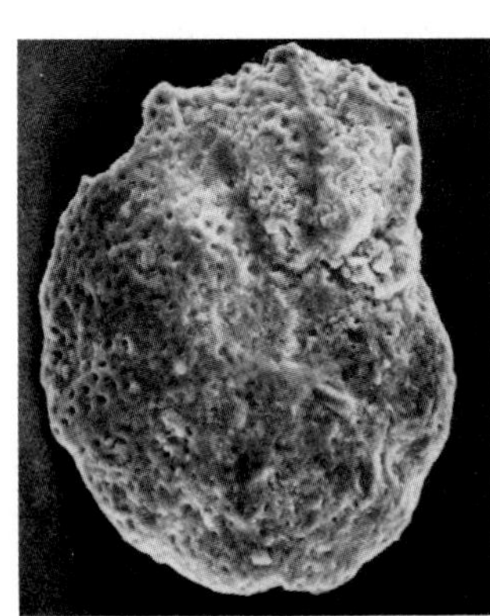
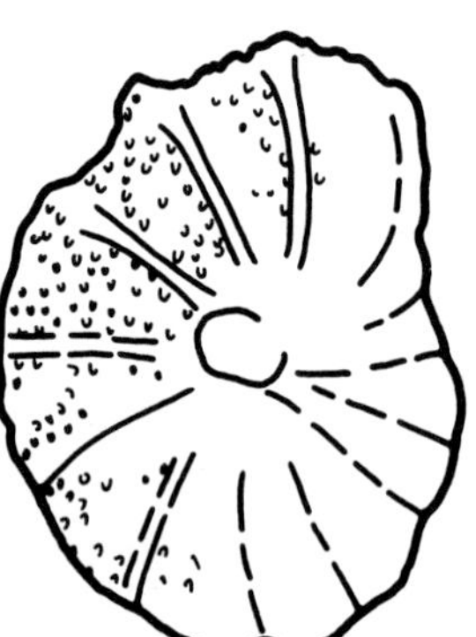

GAVELINELLA KANSASENSIS

FIG. 36.—Common foraminifera from shales immediately above and immediately below the shelf-ridge facies in the Cities Service AB-1A. *Haplophragmoides sp.* and *Spiroplectammina semicomplanata* are common in samples from 9,301 to 9,305 ft (2835–2836 m) and are among the 13 species of foraminifera identified in this interval. *Gavelinella kansasensis* is common in the 9,301–9,305 ft interval and also in a sample at 9,223 ft (2811 m) which is above the sand ridge facies. Also common in the 9,223-ft sample is *Trochamminoides* cf. *T. velascoensis* Cushman and in small numbers *Dorothia smokyenis* Wall, *Gyroidina globosa* var. *orbicella* Bandy and *Lenticulina muensteri* (Roemer), (D. Dailey, pers. commun.) (Compare with Table 6, Tillman and Martinsen, 1984).

ing deposition of the *Shelf Silty Shale Facies* at Hartzog Draw field is indicated by the state of preservation of many of the foraminiferal tests. The larger foraminifera commonly exhibit damage or loss of the final chamber or two, or an abraded periphery may be developed on planispiral tests. An appreciable amount of pyrite as tiny crystals and granules, and as coatings on microfossils and sand clusters, is present in the material recovered for micropaleontologic analysis. In addition, radiolarian tests are invariably pyritized. These aspects may be possible evidence for rapid burial in at least periodically turbulent environments.

Current Flow Direction Analysis

Directional data for Shannon outcrops was discussed by Spearing (1975) and Tillman and Martinsen (1984, p. 114), who recorded very similar southerly mean transport directions for measurements taken from outcrop trough and planar-

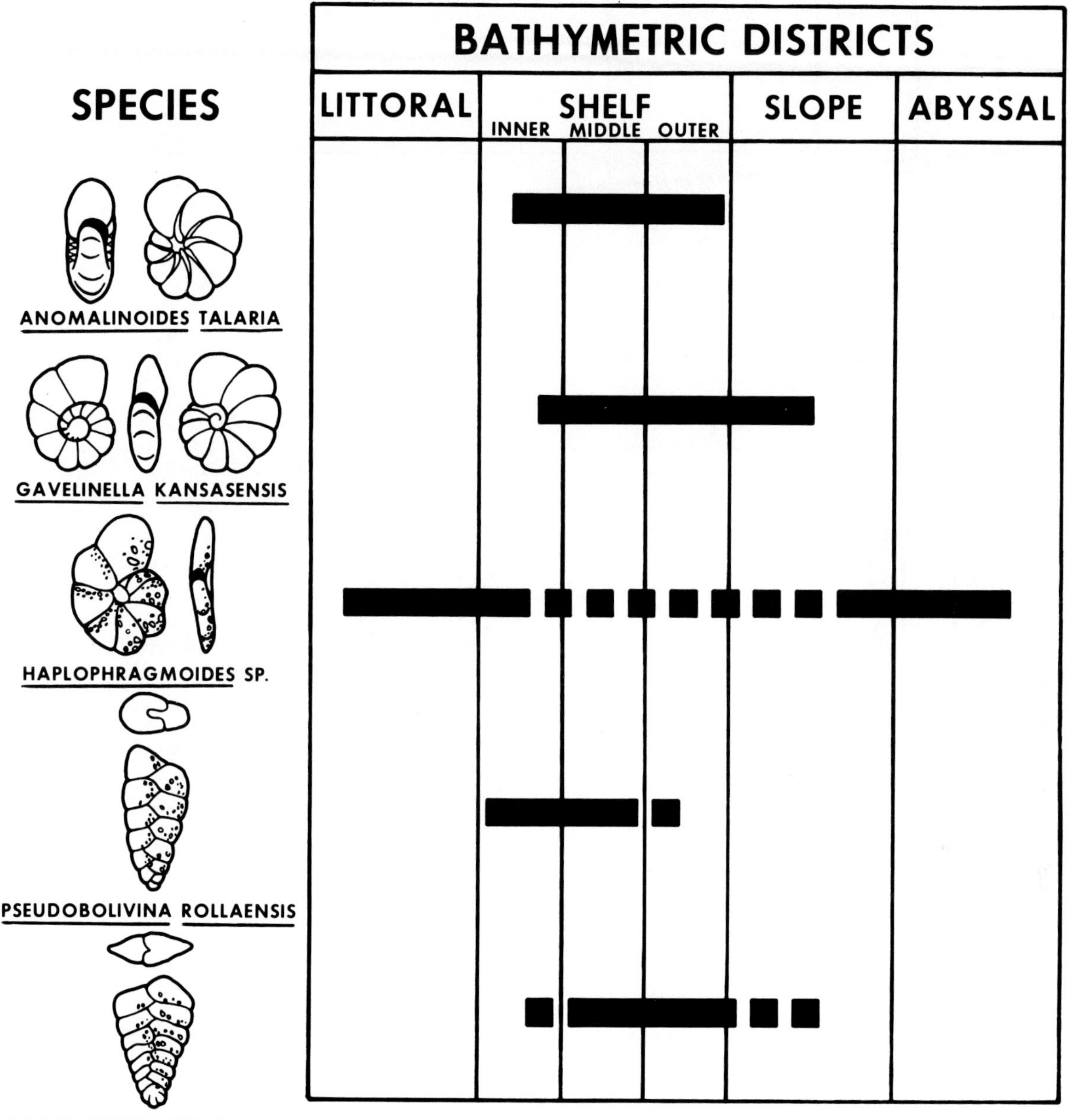

FIG. 37.—Depth zonation of common foraminifera in shales bracketing shelf-ridge sandstones in Cities Service AB-1A. Note that all the Campanian foraminifera identified occur at middle to outer shelf depths. (D. Dailey, pers. commun.).

tabular bedding in the Salt Creek field area of Wyoming. The mean transport directions recorded were 200° ± 15° and 188°, respectively. Analysis by Richard Scott (1978, pers. commun.) and R. Tye and J. Ebanks (*in* Hearn and others, 1984, Fig. 13) using subsurface oriented cores shows a similar mean transport direction.

Tye and Ebanks limited their data mainly to high-angle crossbedding in the reservoir portions of three cores. They determined mean transport directions for three wells in the secondary recovery pilot area (Hearn and others, 1984) in the thickest portion of the field. Mean transport directions in the three wells were 131°, 166° and 169°.

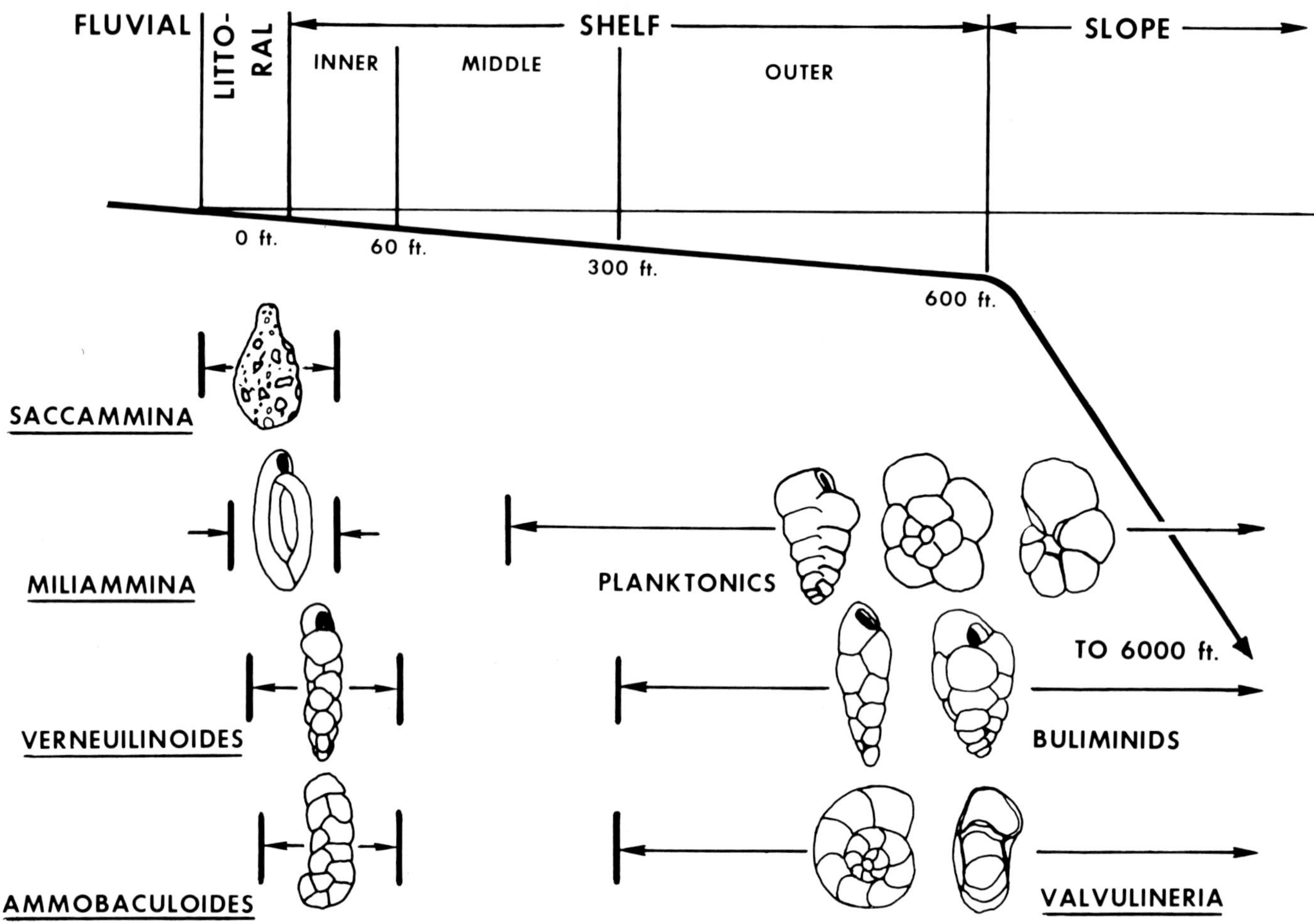

FIG. 38.—Common non-Shannon foraminifera which have been observed in Upper Cretaceous environments of deposition shallower and deeper than that interpreted for the Shannon by Dailey (pers. commun.). The absence of these types of foraminifera in the *Shannon* is supportive evidence for a middle to outer shelf depth of deposition for Hartzog Draw field.

Richard M. Scott analyzed the complete set of crossbed data (n = 132) in the Cities Service Federal AS-1 core (Fig. 11). An average transport direction of 181° was calculated. Two-thirds of the transport directions fall between 155° and 207°. In Scott's work two variations were attempted to remove poorer quality data from the data set: (1) angles less than 10° were removed, and (2) dip directions deemed by the geologist doing the measurements to be of significantly poorer quality were also removed. The first variation only cut the number of points from 132 to 125; the second cut the number from 125 to 119. The mean vector result for the variations was 182° and 181°, respectively, indicating that removal of apparently lower quality data from the core data set had no effect on calculation of the mean transport direction.

Model

The Shannon sandstones in the Powder River Basin of Wyoming were deposited as shelf-ridge complexes on a wide shelf during early Campanian time in the Western Interior Seaway (Fig. 39) stretching from Alaska to Mexico. The wide shelf model proposed by Asquith (1970), in which individual shelf-ridge complexes can be recognized on the shelf and at the shelf-slope boundary, is believed to be applicable for deposition during Shannon time.

The Shannon sandstones in the Hartzog Draw field area were probably deposited in water depths well below effective wave base, at depths as great as 300 ft (91 m; middle to outer shelf) and 100 mi (160 km) east of the contemporaneous shoreline (Fig. 35) following the last stages of the Telegraph Creek-Eagle regression about 80.5 ma (Santonian to Campanian). During the regression, the shoreline moved east 240 mi (384 km) in 5.5 million years at a rate of about 50–70 mi (83.5–116.5 km) per million years (Gill and Cobban, 1973). Based on interpretations of Gill and Cobban (1973), this regression began as early as 83.5 ma and probably ended briefly about 80 ma. The end of the "Shannon regression" was marked by a minor, but rapid,

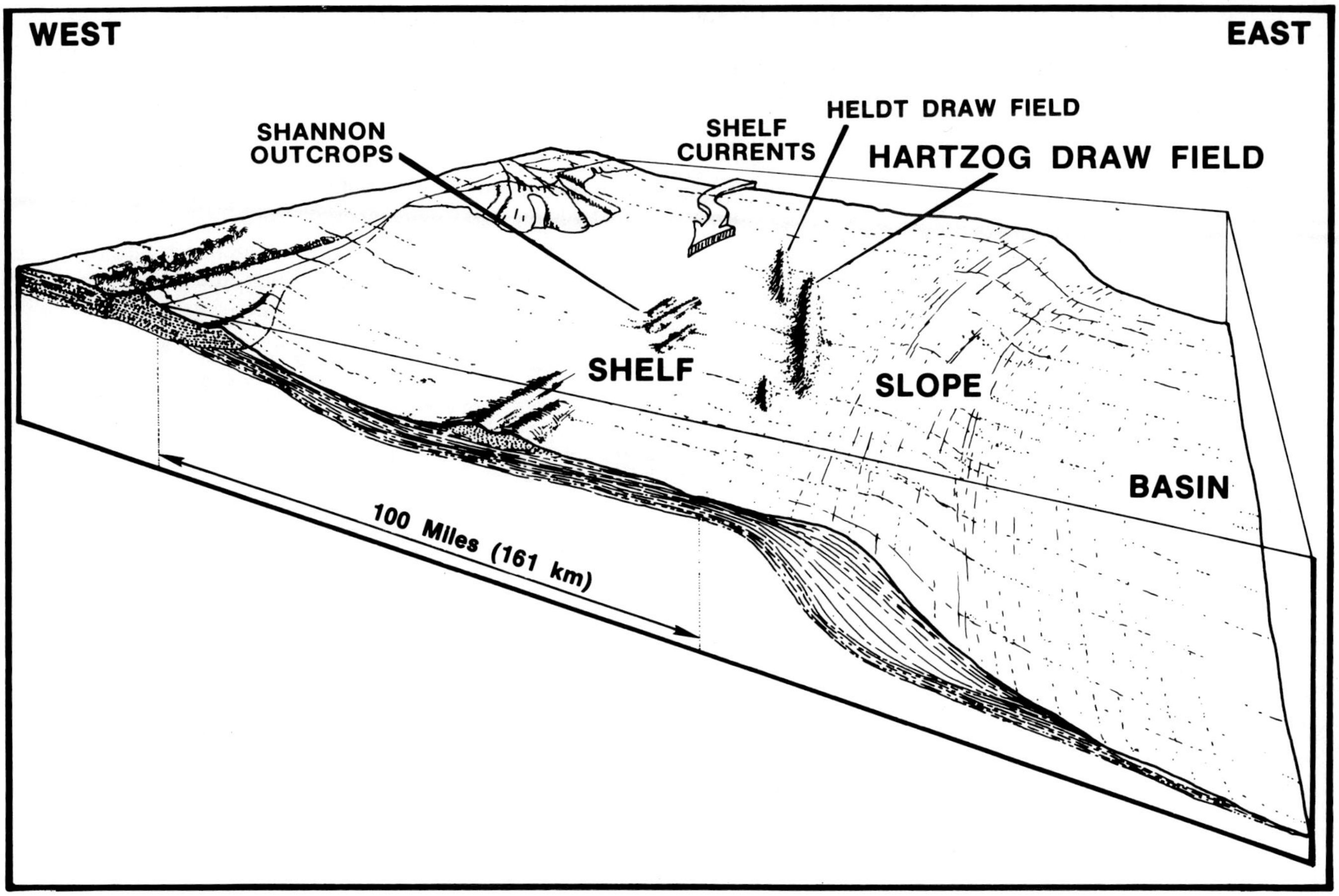

FIG. 39.—Wide shelf on which Hartzog Draw and other Shannon shelf-ridge sandstones were deposited. NNW-SSE-trending subsurface shelf-ridge sandstones are superimposed on a model developed by Asquith (1970). Note that Hartzog Draw field is the largest Shannon shelf-ridge sandstone discovered to date.

transgression which may have lasted less than one million years prior to continued regression.

Combining interpretations related to stratigraphic sequence, lithology, sedimentary structures, micropaleontology (foraminifera) and trace fossils, it is suggested that the Shannon sandstones in Hartzog Draw field were deposited as offshore sand ridges on the middle to outer shelf of the western Pierre-Cody seaway below daily effective wave base at a water depth probably greater than 60 and less than 300 ft (18.2 m, 91 m). The shelf-ridge sandstone complexes are completely enveloped on all sides by fine-grained marine shale deposits.

An isopach map of the Shannon sand zone by Crews and others (1976) (Fig. 40) shows the distribution of Shannon Sandstones. The thicker sandstones (*Central Ridge or Ridge-*

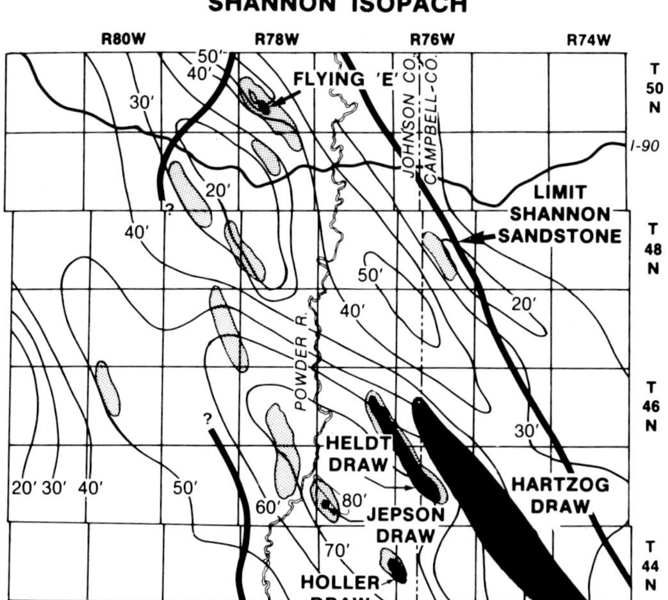

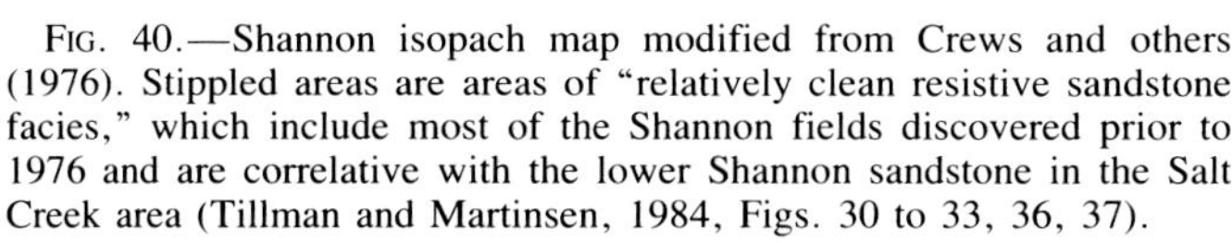

FIG. 40.—Shannon isopach map modified from Crews and others (1976). Stippled areas are areas of "relatively clean resistive sandstone facies," which include most of the Shannon fields discovered prior to 1976 and are correlative with the lower Shannon sandstone in the Salt Creek area (Tillman and Martinsen, 1984, Figs. 30 to 33, 36, 37).

Margin Sandstones) are associated with the thicker overall Shannon zone interval.

The map of Crews and others (1976), as well as other evidence, suggests that across the middle and outer shelf of the western Upper Cretaceous Pierre seaway in Wyoming, there is a series of northwest-trending shelf ridges that extended at least as far west as the Salt Creek area outcrops (Fig. 41). One thing to note is that all the other Shannon fields discovered to date are smaller than Hartzog Draw field. Whether this may be related to Hartzog Draw's being the furthest offshore was not determined.

Heldt Draw field, which lies immediately west of Hartzog Draw field, trends northwest-southeast and also produces from the Shannon, is considerably smaller (Davis, 1976). Fields that produce from the Sussex Sandstone also have a northwest-southeast trend (Fig. 41). House Creek field, a very long Sussex field, was discussed by Berg (1975) and Hobson and others (1982, 1984).

By integration of the vertical sequence relationships observed in surface outcrops (Tillman and Martinsen, 1984)

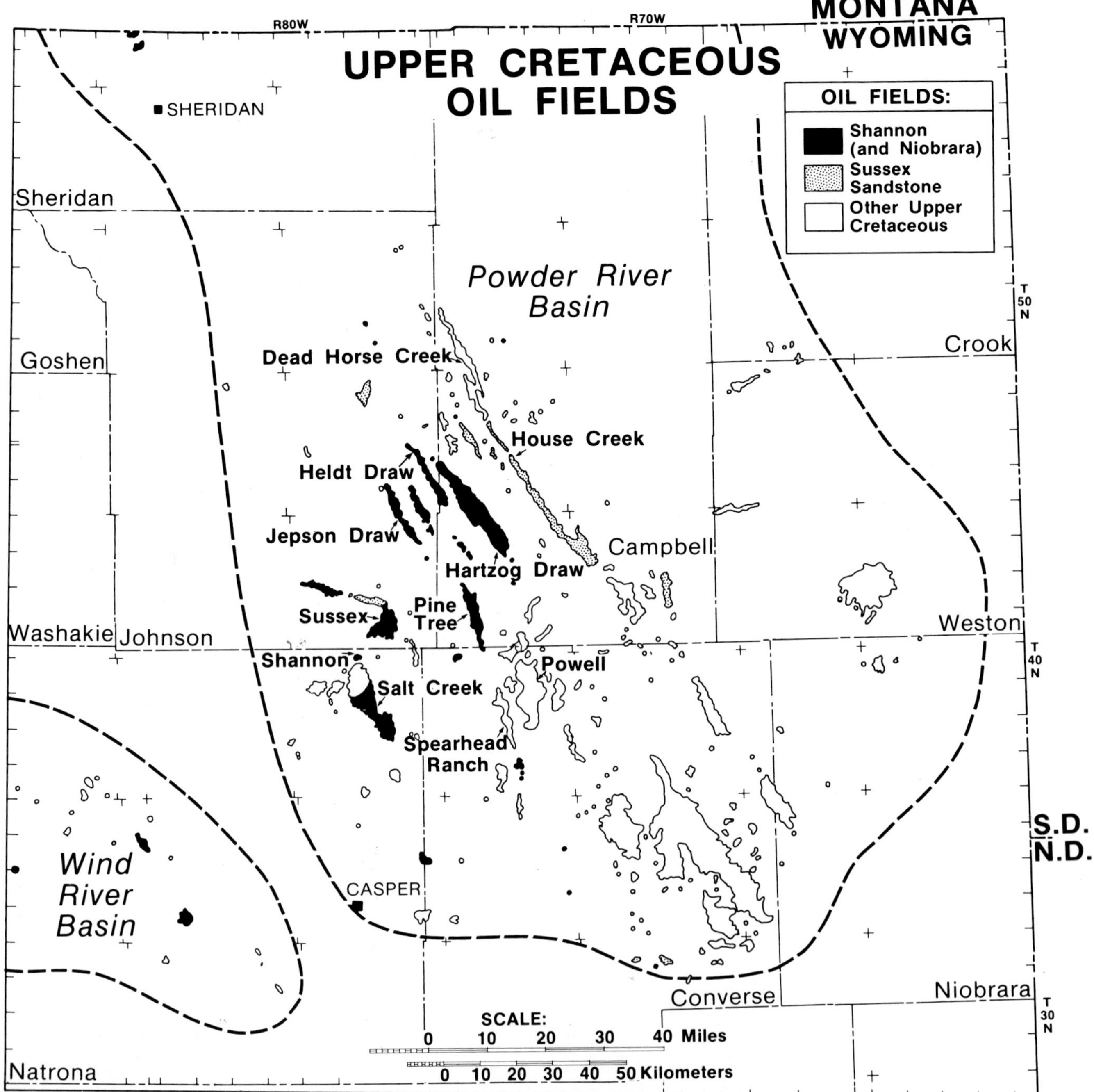

FIG. 41.—Upper Cretaceous oil fields, Powder River Basin, Wyoming. Note the parallelism and elongation of the fields. Among the larger fields are House Creek and Dead Horse Creek fields which produce from the Sussex sandstones.

and subsurface stratigraphic cross sections, a three-dimensional model for Shannon sand-ridge complexes was developed. The model developed for the Shannon Sandstone is very similar to the model developed by Porter (1976) for the Upper Cretaceous Hygiene Sandstone in Colorado. The three-dimensional Shannon self-ridge model is shown in Figure 42. The processes of deposition are summarized as follows.

The base of Shannon shelf-ridge complexes forms gently sloping surfaces at the top of the *Shelf Silty Shale Facies*. A strong bilateral asymmetry is suggested when the base of the crossbedded high-energy facies at Hartzog Draw and Heldt Draw fields are compared (Figs. 30A, 43). A similar asymmetry occurs between Jepson Draw and N. Holler Draw fields (Fig. 30A). Whether this opposite sense of slope at the base of the high-energy part of the complexes occurs in other areas was not determined. Coarser grained deposits built upward from these sloping surfaces. The highest energy depositional currents produced *Central Ridge* and *High-Energy Ridge-Margin Facies*. Prior to and possibly concurrent with deposition of these high-energy sand facies, slightly less competent currents flowed between the bars, producing rippled sandstones and interbedded siltstones and shales of the *Inter-Ridge Sandstone* and *Inter-Ridge Facies (Shaly)*. These less competent currents may also have deposited sediments on the crests of the ridges; however, these sediments may have been eroded shortly after deposition. Short- and long-term fluctuations of near unidirectional currents active in the area created an interfingering of each of these facies. During periods of very slow deposition, silty shale was deposited as drapes over the whole shelf-ridge complex. In the area of the shelf-ridge complex, during periods of slow deposition, burrowing organisms were able to move in and establish themselves well enough so as to thoroughly burrow portions of the previously deposited facies, producing the *Burrowed* or *Bioturbated Shelf Siltstone Facies*. Introduction of strong unidirectional currents into the area, after these periods of relatively slow deposition and burrowing, resulted in the ripping up of thin shale and siderite beds and redeposition of shale and siderite rip-up clasts within the subsequently deposited or, in some areas, contemporaneously deposited *High-Energy Ridge-Margin Facies*.

The vertical sequences of facies observed in cores suggest a single preferred vertical sequence of facies at Hartzog Draw field is unlikely. Instead, several probabilistic sequences are expected. The most likely vertical sequence (Fig. 44) begins with a *Shelf Silty Shale Facies* below the shelf-ridge complex. The most likely sequence from base to top within the shelf-ridge is (1) *Inter-Ridge Facies,* (2) *Low-Energy Ridge-Margin Facies,* (3) *High-Energy Ridge-Margin Facies,* (4) *Central Ridge Facies,* (5) *Shelf Silty-Shale Facies*. A similar, but slightly different sequence seems to be most probable in Shannon outcrops in the Salt Creek

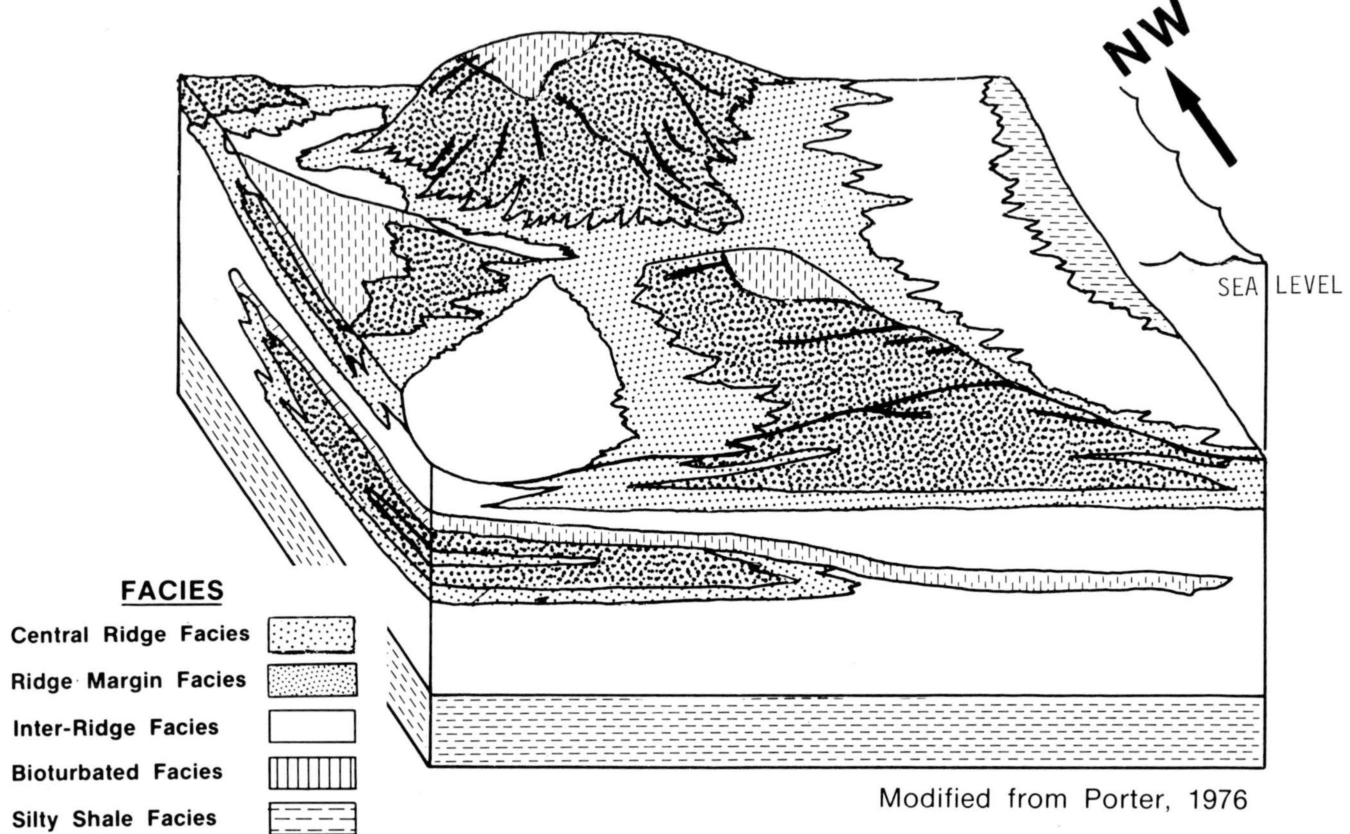

FIG. 42.—Model of facies distributions, Shannon shelf-ridge sandstone complex. This diagram has extreme vertical exaggeration, but the abrupt lateral changes indicated here are substantiated by detailed outcrop study (Tillman and Martinsen, 1984, Figs. 35–37).

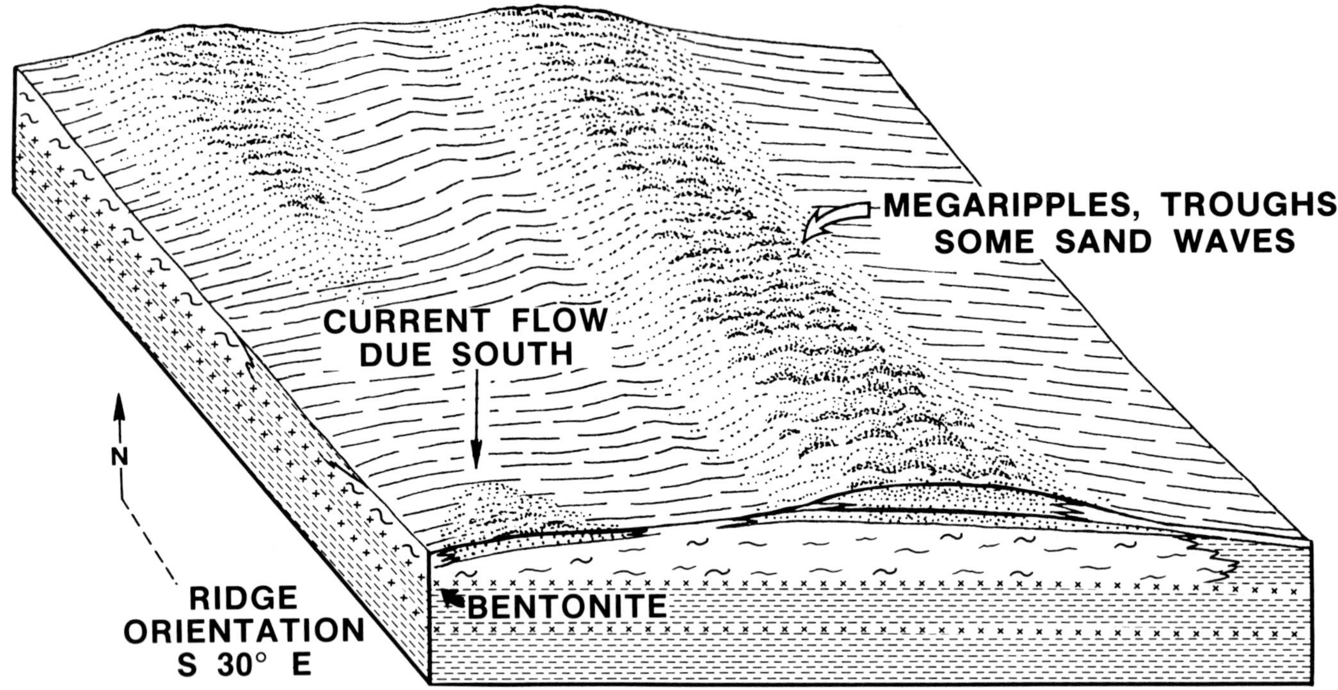

FIG. 43.—Bilateral symmetry of shelf ridges suggested by cross sections through Hartzog and Heldt Draw fields. Note (1) oblique angle of sand ridges relative to transporting currents (south), (2) eastward tilt of base of Hartzog Draw sand ridge complex and westward tilt of base of Heldt Draw field (see also Fig. 30), and (3) parallelism of nearby sand ridges.

field area (Tillman and Martinsen, 1984, Fig. 17). The major difference in vertical sequences between the two areas is the more common occurrences of a ripple-bedded facies directly below the *Central Ridge Facies* in the outcrops; however, note that in outcrop this underlying facies is an *Inter-Ridge Sandstone Facies* which contains very little interbedded shale. This *Inter-Ridge Sandstone Facies* (Fig. 10) is almost totally absent at Hartzog Draw field and in its place is the rippled interbedded shale and sandstone *Inter-Ridge Facies (Shaly)*. A *Bioturbated Shelf Sandstone* occurs below the *Inter-Ridge Sandstone Facies* in outcrops, but it too is virtually absent in Hartzog Draw field cores. The abundance of these two sandstone facies at Salt Creek and their near absence at Hartzog Draw field reinforces the conclusion that something caused the lower portions of the Shannon shelf-ridge complexes to be much sandier in the Salt Creek field outcrop area (see Tillman and Martinsen, 1984, p. 134).

Other differences between the Shannon shelf-ridge deposits in the Salt Creek area and those at Hartzog Draw were discussed by Tillman and Martinsen (1984, p. 134). The major lithologic difference between the shelf-ridge deposits at Hartzog Draw field and those in the Salt Creek area is the greater abundance of shale interbedded with sandstone in the rippled facies at Hartzog Draw field.

Exactly how sands were carried eastward across the Shannon shelf from the western shoreline, which presumably included the Eagle shoreline and delta sandstones in southern Montana, is a mystery. Exactly what the processes were, in time and space, that caused the currents that moved the sand south are also not definitely known. Storm enhancement of shelf subparallel currents is the most likely mechanism of sand movement. The processes involved in moving and depositing the sand, however, must account for the formation of large shelf-ridge complexes containing linearly-trending areas of high-energy crossbedded sandstones as well as elongate to oblate areas of coalescing bars. The high-energy current deposits which form the complexes show relatively small variation from a south transport direction in oriented cores at Hartzog Draw field. Based on subsurface studies (Seeling, 1978; Martinsen and Tillman, 1978; Tillman and Martinsen, 1979) and surface studies (Tillman and Martinsen, 1984), it appears that, once begun, the shelf ridges built upward (rather than laterally) in areas from 0.5 to 3 mi (0.8–4.8 km) wide. Any assignment of processes to the Shannon shelf-ridge complexes must account for a very strong preference for vertical rather than lateral accumulation of thick sections of crossbedded sandstones.

Storm currents superimposed on other bottom currents are believed by us to be the major cause of the nearly shelf-parallel currents which seem to have been responsible for the major shelf-ridge buildups of the Shannon (Fig. 39).

The fact that the sand ridges were at least periodically current-deposited is suggested by the highly irregular periodicity of the bottom current fluctuations. The origin of the "background" bottom currents is conjectural. The bottom currents may have been at least partially tidal in origin; however, differentiation between tidal, geostrophic and storm-driven currents on the *open shelf* is very difficult. Typical criteria used to identify subtidal/tidal current deposits in the nearshore environment, such as herringbone crosslamination, mud couplets and sigmoid bedding, may

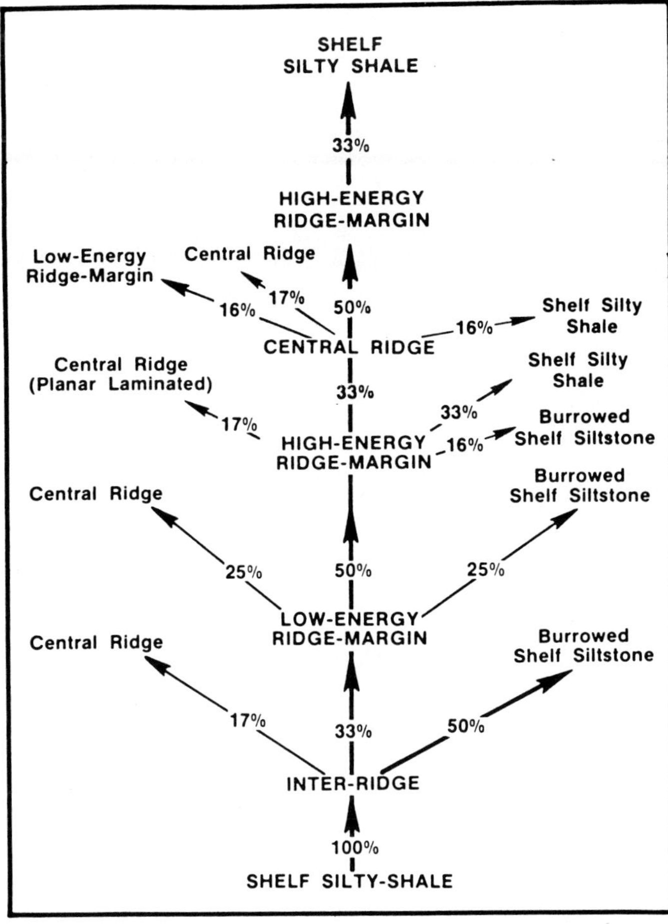

FIG. 44.—Vertical sequence of facies observed in cores at Hartzog Draw field. Individual percentage values may be considered as the probability of finding each specified overlying facies. This figure was developed from Table 3A. See text for discussion. The three cored wells from this paper plus the two from Tillman and Martinsen (1986) were used to compile this figure and Table 3A.

not be typical of tidal current deposits on the open shelf. Swift and Rice (1984) examine in more detail the various aspects of storm and storm-related current deposition in the Western Interior Seaway during the Upper Cretaceous. In subsurface exploration, where logs are the primary geological tool, it may be difficult to separate shelf from shoreline sandstones (Tillman, 1985). A comparison of Shannon-type shelf-sand-ridge and shoreline sandstones is given in Tillman and Martinsen (1984, p. 139).

SUMMARY AND CONCLUSIONS

Hartzog Draw field, located in the Powder River Basin, Wyoming, was discovered in August 1975. It is one of the largest oil fields discovered in the Rocky Mountain province in recent years, with estimates of 350 million barrels of oil initially in place. Field development extends more than 22 mi (35 km) in a northwest-southeast direction and as much as 3-1/2 mi (5.8 km) in width, encompassing in excess of 28,000 productive acres. Development drilling of more than 175 wells on 160-acre spacing has had a better than 95% success ratio, and initial production rates commonly exceed 1,000 barrels of oil per day, with several wells having potential for in excess of 3,000 barrels per day.

The field initially had a solution gas drive and is now under waterflood. There is no water table and no produced formation water. Even zones that calculate water saturations of over 65% from logs do not produce water. Net pay may be based on the porosity, permeability and thickness of the sandstones which produce from individual sedimentary facies (Table A5).

Production at Hartzog Draw is at depths from 9,000 to 9,600 ft (2,727 to 2,910 m). Oil accumulation is stratigraphically controlled, structure having almost no influence on entrapment. The productive interval in the shelf-ridge complex, mostly *Central Ridge* and *High-Energy Ridge-Margin Facies,* is quartzitic and glauconitic, fine- to medium-grained, moderately well sorted, highly trough crossbedded, and is composed of a stacked sequence of sandstones as much as 60 ft (18 m) thick. Sideritic clasts and shale rip-up clasts occur locally in the high-angle trough crossbedded units.

Of the nine facies observed in cores, only two, the *Central Ridge Facies* and the *High-Energy Ridge-Margin Facies,* are consistently high quality reservoirs. Three others, the *Low-Energy Ridge-Margin Facies,* a ripple and trough crossbedded sandstone, the *Central Ridge (Planar Laminated) Facies* and the *Inter-Ridge Facies,* a rippled interbedded sandstone and shale, generally are marginal to very low quality reservoirs.

Shannon Sandstone deposition apparently occurred below effective normal wave base near the middle of the western shelf of the Pierre seaway, more than 100 km from shore in water depths from 60 to 300 ft (18.2–91 m). Recognition of mid- to outer shelf foraminiferal assemblages in shaly portions of the field substantiates the shelf environment of deposition interpretation and an Early Campanian age for the sediments at Hartzog Draw field.

In the reservoir facies, effective porosities average about 12%, and permeabilities average 12 md for the field as a whole. Data from cores indicate that the *Central Ridge* and the *High-Energy Ridge-Margin Facies* have a significantly better average porosity and permeability (15%, 15 md and 14%, 19 md, respectively) than either the *Low-Energy Ridge-Margin Facies* (11%, 11 md) or *Inter-Ridge Facies (Shaly)* (5.2%, 1.2 md). In addition, wells with a thick *Central Ridge Facies* appear to maintain higher reservoir pressure and better production. Net pay thickness closely parallels net sand thickness. Maximum production within the shelf-ridge complex is in the east central area of the field where the *Central Ridge Facies* is thickest. Moderately good production occurs along the northeast flank where the *High-Energy Ridge-Margin Facies* is most abundant.

Recognition of the facies, as well as understanding their distribution and inter-relationships, have been prerequisites to developing primary, secondary and tertiary programs to maximize oil recovery from Hartzog Draw field.

This paper is intended as a geologic and reservoir model for one type of offshore shelf-ridge sandstone. The offshore sand ridge described here is just one of a spectrum of several types that have been alluded to by the senior author

(Tillman, 1985) and others. This documentation of thick reservoir sandstone such as Hartzog Draw field in the middle to outer shelf region should be useful for explorationists. No longer must an explorationist restrict his efforts in exploring in the marine environment to nearshore and deep-water sandstones. The recognition of economic shelf-ridge deposits of this type opens up the whole shelf area for exploration. Not all shelves will have conditions conducive to formation of this type of shelf-ridge sandstone, however, so care should be exercised in trying to adapt all shelves and all offshore shelf-ridge sandstones to this model.

ACKNOWLEDGMENTS

Many co-workers have assisted in this study. Among them were Don Terrell, who initiated the request for analysis of the first core in the field, and V. I. Hill, Jr., who has from the start strongly supported this study and encouraged the coring in the field. Bob Colby and his group provided log calculations. Bob Hunt and Charles Hearn aided in interpretation and discussion of engineering and production data. R. Snyder and R. M. Scott assisted in describing cores. Discussions with John Hobson, Jim Ebanks, Robert Tye, V. Ranganathan and Mike Fowler aided in formulating some conclusions.

In the Cities Service Research Laboratory, Don Dailey's timely analysis and interpretation of the microfauna were very important to the construction of the depositional model for the field. John Hobson, Jim Ebanks, Charles Hearn, Dag Nummedal and Susan Jackson reviewed the manuscript. Fred Mason prepared the cores for examination and did the photography of the cores. R. M. Scott aided in selecting and describing scanning electron microscope photographs and in measuring crossbed orientation in oriented cores. W. Almon supplied X-ray diffraction data. The drafting department at the Research Laboratory also deserves recognition for their cooperation and suggestions which made many of the illustrations more meaningful. Jeannie Sommers prepared a substantial number of the illustrations. Maggie Draughon and Janice Brewer typed the manuscript.

REFERENCES

ASQUITH, D. O., 1970, Depositional topography and major marine environments, Late Cretaceous, Wyoming: American Association of Petroleum Geologists Bulletin, v. 54, p. 1184–1224.

BERG, R. R., 1975, Depositional environments of the Upper Cretaceous Sussex Sandstone, House Creek field, Wyoming: American Association of Petroleum Geologists Bulletin, v. 59, p. 2099–2110.

CREWS, G. C., BARLOW, J. A., Jr., AND HAUN, J. D., 1976, Upper Cretaceous Gammon, Shannon and Sussex sandstones, central Powder River Basin, Wyoming: Wyoming Geological Association, 28th Annual Field Conference, p. 9–20.

DAVIS, M. J., 1976, An environmental interpretation of the Upper Cretaceous Shannon Sandstone, Heldt Draw field, Wyoming: Wyoming Geological Association, 28th Annual Field Conference, p. 125–138.

GILL, J. R., AND COBBAN, W. A., 1966, The Red Bird Section of the Upper Cretaceous Pierre Shale in Wyoming: U.S. Geological Survey Professional Paper 393A, 73 p.

———, AND ———, 1969, Paleogeographic map for latest Eagle time showing position of strandline during range zone of *Baculites sp.* (smooth): U.S. Geological Survey Open-File Report 1969, sheet 2 of 6 paleogeographic maps.

———, AND ———, 1973, Stratigraphy and geologic history of the Montana Group and equivalent rocks, Montana, Wyoming and North Dakota: U.S. Geological Survey Professional Paper 776, 37 p.

HEARN, C. L., EBANKS, W. J., Jr., TYE, R. S., AND RANGANATHAN, V., 1984, Geological factors influencing reservoir performance of the Hartzog Draw Field, Wyoming: Journal of Petroleum Technology, p. 1335–1344.

———, HOBSON, J. P., AND FOWLER, M. L., 1986, Reservoir characterization for simulation, Hartzog Draw field, Wyoming, *in* Lake, L. W., and Carroll, H. B., Jr., eds., Reservoir characterization: Academic Press, Inc., Orlando, Florida, p. 341–372.

HOBSON, J., FOWLER, M. L., AND BEAUMONT, E. A., 1982, Depositional and statistical exploration models, Upper Cretaceous offshore sandstone complex, Sussex Member, House Creek field, Wyoming: American Association of Petroleum Geologists Bulletin, v. 66, p. 689–707.

———, ———, AND TILLMAN, R. W., 1984, Asymmetry of form and lithology, Late Cretaceous shelf sandstone complexes, House Creek and Hartzog Draw fields, Powder River Basin, Wyoming (abs.): Program for Shelf Sands and Sandstones Symposium, Calgary, Alberta p. 43.

HUNT, R. D., AND HEARN, C. L., 1982, Reservoir management of the Hartzog Draw field: Journal of Petroleum Technology, p. 1575–1582.

MARTINSEN, R. S., 1981, Hartzog Draw field, *in* Powder River Basin Oil and Gas Fields, v. I: Wyoming Geological Association, p. 187 and map in pocket.

———, AND TILLMAN, R. W., 1978, Hartzog Draw, new giant oil field (abs.): American Association of Petroleum Geologists Bulletin, v. 62, p. 540.

———, AND ———, 1979, Facies and reservoir characteristics of a shelf sandstone: Hartzog Draw field, Powder River Basin, Wyoming: American Association of Petroleum Geologists Bulletin, v. 63, p. 491.

PORTER, K. W., 1976, Marine shelf model, Hygiene Member of the Pierre Shale, Upper Cretaceous Denver Basin, Colorado, *in* Epis, R., and Weimer, R., eds., Studies in Colorado Field Geology: Professional Contributions of Colorado School of Mines, Golden, CO, p. 251–263.

RANGANANTHAN, V., AND TYE, R. S., 1986, Petrography, diagenesis and facies control on porosity in Shannon Sandstone, Hartzog Draw field, Wyoming: American Association of Petroleum Geologists, v. 70, p. 56–69.

REINECK, H. E., AND SINGH, I. B., 1971, Genesis of laminated sand and graded rhythemites in storm-sand layers of shelf mud: Sedimentology, v. 18, p. 123–128.

RYER, T. A., AND MCPHILLIPS, M., 1983, Early Late Cretaceous paleogeography of east central Utah, *in* Reynolds, M. W. and Dolly, E. D., eds., Mesozoic Paleography of West Central United States: Rocky Mountain Section, Society of Economic Paleontologists and Mineralogists Symposium No. 2, p. 253–272.

SEELING, A., 1978, The Shannon Sandstone, a further look at the environment of deposition at Heldt Draw field, Wyoming: The Mountain Geologist, v. 15, p. 133–144.

SPEARING, D. R., 1975, Shannon Sandstone, Wyoming, *in* Depositional Environments as Interpreted from Primary Sedimentary Structures and Stratification Sequences: The Society of Economic Paleontologists and Mineralogists Short Course No. 2, Dallas: p. 104–114.

SWIFT, D. J. P., AND RICE, D. D., 1984, Sand bodies on muddy shelves: a model for sedimentation in the Western Interior seaway, North America, *in* Tillman, R. W., and Siemers, C. T., eds., Siliclastic Shelf Sedimentation: Society of Economic Paleontologists and Mineralogists Special Publication 34, p. 43–62.

TILLMAN, R. W., 1985, A spectrum of shelf sandstones, *in* Tillman, R. W., Swift, D. J. P., and Walker, R. G., eds., Shelf Sands and Sandstone Reservoirs: Society of Economic Paleontologists and Mineralogists Short Course, p. 1–46.

———, AND MARTINSEN, R. S., 1979, Hartzog Draw field, Powder River Basin, Wyoming, *in* Flory, R. F., ed., Rocky Mountain High: Wyoming Geological Association, 28th Annual Meeting, Core Seminar Core Book, p. 1–38.

———, AND ———, 1984, The Shannon shelf-ridge sandstone complex, Salt Creek anticline area, Powder River Basin, Wyoming, *in* Tillman, R. W., and Siemers, C. T., eds., Siliclastic Shelf Sedimentation: Society of Economic Paleontologists and Mineralogists Special Publication 34, p. 85–142.

———, AND ———, 1986, Shannon Sandstone, Hartzog Draw field core study, *in* Tillman, R. W., Swift, D. J. P., and Walker, R. G., eds., Shelf Sands and Sandstone Reservoirs: Society of Economic Paleontologists and Mineralogists Short Course No. 13, p. 589–656.

———, AND REINECK, 1975, Discrimination of North Sea sand environments with population-derived grain size parameters, *in* Resumes des Publications, IX[ME] Congress International de Sedimentologie, Nice, Theme 6, p. 217–220.

TYE, R. S., RANGANATHAN, V., EBANKS, W. J., JR, 1986, Facies analysis and reservoir zonation of a Cretaceous shelf sand ridge: Hartzog Draw field, Wyoming, *in* Moslow, T. F., and Rhodes, E. G., eds., Modern and Ancient Shelf Clastics: A Core Workshop: Society of Economic Paleontologists and Mineralogists Core Workshop No. 9, p. 169–216.

APPENDIX A

Detailed analyses of Hartzog Draw field cores

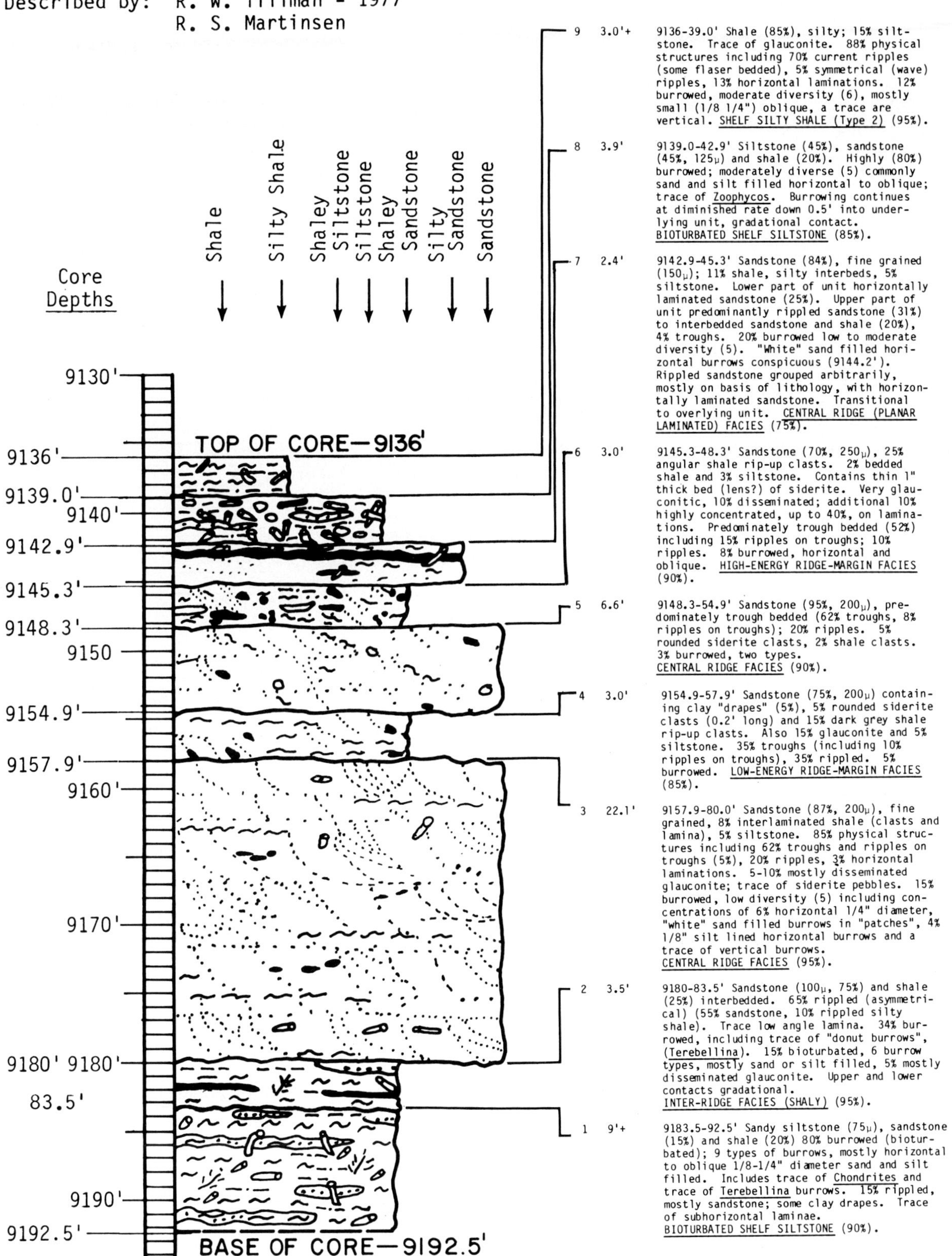

FIG. A1.—Stratigraphic facies description, Cities Service Federal AE-1. Units summarized for this figure from sedimentologic facies descriptions in Figure A2.

CITIES SERVICE NO. 1, FEDERAL AE (5131)
NW SEC. 13 T45N R76W, SHANNON SANDSTONE
HARTZOG DRAW FIELD, CAMPBELL CO., WYOMING

Columnized Data Format: %/Max. set thickness/Additional notes	BIOTURBATED SHELF SILTSTONE (80%) Described by: R. W. Tillman (1977, 1984, 1986) Unit 1 (12.7'+) Core Depth: 9183.8-9192.5' (Log Depth: 9193.8-9202.5')
Lithology	
1. Sandstone (%) (mean grain size, microns)	15%/0.1' (100 microns)
2. Siltstone (%)	65%/0.5'
3. Shale (%); a) Laminated; b) Clasts	20%/0.05'/silty
4. Siderite (%); a) Laminated; b) Clasts	-
5. Glauconite (%); a) Disseminated; b) Laminated	// 3% a
6. Carbonaceous(%); a) Disseminated; b) Laminated	// -
Physical Structures	17%
1. High angle cross-bedding (20°+); a) Troughs; b) Planar-tabular; c) Planar-tangential; d) Curved-tangential (trough)	-
2. Moderate angle cross-bedding (10-20°); a) Troughs; b) Planar-tabular; c) Planar-tangential; d) Curved-tangential (trough)	-
3. Subhorizontal to low angle bedding (<10°); a) Trough; b) Planar-tabular; c) Planar-tangential; d) Curved-tangential; e) Planar	tr/e
4. Horizontal laminations; a) Sandstone; b) Shale	tr/b
5. Rippled; a) Sandstone; b) Shale; c) Current; d) Wave; e) Form	14%
6. Rippled interbedded sandstone and shale	// -
7. Ripples superimposed on troughs	-
8. Reworked: a) By waves & currents; b) Bedding destroyed; c) bioturbated; d) massive; e) soft sediment deformed; f) clasts	tr/a
Biogenic Sedimentary Structures	83%/0.2'
1. Identified burrows; a) *Asterosoma*; c) *Chondrites*; d) "Donut burrows" (*Terabellina*); g) Gastropod tracks; p) Plural curving tubes; s) *Skolithos*; t) *Teichichnus*; th) *Thallasinoides*	tr-c 2%-d
2. Distinct burrows; a) <1/8"; b) 1/8"-1/4"; c) 1/4"-1/2"; d) >1/2"; e) Silt filled; f) Clay filled; g) Sand filled; h) Silt lined; i) Clay lined; j) Spreiten; k) Vertical; l) Oblique; m) Horizontal.*% of burrows in bioturbated in ( ).	3% b,f,k; 2% b,f,l; tr-c,f,k tr-b,f,j; tr-adj; * 2% a,d,k; tr-a,h,l
3. Bioturbated, 75% (+) burrowed (%); a) Distinct burrows (%), b) Mottled (%).	71%/0.2'/b
4. Total interval burrowed (%, footage)	99%
5. Diversity (Number of burrow types); 1-4 low, 4-8 moderate; >8 high.	9, high
Cemented Intervals; a) Siderite; b) Calcite, c) Dolomite	-
Contact Relations (core contacts)	
1. Upper; a) Very sharp (<0.05' transition); b) Sharp (0.05-0.1' transition); c) Transitional (0.1-0.3' transition); d) Gradational (>0.3' transition); e) Contact, erosional (truncated) angular; f) Contact erosional parallel; g) Covered or not covered	d
2. Lower; a), b), c), d), e), f), g)	not cored

Note: Top picked at base of lowest high-angle cross bedded set.

Note: EXPLANATION OF GEOLOGIC CORE CHECKLIST—The following checklists are compiled to allow a systematic detailed description of the lithology, physical structures, and biogenic structures of the various facies observed in the core. The amount of detail included here is not of general interest to most geologists, but the detail is necessary for geologists involved in doing detailed descriptions of cores.

Under "*Lithology*" the percentage values should total 100%. Percentages are based on the percentage of rock *volume* occupied by a feature. Line entries which have a box (with slashes) on the far left are considered to be "additional descriptors" and are not included in the tabulation to total 100%.

"*Physical Structures*" and "*Biogenic Sedimentary Structures*" together total 100%. Individual entries under "Physical" and "Biogenic" total to the percentage indicated in the major heading. Line entries with box (with slashes) on the far left are considered to be additional descriptors and are not included in the tabulation to total 100%.

FIG. A2.—Sedimentologic facies descriptions,

INTER-RIDGE FACIES (SHALY) (95%) Described by: R. W. Tillman	CENTRAL RIDGE FACIES (95%) Described by: R. W. Tillman	LOW-ENERGY RIDGE-MARGIN FACIES (80%) Described by: R. W. Tillman
Unit 2 (3.5') Core Depth: 9180-83.5' (Log Depth: 9190-93.5')	Unit 3 (22.1') Core Depth: 9157.9-9180.0' (Log Depth: 9167.9-9190.0')	Unit 4 (3.0') Core Depth: 9154.9-9157.9' (Log Depth: 9164.9-9167.9')
75%/0.1'/(100 microns)	87%/1.3' (200 microns)	70%/0.2' (200 microns)
2%/interlam. w/shale	5%/0.1'	5%/0.05'
23%/a	7%/0.05'/3% b, 4% a	20%/0.05'/5% a, 15% b
-	tr/0.2'/a, b	5%/0.07'/b
// 5% a	// 5%/0.8'/10% a, 40% on b	// 15% a
// -	// -	// -
66%/0.1'	85%/0.5'	95%/0.5'
-	11%/0.5'/a	-
-	28%	20%/0.5'/a
tr	14%	12%/0.1'
-	3%/0.2'/a	-
65%/0.1'/c, 55% a, 10% b	20%/0.1'/a	38%/0.2'
// 50%/0.15'	// 8%/0.2'	// 40%
-	5%/0.1'	5%
-	4% d	20%/0.15'/15% d, 5% a
34%	15%/0.4'	5%/0.05'
tr-d	tr-d tr-t	-
5 burrow types *	6% c,g,m (0.3', patches) 4% b,e,m *tr-c,g,k	2% c,g,l tr-b,e,m tr-b,e,l * tr-c,g,k
15%	2%/0.2'/w sh. lam.	-
85%, 2.1'	12%/5.1'	8%, 0.24'
6, moderate	5 (low to moderate)	4, low
-	-	-
d	a	c
d	d	b

The percent (by volume) is followed by a / and a value indicating the maximum set thickness for that feature. The area following the second / is for detailed identification of features and notes.

Burrow types which cannot be identified by name are included in *Biogenic Sedimentary Structures*, section 2. The percentage (volume) of burrowing is indicated followed by descriptors concerning size, type of filling, and orientation. The percentage included in ( ) are burrows identified with an otherwise bioturbated interval. The percentage of vertical section burrowed is indicated, and the diversity of burrow types is given.

Upper and lower *Contact Relations* are given at the bottom of each checklist. Lettered descriptors are used to describe the types of contacts.

Cities Service Federal AE-1, Units 1 to 9.

CITIES SERVICE NO. 1, FEDERAL AE (5131)
NW SEC. 13 T45N R76W, SHANNON SANDSTONE
HARTZOG DRAW FIELD, CAMPBELL CO., WYOMING

CENTRAL RIDGE FACIES (90%)

Columnized Data Format:
%/Max. set thickness/Additional notes

Described by: R. W. Tillman

Unit 5
Core Depth: 9148.3-54.9'
(Log Depth: 9158.2-64.9')

Lithology	
1. Sandstone (%) (mean grain size, microns)	95%/0.6' (200 microns)
2. Siltstone (%)	2%/;0.1'
3. Shale (%); a) Laminated; b) Clasts	5%/0.05'/4% b, tr-a
4. Siderite (%); a) Laminated; b) Clasts	5%/0.2'/b
5. Glauconite (%); a) Disseminated; b) Laminated	// 10%/5% a, 30% on b
6. Carbonaceous(%); a) Disseminated; b) Laminated	// -

Physical Structures	97%/0.8'
1. High angle cross-bedding (20°+); a) Troughs; b) Planar-tabular; c) Planar-tangential; d) Curved-tangential (trough)	28%/0.8'/a
2. Moderate angle cross-bedding (10-20°); a) Troughs; b) Planar-tabular; c) Planar-tangential; d) Curved-tangential (trough)	3%/0.3'/a
3. Subhorizontal to low angle bedding (<10°); a) Trough; b) Planar-tabular; c) Planar-tangential; d) Curved-tangential; e) Planar	3%/0.2'
4. Horizontal laminations; a) Sandstone; b) Shale	-
5. Rippled; a) Sandstone; b) Shale; c) Current; d) Wave; e) Form	20%/0.2'
6. Rippled interbedded sandstone and shale	// -
7. Ripples superimposed on troughs	8%/0.2'
8. Reworked: a) By waves & currents; b) Bedding destroyed; c) bioturbated; (d) massive; e) soft sediment deformed; f) clasts	7%/0.1'/a, d

Biogenic Sedimentary Structures	3%/0.1'
1. Identified burrows; a) Asterosoma; c) Chondrites; d) "Donut burrows" (Terabellina); g) Gastropod tracks; p) Plural curving tubes; s) Skolithos; t) Teichichnus; th) Thallasinoides	-
2. Distinct burrows; a) <1/8"; b) 1/8"-1/4"; c) 1/4"-1/2"; d) >1/2"; e) Silt filled; f) Clay filled; g) Sand filled; h) Silt lined; i) Clay lined; j) Spreiten; k) Vertical; l) Oblique; m) Horizontal.*% of burrows in bioturbated in ( ).	2% c,m tr-b,m *
3. Bioturbated, 75% (+) burrowed (%); a) Distinct burrows (%), b) Mottled (%).	-
4. Total interval burrowed (%, footage)	3%, 0.2'
5. Diversity (Number of burrow types); 1-4 low, 4-8 moderate; >8 high.	2, low

Cemented Intervals; a) Siderite; b) Calcite, c) Dolomite	-

Contact Relations (core contacts)	
1. Upper; a) Very sharp (<0.05' transition); b) Sharp (0.05-0.1' transition); c) Transitional (0.1-0.3' transition); d) Gradational (>0.3' transition); e) Contact, erosional (truncated) angular; f) Contact erosional parallel; g) Covered or not covered	d
2. Lower; a), b), c), d), e), f), g)	c

HIGH-ENERGY RIDGE-MARGIN FACIES	CENTRAL RIDGE (PLANAR LAMINATED FACIES (75%)	BIOTURBATED SHELF SILTSTONE (80%)
Described by: R. W. Tillman	Described by: R. W. Tillman	Described by: R. W. Tillman
Unit 6 (3.1') Core Depth: 9145.3-9148.3' (Log Depth: 9155.3-9158.3')	Unit 7 Core Depth: 9142.9-45.3' (Log Depth: 9152.9-55.3')	Unit 8 (3.6') Core Depth: 9139.0-9142.9' (Log Depth: 9149.0-9152.9')
70%/0.2'/(250 microns)	84%/0.3' (150 microns)	10%/0.1' (125 microns)
3%/0.01'	5%/0.1'	25%/0.05'
27%/0.05'/22% b, 5% a	11%/0.1'/10% a, tr-b	65%/0.05'
3%/0.1/b	-	2% b
// 20%/0.2'/10% a, 10% b	// 5%/0.3'/ 4% a, tr-b	// 5% a
// -	// -	// -
92%/0.5'	80%/0.3'	20%/0.1'
27%/0.3'	-	-
10%/0.3'	-	-
30%/0.2'	4%/0.1'/a	-
-	25%/0.5'/a	2%/tr-a, tr-b
10%/0.2'	30%/0.1'	18%/16% b, 2% a
// 2%/0.1'	// 20%/0.2'	// 20%
15%/0.2'	-	-
(20%/0.8'/a)	tr-d	(c)
8%	20%/0.2'	80%
-	-	tr-*Zoophycos* tr-t
5% b,g,m 3% c,g,m *	10% c,g,m at 9144.2' 4% d,g,m tr-b,i 2% b,f,m tr-c,k *	(2% d,e,l) (2% c,e,l) (tr-b,e,m) * (tr-c,g,k)
-	2%/0.1'	73%/2.0'/85% burr.
13%	58%, 1.6'	100%, 3.6'
2	5, moderate	10, high
-	-	-
a	d	b
d	a	d

CITIES SERVICE NO. 1, FEDERAL AE (5131)
NW SEC. 13 T45N R76W, SHANNON SANDSTONE
HARTZOG DRAW FIELD, CAMPBELL CO., WYOMING

Columnized Data Format:
%/Max. set thickness/Additional notes

SHELF SILTY SHALE(TYPE 2) (95%)

Described by: R. W. Tillman

Unit 9 (3.0'+)
Core Depth: 9136-9139.0'
(Log Depth: 9146-9149.0')

Lithology	
1. Sandstone (%) (mean grain size, microns)	tr/0.03' (100 microns)
2. Siltstone (%)	14%/0.1'
3. Shale (%); a) Laminated; b) Clasts	85%/0.1' a, silty
4. Siderite (%); a) Laminated; b) Clasts	-
5. Glauconite (%); a) Disseminated; b) Laminated	// tr
6. Carbonaceous(%); a) Disseminated; b) Laminated	// -

Physical Structures	88%
1. High angle cross-bedding (20°+); a) Troughs; b) Planar-tabular; c) Planar-tangential; d) Curved-tangential (trough)	-
2. Moderate angle cross-bedding (10-20°); a) Troughs; b) Planar-tabular; c) Planar-tangential; d) Curved-tangential (trough)	-
3. Subhorizontal to low angle bedding (<10°); a) Trough; b) Planar-tabular; c) Planar-tangential; d) Curved-tangential; e) Planar	-
4. Horizontal laminations; a) Sandstone; b) Shale	13%/0.15'/8% b, 5% siltstone
5. Rippled; a) Sandstone; b) Shale; c) Current; d) Wave; e) Form	75%/0.1'/70% c flaser, 5% d
6. Rippled interbedded sandstone and shale	// 25%/0.1'
7. Ripples superimposed on troughs	-
8. Reworked: a) By waves & currents; b) Bedding destroyed; c) bioturbated; d) massive; e) soft sediment deformed; f) clasts	- (c)

Biogenic Sedimentary Structures	12%/0.1'
1. Identified burrows; a) Asterosoma; c) Chondrites; d) "Donut burrows" (Terabellina); g) Gastropod tracks; p) Plural curving tubes; s) Skolithos; t) Teichichnus; th) Thallasinoides	-
2. Distinct burrows; a) <1/8"; b) 1/8"-1/4"; c) 1/4"-1/2"; d) >1/2"; e) Silt filled; f) Clay filled; g) Sand filled; h) Silt lined; i) Clay lined; j) Spreiten; k) Vertical; l) Oblique; m) Horizontal.*% of burrows in bioturbated in ( ).	5% b,e,l tr-c,e,k 4% b,e,m * tr-d,e,j,m
3. Bioturbated, 75% (+) burrowed (%); a) Distinct burrows (%), b) Mottled (%).	tr/0.1'
4. Total interval burrowed (%, footage)	70%/2.1'
5. Diversity (Number of burrow types); 1-4 low, 4-8 moderate; >8 high.	6, moderate

Cemented Intervals; a) Siderite; b) Calcite, c) Dolomite	-

Contact Relations (core contacts)	
1. Upper; a) Very sharp (<0.05' transition); b) Sharp (0.05-0.1' transition); c) Transitional (0.1-0.3' transition); d) Gradational (>0.3' transition); e) Contact, erosional (truncated) angular; f) Contact erosional parallel; g) Covered or not covered	not cored
2. Lower; a), b), c), d), e), f), g)	b

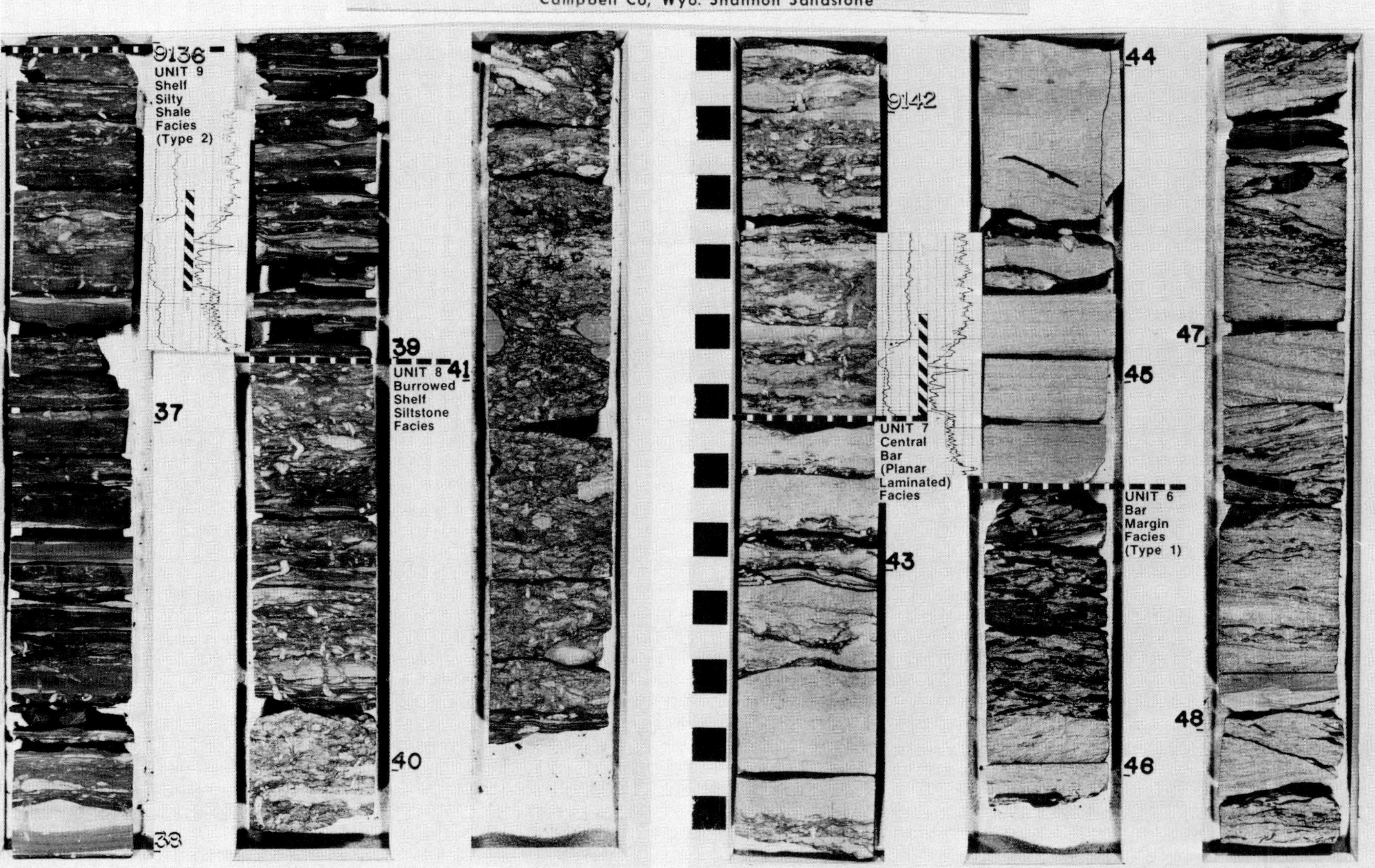

FIG. A3.—Slabbed core, Cities Service Federal AE-1. Unit numbers and facies designations are indicated on photographs. Scale on left is in tenths of feet.

Cities Service Fed. AE-1 (5131)
C,NW,NW Sec. 13 T45N R76W Hartzog Draw Field
Campbell Co, Wyo. Shannon Sandstone

58 60 UNIT 3 Central Bar Facies

61

59

9156

57

53

54

UNIT 4 Bar Margin Facies (Type 2)

51

52

UNIT 5 Central Bar Facies

9149

50

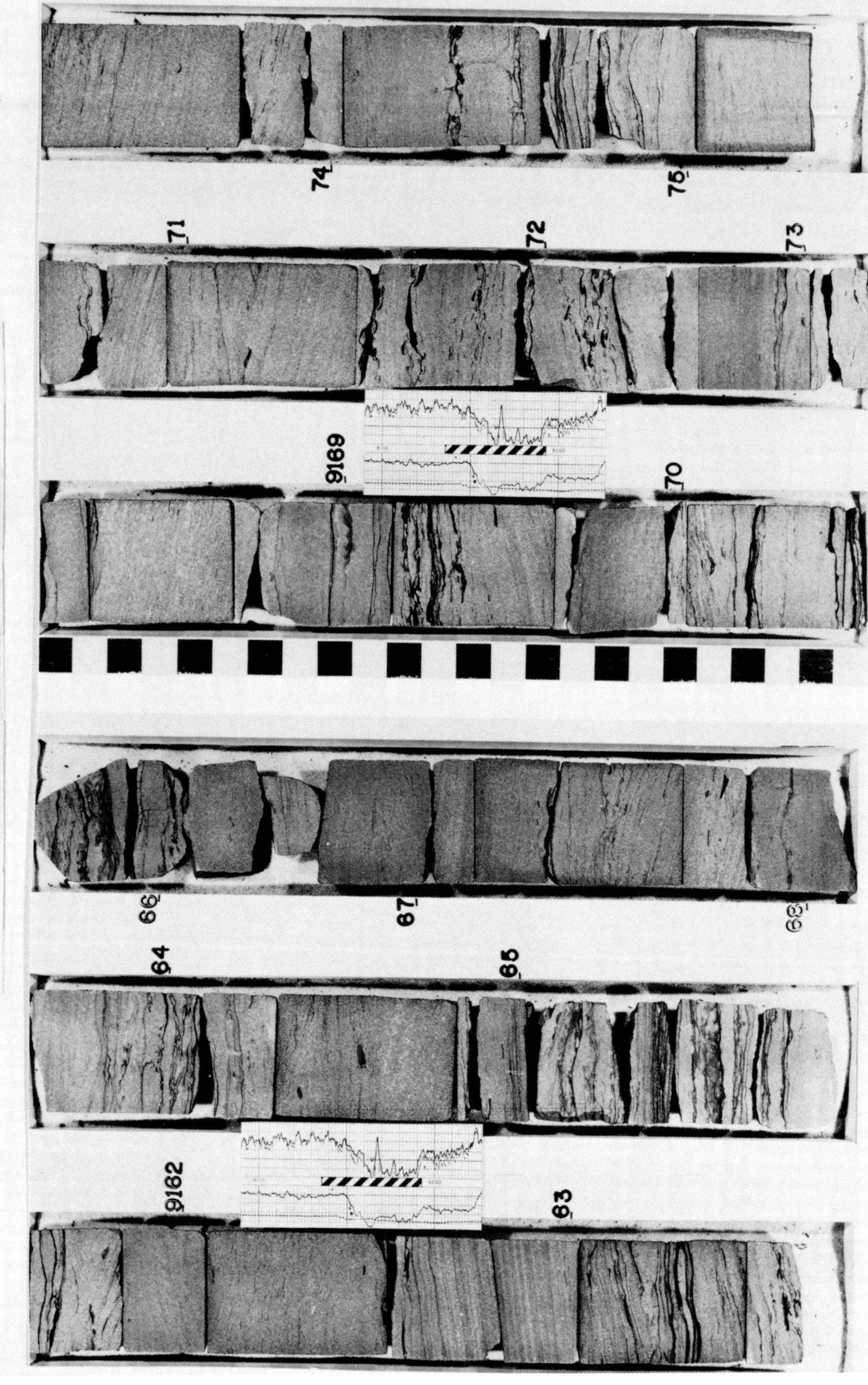
Cities Service Fed. AE-1 (5131)
C,NW,NW Sec. 13 T45N R76W Hartzog Draw Field
Campbell Co, Wyo. Shannon Sandstone
9162
63
64
65
66
67
68
9169
70
71
72
73
74
75

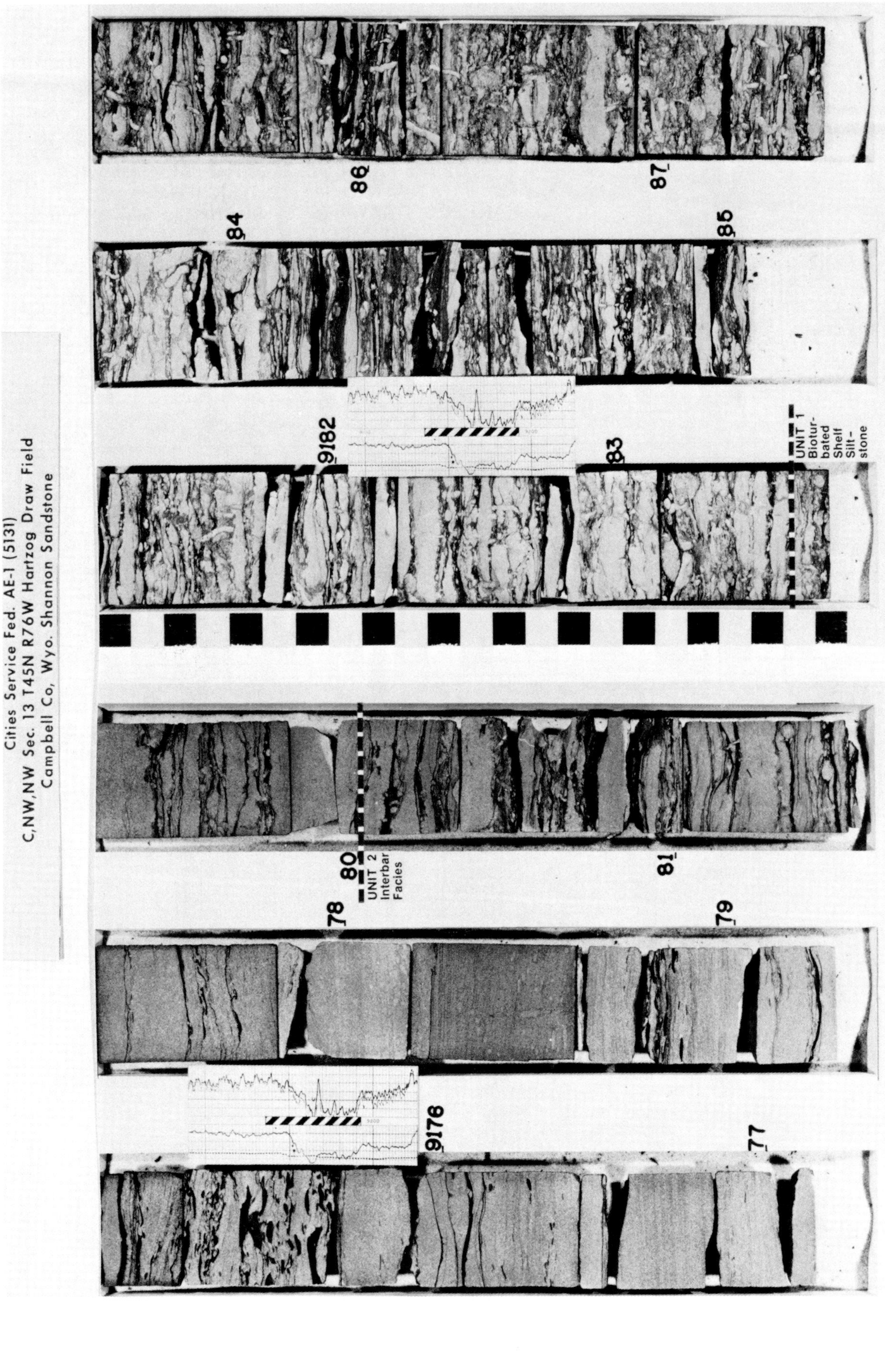

Cities Service Fed. AE-1 (5131)
C,NW,NW Sec. 13 T45N R76W Hartzog Draw Field
Campbell Co, Wyo. Shannon Sandstone

9176
77
78
79
80
UNIT 2
Interbar
Facies
81
9182
83
UNIT 1
Biotur-
bated
Shelf
Silt-
stone
84
85
86
87

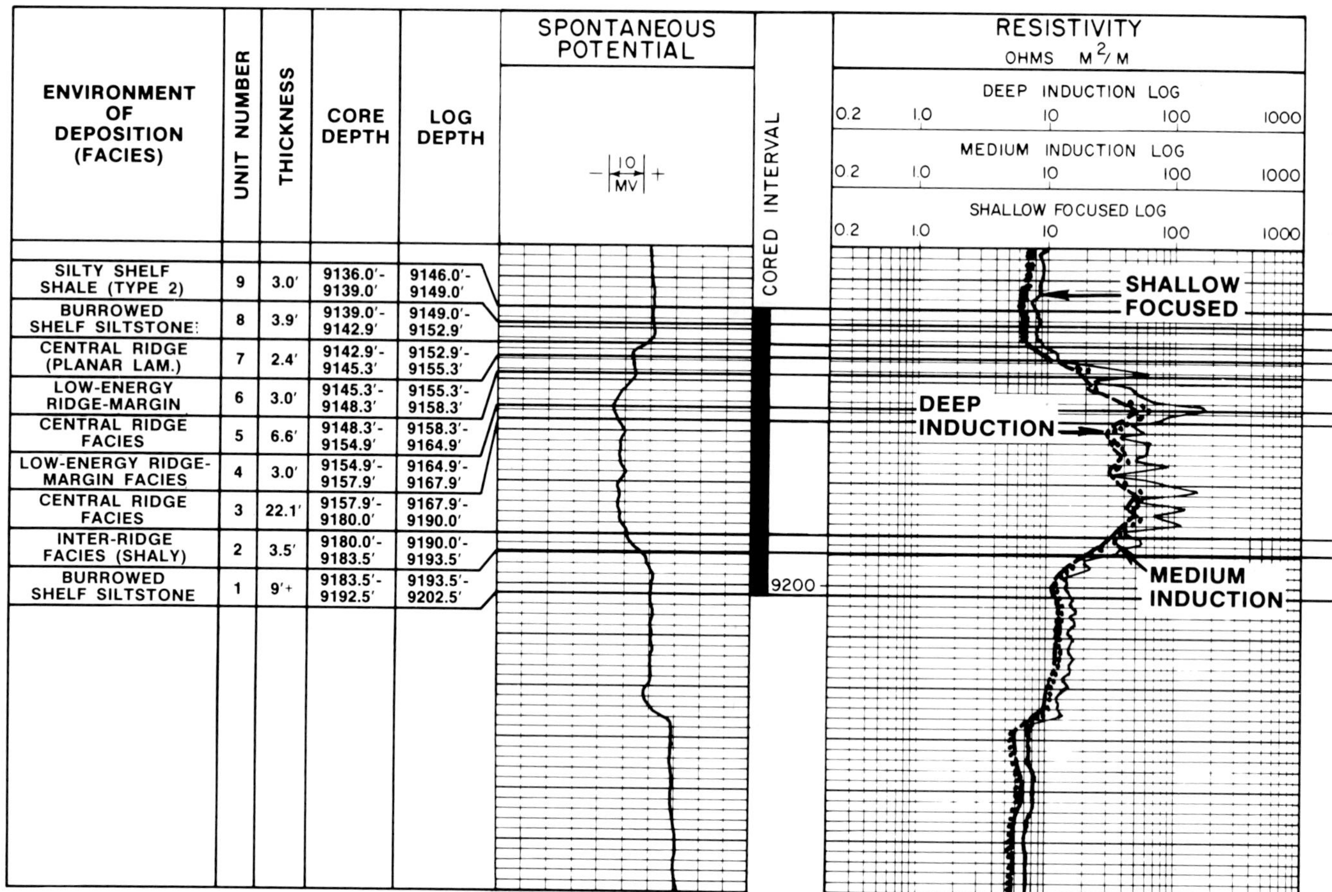

ENVIRONMENT OF DEPOSITION (FACIES)	UNIT NUMBER	THICKNESS	CORE DEPTH	LOG DEPTH
SILTY SHELF SHALE (TYPE 2)	9	3.0′	9136.0′-9139.0′	9146.0′-9149.0′
BURROWED SHELF SILTSTONE	8	3.9′	9139.0′-9142.9′	9149.0′-9152.9′
CENTRAL RIDGE (PLANAR LAM.)	7	2.4′	9142.9′-9145.3′	9152.9′-9155.3′
LOW-ENERGY RIDGE-MARGIN	6	3.0′	9145.3′-9148.3′	9155.3′-9158.3′
CENTRAL RIDGE FACIES	5	6.6′	9148.3′-9154.9′	9158.3′-9164.9′
LOW-ENERGY RIDGE-MARGIN FACIES	4	3.0′	9154.9′-9157.9′	9164.9′-9167.9′
CENTRAL RIDGE FACIES	3	22.1′	9157.9′-9180.0′	9167.9′-9190.0′
INTER-RIDGE FACIES (SHALY)	2	3.5′	9180.0′-9183.5′	9190.0′-9193.5′
BURROWED SHELF SILTSTONE	1	9′+	9183.5′-9192.5′	9193.5′-9202.5′

FIG. A4.—Facies designation on subsurface logs and core gamma log. Note that core depths are 10 ft (3m) shallower than subsurface gamma-

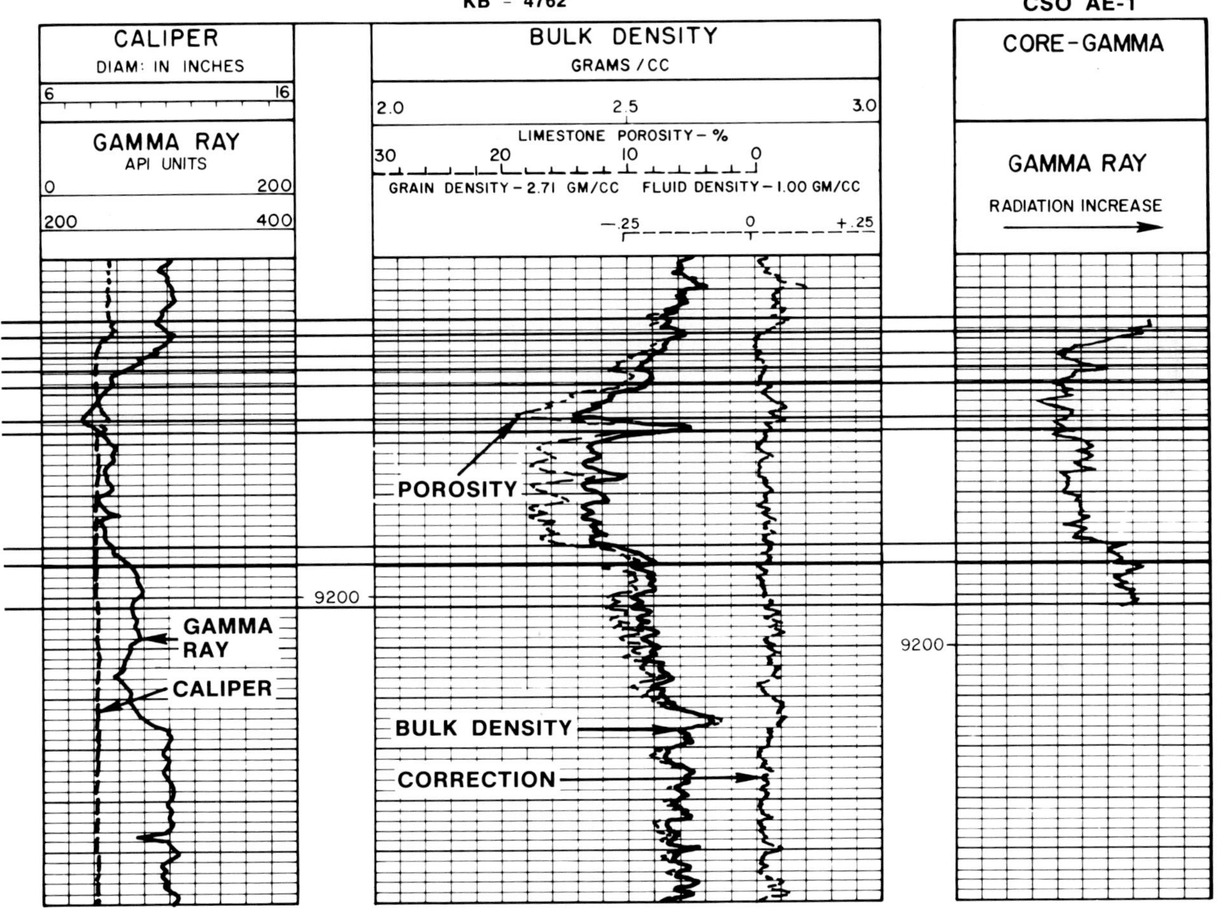

ray log. Facies are designated and logs are corrected to core depths. Units correspond to those described and photographed in Figures A1 and A3.

SOUTHLAND ROYALTY CO., BUD CHRISTENSEN NO. 2 (6343)
SW SEC. 34 T46N R76W, HARTZOG DRAW FIELD
SHANNON SANDSTONE, CAMPBELL CO., WYOMING
CORED INTERVAL 52.9'; 9163-9215.9'

Described by: R. W. Tillman - 1977
R. S. Martinsen

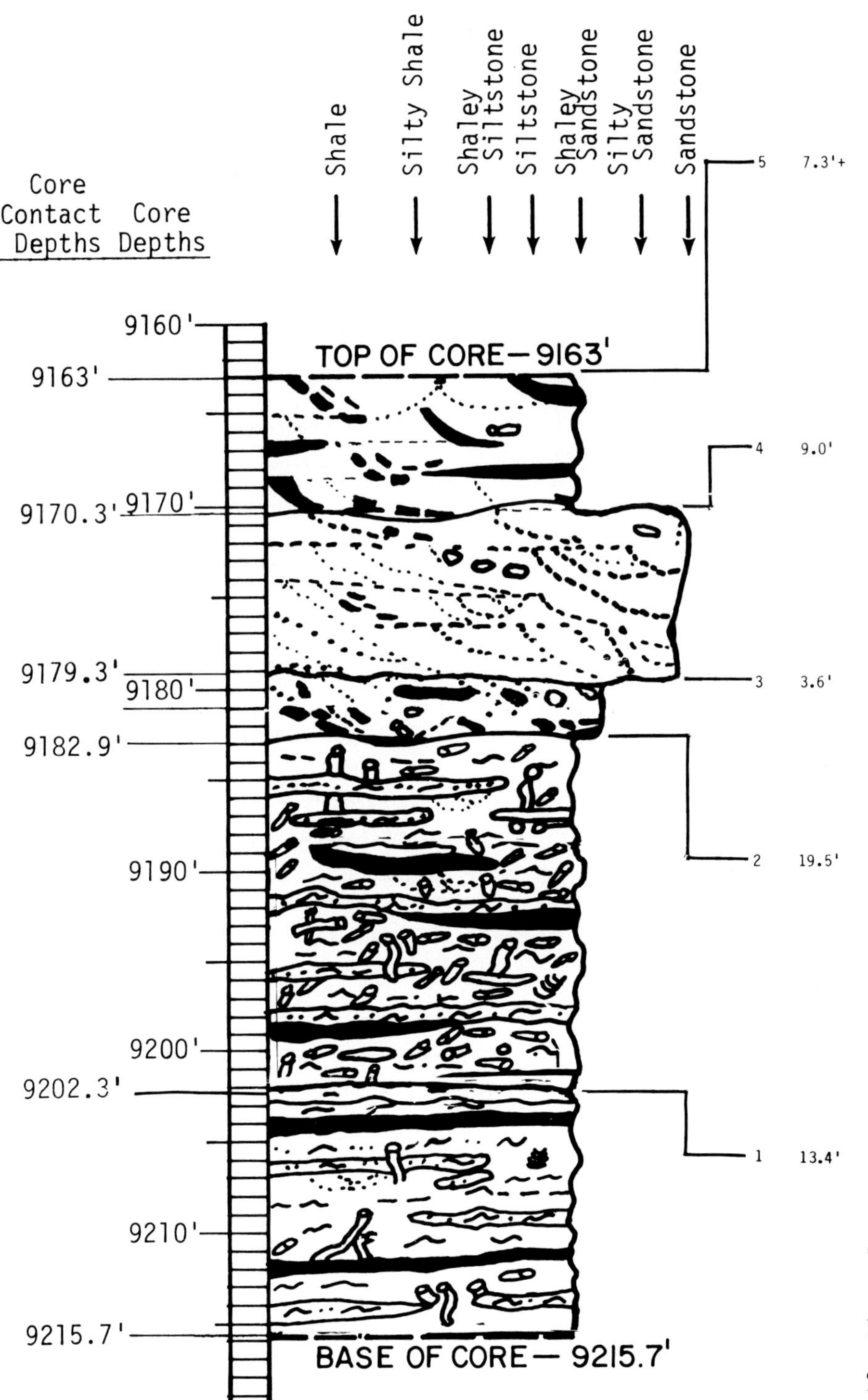

9163-70.3' Sandstone (200μ), 15% glauconite (up to 75% on some lamina). 10% shale beds. Abundant shale clasts (11%), and a trace of siderite clasts. 58% trough cross bedding including 43% ripples on troughs, 24% ripples, 7% horizontal laminations (sandstone and shale drapes). 5% reworked. Burrowing is sparse (6%) and burrows range from 1/8 to 1/2 inch in diameter and are horizontal to oblique. Cored interval probably does not extend to the top of the unit. Lower contact is transitional with underlying unit through about one foot. HIGH-ENERGY RIDGE-MARGIN FACIES (90%).

9170.3-79.3' Sandstone (92%, 200μ). 80% high to low angle trough beds including 7% ripples superimposed on troughs. 15% rippled and 2% reworked (by currents?). Less than 5% glauconite (up to 60% on some lamina). 3% oblate siderite clasts, 2% shale clasts; 2% shale drapes. Burrowing minimal (3%). Only two burrow types; 1/8 to 1/2 inch diameter horizontal to oblique. Maximum bed thickness one foot. Both upper and lower contacts transitional. CENTRAL RIDGE FACIES (95%).

9279.3-82.9' Sandstone (200μ), 15% glauconite. Abundant rip-up clasts (15% shale; 3% siderite). 74% troughs (including 3% ripples on troughs), 10% reworked, 8% horizontal laminations, 8% ripples. Only a trace of burrowing. 2% 1/8-1/2 inch horizontal burrows. Upper contact transitional, lower contact very sharp. HIGH ENERGY RIDGE-MARGIN FACIES (90%).

9182.8-9202.3' Sandstone (150μ), 5% disseminated glauconite, up to 30% on some laminations. 42% shale, mostly rippled. Biogenic structures (82%) dominate. 3% Chondrites, a trace of donut burrows (Terabellina) and Teichichnus. A high diversity of burrow types; predominently horizontal to oblique < 1/4" diameter silt filled, some sand filled. 74% of interval more than 75% burrowed (bioturbated). Some portions "multiply burrowed." Dominent physical structures are ripples (17%); a trace of horizontal laminations. BIOTURBATED SHELF SANDSTONE (90%).

9202.3-9215.7' Interbedded sandstone (60%) and shale (30%); 10% siltstone. Maximum bed thickness 0.15'. Shale mostly as "drape" over ripples. 64% ripples, mostly asymmetrical; a trace of horizontal laminations. 35% burrowed including 8% bioturbated (>75% burrowed) intervals up to 0.7' thick. High diversity (10) of burrow types. Mostly oblique to horizontal silt filled burrows, 8% sand filled. Equal distribution of small < 1/4" and larger 1/4-1/2" diameter burrows. Lower contact not cored, upper contact transitional. INTER-RIDGE FACIES (SHALY) (95%).

Fig. A5.—Stratigraphic facies description, Southland Royalty Company, Bud Christensen No. 2. This well is the offset well to the discovery well for Hartzog Draw field. Units summarized for this figure from sedimentologic facies description in Figure A6.

SOUTHLAND ROYALTY CO., BUD CHRISTENSEN NO. 2
SHANNON SANDSTONE SW SEC. 34 T46N R76W
HARTZOG DRAW FIELD
CAMPBELL CO., WYOMING

Described by: R. Tillman (3/77) (11/84)
R. Martinsen (3/77)

Columnized Data Format: Shannon
%/Max. set thickness/Additional notes

	INTER-RIDGE FACIES (SHALY) Unit 1 (13.4') Core Depth: 9202.3-9215.7' (Log Depth: 9215.3-9228.7')
Lithology	
1. Sandstone (%)	60%/0.15' (175μ)
2. Siltstone (%)	10%/0.1'
3. Shale (%); a) Laminated; b) Clasts	30%/0.1'/a
4. Siderite (%); a) Laminated; b) Clasts	-
5. Glauconite (%); a) Disseminated; b) Laminated	3%/0.1'/a; up to 15% on b
Physical Structures	65%/0.2'
1. High angle cross-bedding (20°+); a) Troughs; b) Planar-tabular; c) Planar-tangential; d) Curved-tangential (trough)	-
2. Moderate angle cross-bedding (10-20°); a) Troughs; b) Planar-tabular; c) Planar-tangential; d) Curved-tangential (trough)	-
3. Subhorizontal to low angle bedding (<10°); a) Trough; b) Planar-tabular; c) Planar-tangential; d) Curved-tangential; e) Planar	-
4. Horizontal laminations; a) Sandstone; b) Shale	tr/0.05'
5. Rippled; a) Sandstone; b) Shale; c) Current; d) Wave	64%/0.4'/a,c
6. Rippled interbedded sandstone and shale	64%/0.4'
7. Ripples superimposed on troughs	-
8. Reworked: a) By waves & currents; b) Bedding destroyed massive; c) soft sediment deformed; d) clasts	-
Biogenic Sedimentary Structures	35%/0.2'
1. Identified burrows; a) *Asterosoma*; c) *Chondrites*; d) "Donut burrows" (*Terabellina*); g) Gastropod tracks; p) Plural curving tubes; s) *Skolithos*; t) *Teichichnus*; th) *Thallasinoides*	tr-c (02.3') tr-d (11.0')
2. Distinct burrows; a) <1/8"; b) 1/8"-1/4"; c) 1/4"-1/2"; d) >1/2"; e) Silt filled; f) Clay filled; g) Sand filled; h) Silt lined; i) Clay lined; j) Spreiten; k) Vertical; l) Oblique; m) Horizontal	20% c,e,l 5% a,e,l 5% b,g,l 5% a,e,m 3% b,e,l tr-a,f,l 3% b,g,m tr-a,i,l
3. Burrowing (non-bioturbated) (%)	27%/0.15'
4. Bioturbated, 75% (+) burrowed (%); % burrowed; a) Distinct, b) Mottled	8%, 80%/0.7'/a
5. Total interval burrowed (%, footage)	97%/13.0'
6. Diversity (Number of burrow types); 1-4 low, 4-8 moderate; >8 high	10, high diversity
Cemented Intervals; a) Siderite; b) Calcite	30% b/sandstone beds
Contact Relations (core contacts)	
1. Upper; a) Very sharp (<0.05' transition); b) Sharp (0.05-0.1' transition); c) Transitional (0.1-0.3' transition); d) Gradational (>0.3' transition); e) Contact,erosional (truncated) angular; f) Contact erosional parallel; g) Covered or not covered	c
2. Lower; a), b), c), d), e), f), g)	-

FIG. A6.—Sedimentologic facies descriptions, Southland

BIOTURBATED SHELF SANDSTONE	HIGH-ENERGY RIDGE-MARGIN FACIES	CENTRAL RIDGE FACIES	HIGH-ENERGY RIDGE-MARGIN FACIES
Unit 2 (19.5')	Unit 3 (3.5')	Unit 4 (9.0')	Unit 5 (7.3')
Core Depth: 9182.8-9202.3'	Core Depth: 9279.3-82.8'	Core Depth: 9170.3-79.3'	Core Depth: 9163-70.3'
(Log Depth: 9195.8-9215.8')	(Log Depth: 9292.3-95.8')	(Log Depth: 9183.3-92.3')	(Log Depth: 9176-83.3')
47%/0.2' (150μ)	79%/0.15'/f-m (200μ)	92%/1.0'/f (200μ)	75%/0.5'/fg (225μ)
11%/0.1'	tr/0.05'	-	4%/0.05'/with sh
42%/0.1'	16%/0.05'/tr-a, 15%b	5%/0.03'/2%a, 3%b	21%/0.1'/10%a, 11%b
-	4%/tr-a, 3%b	3%/0.3'/a, oblate	tr/0.1'/a,1/2" dia.
5%/a; up to 30% on b	15%/0.1'/8% a, 7% b	5%/4% a; tr-b, up to 60%	15%/0.2'/5%a;10%b,75%ctr.
18%	98%/0.15'	97%/0.8'	94%/0.2'
-	-	10%/0.3'/a	10%/0.2'
-	15%/0.1'/a	58%/0.8'/a	45%/0.2'
-	54%/0.15'/a	5%/0.1'/a	5%/0.2'
tr/0.02'/a	8%/0.1'	-	7%/0.1'/shale
17%/0.2'/a,c	5%/0.1'/a	10%/0.1'/a	8%/0.1'
15%/0.2'	3%/0.1'	// 5%/0.1'	// 16%/0.5'
-	5%/0.05'	7%/0.3'	14%/0.2'
-	10%/0.3'/a	2%/0.1'/a	5%/0.5'/compaction
82%/1.5'	2%/0.05'	3%/0.05'	6%/0.08'
2%-c tr-d (00.2') tr-t (93.8')	-	-	-
2% c,e,l 24% a,e,m 10% b,e,l 10% a,e,l 7% b,g,l 3% a,f,l 3% b,g,m 2% a,i,l 15% a,e,l tr-a,f,k	tr - b,l/0.05' tr - c,l/0.05'	2% - b,l/0.05' tr - c,k/0.05'	3% - c,l 2% - b,l tr - c,k
8%/10.1'	2%/0.05'	3%/0.05'	6%/0.08'
74%, 90%/1.5'/a	-	-	-
96%/18.7'	1%/0.1'	1%, 0.1'	5%/0.4'
13, high diversity	2 - low diversity	2 - low diversity	3 (low diversity)
40%/sandstone	10%/b, tr-a	0%	5% b/0.2'
a	b	d	-
c	a	b	d

Royalty Company, Bud Christensen No. 2, Units 1 to 5.

FIG. A7.—Slabbed core, Southland Royalty Company, Bud Christensen No. 2. Unit numbers and facies designations are indicated on photographs. Scale on left is in tenths of feet.

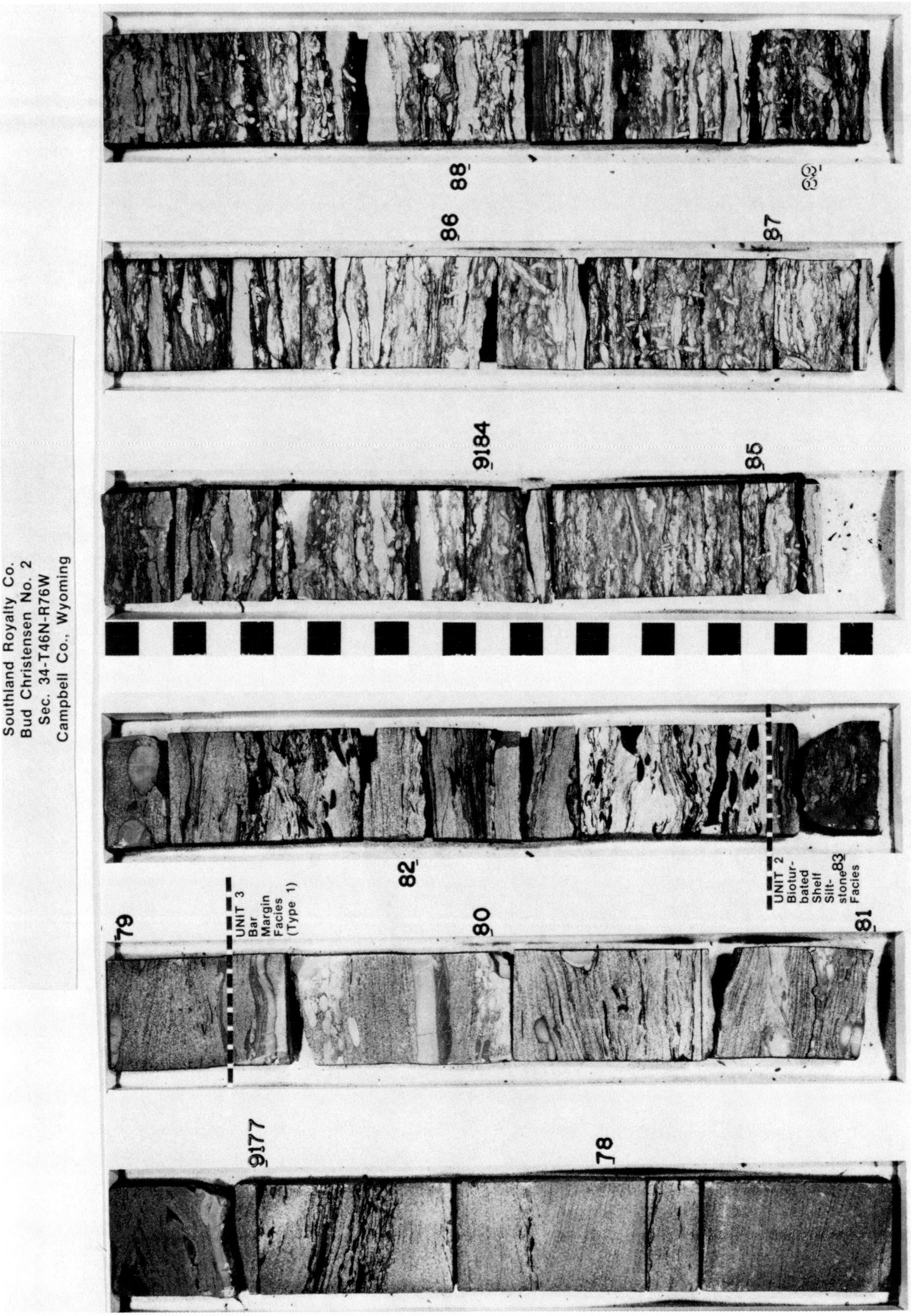

Southland Royalty Co.
Bud Christensen No. 2
Sec. 34-T46N-R76W
Campbell Co., Wyoming

UNIT 3
Bar
Margin
Facies
(Type 1)

UNIT 2
Biotur-
bated
Shelf
Silt-
stone
Facies

9177 78 79 80 81 82 83 9184 85 86 87 88 89

Southland Royalty Co.
Bud Christensen No. 2
Sec. 34-T46N-R76W
Campbell Co., Wyoming
9200
01
98
99
9196
97
94
95
92
93
9190
91

Southland Royalty Co.
Bud Christensen No. 2
Sec. 34-T46N-R76W
Campbell Co., Wyoming
9202
UNIT 1
Interbar
Facies
03
04
05
06
07
08
9209
10
11
12
13
14

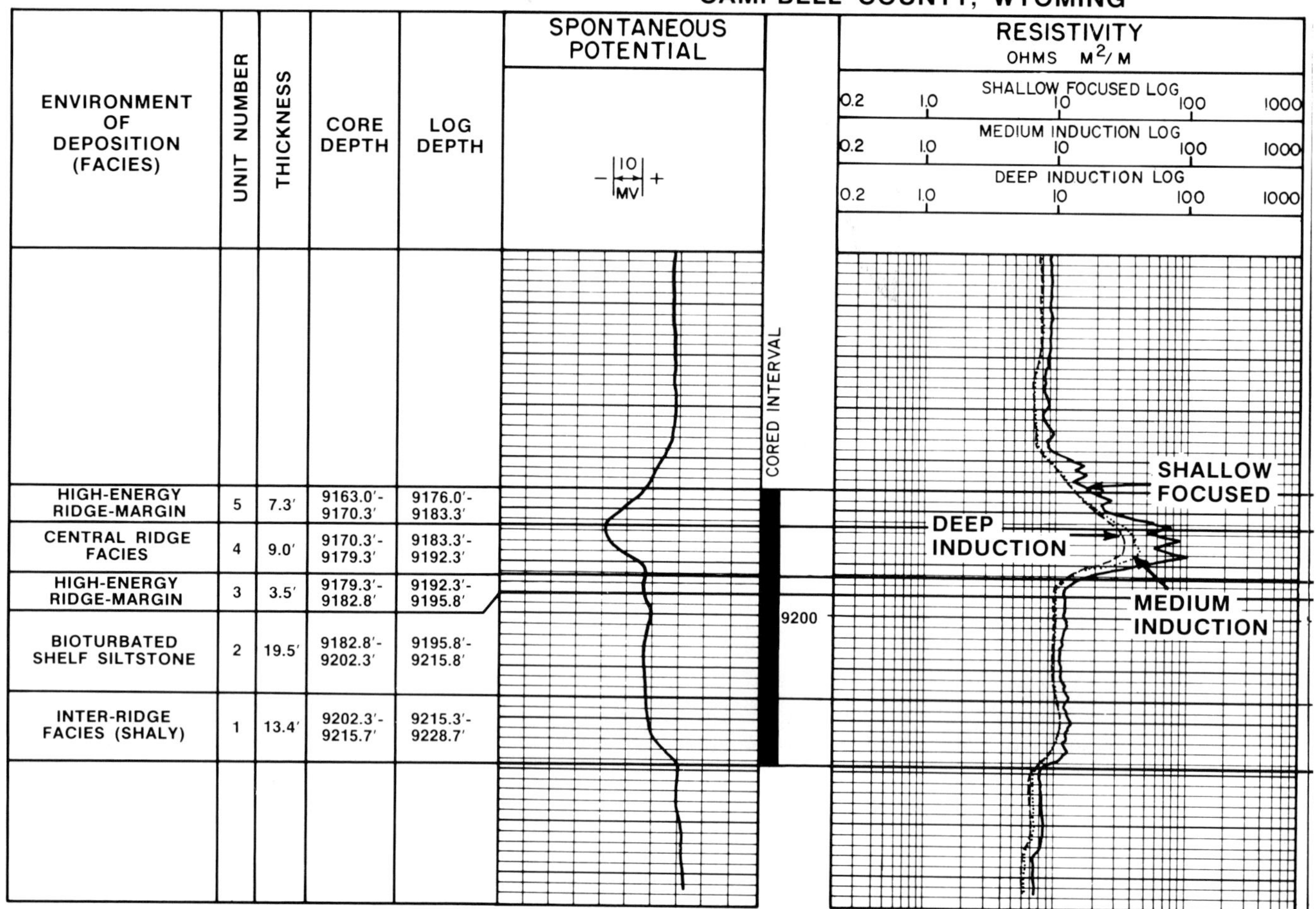

FIG. A8.—Facies designations correlated with subsurface logs and core gamma log, Southland Bud Christensen No. 2. Note that core depths are described and photographed in Figures A5 and A7.

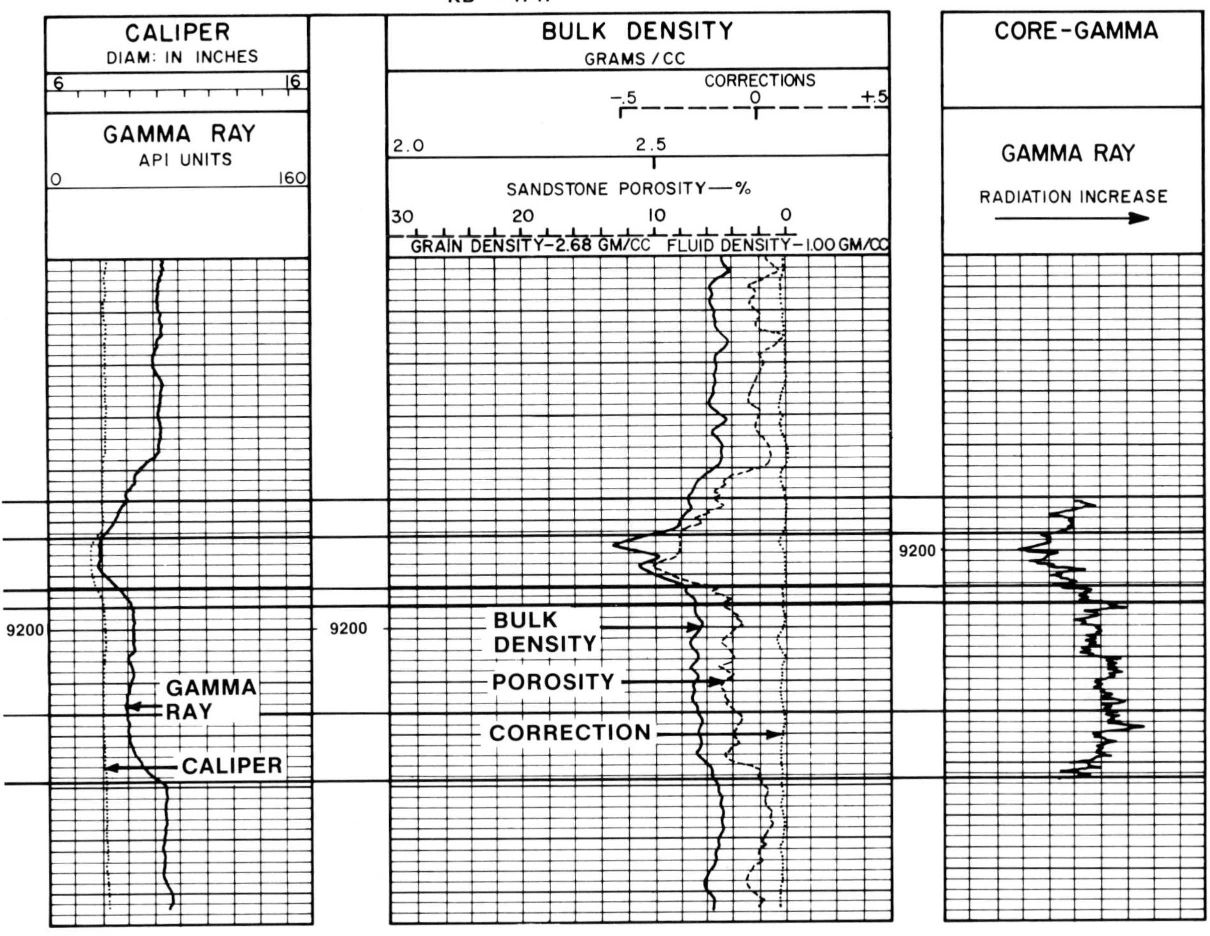

13 ft (4 m) shallower than the subsurface gamma-ray log. Facies and core depths are corrected to log depths. Units described correspond to those

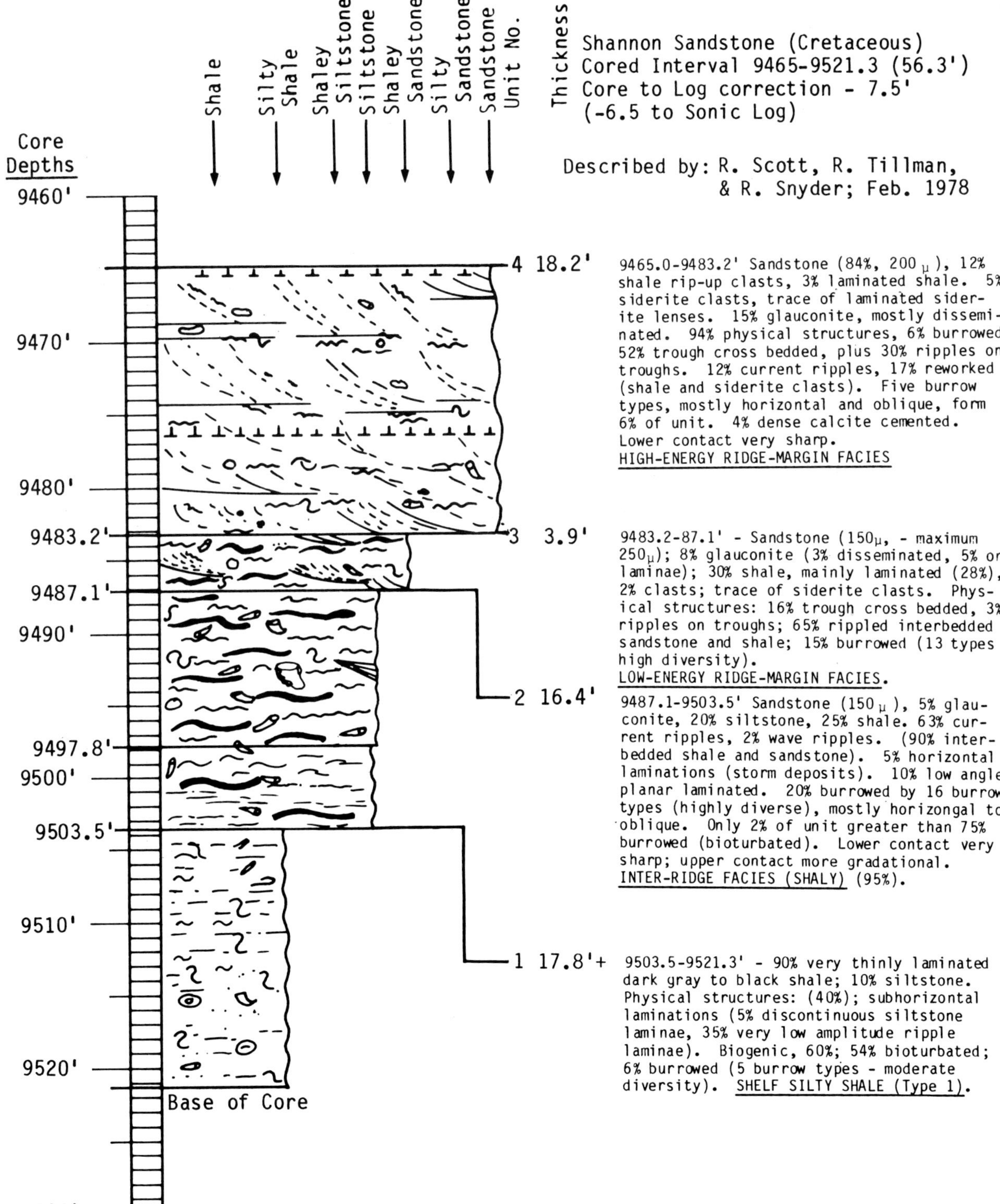

FIG. A9.—Stratigraphic facies description, Cities Service Federal AS-1. Units summarized for this figure from sedimentologic facies description in Figure A10.

CITIES SERVICE FEDERAL AS-1
NW NW Sec. 34 T45N R75W, HARTZOG DRAW FIELD
CAMPBELL CO., WYOMING (SHANNON SANDSTONE)
Cored Interval 9465-9521.3' (56.3')

Columnized Data Format:
%/Max. set thickness/Additional notes

Described by: R. Tillman, 1978
R. Scott

	SHELF SILTY SHALE, BURROWED, 80%
	Unit 1 (17.8') Core Depth: 9503.5-9521.3' (Log Depth:)
Lithology	
1. Sandstone (%)	tr (100μ)
2. Siltstone (%)	10%/0.02'
3. Shale (%); a) Laminated; b) Clasts	89%/ /a
4. Siderite (%); a) Laminated; b) Clasts	-
5. Glauconite (%); a) Disseminated; b) Laminated	-
Physical Structures	40%/0.01'
1. High angle cross-bedding (20°+); a) Troughs; b) Planar-tabular; c) Planar-tangential; d) Curved-tangential (trough)	-
2. Moderate angle cross-bedding (10-20°); a) Troughs; b) Planar-tabular; c) Planar-tangential; d) Curved-tangential (trough)	-
3. Subhorizontal to low angle bedding (<10°); a) Trough; b) Planar-tabular; c) Planar-tangential; d) Curved-tangential; e) Planar	-
4. Horizontal laminations; a) Sandstone; b) Shale	5%/0.01'/b, silt
5. Rippled; a) Sandstone; b) Shale; c) Current; d) Wave	35%/0.02'/b
6. Rippled interbedded sandstone and shale	// 40%/shale and silt
7. Ripples superimposed on troughs	-
8. Reworked: a) By waves & currents; b) Bedding destroyed massive; c) soft sediment deformed; d) clasts	-
Biogenic Sedimentary Structures	60%
1. Identified burrows; a) *Asterosoma*; c) *Chondrites*; d) "Donut burrows" (*Terabellina*); g) Gastropod tracks; p) Plural curving tubes; s) *Skolithos*; t) *Teichichnus*; th) *Thallasinoides*	tr-d
2. Distinct burrows; a) <1/8"; b) 1/8"-1/4"; c) 1/4"-1/2"; d) >1/2"; e) Silt filled; f) Clay filled; g) Sand filled; h) Silt lined; i) Clay lined; j) Spreiten; k) Vertical; l) Oblique; m) Horizontal	tr-a,e,l tr-b,e,l tr-a,e,m tr-a,f,l tr-a,f,m
3. Burrowing (non-bioturbated) (%)	6%
4. Bioturbated, 75% (+) burrowed (%); % burrowed; a) Distinct, b) Mottled	54% (80%)/0.1'/b
5. Total interval burrowed (%, footage)	80%/14.2'
6. Diversity (Number of burrow types); 1-4 low, 4-8 moderate; >8 high	6, moderate
Cemented Intervals; a) Siderite; b) Calcite	-
Contact Relations (core contacts)	
1. Upper; a) Very sharp (<0.05' transition); b) Sharp (0.05-0.1' transition); c) Transitional (0.1-0.3' transition); d) Gradational (>0.3' transition); e) Contact, erosional (truncated) angular; f) Contact erosional parallel; g) Covered or not covered	a, 0.01'
2. Lower; a), b), c), d), e), f), g)	NA

FIG. A10.—Sedimentologic facies descriptions,

INTER-RIDGE FACIES (SHALY)	LOW-ENERGY RIDGE-MARGIN FACIES 75%	HIGH-ENERGY RIDGE-MARGIN FACIES (90%)
Unit 2 (16.4')	Unit 3 (3.9')	Unit 4 (18.2' cored)
Core Depth: 9487.1-9503.5'	Core Depth: 9483.2-9487.1'	Core Depth: 9465.0-9483.2'
(Log Depth: 9479.6-9496.0')	(Log Depth: 9475.7-9479.6')	(Log Depth: 9453.0-9475.7')
55%/0.2' (150μ)	70%/0.4' (150μ; max. 250μ)	84%/0.9' (200μ)
20%/0.1' (75μ)	-	tr/0.01'
25%/0.05'/a	30%/0.25'/28%a, 2%b	15%/0.05'/12%b; 3%a, 0.1'
-	tr//b	7%/0.25'/5%b, 2%a
5%/3%a, 2%b	8%//3%a, 5%b	15%//12%a, 3%a
80%	85%	94%
-	-	7%/0.6'/a
-	16%/0.4'a	38%/0.95'/a
10/0.2'/4%e, storm deposit	-	7%/0.3'/a
5%/0.05'/a, storm deposit	2%/0.01'/b, wavy	-
65%/0.1'/63%a,c,silt;2%d	62%/0.1'/40%a, 24%b	12%/0.3'/10%a, 2%b
// 90%/0.3'	// 65%/0.8'	// 13%
-	3%/0.1'	30%/
-	2%/0.1'/d	17%/0.2'/d, 12% sh, 5% siderite
20%	15%	6%/0.3'
-	-	-
tr-d,g,l 2%-b,e,l tr-d,g,k 2%-b,e,k tr-b,g,k tr-b,g,l tr-b,g,m tr-b,g,m tr-b,g,l tr-a,g,k tr-c,e,l tr-a,g,l tr-c,e,m tr-a,e,m tr-b,e,h,l tr-a,e,l	tr-d,g,m tr-a,g,m tr-c,g,l tr-a,g,m tr-c,g,m tr-a,g,l 2%-b,g,l tr-a,e,m tr-b,g,m tr-a,e,l tr-b,e,l tr-c,g,h,l tr-b,e,m tr-b,e,h,l	- tr-b,g,h,l tr-a,g,k tr-b,g,h,m tr-a,g,h,l tr-a,g,h,m tr-c,g,m
18%/0.01'	14%	6%/0.3'
2%/0.02'	tr	-
93%/15.2'	66%/2.6'	9%/0.4'
16 - high	13, high	5, moderate
-	8%b	4%//b
d, 1.0'	a, 0.04'	NA
a, 0.01'	d, 1.0'	a, 0.04'

Cities Service Federal AS-1, Units 1 to 4.

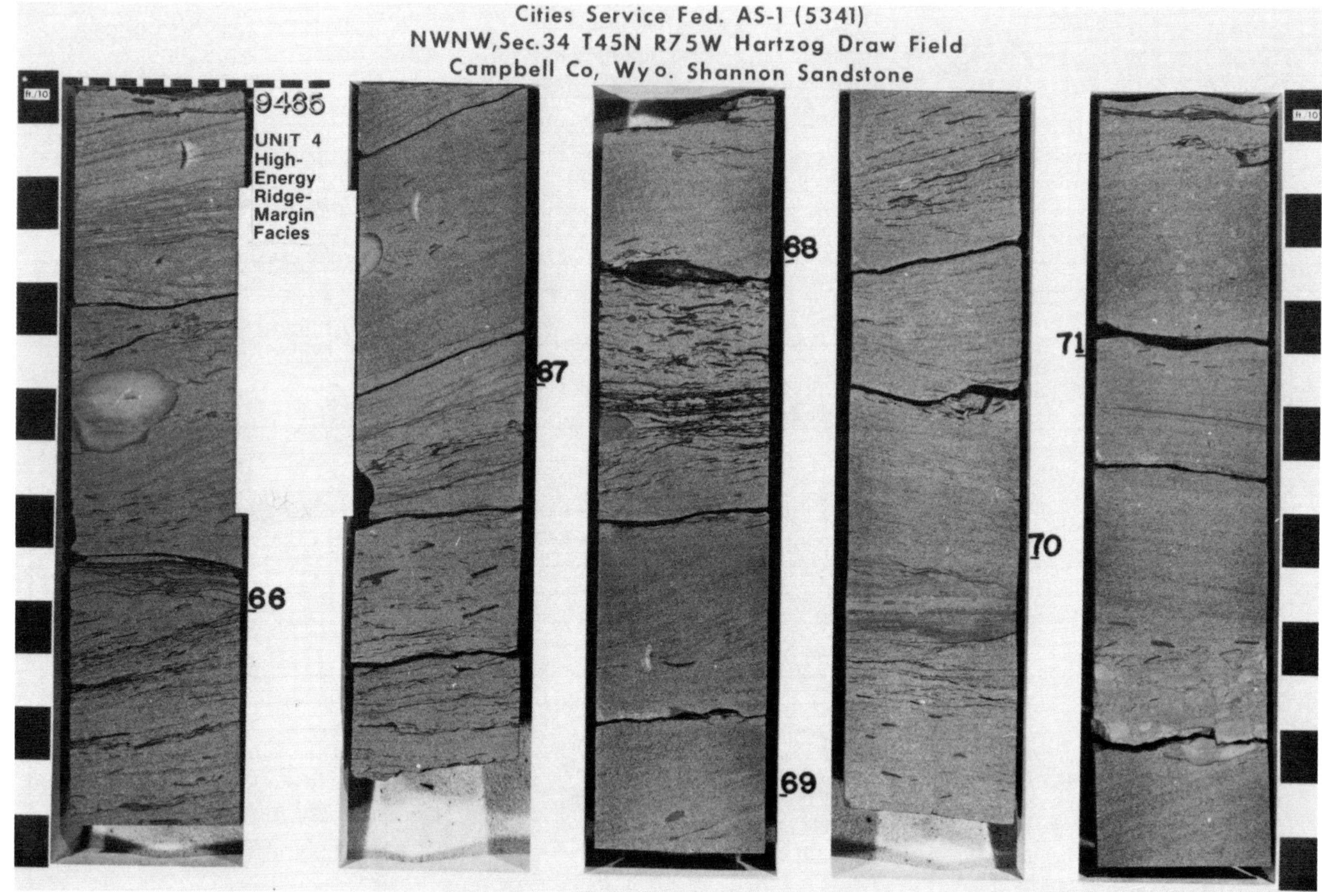

FIG. A11.—Slabbed core, Cities Service Federal AS-1. Unit numbers and facies designations are indicated on photographs. Scale on left is in tenths of feet.

Cities Service Fed. AS-1 (5341)
NWNW,Sec.34 T45N R75W Hartzog Draw Field
Campbell Co, Wyo. Shannon Sandstone
9472
73
74
75
76
77
78
9479
80
81
82
83
84
85
UNIT
3
Low-
Energy
Ridge-
Margin
Facies
CALIPER
9450
9500
CORE (9503.5')
TO GR LOG (9496')
CORRECTION -7.5ft

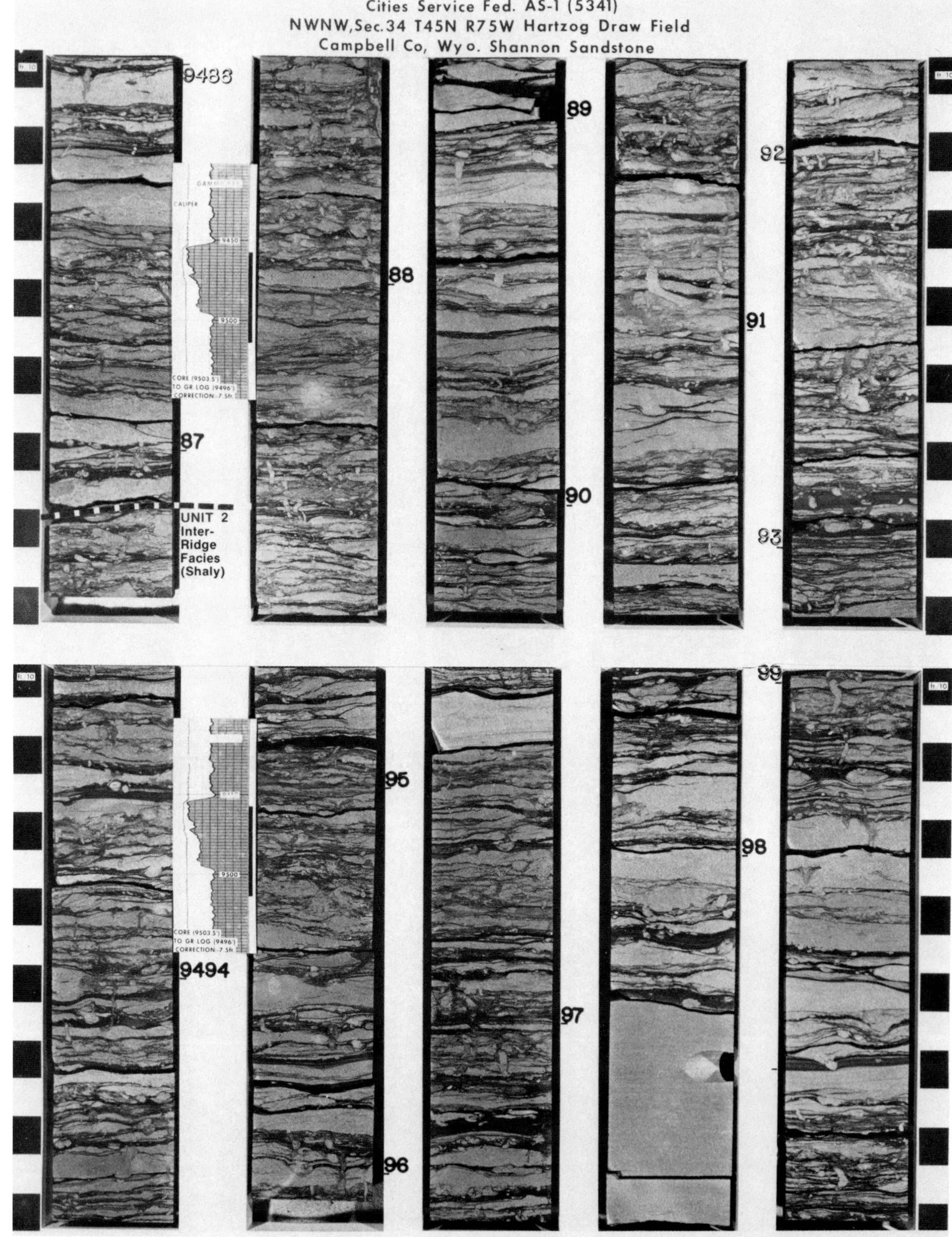
Cities Service Fed. AS-1 (5341)
NWNW,Sec.34 T45N R75W Hartzog Draw Field
Campbell Co, Wyo. Shannon Sandstone
9486
87
88
89
90
91
92
93
UNIT 2
Inter-
Ridge
Facies
(Shaly)
9494
95
96
97
98
99
CORE (9503.5')
TO GR LOG (9496')
CORRECTION -7.5ft
9450
9500

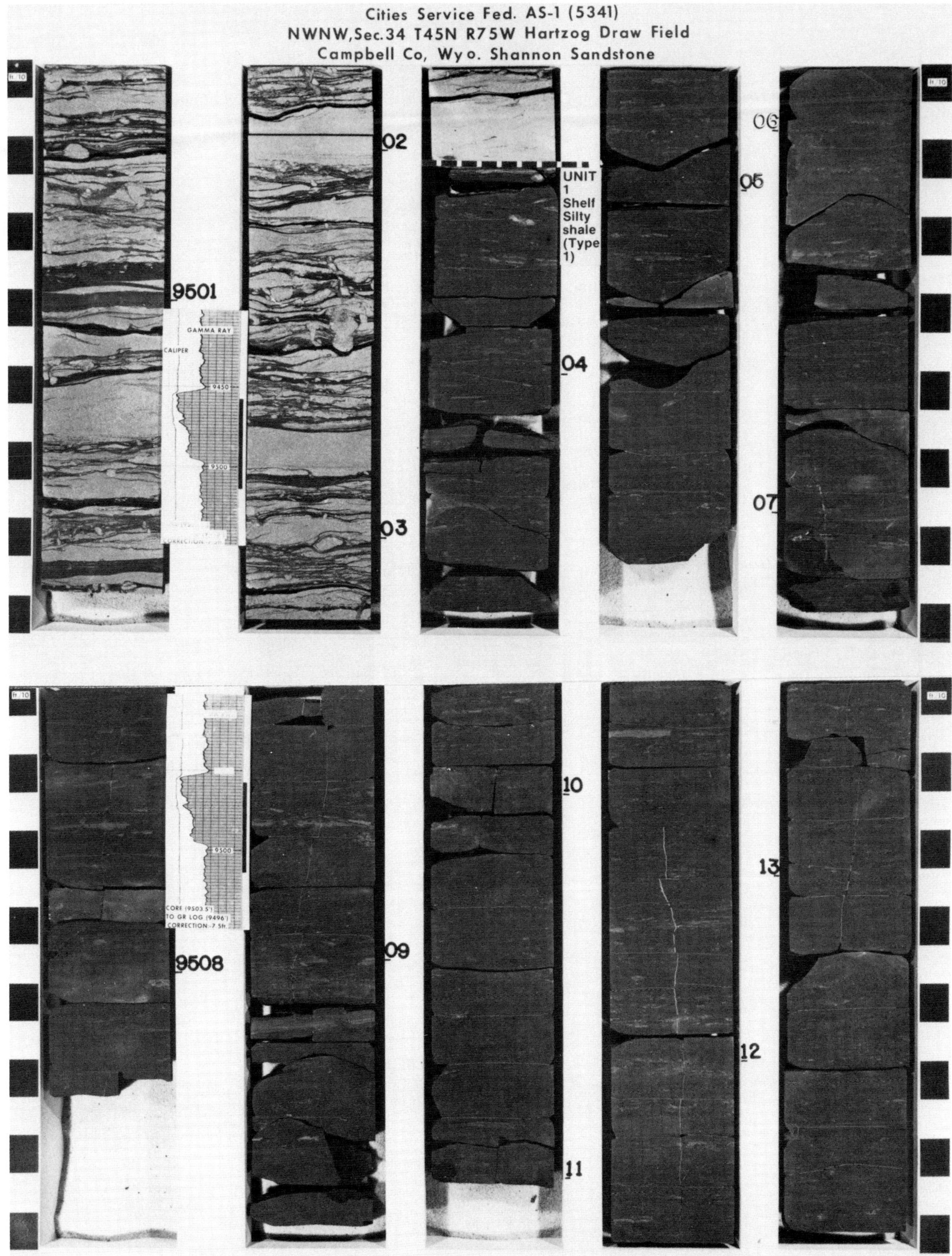

Cities Service Fed. AS-1 (5341)
NWNW,Sec.34 T45N R75W Hartzog Draw Field
Campbell Co, Wyo. Shannon Sandstone
9501
GAMMA RAY
CALIPER
9450
9500
02
03
04
UNIT
1
Shelf
Silty
shale
(Type
1)
05
06
07
9500
CORE (9503.5')
TO GR LOG (9496')
CORRECTION -7.5ft
9508
09
10
11
12
13

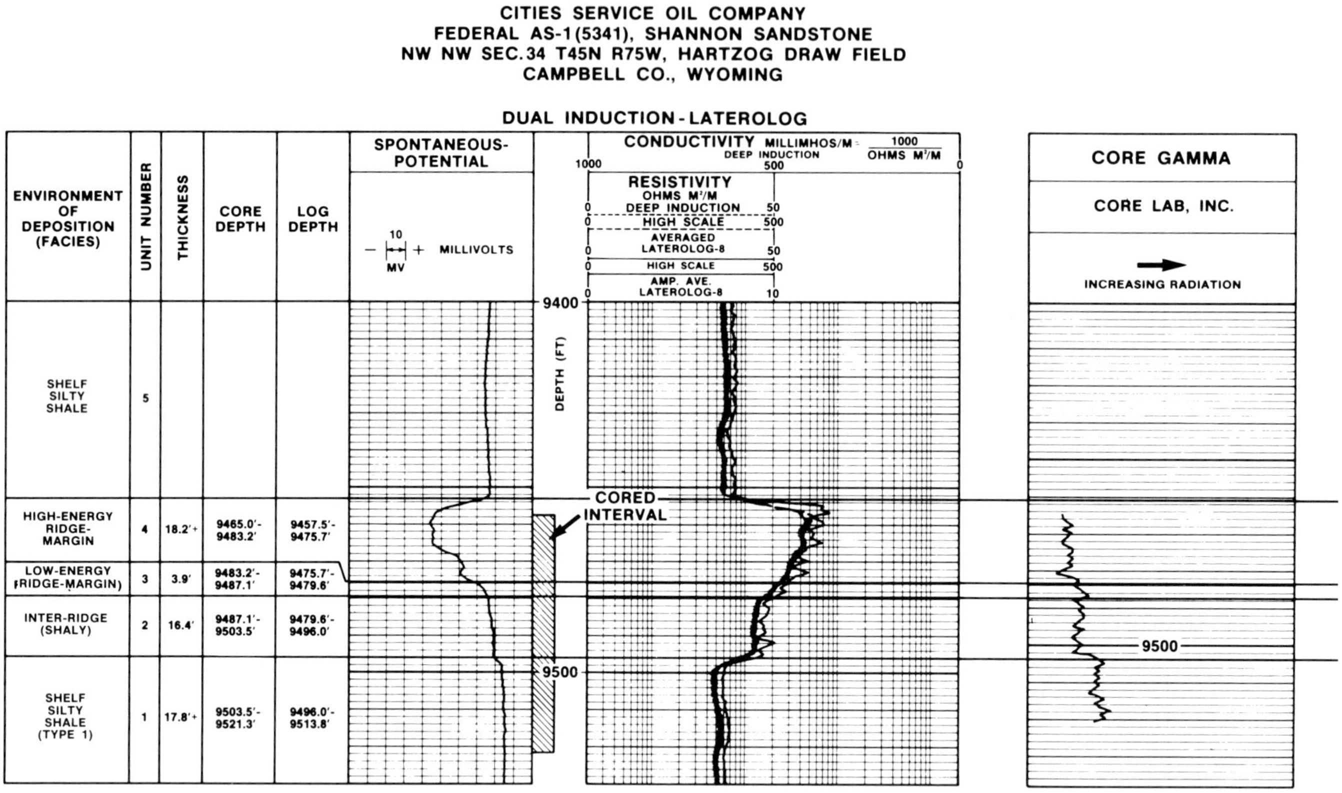

CITIES SERVICE OIL COMPANY
FEDERAL AS-1(5341), SHANNON SANDSTONE
NW NW SEC.34 T45N R75W, HARTZOG DRAW FIELD
CAMPBELL CO., WYOMING

DUAL INDUCTION-LATEROLOG

ENVIRONMENT OF DEPOSITION (FACIES)	UNIT NUMBER	THICKNESS	CORE DEPTH	LOG DEPTH
SHELF SILTY SHALE	5			
HIGH-ENERGY RIDGE-MARGIN	4	18.2'+	9465.0'-9483.2'	9457.5'-9475.7'
LOW-ENERGY (RIDGE-MARGIN)	3	3.9'	9483.2'-9487.1'	9475.7'-9479.6'
INTER-RIDGE (SHALY)	2	16.4'	9487.1'-9503.5'	9479.6'-9496.0'
SHELF SILTY SHALE (TYPE 1)	1	17.8'+	9503.5'-9521.3'	9496.0'-9513.8'

SPONTANEOUS-POTENTIAL
10
− + MILLIVOLTS
MV

CONDUCTIVITY MILLIMHOS/M = 1000 / OHMS M²/M
DEEP INDUCTION
1000 500 0

RESISTIVITY
OHMS M²/M
0 DEEP INDUCTION 50
0 HIGH SCALE 500
AVERAGED
0 LATEROLOG-8 50
0 HIGH SCALE 500
AMP. AVE.
0 LATEROLOG-8 10

DEPTH (FT)
9400
CORED INTERVAL
9500

CORE GAMMA
CORE LAB, INC.
INCREASING RADIATION
9500

FIG. A12.—Facies designations correlated to subsurface logs and core gamma log, Cities Service Federal AS-1. Note that core depths are 7.5 ft in Figures A9 and A11.

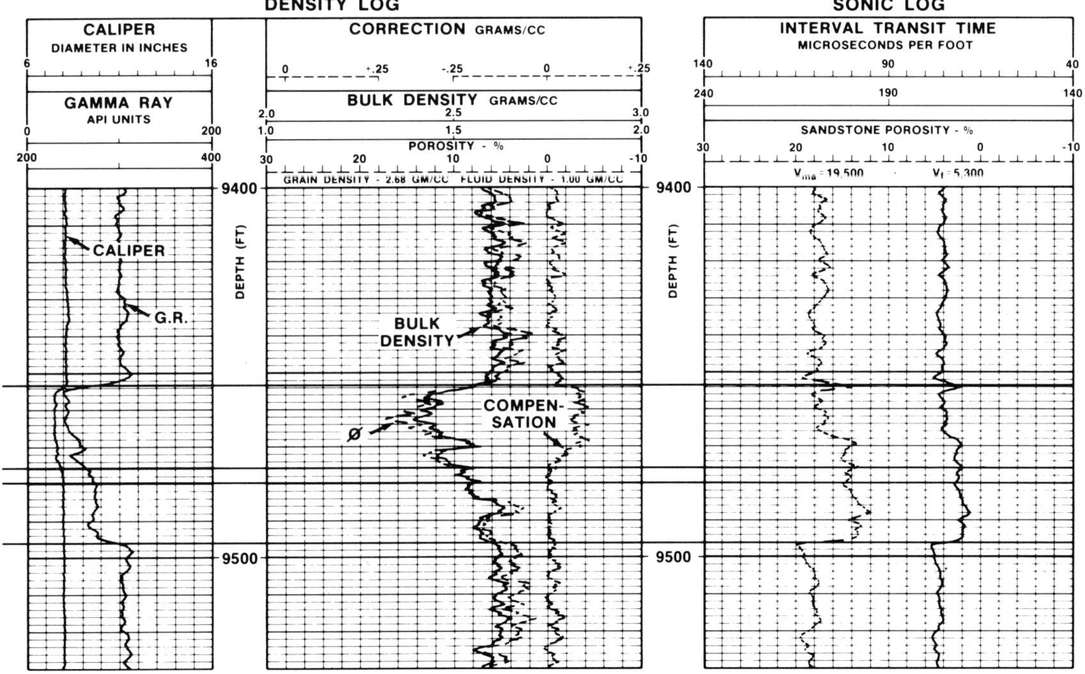

(2.3 m) deeper than log depths. Facies and core depths are correlated to log depths. Units indicated correspond to those described and photographed

TABLE A1.—COMPILATION OF PERCENTAGES OF SEDIMENTOLOGIC FEATURES;

CENTRAL RIDGE FACIES

		Lithology (%)						Physical					
									Cross-Bedded				Hori
Well No.	Unit No.	Sand-stone	Silt-stone	Shale *	*Sid-erite	Glau-conite Avg./Max.	Avg. Grain Size, Microns	Total %	High Angle	Med-ium	Low Angle	Total	Sand-stone
CS AB-1A	4	91	tr	2 1%b	6 b	5/60	250	99	7	38	10	56	tr
South BC-2	4	92	—	5 3%b	3 b	5/60	200	97	10	52	5	67	—
CS AE-1	3	87	5	8 3%b	tr b	5	200	85	11	28	14	53	3
CS AE-1	5	95	2	5 4%b	5 b	10/30	200	97	28	31	3	62	—
CS AK-1	4A	92	—	6 4%b	2 b	10	250	100	11	26	45	84	—
CS AK-1	4B	95	1	4 3%b	1 b	10	200	100	12	26	38	74	—
Mean	N = 6	92%	1%	5% 3%b	3%	8%	215	92	13	33	19	65	0
Range	—	91–97	0–5	2–7 1–4b	1–7	5–15	200–250	85–100	7–28	26–52	3–45	52–84	0–3

*Siderite or Shale: Total; (a) % laminae, (b) % clasts
**Reworked: (a) by waves and currents (default), (d) clasts

CENTRAL RIDGE (PLANAR LAMINATED) FACIES

		Lithology (%)						Physical					
									Cross-Bedded				Hori
Well No.	Unit No.	Sand-stone	Silt-stone	Shale *	*Sid-erite	Glau-conite Avg./Max.	Avg. Grain Size, Microns	Total %	High Angle	Med-ium	Low Angle	Total	Sand-stone
CS AE-1	7	84	5	11 tr-d	—	4/30	150	80	—	—	4	4	25 10° dip

*Siderite or Shale: Total; (a) % laminae, (b) % clasts
**Reworked: (a) by waves and currents (default), (d) clasts (shale and siderite).

CENTRAL RIDGE FACIES AND *CENTRAL RIDGE (PLANAR LAMINATED) FACIES*

Structures (%)					Biological Structures (Burrows) (%)						Contacts		
zontal Laminae		Rip-ples	Ripples on Troughs	**Re-worked a, d	Total %	Iden-tified %	Dis-tinct %	Biotur-bated %	Interval Bur-rowed %	Diver-sity	Transi-tion Thick-nesses	***	
Shale	Total											Upper	Lower
—	tr	5	30	8 7%d	tr	—	tr	—	1	1	0.1′/0.01′	G	T
—	0	10	7	8 6%d	3	—	3	—	1	2	0.4′/0.1′	G	S
—	3	20	5	4 d	15	2 t, te	11	2	12	5	0.1′/0.4′	S	G
—	0	20	8	7 d	3	—	3	—	3	2	1.0′/0.2′	G	T
—	0	2	5	11 6%d	0	—	—	—	0	0	1.0′/1.0′	G	G
—	0	2	5	5 4%d	0	—	—	—	0	0	NA/1.0′	G	—
0	0	10	10	7 7%d	4	0	3	0	4	2	0.5′/0.4′	G	G
0	0–3	2–20	5–30	4–11 4–7d	1–15	0–2	0–11	0–2	1–12	0–5	0.1–1′/0.01–1′	T-G	T-G

Structures (%)					Biological Structures (Burrows) (%)						Contacts		
zontal Laminae		Rip-ples	Ripples on Troughs	**Re-worked a, d	Total %	Iden-tified %	Dis-tinct %	Biotur-bated %	Interval Bur-rowed %	Diver-sity	Transi-tion Thick-nesses	***	
Shale	Total											Upper	Lower
—	25	30	—	tr d	20	10% white horiz.	8	2	58	5	0.7′/0.01′	G	VS

***Core Contact Types: VS = Very Sharp (<0.05′)
S = Sharp (0.05–0.1′)
T = Transitional (0.1–0.5′)
G = Gradational (>0.5′)

TABLE A2.—*HIGH-, LOW-ENERGY RIDGE-MARGIN FACIES*, COMPILATION OF

HIGH-ENERGY RIDGE-MARGIN FACIES

		Lithology (%)						Physical					
								Cross-Bedded					Hori
Well No.	Unit No.	Sand-stone	Silt-stone	Shale *	*Sid-erite	Glau-conite Avg./Max.	Avg. Grain Size, Microns	Total %	High Angle	Med-ium	Low Angle	Total	Sand-stone
CS AB-1A	5	70	4	25 20b	8 b	15/75	250	97	—	20	7	27	tr
South BC-2	3	79	1	16 13b	4 3b	15/70	200	98	—	15	41	56	8
South BC-2	5	60	4	40 16b	1 a	15/75	225	94	10	45	5	47	—
CS AE-1	6	70	3	27 22b	3 b	20/40	250	92	—	27	10	37	—
CS AK-1	3	88	1	9 6d	2 tr-b	15/50	225	97	6	25	34	65	—
CS AS-1	4	84	1	15 12%b	7 5b	15	200	94	7	38	7	52	—
Aver-age	N = 6	75	3	22 15d	5	16	225	96	4	27	16	47	1
Range	—	60–88	1–4	9–40 6–22b	1–8	15–20	200–250	92–98	0–10	15–45	5–41	30–65	0–8

*Siderite or Shale: Total; (a) laminae, (b) clasts.

LOW-ENERGY RIDGE-MARGIN FACIES

		Lithology (%)						Physical					
								Cross-Bedded					Hori
Well No.	Unit No.	Sand-stone	Silt-stone	Shale *	Sid-erite	Glau-conite Avg./Max.	Avg. Grain Size, Microns	Total %	High Angle	Med-ium	Low Angle	Total	Sand-stone
CS† AB-1A	7	45	20	35 20b†	—	15	175	80	—	10	5	15	—
CS AE-1	4	75	5	20 15b	5 b	15	200	95	—	20	12	32	—
CS AK-1	2	65	—	35 5b	18 9b	20	125	90	10	15	10	35	—
CS AS-1	3	70	—	30 2b	tr b	8	150	85	—	16	—	16	—
Mean	N = 4	64	6	30 6b	6 4b	15	165	88	3	15	7	25	0
Range	—	45–75	0–20	20–35 2–20b	1–18 5–96	8–20	125–200	85–95	0–10	0–20	0–12	15–35	0

*Siderite or Shale: Total; (a) laminae, (b) clasts.
†Unit 7 (AB-1A) Transitional to High-Energy Ridge-Margin Facies because of abundant shale clasts.
**Reworked: (a) by waves and currents (default), (d) clasts (shale and siderite).

PERCENTAGES OF SEDIMENTOLOGIC FEATURES

Structures (%)					Biological Structures (Burrows) (%)						Contacts		
zontal Laminae												***	
Shale	Total	Rip-ples	Ripples on Troughs	Re-worked a, d	Total %	Iden-tified %	Dis-tinct %	Biotur-bated %	Interval Bur-rowed %	Diver-sity	Transi-tion Thick-nesses	Upper	Lower
2	3	20	17	30 28d	3	—	3	—	19	1	0.5′ 0.1′	G	G
—	8	5	11	18 16d	2	—	2	—	1	2	0.1′ <0.05′	S	VS
7	7	5	14	21 16d	6	—	6	—	5	3	0.4′ NA	G	—
—	0	10	15	30 25d	8	—	8	—	13	2	0.01′ 1.0′	VS	G
—	—	14	11	7 tr-d	3	—	3	—	2	2	1.0′ 0.04′	G	VS
—	0	12	30	17 d	6	—	6	—	9	5	NA 0.04′	NA	A
1	3	11	15	21 17d	4	0	4	0	8	2	0.4′ 0.3′	G	G
0–7	0–8	5–20	11–17	6–30 1–28d	2–8	0	2–8	0	1–19	1–3	0.01–1.0′ 0.04–1.0′	VS-G	VS-G

Structures (%)					Biological Structures (Burrows) (%)						Contacts		
zontal Laminae													
Shale	Total	Rip-ples	Ripples on Troughs	**Re-worked a, d	Total %	Iden-tified %	Dis-tinct %	Biotur-bated %	Interval Bur-rowed %	Diver-sity	Transi-tion Thick-nesses	Upper	Lower
10	10	34	tr	20 d	20	—	18	2	33	4	0.0.1′ 0.1′	VS	S
—	0	38	5	20 d	5	—	5	—	8	4	0.2′ 0.1′	T	S
—	—	41	30	14 d	10	—	10	—	12	4	0.04′ 0.03′	VS	VS
2	2	61	3	3 d	15	—	14	1	66	13	0.04′ 1.0′	VS	G
3	3	44	10	14 d	13	0	12	1	30	6	0.07′ 0.3′	S	T
0–10	0–10	34–61	1–30	3–20 d	5–20	—	5–18	1–2	8–66	4–13	0.01–0.2′ 0.03–1.0′	VS-T	VS-G

***Core Contact Types: VS = Very Sharp (<0.05′)
S = Sharp (0.05–0.1′)
T = Transitional (0.1–0.5′)
G = Gradational (>0.5′)

TABLE A3.—*INTER-RIDGE FACIES (SHALY)* AND *BIOTURBATED OR BURROWED SHELF SILTSTONE*

INTER-RIDGE FACIES (SHALY)

		Lithology (%)						Physical					
									Cross-Bedded				Hori
Well No.	Unit No.	Sand-stone	Silt-stone	Shale *	*Sid-erite	Glau-conite	Avg. Grain Size, Microns	Total %	High Angle	Med-ium	Low Angle	Total	Sand-stone
CS AB-1A	2A	46	12	42 a	tr	tr 5% in SS	150	63	—	tr	tr	2	tr
CS AB-1A	3A	35	30	45 a	0	3% 10% in SS	100	60	—	—	tr	tr	2
South BC-2*	1	60	10	30 a	0	3%	175	65	—	—	—	—	tr
CS AE-1	2	75	2	23 a		5	100	66	—	—	1	1	—
CS AK-1	1	78	1	21 a		2	140	66	—	1	1	2	1
CS AS-1	2	55	20	25 a		5	150	80	—	—	10	10	5
Avg.	N = 6	58%	13%	31 a		3	135	62	0	0	2	3	2
Range	—	35–78	1–30	21–45		1–5	100–175	60–80	0	0	1–10	1–10	1–5

*Typical interval of IB = 9214–15′

BURROWED (OR BIOTURBATED) SHELF SILTSTONE

		Lithology (%)						Physical					
									Cross-Bedded				Hori
Well No.	Unit No.	Sand-stone	Silt-stone	Shale *	*Sid-erite	Glau-conite Avg./Max.	Avg. Grain Size, Microns	Total %	High Angle	Med-ium	Low Angle	Total	Sand-stone
CS AB-1A	2B	20	35	45	—	2	150	20	—	—	tr	tr	tr
CS AB-1A	3B	25	55	20	—	3	150	30	—	—	—	0	2 silt
CS AB-1A	6	25	35	40	—	tr	125 (175)SS	33	—	—	10	10	2
CS AB-1A	8	15	35	50	tr	3	125	10	—	—	—	0	1
South BC-2	2	47	11	42	—	5	150	18	—	—	—	0	1
CS AE-1	1	15	65	20	—	3	75	17	—	—	1	1	—
CS AE-1	8	10	25	65	2	5	125	20	—	—	—	0	1
Mean	N = 7	22	37	39	tr	3	130	21	—	—	2	2	1
Range		10–25	11–65	20–65	0–2	tr–5	75–150	10–33	—	—	0–10	0–10	1–2

**Number below line is total bioturbated volume. Number above line is burrows so thickly intertwined that not described.

†Number above line is total distinct burrows (some within bioturbated interval). Number below line is distinct burrows outside bioturbated intervals.

FACIES, COMPILATION OF PERCENTAGES OF SEDIMENTOLOGIC FEATURES

Structures (%)					Biological Structures (Burrows) (%)						Contacts		
zontal Laminae													
Shale	Total	Rip-ples	Ripples on Troughs	Re-worked a, d	Total %	Iden-tified %	Dis-tinct %	Biotur-bated %	Interval Bur-rowed %	Diver-sity	Transi-tion Thick-nesses	*** Upper	Lower
4	5	54	tr	—	37	3 a,c,d	26	8	92	10	0.5′ / 0.06′	G	S
4	6	53	tr	—	40	4 c,d,t	7	10	90	10	0.5′ / 0.05′	G	S
—	tr	64	—	—	35	2 d,t	27	8	97	10	0.3′ / NA	T	—
—	—	65	—	—	34	1 t	5	15	85	6	0.5′ / 1.0′	G	G
—	1	63	—	20 a	34	—	22	12	94	13	0.03′ / NA	VS	—
—	5	65	—	—	20	—	18	2	93	16	1.0 ′ / 0.01′	G	VS
1	3	61	0	4	34	2	24	9	92	11	0.5′ / 0.3′		
0–4	1–6	53–65	0–1	0–20	20–40	0–4	5–27	2–15	85–97	6–16	0.03–1.0′ / 0.01–1.0′		

Structures (%)					Biological Structures (Burrows) (%)						Contacts		
zontal Laminae													
Shale	Total	Rip-ples	Ripples on Troughs	Re-worked a, d	Total %	Iden-tified %	Dis-tinct %	**Biotur-bated %	Interval Bur-rowed %	Diver-sity	Transi-tion Thick-nesses	*** Upper	Lower
—	tr	17	tr	—	80	3 c,d	52	70 / 25	100	9	0.05′ / 0.5′	S	G
3	5	26	—		70	7 c,d,t	6	35	99	9	0.01′ / 0.5′	G	VS
8	10	12	tr	—	67	4 c,t	53	10	97	7	0.1′ / 0.5′	S	G
3	4	5	—	tr a	90	tr c	24† / 13	77 / 45	98	6	0.2′ / 0.01′	T	VS
—	1	17	—	—	82	2 d,t	8	74	96	13	<0.05′ / 0.3′	VS	T
tr	tr	15	—	tr a	83	2 c,te	10	71	99	9	0.5′ / —	G	NA
1	2	20	—	2 a	80	tr Zooph	6	73	100	10	0.06′ / 0.7′	S	G
2	3	16	0	0	79	3	20	58	98	9	0.14′ / 0.4′		
0–8	1–10	5–26	0–1	0–2	67–90	1–7	6–53	10–77	96–100	6–10	0.001–.2′ / 0.01–.7′		

TABLE A4.—*SHELF SILTY SHALE FACIES*, COMPILATION

Shelf Silty-Shale (Type 1)

		Lithology (%)						Physical					
									Cross-Bedded				Hori
Well No.	Unit No.	Sand-stone	Silt-stone	Shale	Sid-erite	Glau-conite Avg./Max.	Avg. Grain Size, Microns	Total %	High Angle	Med-ium	Low Angle	Total	Sand-stone
CS AB-1A	1	tr	34	65	—	2	70	65	—	—	—	—	21 silt
CS AS-1	1	1	10	89	—	—	100	40	—	—	—	—	5 silt
Mean	—	1	22	77	0	1	85	53	—	—	—	—	13 silt
Range	—	1	10–34	65–89	—	0–2	70–100	40–65	—	—	—	—	26 silt

Shelf Silty-Shale (Type 2)

		Lithology (%)						Physical					
									Cross-Bedded				Hori
Well No.	Unit No.	Sand-stone	Silt-stone	Shale	Sid-erite	Glau-conite Avg./Max.	Avg. Grain Size, Microns	Total %	High Angle	Med-ium	Low Angle	Total	Sand-stone
CS AB-1A	9	10	40	50	tr d,cmt	tr in ss	100	85	—	—	—	0	10 silt
CS AE-1	9	1	14	85	—	tr	100	88	—	—	—	0	5 silt
Mean	—	6	27	67.5	1	tr	100	86	—	—	—	0	8 silt
Range	—	1–10	14–40	50–85	0–1	0–2	100	85–88	0	0	0	0	5–10 silt

OF PERCENTAGES OF SEDIMENTOLOGIC FEATURES

Structures (%)					Biological Structure (Burrows) (%)						Contacts		
zontal Laminae													
Shale	Total	Ripples	Ripples on Troughs	Reworked a, d	Total %	Identified %	Distinct %	Bioturbated %	Interval Burrowed %	Diversity	Transition Thicknesses	Upper	Lower
34	55	10	—	tr	35	2 d	20	2	80	7	0.06′	S	NA
—	5	35	—	—	60	tr d	6	54	80	6	0.01′	VS	NA
17	30	23	—	tr	46	2	13	28	80	7	0.04′	VS	NA
0–34	5–55	10–35	—	0–1	35–60	1–2	6–20	2–54	80	6–7	0.01–0.04′	S-VS	NA

Structures (%)					Biological Structures (Burrows) (%)						Contacts		
zontal Laminae													
Shale	Total	Ripples	Ripples on Troughs	Reworked a,d	Total %	Identified %	Distinct %	Bioturbated %	Interval Burrowed %	Diversity	Transition Thicknesses	Upper	Lower
15	25	60	—	—	15	tr d	12	2	50	6	0.02′	NA	T
8	13	75 flaser	—	—	12	—	11	tr	70	6	0.06′	NA	S
12	19	68	0	0	14	tr d	11	2	60	6	0.13′	NA	S
8–15	13–25	60–75	—	—	12–15	0–tr	11–12	1–2	50–70	6	0.06–0.02′	NA	S to T

TABLE A5.—PERMEABILITY AND POROSITY VALUES TABULATED FOR INDIVIDUAL FACIES IN HARTZOG DRAW FIELD

Well Unit No.	K	Ø
CENTRAL RIDGE FACIES		
CS AK-1, Unit 4B	40	17.4
CS AK-1, Unit 4A	20.8	14.4
CS AB-1A, Unit 4	2.8	11.9
CS AE-1, Unit 5	10.9	14.2
Mean	14.6	15.3
Maximum	40.0	17.4
Minimum	2.8	11.9
CENTRAL RIDGE (PLANAR LAMINATED) FACIES		
CS AE-1, Unit 7	0.2	6.2
HIGH-ENERGY RIDGE-MARGIN FACIES		
CS AS-1, Unit 4	31.3	14.2
CS AK-1, Unit 3	15.0	14.7
CS AB-1A, Unit 5	0.8	9.9
CS AE-1, Unit 6	10.5	10.3
South BC-2, Unit 5	6.1	8.0
Mean	18.8	12.9
Maximum	31.3	14.7
Minimum	0.8	8.0
LOW-ENERGY RIDGE-MARGIN FACIES		
CS AS-1, Unit 3	2.2	10
CS AK-1, Unit 2	13.4	9.4
CS AB-1A, Unit 7	0.2	7.9
CS AE-1, Unit 4	2.8	5.5
CS AE-1, Unit 3	5.1	12.5
Mean	5.0	11.0
Maximum	13.4	12.5
Minimum	0.2	5.5
INTER-RIDGE FACIES (SHALY)		
CS AS-1, Unit 2	0.16	6.1
CS AK-1, Unit 1	2.2	1.8
CS AB-1A, Unit 3A	0.01	5.1
CS AE-1, Unit 2	2.1	7.7
South BC-2, Unit 1	1.4	6.7
Mean	1.2	5.2
Maximum	2.2	7.7
Minimum	0.01	1.8
BURROWED (AND BIOTURBATED) SHELF SILTSTONE FACIES		
CS AB-1A, Unit 6	0.02	5.4
CS AB-1A, Unit 3B	0.02	6
CS AE-1, Unit 1	0.72	5.9
South BC-2, Unit 2	3.0	7.4
Mean	1.9	6.7
Maximum	3.0	7.4
Minimum	0.02	5.4

TABLE A6.—POROSITY AND PERMEABILITY MEASUREMENTS.

CITIES SERVICE OIL COMPANY, FEDERAL AE-1 (5131), NW NW, SEC. 13 T45N R76W
HARTZOG DRAW FIELD, SHANNON SANDSTONE, CAMPBELL CO., WYOMING

(CORE LABORATORIES, INC.) WHOLE CORE ANALYSIS

Sample Number	Depth (feet)	Perm. to Air		Porosity	Fluid Saturation		Grain Density
		Max.	90 Deg.		Oil	Water	
Unit 9 *SHELF SILTY SHALE FACIES;* (3.0′) 9136.0–39.0′ (9146–49.0′, log depths)							
No values							
Unit 8 *BURROWED SHELF SILTSTONE FACIES;* (3.9′) 9139.0–42.9′ (9149.0–52.9′, log depths)							
No values							
Unit 7 *CENTRAL RIDGE (PLANAR LAMINATED) FACIES;* (2.4′) 9142.9–45.3′ (9152.9–55.3′, log depths)							
No values							
1	9143–44	0.08	*	4.4	36.9	46.1	2.63
2	9144–45	0.34	*	8.0	30.9	19.8	2.65
MEANS		0.21	—	6.2	33.9	32.9	2.64
Unit 6 *HIGH-ENERGY RIDGE-MARGIN FACIES;* (3.0′) 9145.3–48.3′ (9155.3–58.3′, log depths)							
3	9145–46	18.0	9.7	10.4	35.6	24.3	2.73
4	9146–47	13.0	6.9	8.6	24.4	51.1	2.71
5	9147–48	5.1	3.5	11.4	33.8	27.0	2.72
6	9148–49	5.7	5.7	10.8	30.2	26.7	2.68
MEANS		10.5	6.9	10.3	31.0	32.2	2.71
Unit 5 *CENTRAL RIDGE FACIES;* (6.6′) 9148.3–54.9′ (9158.3–64.9′, log depths)							
7	9149–50	8.9	8.3	13.9	28.3	25.6	2.74
8	9150–51	7.9	7.2	11.6	23.2	29.9	2.66
9	9151–52	1.7	1.6	15.4	31.3	16.8	2.66
10	9152–53	13.0	8.8	15.3	34.3	25.7	2.66
11	9153–54	18.0	15.0	17.2	31.0	20.3	2.66
12	9154–55	16.0	16.0	11.8	26.3	27.9	2.69
MEANS		10.9	9.6	14.2	29.1	24.3	2.67
Unit 4 *LOW-ENERGY RIDGE-MARGIN FACIES;* (3.0′) 9154.9–57.9′ (9164.9–67.9′, log depths)							
13	9155–56	4.5	2.1	6.6	24.9	52.9	2.96
14	9156–57	1.6	2.3	2.3	40.0	53.3	2.72
15	9157–58	2.3	2.1	7.6	33.3	34.7	2.70
MEANS		2.8	2.1	5.5	32.7	46.9	2.79
Unit 3 *CENTRAL RIDGE FACIES;* (22.1′) 9157.9–80.0′ (9167.9–90.0′, log depths)							
16	9158–59	2.4	2.3	12.9	30.9	29.5	2.66
17	9159–60	4.5	3.3	12.5	35.1	22.9	2.67
18	9160–61	4.5	3.6	10.0	35.2	23.5	2.67
19	9161–62	4.4	3.5	13.6	33.4	22.2	2.69
20	9162–63	6.2	6.2	13.4	31.2	28.4	2.66
21	9163–64	7.9	4.4	12.8	31.3	23.9	2.71
22	9164–65	4.2	4.2	7.9	26.5	40.4	2.69
23	9165–66	0.5	*	10.3	30.2	30.2	2.66
24	9166–67	7.2	*	13.0	34.9	20.4	2.62
25	9167–68	8.4	7.6	14.8	29.0	25.2	2.69
26	9168–69	6.2	6.1	12.6	36.5	18.3	2.65
27	9169–70	6.1	3.7	12.3	31.4	28.3	2.67
28	9170–71	5.0	3.8	11.4	32.3	25.5	2.69
29	9171–72	10.0	10.0	14.2	29.2	21.3	2.65
30	9172–73	3.7	*	12.2	31.7	26.9	2.64
31	9173–74	7.0	6.3	15.8	38.7	25.4	2.69
32	9174–75	5.3	5.3	13.2	31.9	27.6	2.66
33	9175–76	6.8	5.1	10.1	31.3	27.4	2.68
34	9176–77	4.4	*	14.6	35.1	24.7	2.63
35	9177–78	4.5	3.5	13.4	28.2	26.8	2.67
36	9178–79	1.4	1.4	12.2	23.3	29.6	2.68
37	9179–80	1.1	1.0	10.9	24.5	24.5	2.68
MEANS		5.1	4.1	12.5	31.9	25.1	2.66
Unit 2 *INTER-RIDGE FACIES (SHALY);* (3.5′+) 9180–83.5′ (9190.0–9193.5′, log depths)							
38	9180–81	0.09	*	7.9	35.8	51.1	2.63
39	9181–82	2.8	2.1	7.4	25.6	43.1	2.71
40	9182–83	3.8	2.4	7.7	35.6	27.9	2.70
MEANS		2.1	2.3	7.7	32.3	40.7	2.68

TABLE A6.—CONTINUED.

Sample Number	Depth (feet)	Perm. to Air		Porosity	Fluid Saturation		Grain Density
		Max.	90 Deg.		Oil	Water	
Unit 1 *BIOTURBATED SHELF SILTSTONE;* (9′) 9183.5–92.5′ (9193.5–9202.5′, log depths)							
41	9283–84	1.4	1.3	6.9	30.3	57.7	2.70
42	9184–85	0.91	0.45	5.0	28.3	64.6	2.74
43	9185–86	1.4	0.81	6.2	31.0	55.4	2.72
44	9186–87	0.2	0.1	5.8	28.2	70.4	2.73
45	9187–88	0.1	0.06	5.9	23.9	68.2	2.71
46	9188–89	1.1	0.77	4.9	10.5	83.9	2.70
47	9189–90	0.88	0.82	6.5	29.6	62.4	2.71
48	9190–91	0.26	0.26	5.6	21.7	68.6	2.72
49	9191–92	0.28	0.22	6.1	23.5	70.4	2.73
MEANS		0.72	0.53	5.9	25.2	66.8	2.71

*Sample not suitable for whole core analysis
See also Figures A1 and A2

TABLE A7.—POROSITY AND PERMEABILITY MEASUREMENTS.

SOUTHLAND ROYALTY COMPANY
BUD CHRISTENSEN NO. 2 (6343), SEC. 34-46N-76W
HARTZOG DRAW FIELD, SHANNON SANDSTONE
CAMPBELL CO., WYOMING

WHOLE CORE ANALYSIS
(CORE LABORATORIES, INC.)

Sample Number	Depth (feet)	Perm. to Air		Porosity	Fluid Saturation		Grain Density
		Maximum	90 Deg.		Oil	Water	
Unit 2 *BIOTURBATED SHELF SILTSTONE;* (19.4′) 9182.9–9202.3′ (9295.9–9215.3′; log depth)							
21	9183–84	1.6	0.67	6.9	9.6	88.5	2.73
22	9184–85	0.68	0.55	8.1	3.2	95.0	—
23	9185–86	4.9	3.4	7.8	2.8	92.7	—
24	9186–87	1.6	1.3	7.6	2.7	89.7	—
25	9187–88	1.8	1.4	7.3	3.1	83.3	—
26	9188–89	9.5	1.4	7.5	3.1	92.2	—
27	9189–90	3.0	1.2	8.4	3.5	85.0	—
28	9190–91	15.0	4.9	6.9	3.2	93.8	2.71
29	9191–92	1.5	1.2	7.1	3.2	93.9	—
30	9192–93	4.3	2.9	7.7	2.9	84.7	—
31	9193–94	3.1	1.8	7.1	3.2	85.8	—
32	9194–95	4.6	1.8	7.7	5.8	85.0	—
33	9195–96	0.42	0.38	7.2	3.1	93.2	—
34	9196–97	0.94	0.32	7.8	2.5	78.7	2.70
35	9197–98	0.83	0.51	6.9	2.8	87.1	—
36	9198–99	0.80	0.56	7.5	3.3	91.3	—
37	9199–00	1.3	1.3	7.0	3.0	89.5	—
38	9200–01	0.64	0.62	7.3	3.2	91.7	—
39	9201–02	0.23	0.19	7.6	3.3	88.8	—
MEANS		3.0	1.4	7.4	3.6	88.9	2.71
Unit 1 *INTER-RIDGE FACIES (SHALY);* (13.4′) 9202.3–9215.7′ (9215.3–9228.7′; log depth)							
40	9202–03	7.1	2.60	7.6	3.2	90.9	—
41	9203–04	0.85	0.75	6.2	2.7	87.4	2.71
42	9204–05	1.4	0.64	7.1	2.7	80.3	—
43	9205–06	1.3	0.64	7.0	3.6	92.5	—
44	9206–07	0.29	0.29	4.6	3.2	90.2	—
45	9207–08	0.61	0.32	6.8	3.0	95.5	—
46	9208–09	0.49	0.34	7.7	3.1	91.7	—
47	9209–10	0.76	0.67	7.2	3.3	95.0	—
48	9210–11	1.1	0.36	5.1	4.1	90.6	—
49	9211–12	1.1	0.35	6.4	3.3	95.4	—
50	9212–13	1.3	0.62	7.1	3.6	93.3	—
51	9213–14	1.2	1.10	7.3	11.8	82.3	2.71
52	9214–15	1.2	0.70	6.7	4.4	93.4	—
MEANS		1.4	0.70	6.7	4.0	90.6	2.71

TABLE A7.—CONTINUED.

Sample Number	Depth (feet)	Perm. to Air Maximum	Perm. to Air 90 Deg.	Porosity	Fluid Saturation Oil	Fluid Saturation Water	Grain Density
Unit 5 HIGH-ENERGY RIDGE-MARGIN FACIES; (7.3′) 9163–70.3′ (9176–83.3; log depth)							
1	9163–64	12.0	9.1	7.4	6.3	90.0	—
2	9164–65	12.0	6.6	8.0	19.8	34.0	2.73
3	9165–66	2.0	1.7	9.5	23.5	62.7	—
4	9166–67	5.3	4.2	9.1	33.9	30.3	—
5	9167–68	2.2	1.7	7.1	22.4	61.4	2.67
6	9168–69	8.6	6.7	6.5	10.5	75.0	—
7	9169–70	0.75	*	8.4	13.2	85.2	—
MEANS		6.1	5.0	8.0	18.5	62.6	2.70
Unit 4 CENTRAL RIDGE FACIES; (9.0′) 9170.3–79.3′ (9123.3–92.3′; log depth)							
8	9170–71	4.9	4.6	11.7	36.0	30.7	—
9	9171–72	9.7	9.6	14.0	51.9	29.2	2.71
10	9172–73	8.8	8.4	14.4	35.2	32.6	—
11	9173–74	3.4	2.5	11.5	33.5	33.5	—
12	9174–75	5.5	5.5	11.4	28.0	37.4	2.64
13	9175–76	8.8	8.0	13.5	33.2	23.0	—
14	9176–77	7.9	7.5	10.2	24.6	21.5	—
15	9177–78	6.2	6.2	12.1	37.1	25.8	—
16	9178–79	12.0	11.0	13.5	36.8	25.8	2.64
MEANS		7.5	7.5	12.5	35.1	28.9	2.66
Unit 3 HIGH-ENERGY RIDGE-MARGIN FACIES; (3.6′) 9279.3–82.9′ (9292.3–95.9′; log depth)							
17	9179–80	3.4	3.2	10.1	34.4	20.6	—
18	9180–81	1.9	1.9	9.2	31.5	22.2	—
19	9181–82	2.6	2.6	7.8	29.9	37.4	—
20	9182–83	9.6	7.7	6.6	23.7	53.9	—
MEANS		4.3	3.9	8.4	29.8	33.2	

*Sample not suitable for core analysis
See also Figures A5 and A6

TABLE A8.—POROSITY AND PERMEABILITY MEASUREMENTS.

CITIES SERVICE OIL COMPANY
FEDERAL AS-1 (5341) NE SEC. 29 T45N R75W
HARTZOG DRAW FIELD, SHANNON SANDSTONE
CAMPBELL CO., WYOMING

Core Laboratories, Inc. (Casper, Wyoming)

Sample Number	Depth (feet)	Permeability	Porosity
Unit 4 HIGH-ENERGY RIDGE-MARGIN FACIES (18.2′+); 9465–83.2′ (9457.5–75.7′, log depth)			
44	9465	0.79	7.9
45	9466	94	17.7
46	9467	58	16.2
47	9468	55	16.0
48	9469	12	15.1
49	9470	49	15.4
50	9471	33	17.2
51	9472	41	15.2
52	9473	48	15.7
53	9474	37	16.2
54	9475	9.8	14.2
55	9476	8.6	15.8
56	9477	24	15.9
57	9478	5.7	13.4
58	9479	39	12.8
59	9480	11	14.9
60	9481	54	15.2
61	9482	5.9	10.2
62	9483	9.2	13.0
Mean		31.3	14.2
Unit 3 LOW-ENERGY RIDGE-MARGIN FACIES; (3.9′) 9483.2–87.1′ (9475.7–79.6′, log depth)			
63	9484	6.5	12.8
64	9485	1.8	10.3
65	9486	0.41	8.1
66	9487	0.16	8.8
Mean		2.2	10.0
Unit 2 INTER-RIDGE FACIES (SHALY); (16.4′) 9487.1–9503.5′ (9479.6–96.0′, log depth)			
67	9488	0.22	9.1
68	9489	0.19	5.9
69	9490	0.17	7.3
70	9491	0.35	7.3
71	9492	0.17	5.5
72	9493	0.14	5.3
73	9494	0.56	4.5
74	9495	*	5.6
75	9496	0.03	4.3
76	9497	0.07	6.1
77	9498	*	8.0
78	9499	0.04	5.1
79	9500	0.26	7.3
80	9501	0.05	6.7
81	9502	0.01	5.1
82	9503	0.01	4.7
Mean		0.16	6.1
Unit 1—SHELF SILTY SHALE FACIES (17.8′+) 9503.5–21.3′; (9496.0–9513.8′ log depth)			

No values

See also Figures A9 and A10.

APPENDIX B: UNSOLVED PROBLEMS

Studies done to date on Shannon-type sand bodies still leave many unanswered questions concerning the origin of shelf-ridge sandstone reservoirs. Some of the more important unanswered questions are listed below.

1. How are Shannon age sands moved to a far offshore position for redistribution southward by currents?
2. Was there a delta system in the northwest part of the Powder River basin during Shannon time which yielded sediments to southward-flowing currents (Fig. B1)?
3. What processes, or combination of processes, caused southward flowing currents which were at least periodically strong enough to move medium-grained sand?
4. What processes caused the shelf-ridge complex deposition to be limited to zones 1–3 miles (1.6–4.8 km) wide? Why were the sands concentrated in the areas of the shelf-ridge complexes, while in areas nearby no sands were deposited?
5. Were slightly positive areas on the sea bottom responsible for isolating the shelf-ridge complexes and/or individual bars which make up the shelf-ridge complex?
6. Is the position of Hartzog Draw field near the inferred shelf slope boundary significant in causing it to be a somewhat larger shelf-ridge complex?
7. Could the sand deposited at Hartzog Draw be the result of superimposed, mildly to strongly catastrophic events, such as major storm currents superimposed on some type of contour currents? Can the effects of tidal currents be discounted a hundred miles offshore?
8. What controlled the recurrence of stacked generally coarsening-upward sequences on top of each other at Hartzog Draw? Why didn't the shelf-ridge complex build laterally rather than upward once local topographic relief had been established?

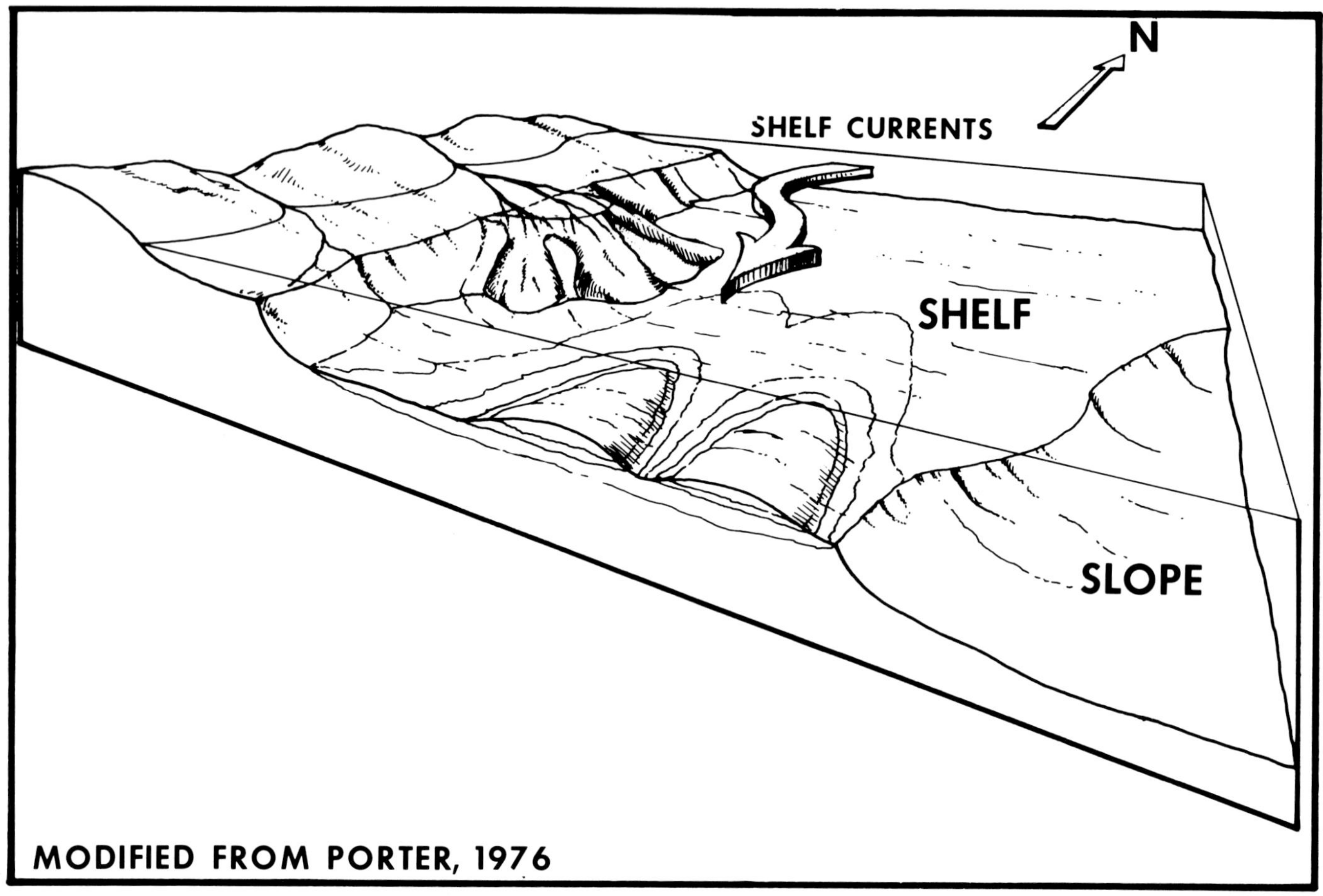

FIG. B1.—Hypothetical model indicating possible source (delta) of sands which initially were distributed over wide areas of the shelf and were then carried south to be redeposited as shelf ridges such as Hartzog Draw field.

9. Why does the steepest, elongate side of the bar (NNE side) apparently face obliquely into the currents that deposited the shelf-ridge?
10. What processes were effective in causing a thicker *Inter-Ridge Facies (Shaly)* to be built in the northerly and westerly portions of the field *prior* to deposition of the high-angle crossbedded reservoir sandstones?
11. Do the stacked genetic features that form the shelf-ridge complex have sufficiently impermeable boundaries to affect initial or enhanced recovery production?
12. How much time is reflected in the accumulation of a shelf-ridge such as Hartzog Draw?
13. Within each of the genetically defined facies, how much effect do secondary diagenetic changes have in restricting lateral and vertical flow of oil?
14. Knowing that sand ridges of the scale of Hartzog Draw field are present in largely shale sections, such as the Cody and Pierre Shales of the Powder River, Wind River, and Hanna and Southern Williston basins, how does one predict the location of the ridge complexes, and still further, how does one delineate those sand ridges which will yield oil?

DIPMETER INTERPRETATION OF TURBIDITE-CHANNEL RESERVOIR SANDSTONES, INDIAN DRAW FIELD, NEW MEXICO

SANDRA PHILLIPS

Department of Geology, Texas A & M University, College Station, Texas, 77843

ABSTRACT: Detailed stratigraphic interpretation of high-resolution dipmeter logs can provide important information concerning the geometry and distribution of reservior sandstones. Stratigraphic dip data were correlated with primary rock properties observed in cores and borehole-log data to define the internal geometry of turbidite-channel sandstones in the Cherry Canyon Formation at Indian Draw field, New Mexico. Characteristic dip patterns allowed the delineation of erosional unconformities, channel sequences, slump faulting, contorted and massive bedding, and sedimentary drape.

The erosional unconformity which marks the base of the Indian Draw channel exhibits a characteristic dip pattern consisting of an abrupt change in the trend of dip magnitude and dip azimuth across the unconformity. Slump faults exhibit an abrupt increase in dip with depth over a small interval and an associated progressive dip azimuth rotation approaching the fault. Contorted beds show a random dip pattern, often marked by poor quality, high magnitude dips. Massively bedded sandstones lack computed dips and sedimentary drape patterns typically consist of a decrease in dip upward within basinal deposits overlying a sandstone.

Detailed mapping of the reservoir sandstones indicates deposition as stacked, laterally discontinuous lenses within the previously eroded channel. Direction of sedimentary drape over sandstone lenses can be used to map their trends. Channel-fill lenses are 5 to 30 ft (1.5 to 9.1 m) thick, are elongated parallel to depositional dip and have a sinuous geometry. Such turbidite channel deposits can be anticipated to form complex multilayered reservoirs, consisting of a series of isolated sandstone lenses of restricted areal extent.

INTRODUCTION

High-resolution dipmeter logs have been used extensively in petroleum exploration as a tool for interpreting structural settings for many years. Only recently, however, has the potential for utilizing dipmeters as a sedimentologic and stratigraphic tool begun to be recognized (Pirson, 1977). Interpretation of sedimentary dips from dipmeter logs can provide information indicative of sediment transport direction, depositional environment (Gilreath and others, 1969) and sandstone geometry. Dipmeter log patterns alone, however, cannot provide a unique solution for the depositional environment since similar log profiles can be produced by sequences from several different environments.

Integration of conventional core data with stratigraphic dip data provides a more reliable basis for interpreting depositional environment, sandstone geometry and reservoir trends. This approach has been utilized in the present paper to characterize the Cherry Canyon sandstones at Indian Draw field, New Mexico (Fig. 1).

These sandstones form a complex multilayered reservior. Characteristic diplog patterns define such features as erosional unconformities, channel sequences, slump faulting, contorted and massive bedding, and sedimentary drape. It is demonstrated that the diplog patterns can be used to map the distribution and individual geometry of the channel-fill sandstones which comprise the reservoir complex.

GEOLOGIC SETTING

Guadalupian (Middle Permian) Cherry Canyon sandstones of the Delaware Mountain Group occur as deep-water facies within a thick clastic sequence deposited in the Delaware basin of West Texas and southeast New Mexico (King, 1948; Newell and others, 1953; Meissner, 1972; Fig. 2). Cherry Canyon sandstones are underlain and overlain by similar deep-water facies of the Brushy Canyon and Bell Canyon formations.

Present Address: ARCO Oil and Gas Company, Exploration Research, 2300 W. Plano Pkwy., Plano, Texas, 75075.

Indian Draw field is located in the southeast part of Eddy County, New Mexico (Fig. 1). The Cherry Canyon section in the Indian Draw area consists of very fine-grained, quartzose sandstones interbedded with dolomitic limestones, siltstones and shales. Within the field, the section is subdivided, in ascending order, into the following six intervals (Fig. 3); the "Big Dolomite," "TG Limestone," "A_1," "A_2" and "A_3" sandstone intervals (collectively termed the "Indian Draw Zone") and the "Two-Finger Limestone." The "A_3" interval produces oil at Indian Draw field when porosities are greater than 20%. Thicknesses of the reservoir sandstones range from less than 10 ft (3 m) to as much as 33 ft (10 m).

The discovery well for the field, Amoco Production Old Indian Draw Unit 1 (Fig. 4), was completed in December 1973. Initial production was 108 barrels (17 m^3) of oil per day plus 5 barrels (1 m^3) of water from 3,260 to 3,290 ft (994 to 1,003 m) in the upper Cherry Canyon, Delaware sandstone (International Oil Scouts Association, 1974). A structure contour map on the base of the "Two-Finger Limestone" shows regional dip to the east-southeast (Fig. 4). Drilling outside the Indian Draw area encountered thin, tight sand, delineating the western boundary of the field as a stratigraphic pinchout (Cromwell, 1979). An erratic distribution of oil and water within the "A_3" pay sand interval is similar to other Delaware Basin fields and indicates that sandstones occur as a complex of lenses within linear trends (Grauten, 1979).

CHARACTERISTICS OF THE CHERRY CANYON SANDSTONES

A total of 168 ft (51 m) of core was examined from the Amoco Old Indian Draw Unit 4, 15 and 35 wells (Fig. 4). The cores were slabbed, and sedimentary structures and gross compositional characteristics were described, sketched and photographed. Whole-core porosities and permeabilities were measured by Core Laboratories, Inc. Gamma-ray/sonic logs were correlated with cores to correct for depth differences. Dipmeter logs were then correlated to the gamma-ray/sonic logs and to the cores. Distinctive log characteristics cor-

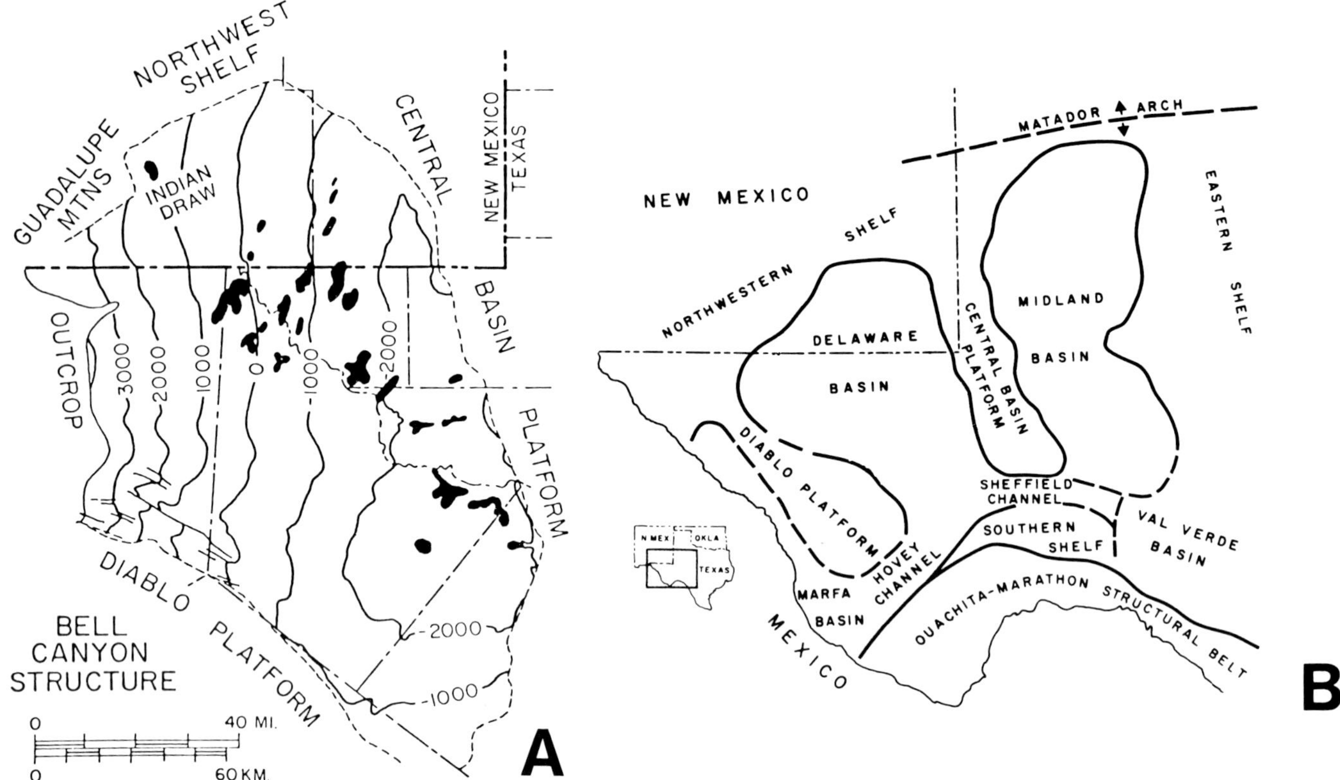

FIG. 1.—(A) Structure map of the Delaware basin, West Texas and southeast New Mexico, showing structure on top of the Bell Canyon sandstone and location of Indian Draw field. Depths contoured are based on a datum of mean sea level (subsea depths). Contour interval is 1,000 ft (300 m). Stratigraphic oil fields in the upper Delaware Mountain Group are shown in black (modified from Berg, 1979). (B) Regional map of New Mexico and Texas showing Permian tectonic elements and location of Delaware basin, detailed in [A] (from Williamson, 1979).

responding to specific lithologies as identified from cores were used as a basis for extrapolating to uncored wells. Gamma-ray/sonic logs from wells in the study area were also used to construct structure contour maps, isopach maps and geologic cross sections. In the construction of net sandstone isopach maps, an economic cutoff of 82 microsec/ft was used, which is equivalent to about 20% porosity (modified from original cutoff values suggested by Cromwell, 1979).

Cherry Canyon sandstones observed in cores are typically very fine-grained, of variable thickness and interbedded with thin siltstones and shales. The cored intervals show ordered, repetitive sequences of bedding types with characteristic sedimentary structures which constitute "bedsets." A complete ordered sequence, when present, consists of the "A" through "E" divisions of Bouma's turbidite sequence (Bouma, 1962; Fig. 5). In addition to the "A" through "E" divisions, a bioturbated sandstone unit (arbitrarily designated "F" in Fig. 5) is also found in the Cherry Canyon section.

Bedset types observed in cores in order of their most common occurrence are (1) incomplete AE bedsets (Figs. 5A, B, C); (2) slightly more complete ACE, and rarely ABCE or ADE, bedsets (Figs. 5D, E) missing one or more of the intermediate BCD divisions; and (3) complete ABCDE bedsets containing all the turbidite divisions. Dominant bedset type in the three cores is the AE type.

Cursory thin section examination indicates the sandstones are feldspathic sublitharenities (Folk's 1974 classification), with an average detrital composition of 69% monocrystalline quartz, 15% feldspar, 11% rock frag-

FIG. 2.—Diagrammatic cross section from the northwest shelf (north) to the Delaware basin (south) showing stratigraphic relationships in the Middle Permian Guadalupian Series (from Berg, 1979).

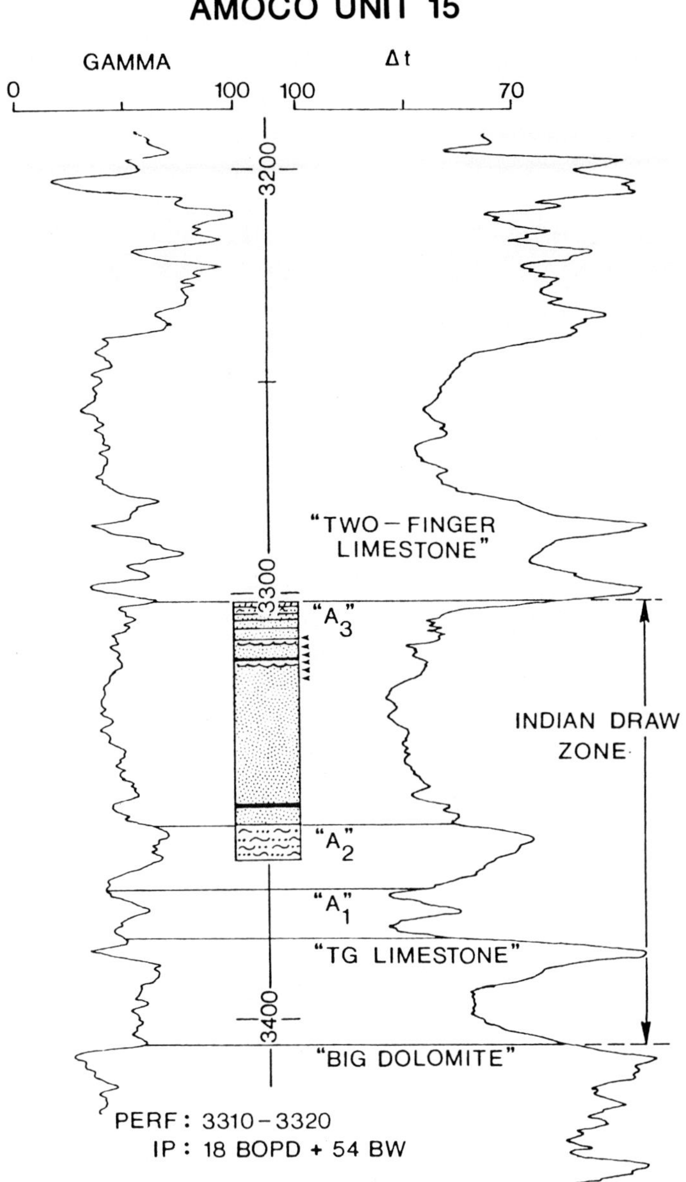

FIG. 3.—Gamma-ray and sonic-log characteristics of the Cherry Canyon section, Amoco Unit 15, Indian Draw field. Diagram shows the informal stratigraphic units used in mapping the field. The interval from the top of the "Big Dolomite" to the base of the "Two-Finger Limestone" is designated the "Indian Draw Zone." Log depths are in feet. Core corrected down 6 ft (1.8 m) to log.

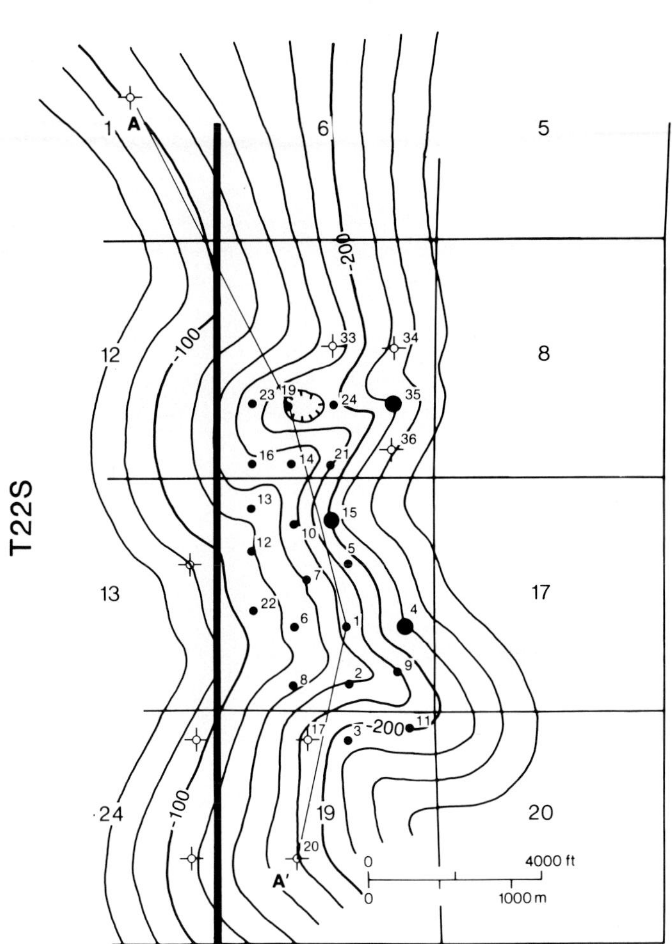

FIG. 4.—Structure map on the base of the "Two-Finger Limestone," Indian Draw field (modified from Cromwell, 1979). Depths contoured are based on a datum of mean sea level (subsea depths). Contour interval is 20 ft (6.1 m). Cores were examined from Amoco Unit 4, 15, and 35 (larger circles). The discovery well for the field is the Amoco Unit 1 well indicated on cross section A-A′. Cross section A-A′ is shown in Figure 8. All wells in the field are Amoco wells.

ments, and 3% matrix. Matrix is composed largely of clay minerals. Cement averages 20% of the total rock composition and consists of carbonate (14%) and silica (6%) as grain overgrowths.

Porosity and permeability (Table 1) exhibit an erratic distribution which cannot be related to either bedding character of facies units or internal variations in composition and texture. In the Amoco Unit 15 core, for example, an average permeability of 10 md and an average porosity of 17.4% were observed in the uppermost thin-bedded AE units (3,300–3,304 ft; 1006–1007 m) (Fig. 6). In the section immediately below (3,305–3,319 ft; 1007–1012 m), which includes a thinner ACE turbidite bedset and the upper part of the underlying thicker ACE channel turbidite, average permeability is 28 md and average pososity is 22.5%. The lower half of the thick channel turbidite shows a gradual decrease in permeability to an average of only 5 md and porosity of 22.6%. This lack of a systematic variation in porosity and permeability within individual bedsets corresponding to textural changes appears to be due to (1) the patchy distribution of calcite cement and (2) authigenic alteration of plagioclase feldspar to clay.

DEPOSITIONAL ENVIRONMENT

The sandstones at Indian Draw field are located about 10 mi (16 km) downdip from the approximate edge of the

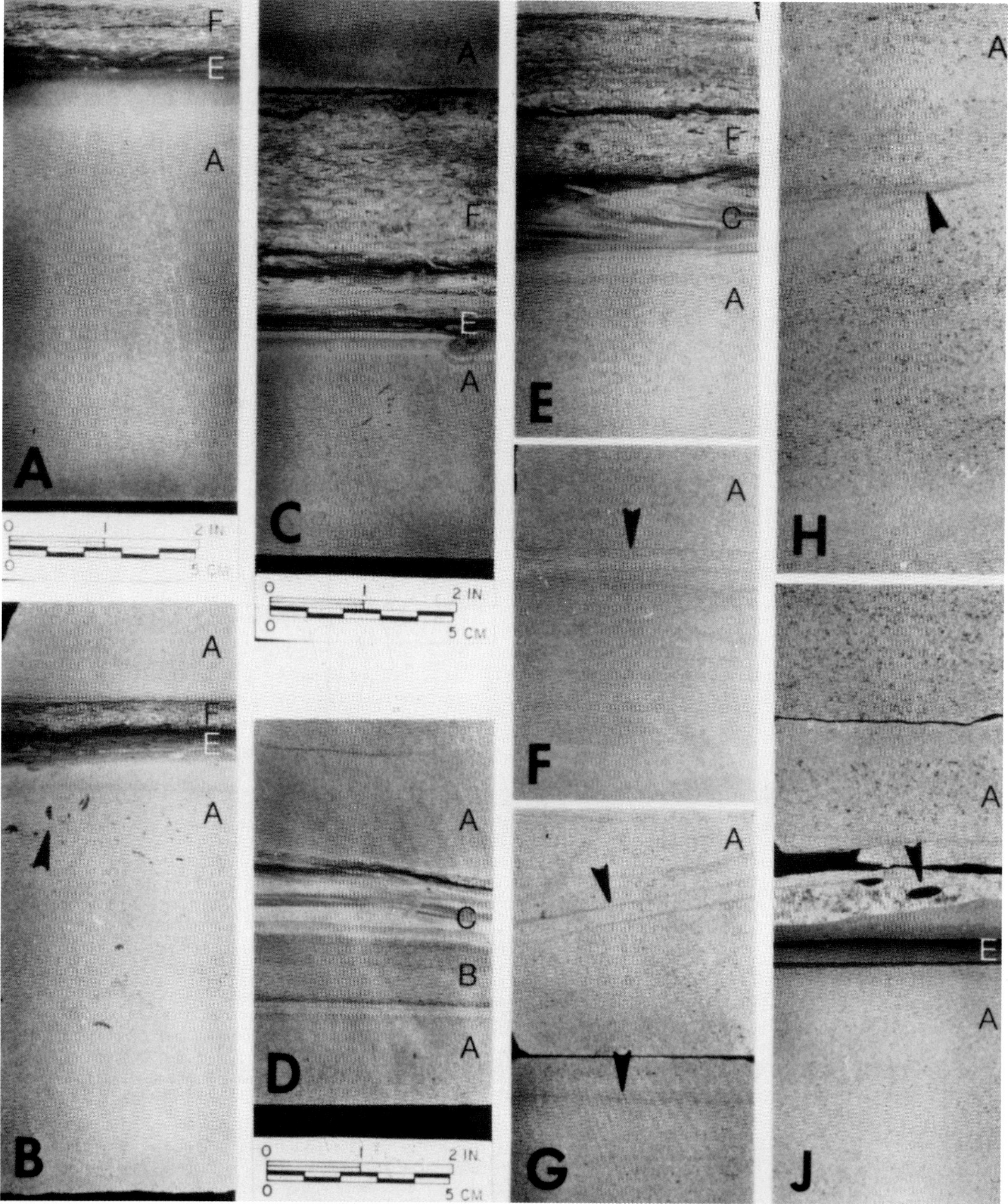

FIG. 5.—Sedimentary structures in Amoco Unit 15 core, Indian Draw field (R. R. Berg, 1984, pers. commun.). Photographs are identified by bold-face letters in lower left corner; turbidite bedset divisions are designated by small letters at right-hand margins. (A) Turbidite bedset of the AE type showing massive A sandstone overlain by thin E shale and succeeded by highly bioturbated sandstone (F); 3,299.0 ft (1,005.5 m). (B,C) Turbidite bedset sequence composed of massive A sandstone overlain by thin E shale, succeeded by highly bioturbated sandstone (F) and overlain

TABLE 1.—AVERAGE POROSITIES AND PERMEABILITIES OF CHERRY CANYON SANDSTONES AT INDIAN DRAW FIELD, EDDY COUNTY, NEW MEXICO

Well	Cored Interval ft	Samples	Porosity[a] Average	Porosity[a] Range	Permeability[b] Average	Permeability[b] Range
Amoco Unit 4	3,308–3,354	92	22.4	10.6–27.4	15.0	0.1–75.0
Amoco Unit 15	3,302–3,346	53	21.2	5.5–25.9	9.0	0.1–100.0
Amoco Unit 35	3,330–3,362	37	22.3	5.9–24.6	21.0	0.1–51.0
Average			22.1		15.0	

[a]Values in percent, from commercial core analyses.
[b]Values in millidarcys, from commercial core analyses.

Guadalupian reef front, indicating a deep-water depositional setting, as first proposed by Cromwell (1979). This, plus the repetitive, ordered sequences of Bouma division sedimentary structures, sharp basal contacts, shale clasts and soft-sediment deformation features (Fig. 5), indicate the Cherry Canyon sandstones were deposited by turbidity currents.

At Indian Draw, the thickness of the interval from the base of the "Two-Finger Limestone" to the top of the "Big Dolomite" (Fig. 3) increases rapidly from 49 ft (15 m) in the Amoco Unit 33 dry hole to over 120 ft (37 m) within the field (Fig. 7A). The "Two-Finger Limestone" and "Big Dolomite" units can be correlated outside the field, but the equivalent reservoir sandstone interval is thin and non-porous and consists of fine-grained deposits. Also, correlation of the section illustrates truncation of the local stratigraphic markers within widespread basinal sediments (Fig. 8). This evidence implies that the sandstones of the Indian Draw Zone were deposited inside an erosional channel. Outcrop studies of the Delaware Mountain Group have documented erosional channels filled with massive sandstone (Jacka and others, 1968; Jacka, 1979; Harms, 1974; Williamson, 1977, 1979). These channels are on the order of 0.6 mi (1 km) or more in width and as much as 100 ft (30 m) deep. Erosional channels can also be demonstrated for the upper Bell Canyon section in the subsurface (Weinmeister, 1978; Berg, 1979), which contain a stratigraphic sequence similar to that observed in the Indian Draw Zone. The pelagic limestones of the "Big Dolomite" were deposited within the channel as a uniform mantle following erosion of the channel (Cromwell, 1979) and prior to infill by turbidite sandstones. This "Big Dolomite" fill is similar to the shale units which mantle the bases of erosional channels in the Bell Canyon section (Berg, 1979).

Berg's (1979) model of the geometry and bedset associations of erosional channel-turbidite-fill deposits (Fig. 9),

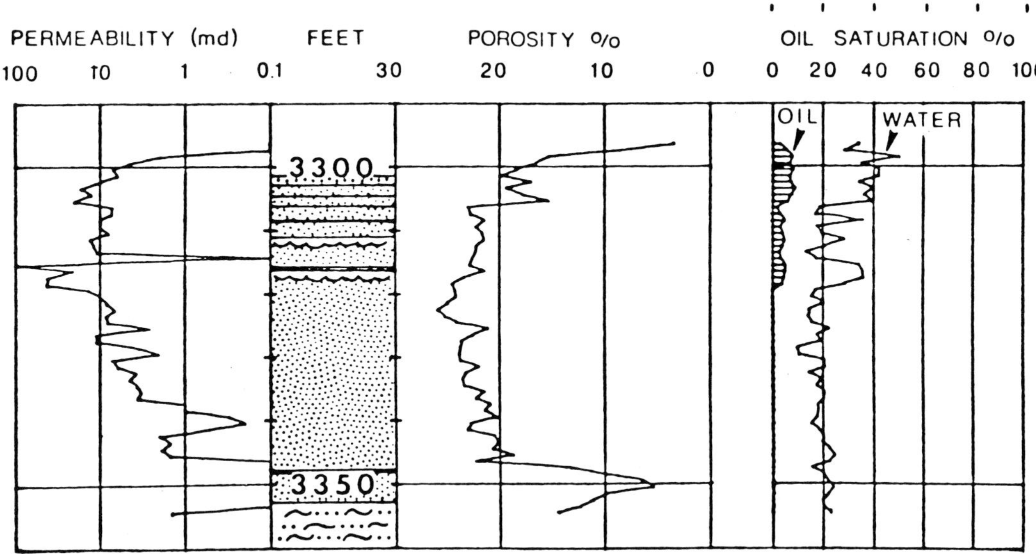

FIG. 6.—Core analysis plot showing porosity, permeability, and fluid saturation for Cherry Canyon sandstones, Amoco Unit 15, Indian Draw field.

in sharp contact with massive A sandstone; 3,300.1 (1,005.9 m), 3,302.3 ft (1,006.5 m). (D) Upper part of ABCE turbidite and overlying massive A sandstone; 3,304.5 ft (1,007.2 m). (E) Upper part of thick, channel turbidite composed largely of massive A sandstone; thin CE divisions overlain by highly bioturbated sandstone (F) at top of section; 3,308.6 ft (1,008.5 m). (F,G) Massive A sandstone containing indistinct horizontal to inclined laminae (arrows); 3,310.0 (1,008.9 m), 3,341.0 ft (1,018.3 m). (H,J) Massive A sandstone containing thin zone of convergent laminae (H, arrow) and minute shale clasts in basal part of A division that gives the sandstone a speckled appearance. Lower part of the massive sandstone has a sharp basal contact with lower shale clasts (J, arrow) overlying the top part of AE turbidite bedset; 3,341.5 (1,018.5 m), 3,342.0 ft (1018.6 m).

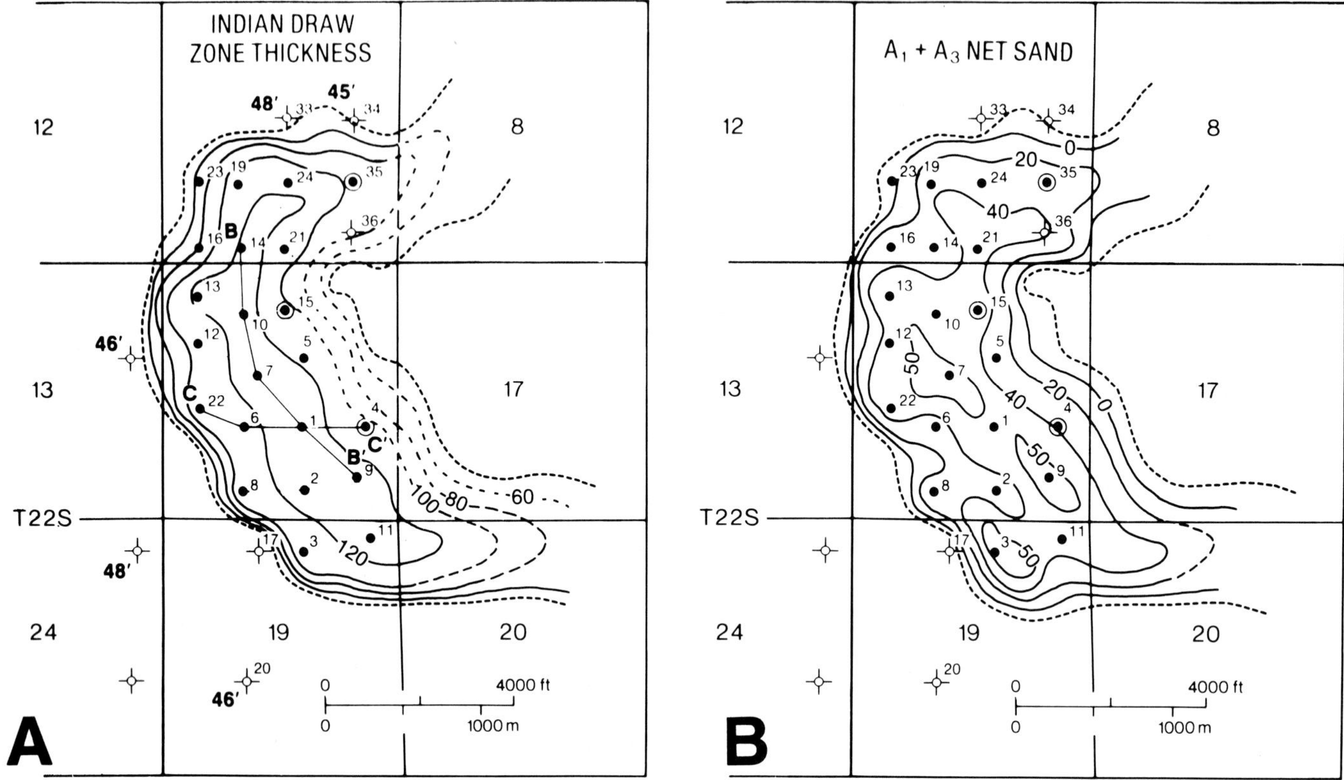

FIG. 7.—Isopach maps of Indian Draw field. (A) channel-fill isopach showing thicknesses of the Indian Draw Zone interval, between the top of the "Big Dolomite" and the base of the "Two-Finger Limestone." Contour interval is 20 ft (6 m). Cross sections B-B′ and C-C′ are shown in Figures 15 and 16. (B) Net "A_1" sandstones plus net "A_3" sandstones showing non-uniform (Fig. 3) distribution of channel-fill sandstones. Contours are in feet.

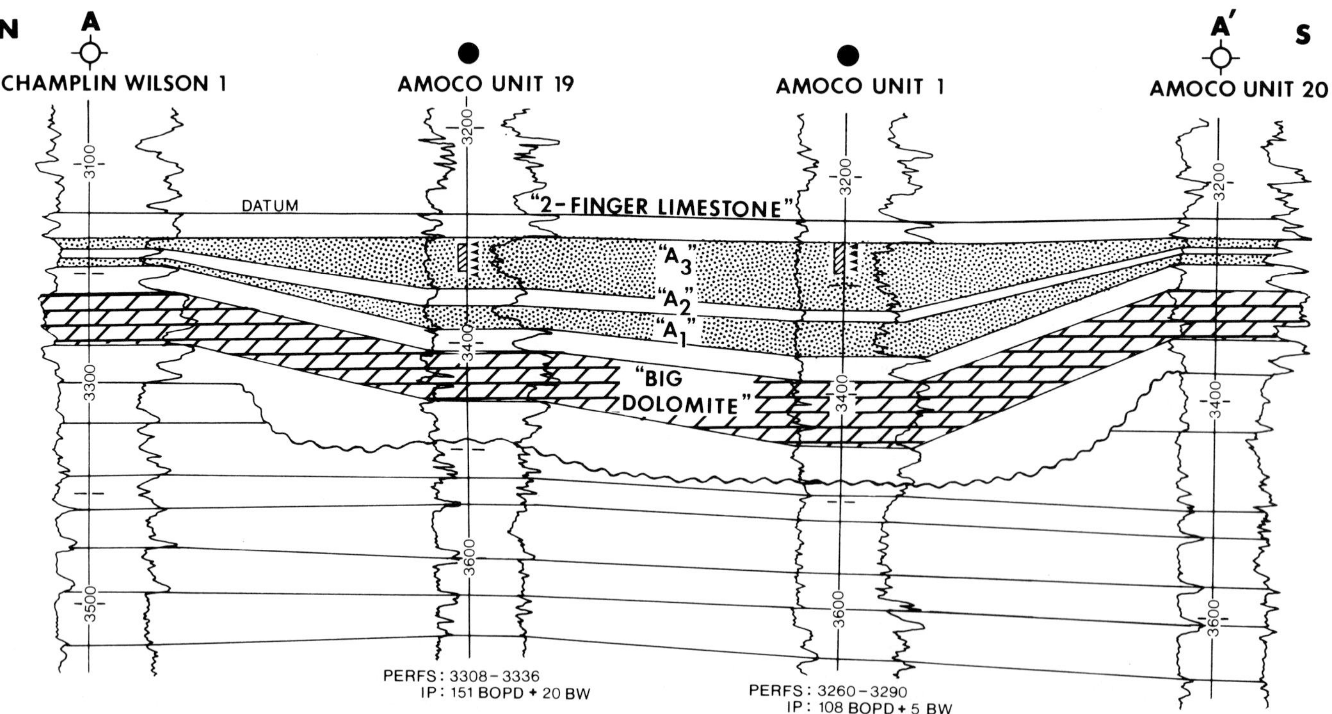

FIG. 8.—Stratigraphic cross section A-A′ showing correlation of the Cherry Canyon section in Indian Draw field. Correlations illustrate truncation of local stratigraphic markers and rapid thickening of the Indian Draw Zone within the field. No horizontal scale (modified from Cromwell, 1979). Location of cross section shown in Figure 4.

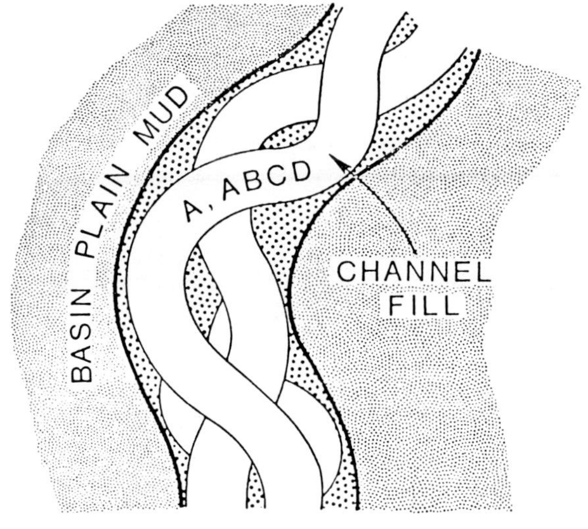

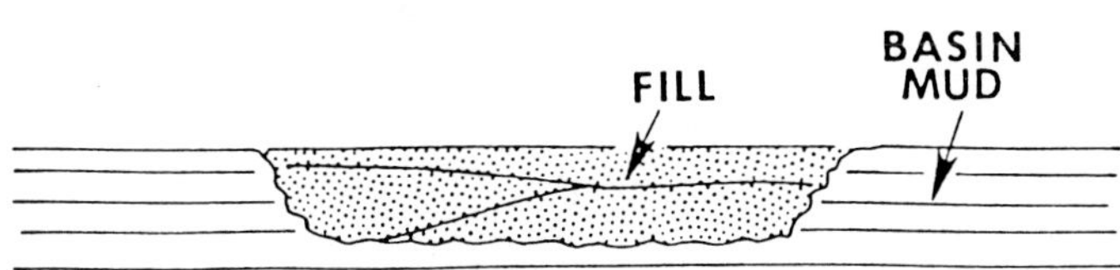

FIG. 9.—Idealized depositional model and diagrammatic cross sectional view of erosional channel which was subsequently filled with turbidites (R. R. Berg, pers. commun.).

probably is applicable in gross aspect to the Indian Draw field sequence. The Indian Draw channel probably developed during a period when sea level was regionally lowered, and shelf-derived sand carried by density currents eroded into basin-plain sediments. Studies of erosional channels in the upper Bell Canyon at Paduca and El Mar fields (Weinmeister, 1978; Berg, 1979) indicate a difference in timing and discharge between the flows which eroded the major channels and those which followed and filled the channels. Berg (1979) concluded that the flows which eroded the Bell Canyon channels were high discharge flows scouring channels of low sinuosity. Subsequent flows responsible for deposition of sandstone were of lower discharge and resulted in a more highly sinuous pattern for the channel-fill deposits. Meandering channel systems comparable to mature fluvial systems have been observed on modern submarine fans such as the Amazon (Damuth and others, 1983) and Mississippi fans (Pryor and others, 1983). In addition, studies of sandstone deposition in modern submarine canyons have concluded that deep-water erosional and depositional processes can reproduce the external geometry of fluvial systems despite differences in actual flow mechanisms.

DIPMETER INTERPRETATION

Stratigraphic and structural dip data were compiled from Schlumberger HDT logs available for 26 wells in the Indian Draw field. The processing parameters used in computer correlation of 22 of the surveys were (1) correlation length: 4 ft (1.2 m); (2) step length: 2 ft (0.6 m); and (3) search angle: 60°. The remaining four surveys used processing parameters of 50% of the values of each of the parameters listed above. Basic dip data from survey computer printouts were coded and tabulated on an interpretive basis in order to utilize only the most reliable and best quality data and to differentiate dips influenced by structural deformation from those reflecting sedimentary features. Values of mean dip magnitude and mean dip azimuth were calculated from selected groups of dips which exhibited the greatest degree of consistency for the various stratigraphic zones. Dips with magnitudes of less than 1° or a low quality index were omitted from the calculation of mean values.

Stratigraphic dip data were correlated with rock properties observed in the cores and borehole-log data to define the Indian Draw channel configuration and internal morphology of the turbidite-channel sandstones. The inferred presence of a scoured channel based on cross sections (Fig. 8) was confirmed by the dipmeter logs. The dipmeter from the Amoco Unit 19 well (Fig. 10) illustrates a characteristic abrupt change in dip and azimuth patterns which is interpreted to be the erosional unconformity which marks the base of the channel. Dips of 6° to 9° above this unconformity have a notably consistent, unimodal azimuth trend (southeast) and increase in magnitude with depth. Dips below the unconformity also have a relatively uniform azimuth trend, but the trend is west-southwest with dips of 2° to 4°. In many of the other wells the unconformity is less sharply defined but can still be recognized by the same general characteristics.

The dipmeters from the Amoco Unit 19 and 22 wells illustrate another dipmeter pattern which can be related to the thin-bedded channel-fill sequence. On both logs (Fig. 10), a successive decrease in dip magnitude through an interval of over 100 ft (30 m) or more of section is observed above the unconformity, whereas the corresponding dip azimuths maintain a relatively consistent trend. Frequently, near the edges of large channels, an increase in dip with depth is observed as the bottom of the channel is approached. The dips at the top of the channel sequence are often of very low magnitude and of variable azimuth, reflecting the final filling of the channel before abandonment.

Channel configuration within the field can also be estimated from dipmeter logs. This is because the Big Dolomite unit and underlying pelagic carbonates mantle the channel surface, so dip azimuths and magnitudes in these sedimentary units should mimic those of the underlying channel walls. Interpretation of the dip of the channel surface (as determined by dips in the "Big Dolomite" and underlying basal channel-fill sediments) in the Amoco Unit 19 (3,454–3,474 ft; 1,053–1,059 m) and 22 (3,370–3,390 ft; 1,027–1,033 m) wells (Fig. 10) indicates that the western channel wall has a slope of approximately 8° to 10°. In the Amoco Unit 23 well near the northwestern channel margin, a channel slope of 4° to 6° is indicated, and associated dip azimuths imply that the channel axis lies to the southeast.

The slope of the eastern channel edge appears to be gentler than that of the western edge and its position much

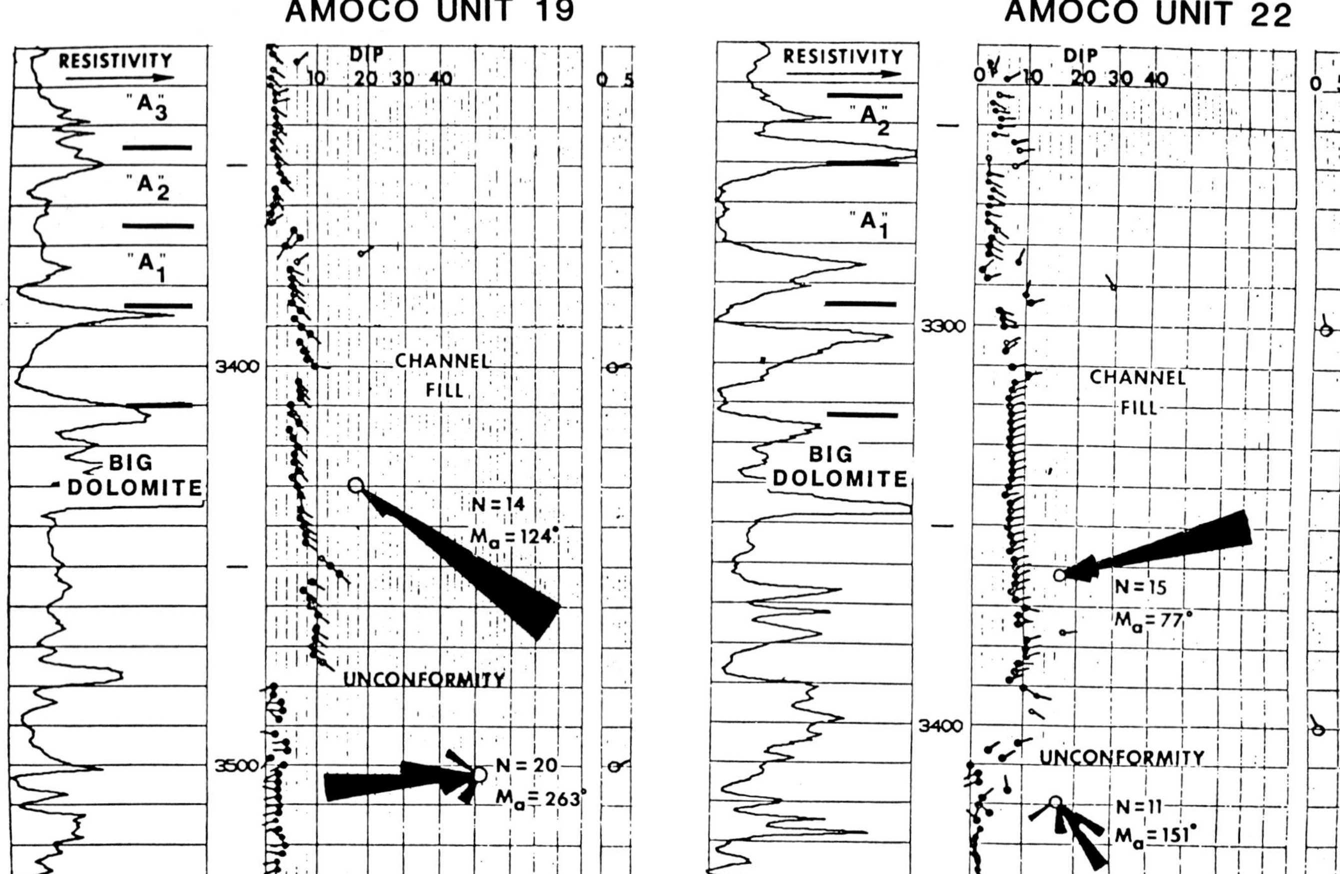

FIG. 10.—High-resolution dipmeter logs from Amoco Unit 19 and 22 wells illustrating the dipmeter response pattern associated with an erosional unconformity at the base of the channel at Indian Draw field. Computer processing parameters used were correlation length 4 ft (1.2 m), step length 2 ft (0.6 m) and search angle 60°. Azimuth frequency plots show the distribution of dip directions above and below the unconformity.

more difficult to define. The reservoir is bounded on the east by an oil-water contact; thus, the eastern limits of the channel were never fully defined by drilling. The existence and approximate location of the eastern edge, however, is confirmed by the reversal of dip from the regional trend in the "Big Dolomite" unit which mantles the base of the channel. In the Amoco Unit 21 well (3,422–3,436 ft; 1,043–1,047 m) (Fig. 11), as well as in Amoco Unit 15 (3,412–3,424 ft; 1,040–1,044 m), the dips in the "Big Dolomite" unit exhibit a reversal of direction from the regional southeast dip, which implies that the center of the channel is to the southwest.

Slumped and contorted bedding often take the form of multiple normal faults. This is recognized on dipmeter logs by an abrupt increase in dip magnitude with depth over a small interval and an associated progressive dip azimuth rotation approaching the fault (Fig. 12). Slump faults associated with the Indian Draw sandstones are typically small-scale features which rarely exhibit dips over 30°. Also, dip magnitude and azimuth of contorted beds are usually extremely erratic, and recorded dips are often of poor quality (open circles; Fig. 12). The pattern is completely random, and dips may exceed 30° in magnitude.

Two sedimentary features, massively bedded sandstones and sediment drape, played a major role in the mapping of reservoir geometry, discussed below. The massive sandstones of the channel-fill sequence generally exhibit the best reservoir characteristics. Sediment drape over the A-dominated bedset units provides important details for contouring sandstone trends as well as delineating multiple sandstone units. Both of these features exhibit characteristic diplog patterns. The dipmeter from the Amoco Unit 15 well illustrates the typical dip patterns observed and their relation to the core lithology (Fig. 13).

The most common dipmeter response to massive sandstones is the absence of computed dips. Since no stratification is present, the dipmeter reads no change in resistivity and no dips can be computed. In some cases indistinct laminae, usually high-angle, may be present in the units. The diplog may then display erratic dips, often of high magnitude and poor quality. Sediment drape over these massive sandstones may be observed in overlying shale, bioturbated sandstone and shale (F), or thinly interbedded sandstone and shale (AE turbidite bedsets). The typical drape pattern is a decrease in dip upward, usually with a relatively consistent azimuth trend. In the Amoco Unit 15 well, the thin

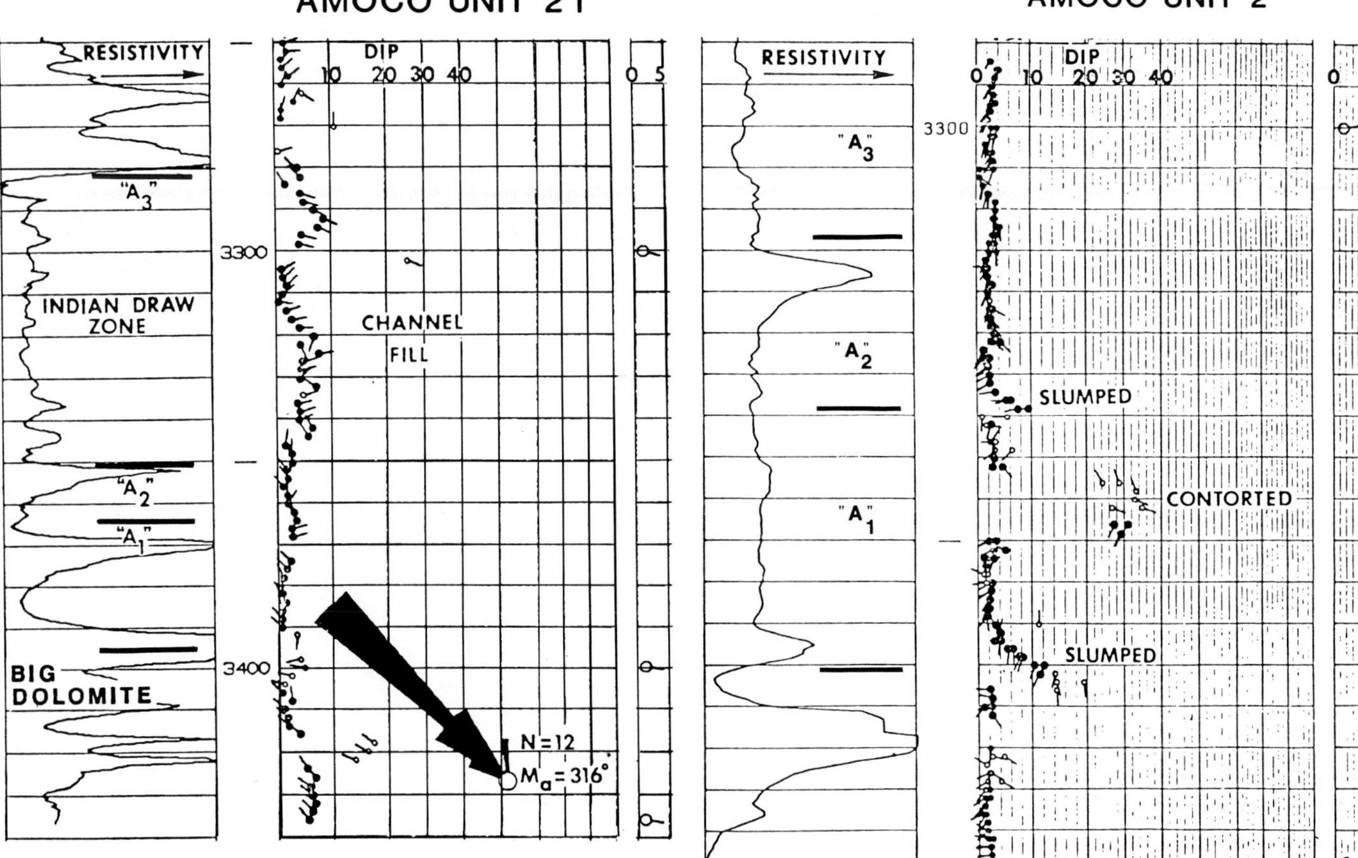

FIG. 11.—Dipmeter log and azimuth frequency plot of the Cherry Canyon section at Indian Draw field illustrating channel-edge related characteristics observed in Amoco Unit 21 well located near the eastern channel margin. Computer processing parameters used were correlation length 4 ft (1.2 m), step length 2 ft (0.6 m) and search angle 60°.

FIG. 12.—High-resolution dipmeter log of Amoco Unit 2 well illustrating characteristic dip patterns exhibited by slumped beds and contorted beds, Indian Draw field. Computer processing parameters used were correlation length 2 ft (0.6 m), step length 1 ft (0.3 m), search angle 30°.

AE units above the thick, massive sandstone form a sediment drape dipping to the southeast (Fig. 13). In general, the direction of drape is perpendicular to the sand body trend, and points away from the thickest part of the underlying sand body.

In some instances, basin-plain deposits overlying a sandstone lens can become unstable and slump rather than forming a drape. In many cases, there appears to be a correspondence between lithology and sediment behavior. Where clay or shale is the principal sediment overlying a sandstone lens, the sediment tends to form a drape conforming with topography. In contrast, when thinly interbedded siltstone and shale overlie a sandstone lens, the sediment frequently exhibits characteristics of slumping. This slumping may suggest location on the flank of the sand body. As discussed above, slump patterns can be identified from the dipmeter logs. These patterns can be used in conjunction with drape patterns to map individual sand body trends. The diplog from the Amoco Unit 2 well illustrates slumping interpreted to be near the flanks of sandstone lenses in the "A_1" interval in that well.

General dipmeter patterns discussed above and their relative positions with respect to channel geometry are summarized in Figure 14.

RESERVOIR GEOLOGY AND DEPOSITIONAL PATTERNS

The Indian Draw reservoir sandstones were deposited in a meander loop of a sinuous submarine erosional channel (Figs. 8, 9). Cross sections through the channel in the field area reveal a maximum total channel fill of approximately 1 mi. (1.6 km) in width and 220 ft (67 m) deep (Figs. 8, 15, 16). Local stratigraphic markers were truncated and the paleotopography developed prior to deposition of the Indian Draw sands. The first sediments to be deposited in the channel were pelagic limestones which uniformly blanket the geomorphic surface (Cromwell, 1979). These carbonates average 70 to 90 ft (21 to 27 m) in thickness throughout the local area. Since they are uniformly distributed, the channel configuration was not altered by their deposition, retaining its approximate initial physiography until deposition of the turbidite sandstones. An isopach of the Indian Draw Zone interval above the basal carbonates (Fig. 7A) shows a rapid increase in sediment thickness perpendicular to the channel edges.

Sandstones within the channel are divided into the lower "A_1" sandstones and the upper "A_3" sandstones (Fig. 3), which are separated by a zone of bioturbated sandstone and shale ("A_2"). Total thickness of "A_1" + "A_3" sandstones

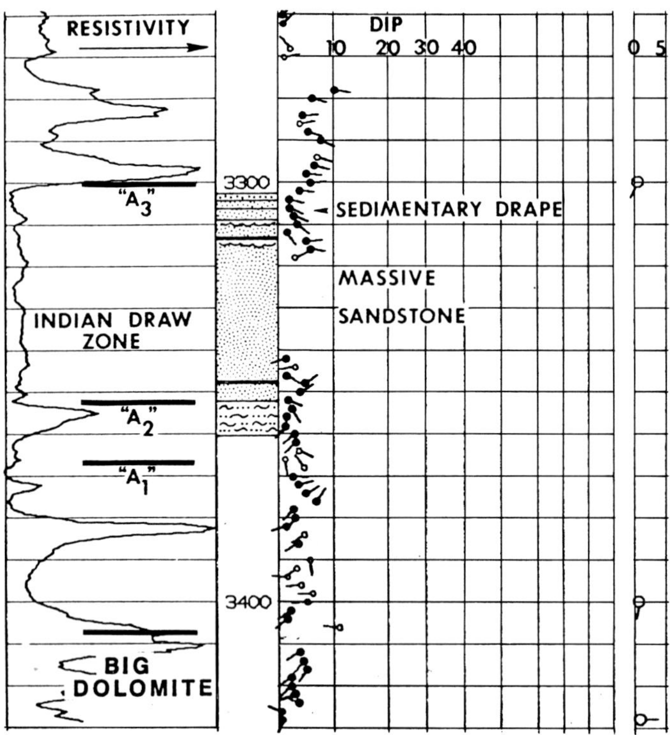

FIG. 13.—Characteristic dip patterns associated with massive sandstones and sedimentary drape, Amoco Unit 15 well, Indian Draw field. Computer processing parameters used were correlation length 4 ft (1.2 m), step length 2 ft (0.6 m) and search angle 60°.

(Fig. 7B) indicates that filling of the channel did not take place uniformly. This appears to be due to the initial asymmetric cross sectional geometry of the channel and a change in depositional patterns as the channel was progressively filled. Channel asymmetry, characterized by a steeper western flank, is suggested by diplog patterns in the "Big Dolomite" unit, previously discussed, as well as channel-fill isopachs (Fig. 7A), which exhibit rapid thinning along the western channel margin in contrast to a more gradual thinning along the eastern edge.

Cross sections through the field (Figs. 15, 16) illustrate the lenticular nature of the channel-fill sandstones and associated shales. Although sandstones in both the "A_1" and "A_3" intervals have a lenticular geometry, there appears to be a marked difference in their respective distributions. Distribution of the sediments deposited within the "A_1" interval must have been controlled by the physiography of the channel bottom and the hydraulic energy of the intermittent turbidity flows. Sands were deposited by successive turbidity flows, forming a series of stacked, lenticular sandstone bodies.

Initial deposition was confined to the pre-existing physiographic low adjacent to the western channel margin (Fig. 17A). Geometry of the "A_1" sandstones was mapped from analysis of drape patterns on diplogs, as previously described. An isopach of "A_1" interval net sandstone (Fig. 17A) and a net sandstone isopach of the individual sandstone lenses (Fig. 17B) illustrate the sinuous geometry of the sandstones as well as their concentration along the steeper western part of the channel. The "A_1" lenses are typically 5 to 10 ft (1.5 to 3 m) in thickness. Strike and dip symbols on Figure 17B indicate the direction of drape.

Following the period of active turbidite deposition of the "A_1" interval, sediment influx diminished. Fine-grained basin-plain sediments of the "A_2" interval were deposited from suspension blanketing the underlying "A_1" sandstones. Direction of drape patterns in "A_2" sediments was used in conjunction with stratigraphic correlations to map the morphology of the "A_1" sandstones. As the "A_2" sediments progressively filled the channel, paleotopographic relief of the channel bottom was gradually diminished. This apparently created a fairly smooth geomorphic surface and probably lowered the gradient, resulting in a somewhat different depositional pattern for the succeeding "A_3" sandstones.

Differences between the distribution of "A_1" and "A_3" sandstones reflects the change in depositional patterns as the channel was progressively filled. Total net "A_3" sandstone thicknesses (Fig. 18A) range from 0 to 44 ft (0 to 13 m) and show an irregular, arcuate composite geometry. As with the "A_1" sandstones, however, detailed mapping of individual sandstone bodies and diplog patterns of drape illustrate their depositional pattern and spatial relationships much more clearly. An isopach of selected "A_3" sandstone lenses (Fig. 18B) reveals a series of sandstone bodies which successively offlapped toward the south-southwest with time. The maximum thickness of any apparently continuous sandstone body is the 33-ft-thick (10 m) bed observed in the Amoco Unit 15 core. Sandstone lenses are elongated parallel to depositional dip and exhibit overlapping relationships in cross section, which reflect the timing and laterally prograding style of sand deposition.

Deposition of the *e, f,* and *g* sandstones (Fig. 18B) proceeded laterally from the inside of the meander loop to form a series of broadly curved, lens-shaped bodies that tended to be progressively displaced in a downstream direction similar to fluvial point bars. After each turbidity flow ceased, very fine sediment was deposited from suspension to form a "clay drape" over the entire channel. This "clay drape" is observed on logs as a shale break between individual sandstones. Where thinly interbedded sandstone and shale is the principal deposit overlying a sandstone lens, the sediment frequently exhibits diplog characteristics of slumping.

After a number of flows had progressively filled the channel with "A_3" deposits, succeeding flows were diverted to another location. The abandoned flow path was filled by thinly interbedded sandstones, siltstones and shales of the channel-fill facies. This final channel fill ranges from 6 to 14 ft (1.8 to 4.3 m) in thickness and forms an irregular, highly sinuous deposit around the outside of the meander loop (Fig. 19). It may be composed of CDE turbidite sequences that occur without a massive AB unit.

Following the clastic influx of the "A_3" interval, the uni-

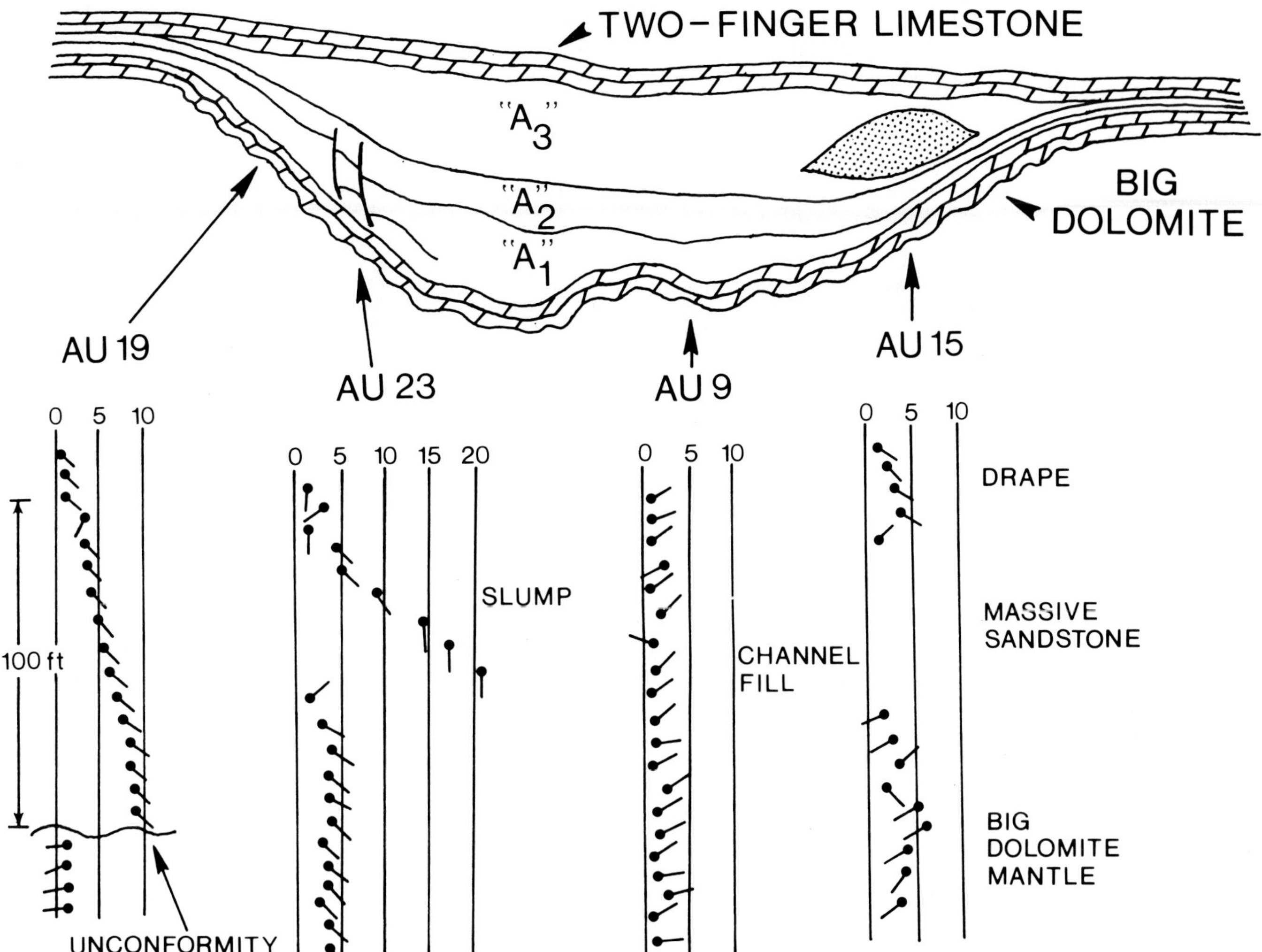

FIG. 14.—Schematic diagram illustrating the distribution of typical structural and stratigraphic features across the erosional channel at Indian Draw field and their characteristic dipmeter patterns. Amoco Unit 19 diplog shows abrupt change in trend of dip magnitude and dip azimuth, marking erosional unconformity at base of channel sequence. Amoco Unit 23 diplog illustrates small-scale abrupt increase in dip with depth typical of slumping. Amoco Unit 9 diplog shows relatively consistent dip azimuths and low dip magnitudes in channel-fill sediments near center of channel. Amoco Unit 15 diplog illustrates (1) typical drape pattern of fine-grained sediments consisting of small-scale decrease in dip upward, (2) lack of computed dips commonly observed in massive sandstone intervals, and (3) conformance of dips in pelagic "Big Dolomite" unit to channel topography.

formly thin limestones of the "Two-Finger" were deposited over a large area of the Delaware basin. These limestones mark the top of the erosional channel sequence at Indian Draw field. A diagrammatic cross section illustrating the stratigraphic distribution of the turbidite deposits is shown in Figure 20.

CONCLUSIONS

The Cherry Canyon sandstones at Indian Draw field were deposited downdip from the Guadalupian reef front in the Delaware basin in relatively deep water. Interpretation of primary rock properties substantiates earlier conclusions (Cromwell, 1979) that these sandstones were deposited by turbidity currents.

Correlation of borehole logs through the field area reveal that the Cherry Canyon sandstones occur as stacked, laterally discontinuous sandstone bodies within an erosional channel. Geometry and distribution of the individual sandstone lenses within the field illustrate the complex, multilayered geometries of erosional channel deposits. The channel is asymmetric in cross section, having a relatively steep western edge and a more gently sloping eastern edge. The first sediments deposited in the channel were pelagic carbonates which uniformly mantled the channel topography.

Distribution of the first clastics deposited within the channel ("A_1" sandstones) was controlled by physiography of the channel bottom and hydraulic energy of the intermittent turbidity flows. These sandstones occur as stacked, isolated lenses which have sinuous geometry, elongated parallel to depositional dip.

Fine-grained sediments of the overlying "A_2" interval were deposited uniformly and formed a low-gradient, smooth

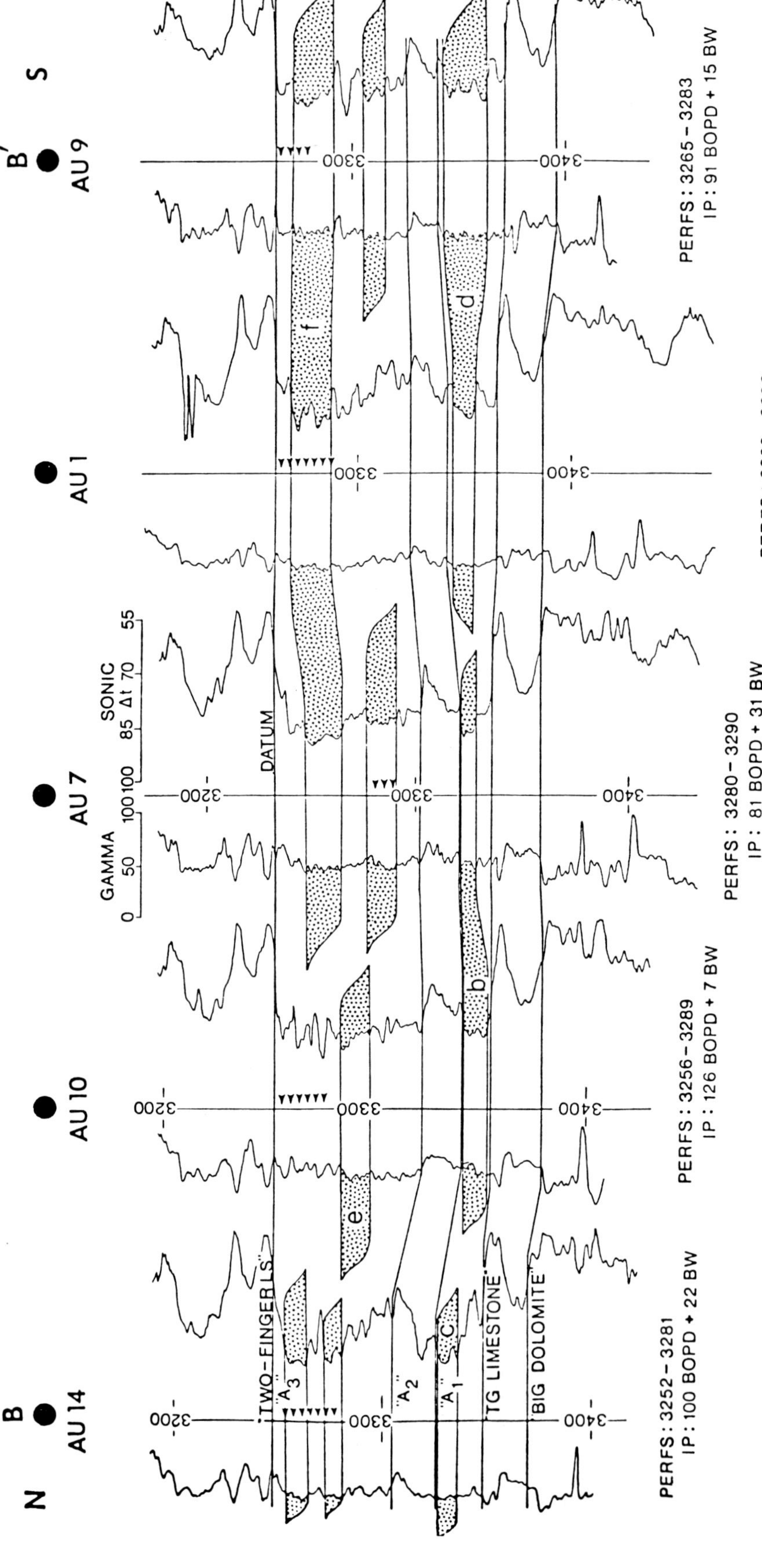

FIG. 15.—Stratigraphic cross section B-B′ illustrating the lateral variation and lenticular geometry of the channel-fill sandstones. Section shows the relationship of sandstone lenses *b*, *c*, *d*, *e*, and *f* shown also in Figures 17B and 18B. No horizontal scale. Location of cross section is shown in Figure 7.

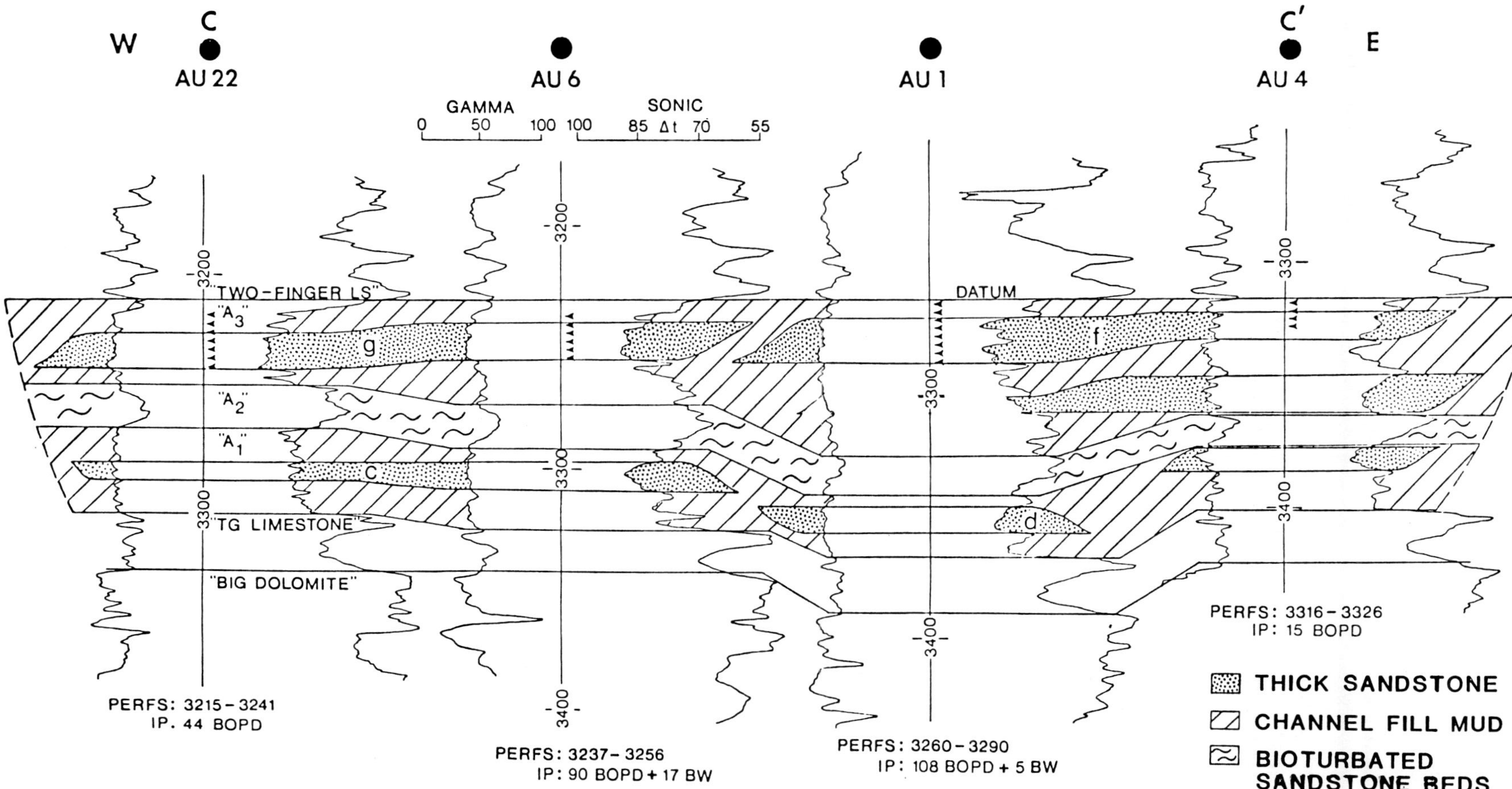

FIG. 16.—Stratigraphic cross section C-C′ illustrating sandstone lenses, channel-fill facies and the bioturbated fine-grained facies ("A_2"). Section shows rapid lateral facies changes and tthe relationship of sandstone lenses *c*, *d*, *f*, and *g* shown also in Figures 17B and 18B. No horizontal scale. Location of cross section is shown in Figure 7.

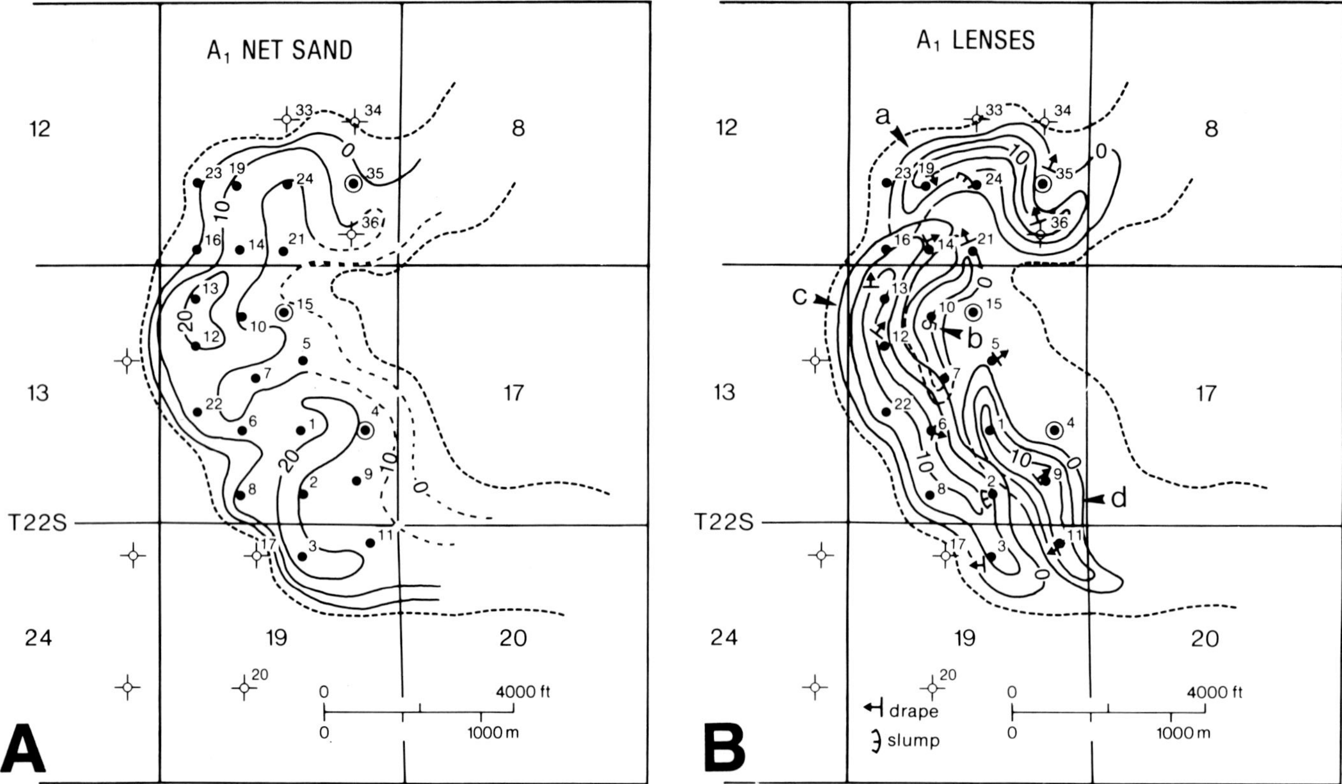

FIG. 17.—Isopach maps of "A_1" interval sandstones illustrating (A) total net sandstone, showing sinuous composite geometry of the numerous stacked, lenticular sandstones. Contour interval is 10 ft (3 m). (B) Isopach of "A_1" lenses shows the geometry and distribution of sandstone lenses *a, b, c,* and *d*. Individual lenses are labeled on map. Contour interval is 5 ft (1.5 m).

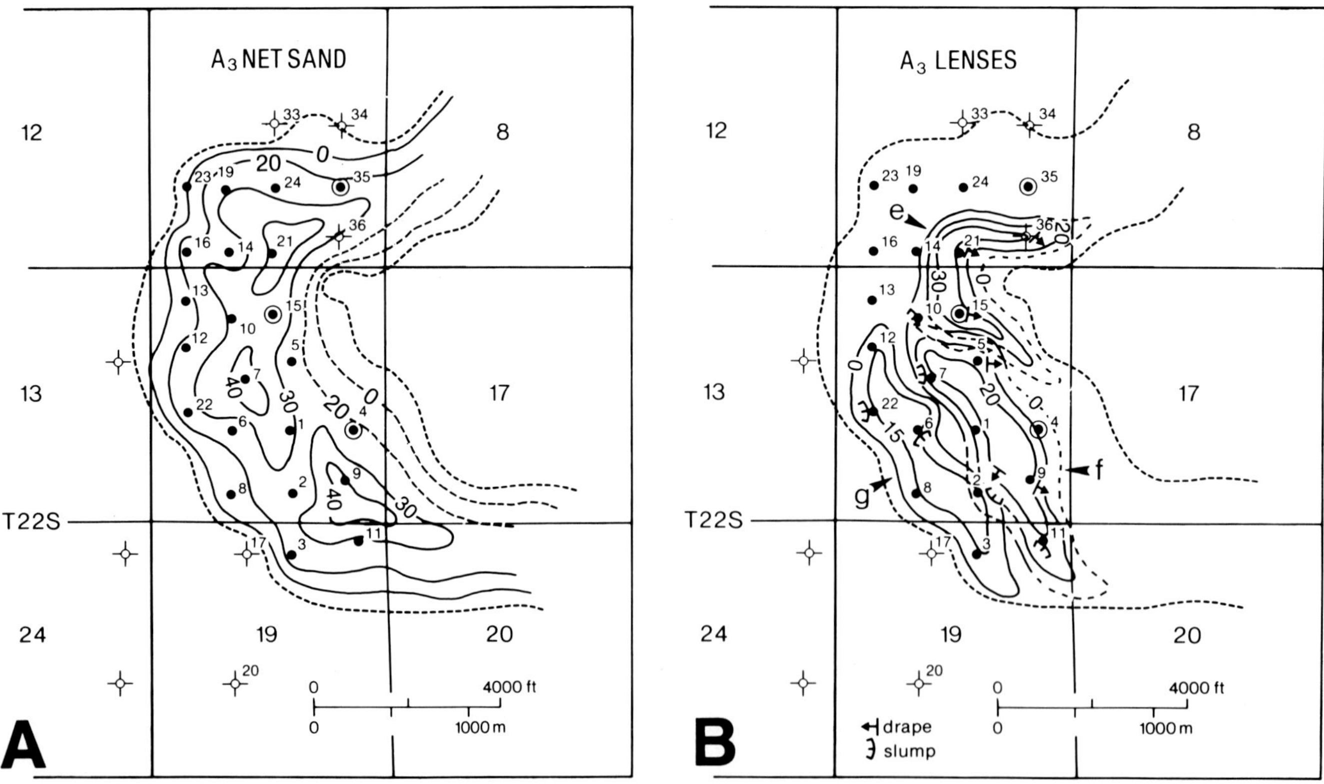

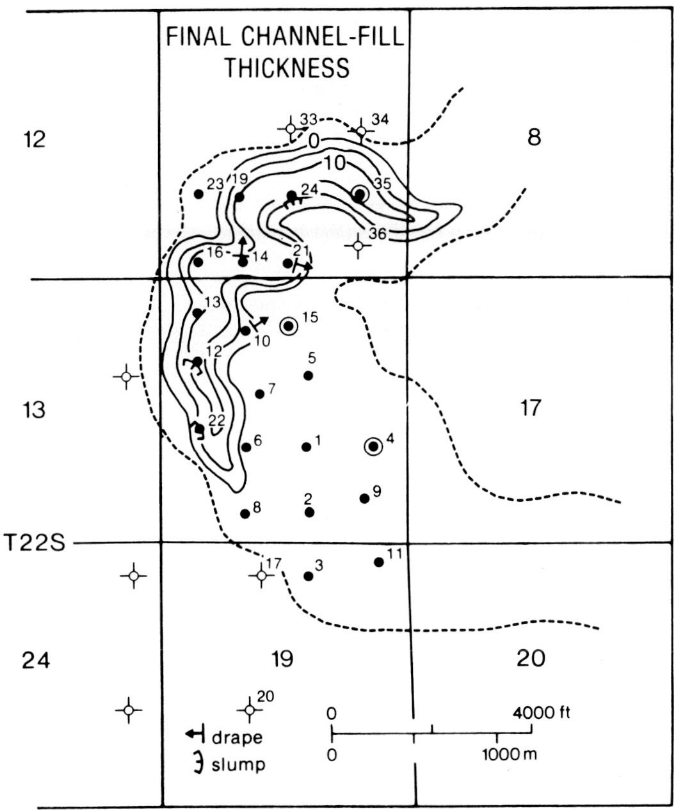

FIG. 19.—Isopach of final channel fill showing the highly sinuous geometry of the thinly interbedded sandstones, siltstones and shales. Contour interval 5 ft (1.5 m).

geomorphic surface on which the "A_3" sandstones were deposited. Depositional patterns of "A_3" sandstones are similar to those of point bars in meandering streams in that they form a series of successive, offlaping lenses deposited in a laterally prograding fashion with progressive shifting of deposition down the channel axis. Optimum reservoir properties appear to be present in the "A_3" channel sandstones. Production from any individual sandstone lens can be expected to be limited by the restricted areal extent of the sandstones.

ACKNOWLEDGMENTS

I thank Amoco Production Company and Bass Enterprises Production Company for providing the financial support for this research, as well as the cores, well logs, core analyses and field information upon which the study is based. Additional funding was provided by a graduate fellowship from the Fannie and John Hertz Foundation, for which I am most grateful. Thanks also to David W. Cromwell for his contribution made by the initial study of the field and for graciously sharing his ideas with me. Petrographic analyses were made by Wayne M. Ahr and David G. Kersey. I am greatly indebted to Robert R. Berg, who aided in selection of the topic, obtained financial support for the study and served as research advisor. He also contributed his original core photographs and descriptions to the study. I also express sincere thanks to Roger Slatt, who greatly improved the manuscript with comments and editorial suggestions.

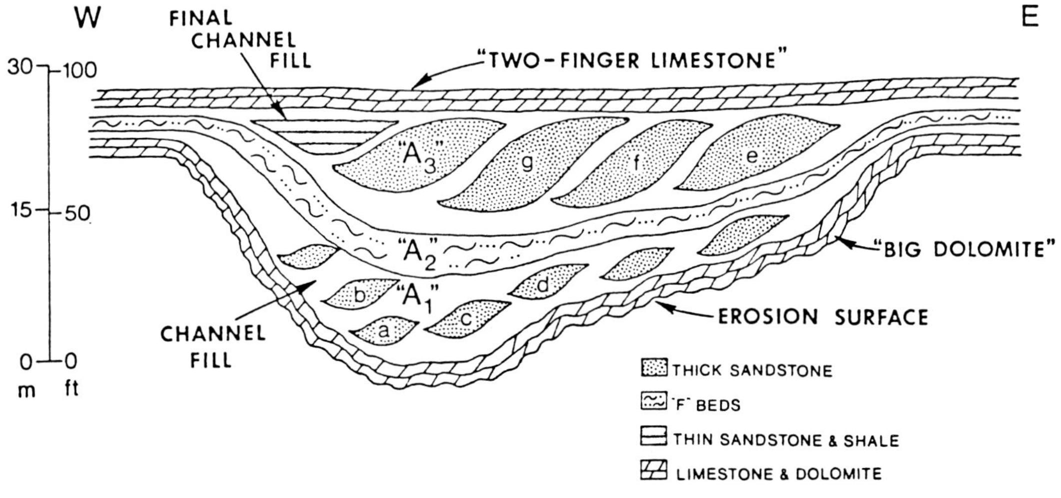

FIG. 20.—Diagrammatic cross section illustrating the size, shape and distribution of deposits within the erosional channel at Indian Draw field. "A_1" sandstones occur as small, stacked, isolated lenses with their greatest depositional concentration in the physiographic low adjacent to the western channel margin. "A_3" sandstones resemble point bar deposits, forming a series of successive offlaping lenses deposited in a laterally prograding fashion.

FIG. 18.—Isopach of "A_3" interval sandstones illustrating (A) total net sandstone, showing sinuous composite geometry of stacked, lenticular sandstones. (B) Isopach of selected "A_3" lenses shows the geometry and distribution of sandstone lenses *e*, *f*, and *g*. Individual lenses are labeled on map.

REFERENCES

Berg, R. R., 1979, Reservoir sandstones of the Delaware Mountain Group, southeast New Mexico, *in* Sullivan, N. M., ed., Guadalupian Delaware Mountain Group of West Texas and Southeast New Mexico: Society of Economic Paleontologists and Mineralogists, Permian Basin Section Guidebook, Publication 79-18, p. 75–95.

Bouma, A. H., 1962, Sedimentology of Some Flysch Deposits: A Graphic Approach to Facies Interpretation: Elsevier, Amsterdam, 168 p.

Cromwell, D. W., 1979, Indian Draw Delaware field: a model for deeper Delaware sand exploration, *in* Sullivan, N. M., ed., Guadalupian Delaware Mountain Group of West Texas and Southeast New Mexico: Society of Economic Paleontologists and Mineralogists, Permian Basin Section Guidebook, Publication 79-18, p. 142–152.

Damuth, J. E., Venkataratham, K., Flood, R. D., Kowsmann, R. O., Monteiro, M. C., Gorini, M. A., Palma, J. J. C., and Belderson, R. H., 1983, Distributary channel meandering and bifurcation patterns on the Amazon deep-sea fan as revealed by long-range side-scan sonar (GLORIA): Geology, v. 11, p. 94–98.

Folk, R. L., 1974, Petrology of Sedimentary Rocks: Hemphill, Austin, 182 p.

Gilreath, J. A., Healy, J. S., and Yelverton, J. N., 1969, Depositional environments defined by dipmeter interpretation: Gulf Coast Association of Geological Societies Transactions, v. 19, p. 101–111.

Grauten, W. F., 1979, Fluid relations in Delaware Mountain sandstone, *in* Sullivan, N. H., ed., Guadalupian Delaware Mountain Group of West Texas and Southeast New Mexico: Society of Economic Paleontologists and Mineralogists, Permian Basin Section Guidebook, Publication 79-18, p. 191–204.

Harms, J. C., 1974, Brushy Canyon Formation, Texas: a deep-water density current deposit: Geological Society of America Bulletin, v. 85, p. 1763–1784.

International Oil Scouts Association, 1974, International Oil and Gas Development Yearbook, v. 44, part 1, p. 200.

Jacka, A. D., Beck R. H., St. Germain, L. C., and Harrison, S. C., 1968, Permian deep-sea fans of the Delaware Mountain Group (Guadalupian), Delaware basin, *in* Sullivan, N. H., ed., Guadalupian Facies, Apache Mountains Area, West Texas: Society of Economic Paleontologists and Mineralogists, Permian Basin Section, Publication 68-11, p. 49–90.

———, 1979, Deposition and entrapment of hydrocarbons in Bell Canyon and Cherry Canyon deep-sea fans of the Delaware basin, *in* Sullivan, N. H., ed., Guadalupian Delaware Mountain Group of West Texas and Southeast New Mexico: Society of Economic Paleontologists and Mineralogists, Permian Basin Section Guidebook, Publication 79-18, p. 104–120.

King, P. B., 1948, Geology of the southern Guadalupe Mountains, Texas: U.S. Geological Survey Professional Paper 215, 183 p.

Meissner, F. F., 1972, Cyclical sedimentation in mid-Permian strata of the Permian basin (abs.), *in* Cyclical Sedimentation in the Permian Basin: West Texas Geological Society Symposium, p. 203–222.

Newell, N. D., Rigby, J. K., Fisher, A. G., Whitman, A. J., Hickox, J. E., and Bradley, J. S., 1953, The Permian Reef Complex of the Guadalupe Mountains Region, Texas and New Mexico: Freeman and Co., San Francisco, 236 p.

Pirson, S. J., 1977, Continuous dipmeter as a sedimentation tool, *in* Geologic Well Log Analysis, 2nd Edition: Gulf Publishing Co., Houston, p. 130–162.

Pryor, D. B., Adams, C. E., and Coleman, J. M., 1983, Characteristics of a deep-sea channel on the middle Mississippi fan as revealed by a high resolution survey: Gulf Coast Association of Geological Societies Transactions, v. 33, p. 389–394.

Weinmeister, M. P., 1978, Origin of upper Bell Canyon reservoir sandstones (Guadalupian), El Mar and Paduca fields, southeast New Mexico and West Texas: Unpublished M.S. Thesis, Texas A & M University, 96 p.

Williamson, C. R., 1977, Deep-sea channels of the Bell Canyon Formation of the Delaware basin, Texas-New Mexico, *in* Sullivan, N. H., ed., Upper Guadalupian Facies, Permian Reef Complex, Guadalupe Mountains, New Mexico and Texas: Society of Economic Paleontologists and Mineralogists, Permian Basin Section, Publication 77-16, p. 409–431.

———, 1979, Deep-sea sedimentation and stratigraphy, Bell Canyon Formation (Permian), Delaware basin, *in* Sullivan, N. H., ed., Guadalupian Delaware Mountain Group of West Texas and Southeast New Mexico: Society of Economic Paleontologists and Mineralogists, Permian Basin Section Guidebook, Publication 79-18, p. 39–74.

CARBONATE RESERVOIR DESCRIPTION

DANIEL JARDINE AND JOHN WILLIAM WILSHART
Esso Resources Canada Limited 339 50th Avenue SE Calgary, Alberta, Canada

ABSTRACT: Carbonate reservoirs are characterized by extremely heterogeneous porosity and permeability. These heterogeneities are caused by the wide spectrum of environments in which carbonates are deposited and by subsequent diagenetic alteration of the original rock fabric. Pore systems range from thick, vuggy reservoirs in the coarse-grained skeletal-rich facies of the reef margin or platform margin to highly stratified, often discontinuous, reservoirs in the reef interior, platform interior, and nearshore facies.

Eight fields from the western Canada basin have been selected to illustrate the variety of reservoir configurations which can be found. Boundary Lake field and Steelman field are thinly bedded sequences of cyclic nearshore carbonates which are best suited to pattern waterflood. Quirk Creek field is a thick, overthrusted open platform gas pool in which fractures associated with the leading edge of the structure provide high deliverability in an otherwise low permeability reservoir. Clarke Lake field is a fractured, dolomitized platform margin gas reservoir which is highly susceptible to water channeling. Golden Spike, Judy Creek, Norman Wells, and Redwater fields are reef bioherms which display the classic reservoir configuration of highly permeable reef margin facies enclosing strongly stratified reef interior sediments. With increasing reef size, the relative volume occupied by oil becomes smaller and the aquifer becomes larger and more effective.

Golden Spike field was entirely filled with oil when discovered and early production was characterized by rapidly declining pressure. An initial gas flood depletion plan was followed by a gravity-controlled, gas-driven miscible flood. This was very effective until the miscible bank was broken up and dispersed by several thin impermeable beds in the reef interior, which proved to be barriers to vertical flow.

Judy Creek field had a small ineffective original aquifer and was produced initially by a downdip peripheral waterflood. This ultimately proved to be ineffective in repressuring the discontinuous porosity in the reef interior and was supplemented with a pattern waterflood.

Depletion plans for these pools require detailed, integrated geological engineering studies to develop accurate reservoir models. These models are then used in computer simulation studies to forecast reservoir behavior under various depletion methods and to assist in ongoing reservoir surveillance studies.

INTRODUCTION

Carbonates make up about 20% of the world's sedimentary rocks and contain 40% of the world's major oil pools. The western Canada basin contains 2.5×10^9 m^3 (16 billion BBL) of oil, of which 70% is reservoired in carbonates. These reservoirs range in age from the Devonian to the Triassic and represent almost the entire spectrum of carbonate types. Production history from these reservoirs extends for periods of as many as 30 years—at first under severe restrictions caused by limited market demand and later at maximum rates because of a shortfall in supply. Canadian oil and gas pools, therefore, provide excellent examples of the variety and variability of carbonate reservoir types and their producing characteristics under various depletion strategies.

Part I-A of this paper describes, in very general terms, the geometry, depositional environment, and porosity characteristics of various carbonate facies and the diagenetic influences that can modify the original textures and alter the reservoir quality. To relate these principles to actual cases, six typical carbonate pools are described and schematically illustrated in part I-B. Each of these pools has unique reservoir characteristics that require detailed integration of the geological-engineering disciplines for optimum development. In Part II, two additional pools, the Golden Spike and Judy Creek reef reservoirs, are described in detail to illustrate how integrated geological-engineering studies can lead to better understanding of reservoir behavior and to changes in operating practices.

PART I-A—DEPOSITIONAL SETTINGS, POROSITY AND GEOMETRY

Depositional Settings

Carbonate rock types illustrated schematically in Figures 1, 2, and 3 are found in a wide variety of depositional settings. These settings range from reefs covering only a few hectares to carbonate platforms (shelves) and offshore banks that often extend over thousands of square kilometers. The geometry and relationships to the non-reservoir rocks associated with them are controlled by the local and regional environments of deposition.

Figure 1 illustrates an environmental model of a common carbonate setting characterized by an abrupt change in bathymetry between a shallow-water shelf and a deeper water basin. The depositional environments that arise from this geometry are (a) nearshore, fringing the coastal plain; (b) platform interior; (c) platform margin; (d) platform slope or forereef; and (e) moderate to deep-water basin. Note that the scale of this model may vary considerably, both horizontally and vertically.

(a) At the edge of the coastal plain, lime sands often form beaches along the shoreline and may give way landward to supratidal mudflats characterized by algal-encrusted laminites. In arid climates sabkha conditions develop, in which the mudflat carbonates are associated with thin beds of evaporites.
(b) Platform interior sediments are usually low-energy lime muds, but small reef bioherms, patch reefs, and lime sand shoals may be present, if the platform is not completely restricted and is open to marine tidal, wave, and current action. In some instances an open platform may develop extensive deposits of bedded reef-building organisms. These give rise to porous tabular deposits often referred to as reef biostromes.
(c) The platform margin is characterized by shoals of lime sand (calcarenites) composed of skeletal fragments or ooid grains deposited in a high-energy, agitated environment. Where the platform margin is steep, a barrier reef composed of colonial and encrusting organisms may

CARBONATE ENVIRONMENTS

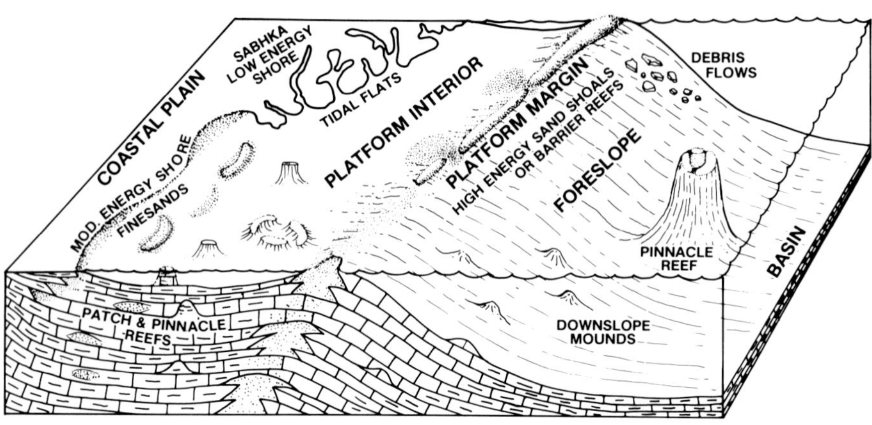

FIG. 1.—Schematic carbonate environment model.

form, especially on the windward side.

(d) Platform slope and forereef environments are characterized by a gradual decrease in grain size downslope. Sediment types are to a large extent dependent on the steepness of the seafloor slope and the nature of the platform margin. Talus slopes, lime sands, muds, mud mounds, and downslope pinnacle reef bioherms can be characteristic under various conditions.

(e) Basinal carbonates are usually lime muds which do not form reservoir rocks. In special instances thick beds of chalks which may be reservoirs are present. Deep-water chalks are formed largely by the accumulation of the skeletons of coccoliths (pelagic algae). Coccoliths only exist in parts of marine basins that are far removed from a source of terrigeneous clastic sediments. Chalks form thick, texturally homogeneous deposits and are represented in the geological record from the Triassic to the Recent.

A similar model can be used to represent atoll-type reef bioherms. In this case the cross section model would show a degree of bilateral symmetry with lagoonal lime muds and lime sands in the reef interior, contained by organically-bound reef-margin sediments with relatively steeply-dipping foreslope merging into starved basin deposits.

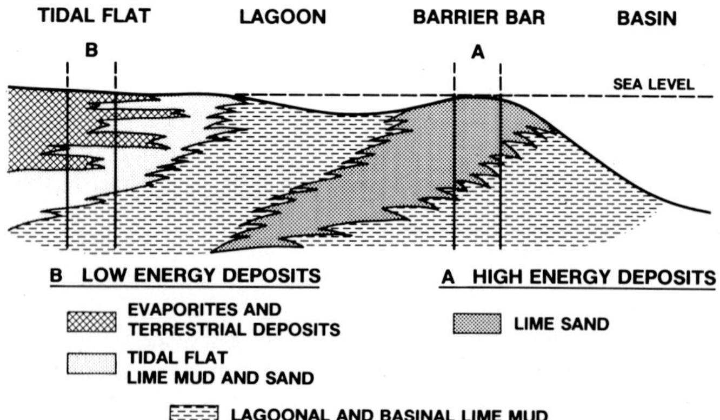

FIG. 2.—Cross section showing high- and low-energy facies.

The other end member of carbonate environmental models is one in which there is no break in slope and the sea bottom uniformly deepens from the shoreline into the basin (ramp). In this case the highest energy conditions exist along the shoreline, where lime sands are deposited, and muddy sediments are deposited offshore.

It should be stressed that the simple model shown here will rarely exist in the subsurface due to the repeated rises and falls in relative sea level during geologic time. During periods of sea-level rise (transgression), deeper water sediments will bury the pre-existing shallow-water sediments. When sea level remains relatively constant, shallow-water sediments will prograde over the deep-water deposits. If sea level falls (regression), carbonate deposition may be relocated a considerable distance basinward. Slow, long-term rises in sea level allow the development of thick, extensive mud-dominated platforms, commonly referred to as shelves or banks. Rapid rises in sea level often drown the platform sediments, but if carbonate sedimentation is re-initiated, thick reef bioherms or offshore banks are commonly developed. These provide excellent reservoirs.

During periods of arid climates, periodic desiccation of the basin gives rise to cyclical deposits of anhydrites, salts, and carbonates. Evaporites are very important in providing good reservoir seals.

The cross section in Figure 2 illustrates the concepts of low- and high-energy regimes on a carbonate platform. High-energy carbonate sands form as an offshore bar or as a mainland beach, where the coast is exposed to the open sea. Carbonate muds are deposited in low-energy, shallow-water lagoons behind a barrier or in deep-water basinal environments.

The nature of the shoaling upward sequence at a given locality depends upon its position on the platform. In the more open marine settings shoaling-upward sequences grade from deeper water muddy carbonates below to shallower water lime sands above (Fig. 2, stratigraphic column A). In protected landward positions on the platform, the shoaling upward sequence consists entirely of muddy carbonates representing lagoonal, intertidal and supratidal environments (Fig. 2, stratigraphic column B). If circulation is suf-

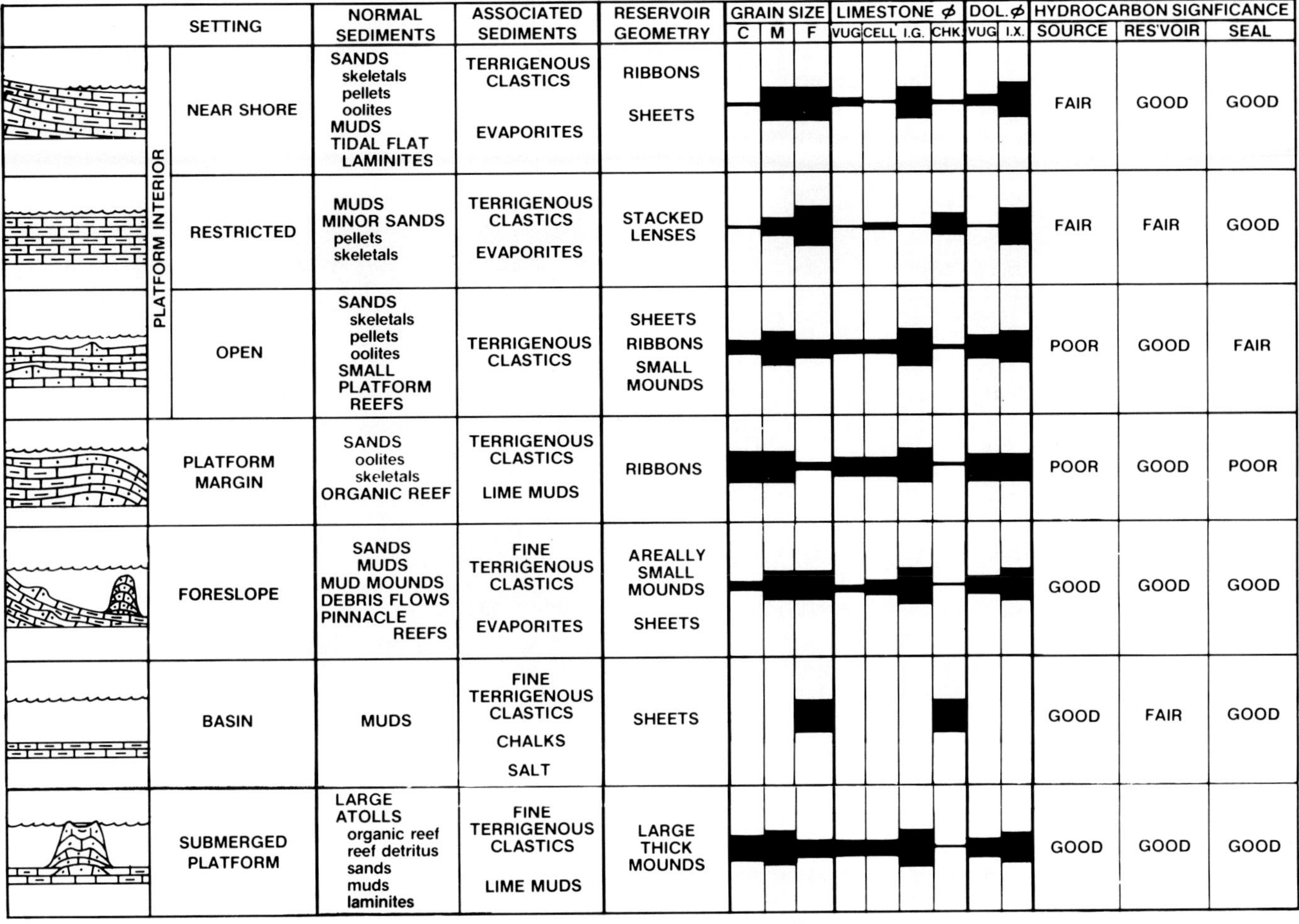

SYNOPSIS OF CARBONATE RESERVOIRS

	SETTING	NORMAL SEDIMENTS	ASSOCIATED SEDIMENTS	RESERVOIR GEOMETRY	HYDROCARBON SIGNFICANCE SOURCE	RES'VOIR	SEAL
PLATFORM INTERIOR	NEAR SHORE	SANDS skeletals pellets oolites MUDS TIDAL FLAT LAMINITES	TERRIGENOUS CLASTICS EVAPORITES	RIBBONS SHEETS	FAIR	GOOD	GOOD
PLATFORM INTERIOR	RESTRICTED	MUDS MINOR SANDS pellets skeletals	TERRIGENOUS CLASTICS EVAPORITES	STACKED LENSES	FAIR	FAIR	GOOD
PLATFORM INTERIOR	OPEN	SANDS skeletals pellets oolites SMALL PLATFORM REEFS	TERRIGENOUS CLASTICS	SHEETS RIBBONS SMALL MOUNDS	POOR	GOOD	FAIR
	PLATFORM MARGIN	SANDS oolites skeletals ORGANIC REEF	TERRIGENOUS CLASTICS LIME MUDS	RIBBONS	POOR	GOOD	POOR
	FORESLOPE	SANDS MUDS MUD MOUNDS DEBRIS FLOWS PINNACLE REEFS	FINE TERRIGENOUS CLASTICS EVAPORITES	AREALLY SMALL MOUNDS SHEETS	GOOD	GOOD	GOOD
	BASIN	MUDS	FINE TERRIGENOUS CLASTICS CHALKS SALT	SHEETS	GOOD	FAIR	GOOD
	SUBMERGED PLATFORM	LARGE ATOLLS organic reef reef detritus sands muds laminites	FINE TERRIGENOUS CLASTICS LIME MUDS	LARGE THICK MOUNDS	GOOD	GOOD	GOOD

FIG. 3.—Synopsis of carbonate environments.

ficiently restricted and the climate is sufficiently arid, evaporites are deposited subaqueously in the lagoons or subaerially on the tidal flats (sabkhas).

Porosity and Geometry

Figure 3 summarizes the important characteristics of each of the depositional settings shown on the model of carbonate environments (Fig. 1). The normal carbonate sediments are listed, together with other sediment types that are commonly associated with them, the geometry of the reservoirs that can be expected and the relative abundance of coarse-, medium-, and fine-grained sediment. The porosity columns suggest the relative abundance of the main limestone porosity types—vuggy, intraparticulate, intergranular, and chalky. When the sediments are dolomitized, the porosity types become vuggy or intercrystalline. Source, reservoir, and seal parameters are characteristic of the specific environments only and should not be construed as being representative in a given sequence of sedimentary cycles (Fig. 3). For example, although open platform sediments may not contain good seals, they are frequently overlain by restricted platform muds or by evaporites which can provide excellent seals for the open platform reservoir rocks. Similarly, platform margins do not contain good source beds or seals, but they may be deposited on deeper water sediments or be covered by deeper water muds which can provide both seals and source for hydrocarbons.

Nearshore sediments.—

Nearshore sediments can be sands or muds or both. Reservoir geometry is ribbonlike for shoreline sediments, but progradation can produce sheetlike reservoirs. Frequently, there are a number of very thin reservoirs superimposed on each other. Grain size is medium to fine and intergranular porosity is most prevalent. With dolomitization, microvuggy and fine intercrystalline porosity is developed. Dolomitized or fractured algal flat laminites or shoreline sands often form good reservoirs.

Restricted platform sediments.—

The restricted platform contains mainly muddy sediments of predominantly fine-grained texture. Chalky porosity is sometimes developed, but these sediments are usually not very effective reservoirs unless they have been dolomitized. Since porosity is usually only locally developed, a number of stacked, but discontinuous, reservoirs is characteristic. Pores are very small and permeability is low.

Open platform sediments.—

Open platform sediments can be heterogeneous mixtures of muds, sands, and small platform reefs, and grain size is quite variable. Reservoir geometry tends to be sheetlike; vuggy, cellular, and intergranular porosity is common. Small to large vugs are often developed, but intergranular and intercrystaline porosity is prevalent. All platform reservoirs are well stratified and poor vertical communication is normal. Horizontal permeability may locally be very good.

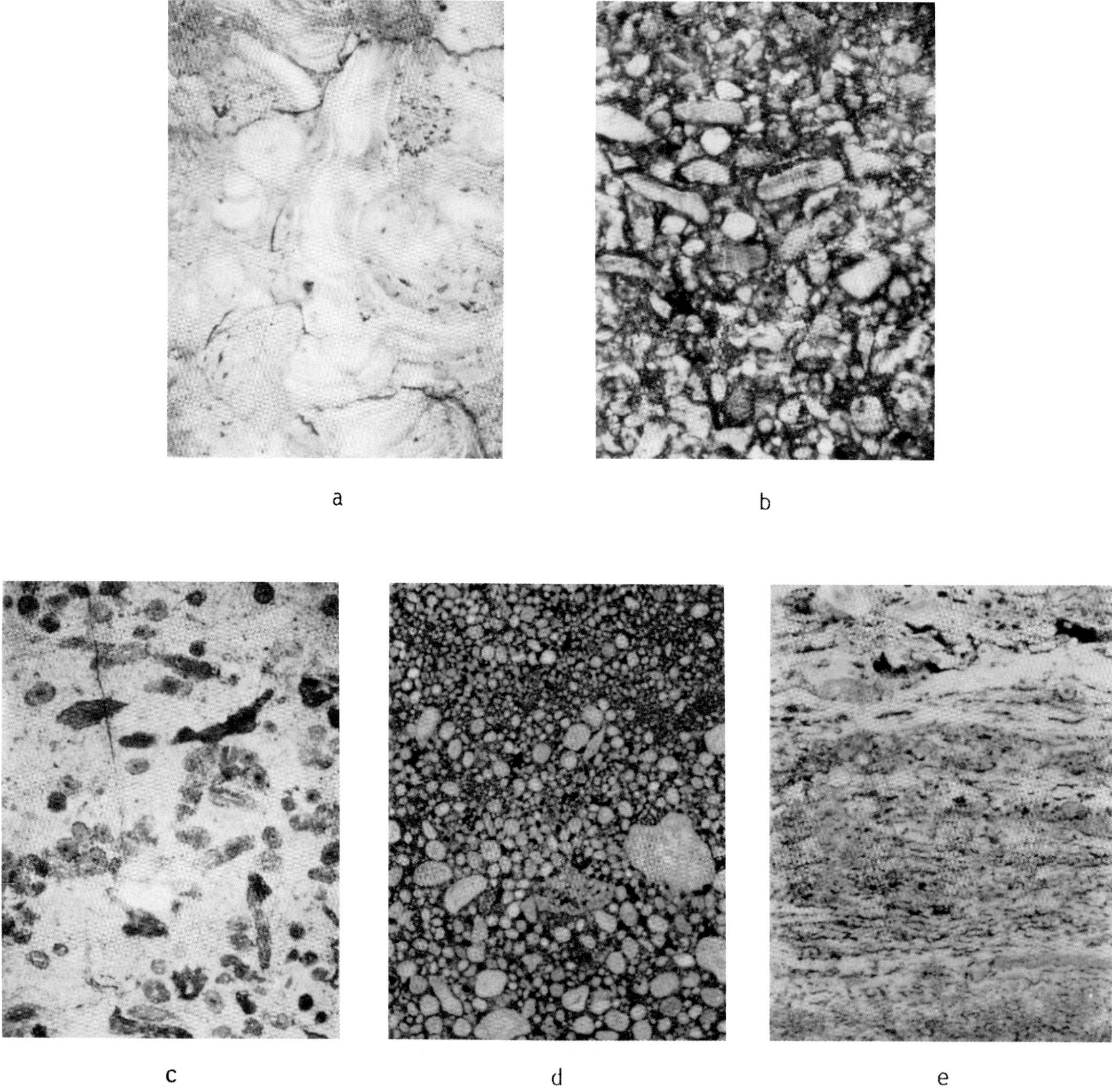

FIG. 4.—Examples of types of deposition: (a) Reef bioherm—stromatoporoid organic-reef framework typical of bioherm margins. Organic-cell porosity predominates. Porosity 14 percent; permeability (k90) 35 md. (b) Platform margin. Skeletal conglomerate composed of stromatoporoid and coral fragments. The original intergranular and organic-cell porosity has been destroyed by calcite cementation. Porosity <1 percent; permeability (k90) <0.001 md. (c) Platform (bank). Micritic-skeletal (amphiporid) limestone. Interior portions of banks characteristically have a high proportion of micrite (lime mud). Porosity in this sample is mainly of the organic-cell and chalky types. Porosity 3 percent; permeability (k90) 700 md. (d) Open platform. Oolitic-lump limestone with excellent intergranular porosity. Porosity 18 percent; permeability (k90) 700 md. (e) Nearshore. Algal laminite typical of the intertidal zone. The vuggy (birdseye) porosity is distinctly layered. Porosity 9 percent; permeability (k90) 200 md. (After Jardine and others, 1977.)

Platform margin sediments.—

Platform margin sediments are mainly sands or organic reefs, and the grain size is coarse to medium since the fines are winnowed out in this high-energy environment. Reservoirs are ribbonlike and frequently thick. Excellent vuggy, cellular, and intergranular porosity is commonly developed. When dolomitized, the porosity is often composed of large vugs and medium to coarsely intercrystalline pore networks. The rocks are poorly bedded and both vertical and horizontal permeability can be well developed.

Foreslope sediments.—

Foreslope sediments tend to be highly variable. They may be mainly muds or they may contain a high proportion of coarse material. There is a decrease in grain size and, consequently, a decrease in porosity and permeability in a downslope direction. Downslope organic reefs are characteristically small in area but may be quite thick and provide excellent reservoirs.

Basinal sediments.—

Basinal sediments are normally non-porous, but they often provide excellent source beds. If chalks are developed, their uniformity and purity give rise to a sediment with exceptionally high initial porosity, but this is dramatically reduced during burial and compaction. Although porosity in chalk may remain relatively high, the permeability may in some reservoirs be only a few millidarcies because of the very fine grain size.

The large atoll-like reef bioherms that sometimes develop on a submerged platform are each microcosms of the carbonate environmental model. Typically, they display an interior restricted-type facies of highly stratified muds and sands which are discontinuous reservoirs with poor vertical communication. The reef margins are composed of reef framework and associated detritus which form excellent, thick reservoirs with good lateral and vertical communication. Dolomitization usually improves permeability and reservoir continuity between the various reef facies. Examples of rock types are illustrated by the core photographs in Figure 4.

Secondary Porosity

Diagenetic alteration is the rule rather than the exception in carbonates, and these processes frequently modify or obscure the primary depositional porosity characteristics. The five processes of porosity alteration that commonly occur are noted in Figure 5 and illustrated in Figure 6. Dissolution (Fig. 6a) generally has a favorable effect on reservoir quality because it tends to improve porosity and markedly enhances permeability.

Dolomitization can play a dual role. Generally, it will improve a reservoir by increasing the pore size (Fig. 6b); at other times, porosity may be destroyed when dolomitization creates a dense interlocking crystal mosaic. Dolomitization preferentially affects the fine-grained matrix material and may be accompanied by solution of the larger skeletal grains to form vugs.

PROCESS	FAVORABLE EFFECTS	UNFAVORABLE EFFECTS
LEACHING	INCREASE ⌀ & K	
DOLOMITIZATION	INCREASE K	CAN ALSO DECREASE ⌀ & K
FRACTURING JOINTS BRECCIA	INCREASE K	INCREASE CHANNELING
RECRYSTALLIZATION	MAY INCREASE PORE SIZE & K	DECREASE ⌀ & K
CEMENTATION BY CALCITE DOLOMITE ANHYDRITE PYROBITUMEN SILICA		DECREASE ⌀ & K

FIG. 5.—Summary of diagenesis and secondary porosity.

The influence of dolomitization on reservoir properties has been evaluated empirically by comparing undolomitized with dolomitized reefs in western Canada. Limestone reefs have an average porosity of 8 percent and an average permeability of 68 md, whereas dolomitized reefs have an average porosity of 9 percent and a permeability of 800 md. These comparisons indicate that, although dolomitization may not change porosity significantly, it can increase permeability dramatically. This increase can be attributed to better development of solution vugs in dolomites and to fracturing, which is more intense in dolomites because of their more brittle nature.

Fracturing (Fig. 6c), which is most common in tectonically active areas, can create permeability in rocks where none previously existed and can form a pathway for leaching or cementing solutions. Fracturing often significantly increases the reservoir producibility, but from an operational standpoint, it can also result in high-water rates because of channeling.

Recrystallization or micrite enlargement is generally a reservoir-enhancing process that produces larger calcite crystals and a "chalky" texture (Fig. 6d). Reservoirs with chalky textures often have high porosity that is easily recognized on wireline logs, but very low permeability that is not apparent without cores or production tests.

There are many different types of cements that can destroy porosity. Even small amounts of certain cements such as pyrobitumen can seriously affect productivity because of a tendency to occupy the narrow connections between pores, thereby reducing the permeability. An example of cementation by calcite is shown in Figure 6e.

In summary, the predominant feature of carbonate porosity is heterogeneity. Pore systems in silicic sandstones are often relatively simple, being mainly intergranular and controlled by the size and shape of the grains and the amount of matrix and secondary cement. Carbonate pore systems, on the other hand, are usually more complex. The primary intergranular porosity is more variable because of the diversity of grain size and shape; in addition, the skeletal grains themselves often have intraparticle cellular porosity characteristic of the original organism. Because of the suscep-

FIG. 6.—Examples of types of porosity: (a) Leached fossil casts in a skeletal micritic limestone. Some types of skeletal material are much more susceptible to leaching than others because of original differences in mineralogy. Porosity 9 percent; permeability (k90) 72 md. (b) Dolomitized skeletal micritic limestone with leached fossil (amphiporid) casts. Secondary vuggy and intercrystalline porosity predominate. Porosity 16 percent; permeability (k90) 300 md. (c) Fracture porosity in a dolomitized forereef breccia. The fractures have been partly infilled with white coarsely crystalline dolomite cement. Two distinct fracture systems of different ages are evident. Porosity 3 percent; permeability (k90) 30 md. (d) Recrystallization in a skeletal-pellet-micritic limestone. The darker layers are relatively unaltered, non-porous micrite (lime mud). Recrystallization of the micrite to a coarser, porous, chalky texture (light patches) is confined to the skeletal (amphiporid)-rich layers where some primary organic-cell porosity is still present. Porosity 12 percent; permeability (k90) 15 md. (e) Cementation in a skeletal (stromatoporoid) conglomerate. All the originally intergranular porosity has been completely infilled with white coarsely crystalline calcite. The primary organic-cell porosity in the stromatoporoid fragments also has been destroyed as a result of calcite cementation. Porosity <1.5 percent; permeability (k90) <0.01 md. (After Jardine and others, 1977.)

tibility of carbonates to dissolution and replacement, the original pore systems are commonly modified by post-depositional diagenetic processes to even more complex pore networks.

PART I-B—THE GEOLOGY AND POOL DEVELOPMENT OF SIX CARBONATE POOLS

The heterogeneous rock fabric, which results from the variety of depositional environments and the subsequent

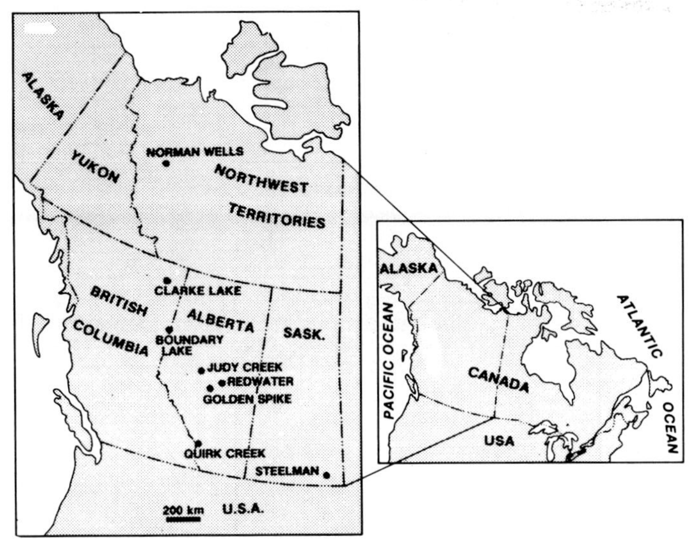

FIG. 7.—Geographic distribution of selected carbonate pools, western Canada.

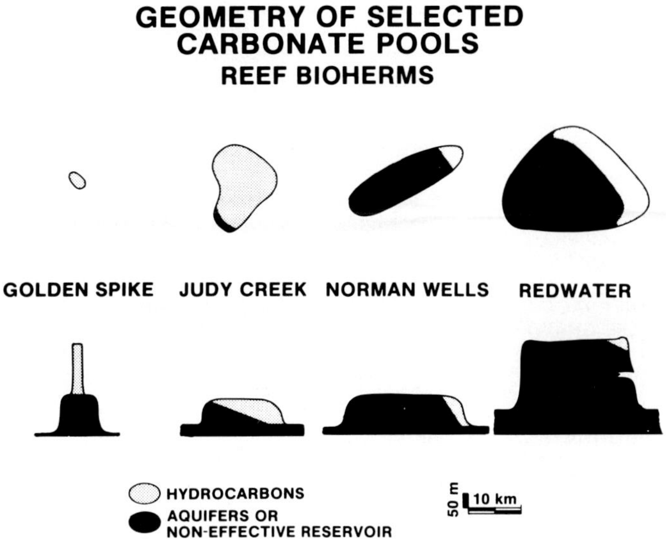

GEOMETRY OF SELECTED
CARBONATE POOLS
REEF BIOHERMS

GOLDEN SPIKE JUDY CREEK NORMAN WELLS REDWATER

HYDROCARBONS
AQUIFERS OR
NON-EFFECTIVE RESERVOIR
50 m
10 km

FIG. 9.—Geometry of selected western Canada carbonate reservoirs—reef pools.

diagenetic alteration described above, strongly affects the performance behavior of carbonate reservoirs and, therefore, the production practices used in their development and depletion. This cause-and-effect relationship can be demonstrated by describing six typical oil and gas pools from western Canada. The geographic location of these pools is indicated on the map in Figure 7.

Platform pools (Fig. 8) and reef pools (Fig. 9) have differing plan views and cross sections. In order to facilitate comparison, each of the pools is shown on the same scale and with the regional southwesterly dip removed. The large areal extent, amoeboid outline, and sheetlike nature of the platform interior pools (Boundary Lake and Steelman), and the ribbonlike outline and thick buildup of the platform margin pool (Clarke Lake) are shown in Figure 8.

Reefs may be listed in order of size (Fig. 9). Features that all have in common include an underlying shallow-water platform and a thick buildup of reefal carbonates encased in terrigenous shales and muddy limestones. Small reefs tend to be filled with oil, but with increasing size the relative volume occupied by oil becomes smaller. Comparative data tables for these pools are listed in Figures 10 and 11.

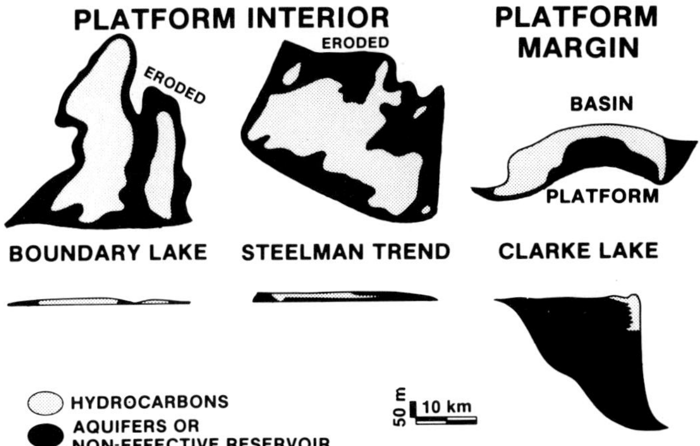

FIG. 8.—Geometry of selected western Canada carbonate reservoirs—platform pools.

Boundary Lake Field

Boundary Lake (Fig. 12) produces oil from Triassic nearshore carbonates. The reservoir covers an area of 310 km^2 (120 mi^2) but is only 10 m (30 ft) thick. This pool is a combination stratigraphic-structural trap. Reservoir beds dip gently to the south and terminate updip at an unconformity where they are sealed above by tight siltstone beds. The base seal is an anhydrite that is also truncated by the unconformity.

There is no water leg present and the pool is terminated downdip by the disappearance of effective porosity. A small gas cap is present in a few structurally high wells.

The reservoir occurs within a cyclic sequence of shallow-water evaporitic carbonates consisting of thinly interbedded micrites and calcarenites. The supratidal algal beds have been dolomitized selectively and form the best reservoirs. As many as five separate porous zones can be stacked within a net pay section of 9 m (28 ft). The thickness of the total unit at any locality is determined by the amount of relief on the unconformity surface.

Boundary Lake has been on production since 1961 and was placed under waterflood operation 3 years after discovery. The period before waterflooding was characterized by high production rates, progressively increasing gas/oil ratios, and rapidly declining reservoir pressure. Continued production under natural depletion would have resulted in an ultimate oil recovery of only 12 percent.

The Boundary Lake type of reservoir, consisting of thin, but continuous, porous zones confined above and below by dense beds, lends itself to pattern waterflooding. Inverted

	BOUNDARY LAKE	STEELMAN	QUIRK CREEK	CLARKE LAKE
DEPOSITIONAL SETTING	NEAR SHORE	NEAR SHORE	OPEN PLATFORM	PLATFORM MARGIN
AGE	TRIASSIC	MISSISSIPPIAN	MISSISSIPPIAN	DEVONIAN
ROCK TYPE	DOLOMITE	LIMESTONE	LIMESTONE / DOL.	DOLOMITE
NUMBER OF WELLS	300	944	15	51
AREA OF POOL (sq. km.)	310	293	29	115
CARBONATE THICKNESS (m.)	10	24	245	300
TRAP FILL (%)	100	90	80	60
AQUIFER	NONE	INEFFECTIVE	SMALL	SMALL
AVERAGE DEPTH (m.)	1310	1400	1920	1920
OIL GRAVITY (API)	36	38	(GAS)	(GAS)
MAXIMUM PAY (m.)	9	18	100	107
AVERAGE POROSITY (%)	22	17	8	7
AVERAGE MATRIX PERM. (K 90 mD)	30	10	3	50
CONNATE WATER (SW %)	10	35	20	16
HYDROCARBONS-IN-PLACE	91 x 10^6m^3	101 x 10^6m^3	14 x 10^9m^3	62 x 10^9m^3
DEPLETION-NATURAL	SOLUTION GAS	SOLUTION GAS	GAS EXPANSION	GAS EXPANSION
DEPLETION-ENHANCED	WATER FLOOD	WATER FLOOD	NONE	NONE
APPROX. RECOVERY FACTOR (%)	35	45	85	70
PRODUCTION THRU 1980	22 x 10^6m^3	34 x 10^6m^3	6 x 10^9m^3	28 x 10^9m^3

FIG. 10.—Comparative pool data—platform pools.

nine-spot patterns as close as 32 ha (80 acres) have been implemented and will increase the ultimate recovery to about 35 percent of the oil-in-place. Detailed correlation and mapping of the individual porous zones is a prerequisite for assuring that every zone is being waterflooded effectively.

Steelman Trend

Even a cursory comparison of the Steelman trend (Fig. 13) with Boundary Lake will indicate a pronounced similarity between these major oil pools, although they are of different geologic age. Both cover large areas, have irregular outlines, contain thin reservoirs, and are intimately associated with evaporites. Both reservoirs are truncated and sealed updip by an unconformity, although most of the porous horizons at Steelman terminate updip before reaching the unconformity surface. Steelman has a downdip aquifer, but this is only locally porous and it is ineffective.

The basal portion of the Steelman reservoir section over most of the trend is anhydrite, and the carbonate unit that contains most of the reservoir is only 10 m (30 ft) thick. An algal laminite-oolite-pisolite reservoir is present locally at the downdip end of the pool. Elsewhere the reservoir beds were deposited as cyclical interbeds of lime muds and lime sands. The sands varied from very fine-grained and well sorted to coarse-grained, poorly sorted and muddy. They were laid down as current-sorted bar-like deposits with their long axes approximately at right angles to the regional depositional strike.

The key to understanding the distribution of an effective

	GOLDEN SPIKE	JUDY CREEK	NORMAN WELLS	REDWATER
DEPOSITIONAL SETTING	REEF	REEF	REEF	REEF
AGE	DEVONIAN	DEVONIAN	DEVONIAN	DEVONIAN
ROCK TYPE	LIMESTONE	LIMESTONE	LIMESTONE	LIMESTONE
NUMBER OF WELLS	55	220	60 → 250	960
AREA OF REEF (sq. km.)	6	120	140	520
AREA OF POOL (sq. km.)	6	116	16	152
REEF BUILDUP (m.)	170	65	110	245
TRAP FILL (%)	100	75	10	5
AQUIFER	NONE	SMALL	INEFFECTIVE	LARGE
AVERAGE DEPTH (m.)	1730	2640	400	975
OIL GRAVITY (API)	37	41	39	36
MAXIMUM PAY (m.)	170	65	110	60
AVERAGE POROSITY (%)	9	9	10	7
AVERAGE MATRIX PERM. (k 90 mD)	100	45	4	100
CONNATE WATER (SW %)	11	16	10	25
ORIGINAL OIL-IN-PLACE (10^6m^3)	51	132	100	207
DEPLETION-NATURAL	SOLUTION GAS / GRAVITY	SOLUTION GAS	SOLUTION GAS	WATERDRIVE
DEPLETION-ENHANCED	MISCIBLE / GAS FLOOD	WATERFLOOD	WATER FLOOD	WATER INJECTION
APPROX. RECOVER FACTOR (%)	65	45	40	65
PRODUCTION THRU 1980 (10^6m^3)	27	39	4	114

FIG. 11.—Comparative pool data—reef pools.

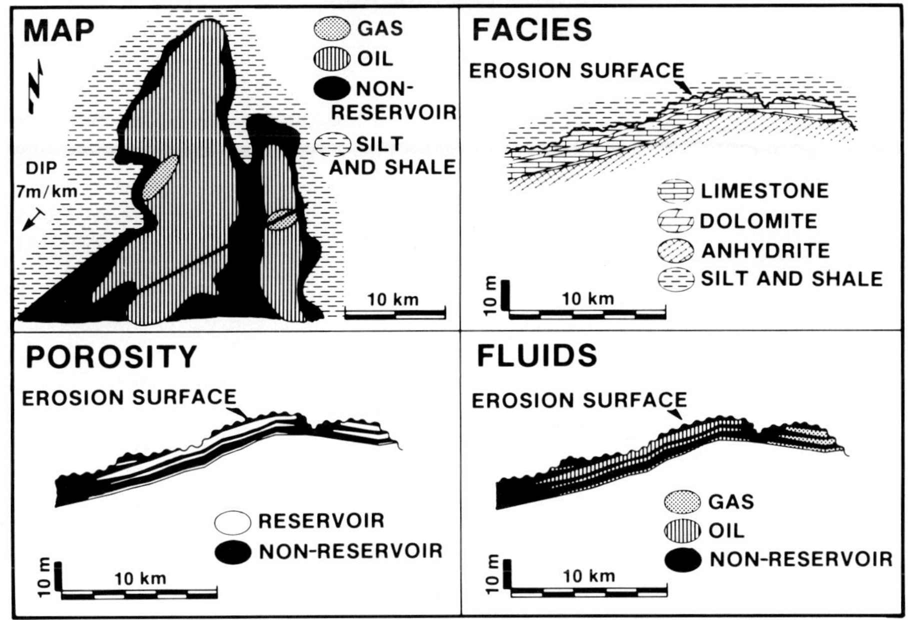

FIG. 12.—Boundary Lake oil pool. Map and cross sections illustrating lithofacies, porosity, and fluid distribution. (After Jardine and others, 1977.)

reservoir lies in the nature of the matrix between the sand grains. If the matrix is less than 0.01 mm in diameter, the rock is an ineffective reservoir, although leached microvugs may be visible and the measured porosity is high. Where the matrix has been recrystallized (a process which is also known as chalkification or micrite enlargement) and partly dolomitized, permeability is increased, connate water saturation is decreased, and the rock becomes an effective oil reservoir. Accurate reservoir description requires mapping the distribution of the original depositional facies and the subsequent reservoir facies that result from diagenetic modification of the original textures.

A pilot waterflood was installed at Steelman 3 years after discovery and full-scale waterflooding began 3 years later. As expected, the prewaterflood production was characterized by declining oil rates and increasing gas/oil ratios.

Steelman contains a series of joint-type fractures oriented in a northeast-southwest direction and the inverted 5-spot

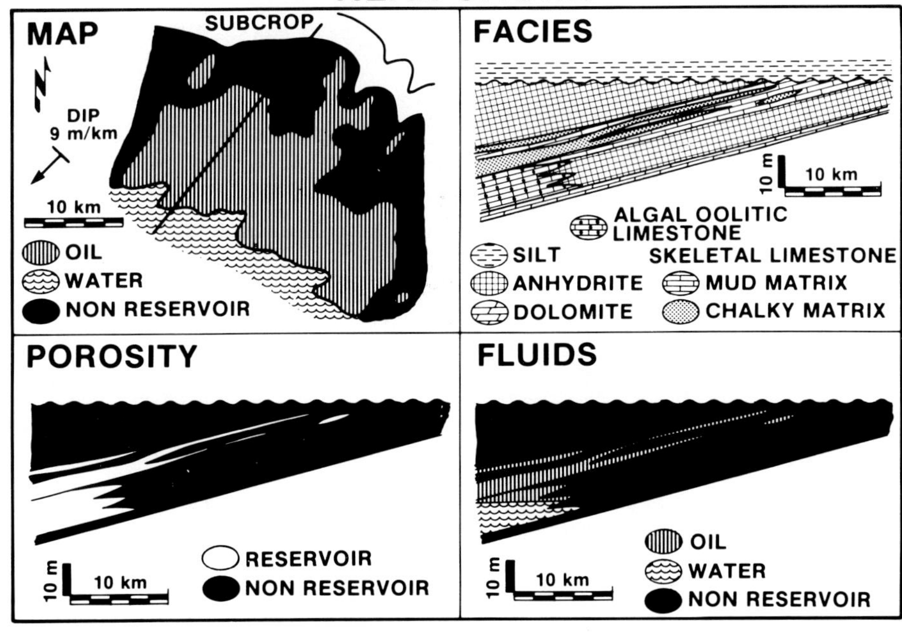

FIG. 13.—Steelman oil pool. Map and cross sections illustrating lithofacies, porosity, and fluid distribution.

waterflood pattern was oriented normal to this trend. No premature water breakthrough has been noted, and the current water-oil ratio is about 3 m^3/m^3 (3 bbl/BBL) after 25 years of sustained production.

Quirk Creek Field

Quirk Creek field (Fig. 14) is one of a series of major gas pools contained within the foothills overthrust belt of western Canada. Vertical displacement along these thrusts ranges to as much as 2,000 m (6,000 ft), and horizontal movement can be from 1 to 30 km (<1 mi to 20 mi) in an eastward direction.

Thrusting sometimes involves only a single, relatively simple plate, but more commonly, several plates are stacked and the reservoirs may be found in repeated sections. Thick Devonian, Mississippian, and Triassic platform carbonates are involved in the thrusts and form important reservoirs for sour gas.

Quirk Creek is the uppermost plate in a stack of at least three major thrusts with several imbricates and produces gas from Mississippian carbonates. The pool is 15 km (10 mi) long and about 2 km (over 1 mi) wide. The updip limit of the pool is a major sole fault, and the downdip extension on the back limb of the thrust is limited by water. About 400 m (1,300 ft) of gas leg is present between the crest of the structure and the water line. There are several imbricates near the front edge of the Quirk Creek plate, but these are difficult to define precisely by the seismograph and only one is shown in Figure 14.

The reservoir unit was deposited as lime muds with subsidiary lime sands in a series of thin bars which shifted vertically with time, producing a series of thin, stacked lenticular reservoirs with poor horizontal and vertical continuity. The sands and muds were partially dolomitized, and porosity is mainly microvugs and fine intercrystalline pores. The reservoir has an average porosity of 8 percent, but permeability is very low as the large grains were left undolomitized or the vugs became filled with calcite; however, fractures associated with the thrusts have improved reservoir communication and have markedly improved the effective permeability. Fractures are particularly prevalent

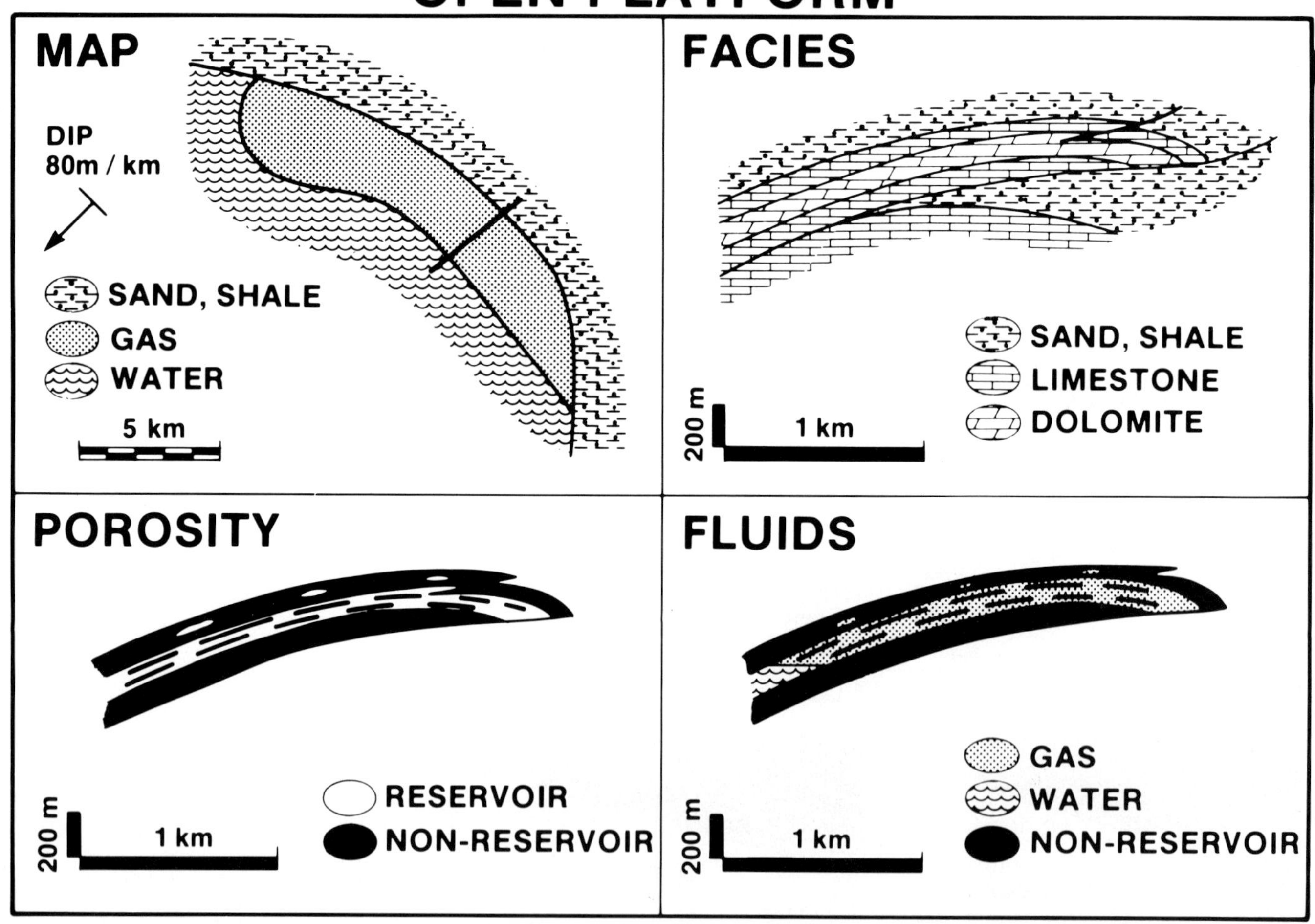

FIG. 14.—Quirk Creek gas pool. Map and cross sections illustrating lithofacies, porosity and fluid distribution.

near the leading edge of the thrust plate and in the imbricate zones. Wells drilled between the crestal portion of the structure and the front edge of the sole fault have deliverabilities of as much as 425 × 10^3 m^3/day (15 MMCF/D), or about three times the productivity of wells on the back limbs of the structure. There is little evidence of water channelling at Quirk Creek, or, indeed, in any of the foothill structures, as the water zones are usually not in direct vertical communication with the gas zones.

Clarke Lake Field

Clarke Lake is an example of a platform margin reservoir. The cross section and the pool outline on the map (Fig. 15) show that Clarke Lake is one of several gas pools located at the edge of a very extensive carbonate platform about 300 m (1,000 ft) thick. (Gas pools in the basin are associated with downslope reef bioherms.) The pool is sealed at the top by shale and at the base by non-porous dolomite. The lateral beds are basinal shales and non-porous platform interior limestones which have been plugged with sparry calcite.

Local dolomitization has occurred along selective zones at the platform margin and along linear trends shelfward of the platform edge. The dolomitized belts of porosity are narrow and difficult to map, resulting in a poor exploration and delination drilling record. The ratio of successful wells to dry holes is about 1:1. Maximum pay is 107 m (350 ft) and the gas is underlain by a local aquifer. Some of the better wells can deliver as much as 850 × 10^3 m^3/day (30 MMCF/D). The field-producing gas-water ratio is about 2,800 m^3/m^3 (65 barrels/MMCF), and there has been a relatively slow pressure decline from the original 20,000 kPa (2,900 psi) to about 13,500 kPa (2,000 psi) after 17 years and over 28 × 10^9 m^3 (over 1 TCF) of gas production.

This pool has two very distinct reservoir components. The host rock is composed of a low permeability dolomite which is intersected by a network of highly permeable secondary dolomites associated with a complex of fractures. The following production characteristics result from this combination.

(1) Initial production rates are very high as gas is produced from the fractures. Rates decline rapidly when the fracture gas is exhausted and the flow becomes dependent on the low-matrix permeability of the host rock.

(2) As productivity declines, water is channelled from the underlying aquifer. This occurs when it becomes as easy for water to flow up the permeable channels as it is for the gas to escape from the poorly porous matrix. To help alleviate this problem, most wells are drilled only into the top 3–7 m (10–20 ft) of the reservoir.

(3) This reservoir probably will be abandoned at a relatively high pressure—about 10,000 kPa (1,500 psig) as productivity declines and water production exceeds the economic limit. It should be noted that the conventional pressure/production pilot will indicate substantially higher reserves than will actually result.

Redwater Field

Redwater field (Fig. 16) is a very large reef bioherm which rests on a drowned carbonate platform and is enclosed in basinal shales. The reef covers 520 km^2 (200 mi^2) and is 245 m (800 ft) thick. In cross section, the reef has the typical appearance of an atoll with an exterior margin facies of an organic reef with its associated skeletal detritus and an interior lagoonal and intertidal to supratidal facies of algal limestones and finer detritus.

From the point of view of pore volume and continuity,

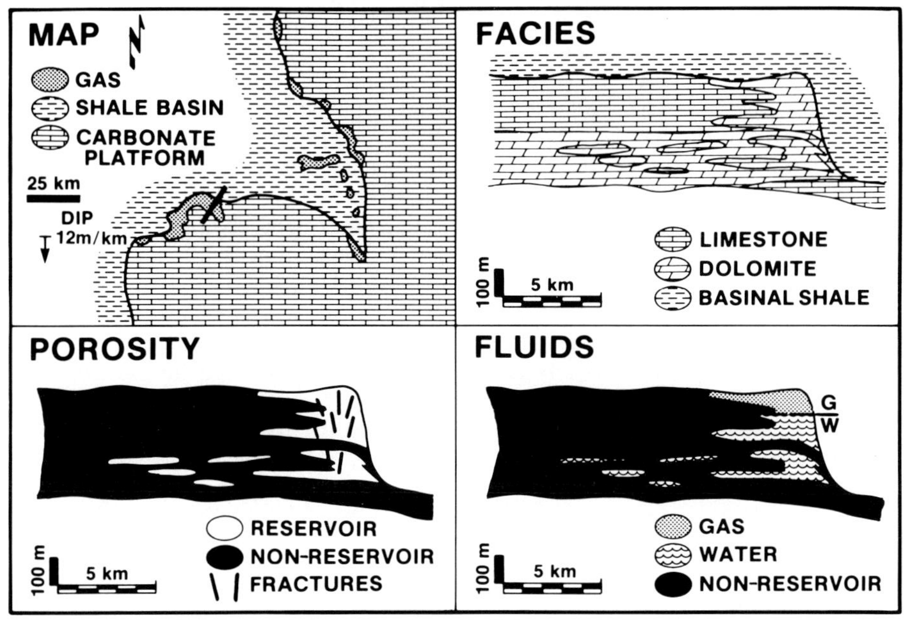

FIG. 15.—Clarke Lake gas pool. Map and cross sections illustrating lithofacies, porosity, and fluid distribution. (After Jardine and others, 1977.)

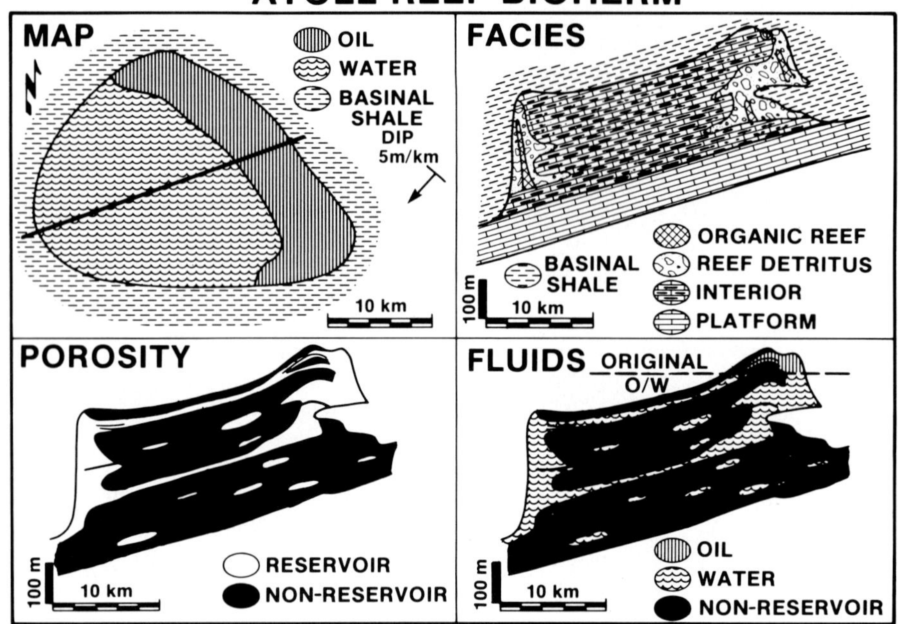

FIG. 16.—Redwater oil pool. Map and cross sections illustrating lithofacies, porosity, and fluid distribution. (After Jardine and others, 1977.)

the best reservoir is in the organic facies of the reef margin and in the high forereef sediments. The fine-grained interior lime mudstones are highly stratified and have lower and less continuous porosity. Oil is pooled at the updip edge of the reef and is present in both the margin and interior facies. Since 95 percent of the reef volume is occupied by water, and the reef is also connected to a large regional aquifer, the oil pool has a strong natural waterdrive.

The geologic description of the reef complex into its organic reef, forereef, and backreef systems is a key input into the definition of the flow patterns in both the aquifer and the oil zones. Production strategies, in terms of oil well recompletions and location of water injectors, has depended greatly on the geologic modelling of the barrier systems within the reef margin and backreef components.

Until recently, water disposal into the aquifer was limited to produced water, but there are now several water-injection wells supplementing the strong natural waterdrive. Supplemental water injection is aimed at raising the reservoir pressure to improve well productivity. Detailed stratigraphic studies were necessary to establish the optimum placement of the water injection wells.

Normal Wells Field

Oil was discovered at Norman Wells in 1970, but the remote location of the pool (Fig. 7) has precluded full development and the oil is presently refined on site to serve a local northern Canada market. The pool produced only 4 percent of its original oil-in-place of 100×10^6 m^3 (over 600 million BBLS) under primary solution gas drive. With the increase in oil prices which occurred in the late 1970s, southern markets became more economically attractive. About 190 new wells were drilled, and a 900-km (450 mi) pipeline was constructed to move the oil to existing facilities in northern Alberta.

Oil is pooled in the updip end of a linear reef bioherm which is entirely encased in shale (Fig. 17). The reef grew in two major cycles, each of which is typified by open marine facies at the base succeeded by restricted interior facies as the reef grew to sea level and established a peripheral organic reef with a central lagoon.

Porosity and permeability vary according to the facies type. Many of the larger primary porosity voids have been filled with sparry calcite, however, and there has been extensive microleaching of the matrix, the smaller grains and some of the skeletal grains, to produce a chalky porosity throughout the reservoir. These diagenetic effects have tended to narrow the normal distinction between reef margin and reef interior reservoir characteristics and produce a reservoir which is slightly more porous than the average reef but has very poor overall permeability.

Fracturing is an important characteristic of the Norman Wells pool and detailed fracture orientation studies have recently been completed. The paper by Irish and Kempthorne in this volume discusses this field in more detail. Fractures were mapped in adjacent outcrops; an areal photolineament study was prepared and regional orientations from Landsat data were compiled. Oriented cores were cut in six wells and another well was drilled through the reservoir at an angle of 70° to the vertical for a distance of 250 m (800 ft). Well testing involved 25 pressure buildup tests and water injectivity and interference tests. All of these data are in substantial agreement and indicate a preferred fracture orientation in a NNE-SSW direction with a secondary axis 90° to this trend. Preferred directional permeability ratios range from around 8:1 to 2:1 and indicate that the fractures decrease in effectiveness with increasing depth.

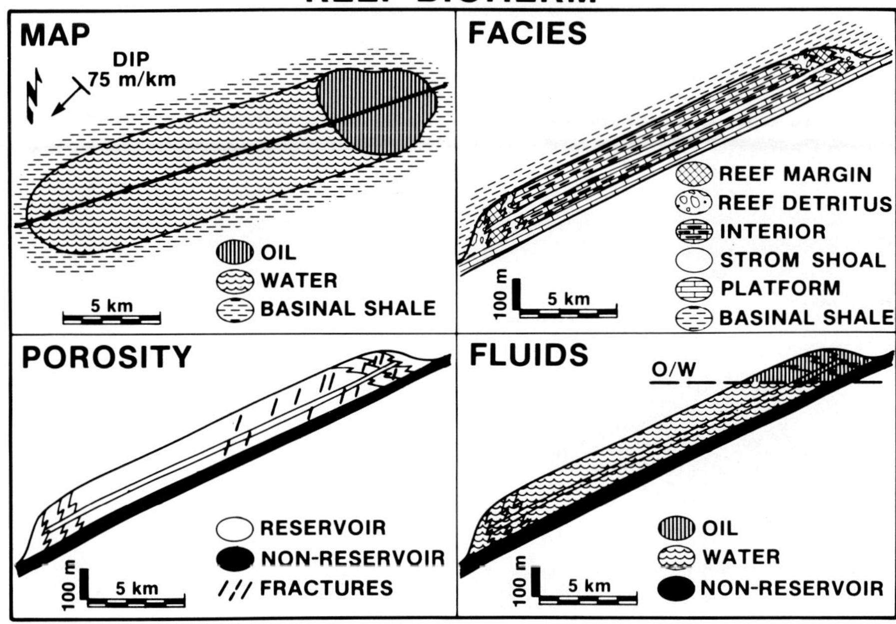

FIG. 17.—Norman Wells oil pool. Map and cross sections illustrating lithofacies, porosity, and fluid distribution.

Computer simulation modelling of the reservoir characteristics, including various ratios of directional permeability, indicated that a five-spot pattern with a 2:1 aspect ratio, elongated along a N30°E direction, is the optimum well configuration. Optimum economic spacing is suggested as 6.2 ha (15 acres)/well and ultimate recovery will be in the range of 40–45 percent of the oil-in-place.

The oil pool lies directly under the very large MacKenzie River (Fig. 18), and the 60 wells drilled to date have been located on the river bank and on two large islands. Since the river freezes over in winter and year round operations are required, the wells are drilled and operated from pads or from six permanent artificial islands. To maintain the required well spacing, most of the newer wells were directionally drilled at angles as high as 70° from vertical. A peak oil rate of 4,000 m^3/d (25,000 BBLS/D) was expected about 5 years from startup of the project.

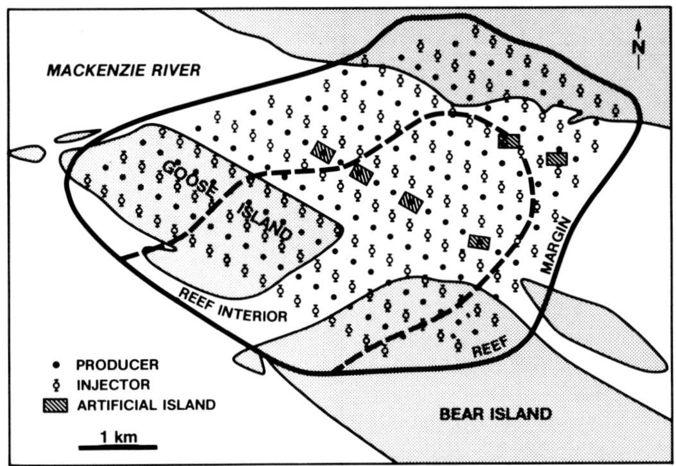

FIG. 18.—Norman Wells. Idealized well locations. (After Kempthorn and Irish, 1980.)

The development plan for Norman Wells provided an excellent example of the need for accurate reservoir description and modelling, especially of the fracture system, in order to evaluate and optimize future pool depletion strategies.

Conclusions

The pools we have just described display pore systems and other geologic characteristics that are unique to each. Pool depletion strategies must be tailored to the natural producing characteristics in order to optimize the performance behavior of each reservoir.

CARBONATE RESERVOIR DESCRIPTION—PART II; RESERVOIR DESCRIPTION AND POOL BEHAVIOR OF CARBONATES

Introduction

In Part I of this paper we described the depositional setting, porosity distribution, and geometry of carbonate reservoirs as illustrated by six western Canada oil and gas pools. In Part II we will use the reservoir description and production behavior of two oil fields, Golden Spike and Judy Creek, to show how the producing characteristics are intimately influenced by the internal reservoir heterogeneities and how detailed reservoir description and modelling can lead to modification of pool depletion strategies.

Golden Spike Field

Golden Spike field (Fig. 19) is a small reef bioherm which covers an area of 6 km^2 (2 mi^2), but has a buildup of 170

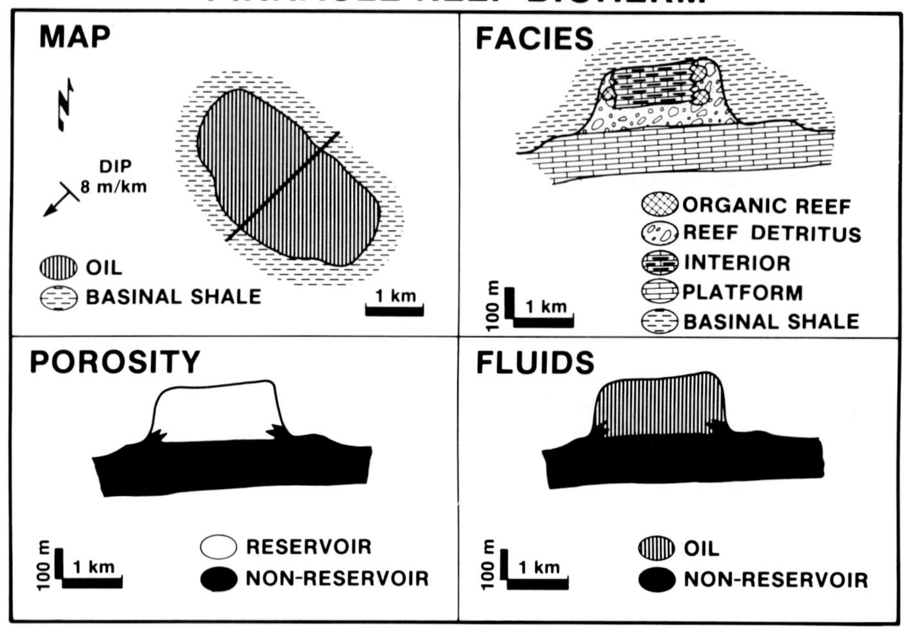

FIG. 19.—Gold Spike oil pool. Map and cross sections illustrating lithofacies, porosity, and fluid distribution.

m (550 ft), all of which was originally filled with oil (Fig. 11). The reef rests on a lower reef buildup of impermeable carbonates which form an effective base seal; an envelope of basinal shale encloses the rest of the reef. Even with this small size, the upper part of the reef is clearly differentiated into an exterior margin organic reef with its associated debris and an interior, strongly bedded facies of calcarenites and supratidal laminites. A few thin interbeds of non-porous lime mudstone are present in the reef interior sediments; otherwise the entire complex is porous.

At the time of discovery in 1949, the Golden Spike reef had an oil zone averaging 150 m (480 ft) with no free gas or water present. Since the primary producing mechanism was solution gas drive supplemented by gravity drainage, pressure decline was rapid in the initial years of production. Pressure maintenance by gas injection into the top of the reservoir was initiated in 1953, and by 1964 a secondary gas cap occupied the top 37 m (120 ft) of the reservoir. Performance to that date suggested a flushing efficiency of about 75 percent and a relatively homogeneous reservoir. At the same time production from the pool was severely prorated and an inexpensive supply of LPG solvent could be generated by cycling some of the reservoir crude oil, stripping off the light ends, and reinjecting the surplus dead oil. By emplacing the solvent as a bank or layer between the secondary gas cap and the oil zone, a vertical, gravity-controlled miscible flood was installed. The expectation was that the solvent bank, driven by continued gas injection, could be displaced downward through the oil zone, "dry cleaning" the oil from the reservoir as it went. The anticipated recovery from this process was 95 percent of the oil-in-place. The solvent bank contained 6 mol percent C_2-C_4 and was sized at 7 percent hydrocarbon-volume.

When the miscible flood was initiated (Fig. 20), there were 14 wells in the pool, including producers and various injectors for gas, solvent, and recycled dead crude. The low allowables could be more than met by the highly productive wells then in existence. Only eight of these wells had completely penetrated the oil zone or about one well per 65 ha (160 acres). By 1975 there were 52 wells in the pool, or about one control well per 10 ha (25 acres). Most of the infill wells were drilled in the early 1970s to meet and maintain the greatly increased allowables at that time.

During the first few years, the movement of the gas contact and the solvent bank were monitored with neutron logs

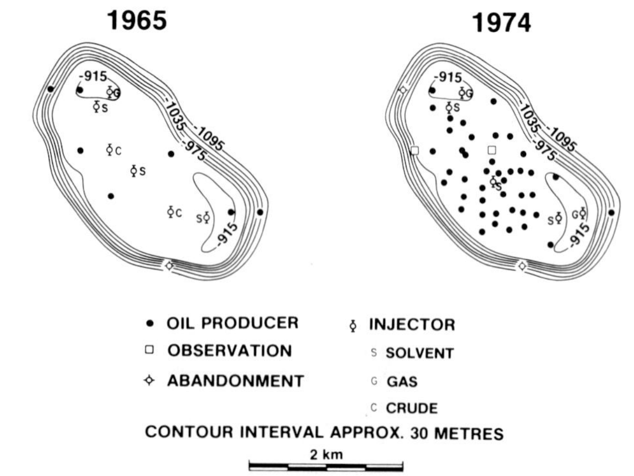

FIG. 20.—Golden Spike. Structure on reef and well status, 1965 and 1974. (After Reitzel and Callow, 1977.)

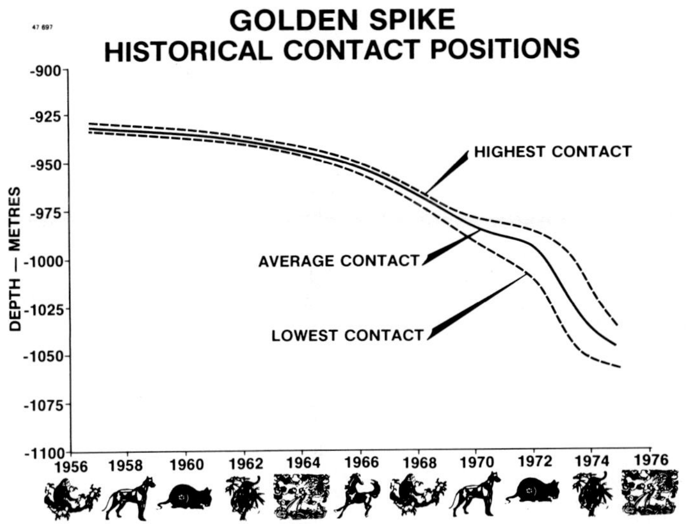

FIG. 21.—Golden Spike. Historical contact measurements. (After Reitzel and Callow, 1977.)

and by sampling through windows in the observation wells. Performance was excellent, as indicated by the gradual lowering of the contacts (Fig. 21). Starting in 1971, some distortion of the oil contact was noted, and this became progressively worse as increased production rates accelerated the gas advance. By 1973 the distortion in the contact had reached 130 ft (40 m).

The new infill wells, which had been required to meet the greatly increased production rates, provided much greater detail of the reef interior, and it was possible to map a number of thin shales and lime mudstones which obviously were acting as barriers to vertical flow (Fig. 22). Much of the distortion of the oil contact was caused by a 3- to 5-m (10–15 ft) impermeable zone which is referred to as the main barrier.

A cross section (Fig. 23) of the pool shows the observed fluid distribution at the 1973 survey. Note that at this time the gas cap was beginning to underrun the main barrier (indicated by the black line), bypassing some 1.4×10^6 m^3 (8.5 million BBL) of oil and about half of the injected solvent bank. Analyses of fluid samples indicated that the remainder of the solvent had by this time dispersed into the gas cap. A year later, in 1974, gas had completely underrun the main barrier, completely bypassing oil on top of the main barrier. Minor amounts of oil were also noted above other barriers of lesser extent, and there was no pure solvent left in the reservoir. The pool had reverted to a gas cap drive again, and gas injection was terminated in 1975. Based on the additional control now available from the infill wells, a detailed reservoir description and computer simulation study was initiated in 1974.

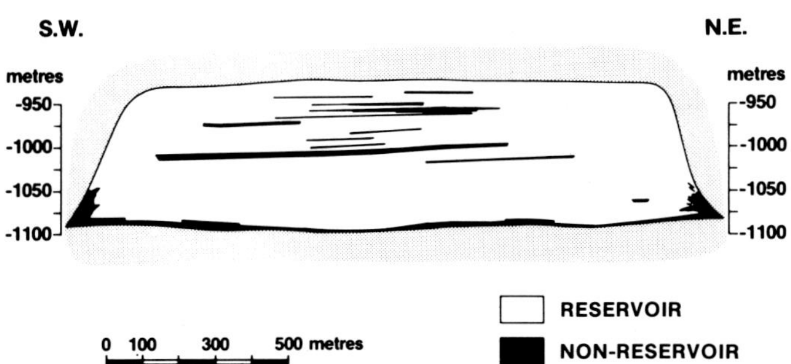

FIG. 22.—Golden Spike. Cross section showing barriers to vertical flow. (After Reitzel and Callow, 1977.)

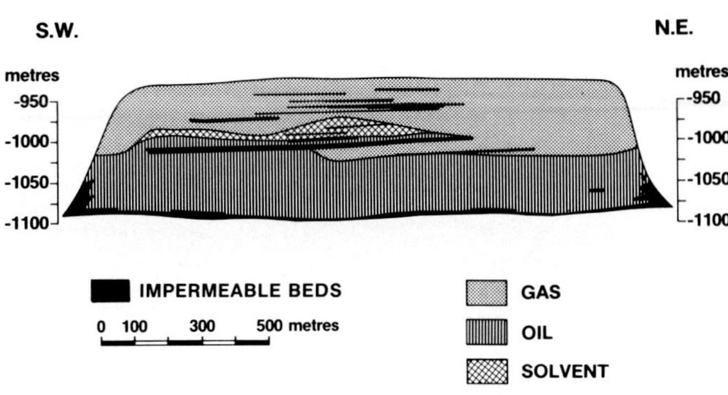

FIG. 23.—Golden Spike. Cross section showing observed fluid distribution in 1973. (After Reitzel and Callow, 1977.)

Geological reservoir description.—

The spatial relationships of the environmental facies were delineated (Fig. 24). The lower 60 m (200 ft) of the reservoir is a shallow-shoal facies deposited in intermediate water depth. Lime mudstones containing tabular stromatoporoids form the basal 15 m (50 ft) of the shoal complex. Average porosity is 5 percent, but porosity becomes ineffective along the margins. Biostromal-type patch reefs and reef detritus make up the rest of the shoal facies. Porosity averages 12 percent and is again reduced along the margins by an increase in carbonate mud.

The upper 115 m (375 ft) of the reef displays the typical atoll reef bioherm facies distribution. A peripheral rim of organic reef and reef detritus encloses a lagoonal, restricted-facies reef interior. Reef margin sediments tend to be highly variable and represent a heterogeneous, poorly bedded sequence averaging 15 percent porosity near the crest edge but becoming muddier and less porous down the foreslope, so that the porosity decreases to less than 7 percent at the toe of the reef.

Sediments of the reef interior are largely calcarenites of stromatoporoid debris averaging 15 percent porosity. These give way upward to low-energy, strongly bedded, algal laminites. The laminites have a porosity of 7 percent, but often contain thin, impermeable lime mudstones or green shale interbeds. The impermeable beds vary from 2 ha (5 acres) to more than 240 ha (600 acres) in extent. Near the base of the supratidal sequence, there is a bed of imperme-

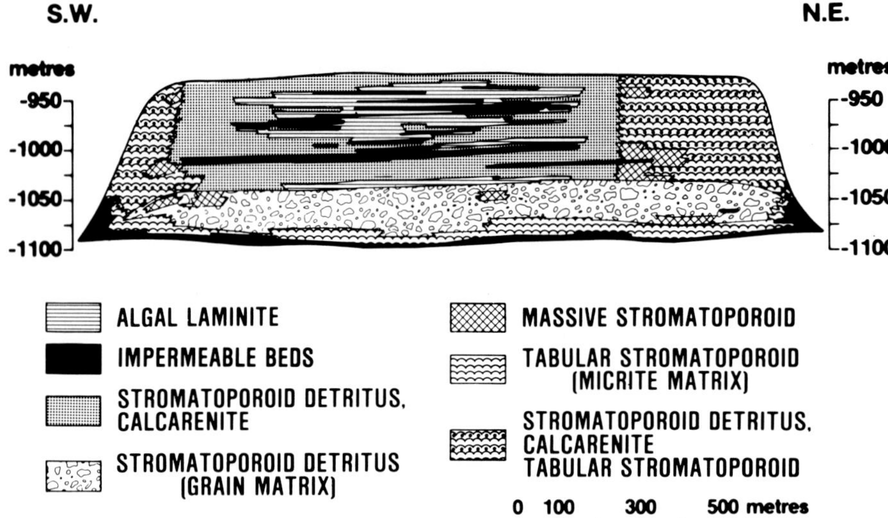

FIG. 24.—Golden Spike. Cross section showing distribution of lithofacies. (After Reitzel and Callow, 1977.)

able lime mudstone 3–5 m (10–15 ft) thick (the main barrier), which occupies most of the reef interior. It is not present in the organic reef or forereef beds. This zone represents a time of slight sea-level drop and subaerial erosion.

Permeability barriers in the reef interior, which were not apparent or significant in the early wells, later became a significant component of the reservoir description. These barriers to vertical flow, together with porosity and permeability as determined from the geological model, were incorporated into a two-dimensional cross section computer model which was used to determine the behavior of the pool.

The computer model was matched with pool performance by comparing the calculated and the observed fluid distribution with time. The behavior of the miscible fluids could not be incorporated directly into the model; therefore, oil, solvent, and gas were treated as three separate fluids that did not mix. To account for the additional recovery contributed by the miscible scheme, relative permeability curves that permitted 100% recovery of oil by solvent and 100% recovery of solvent by gas, were used, in addition to the conventional gas-oil relative permeability curves.

Computer simulation results.—

Various bases were run to test the sensitivity of pool performance to changes in reservoir parameters. The key controlling parameter was found to be horizontal permeability, since it affected the degree of warping of the contact and the amount of oil bypassed on the barriers. An excellent match to observed behavior was obtained by increasing vertical core permeabilities by a factor of two. In reefal carbonates, permeability values measured in cores may not be entirely representative of the effective permeability away from the wellbore due to the heterogeneous nature of the rock and the small size of the sampling.

The 1975 fluid distribution (Fig. 25), as predicted by the model, compares favorably with the field observations. The model indicates the presence of thin oil zones on most of the barriers, and many of these have been confirmed by analysis of contact logs or by testing. (For simplicity, most of the thin oil zones have been omitted from Fig. 25.) The simulation model also indicates that from 1967 through 1972 there was essentially complete oil recovery from the zone swept by the solvent bank. In total, 4.8×10^6 m^3 (30 million BBL) of solvent injected into the pool recovered an additional 1.6×10^6 m^3 (10 million BBL) of oil. Poor performance after 1972 was caused by reduced sweep efficiency due to the barriers and the small bank size. Dispersion of solvent into the gas cap resulted in a further reduction in flushing efficiency and led to a total loss of the solvent bank after 1973.

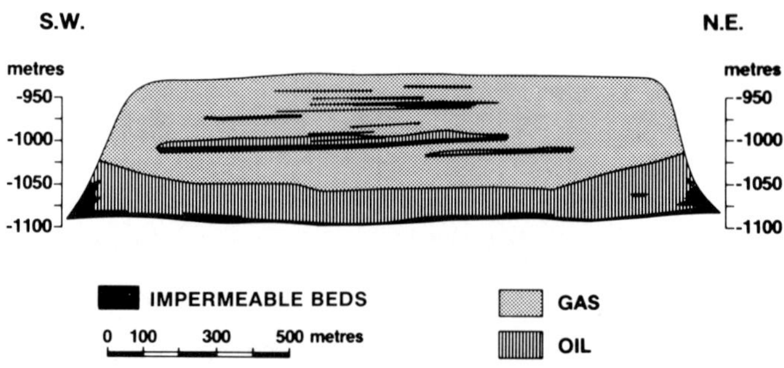

FIG. 25.—Golden Spike. Cross section showing model-predicted fluid distribution in 1975. (After Reitzel and Callow, 1977.)

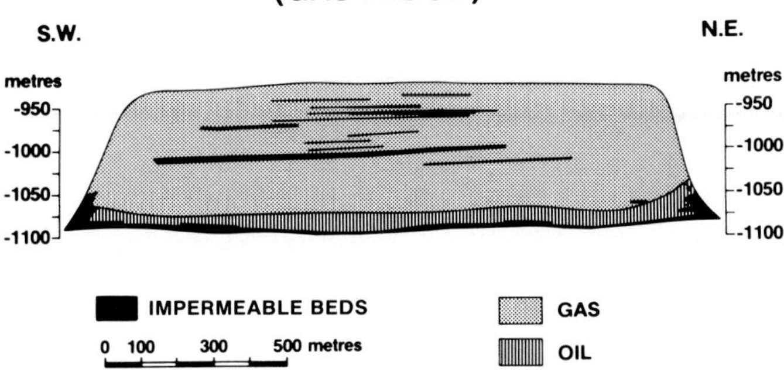

FIG. 26.—Golden Spike. Cross section showing model-predicted fluid distribution in year 2020 for gas flood. (After Reitzel and Callow, 1977.)

Future recovery will be by gas flood, and the model-predicted fluid distribution for the year 2020 (Fig. 26) indicates that oil drainage from the barriers will be almost complete, resulting in an ultimate recovery of about 65 percent of the original oil-in-place. Golden Spike field closely approaches an ideal "homogeneous" reservoir, yet the presence of a few thin, impermeable beds, aggregating only a small fraction of the total reservoir, plays a vital role in the production behavior of the pool.

Judy Creek Field

The Judy Creek reef (Fig. 27) occupies 120 km^2 (47 mi^2) and is 65 m (220 ft) thick from reef top to the platform. The reef rests on a nonporous platform and is enclosed within a sequence of interbedded shales and basinal limestones.

Porosity is best developed in the organic reef and reef detritus of the perimeter. Very good porosity also occurs in a zone of reef detritus across the top of the reef. In the stratified reef interior, porosity is discontinuous and occurs as patches and lenses. About 75 percent of the reef is filled with oil, and the downdip portion of the pool is water-bearing, but this aquifer is too small to be effective.

Judy Creek was discovered and put on production in 1959. By 1962, a peripheral waterflood had been installed to arrest the rapidly declining pressure. As pool production increased from about 4,800 m^3/d (30,000 B/D) in the late 1960s to 11,000 m^3/d (70,000 B/D) in the early 1970s, however, extreme pressure gradients developed. By early 1973, a gradient of nearly 14,000 kPa (2,000 psi) existed across the pool. At that time, reservoir pressure in the vicinity of the injection wells was about 7,000 kPa (1000 psi) above original reservoir pressure and some internal areas of the pool were below the bubble point. These factors indicated the ineffectiveness of the peripheral waterflood. Obviously, there were permeability barriers that did not allow the pressure maintenance scheme to function as expected, and detailed reservoir description studies were initiated in order to determine appropriate remedial action.

Detailed lithofacies.—

The cross section through Judy Creek reef (Fig. 28) demonstrates the complexity of lithofacies within the field. For simplicity, 11 distinct carbonate facies have been reduced to seven in this illustration. The platform beneath the reef has been divided into two sequences of non-porous limestones designated as the S1 and S2 zones. Above the platform, the reef buildup is subdivided into three stratigraphic units representing three distinct periods of reef growth and designated as the S3, S4, and S5 zones.

The S3 zone consists of a narrow, peripheral rim of organic reef with an interior lagoon of reef detrital carbonates

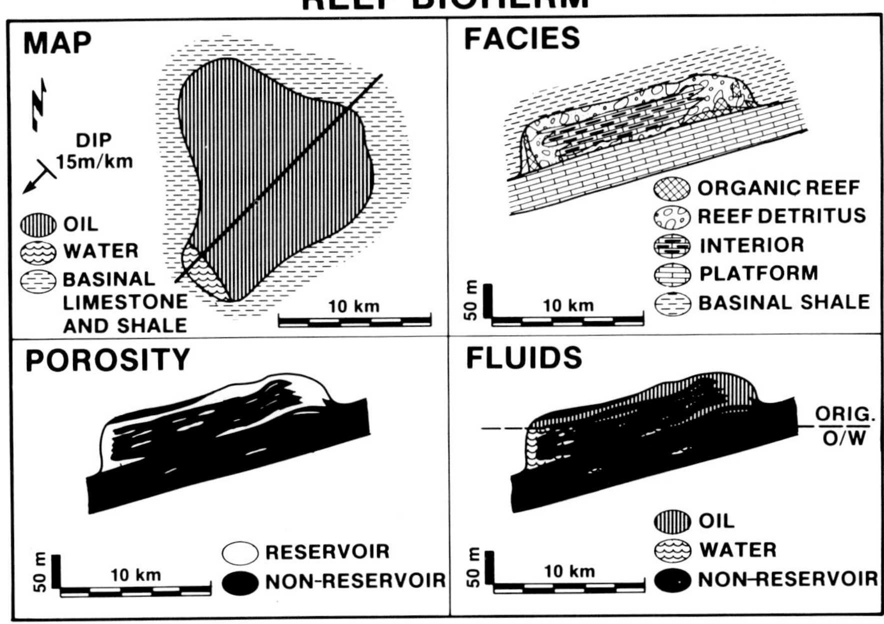

FIG. 27.—Judy Creek oil pool. Map and cross sections illustrating lithofacies, porosity, and fluid distribution. (After Jardine and others, 1977.)

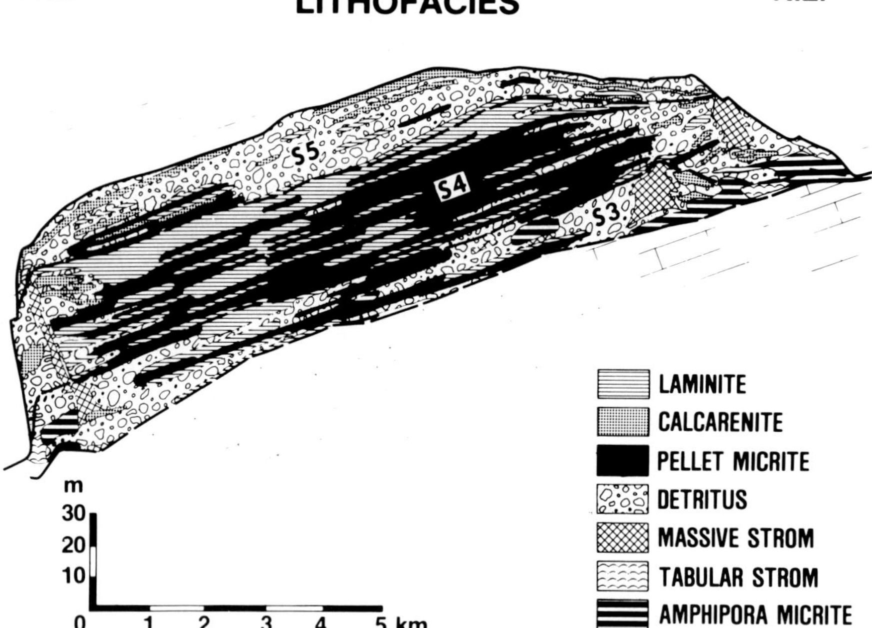

FIG. 28.—Judy Creek. Cross section showing distribution of lithofacies. (After Jardine and others, 1977.)

and lime mudstones. The S4 zone has a thick organic reef buildup and an extensive interior facies of interbedded porous and non-porous limestones. The uppermost zone, the S5, has no organic reef framework. It consists predominantly of coarse conglomeratic organic detritus and calcarenites.

Even on this simplified cross section, the distribution of facies exhibits an extremely complex stratification and lensing of rock types, particularly in the interior portion of the S4 zone.

Porosity distribution.—

The classification of porosity types within the Judy Creek reservoir may be summarized (Fig. 29). A cross plot of horizontal permeability (k_{90}) *versus* porosity resulted in a simpler subdivision of the environmental facies into three

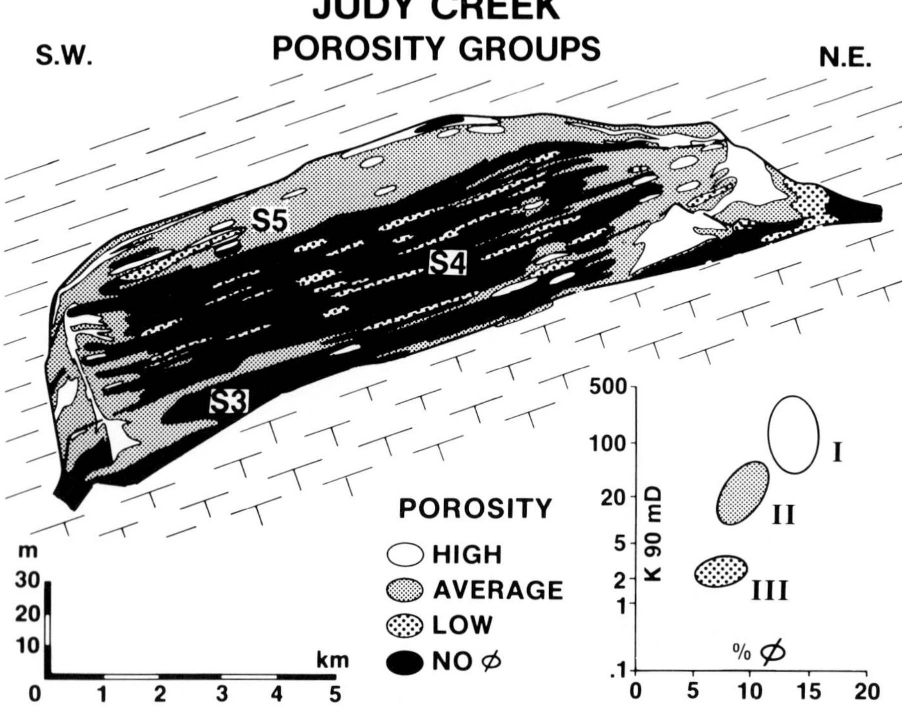

FIG. 29.—Judy Creek. Cross section showing distribution of three facies-controlled porosity groups. (After Jardine and others, 1977.)

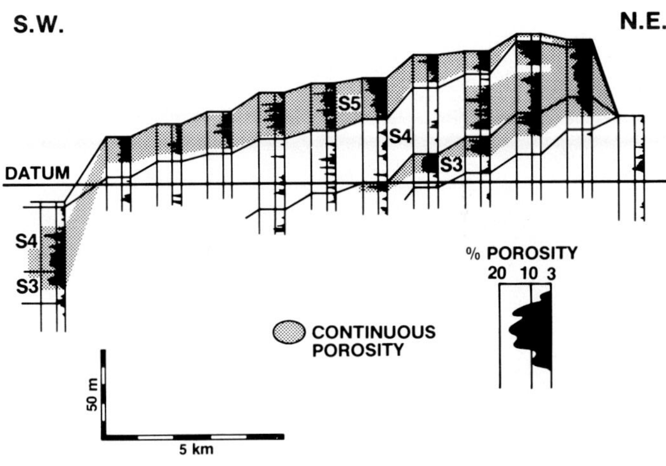

FIG. 30.—Judy Creek. Cross section illustrating continuous and non-continuous porosity. (After Flewitt, 1975.)

reservoir families, illustrated in the inset.

Type I reservoirs occur in the organic reef and shallow-shoal facies and consist of reef framework limestone and associated reef detrital limestone. These are the best reservoirs, having an average porosity of 12.5 percent and an average permeability (k_{90}) of 170 md.

Type II reservoirs are found in shoals in the fore- and backreefs where reef detrital limestone and algal laminites were deposited. Porosity averages 9.5 percent and average permeability is 40 md.

Type III reservoirs occur in the deeper water facies of the forereef and in the interior lagoonal facies. Rock types contain organic debris and pellets in a lime mud matrix. Average porosity is 6.5 percent and average permeability is 3 md. Note that the major portion of the interior lagoonal facies of the S3 and S4 zones is composed of non-porous lime mudstones.

Type I reservoirs provide vertically continuous porosity around the perimeter of the reef and laterally continuous porosity across the top. The interior of the reef is characterized by extreme stratification that results in thin porous beds that tend to be discontinuous, both laterally and vertically. These beds consist mainly of Type II and Type III reservoirs.

A grid of 47 sonic-log cross sections encompassing every well drilled was used to correlate porous beds and non-porous beds throughout the pool. A southwest-to-northeast section (Fig. 30) illustrates how the correlation was facilitated by using the facies descriptions and the simplified quantitative porosity profiles. Porous intervals which are stippled in the figure are interpreted to be in continuity either vertically or horizontally. These occur primarily in the reef perimeter and in the S5 zone. Discontinuous porous beds (not stippled) occur in the interior deposits of S3 and S4 zones. A 3% porosity cutoff was used as the lower limit for effective reservoir.

Net porosity maps.—

From the cross section grid, net porosity maps were prepared for each zone. Perhaps the most useful maps, in terms of understanding fluid flow, are those that delineate areas of continuous porosity. The stippled areas in Figure 31 indicate that the uppermost zone, the S5, has continuous porosity across the top of the reef. The S3 and S4 zones contain continuous porosity only around the periphery of the reef.

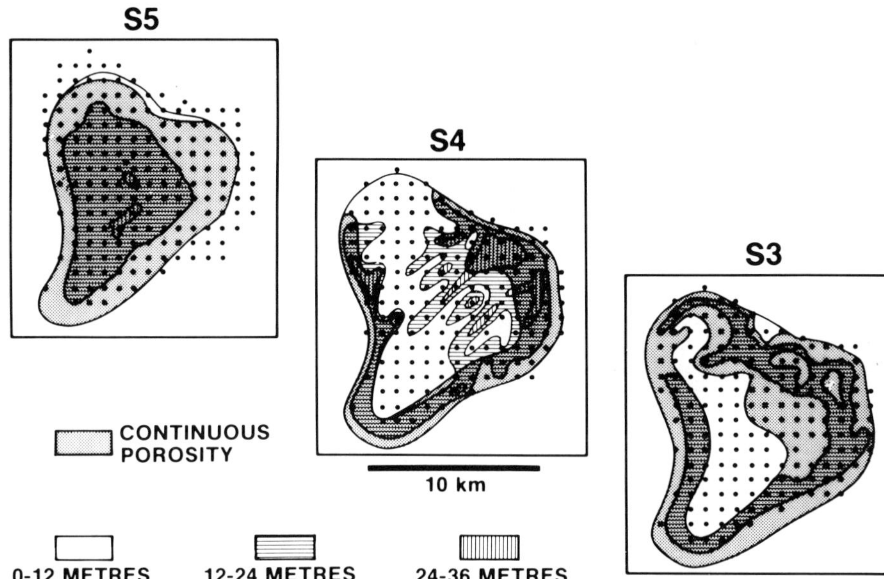

FIG. 31.—Judy Creek. Maps showing isopach net porosity in S3, S4, and S5 zones. (After Flewitt, 1975.)

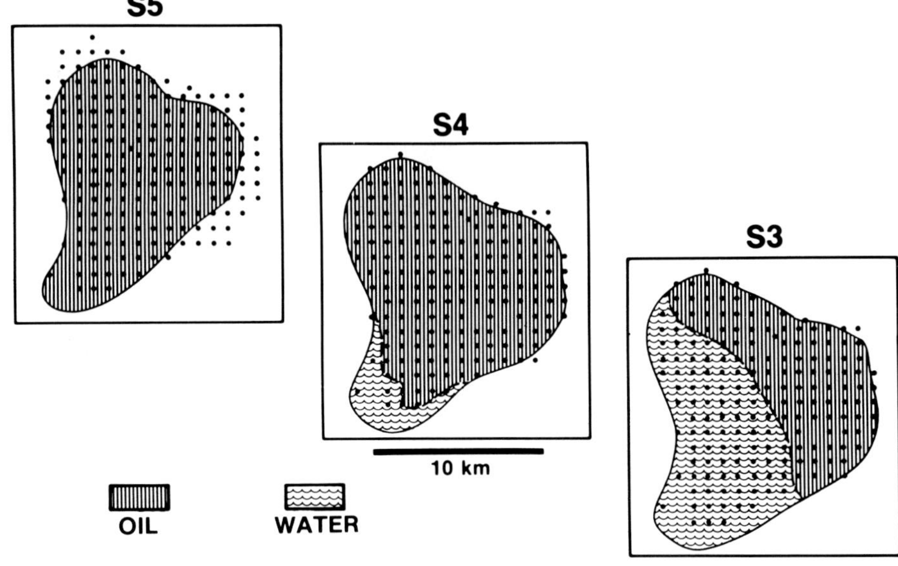

FIG. 32.—Judy Creek. Maps showing original distribution of water and oil in S3, S4, and S5 zones. (After Flewitt, 1975.)

Fluid distribution.—

The original fluid distribution within the S3, S4, and S5 zones at the time of discovery was documented (Fig. 32). The S5 was essentially water-free; the S4 had a limited downdip water leg and the S3 was about half filled with water.

The pressure maintenance scheme, which was initiated in 1962, concentrated water injection in the downdip periphery of the reef, principally into the S3 and S4 porosity. The water movement within the reef over two time periods was documented (Fig. 33): the first period extended from the start of production to 1972, and the second period is from 1972 to 1974. Injected water tended to follow the highly permeable reef rim, but the relatively impermeable interior facies prevented fluid movement to the interior of the reef. In effect, the internal regions of the S4 and S5 zones were not being swept by injection water.

The cross section in Figure 34 illustrates this discontinuity more strikingly. By 1976, water had advanced far updip in the S3 and in the upper part of the S5, which contains the highest continuous porosity in the reef. Here,

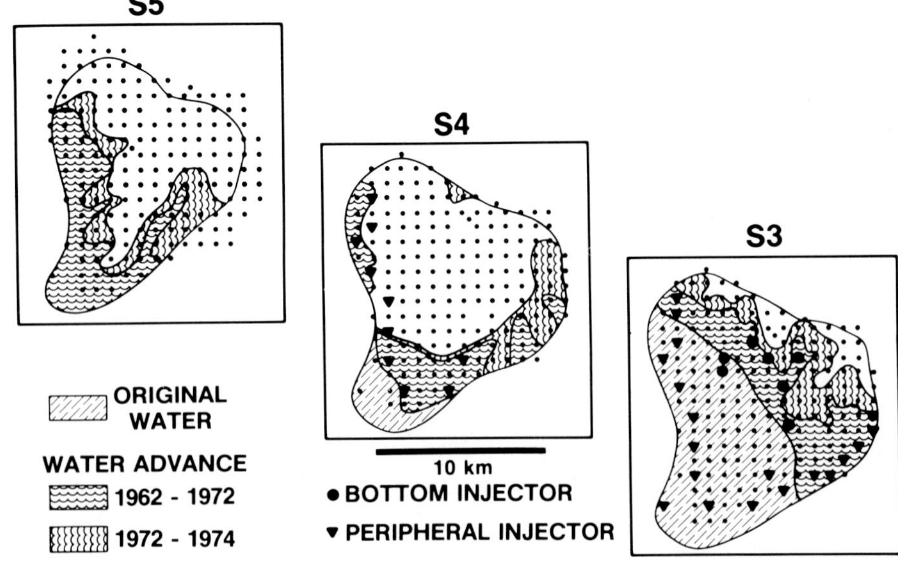

FIG. 33.—Judy Creek. Maps showing rate of water advance in S3, S4, and S5 zones under peripheral and bottom waterflood. (After Flewitt, 1975.)

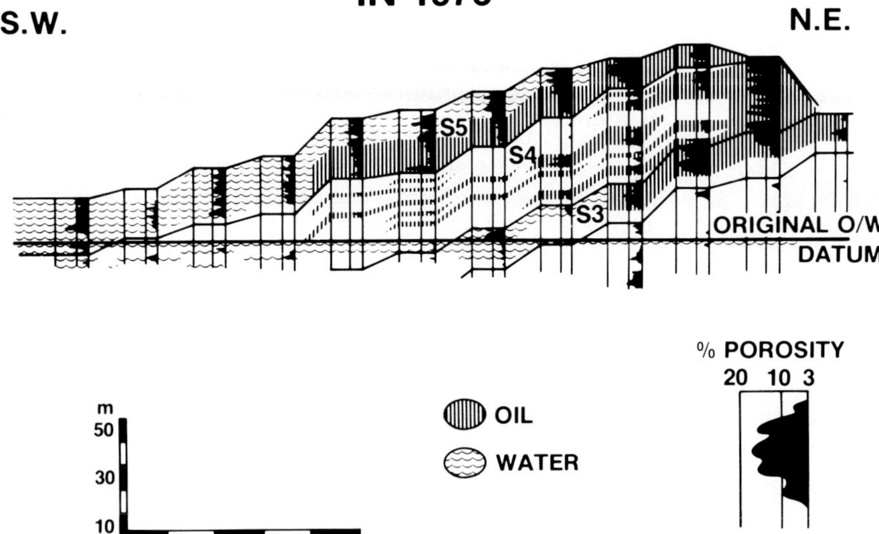

FIG. 34.—Judy Creek. Cross section showing differential water advance in 1976 of water injected downdip. (After Flewitt, 1975.)

the water had overriden the lower part of the S4, which has somewhat lower permeability. Sandwiched between these two tongues of advancing water were several zones of discontinuous porosity in the S4 which had been bypassed and still contained oil.

Pressure distribution.—

Pool isobaric maps were constructed for March 1974 and October 1975 (Fig. 35). The most prolific producing wells, located in the northeastern area of the pool and completed in the S5 zone, were experiencing severe pressure drawdowns. This area could communicate with the injectors only through the organic framework around the outside edge of the reef. Although several saltwater disposal wells were completed in the S3 in this area, injection was not effective in repressuring the S5 because of the intervening impermeable S4 sections. In addition to the poor pressure-maintenance performance of the S5, many potentially productive

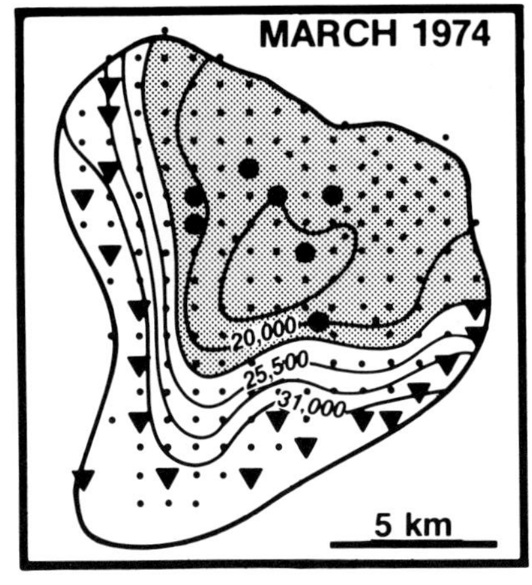

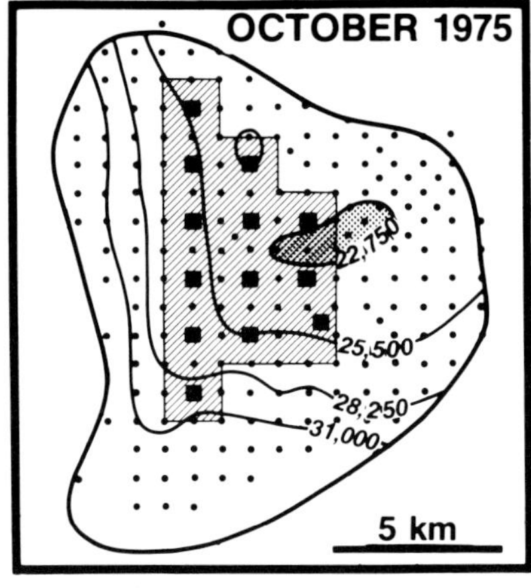

FIG. 35.—Judy Creek. Pool isobaric maps, 1974/75, showing result of conversion from peripheral to pattern flood. (After Flewitt, 1975.)

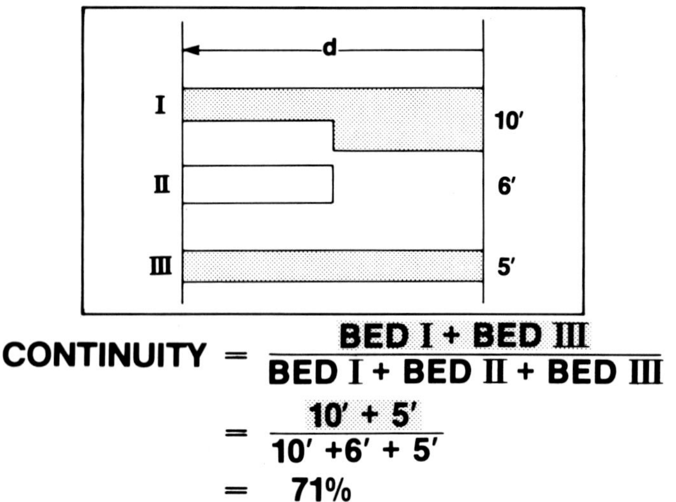

FIG. 36.—Judy Creek. Concept of porosity continuity and example of continuity calculation between well pairs. (After Delaney and Tsang, 1981.)

porous beds within the S4 had been bypassed. They were not being drained through perforations or being pressure-maintained by the injectors in the S3.

Revised depletion plan.—

The reservoir model study and the reservoir simulator studies established two main facts: (1) the best way to recover the oil in the S5 was through pattern flood, and (2) the discontinuous porous beds in the interior of the S4 were not being waterflooded effectively.

The pool depletion strategy was then revised and included the following steps:

(1) Thirteen wells in the central part of the pool were converted to water injection wells, and water was injected into all porous intervals in the S3, S4, and S5 zones. In addition, all producing wells in the pattern waterflood area were reperforated in all zones, and high-volume pumps were installed.

(2) Wells behind the flood front, and previously deemed to be flooded out (although not all porous zones had been opened up), were recompleted and reactivated. Many reactivated wells are currently producing more than 16 m^3/d (100 B/D), and these are expected to recover more than 1.6×10^6 m^3 (10 million BBLS) of oil. Reactivation of an additional 20 wells was planned.

The pattern waterflood, which was placed in operation in 1974 (Fig. 35), shows the dramatic improvement in reservoir pressure approximately 1 year later. After the reservoir pressure was brought under control, injected and produced fluid balances in all zones were maintained at approximately the original reservoir pressure of 24,000 kPa (3,500 psig). Water volumes to the peripheral injectors were also cut back to prevent overpressuring in these areas.

Continuity studies.—

Additional reservoir continuity studies were undertaken in the late 1970s. These were designed to analyze current production performance, including the expected results from infill drilling, and to investigate the feasibility of a miscible flood tertiary recovery scheme. Statistical techniques for quantifying reservoir continuity have been described in the literature (Ghauri and others, 1974; Stiles, 1976).

A simplified geological model can be developed to show the porous beds between a pair of wells (Fig. 36). In practice, a greater degree of intertonguing could be expected, but this would be very difficult to represent mathematically. The distance between the wells is represented by '*d*'. Beds I and III are continuous and Bed II is discontinuous. Continuity between any two wells is defined as the ratio between the summation of the continuous pore volume and the total pore volume. The relationship of volume-to-bed thickness may be designated by a formula (Fig. 36). Note

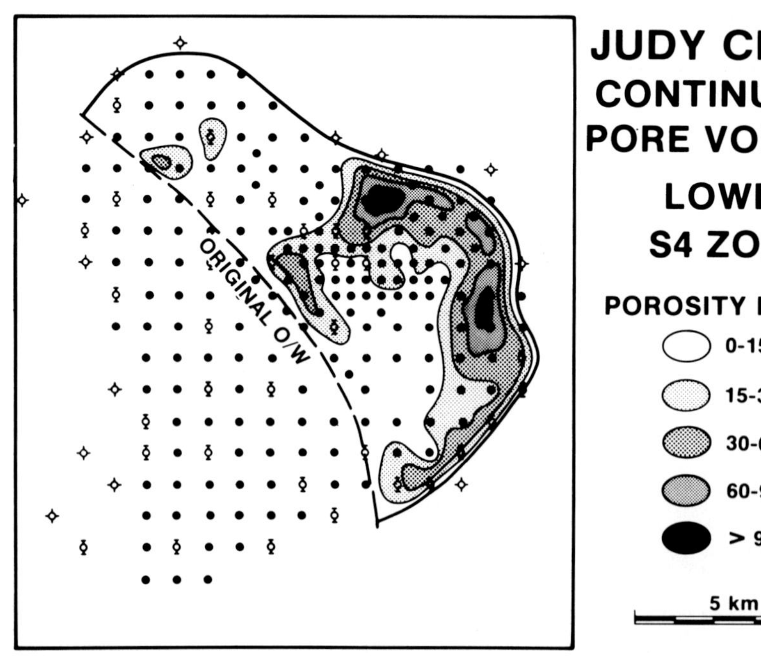

FIG. 37.—Judy Creek. Continuity map of lower S4 zone. (After Delaney and Tsang, 1981.)

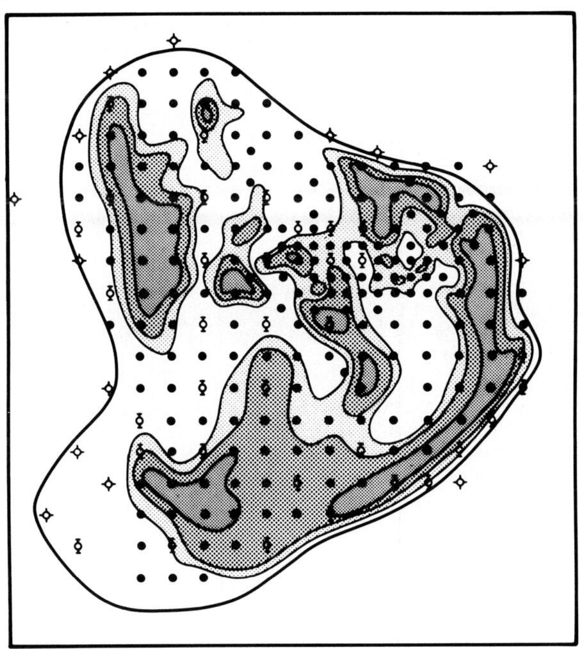

FIG. 38.—Judy Creek. Map showing continuous pore volume, lower S4 oil zone. (After Delaney and Tsang, 1981.)

that the thicker side of Bed I—3 m (10 ft)—is used in the formula, but any fraction of this could be used in the calculation of continuous volume. Analysis at Judy Creek suggested this refinement was unnecessary.

Porosity data, derived by computer analysis of all logs run during the life of the pool (sonic, neutron, and cased-hole CNL) and cross-checked with core analysis, were input into the computer model, together with carefully selected time-stratigraphic geological markers. These provided a detailed porosity analysis of six reservoir zones—the S3, the lower, middle, and upper S4, and the lower and upper S5, each subdivided into 0.6-m (2 ft) slices.

The computer was used to generate quantitative porosity cross sections and graphs of continuity *versus* interwell distance. These were useful in calculating the potential for various possible infill drilling locations. Continuity maps were also mathematically derived and an example of the lower S4 is shown in Figure 37. By cross-multiplying the continuity with the total pore volume, continuous pore-volume maps were also generated (Fig. 38). This type of map, when compared with the estimated flood front for this zone, can be used to monitor behavior and pinpoint potential anomalies for further study.

The large complex data base now available for Judy Creek allows for the detailed analysis of geologically well described, individual surveillance areas which approximate the existing waterflood patterns. Recovery prediction methods can then be applied to each area to calculate its future performance under various depletion strategy alternatives.

CONCLUSIONS

(1) Carbonate reservoirs commonly display a high degree of heterogeneity both laterally and vertically. This is caused by the wide variety of environments in which the rocks were deposited and by the susceptibility of the rocks to diagenetic alteration.

(2) Reservoir simulation models are necessary for complete understanding of the producing characteristics of complex reservoirs. A key input to these models must be provided by the geologist in the form of the vertical and lateral distribution of the reservoir rock types, reservoir rock characteristics, reservoir continuity, and barriers to flow.

(3) Reservoir description is not static, but must be upgraded continuously as depletion of the reservoir proceeds and as new data become available. Performance of individual wells must be monitored to determine fluid migration paths and their effect on pressure patterns within the reservoir. Simulation models should be used to determine the future behavior of the pool under various depletion programs.

(4) For most effective reservoir management, a multidisciplinary holistic approach is essential. The study team should include reservoir geologists, geophysicists, petrographers, and petrophysicists, as well as reservoir, production, and completion engineers.

(5) Optimized depletion plans translate into greater profitability for the operator and more efficient utilization of our increasingly valuable energy resources.

ACKNOWLEDGMENTS

The authors thank Esso Resources Canada Limited and the Society of Petroleum Engineers for permission to publish this paper. Grateful thanks are also due R. P. Glaister for supplying and describing the core photographs of Figures 4 and 6. The material for this paper has been adapted freely from the files of Esso Resources Canada Limited (formerly Imperial Oil Limited) and from various papers prepared by members of Esso Resources Canada Limited staff.

The authors are deeply in debt to the organization and to the numerous authors listed for their gracious permission to allow much of their material to be transcribed and for the help of many members of Esso Resources Canada Limited staff in the preparation of this paper. In particular, the stimulating discussions pertaining to characteristics of carbonate environments with Messrs. J. C. Wendte, R. P. Glaister and F. A. Stoakes must be recognized. Last and not least, the authors gratefully acknowledge the assistance of Esso Resources Canada Limited drafting staff for the preparation of the illustrations.

REFERENCES

DELANEY, R. P., AND TSANG, P. B., 1981. Reservoir continuity study at Judy Creek. A synergy study: 32nd Annual Technical Meeting of the Petroleum Society of Canadian Institute of Mining and Metallurgy Calgary, Alberta, Preprint 81-32-5, p. 38–44.

FLEWITT, W. E., 1975, Refined reservoir description to maximize oil recovery: Paper presented at Society of Professional Well Log Analysts Sixth Annual Logging Symposium, New Orleans, Louisiana, U.S.A., 22 p.

GHAURI, W. K., OSBOURNE, A. F., AND MAGNUSON W. L. , 1974, Changing concepts in carbonate water flooding, West Texas Denver Unit Project—an illustrative example: Journal of Petroleum Technology, v. 26, p. 595–606.

JARDINE, D. J., ANDREWS, D. P., WISHART, J. W., AND YOUNG, J. W., 1977. Distribution and continuity of carbonate reservoirs: Journal of Petroleum Technology, v. 29, p. 873–885.

———, AND WILSHART, J. W., 1982. Carbonate reservoir description: International Petroleum Exhibition and Technical Symposium of the Society of Petroleum Engineers, Bejing, China, March 18–26. SPE Preprint 10010, p. 43–110.

KEMPTHORNE, R. H., AND IRISH, J. P. R., 1980. Norman Wells—a new look at one of Canada's largest oil fields: 55th Annual Fall Technical Conference and Exhibit of Society of Petroleum Engineers of American Institute of Mining, Metallurgical and Petroleum Engineers, Dallas, Texas, SPE Preprint 9477, 13 p.

REITZEL, G. A., AND CALLOW, G. O., 1977. Pool description and performance analysis leads to understanding Golden Spike's miscible flood: Journal of Petroleum Technology, v. 29, p. 867–872.

STILES, L. H., 1976. Optimizing water flood recovery in a mature waterflood, the Fullerton Clearfork Unit location: 51st Annual Fall Meeting. Society of Petroleum Engineers of American Institute of Mining, Metallurgical and Petroleum Engineers, New Orleans, Louisiana, LA, SPE Preprint 6198, 12 p.

THE STUDY OF NATURAL FRACTURES IN A REEF COMPLEX, NORMAN WELLS OIL FIELD, CANADA

J. P. R. IRISH AND R. H. KEMPTHORNE[1]

Esso Resources Canada Ltd., Esso Plaza, 237 4th Avenue SW, Calgary, Canada T2POH6

ABSTRACT: Hydrocarbons at Norman Wells, Northwest Territories, Canada, are pooled in the Devonian Kee Scarp Formation limestone reef buildup which is 15 mi (25 km) long and 5 mi (8 km) wide. The reef complex began to develop during Upper Middle Devonian time on shale banks of the Hare Indian Formation. Minor variations in sea level localized major reef growth on a thin, open marine platform along basinward edges of the Hare Indian shales. Reef growth proceeded during two major shoaling-upward cycles and was terminated by the advancing clastic sediments of the Canol and Imperial Formations, which serve as the top seal for the reefal buildups. The Kee Scarp reef at Norman Wells attains a maximum thickness of 535 ft (160 m) with a reservoir thickness of 360 ft (110 m). The formation has been divided into four zones, named in ascending order: K1, K2, K3, and K4. The K1 zone is interpreted as platform, and the K2–K4 zones as reefal buildup.

In 1920, Imperial Oil Limited drilled the discovery well for the Norman Wells field near oil seepages at Norman Wells 900 miles (1,500 km) north of Edmonton, Alberta. Oil is stratigraphically trapped in the extreme updip end of the reef buildup. There is no gas cap, and a limited water leg provides no fluid pressure support to assist in producing the oil.

The major development plan for producing a portion of the remaining oil at Norman Wells field is discussed in this paper. The plan is based on a recent synergistic geological and engineering study of the reservoir. Information from six oriented cores, well tests, fluid and pressure surveys, outcrop studies, and surface photos has shown that natural vertical fractures present in the reservoir have a strong N30° E orientation. The proposed development plan will make use of the geological and engineering model as well as the directional permeabilities resulting from natural fractures.

INTRODUCTION

Oil sands and hydrocarbon seepages along the MacKenzie River in the Northwest Territories have been reported since the late 1700s by explorers and fur traders. In 1920, Imperial Oil Limited drilled the discovery well near oil seepages at Norman Wells 900 mi (1,500 km) north of Edmonton, Alberta (Fig. 1). The remote location of the field delayed development until the early 1940s when, under the Canol Agreement, the United States Army, Canada, and Imperial Oil drilled a total of 60 wells, expanded the existing refinery and constructed a road and pipeline to Whitehorse 620 mi (1,000 km) to the southwest. Pool production peaked at 4,400 B/D (700 m^3/d) in 1944. The pipeline to Whitehorse was dismantled after 1945, but the pool has continued to supply northern Canada with oil. Production increased to 3,000 B/D (480 m^3/d) by 1982.

Total pool production to the end of 1980 under primary solution gas drive was less than 4% of the estimated 630 million bbl (100×10^6 m^3) original oil-in-place. The MacKenzie River overlies the bulk of the reef reservoir and, since the reservoir is relatively shallow, at a depth of 1,000 to 1,600 ft (300 to 500 m), development has been limited to three producing areas known as the Mainland, Goose Island and Bear Island (Fig. 2). Typical well oil production rates have declined by 50% to a range of 60 to 500 B/D (10 to 80 m^3/d). The present gas-oil ratios (GOR) have escalated to between 2,000 to 4,000 cu ft/bbl (350 to 700 m^3/m^3) from a solution GOR of 335 cu ft/bbl (60 m^3/m^3).

Activity has increased significantly since 1979 with the drilling of seven delineation wells from grounded river ice platforms and land areas, seven highly deviated wells for water injection and production, and one "horizontal" well intersecting 4,000 ft (1,200 m) of reservoir rock from the Mainland to the central pool area under the river.

A line drive waterflood employing 13 wells was started in 1980 for the Mainland area, and an extensive monitoring system has been installed to measure fluid production and injection, gas-oil ratios, and pressures for individual wells and zones (Kempthorne and Irish, 1980). In March 1980 an application was made to the Canadian Federal Department of Indian Affairs and Northern Development for approval to expand the Mainland waterflood to a fieldwide pattern waterflood system. Government approval to proceed was received in 1981, and a 560-mi (900 km) pipeline was started in 1983 for pool expansion start-up in 1985.

RESERVOIR GEOLOGY

Introduction.—

Since the Norman Wells pool is still in a stage of development (Table 1), it is important to recognize that additional data and more detailed work may modify some of the descriptions and interpretations reported in this paper. Although the core and borehole log information used in this study was of good quality and was well distributed across the pool, the details of this assessment must be considered preliminary.

The following description of the vertical and lateral changes within the limestone rocks that compose the reef uses a classification by Embry and Klovan (1971), which is included for reference in Figure 3. These authors expanded the widely used classification by Dunham (1962) to differentiate further between organically bound and non-organically bound limestones. Part of this classification modification recognizes that limestone conglomerates are important rock types. These conglomerates are defined as rocks containing greater than 10% > 2-mm components and are separated into floatstone and rudstone. In a rudstone, greater than 2-mm particles form the supporting framework, whereas in a floatstone they "float" in the finer grained matrix.

Stratigraphy.—

The Kee Scarp Formation developed in Upper Middle Devonian time on shale banks of the Hare Indian Formation

[1]Present address, Petro Canada Resources, Calgary, Alberta, Canada.

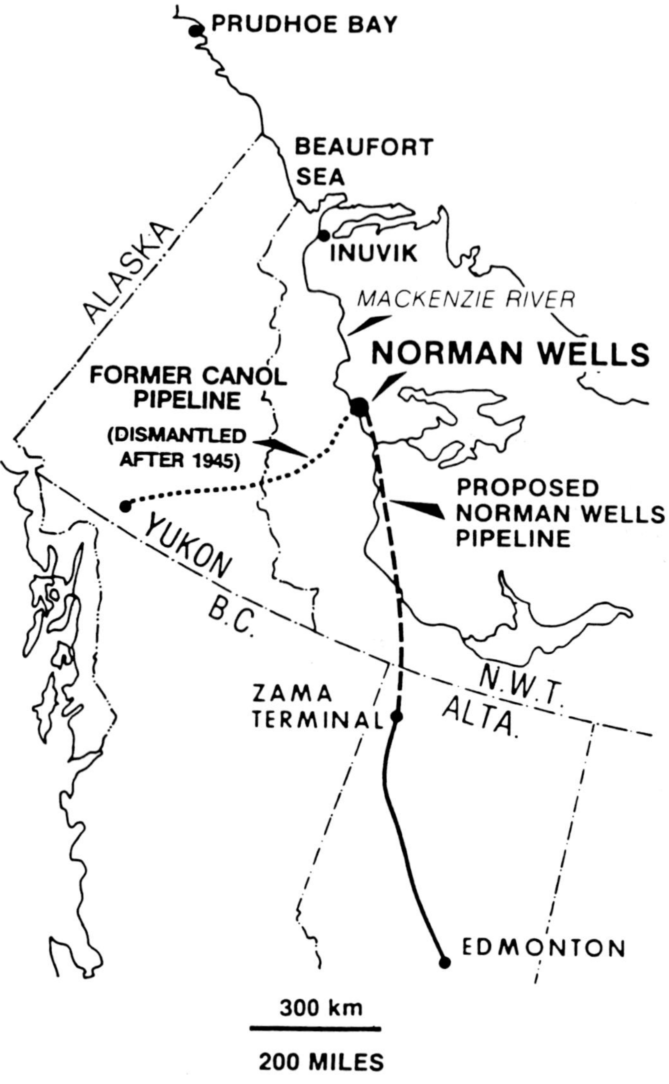

FIG. 1.—Norman Wells project location.

(Fig. 4). At the formation Norman Wells has been divided into four zones, named in ascending order: K1, K2, K3, K4. These are illustrated on a type log of the pool in Figure 5. Shaded portions on the sonic log indicate effective porosity. The Kee Scarp reef at Norman Wells is as thick as 525 ft (160 m) and the reservoir attains a thickness as great as 360 ft (110 m). Regional structural dip is 4-1/2° to the southwest. The K1 zone represents the platform, and the K2–K4 zones represent reefal buildup.

The reefal platform or K1 zone is characterized by more open marine conditions and has a higher argillaceous content than the overlying carbonates. It is generally described as a rudstone or less frequently a floatstone with a medium to dark brown skeletal packstone-to-wackestone matrix. There is a gradation upward from rudstones and floatstones dominated by stubby finger corals at the basal or deepest part of the platform to a rudstone dominated by cylindrical stromotoporoids at the top. Encrusting and tabular corals are also numerous in the coral-prone basal portion. Stylolitization and burrowing are common throughout the whole interval. Sections of crinoid columnals and brachipods are minor but always present.

Porosity development within the platform is very poor except for one 15–30 ft (5–10 m) widespread, uniform, porous interval at its top. Here, alteration was sufficient to create porosities of as much as 10%, although permeabilities remain very low. The rocks in this interval are packstones, and the porosity increase can be attributed to an increase in grain content in the matrix. The platform averages 100 ft (30 m) in thickness.

The reefal buildup consists of two major cycles. Each cycle was caused by a relative rise in sea level, and most of the sedimentation in each cycle took place after the rapid rise which initiated the cycle. The recognition of these and other minor cycles in core and how they are reflected on borehole logs formed the basic framework for the zonation of the reef. The first major cycle, or K2 zone, attains an average thickness of 165 ft (50 m); the second cycle encompasses the K3 and K4 zones and is 195 ft (60 m) thick. Each cycle has a lower portion of stromatoporoid thicket, resulting from widespread open marine conditions, and an upper section with distinct lateral zonation including lagoon, reef flat sand, reef margin, and foreslope sediments.

The K4 is composed of 33 ft (10 m) of reef deposits and is distinguished from the underlying K3 by the significantly higher primary porosities and correspondingly higher permeabilities.

The zone of stromatoporoid thicket marks the first accumulation of the shoaling-upward reef growth cycle after a deepening phase. It is a light to dark brown cylindrical stromatoporoid rudstone with a packstone-to-grainstone matrix (Fig. 6). The cylindrical stromatoporoids at the base of the first cycle are distinctly smaller than those in the second growth cycle, possibly indicating relatively deeper conditions. In addition, there is a finer micritic matrix which allows only minor alteration and therefore low porosity at this level. Crinoids are present along with minor tabular and irregular stromatoporoids, corals and bivalves. Alteration is variable but is usually extensive in the matrix and cylindrical stromatoporoids. All primary porosity has been filled with sparry calcite cement.

Lagoonal sediments are mainly *Amphipora* rudstones and floatstones with a light to medium brown wackestone-to-packstone matrix. The grains are fine, skeletal and occasionally pelletal. *Amphipora* stromatoporoids are common and locally abundant (Fig. 7). Some bulbous and cylindrical stromatoporoids, gastropods and bivalves are present. Other features which are present include calcispheres and algal coatings on stromatoporoids. Alteration ranges from a variable alteration of the matrix, giving a light and dark brown mottled appearance with no alteration of the *Amphipora,* to an extensively altered matrix with partly altered *Amphipora*. Sparry calcite cement fills all original shelter and interparticulate porosity. Wispy stylolites are common in this facies as well as other facies.

Some lagoonal sediments culminate in the deposition of a tidal flat facies. These rocks are generally pelletal mudstones to packstones. Stromatolitic algal laminations and laminar, tubular, and irregular fenestrae are common features and indicate deposition near mean upper tide level.

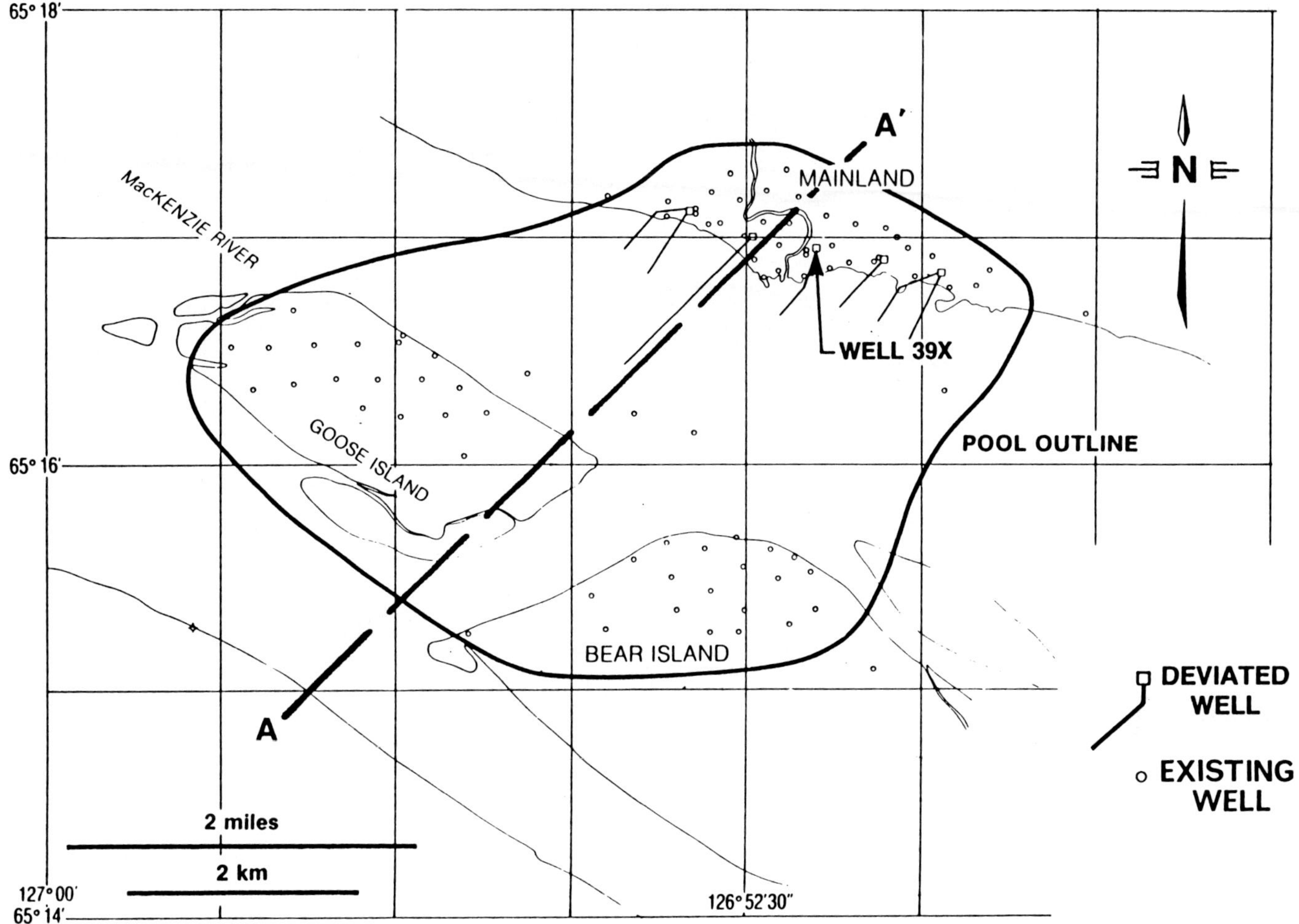

FIG. 2.—Normal Wells pool outline, Northwest Territories, Canada.

The fenestrae are invariably filled with sparry calcite (Fig. 8).

The reef flat carbonate sand facies is one of the better reservoir rock types in the reef comlex. It is typically a light brown, medium-grained, skeletal, grainstone to packstone. Grain size varies locally from fine to coarse sand. Some large grains such as cylindrical and bulbous stromatoporoids, *Amphipora* and brecciated stromatoporoid fragments can be present. There is extensive alteration of the small grains and micrite and variable alteration of the large stromatoporoids. Generally, sparry calcite fills the primary pore space; however, some primary intergranular porosity is preserved in some instances, especially in the upper portions of the reef complex.

Reef margin rocks are dominated by interlayered and thick

TABLE 1.—NORMAN WELLS RESERVOIR CHARACTERISTICS

Average Reservoir Depth (m)	400
Average Horizontal Permeability from core (md)	110
Average Porosity (%)	9.8
Reservoir Temperature (°C)	21
Initial Reservoir Pressure, kPa (−345m ss)	4,900
Bubble Point Pressure (kPa)	4,700
Pour Point (°C)	−40
Crude Gravity (°API)	38.5
Viscosity of Crude (21° C, 4,900 kPa) kPa.S	1.32
Average Horizontal Permeability from Core (md)	4
Average Vertical Permeability from Core (md)	1
Connate Water Saturation (%)	10
Residual Oil Saturation (%)	31
End-point Water Relative Permeability	0.4
Critical Gas Saturation (%)	2

CLASSIFICATION OF CARBONATE ROCK

<table>
<tr><td colspan="6">ALLOCHTHONOUS LIMESTONES ORIGINAL COMPONENTS NOT ORGANICALLY BOUND DURING DEPOSITION</td><td colspan="3">AUTOCHTHONOUS LIMESTONES ORIGINAL COMPONENTS ORGANICALLY BOUND DURING DEPOSITION</td></tr>
<tr><td colspan="4">LESS THAN 10% >2mm COMPONENTS</td><td colspan="2">GREATER THAN 10% > 2mm COMPONENTS</td><td rowspan="4">BY ORGANISMS WHICH ACT AS BAFFLES</td><td rowspan="4">BY ORGANISMS WHICH ENCRUST AND BIND</td><td rowspan="4">BY ORGANISMS WHICH BUILD A RIGID FRAMEWORK</td></tr>
<tr><td colspan="3">CONTAINS LIME MUD (<.03 mm)</td><td>NO LIME MUD</td><td rowspan="3">MATRIX SUPPORTED</td><td rowspan="3">> 2mm COMPONENT SUPPORTED</td></tr>
<tr><td colspan="2">MUD SUPPORTED</td><td colspan="2" rowspan="2">GRAIN SUPPORTED</td></tr>
<tr><td>LESS THAN 10% GRAINS (>.03mm <2mm)</td><td>GREATER THAN 10% GRAINS</td></tr>
<tr><td>MUD-STONE</td><td>WACKE-STONE</td><td>PACK-STONE</td><td>GRAIN-STONE</td><td>FLOAT-STONE</td><td>RUD-STONE</td><td>BAFFLE-STONE</td><td>BIND-STONE</td><td>FRAME-STONE</td></tr>
</table>

FIG. 3.—Modified Dunham carbonate rock classification (Embry & Klovan, 1971).

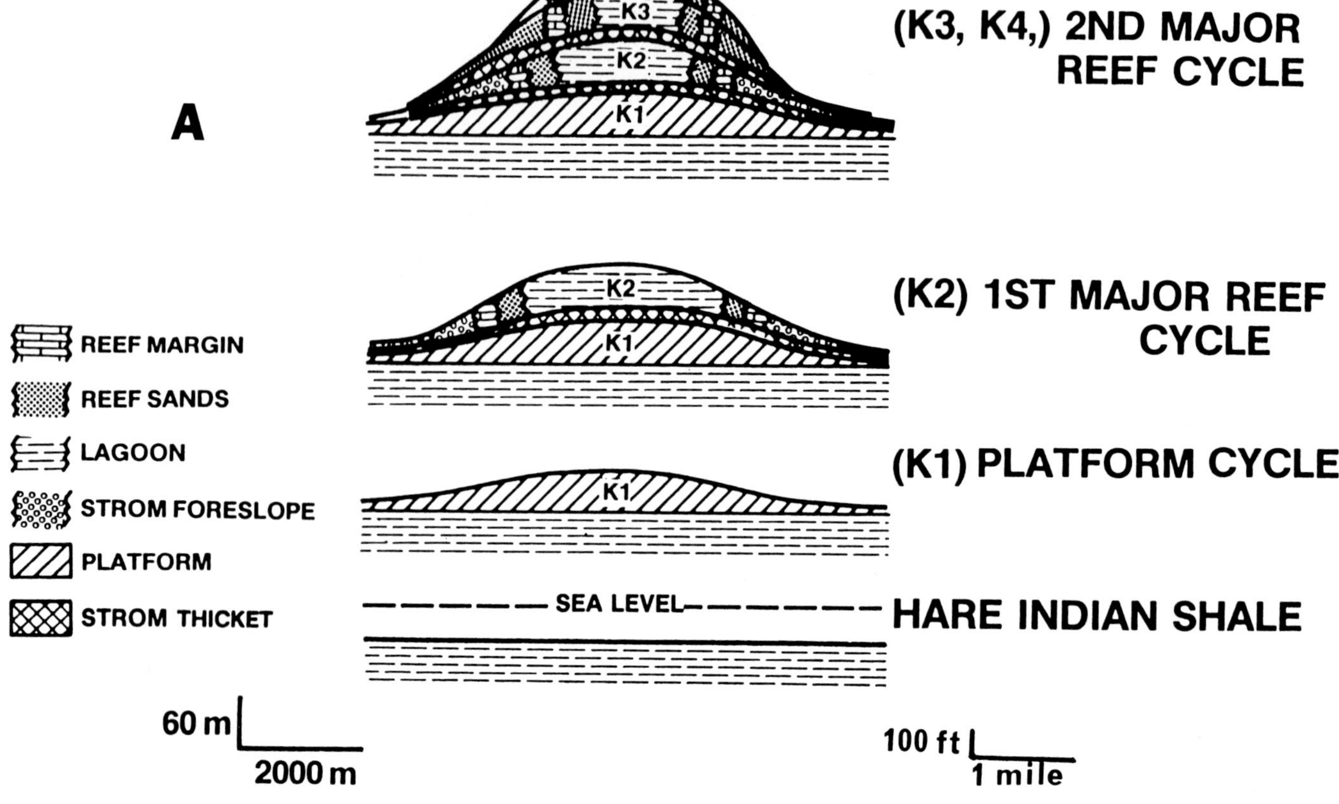

FIG. 4A.—Schematic reef evolution.

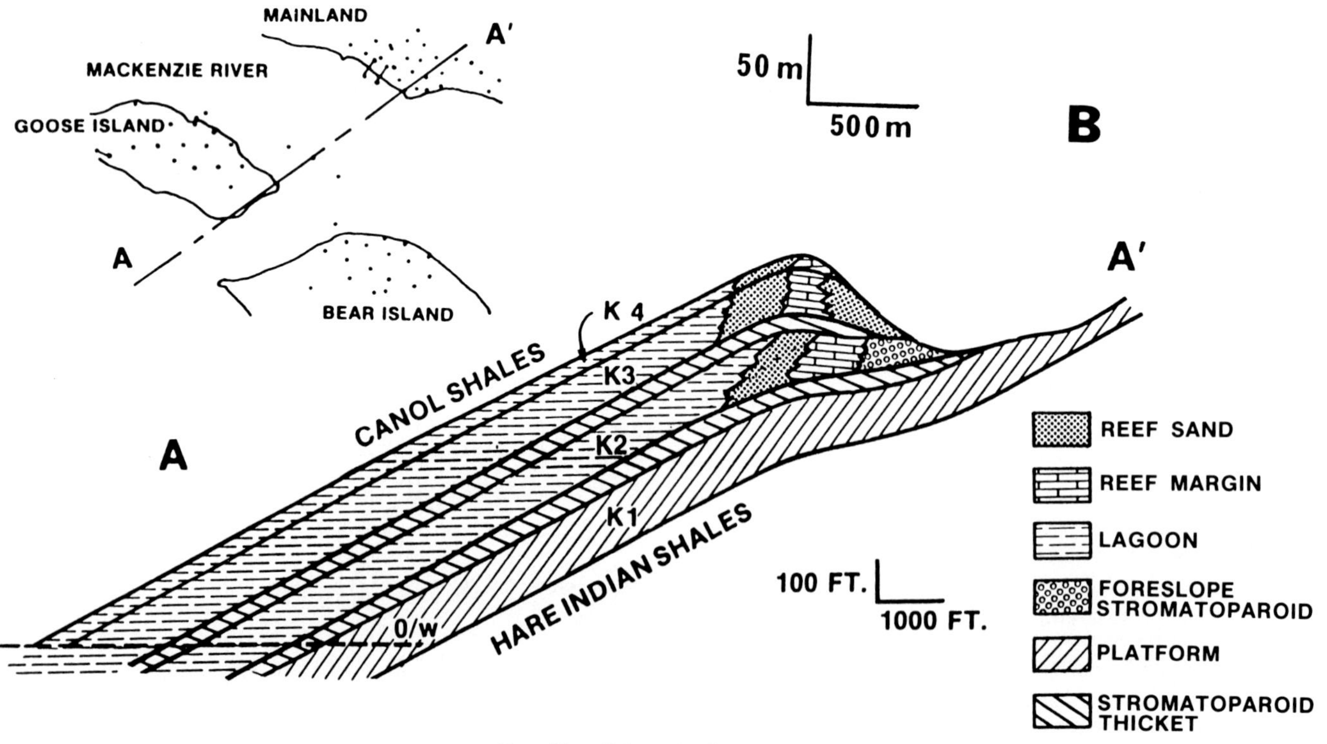

FIG. 4B.—Schematic dip section.

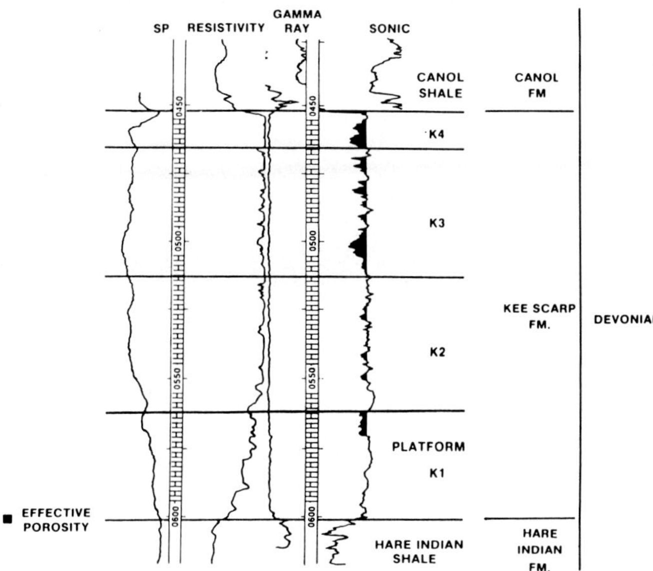

FIG. 5.—Reservoir zonation—type log.

tabular stromatoporoid boundstones as much as 25 cm in thickness (Fig. 9) and brecciated stromatoporoid rudstones. Cylindrical stromatoporoids are common, and corals and crinoids are minor constituents. The matrix is a light brown skeletal grainstone or packstone with only very minor micrite. Again, sparry calcite fills most primary interparticle and growth framework porosity. Occasionally, some primary cellular porosity is present in the more massive stromatoporoids. There is extensive alteration of small grains and stromatoporoids.

Two general rock types typify the foreslope facies, although many variations of the two are found. The first is a medium brown to dark grey, cylindrical and tabular stromatoporoid rudstone (Fig. 10). Irregular, bulbous and wafer stromatoporoids occur as well as stromatoporoid breccia. Minor tabular stromatoporoid boundstone is found in scattered places. The matrix can vary from a grainstone to a packstone. In the lower parts of the foreslope corals, rudstones appear more frequently. They include finger corals (*Thamnopora*) and horn corals as well as some encrusting, tabular and irregular corals. Mixed with the corals are scattered cylindrical stromatoporoids, bivalves and crinoid fragments. Skeletal sands form the second general rock type in this facies (Fig. 11). They are light brown grainstones to packstones with scattered cylindrical and irregular stromatoporoids and minor crinoids. Sparry calcite has filled most of the original foreslope pore space, but occasionally interparticle porosity is preserved.

Porosity and permeability.—

The Norman Wells reservoir has very low matrix permeability. Whereas porosity can reach 24%, permeability is generally restricted to 10 md or less. Crossplots of core data show that for given porosity, horizontal permeability is two to four times as great as vertical permeability. Extensive microleaching of the matrix, small grains, and stromatoporoids has created a fine intercrystalline ("chalky") porosity throughout most of the reservoir. Primary porosity is less extensive because initial voids are largely cemented by sparry calcite. Where primary porosity does occur, it is important as it is accompanied with as much as fivefold increases in permeability, particularly in the structurally higher portions of the reef buildup.

Porosity and permeability also vary with facies type, but the boundaries of such differences overlap and are indistinct. In general, the margin area, which includes carbonate sands, reef framework and foreslope, exhibits higher porosity and permeability than the lagoonal and more open marine sediments.

Fracturing.—

The presence of fractures at Norman Wells has been known since the discovery well produced oil from fractured shales overlying the reef reservoir. Subsequent tests have substantiated the presence and significance of the fractures. A

FIG. 6.—Slabbed core photo from zone of stromatoporoid thicket showing cylindrical stromatoporoids with extensive alterations.

FIG. 7.—Lagoonal sediment with abundant *Amphipora*. Mottled appearance is due to variable alteration of wackestone matrix.

FIG. 8.—Tidal flat facies. Laminar and regular fenestrae are filled with sparry calcite.

1969 water injectivity test on Goose Island well #19 indicated a 30:1 directional permeability. A core of the complete reef section from the first well in the 1978–79 delineation program showed numerous open and cemented vertical or near vertical fractures, one of which had a width of 0.4 in. (10 mm) and a height of 16 ft (5 m). Observations of fractured reservoirs by Carnes (1966), Felsenthal and Ferrell (1971), Aguilera and Poolen (1977), and others have shown that the presence of these fractures will have a significant impact in any recovery scheme planned for this pool. Therefore, a concerted effort was mounted to gather as much information about fractures in the reservoir as possible. Figure 12 summarizes the results of the various methods as discussed below.

Oriented cores, in most cases covering the full reef section, were taken on six wells in the pool. The orientation of both open and calcite-infilled fractures was measured and found to align in a northeasterly direction. A secondary but generally less prevalent direction was also present at about 90° to the major axis. A composite of all the observations of vertical fractures, a total of 199 readings, is illustrated in Figure 13. These data indicate a mean orientation of N26°E and a standard deviation of ±21°.

As part of a pattern waterflood for the Mainland area of the pool, seven deviated wells were drilled at angles of as much as 70° from the vertical. One of the wells, 39X, was cored through the reservoir for a total length of 820 ft (250 m). This allowed the unique study of vertical fractures in two dimensions to be made from the same well. The observations are shown in Figure 14, which provides a visual impression of the degree of fracturing along the wellbore. It also shows that zones of sonic log cycle skipping can be matched to higher concentrations of fractures in the core. Zones of open fractures can be matched to slow travel times

FIG. 9.—Thick tabular stromatoporoid boundstone of reef margin facies. Alteration is extensive.

on the sonic log, whereas zones of cemented fractures correspond to fast travel times on the sonic log.

A photolineament study by E. A. Babcock (pers. commun.) showed close agreement with the combined oriented core results (Fig. 12). The study defined fracture directions in bedrock formations through the mapping of lineaments on aerial photographs. The technique has been shown to be a useful tool for predicting the strike of joints or faults by Alpay (1973), Babcock (1974), and Komar and others (1976). The large amount of outcrop in the study area made lineament mapping a well suited method for determining joint strikes in the Norman Wells area. A number of these outcrops of the Kee Scarp Formation within 3 mi (5 km) of the reservoir were inspected for fracture orientation and dip. Vertical fractures trending northeast with a secondary axis 90° to this direction were prevalent. Measurements of lineaments on the regional surface topography, evident from Landsat photographs, also confirmed the presence of a dominant orientation, roughly N30°E.

Two factors which may make the Kee Scarp at Norman Wells particularly susceptible to fracturing are its shallow depth and its fine grain size. Stearns and Friedman (1972) and others report that limestone at shallow depths has a low ductility and responds to stress in a more brittle manner than limestone at greater depths. The reservoir at Norman Wells has also been subjected to alteration by microleaching, which has created a fine intercrystalline porosity and grain size. Studies by Hugman and Friedman (1979) have indicated that microcrystalline limestones are stronger and more brittle than coarser grained limestones, making them more susceptible to fracturing.

The consistency of the orientation of the fracturing in the Normal Wells area suggests that the dominant cause of the

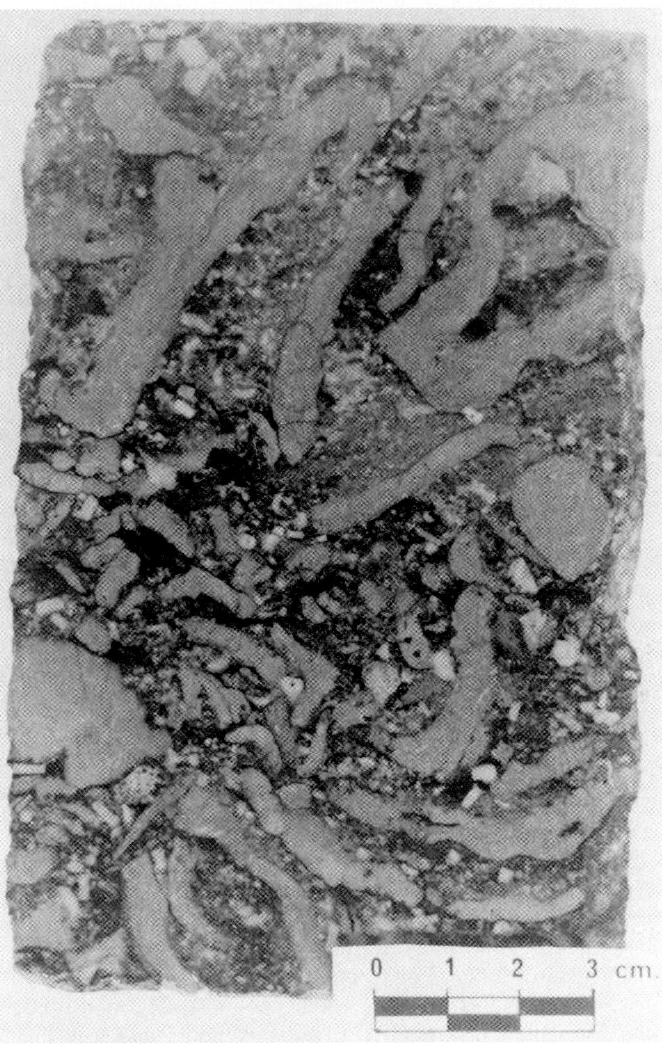

FIG. 10.—Tabular stromatoporoid rudstone from foreslope facies. Crinoid fragments and corals are also present.

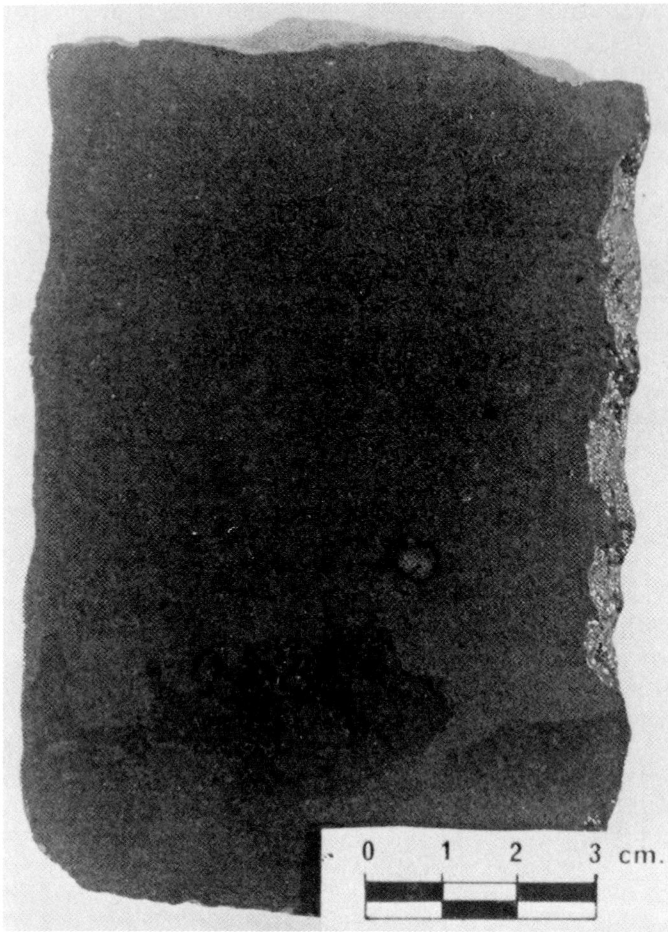

FIG. 11.—Skeletal grainstone from foreslope sand facies. Alteration is extensive.

fractures is a regional stress system. Local structure is not apparent within or adjacent to the reservoir and would not be a major control on the fracture patterns.

Analysis of 25 pressure buildup tests for the Mainland area indicates an average radial permeability of 8.6 md. Combining the average matrix permeability of 2.8 md with the average radial permeability and converting from a radial model to an elliptical model, an average directional permeability ratio of 5:1 results. A similar analysis for Goose and Bear Island indicated ratios of 2.5 and 1.2, respectively. The reduction in fracture significance for the Goose and Bear Island portions of the pool is felt to be due to the increased confining stress as the reservoir depth increases from 1,150 to 1,640 ft (350 m to 500 m). Fractures logged in the deeper parts of the reservoir were commonly hairline to 0.04 in. (1 mm) in width, whereas those from the Mainland area ranged as much as 0.4 in. (10 mm) in width. This feature of increasing fracture width with decreasing reservoir depth could be a result of the unloading of the rock due to a reduction in overburden, allowing fractures to open wider under a less effective confining stress. Unloading also provides a reason for the secondary fracture direction being more prevalent in the surface measurements and less predominant at depth in the reservoir.

Pulse testing and interference testing are common methods of determining directional permeabilities. A water injectivity test, naptha gravity survey and an interference test were used to help delineate the orientation of the preferred permeabilities due to fracturing (Fig. 12). A water injectivity test indicated a directional permeability ratio of 8:1, with the orientation between north and N45°E. The naptha gravity survey was a mapping of the reproduced naptha by-product conserved over the years by reinjection into the reservoir on the Mainland. The fluid eventually was produced on adjacent wells and resulted in an increase in API gravity. Wells along a N30°E trend from the injector produce at a higher API gravity with naptha production decreasing as the distance from this trend increases. An interference test on Goose Island in 1979 indicated a 6:1 directional permeability ratio with a preferred orientation of about north-northeast. More definitive results cannot be expected from current well testing because low reservoir pressures have resulted in high gas saturations that tend to dampen the pressure response.

RESERVOIR MODEL STUDY

The approach used to forecast the total pool production performance has been to model two relatively small but representative areas of the pool, each covering about 32 acres (13 ha) or roughly 1% of the total area. A detailed three-dimensional, three-phase reservoir simulator was developed for each of these areas, and several key parameters were varied as sensitivities to the initial assumptions regarding reservoir conditions. The use of a small pattern model has certain advantages over the gross total pool models in that a more detailed reservoir description can be studied at a lower cost.

The fluid and rock properties used in the model are summarized in Table 2. Model dimensions and the grid systems are shown in Figure 15. The model dimensions relative to the entire pool are shown in Figure 16, as well as the areal extent of the two main areas—reef margin and reef interior. The key parameters in the model were pattern and well spacing and directional permeability.

For the purposes of reservoir modelling, the two main areas, "reef margin" and "reef interior," were divided into five layers with differing porosity and permeability characteristics, as illustrated in Figure 17. Conversion values for the metric system are given in Table 2. The five layers are a simplification and grouping of core-derived porosities and permeabilities taken from wells representative of each main area. Although a more detailed examination of the core porosities and permeabilities indicate a high degree of stratification and variability within each layer, computer reservoir simulation would become prohibitively expensive if they all were incorporated in the simulation model. Five layers were chosen as the base case, with sensitivities of 18 to 24 layers also being tested.

Both idealized geological models deal only with what is considered the reservoir section of the Kee Scarp, which sits above the tight section immediately overlying the K1 zone or platform.

In general, the layers conform to the established strati-

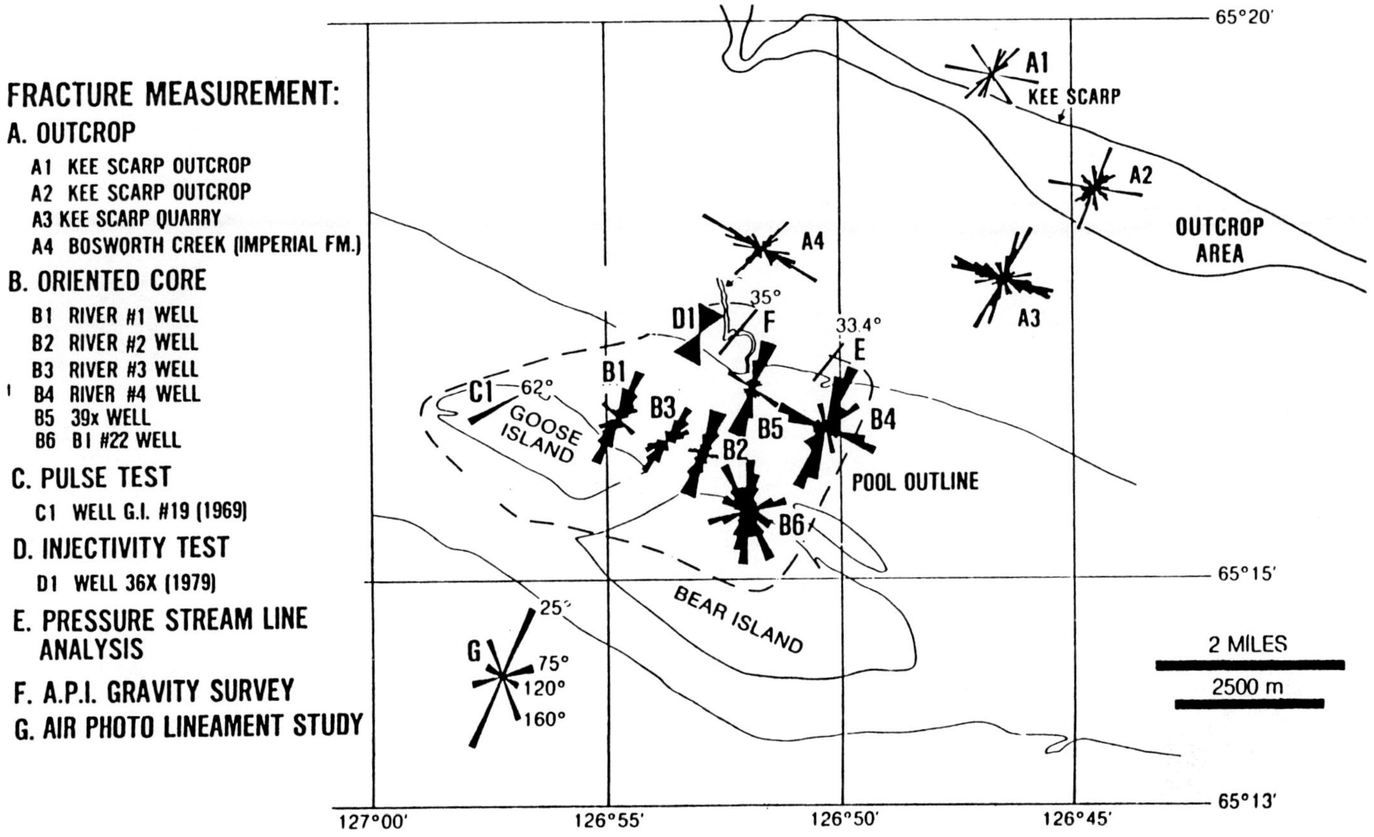

FIG. 12.—Summary of fracture orientation measurements.

graphic breakdown, although the K2 and K3 zones have been subdivided further for modeling purposes. Each of the layers will have lateral variations in facies and reservoir potential, but since the pattern model used was small (13 ha), these variations were not incorporated. The following

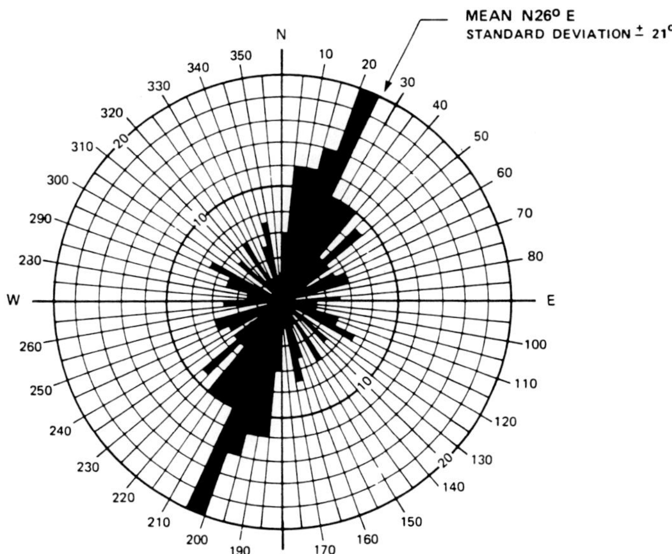

FIG. 13.—Composite vertical fracture orientations from oriented core. Includes 707 m of core and 199 observations from six wells.

descriptions of the layers are brief, since the details of the different environmental facies were discussed in the section on stratigraphy. The degree of alteration is noted in each case, since it is a major control on the porosity and permeability.

Layer One in both models represents a widespread cylindrical stromatoporoid thicket or foreslope-like facies resulting from an increase in water depth creating more open marine conditions across the entire reef complex. The base of this zone roughly marks the beginning of the first major cycle of reef growth (K2 zone). In the reef margin model, alteration was extensive, and in the reef interior model it was moderate to extensive.

Layer Two in both models is dominantly a lagoonal environment with associated thin supratidal facies. Alteration is moderate.

In the reef margin example, the third layer is another foreslope facies but with only partial alteration. Layer Three in the reef interior model is also foreslope but it has extensive alteration. The base of this layer marks the beginning of the second major cycle of reef growth (K3 zone).

Layer Four is a continuation of the foreslope environment in the reef margin area, but the alteration is now extensive. In the reef interior, it is dominantly lagoonal with some supratidal deposits and only partial alteration of the matrix and stromatoporoids.

The final layer, Layer Five, is reef margin facies in the margin model. It is extensively altered and has some pre-

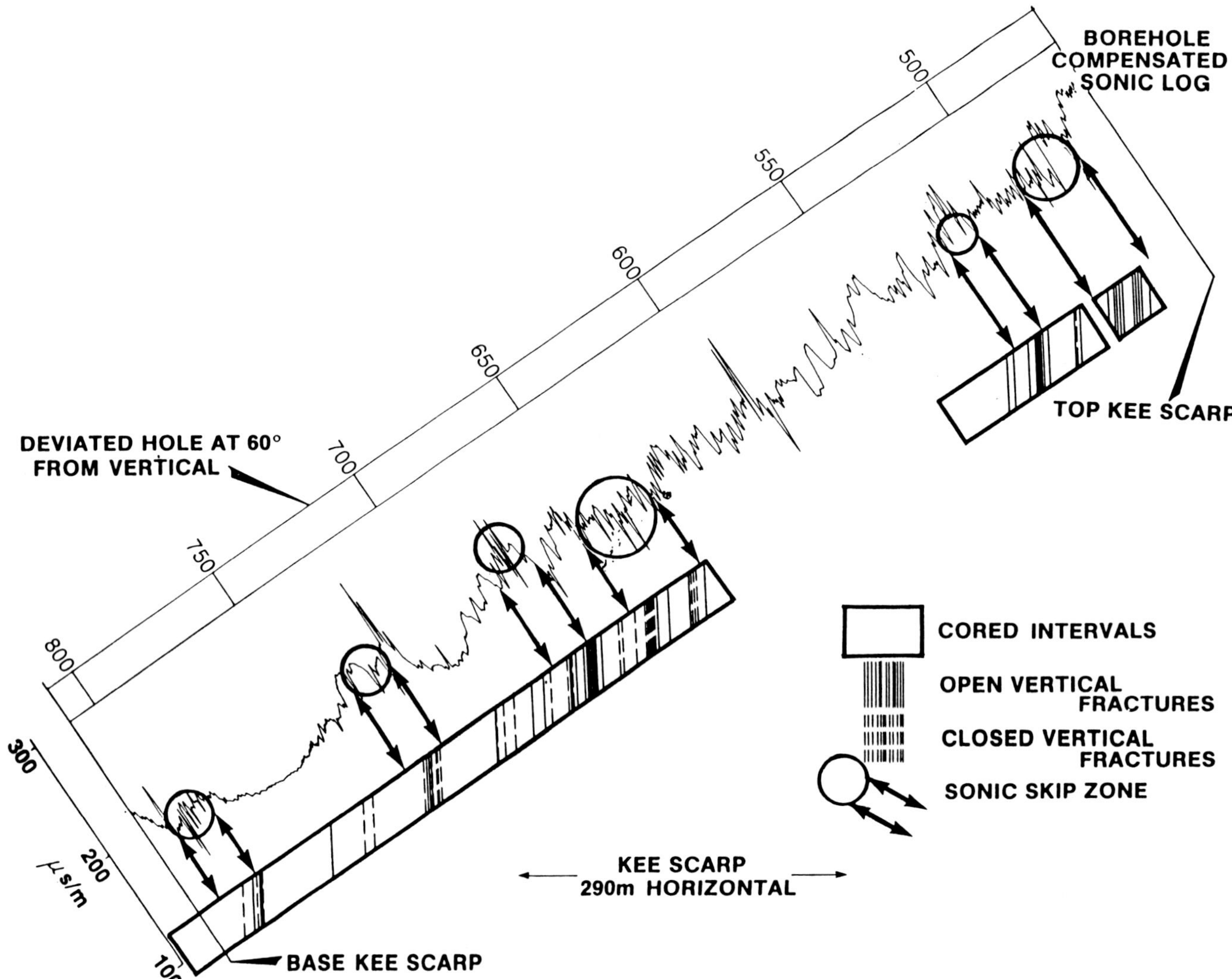

FIG. 14.—Fracture frequency and sonic log response from deviated well '39X'.

served primary porosity in the form of cellular porosity within stromatoporoids. In the reef interior model, it is a combination of backreef sand flat facies with lesser lagoonal facies. In the reef sands, the grain matrix is well altered, but the larger stromatoporoid grains are not altered. Minor intergranular primary porosity is present in the reef sands. Layer Five equates to the K4 zone of the stratigraphic breakdown.

BASE CASE

When the geological and reservoir engineering descriptions are combined with generally accepted production engineering practices into the reservoir simulation model, forecasts of pool production performance can be made.

In the base case, Figure 18 illustrates the oil production, water-oil ratio (WOR), and gas-oil ratio (GOR) as a function of the original oil-in-place for both reef margin and reef interior production wells. Before water breakthrough occurs, the oil production rates stabilize at about 630 B/D (100 m^3/d) and 235 B/D (37 m^3/d) for the reef margin and reef interior wells, respectively. At a WOR of 25, the cumulative oil production is 62 and 60% of the oil-in-place for the reef margin and reef interior wells, respectively. It should be stressed that these model results are usually optimistic, do not fully reflect expected average well performance, and are best used for comparative purposes between variations in parameters. Downgrading or reducing model results to forecast well production rate and pool ultimate oil recovery will be discussed later.

TABLE 2.—SI METRIC CONVERSION FACTORS

°API 141.5/(131.5 + °API)		= g/cm^3
bbl × 1.589 873	E−01	= m^3
acre × 4.046 856	E−01	= ha
ft × 3.048	E−01	= m
mile × 1.609344	E+00	= km
°F(°F − 32)/1.8		= °C
psi × 6.894 757	E+00	= kPa
cp × 1.0	E−03	= Pa.S

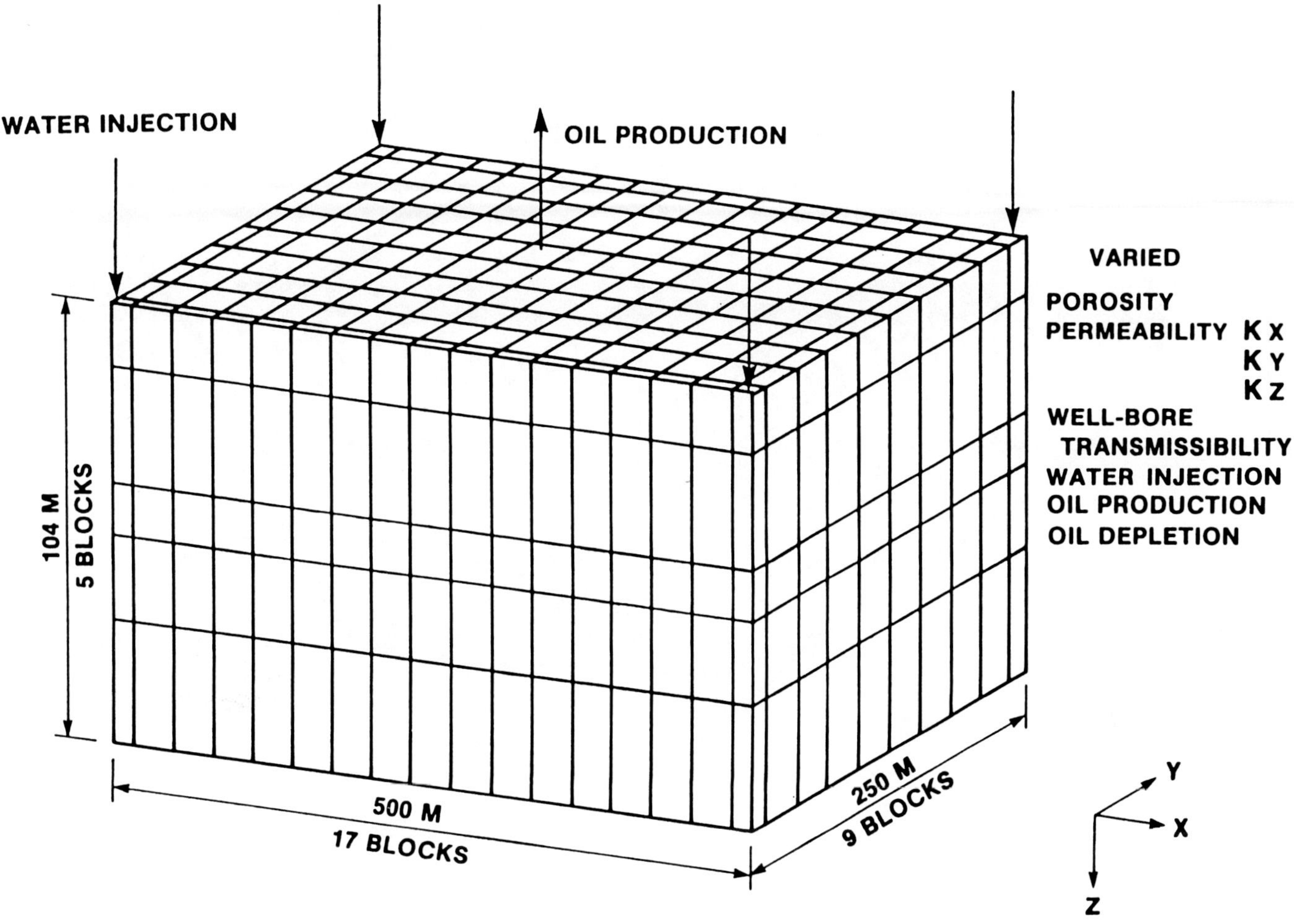

FIG. 15.—Normal Wells 3D-3-phase reservoir model.

Pattern and well spacing effects.—

Various patterns employing a one-to-one injector-to-producer ratio were examined with different well spacings and aspect (pattern length-to-width) ratios. Previous model studies of the pool concluded that fewer water injectors would not adequately replace the production voidage, and well spacing of 16 acres/well (6.2 ha/well) was recommended to achieve a suitable production rate to justify economically a major transportation system. A comparison of the various alternative well patterns and well spacing showed that for a given set of geological parameters, the total pool production was nearly proportional to the number of well patterns. Unadjusted model results indicated that oil production rates of a per well basis would generally fall within a 20% range of each other, and ultimate oil recovery at a 25:1 water-oil ratio would be 60 ± 2% of the original oil-in-place.

The most significant parameter was the orientation and length of the major axis of the pattern. As the directional permeability ratio increased, it became evident that significantly higher production rates from a unit area could be achieved as the distance between rows of water injectors decreased. In general, the maximum oil rate would occur when the pattern aspect ratio was equal to the square root of the directional permeability ratio.

Results of directional permeability variations.—

Since all of the cores taken to date from the pool have shown signs of natural vertical open and closed fractures, an attempt was made to model the pool performance of a finely fractured reservoir system. A review of the existing techniques available for modeling fractured reservoirs showed that they have been developed primarily for massively fractured reservoirs where the width between these fractures is great (approximately 30 ft, 10 m, or more; Mattax and Kyte, 1962; Yamanota and others, 1971; Kleppe and Morse, 1974; Saidi, 1975; and Rossen, 1977).

As a result of the fracture studies to date, it can be concluded that the fracture spacing in the Norman Wells pool is on the order of 0.3 to 6 ft (0.1 to 2.0 m) and with this density of fracturing, the use of a double-porosity fracture model would be prohibitive due to the large number of blocks required. An acceptable alternative is to incorporate an increase in the average block permeability in the direction of the fracture. Ratios of Kx:Ky of 1, 2, 4 and 10 were studied.

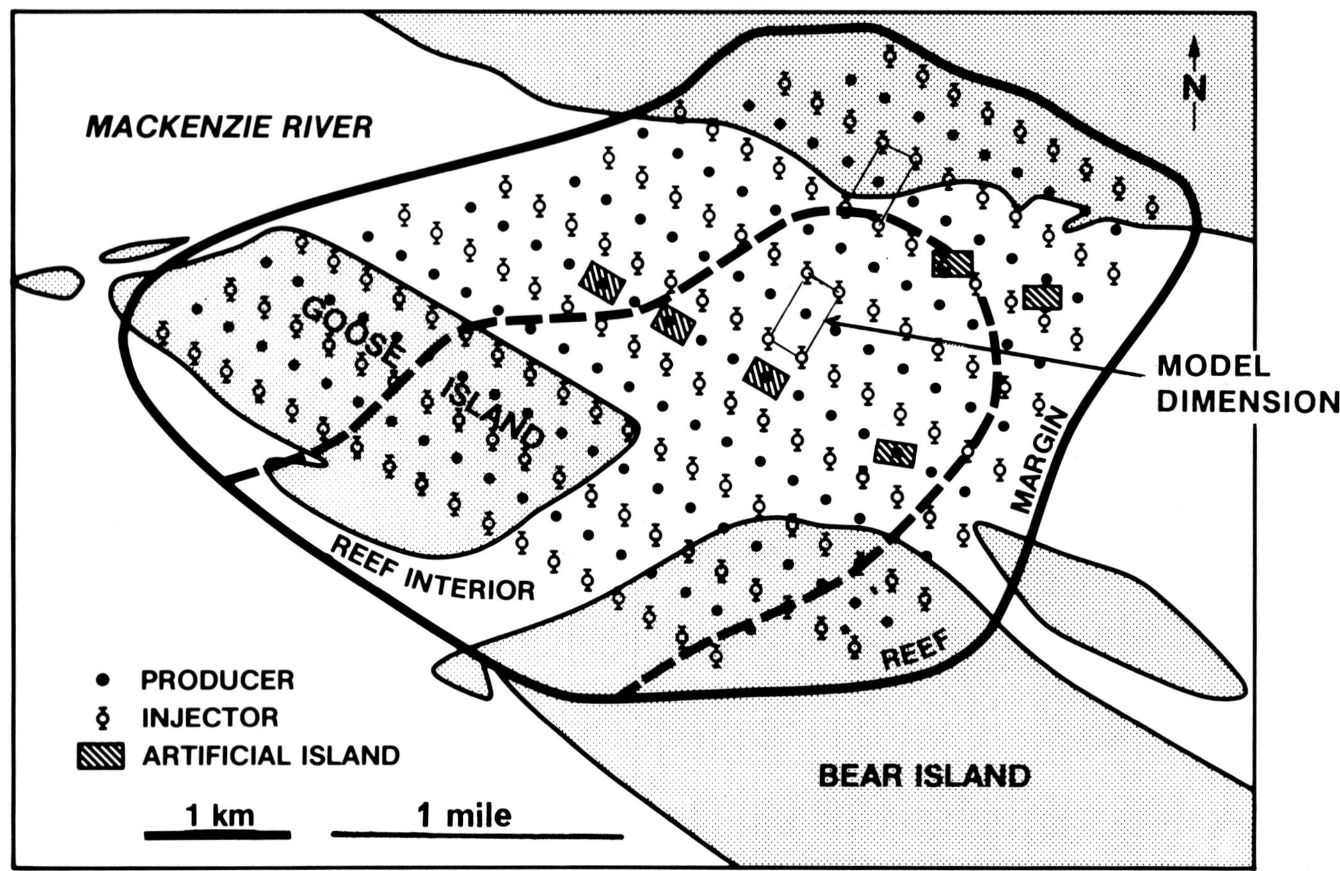

FIG. 16.—Idealized well locations and model dimensions.

Figure 19 illustrates the effect of increasing the permeability in one direction for the reef margin model. Stabilized oil rates were essentially directly proportional to the square root of the directional permeability ratio, although ultimate oil recoveries were basically identical for all cases. It is evident that knowledge of the permeability ratio is critical to the pool production forecast.

Proposed pool development.—

As a result of the geological and reservoir model study, the proposed well pattern is an elongated five-spot with the major axis along the observed vertical fracture direction of N30°E. A 2:1 aspect ratio has been chosen, based on the evidence that the average directional permeability existing in the field is two or more times as great along the major fracture trend as it is at right angles to this trend. A one-to-one injector-to-producer ratio is required to maintain adequate voidage replacement.

Using a 16 acre/well (6.2 ha/well) spacing, which is the optimum economic spacing, a total of as many as 180 new wells will be drilled in addition to the 60 existing wells capable of production. All of the new wells will be drilled from pads or artificial islands to allow year-round operation. Since half of these wells will be producers and half injectors, ultimate oil production from each producer is expected to vary from 0.75 to 3.7 million bbl (0.12 to 0.6×10^6 m^3), depending on the oil-in-place within the confined pattern. Denser well spacing is inefficient as the incremental investment does not provide adequate economic return. Ultimate pool recovery is described in the next section.

Total pool production forecast.—

Using the most representative reservoir characteristics for the reef margin and interior models, the predicted recovery averaged 60% of the original oil-in-place. Reef margin recovery was slightly better than reef interior recovery. Reservoir simulation models cannot incorporate all the factors that affect ultimate oil recovery and some downgrading or reduction of recovery from the models is appropriate, particularly in Norman Wells where pool depletion is at an early stage and calibration of the model to pool waterflood performance is not possible. Reliable reservoir performance predictions can be obtained from the model studies, provided adjustments are made to incorporate reservoir heterogeneities such as reservoir stratification, vertical fractures and reef edge accessibility. At Norman Wells, these factors led us to reduce ultimate oil recovery from the model predictions of 60% down to 43% of original oil-in-place as follows:

(1) Sensitivities to increased reservoir stratification by the use of more representative 18- and 24-layer models

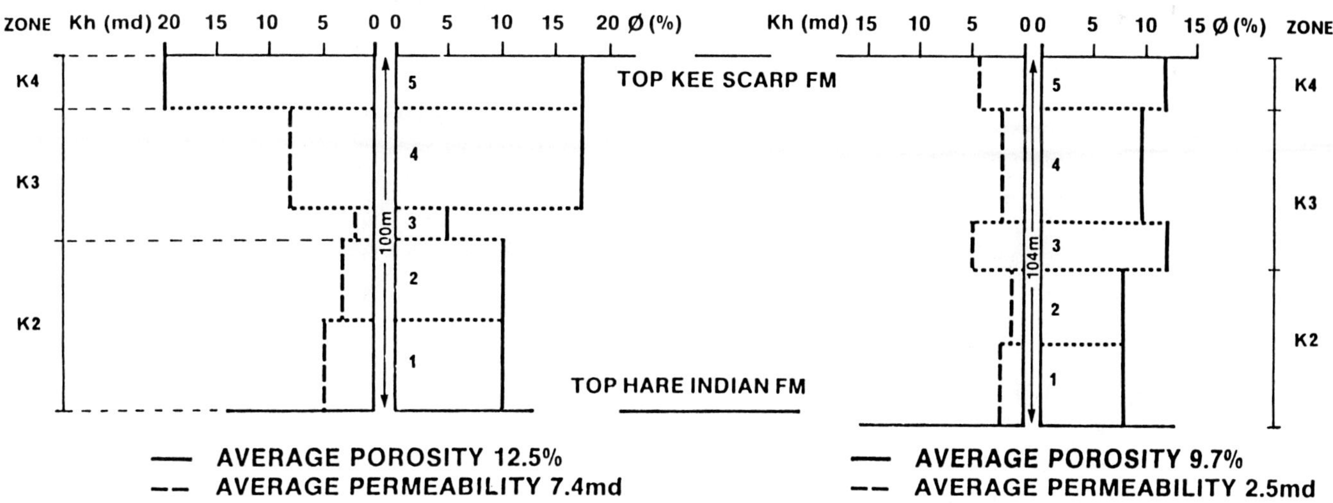

FIG. 17.—Idealized geological model.

with varying permeabilities (rather than five layers) showed that the ultimate oil recovery at equivalent end-point water-oil-ratios would be reduced to 55% from 60% of the original oil-in-place.

(2) Normal drilling inaccuracies in highly deviated wells could result in wellbores that cross the fracture trend, leading to large volumes in the reservoir surrounding the injector and producer locations that have pres-

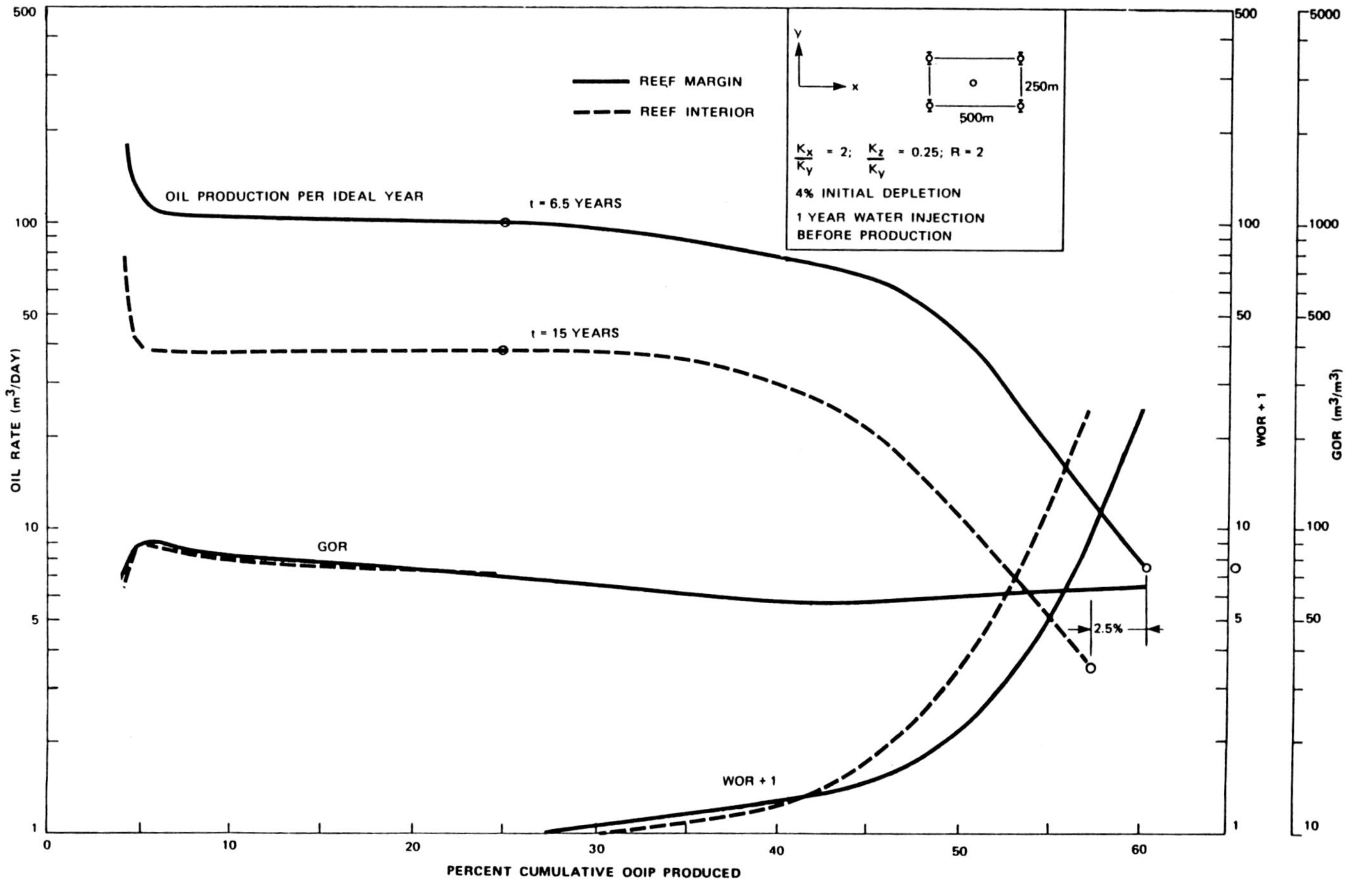

FIG. 18.—Base case comparison of 'reef margin' to 'reef interior' production characteristics.

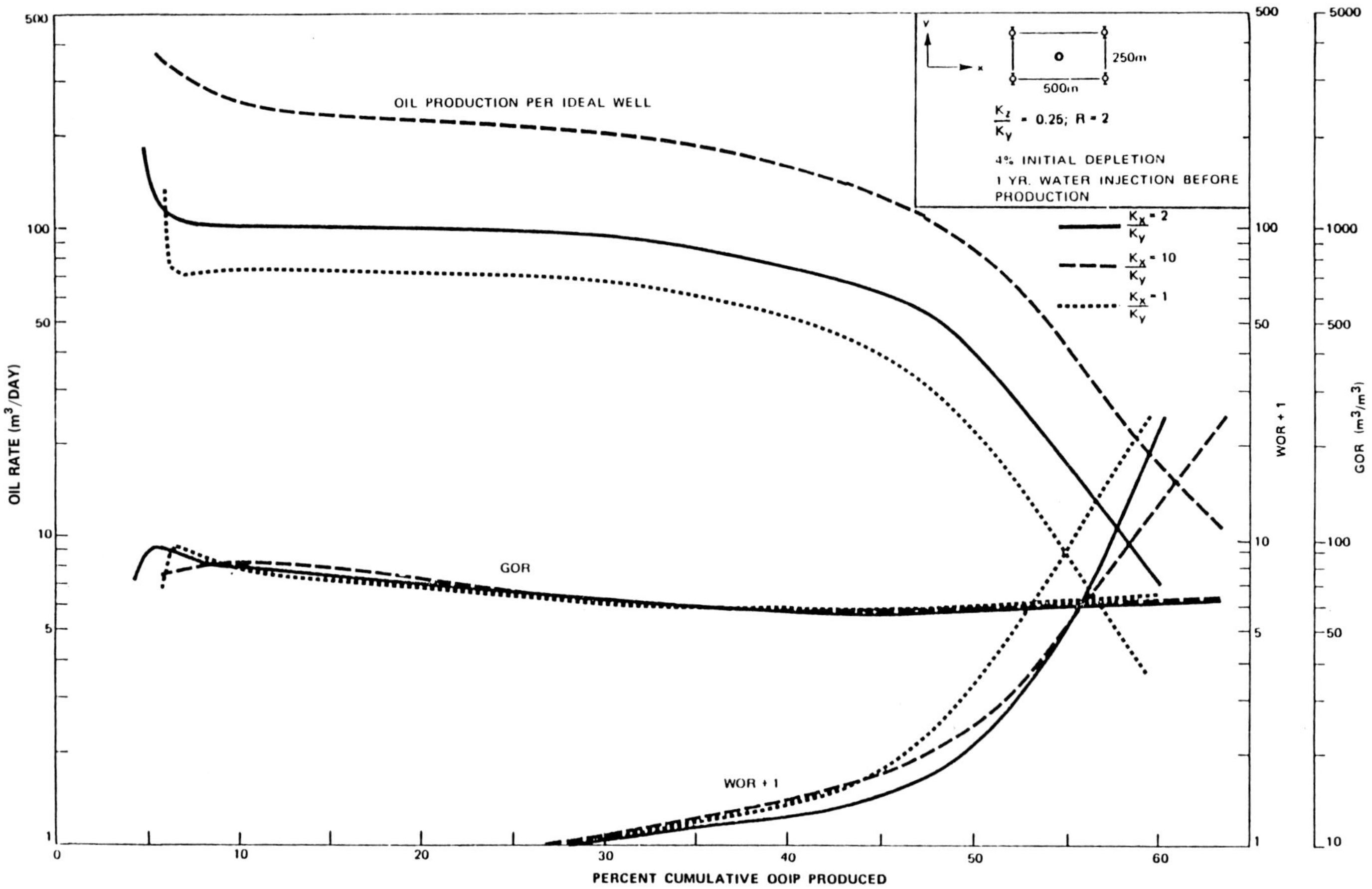

FIG. 19.—Effect of increasing the directional permeability for the reef margin, Kx/Ky.

sures similar to the wellbore. Since fluid flow requires that there be a pressure differential, fluids within these volumes will be subjected to a limited driving force and recovery efficiency will be reduced. These volumes were estimated to amount to 4.5% of the original oil-in-place, equivalent to reducing the previous 55% value to 52.5%.

(3) Since the field observations of oriented core, well interference testing and gravity surveys have shown that there will be variations in the natural fracture orientation, some well patterns will not be oriented exactly parallel with the actual fracture direction. Using statistical analysis of the core fracture observations, it is possible to estimate that this pattern orientation effect will lead to unswept volumes of about 8% of the original oil-in-place, reducing the previous 52.5% value to 48%.

(4) Additional losses for poor reef edge and K1 zone recovery due to well inaccessibility led to an estimated total pool recovery of 43% of the original oil-in-place, or 271 million bbl (43×10^6 m^3).

In arriving at a total pool production forecast of oil, gas and water injection and production, the 3D-3-phase model results discussed in the previous section were applied as follows:

(1) The "reef margin" and "reef interior" base case model runs were assigned a weighting factor of unity. Pore-volume weighting factors were assigned to all reef margin wells and reef interior wells on an individual basis.

(2) The number of "net" reef margin and reef interior wells were calculated by summing the individual well weighting factors.

(3) The total pool production forecast was determined by multiplying the base case model production forecast for the individual reef margins and reef interior wells by the number of "net" reef margins and reef interior wells.

The production forecast generated above was downgraded by the following operational experience factors:

—well production efficiency	15%
—field equipment downtime	10%
—early water breakthrough	5%
Total	30%

where well production efficiency is defined as the ratio of the actual well production rate to its test rate, field equipment downtime refers to planned and unplanned well shut-ins, and the early water breakthrough adjustment accounts for possible breakthrough from small, high permeability zones or through connecting fractures which are not incorporated explicitly in the simulation models.

Water injection was in the long-term forecast to equal 130 to 150% of the stock-tank oil production. A peak rate of 25,000 B/D (4,000 m^3/d) of oil at a 400 cu ft/bbl (70 m^3/m^3) gas-oil ratio can be expected within 5 years or earlier, depending on the time required for repressurizing the reservoir and pool expansion start-up.

Oil production will start to decline within 10 years when water breakthrough occurs on a total pool basis. Gas production and water injection will be higher at the start of the operation. Natural gas liquids will be processed from the produced gas.

CONCLUSION

Relatively stable oil prices have made it attractive to plan for the total pool development of Norman Wells. An extensive geological and engineering study conducted over a period of 2 years had indicated that 43% of 630 million bbl (100 × 10^6 m^3) original oil-in-place can be recovered by pattern waterflooding and a peak pool rate of 25,000 B/D (4,000 m^3/d) can be expected. This will require the drilling of upward of 180 new highly deviated wells from the existing land mass and six artificial islands in the MacKenzie River. An elongated five-spot on 16 acres/well (6.2 ha/well) spacing will take advantage of the strongly oriented natural fracture systems identified in the pool.

This study indicates that the understanding of fractures will play an important role in the future pool production performance. Tools such as oriented core, well testing fluid and pressure surveys have been used in assessing fracture orientation.

ACKNOWLEDGMENTS

The authors thank Esso Resources Canada Limited for permission to publish this paper. Special thanks are due to B. A. Dawe, J. R. Bacon, R. E. Francis, Dr. J. Wendte, and Dr. R. Vierbuchen for their valuable assistance in the study of the pool, and to J. D. McFarland, B. G. Pryce and O. Biberdorf for their review of the paper. B. P. Tsang coordinated several aspects of the paper.

REFERENCES

Aguilera, R., and Van Poolen, H. K., 1977, Current status on the study of naturally fractured reservoirs: Log Analyst, v. 18, p. 3–23.

Alpay, A. O., 1973, Application of aerial photographic interpretation to the study of reservoir natural fracture systems: Journal of Petroleum Technology, v. 25, p. 37–45.

Babcock, E. A., 1974, Jointing in central Alberta: Canadian Journal of Earth Sciences, v. 11, p. 1181–186.

Carnes, P. S., 1966, Effects of natural fractures on directional permeability in waterflooding: Society of Petroleum Engineers, SPE Paper 1423, Seventh Biennial Secondary Recovery Symposium, Wichita Falls, Texas, 24 p.

Dunham, R. J., 1962, Classification of Carbonate Rocks According to Depositional Texture: American Association of Petroleum Geologists Memoir 1, p. 108–121.

Embry, A. F., and Klovan, J. E., 1971, A Late Devonian reef tract on northeastern Banks Island, N.W.T.: Bulletin of Canadian Petroleum Geology, v. 19, p. 730–781.

Felsenthal, M., and Ferrell, H. H., 1971, Factors that can be optimized in waterfloods of the low permeability reservoir: Journal of Petroleum Technology, v. 23, p. 727–730.

Hugman, R. H. H., and Friedman, M., 1979, Effect of texture and composition on mechanical behavior of experimentally deformed carbonate rocks: American Association of Petroleum Geologists Bulletin, v. 63, p. 1478–1489.

Kempthorne, R. H., and Irish, J. P. R., 1980, Norman Wells—a new look at one of Canada's largest oilfields: Society of Petroleum Engineers SPE Paper 9477, 55th Annual Fall Technical Conference and Exhibit of the Society of Petroleum Engineers, Dallas, Texas, 13 p.

Kleppe, J., and Morse, R. A., 1974, Oil production from fractured reservoirs by water displacement: Society of Petroleum Engineers, SPE Paper 5084, Society of Petroleum Engineers 49th Annual Meeting, Houston, 20 p.

Komar, G. A., Overbey, W. K., Jr., and Watts, R. J., 1976, Prediction of fracture orientation from oriented cores and aerial photos in Sand Draw field, Wyoming: Morgantown Energy Research Center Report TRR-76/4, U.S. Department of Energy, 16 p.

Mattax, C. C., and Kyte, J. R., 1962, Imbibition oil recovery from fractured, water drive reservoirs: Society of Petroleum Engineers Journal, v. 2, p. 177–184.

Rossen, R. H., 1977, Mathematical simulation of naturally fractured reservoirs with semi-implicit source terms: Society of Petroleum Engineering Journal, v. 17, p. 201–210.

Saidi, A. M., 1975, Mathematical simulation model describing Iranian fractured reservoirs and its application to Haft Kel field: Proceedings, 9th World Petroleum Congress, Tokyo, p. 209–220.

Stearns, D. W., and Friedman, M., 1972, Reservoirs in fractured rock, stratigraphic oil and gas fields: American Association of Petroleum Geologists Memoir 16, p. 82–106.

Yamanoto, R. H., Ford, W. T., and Bouleguira, A., 1971, Compositional reservoir simulator for fissured systems—the single block model: Society of Petroleum Engineers Journal, v. 11, p. 113–128.

PART II
SHORELINE, DELTAIC AND FLUVIAL RESERVOIRS

SEDIMENTOLOGY OF THE McMURRAY FORMATION ON THE SANDALTA PROJECT STUDY AREA, NORTHERN ALBERTA, AND IMPLICATIONS FOR OIL SANDS DEVELOPMENT

JAMES A. RENNIE
Gulf Canada Resources Inc., Calgary, Alberta

ABSTRACT: Gulf Canada Resources has carried out extensive evaluation studies in the Athabasca oil sands deposit in northern Alberta. The Sandalta project near Fort McMurray was aimed at developing an open-pit mining operation to recover oil from the oil-saturated sands in the McMurray Formation.

The McMurray Formation is the result of a gradual Lower Cretaceous marine transgression into a well defined basin. A sequence of fluvial sands followed by estuarine sediments and subsequent marginal marine sediments is widespread. Four main depositional environments have been recognized for the McMurray Formation in the area. Within the Sandalta project itself, two of these depositional environments account for the bulk of the sediments present and virtually all of the oil-bearing sands. Three-dimensional sedimentological studies have shown that, of the 22 facies identified in these two environments, only three of the sand facies (tidal channel, distributary channel and fluvial-estuarine sand) have any appreciable lateral extent and consistency in oil saturation. These three sands are therefore the prime targets for exploration and development in the study area, whether through surface mining or *in situ* recovery techniques. This model can be used for most of the Athabasca deposit with modifications for local conditions. Conventional techniques of outlining areas for development based on economic surface mineability factors or isopach maps will be misleading if the variability of the pay zones themselves is not accounted for by studying the depositional environments. A three-dimensional computer block model can be used to store the facies data from numerous core holes. This computer data base then becomes a valuable tool for mapping the lateral variations in the subsurface geology, prospect evaluation for development potential, and conceptual mine planning.

INTRODUCTION

The Athabasca deposit constitutes an immense resource potential containing more than 87 billion m^3 of heavy oil or bitumen and covering more than 36,000 km^2. The percentage of the gross resource recoverable by existing technology has yet to be determined. A workable geological model that identifies prime targets for exploration is a necessary tool for effective exploration and development of such a large resource.

The Sandalta project study area is within the portion of the Athabasca oil sands deposit that is amenable to surface mining (Figs. 1, 2). The project was proposed as a surface mining oil sands operation to develop the near-surface oil-saturated McMurray sands. Most of the Athabasca oil sands deposit is too deep for surface mining, and the viscous oil (API gravity 7 to 10) must be recovered by *in situ* heating techniques. Gulf Canada Resources has carried out extensive studies of the sedimentology of the Athabasca deposit with the aim of developing the oil sands through both surface mining and *in situ* enhanced recovery. The scope of this paper is to outline how the detailed sedimentology studies impact exploration and development strategy.

This paper summarizes the depositional environment in a specific area that has a geologic history similar to that of most of the Athabasca deposit and therefore has widespread applicability. The geological model works well in the surface mineable area of the deposit and may be extended to most areas within the Athabasca deposit with some modifications for local conditions. This is discussed in the final section.

In oil sands evaluation work, much emphasis in the past has been placed on oil saturations determined from analyses of frozen core. Oil contents are measured by separating the oil, water, and solids in laboratory apparatus. Oil contents are generally expressed as bulk weight percent, and an economic cutoff grade is often used to define the net pay sands. Pay maps based on these measured oil contents or on oil saturations derived from well logs are often used to outline promising areas for further exploration and development. This approach by itself can be misleading, as will be shown in the section on Implications for Oil Sand Development.

Because of the present economic conditions for oil sands development, the emphasis in exploration is placed on finding and developing thick rich oil sands that have good lateral continuity. Chasing narrow channel sands or sand units with variable shale content and porosity is not economically feasible, and this is where the sedimentologist has a major impact on the exploration and development programs. By comparison with conventional oil development, economical oil sands development is more strongly dependent upon a detailed understanding of the depositional environments which created the reservoir sands.

The study area and scope of investigation.—

The Sandalta project is located within the surface mineable portion of the Athabasca oil sands close to two existing oil sands plants, Syncrude and Suncor, both of which are open-pit mines (Figs. 1, 2). This paper summarizes 2 years of work examining 400 drill holes on the Sandalta lease, 200 of which were fully cored, another 725 drill holes for regional control, three-dimensional block model cross sections, outcrop sections on the Athabasca River and tributaries, and the two operating open-pit mines. The area fortunately has a large number of excellent outcrops to study the lateral extent and three-dimensional relationships of facies identified in core studies. In addition, the two open-pit mines offer an excellent opportunity to see facies units that are not well exposed in natural sections and to compare exposed mine sections to drill hole data. As a participant in both the Syncrude and Sandalta projects, Gulf Canada Resources has had access to detailed core data that is not available in the public record.

This study differs in approach from previous studies of the McMurray Formation in the mineable area in that it combines data from the very detailed scale to the regional scale in an attempt to construct a workable predictive geo-

FIG. 1.—Alberta's oil sands.

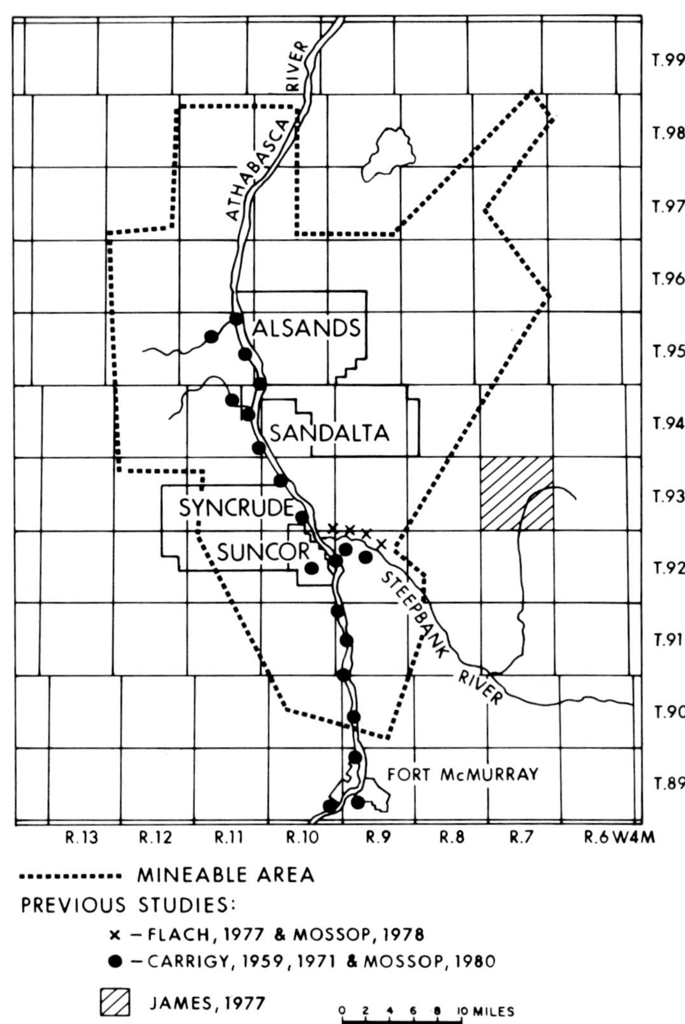

FIG. 2.—Location of study area.

logical model that is valid for more than a confined study area. There have been previous excellent studies (Mossop, 1978, 1980; Flach, 1977; James, 1977) that postulate geological models for a particular study area. The uniqueness of the study area relative to the regional picture, however, was not always recognized. In part this was due to the confidential nature of some of the detailed core data in this area. Fortunately, much of this data was available to the author for the present study. In the case of previous studies based largely on outcrops, the tendency of the sand units to crop out much more frequently than the shale units creates a problem by offering a biased sampling of the various facies. This generates the inevitable assumption that the McMurray Formation is mainly oil sand. In fact, as shown by the widespread core hole data, the McMurray Formation is mainly shale and siltstone with localized sand buildups caused by repetitive scour-and-fill cycles. More regional studies by Carrigy (1959, 1966, 1971) were also strongly influenced by the excellent outcrop exposures on the Athabasca River and tributaries near Fort McMurray. These outcrops are actually representative of only a few of the many sedimentary facies present in the McMurray Formation.

It should be mentioned that these previous studies are not incompatible with the model presented here. In fact, Mossop's and Flach's channel sands from the Steepbank River sections (Fig. 2) correlate with the tidal channel and dis-

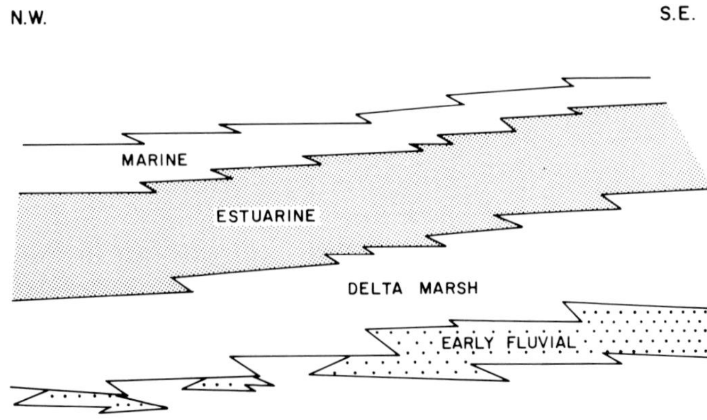

FIG. 3.—Schematic of McMurray Formation interval for the Athabasca region.

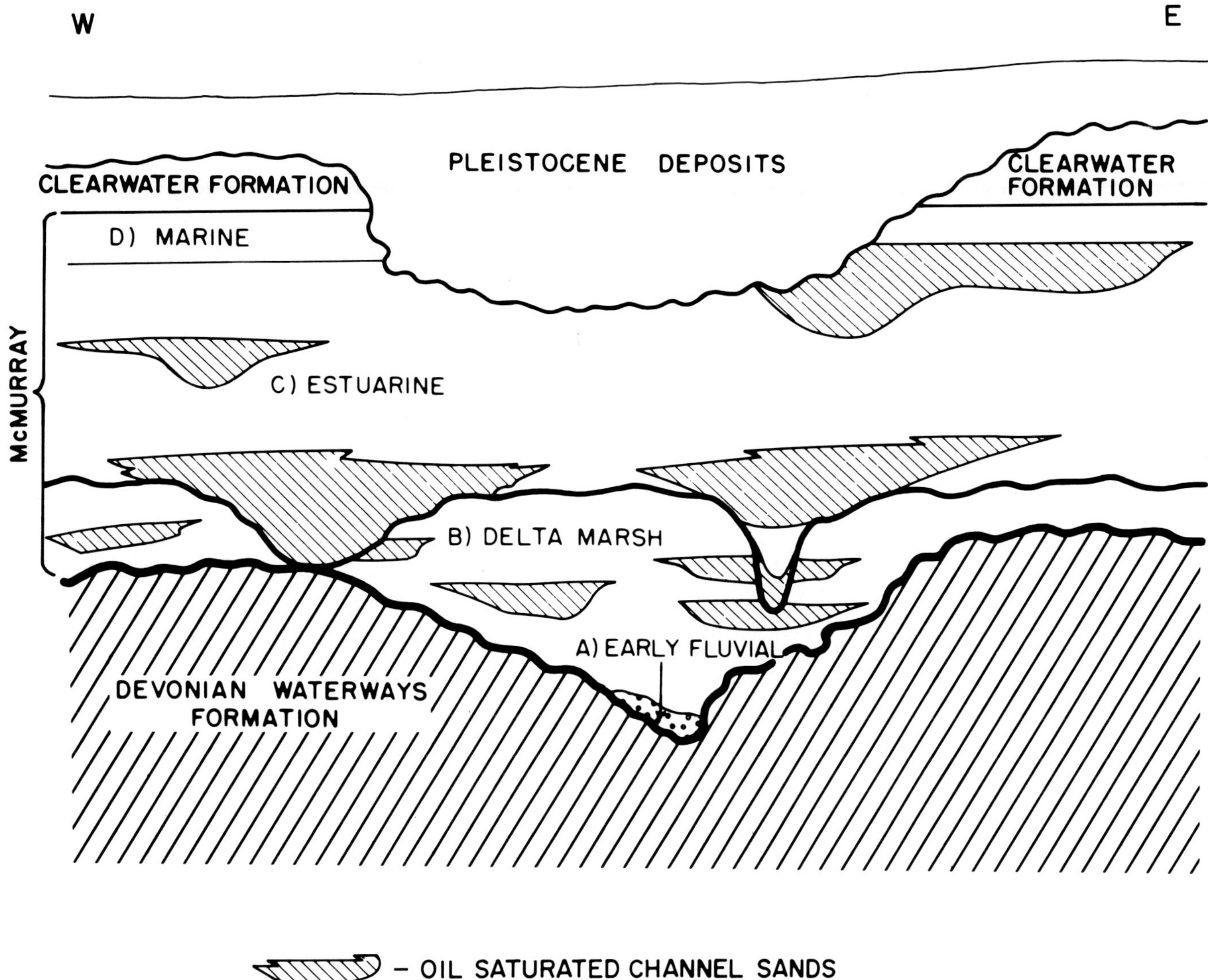

FIG. 4.—Relationship of the four main depositional environments within the McMurray Formation in the study area.

tributary channel facies of the estuarine and the delta marsh environments described herein, i.e., they are facies units within larger scale depositional environments and are a small part of an immense drainage system.

Similarly, the area studied by James (Fig. 2) is relatively small. The facies units encountered in the 14 core holes in that study area are a small sampling of the many facies units in the McMurray Formation. Each facies unit described in that study can be found developed to a lesser or greater extent throughout the region.

The Sandalta lease itself shows a stronger development of tidal flat facies and a lesser development of the fluvial facies compared to some other areas. This has been recognized through the regional studies carried out and is reflected in the regional model presented herein.

The purpose of this paper is not to present detailed facies descriptions but to focus on some of the three-dimensional relationships of the important facies. These spatial relationships can then be used to recognize and map the best targets for exploration and development.

McMURRAY FORMATION OVERVIEW

The McMurray Formation is part of the Cretaceous Mannville Group that is widespread in Alberta. In the study area it is underlain by Devonian carbonates and overlain by Cretaceous Clearwater shales. The McMurray Formation is basically the sediment infill of a pre-existing basin that was developed during pre-Cretaceous erosion of the underlying Paleozoic strata. Widespread salt solution and minor recurring faulting (both pre-Cretaceous and Cretaceous) helped to shape the basin and exerted considerable control on the pattern of sedimentation in the overlying McMurray sediments. This geological history has been well documented in previous studies (McPherson and Kathol, 1977; Stewart, 1963).

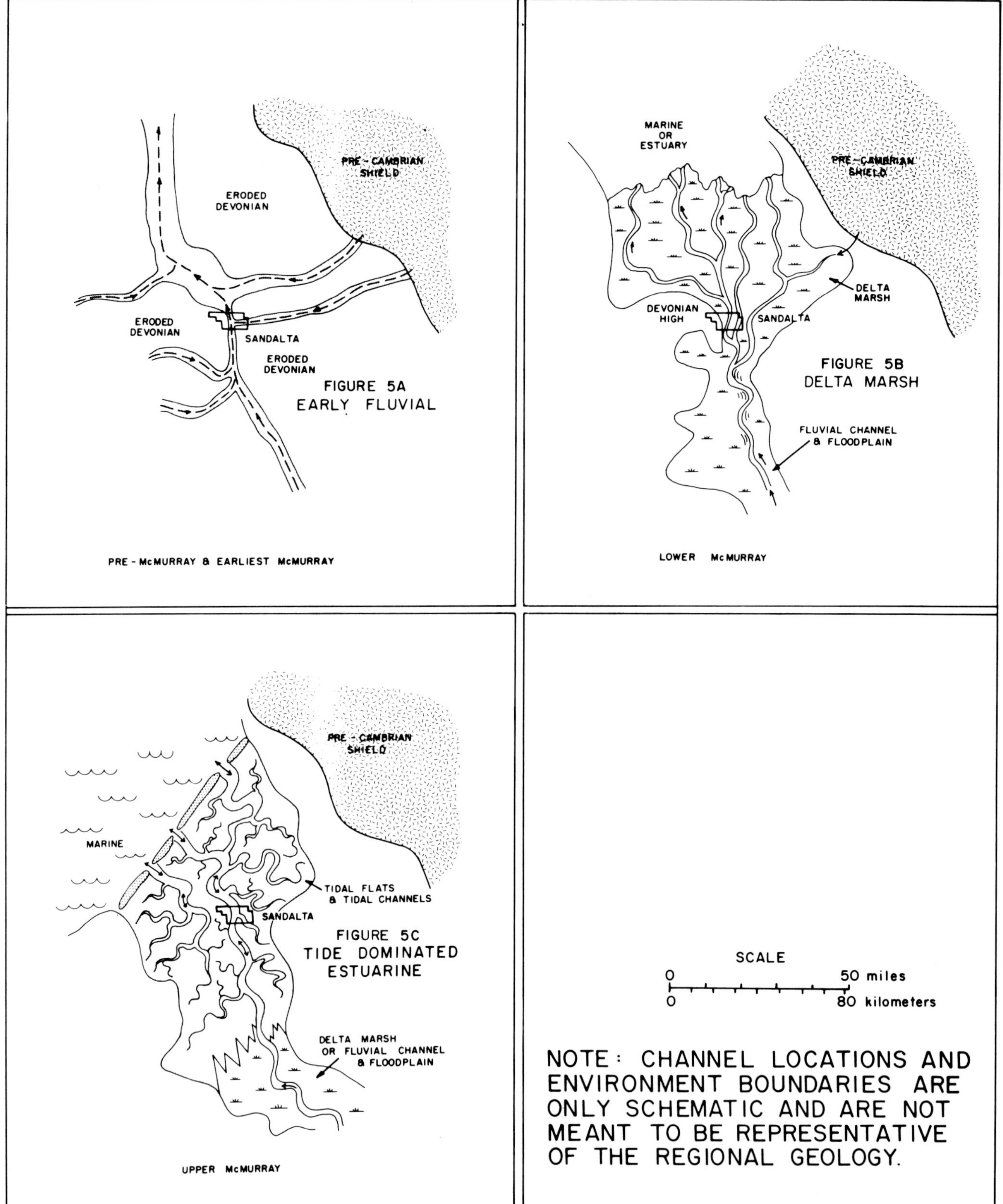

FIG. 5.—Schematic representation of the early fluvial, delta marsh, and estuarine environments in the McMurray Formation, Sandalta study area.

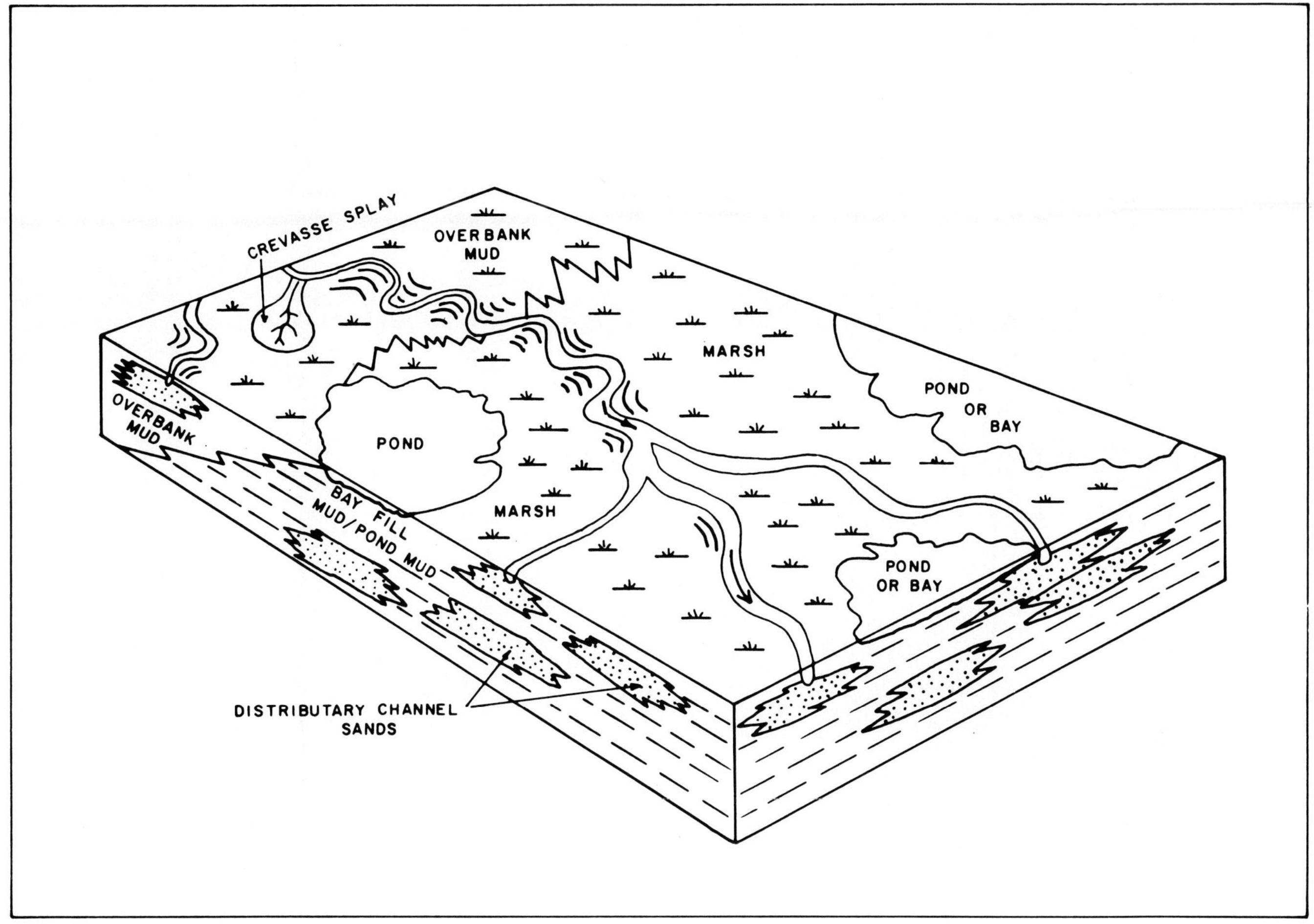

FIG. 6.—Delta marsh environment.

Four main depositional environments have been recognized by the author in the region for the McMurray sediments: early fluvial, delta marsh, estuarine, and marine. These are illustrated schematically in Figure 3. It should be stressed that, due to original sedimentary processes or subsequent scouring by later events, one or more of the four depositional environments is often missing from the sedimentary column at any given location. Palynological analyses carried out for Gulf Canada Resources suggest that the marine transgression in the basin was very slow and initially came from the northwest, so that the estuarine "Upper McMurray" delineated by a geologist in the northwestern area of the basin might be time-equivalent to the delta marsh "Lower McMurray" described by a geologist in the southeastern part of the basin.

Figure 4 illustrates the relationship between the four main depositional environments on the Sandalta project and immediately surrounding area. The overall sedimentary sequence is the result of a slow marine transgression, interrupted by occasional regressive periods that are responsible for most of the preserved sandy sediments. Generally, the early fluvial environment is not well developed, being confined to narrow drainage channels on the underlying Paleozoic surface (Fig. 5A). A marine transgression is hypothesized to account for the abrupt change from the localized high-energy environment of the early fluvial unit to a laterally more extensive complex of lower energy sedimentary units in the delta marsh environment (Fig. 5B). The delta marsh group of sediments is widespread in the study area, but is preferentially located in the broad lows of the underlying Paleozoic surface. Further marine transgression resulted in deposition of the estuarine group of sediments (Fig. 5C).

The estuarine sediments overlie the delta marsh sediments, and a sharp scour surface with considerable relief (as much as 60 m) separates the two environments. In fact, in some places scouring tidal channels from the estuarine environment have completely eroded out the delta marsh sediments. This scour surface is illustrated schematically in Figure 4 (and is shown in more detail in Figs. 20 and 21).

Continuing transgression of the Cretaceous Sea resulted in deposition of the marine sediments. These sediments were generally eroded out in the study area during the Pleistocene. Marine sediments are present to the northwest of the

FIG. 7.—Bay-fill sequence, pond mud facies. This interval is mainly clay-rich mud. Note the occasional coal layers and extensive vertical root traces at the top of the second core box from the right. Top of the cored interval is the upper right corner. Gulf Sandalta CH 184, 12-36-94-10W4, Box 36-33.

Sandalta lease, whereas to the east it appears that true marine conditions did not exist until deposition of the overlying Clearwater Shale.

In the study area, the estuarine and delta marsh groups contain the bulk of the oil-bearing sands. The following sections examine these two environments, their respective styles of sedimentation and the implications for oil sand development.

FIG. 8.—Overbank mud facies exhibiting low-angle laminations. This core is mainly silt with very thin laminations of fine sand. Gulf Sandalta CH 102, Box 79.

FIG. 9.—Distributary channel sand facies, showing medium-angle crossbedding and high oil saturation. Width of core is 6 cm.

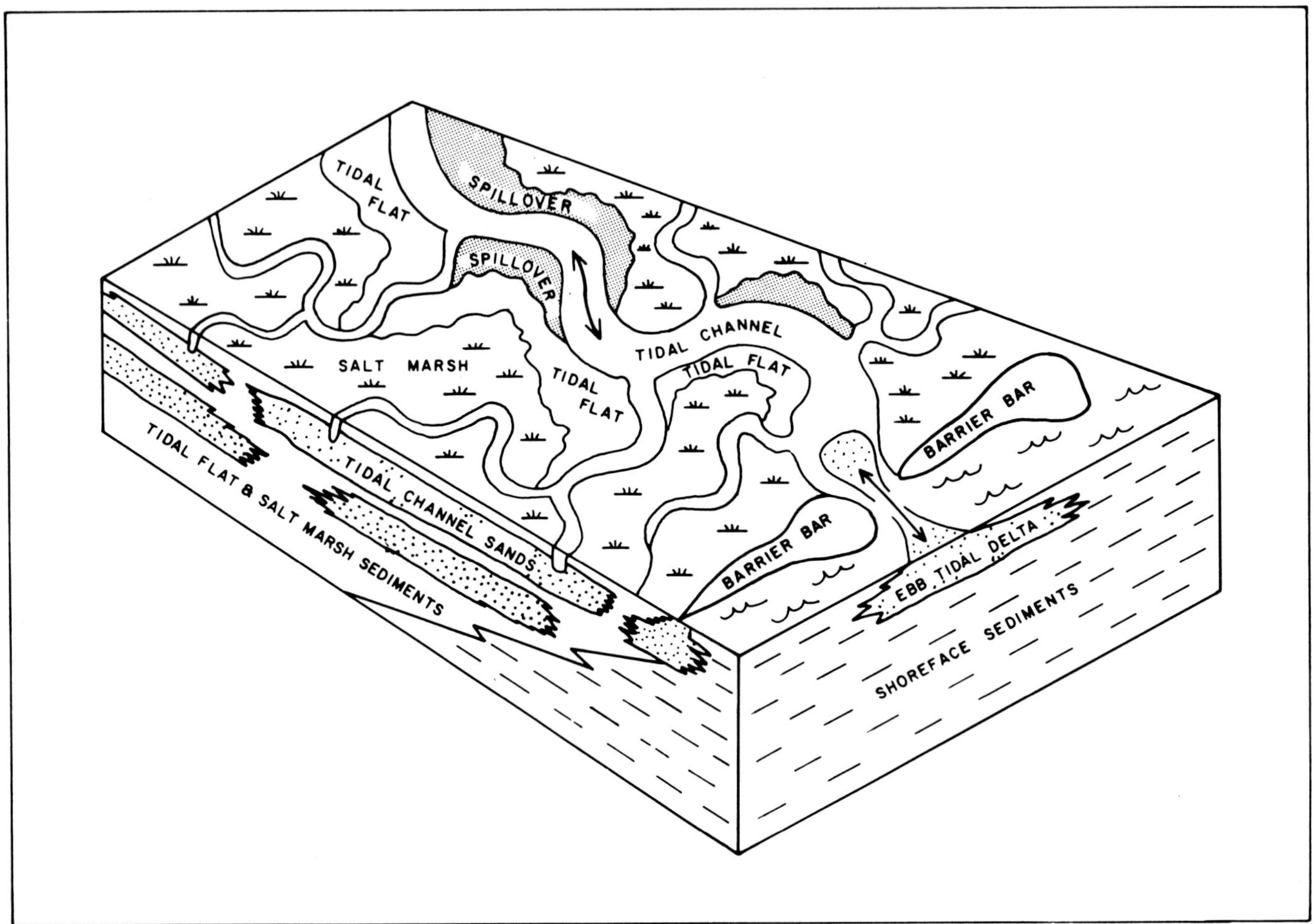

FIG. 10.—Tide-dominated estuarine environment, mesotidal model (modified from Hayes, 1976).

Delta marsh environment.—

The delta marsh group of sediments is widespread, although the thickest accumulations are preferentially located over the Paleozoic lows. Sedimentary features and palynology suggest a deltaic environment. More specifically, the sediments are representative of the outer delta plain, where occasional saltwater flooding can occur and where distributary channels have low sinuosity (Fig. 6). The overbank muds and mud drapes in the distributary channel sediments carry a large variety of pollen, with cypress and fern species being very common. The bay-fill muds have a few algal species and dinoflagellates in addition to the pollens, indicating occasional saltwater flooding. The deltaic system was apparently prograding seaward during this time period. Fluvial sands and overbank muds are more common toward the top of the sequence, whereas near-marine indicators become less common.

Common facies are: (1) bay-fill or pond muds (Fig. 7), typically grey, clay-rich muds with occasional root structures and thin coals. Often a complete bay-fill sequence is preserved, grading from clay at the base upward to silt, rooted silt and finally a thin coal cap. (2) Overbank muds are mainly laminated silt and clay with some thin sand beds and coals (Fig. 8). (3) Cutting through these muddy sediments are distributary channel sands, typically crossbedded fine to medium sands with occasional coarse sand beds (Fig. 9). All sand facies in these core photographs are oil-saturated and appear black, whereas the muds are light colored.

There are other minor facies present in the delta marsh environment (see Fig. 22). Although they do not represent a very significant portion of the sedimentary record in the Sandalta area, some of these facies are more common in other local areas of the Athabasca deposit and are thus included in the regional model.

The distributary channel sands in the Sandalta area tend to be oriented in a north-south direction with flow to the north. Since the outer delta distributary channels were periodically relocated by channel abandonment and not by gradual lateral migration, these sands are thick but narrow and elongated. Whereas they have excellent reservoir characteristics with good oil saturation, their narrow width is a limiting factor on their potential for development. It is not uncommon for an exploratory core hole to intersect a 30-m thickness of oil-saturated channel sand in a channel that is less than 400 m wide. Smaller channels, 2 m thick and 15 m wide, have been observed in outcrop. Further

	SAMPLES							
TERRESTRIAL SPECIES	A	B	C	D	E	F	G	H
INAPERTUROPOLLENITES	61	51	66	44	178	82	71	64
ALISPORITES BILATERALIS	95	34	24	48		21	20	24
BROKEN BISACCATES	35	41	14	25		33	18	46
CLASSOPOLLIS	15	6	4	9		3	49	27
SPHERIPOLLENITES/EXESIPOLLENITES	31	8	14	8		7	12	13
DELTOIDOSPORA HALLI	4	7	8	4	2	8	8	2
CYATHIDITES MINOR	24	5	11	12		1	12	6
IMPARDECISPORA PURVERULENTUS	1							
PODOCARPIDITES CANADENSIS	9		6			4	3	1
CICATRICOSISPORITES AUSTRALIENSIS	5		6				6	4
PERINOPOLLENITES ELATOIDES	5		4	1		3		
DICTYOTRILETES GRANULATUS	1							
BIRETISPORITES POTONIAEI	1						1	1
OSMUNDACIDITES WELLMANII	5	5	1			2	2	
CICATRICOSISPORITES MOHRIOIDES	2		11				11	1
IMPARDECISPORA MARYLANDENSIS	3							
CEREBROPOLLENITES MESOZOICUS	2		1	2		2	2	1
PITYOSPORITES ALATIPOLLENITES	8	4	3	2		3	1	
PETITRILETES RETICULUMSPORITES	1							
UNDULATISPORITES PANNUCEUS	1							
CICATRICOSISPORITES SP.	3		5	2		3		2
PLEURICELLAESPORITES PSILATUS	1		1					
LEPTOLEPIDIES VERRUCATUS	2					2	35	2
PEROMONOLETES ALLENSIS	1	1	1					
IMPARDECISPORA TRIORETICULOSUS	5			1				
FORAMINISPORIS WONTHAGGLENSIS	1	1						2
PRISTINUSPOLLENITES MICROSACCUS	1					2	1	1
PARVISACCITES RADIATUS	1	1	1			3		
BACULATISPORITES COMAUMENSIS	3	2	1			1	1	
CEDRIPITES CRETACEUS	1		1					
APPENDICISPORITES POTOMACENSIS	?1	1						1
GLEICHENLIDITES SENONICUS	1		2	1		1		2
RETITRILETES AUSTROCLAVATIDITES	1	2		1		2		
FORAMINISPORIS ASYMMETRICUS	1			1				
PODOCARPIDITES BIFORMIS	1		2					
TRIPOROLETES RETICULATUS	1		1				?1	
TIGRISPORITES RETICULATUS	1							
LAEVIGATOSPORITES OVATUS	1		1			1		
CINGUTRILETES POCOCKII	1		1	1				
CONCAVISSIMISPORITES ASPER	2		1			1		
CONCAVISSIMISPORITES PUNCTATUS	1							
RETITRILETES MARGINATUS		1						
CIRCULINA PARVA		1						
CONTIGNISPORITES COOKSONITES		1						
CHOMOTRILETES MINOR		1						
CICATRICOSISPORITES IMBRICATUS		1						
CICATRICOSISPORITES HALLEL		1						
CEDRIPITES CANADENSIS		2	1	1		2		1
CICATRICOSISPORITES PSEUDOTRIPARTITUS		2						2
TIGRISPORITES SCURRANDUS		2						
ARAUCARIACIDITES AUSTRALIS		2		2				
TRILOBOSPORITES DOMITUS		?1						
ALISPORITES MINUTUS		1						
DENSOISPORITES MICRORUGULATUS			1					
PLICATELLA CRIMENENSIS			1				1	
CICATRICOSISPORITES MINOR			4	1				1
TODISPORITES MINOR			1	1		1	1	
CICATRICOSISPORITES ANNULATUS			1					
PILOSISPORITES VERUS			1	1				
PILOSISPORITES TRICHOPAPILLOSUS			1				1	
KUYLISPORITES LUNARIS			1					
PITYOSPORITES CONSTRICTUS			1					
HYMENOZONOSPORITES MEXOZOICUS				2				
CYCADOPITES FORMOSUS				3				
PLICATELLA JANSONII				?1				
DISTALTRIANGULATISPORITES PERPLEXUS				1				
CICATRICOSISPORITES AUGUSTUS				1			1	1
CONCAVISSIMISPORITES VARIVERRUCATUS				1				
CICATRICOSISPORITES VENUSTUS				1			2	3
VITREISPORITES PALLIDUS				1				
EQUIETOSPORITES MULTICOSTATUS				1				
ALISPORITES GRANDIS						1		
DELTOIDOSPORA PSILOSTOMA							1	
OBTUSISPORIS CANADENSIS							2	
AEQUITRIRADITES SPINULOSUS							2	
RETITRILETES CLAVATIDITES							1	
KLUKISPORITES PSEUDORETICULATUS							1	3
IMPARDECISPORA MINOR								1
SCHIZOPHACUS PARVUS								1
ALGAE								
VERYHACHIUM SP.	X							
MICRHYSTRIDIUM STELLATUM					X		X	
CLEISTOSPHAERIDIUM SP.						?		
CANNINGIA CF. C. ATTADALICA							X	

FIG. 11.—List of pollen species for individual samples from the delta marsh environment. Note: x = less than one grain present in a total 200-grain count after averaging from a larger grain count.

upstream on this model, however, we would expect to find more laterally extensive point bar sands, and this seems to occur toward the southeastern (upstream) portion of the Athabasca deposit and locally where tributary rivers entered the basin.

Estuarine environment.—

The estuarine group of sediments was dominated by tidal processes, and the style of sedimentation was much different from that observed in the delta marsh environment. Based on core studies, a mesotidal model has been selected (Fig. 10). The sand bodies tend to be more laterally extensive due to active lateral migration of the meandering tidal creeks and channels. Although individual tidal channel sands can occur as discrete narrow channels, the repetitive scour-and-fill cycles and laterally migrating channels have resulted in deposition and preservation of some multicycle sand bodies 400 m to over 3,000 m wide and several kilometers in length. These large sand bodies are potentially valuable targets for development of the oil sand resource they contain.

Sedimentary features are dominated by small-scale bidirectional flaser bedding with extensive burrowing by small organisms in the lower energy environments. The study area is located considerably inland from the barrier islands in the block model shown in Figure 10. Palynological analyses indicate a brackish water environment with no indication of open marine environment. The pollen species from samples in the delta marsh environment are very similar to the species in the estuarine sediments. The only indication of more estuarine or marine influence in the estuarine sediments is the slight increase in algae and Inaperturopollenites combined with a decrease in Cyathidites (Figs. 11, 12), a trend that commonly occurs in the McMurray Formation as marine influences increase. There is more variation in pollen assemblages between individual high-energy and low-energy facies than between the two major environments. This is to be expected since the pollen is subject to the same sorting mechanisms of currents and waves as the enclosing sediments.

The base of the estuarine group is usually marked by a sharp scour surface. Palynological studies have not indicated a significant hiatus associated with this contact, at least on the time scale recognizable by palynologists. The scouring of tidal channel sands into underlying sediments is common in mesotidal environments and can occur over short periods of time. This situation is very similar to that of some areas on the present South Carolina coastline (Wojtal and Moslow, 1980; Duc, 1981).

Common facies in the estuarine group of sediments are: (1) tidal flat sands and muds exhibiting flaser bedding and moderate to intense bioturbation, (Figs. 13–16). (2) Salt marsh muds are also common and are typically thin waxy, featureless muds with occasional root structures. (3) These sediments are scoured into by tidal channel sands (Fig. 17) which are finer grained, better sorted and have more small-

TERRESTRIAL SPECIES	J	K	L	M	N	O	P	Q	R	S	T
	SAMPLES										
INAPERTUROPOLLENITES SP.		77	78	79	110	140	113		97	206	99
ALISPORITES BILATERALIS		55	39	44	131	6	66		17	75	68
BROKEN BISACCATES		25	60	23	37	13	25		39	15	53
CLASSOPOLLIS CLASSOIDES		12	8	12	5	8	14		24	37	20
SPHERIPOLLENITES/EXESIPOLLENITES		12	7	9	14	18	11		15	41	23
DELTOIDOSPORA HALLI		4	8	4	13	3	12		7	5	7
CYATHIDITES MINOR		6	8	3	15	1	10		8	13	11
OSMUNDACIDITES WELLMANII		4	1	1	9	2	4		5	6	4
PRISTINUSPOLLENITES INCHOATUS		1									
CICATRICOSISPORITES SP.		4	1	4	1	1	1		2	4	
CICTRICOSISPORITES VENUSTUS		2									
CICATRICOSISPORITES POTOMACENSIS		2								3	
PERINOPOLLENITES ELATOIDES		3				3	4			9	4
CICATRICOSISPORITES HUGHESI		2									
CICATRICOSISPORITES HALLEL		2	2						1		
CICATRICOSISPORITES ANNULATUS		2	1	1						1	
TRIPOROLETES SIMPLEX		?2					?1				
LAEVIGATOSPORITES OVATUS		1	1	1	1	1	1		1	5	2
CEREBROPOLLENITES MESOZOICUS		2	2		4	2	1		1	2	1
CICATRICOSISPORITES MOHRIOIDES		3	5		1					1	3
CYCADOPITES FORMOSUS		2			5		3		2	15	3
CICATRICOSISPORITES AUSTRALIENSIS		1		2	1		2		2	2	3
PODOCARPIDITES CANADENSIS		4	3	2	7	1	3			3	9
PODOCARPIDITES BIFORMIS		2		2	1						4
CICATRICOSISPORITES MINOR		1					2			1	
CICATRICOSISPORITES AUGUSTUS		1		1					1		
DISTALTRIANGULATISPORITES PERPLEXUS		1	1								
RETITRILETES AUSTROCLAVATIDITES		2			1						
CEDRIPITES CANADENSIS		1	2	1	2		1				
CONCAVISSIMISPORITES ASPER		1		1	1						
PARVISACCITES RADIATUS			2		2	1	2		1	2	
PLEURICELLAESPORITES PSILATUS			1			3					
CONCAVISSIMISPORITES PENOLAENSIS			1			?1					
PITYOSPORITES CONSTRICTUS			2			1					1
COPTOSPORA RETICULATA			?1								2
BIRETISPORITES POTONIAEI			2							2	
RETITRILETES CLAVATIDITES			1								
CORONATISPORA VAIDENSIS			?1								1
EUCOMMLIDITES MINOR			1								
UNDULATISPORITES PANNUCEUS			1								
FORAMINISPORIS ASYMMETRICUS			1	1	1		1		1		7
PITYOSPORITES ALATIPOLLENITES			1		2		8		1	2	4
BACULATISPORITES COMAUMENSIS			1						2		3
PILOSISPORITES VERUS			1				1				
APPENDICISPORITES ERDTMANIL				1							
DENSOISPORITES VELATUS				2							1
PILOSISPORITES TRICHOPAPILLOSUS				1							2
LARICOIDITES MAGNUS				1	1	1					1
CICATRICOSISPORITES PURBECKENSIS				?1							
FORAMINISPORIS WONTHAGGLENSIS				1							1

RETITRILETES RETICULUMSPORITES	1							
PILOSISPORITES ERICUS	1							
GLEICHENLIDITES SENONICUS	1	5		2		1	2	2
CIRCULINA PARVA	1			4		1		
OBTUSISPORIS CANADENSIS		1						
SCHIZOPHACUS PARVUS		1						
AEQUITRIRADITES SPINULOSUS		2		1				2
IMPARDECISPORA TRIORETICULOSUS		1						
PODOCARPITITES MULTESIMUS		1						
DELTOIDOSPORA DIAPHANA		1		1				
LEPTOLEPIDITES VERRUCATUS		1				1		
PEROMONOLETES ALLENSIS		8	3	3		3	17	
TIGRISPORITES RETICULATUS		1						
TIGISPORITES SCURRANDUS		2		1				
DELTOIDOSPORA PSILOSTOMA		1						
FORAMINISPORIS DAILYI			1					
DENSOISPORITES MICRORUGULATUS			1					
PRISTINUSPOLLENITES MICROSACCUS			2				1	
CINGUTRILETES POCOCKIL			2					
PUNCTATOSPORITES SCABRATUS			1					
CEDRIPITES CRETACEUS			1					3
ARAUCARIACIDITES AUSTRALIS			1					
ALISPORITES MINUTUS								
CONCAVISSIMISPORITES PARKINII				2		1		
JANUASPORITES SPINIFERUS							1	
CICATRICOSISPORITES UNDATUS				1				
CYATHIDITES AUSTRALIS				3				
CALLIALASPORITES DAMPIERI				?1			1	
COOKSONITES VARIABILIS				1			1	
APPENDICISPORITES POTOMACENSIS				1				
TODISPORITES MINOR				1		1		2
NEORAISTRICA TRUNCATA				1				
CERATOSPORITES POCOCKIL				1				
ANTULSPORITES DISTAVERRUCOSUS							1	
DELTOIDOSPORA JUNCTA							1	
VITREISPORITES PALLIDUS							1	
KLUKISPORITES PSEUDORETICULATUS							1	
MICRORETICULATISPORITES UNIFORMIS							1	
APPENDICISPORITES CRISTATA								1
CERATOSPORITES EQUALIS								1
DICTYOTRILETES GRANULATUS								1
IMPARDECISPORA PURVERULENTUS								1

ALGAE

MICRHYSTRIDIUM STELLATUM	1			x				1
MICRHYSTRIDIUM SP.	x				x			
CLEISTOSPHAERIDIUM POLYPES	x		x	?				1
LEPTODINIUM - MENDICODINIUM	x			?				?
PERIDINIALES	x							?
MUDERONGIA ASYMMETRICA		1			x			
BAITISPHAERIDIUM CRAMERI		?			x			
VERYHACHIUM LAIRDI					x			
VERYHACHIUM RHOMBOIDIUM					x			

FIG. 12.—List of pollen species for individual samples from the estuarine environment.

scale complex bedding (e.g., ripple-laminated sands with mud flasers) than the fluvial distributary channel sands. More detailed descriptions of the channel sands are given in Table 1. (4) The tidal channel sands in some instances contain concentrations of channel breccia (Fig. 18), where blocks of salt marsh and tidal flat sediments have fallen into undercutting tidal creeks and were quickly buried by sand. (5) On channel margins spillover sands are common. The spillover sands are composed of repetitive cycles of high-energy plane-bedded sands with cycles of low-energy bioturbated muds (Fig. 19). This is interpreted to be the result of repeated storm cycles that pushed previously deposited sand from the channels back into the tidal flat environment, similar in process to levee sediments deposited on the margins of freshwater channels. These spillover sands shale out in short distances away from the channel margin, and thus it is important to differentiate these sands from the tidal channel sands. The repetitive cycles of high- and low-energy environments are the key and they can only be recognized from core material. Figures 17 and 19 illustrate the difference between spillover sands and tidal channel sands. The log signatures of tidal channel sands with mud flasers are identical to the signatures of spillover sands, but their relative value for oil sands development is vastly different. This is discussed in the following section on Implications for Development.

Other minor facies and subfacies in the estuarine environment are constituents of the overall sedimentary picture in the Sandalta area, but some are much more prevalent in other local areas and thus are included in the facies chart (Fig. 22).

Spatial relationships of the sedimentary facies units.—

The entire McMurray sequence is characterized by repetitive scour-and-fill cycles, so that within a major depositional environment sequence, a particular facies may occur any number of times in the stratigraphic column. Only the boundaries between the main depositional environments have any time connotation.

Figures 20 and 21 are simplified facies cross sections. Many of the thinner facies have been omitted. Since both the distributary channels and major tidal channels trend north-south on the Sandalta lease, there is good correlation parallel to this trend as shown in Figure 20. Many of the nat-

FIG. 13.—Lower tidal flat facies; flaser bedding is the dominant feature. Gulf Sandalta CH 178, Box 22.

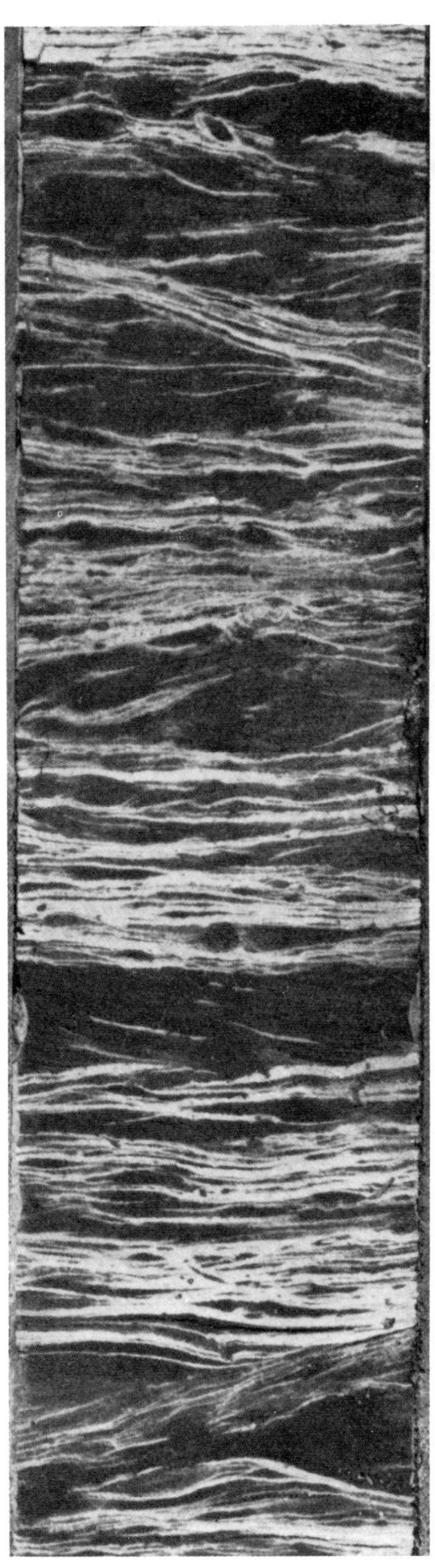

FIG. 14.—Upper tidal flat facies; flaser bedding is the dominant feature. Gulf Sandalta CH 100, Box 30. Width of core is 6 cm.

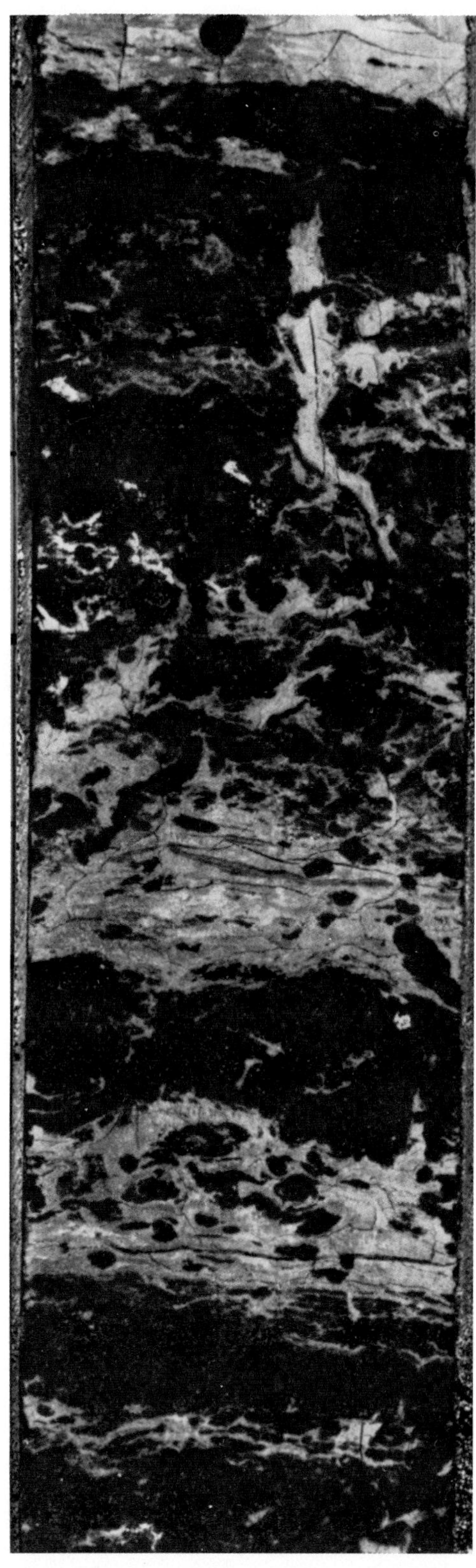

FIG. 15.—Lower tidal flat facies, extensively borrowed. Gulf Sandalta CH 137, Box 25. Width of core is 6 cm.

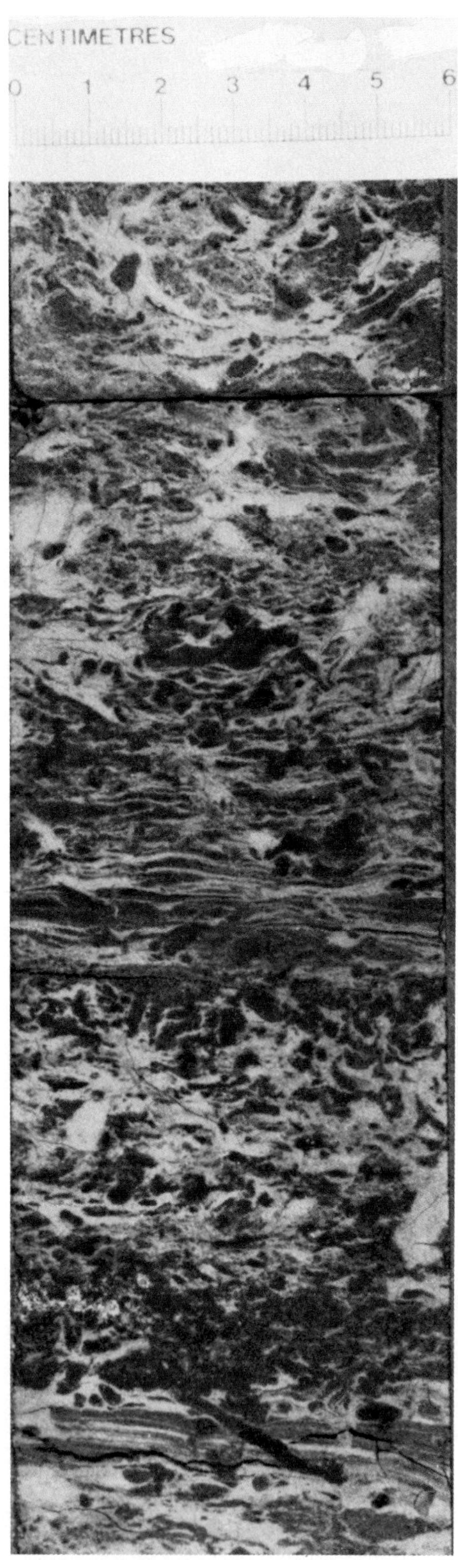

FIG. 16.—Upper tidal flat facies, extensively burrowed. Gulf Sandalta CH 179, Box 21.

FIG. 17.—Tidal Channel Sand facies. The bedding is mainly bidirectional ripple laminations. Note the occasional thin mud flaser. Gulf Sandalta CH 101, Box 7-6. Top of cored interval is the upper right corner.

ural outcrop exposures exhibit this good correlation because present incised creek channels tend to follow the pre-existing McMurray channel sands. The east-west cross section in Figure 21 shows very little correlation even with drill holes 400 m apart. The grey shaded areas represent the best reservoir sands, whereas the dotted shaded areas are shalier sands or sands that are known to be erratic in oil saturation and are laterally discontinuous. Note that some of the fluvial channel sands in the lower portion of the McMurray Formation are water-saturated. It is obvious from Figure 21 that net sand isopach maps would not adequately represent the trend in the pay zones in this scour-and-fill sequence.

To circumvent this problem on the Sandalta lease, a three-dimensional facies block model was built and stored on computer. Medsystem 10, a mine planning block model offered by Mintec, Inc., was selected for this purpose. In order to build such a block model, a grid of east-west and north-south cross sections 400 m apart were hand drawn and then digitized. The computer was then allowed to extrapolate facies boundaries in a north-south direction between east-west digitized cross sections, i.e., in the direc-

TABLE 1.—COMMON CHARACTERISTICS OF THE THREE PRIME TARGET SAND FACIES

Facies	Common Characteristics
Tidal Channel Sand	—Occurs above the scour base of the estuarine environment. —Very fine- to fine-grained, well sorted, highly quartzose (95% to 98% quartz). —Small-scale bidirectional ripple cross laminations form the dominant bedding and are occasionally interrupted by mud flasers or zones of channel breccia having angular clasts of saltmarsh mud or tidal flat sediments. —Oil saturation is high and relatively consistent. —Fines content (silt and clay) averages 12%. —30% to 36% porosity.
Fluvial-Estuarine Sand	—Sand bodies are large-scale sand waves or mega-ripples pushed into the estuarine environment by rivers at flood stages. —Very coarse sand-to-gravel sizes, angular, poor to moderate sorting, highly quartzose. —Large-scale crossbedding or massive appearance. —Generally found near the base of the estuarine sequence, scoured into other estuarine sediments. —Fines content (silt and clay) less than 7%. —29% to 32% porosity.
Distributary Channel Sand (Fluvial Channel)	—Occurs throughout the delta marsh sequence, scoured into pond mud or overbank mud facies. —Large-scale crossbedding is the dominant bedding type; occasional graded-tabular crossbedding occurs. —Fine- to medium-grained, moderate- to well-sorted, highly quartzose sand with occasional coarse sand beds. —Scattered thin clay drapes on reactivation surfaces. —Scattered coal fragments and carbonaceous laminae. —High oil saturation above the oil/water contact, but water-saturated with no residual oil staining below the contact. —More erratic fines content than tidal channel or fluvial-estuarine sands. —27% to 34% porosity.

FIG. 18.—Channel breccia facies. Angular clasts of saltmarsh mud and tidal flat sediments are incorporated into a clean oil-saturated channel sand matrix. Gulf Sandalta CH 177, 16-34-94-10W4.

tion parallel to preferred orientation of the channels. The blocks in the model (1.2 million blocks, 100 m × 100 m × 1.5 m thick) were then assigned codes for facies type, fines content and expected range of oil contents. Following a standard geostatistical evaluation approach, oil contents were extrapolated by Kriging (David, 1976) core analyses values within facies units from well data in a north-south oriented elliptical envelope. These values usually fell within

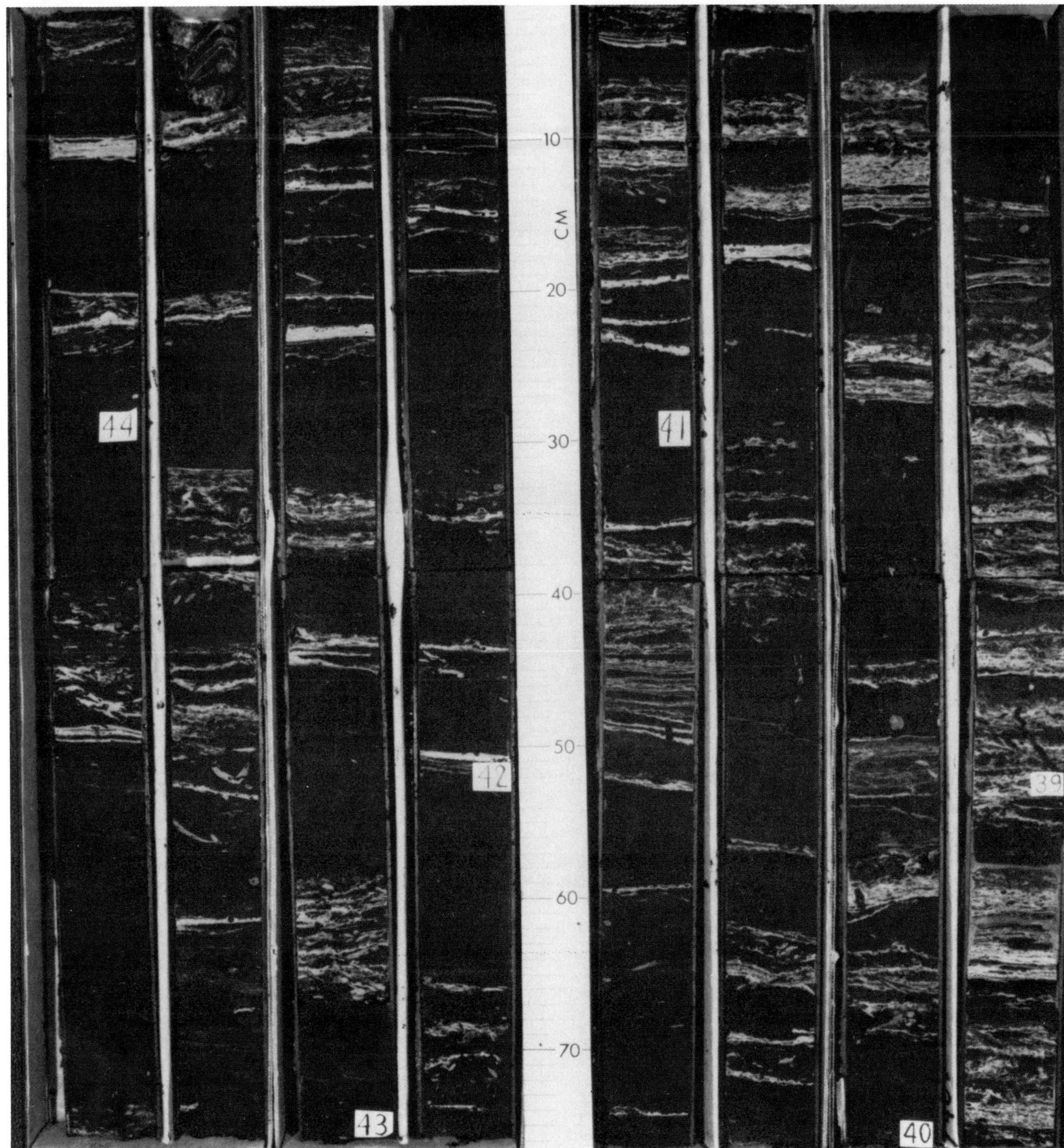

FIG. 19.—Spillover sand facies. Note the repetitive cycles of clean sand (in this case high-energy plane beds) with cycles of lower energy, burrowed, muddy tidal flat sediments. Compare the differences in sedimentary features with Figure 17, tidal channel sand. Gulf Sandalta CH 191, 10-32-94-9W4, Box 20-17. Top of the cored interval is the upper right corner.

the range expected for the facies in any particular block. Where there was any discrepancy, the facies data would overrule the Kriged values.

This proved to be a very useful tool for effectively displaying the data in either vertical cross sections or horizontal slices representing mine benches. For evaluating prospects for open-pit mining potential, the computer block model is a necessity to evaluate the effects of various cutoff

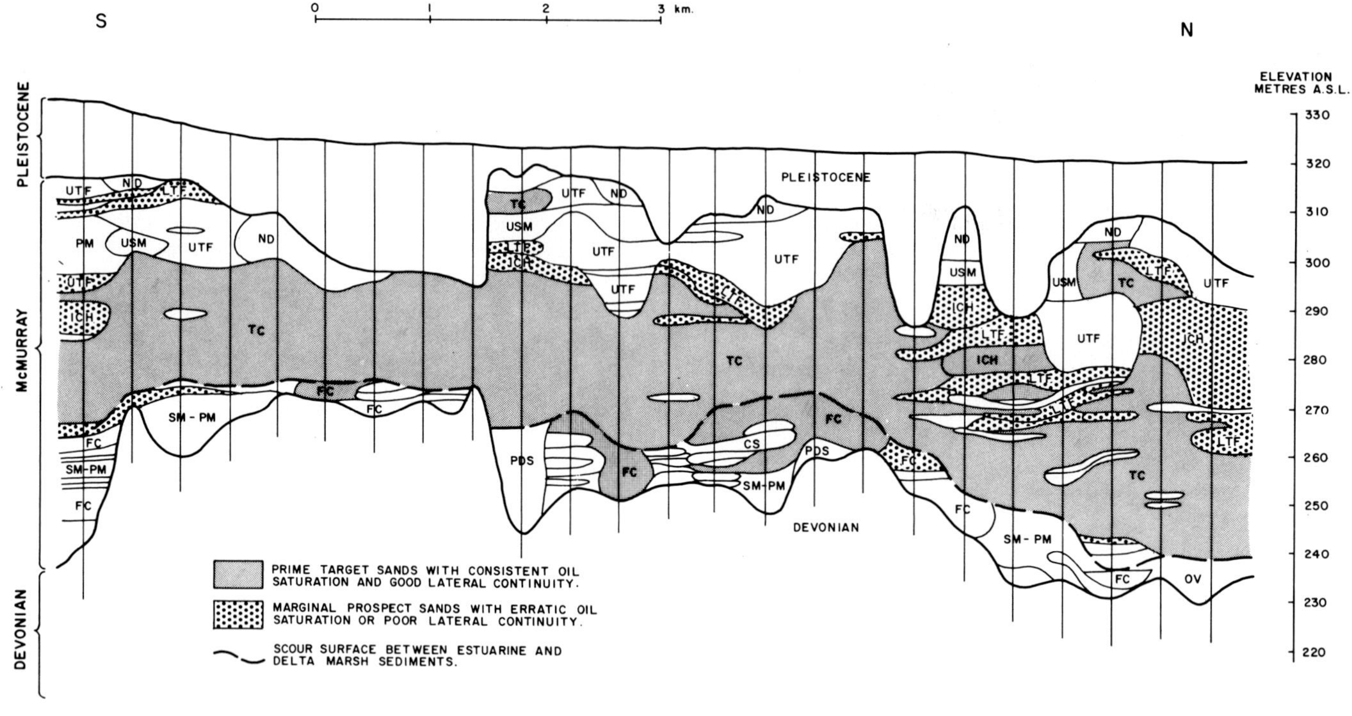

LEGEND		
Depositional Environment	Symbol	Facies
	ND	No Data
ESTUARINE	PM	Prodelta Mud
	USM	Upper Salt Marsh
	ICH	Spillover
	AC	Abandoned Tidal Channel
	TC	Tidal Channel
	UTF	Upper Tidal Flat
	LTF	Lower Tidal Flat
	TM	Tidal Mud
	CB	Channel Breccia
	FES	Fluvial Estuarine Sand
		Widespread Scour
DELTA MARSH	AC	Abandoned Distributary Channel
	OV	Overbank Muds
	FC	Fluvial or Distributary Channel
	CS	Crevasse Splay
	SM – PM	Pond Mud
	FB	Fluvial Breccia
FLUVIAL	EF	Early Fluvial
	PDS	Post-Depositional Slump
DEVONIAN		

FIG. 20.—North-south cross section of sedimentary facies in McMurray Formation, Sandalta project.

parameters or equipment selectivity on scheduling and overall economics of the various prospects.

IMPLICATIONS FOR DEVELOPMENT

The facies chart in Figure 22 lists the 22 facies identified in the delta marsh and estuarine sediments. The three sand facies (tidal channel, fluvial-estuarine and distributary channel) shown in the shaded bars are the only sands with any lateral continuity and consistency in reservoir quality and oil saturation. They are thus the prime targets for development. Typical characteristics of these sands are listed in Table 1.

In the mineable area the marginal sands are considered by the author to be marginal "pay," which is only added to potential reserves if encountered in pursuit of the better sands. If the marginal sands overlie economically mineable prime target sands, they will be mined while attempting to recover the higher grade sands, but a mine plan should not be dependent on these marginal sands because they are unreliable in reservoir quality and lateral continuity. The tidal flat sands have erratic sand and oil contents and can shale out rapidly. The most potentially misleading facies is the spillover sand. It can have oil contents as high or higher than the channel sands, but it pinches out quickly, making it a poor target for development. It is thus important to obtain core material and correctly delineate the sedimentary facies. Core analysis results, or log-derived oil saturations alone, can be misleading because the spillover sands will only be identified as rich oil sands and not differentiated from tidal channel sands. The oil contents expressed as weight percent bitumen in Figure 22 are the measured values from core analysis. Note that there is considerable overlapping of the range of values for some facies units. It would

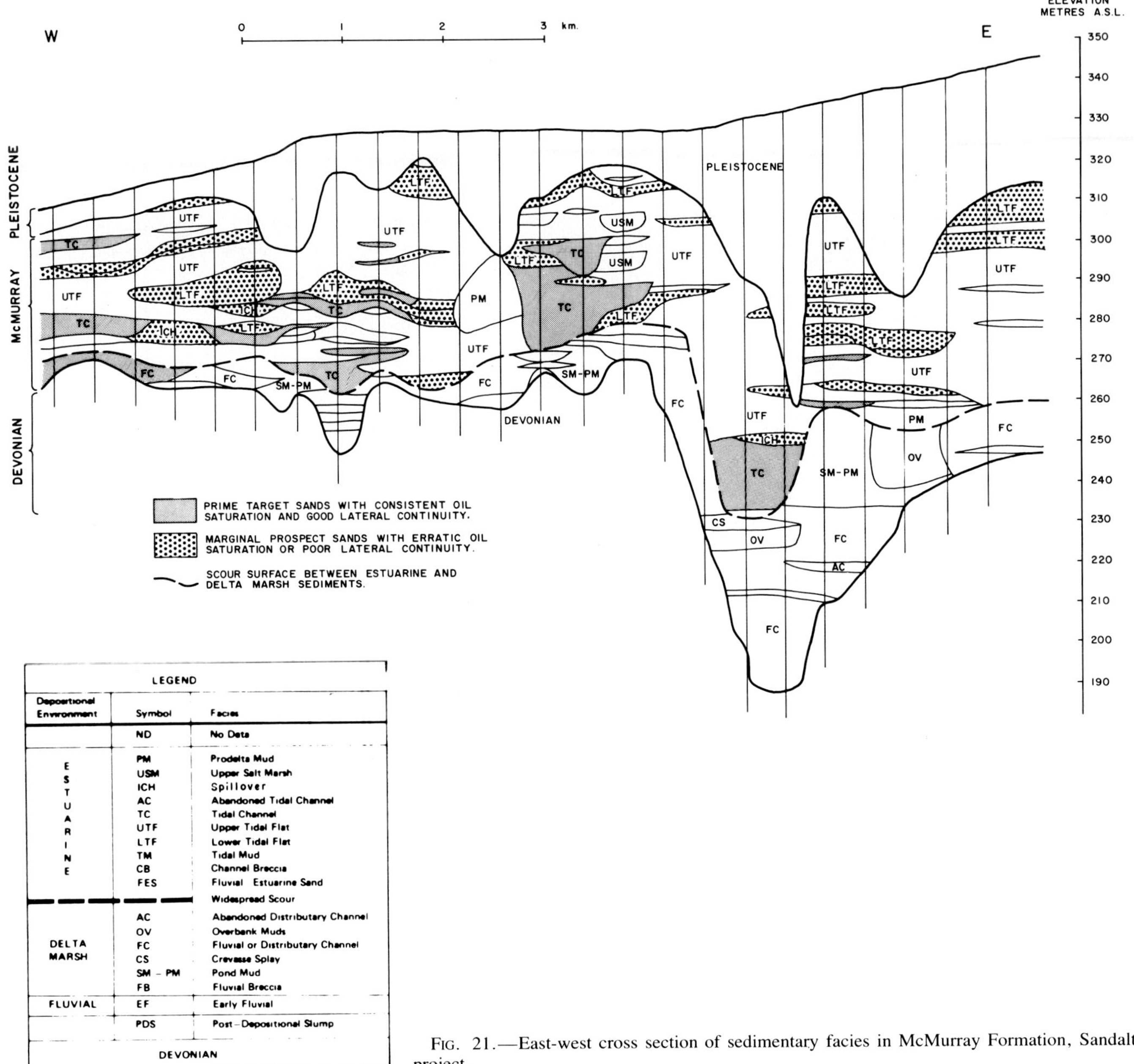

LEGEND

Depositional Environment	Symbol	Facies
	ND	No Data
ESTUARINE	PM	Prodelta Mud
	USM	Upper Salt Marsh
	ICH	Spillover
	AC	Abandoned Tidal Channel
	TC	Tidal Channel
	UTF	Upper Tidal Flat
	LTF	Lower Tidal Flat
	TM	Tidal Mud
	CB	Channel Breccia
	FES	Fluvial Estuarine Sand
		Widespread Scour
DELTA MARSH	AC	Abandoned Distributary Channel
	OV	Overbank Muds
	FC	Fluvial or Distributary Channel
	CS	Crevasse Splay
	SM - PM	Pond Mud
	FB	Fluvial Breccia
FLUVIAL	EF	Early Fluvial
	PDS	Post-Depositional Slump
DEVONIAN		

FIG. 21.—East-west cross section of sedimentary facies in McMurray Formation, Sandalta project.

be impossible to distinguish spillover sands from tidal channel or fluvial channel sands based on weight percent bitumen alone, yet in the past it was standard practice to use the weight percent bitumen figures as the only variable describing the quality of the pay zone in evaluation mapping.

In evaluation of mineable oil sands, it unfortunately has been common industry practice to outline areas for further exploration or development based on waste/ore ratio maps or economic mining factors. In the case of *in situ* oil sand development schemes, net pay isopach maps are commonly used. These two approaches can be misleading because they ignore the valuable information to be gained by taking a closer look at the depositional environment of the pay zone itself. A graphic case history follows:

Figure 23A shows an area within the Sandalta lease which was outlined for potential development by conventional mapping techniques with drill hole control of 4 to 5 drill holes per square mile. The pay zone itself, however, is mostly spillover sands and tidal flat sands, both of which are erratic in sand content and lateral extent. Figure 23B shows that, as expected for these facies, the subsequent development drilling to 12 drill holes per square mile drastically changed the outline of the potential area. The value of this area is very much decreased based on the additional data. The sand distribution is too erratic for successful surface mining. The outline would probably change even more with increased drilling, and the area would then also have little value for *in situ* recovery schemes. In the existing open-pit

DEPOSITIONAL ENVIRONMENT	FACIES	SUBFACIES	LITHOLOGY	WEIGHT % BITUMEN
	PRODELTA MUD		SILT + CLAY	0% – 2%
	SALT MARSH		CLAY	0
	SPILLOVER		CYCLES OF SAND WITH SILT + CLAY	9% – 13%
	ABANDONED CHANNEL		SILT, CLAY + SAND	0% – 2%
	TIDAL CREEK TIDAL CHANNEL	TIDAL CHANNEL T C + BRECCIA	SAND + MUD FLASERS	10% – 17%
	CHANNEL BRECCIA		SAND WITH CLASTS OF SALT MARSH OR TIDAL FLAT.	4% – 14%
ESTUARINE	TIDAL FLAT SEDIMENTS UTF = UPPER TIDAL FLAT LTF = LOWER TIDAL FLAT	TIDAL MUD	CLAY	0
		UTF INTERBEDDED	SILT + SAND	1% – 7%
		UTF BURROWED	SILT + SAND	1% – 7%
		UTF MIXED	SILT + SAND	1% – 7%
		LTF INTERBEDDED	SAND + SILT	3% – 10%
		LTF BURROWED	SAND + SILT	3% – 10%
EROSIONAL UNCONFORMITY		LTF MIXED	SAND + SILT	3% – 10%
	FLUVIAL ESTUARINE SAND		COARSE SAND	5% – 15%
DELTA MARSH	ABANDONED CHANNEL		SILT, CLAY, SAND	0% – 2%
	OVERBANK MUDS	OVERBANK MUD COAL	SILT + CLAY, COAL	0% – 3%
	FLUVIAL (DISTRIBUTARY) CHANNEL	FLUVIAL SAND FLUVIAL BRECCIA	SAND	4% – 16%
	CREVASSE SPLAY		SAND + DEBRIS	0% – 7%
	POND MUD		CLAY	0
FLUVIAL	EARLY FLUVIAL		IMMATURE SAND & GRAVEL	0% – 4%

FIG. 22.—Sedimentary facies, McMurray Formation, Sandalta project. Note: Within a particular depositional environment, the individual facies can occur in any order or can be repeated a number of times. Only the boundaries between the main depositional environments carry any time connotation.

mines, correlation for more than 100 m horizontally in these tidal flat and spillover sand facies units is often impossible.

In contrast, Figures 24A and 24B illustrate an area within the Sandalta lease where the pay zone is mainly tidal channel sand, one of the three primary targets identified for development. In this case, as expected for this facies, the outline and overall reserves did not change significantly even after development drilling to 16 core holes per square mile.

The above case history illustrates how recognition of facies through core studies and delineation of prime target sands early in the exploration game can save a lot of disappointment and wasted development drilling costs. This allows the operator to concentrate on the best areas.

Whereas this model has proved to be very useful for exploration and development in the study area and immediate vicinity, modifications are necessary for other areas of the Athabasca deposit where some facies are more widespread and others are absent. For example, local thick pockets of good point bar sands at the top of the delta marsh sedimentary sequence are found along the margins of the Athabasca deposit, where tributary rivers have supplied more sand. Also, fluvial-estuarine sands are not very common on the Sandalta project area, but extensive fluvial-estuarine sands have developed in the estuarine sediment group elsewhere where major tributaries have contributed coarse sand during floods. In the study area, erosion during the estuarine pe-

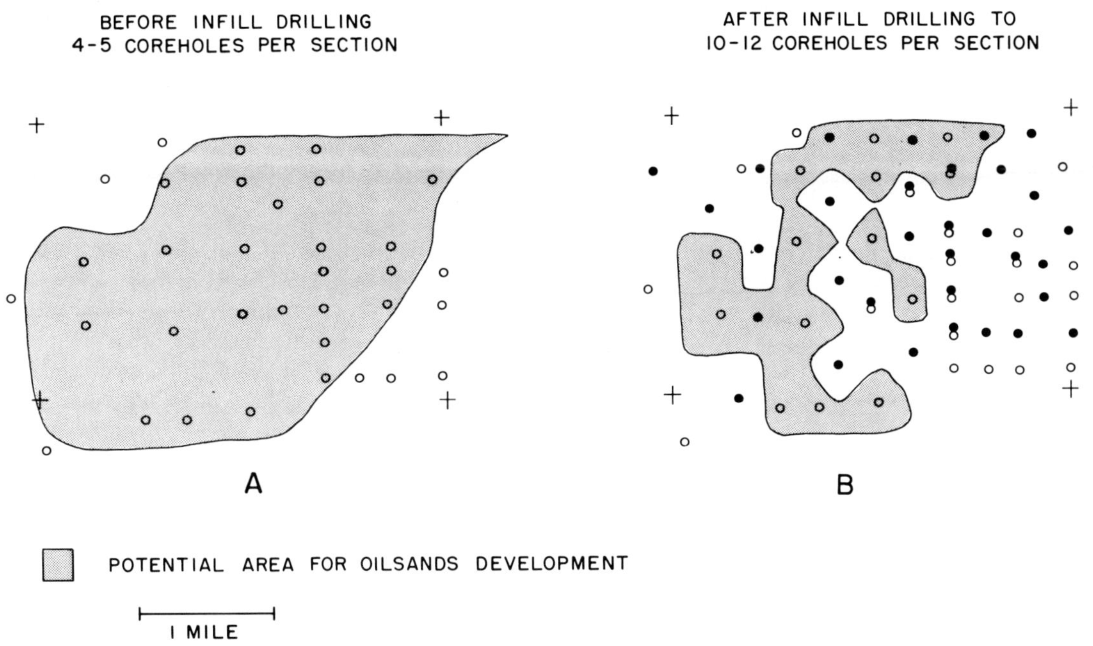

BEFORE INFILL DRILLING
4-5 COREHOLES PER SECTION
AFTER INFILL DRILLING TO
10-12 COREHOLES PER SECTION
A
B
POTENTIAL AREA FOR OILSANDS DEVELOPMENT
1 MILE

FIG. 23.—Pay zone is mainly spillover sands and tidal flat sands.

BEFORE INFILL DRILLING
4-5 COREHOLES PER SECTION
AFTER INFILL DRILLING TO
12-16 COREHOLES PER SECTION
A
B
POTENTIAL AREA FOR OILSANDS DEVELOPMENT
1 MILE

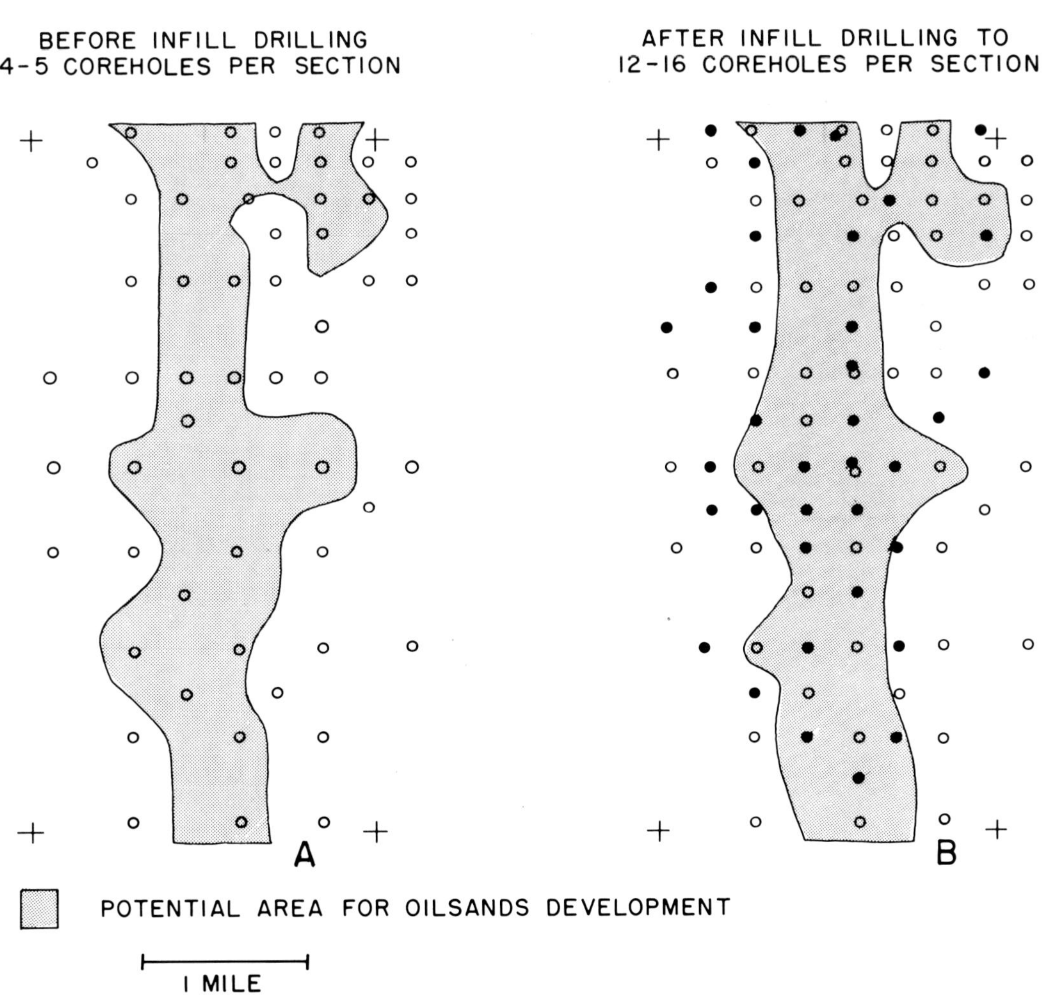

FIG. 24.—Pay zone is mainly tidal channel sands.

riod has removed many meters of progradational fluvial sand from the top of the delta marsh sequence. Immediately to the north of the study area, these sediments have been preserved, apparently because the underlying Paleozoic topography was not as confining and thus the tidal currents were not as strong. Other similar occurrences of the progradational fluvial sand can be expected locally within the Athabasca deposit.

In the regional study area, the later McMurray marine sediments are not well preserved due to subsequent erosion during the Pleistocene. In places where erosion has been less severe, the marine sediments contain some marginal prospect shoreface sands that could conceivably add to total reserves if encountered in pursuit of the three prime target sands.

Although the early fluvial sands are generally not oil-bearing in the study area, they are both oil-bearing and laterally more extensive in parts of the southeast portion of the Athabasca deposit. A study similar to this one may help to identify individual facies that are the prime targets for exploration in that area.

CONCLUSIONS*

An overall transgressive sequence during the Early Cretaceous was responsible for the basin-filling McMurray Formation sediments. The McMurray Formation is mainly shale with localized buildups of sand caused by repetitive scour-and-fill cycles. Four stages of sedimentation have been recognized and are called early fluvial, delta marsh, estuarine and marine. In the study area only the delta marsh and estuarine environments contain significant oil-bearing sands.

Of the 22 facies identified in the delta marsh and estuarine sediments, only three sand facies have any appreciable lateral extent and consistency in oil saturation. They are the tidal channel sands, fluvial-estuarine sands and distributary channel sands. These sands are therefore considered to be the prime targets for exploration and development. Other oil-bearing facies should be considered marginal pay that might contribute to overall reserves in a mining project only if they are encountered in pursuit of the three prime target sands.

Conventional mapping techniques, such as waste/ore ratios or isopach maps, are often misleading in this scour-and-fill sequence. In addition, any mapping based solely on oil contents from core analysis or logs will be misleading. Computer block models are most useful in mapping the spatial distribution of the various facies and in evaluating prospects for development potential.

Detailed facies studies can help immensely in the exploration and development of the Athabasca oil sands by identifying the primary sand targets. This can save a lot of time and development drilling costs by allowing the operator to concentrate on the sands that are most reliable in lateral extent and reservoir quality.

It is also recognized that the overall transgressive sedimentary sequence described herein was repeated in similar fashion with some modifications in many other areas during the Cretaceous, particularly in other Mannville sediments in Alberta. This type of detailed study could be useful in understanding the geology and identifying the main targets for exploration or development in those areas.

*AUTHOR'S NOTE: After this paper was completed, additional drilling of 62 core holes was carried out in the winter of 1984–85 to prove a potential mine area within the tidal channel sand body shown in Figure 24. In addition, 39 km of refraction seismic and electromagnetic profiles were obtained. The results of both the drilling and the geophysics show that the tidal channel sand body has a high degree of lateral continuity and consistent high reservoir quality. This followup work further illustrates the value of using facies studies to select the best prospects for development drilling.

ACKNOWLEDGMENTS

The author acknowledges Gulf Canada Resources, Inc. for permission to publish these data, Elliot Burden for carrying out the palynological analyses, and Ross Lennox for his helpful review of the text. G. A. Stewart provided some helpful analysis of sedimentary units early in the study of the Sandalta core data. I also express thanks to the mine personnel at the Suncor and Syncrude mines for their assistance in examining the mine sites.

REFERENCES

CARRIGY, M. A., 1959, Geology of the McMurray Formation: Research Council of Alberta, Memoir I, Part III, 130 p.

———, 1966, Lithology of the Athabasca oil sands: Research Council of Alberta, Bulletin No. 18, 48 p.

———, 1971, Deltaic sedimentation in Athabasca tar sands: American Association of Petroleum Geologists Bulletin, v. 55, p. 1155–1169.

DAVID, M., 1976, The practice of Kriging, *in* Guarascio, M., and others, eds., NATO Advanced Study Institute, Advanced Geostatistics in the Mining Industry: Reidel, Dordrecht, Holland, 1976, p. 31–48.

DUC, A. W., 1981, Back-barrier stratigraphy, Kiawah Island, South Carolina: Unpublished M.S. Thesis, University of South Carolina, 253 p.

FLACH, P. D., 1977, A lithofacies analysis of the McMurray Formation, lower Steepbank River, Alberta: Unpublished M.S. Thesis, University of Alberta, 139 p.

HAYES, M. O., 1976, Terrigenous clastic depositional environments, some modern examples: Technical Report No. 11-CRD, Coastal Research Division, Department of Geology, University of South Carolina, 4th Edition (1980), p. I-50 to I-109.

JAMES, D. P., 1977, The sedimentology of the McMurray Formation, east Athabasca: Unpublished M.S. Thesis, University of Calgary, 198 p.

MCPHERSON, R. A., AND KATHOL, C. P., 1977, Surficial geology of potential mining areas in the Athabasca oil sands region: Alberta Research Council Open-File Report 1977-4, 185 p.

MOSSOP, G. D., 1978, Epsilon cross-strata in the Athabasca oil sands (abs.), *in* Miall, A. D., ed, Fluvial Sedimentology: Canadian Society of Petroleum Geologists Memoir 5, p. 854–855.

———, 1980, Facies control on bitumen saturation in the Athabasca oil sand, *in* Miall, A. D., ed., Fluvial Sedimentology: Canadian Society of Petroleum Geologists Memoir 5, p. 609–632.

STEWART, G. A., 1963, Geological controls on the distribution of Athabasca oil sand reserves, *in* Carrigy, M. A., ed., The K. A. Clark Volume: Alberta Research Council Information Series No. 45, 1963, p. 15–26.

WOJTAL, A. M. AND MOSLOW, T. F., 1980, Stratigraphy of barrier and back-barrier facies, Kiawah Island, South Carolina, *in* Frey, R. W., ed., Coastal Environments of Georgia and South Carolina: Geological Society of America Field Trip Guide Book, v. II, p. 289–310.

DEPOSITIONAL MODELLING OF THE UPPER MANNVILLE (LOWER CRETACEOUS), EAST CENTRAL ALBERTA: IMPLICATIONS FOR THE RECOGNITION OF BRACKISH WATER DEPOSITS

DARYL M. WIGHTMAN[1], S. GEORGE PEMBERTON[2] AND CHAITANYA SINGH[1]

[1]Alberta Geological Survey, Alberta Research Council, P.O. Box 8330, Station F, Edmonton, Alberta Canada, T6H 5X2; [2]Department of Geology, University of Alberta, Edmonton, Alberta, Canada, T6G 2E3

ABSTRACT: Thick (15 to 35 m, 49 to 115 ft) sandstones in the upper Mannville (Aptian to early Albian) in east central Alberta have several genetic origins. In the western part of the study area, thick sandstones have formed by the stacking of fluvial channel deposits. The variable thickness and irregular distribution of the sandstones on cross sections support the presence of multiple paleochannels. Oil saturation and reservoir pressure data corroborate this interpretation in the Hairy Hill area.

In the eastern part of the study area, sandstones of similar thickness have formed by an amalgamation of channel and marine shoreline facies and by the stacking of crevasse splay deposits in brackish water bays. Ichnological and palynological data confirm that the upper Mannville becomes more marine in the eastern part of the study area. It is concluded that during upper Mannville time, the streams in the west were flowing into a marine embayment which lay to the east.

Brackish water deposits have predictive value, as they signal the change from continental to marine sedimentation. Based on the upper Mannville, several criteria have been proposed for the interpretation of these environments in core. The presence of syneresis cracks, pyrite and siderite and a limited assemblage of certain trace fossils and marine microplankton together indicate a marginal marine setting. These criteria are complementary and represent a multidisciplinary approach which the authors advocate for the recognition of paleoenvironments.

INTRODUCTION

In a series of recent papers (Putnam, 1980, 1982a, 1982b, 1983; Putnam and Oliver, 1980; Smith and Putnam, 1980) the upper Mannville of east central Alberta has been interpreted as an anastomosed fluvial deposit. The model was proposed to explain the occurrence of unusually thick (15 to 35 m, 49 to 115 ft) ribbon-shaped sandstones. One of the present authors (DMW) discussed the original paper by Putnam and Oliver (1980) and rejected the anastomosed fluvial interpretation for various reasons (Wightman and others, 1981). The purpose of this paper is to propose an alternate interpretation. In addition, during the course of the study, the advantages of using several types of data to recognize brackish water deposits became apparent. The second part of the paper is an outgrowth of this approach.

REGIONAL GEOLOGY

The Lower Cretaceous Mannville Group of Aptian to early Albian age (Stelck and Kramers, 1980) is a siliciclastic sequence composed mainly of sandstone and is economically important because of reserves of heavy oil and gas. In the study area (Fig. 1) it is 125 to 225 m (410 to 738 ft) thick, dips approximately 1.9 m/km (10 ft/mi) to the southwest (Orr and others, 1977) and is encountered at depths of 350 to 850 m (1,148 to 2,789 ft). The Mannville Group overlies a regional unconformity developed on Devonian carbonates, primarily of the Beaverhill Lake Group, and is disconformably overlain by the Joli Fou Formation of early late Albian age (Stelck, 1958, 1975; Caldwell and others, 1978).

Putnam (1982a) reviewed the stratigraphy of the Mannville Group in the Lloydminster area and his informal subgroups of lower, middle and upper are utilized in this paper (Fig. 2). Precise correlation of the upper Mannville to stratigraphic units in the United States is uncertain. Christopher (1984) correlates the Mannville Group to the Lakota, Fuson and Fall River Formations in the Williston basin, whereas Vuke (1984) correlates it to the Lakota and Fall River Formations in western North Dakota. The lower Mannville (Dina-McMurray Formations) is continental in origin and largely filled topographic lows developed on the sub-Cretaceous unconformity (Williams, 1963; Jardine, 1974). The lower Mannville probably represents the earliest response to a rising sea level to the north, as marine deposits (Cummings-Clearwater Formations) of the boreal Clearwater sea overlie the Dina-McMurray Formations in much of central and northern Alberta. The middle Mannville consists of several transgressive/regressive cycles deposited in the Clearwater sea (Vigrass, 1977; Putnam, 1982a). The Sparky Formation, the last regressive sequence in the middle Mannville, is capped by a coal over most of the study area. Thus, continental conditions prevailed over east central Alberta at the end of middle Mannville time.

Various depositional interpretations have been proposed for the upper Mannville with some authors favoring a continental origin (Williams, 1963; Vigrass, 1977), whereas others include marine and nonmarine environments (Orr and others, 1977) in the interval. Putnam and Oliver (1980) interpreted the upper Mannville as an anastomosed fluvial deposit over a large area (approximately 200 × 300 km, 124 × 186 mi) of east central Alberta. They suggested that nearshore marine deposits may be found to the north of their study area (north of Township 64). In two subsequent publications (Putnam, 1982a; Putnam, 1983) the anastomosed fluvial system was extended about 100 km (62 mi) eastward into Saskatchewan. Lentin and Putnam (1981), however, presented data which are contrary to this depositional model as they pointed out that there are marine microplankton (acritarchs) in the upper Mannville. Lentin and Putnam (1981) postulated the existence of extensive salt-water ponds and brackish lakes to explain the presence of the acritarchs. Our study area is situated in the central part of Putnam and Oliver's (1980) anastomosed system and well to the south of their area of marine deposition.

Contribution No. 1457, Alberta Research Council

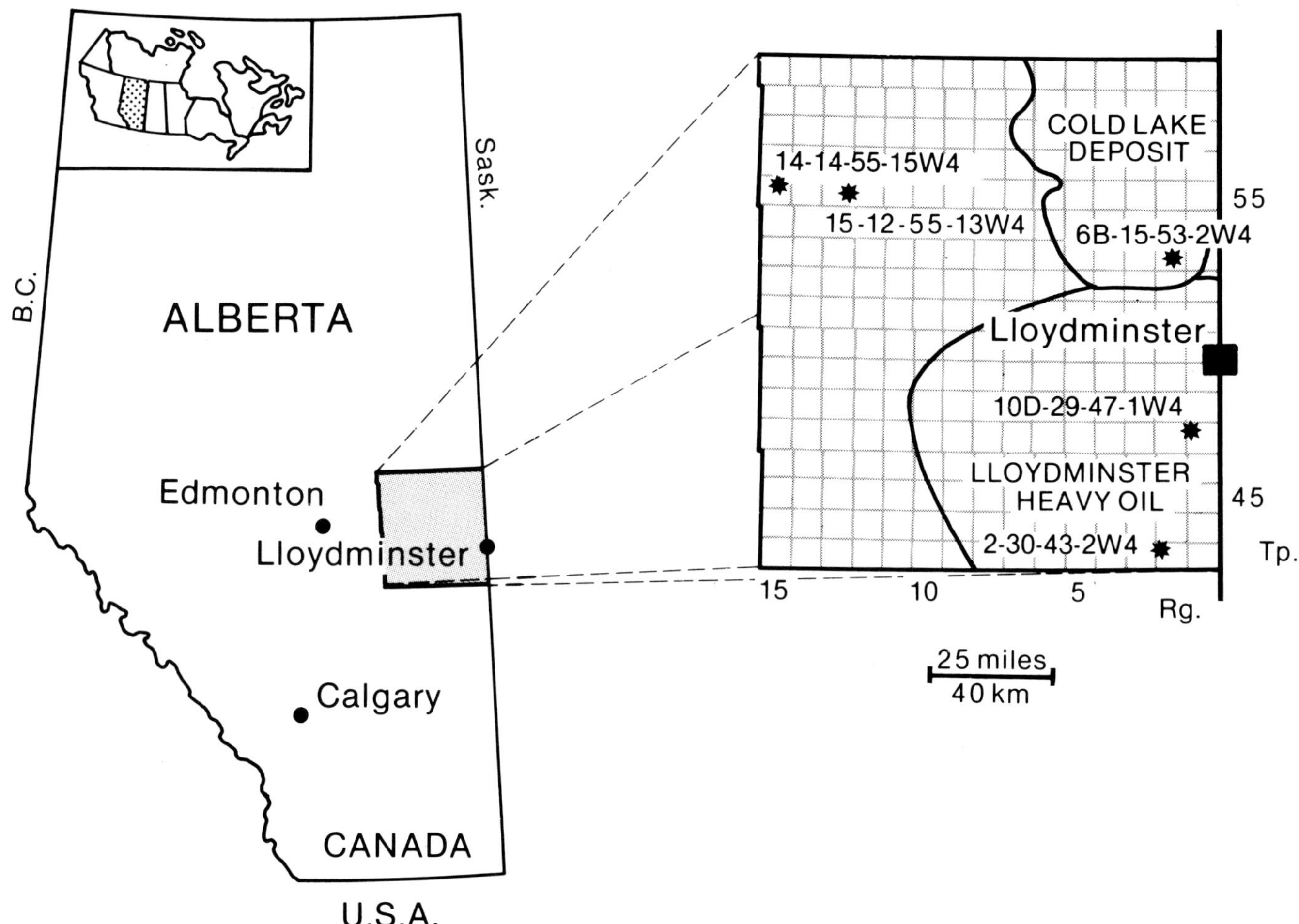

FIG. 1.—Location of study area, eastern Alberta, Canada.

METHOD OF STUDY

Several cross sections were constructed in the region bounded by Townships 47 to 55, Ranges 1 to 15W4, and 10 cores were examined within the area. The core descriptions included analysis of trace fossils and samples were taken for palynology. X-ray diffraction was used to identify clay mineralogy, pyrite and siderite, whereas commercial laboratories measured oil saturations and permeabilities. Westmin Resources kindly provided the reservoir pressure data in the Hairy Hill area.

	Cold Lake Area		Lloydminster Area
	Joli Fou Fm.		Joli Fou Fm.
Mannville Group	Upper Grand Rapids Fm.	Upper	Colony Fm.
			McLaren Fm.
	Lower Grand Rapids Fm.		Waseca Fm.
		Middle	Sparky Fm.
			General Petroleums Fm.
	Clearwater Fm.		Rex Fm.
			Lloydminster Fm.
			Cummings Fm.
	McMurray Fm.	Lower	Dina Fm.
	Devonian		Devonian

FIG. 2.—Mannville stratigraphy, east central Alberta (modified after Putnam, 1982a; 1982b).

DESCRIPTION AND INTERPRETATION OF FACIES

Three cores from the study area are discussed in detail. Two of the cores (15-12-55-13W4 and 10D-29-47-1W4) lie within the fluvial channel sandstones as mapped by Putnam and Oliver (1980) and the third (6B-15-53-2W4) is situated in the interchannel area. For each core, the facies are described, interpreted, and the results summarized in the subsections entitled Conclusions. Grain size data for core 6B-15-53-2W4 have been visually estimated, whereas data for the other two cores were obtained by sieving (the sandstones are essentially unlithified). The significance and implications of the facies work are given in the discussion at the end of each section.

Well 15-12-55-13W4

A thick sandstone (523 to 558 m, 1,716 to 1,831 ft) occurs in the upper Mannville of well 15-12-55-13W4 and is cored in its entirety (Fig. 3). The sandstone overlies a thin coal, as many of the other thick sandstones do (interpreted from geophysical logs).

Crossbedded sandstone facies (544 to 558 m, 1,785 to 1,831 ft).—

This facies is mainly fine-grained (Figs. 4, 5) and contains thick (as much as 75 cm, 30 in.) sets of high-angle crossbeds with intervening zones of thinner crossbeds, scour-and-fill structures, ripple cross-stratification and beds that appear massive. In five of the sets of crossbeds, the angle of the foreset laminae is almost constant from the lower to the upper part, indicating that the foresets are planar. The lower contacts of these laminae are angular to tangential. One set of crossbeds has concave foresets (Pettijohn and others, 1973) as the laminae steepen upward. Reactivation surfaces are not evident in the crossbeds. In one interval, several sets of crossbeds are separated by thin interbeds of very fine-grained sandstone which are ripple cross-stratified; some of the laminae appear to dip in the direction opposite to that of the crossbeds (Fig. 6). A heavy oil/water contact occurs at 552 m (1,811 ft), which is visible in the core.

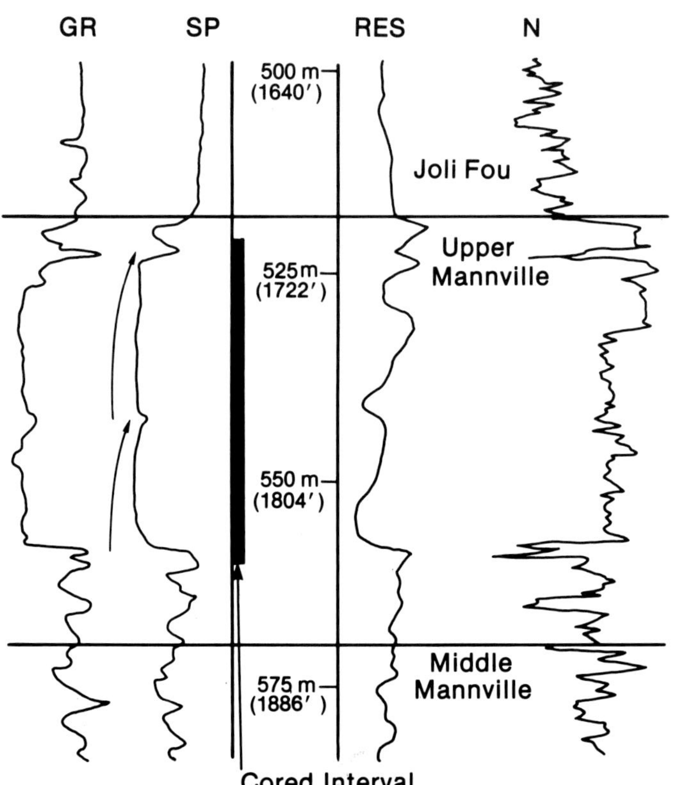

FIG. 3.—Geophysical logs and cored interval for 15-12-55-13W4.

Rippled sandstone facies (542 to 544 m, 1,778 to 1,785 ft).—

At 544 m (1,785 ft), four distinct changes occur in the sandstone: (1) it decreases to very fine-grained (Fig. 5), with a concomitant increase in silty matrix. A few beds have little or no oil saturation because of the high matrix content and appear white rather than dark brown. (2) Shale clasts and thin interbeds of shale (too thin to be plotted in Fig. 5) are present in the sandstone (Fig. 7). The thickest (25 cm, 10 in.) shale bed occurs at the top of the interval. (3) The sedimentary structures decrease in size to ripple cross-stratification; and (4) climbing ripples are present (Fig. 7). Some of the shales contain trace fossils (*Planolites* and possibly *Palaeophycus*) which are not diagnostic of marine or nonmarine environments. Marine microplankton (Table 1) have been recovered from the thickest shale, however, indicating that there was at least some saline influence during clay deposition (Fig. 8).

Interpretation.—The features of the crossbedded sandstone facies (544 to 558 m, 1,785 to 1,831 ft) are typical of the lower part of fluvial channel deposits. The thick sets of crossbeds indicate that strong currents were forming relatively large bedforms (75 cm or 30 in. is a minimum height for one of the bedforms) in the deeper part of the channel. The absence of reactivation surfaces is consistent with relatively steady unidirectional flow (Allen, 1980). Thin, ripple cross-stratified interbeds that separate sets of crossbeds represent back flow and co-flow ripples (Reineck and Singh, 1980) that formed in front of the bedform (Fig. 9). Relatively stable separation eddies, which are more characteristic of straight crested bedforms, are required for these ripples to form (Collinson and Thompson, 1982).

The features of the rippled sandstone facies (542 to 544 m, 1,778 to 1,785 ft) are interpreted as representing the top of a fluvial sequence (Allen, 1970; Harms and others, 1982). The presence of climbing ripples is consistent with this interpretation, as they generally form in environments such as levees or crevasse splay deposits (Horne, 1979), in late-stage active channel infills (Horne and others, 1978) or on the upper parts of point bars (Reineck and Singh, 1980).

Thus, the crossbedded and rippled sandstone facies combine to produce a single, fining-upward unit that is interpreted as a fluvial channel sequence. The brackish water shale at the top of the sequence indicates that the stream was flowing into a nearby marine body of water and during abandonment, saline water flooded the channel.

Crossbedded sandstone facies (526 to 542 m, 1,726 to 1,778 ft).—

A second crossbedded sandstone facies is initiated at 542 m (1,778 ft) and overlies the shale of the underlying facies with a sharp contact. The angle of the foreset laminae was measured in six sets of crossbeds; it is relatively constant for three sets (planar foresets), whereas the angle steepens upward (concave foresets) in the others. Reactivation surfaces are not apparent and neither are the thin rippled interbeds that occur in the lower crossbedded facies. In one set of crossbeds, soft sediment deformation has produced recumbent folds. Of special significance is the fact that in

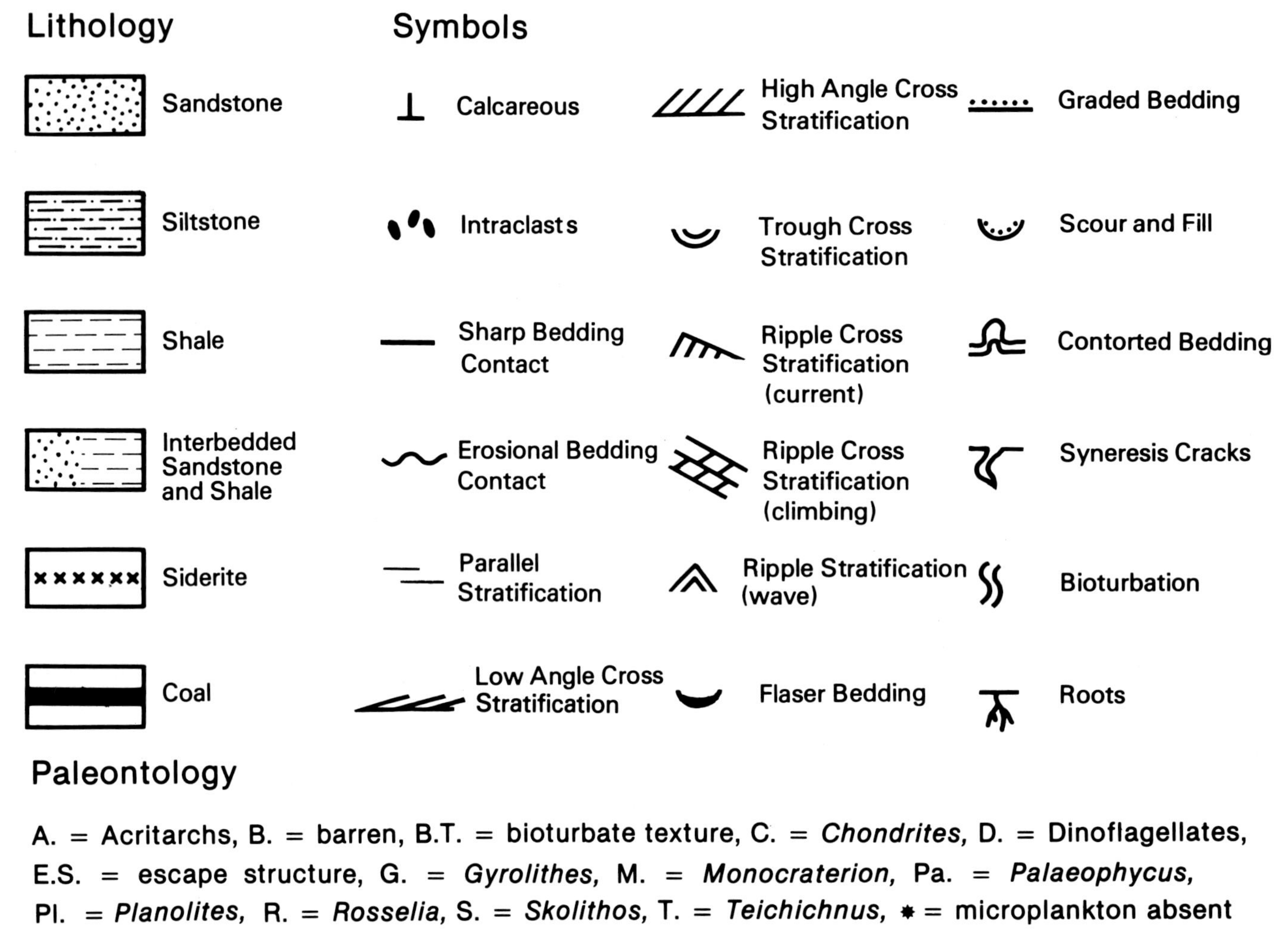

Lithology

Sandstone

Siltstone

Shale

Interbedded Sandstone and Shale

Siderite

Coal

Symbols

Calcareous

Intraclasts

Sharp Bedding Contact

Erosional Bedding Contact

Parallel Stratification

Low Angle Cross Stratification

High Angle Cross Stratification

Trough Cross Stratification

Ripple Cross Stratification (current)

Ripple Cross Stratification (climbing)

Ripple Stratification (wave)

Flaser Bedding

Graded Bedding

Scour and Fill

Contorted Bedding

Syneresis Cracks

Bioturbation

Roots

Paleontology

A. = Acritarchs, B. = barren, B.T. = bioturbate texture, C. = *Chondrites,* D. = Dinoflagellates, E.S. = escape structure, G. = *Gyrolithes,* M. = *Monocraterion,* Pa. = *Palaeophycus,* Pl. = *Planolites,* R. = *Rosselia,* S. = *Skolithos,* T. = *Teichichnus,* * = microplankton absent

FIG. 4.—Legend for Figures, 5, 15 and 19.

the basal 3 m (10 ft) of this facies, the water saturation is 87 to 97% (Fig. 5) so that water-saturated sandstone occurs between two oil-saturated zones. Water saturation can be caused by a high matrix content, but this is not the case here as there is no reduction in porosity (not plotted, Fig. 5), grain size or permeability (Fig. 5). At 527.5 m (1,731 ft), the oil saturation again decreases, but this is due to the gas cap which is produced in this well.

Rippled sandstone facies (523 to 526 m, 1,716 to 1,726 ft).—

The crossbedded sandstone facies is again overlain by the rippled sandstone facies, which, in this interval, contains no shale interbeds. Oil saturation is low because of the high gas saturation.

Interpretation.—The crossbedded and rippled sandstone facies are interpreted as a single genetic unit and represent a second fluvial channel sequence (523 to 542 m, 1,716 to 1,778 ft). The presence of rootlets (Fig. 5) in the rippled sandstone facies corroborates its interpretation as the upper part of a fluvial sequence.

Gray silty shale facies (521.5 to 523 m, 1,711 to 1,716 ft).—

A gray, silty shale caps the second fluvial deposit. The shale coarsens upward into a thin sandstone (518 to 521.5 m, 1,699 to 1,711 ft) of which the lower 50 cm (not shown, Fig. 5) is contained in the core. Trace fossils present in the shale are non-diagnostic but marine acritarchs (Table 1) are relatively common (Fig. 8); one species of dinoflagellate is also present. Trace fossils (*Monocraterion, Palaeophycus, Planolites, Skolithos*) in the sandstone indicate a marine depositional environment, which is consistent with the palynological data from the underlying shale.

Interpretation.—The shale was deposited in brackish water during the abandonment phase of the channel, similar to that of the shale at the top of the first fluvial sequence. The marine influence is slightly stronger in this interval, however, as evidenced by the greater diversity and abundance of microplankton. This may be due, in part, to preservation, as the shale at the top of the first sequence is thin (part of it may have been removed), and marine micro-

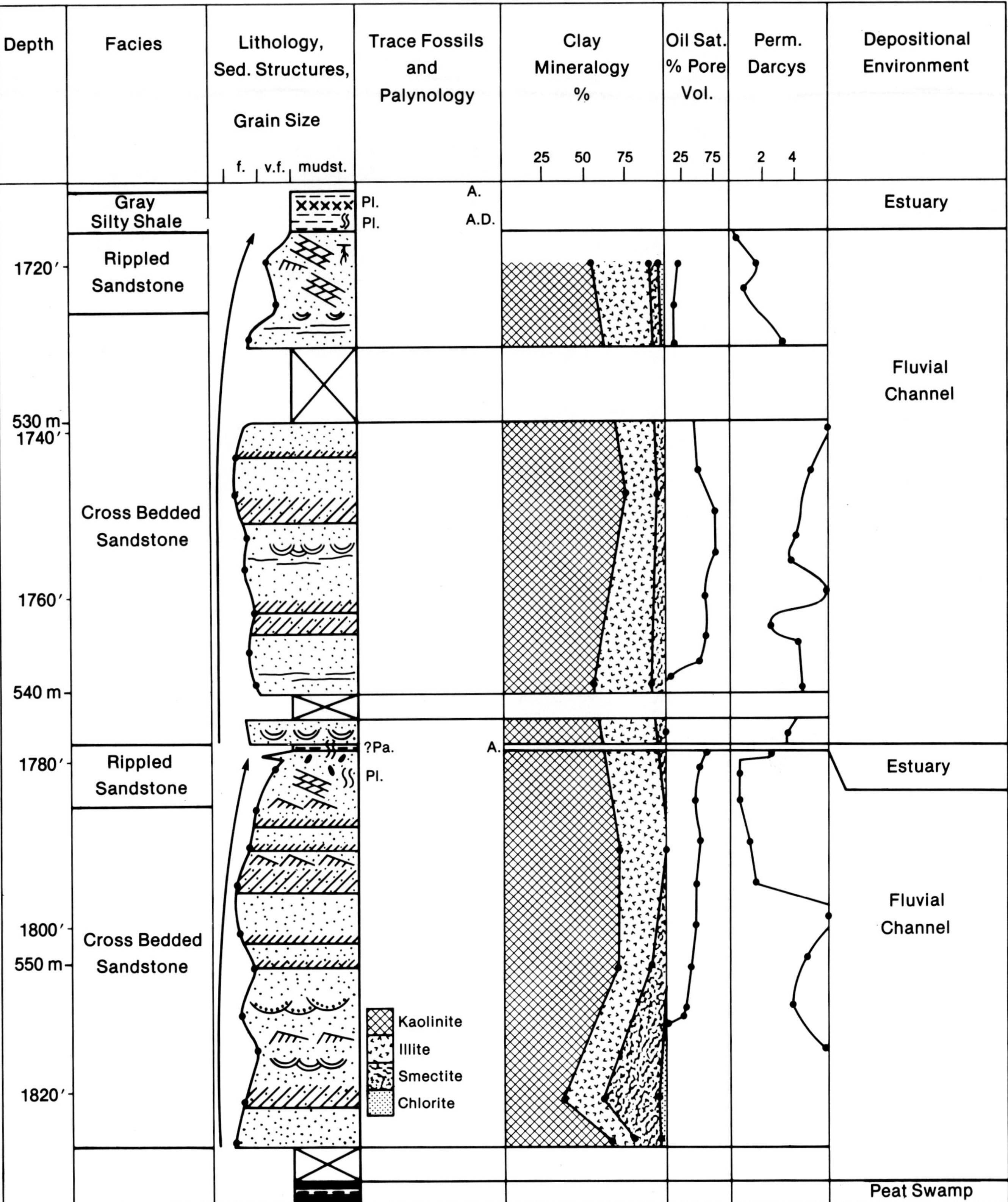

FIG. 5.—Core-derived data and sedimentologic description for well 15-12-55-13W4. Legend in Figure 4.

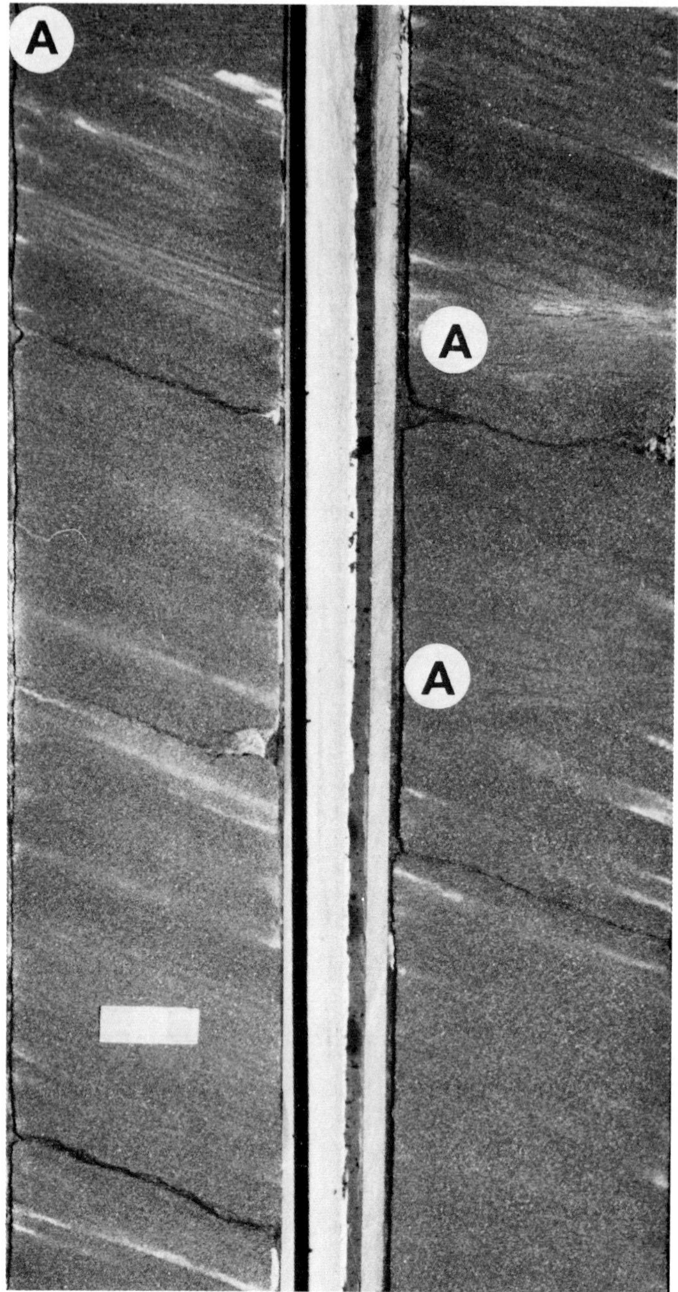

FIG. 6.—Crossbedded sandstone facies. Thick sets of high-angle crossbeds with thin, rippled interbeds (A). Dark color is due to oil staining. 15-12-55-13W4, depth approximately 546 m (1,791 ft). Scale in all photographs is 3 cm.

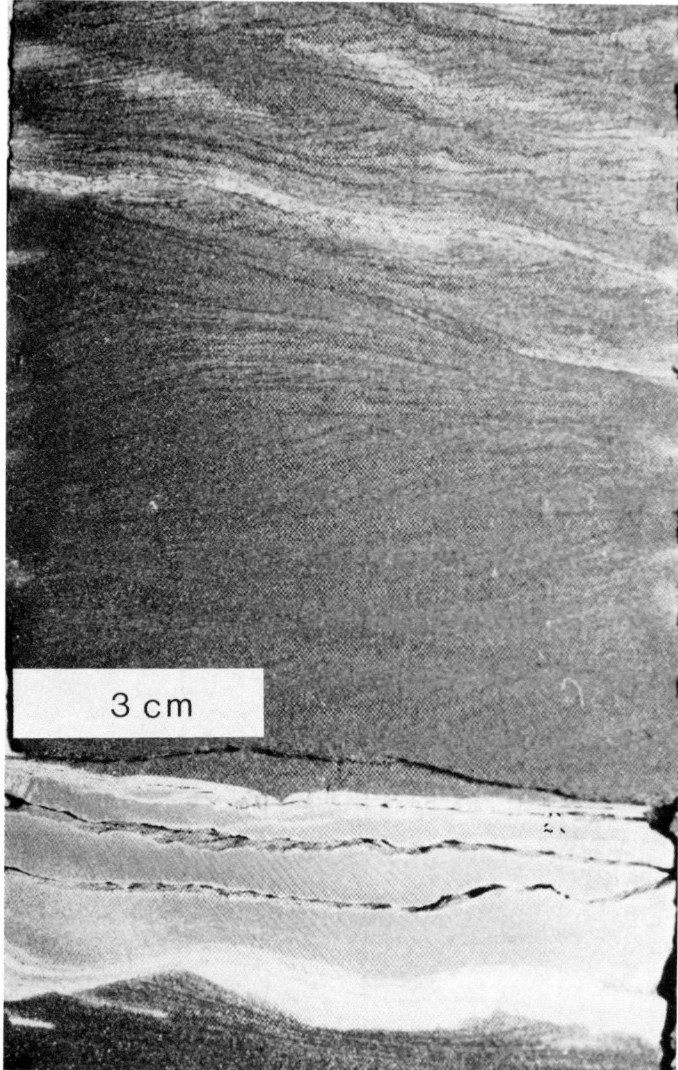

FIG. 7.—Rippled sandstone facies. Climbing ripple stratification overlying a shale bed. 15-12-55-13W4, depth approximately 543 m (1,781 ft.).

plankton are most abundant in the upper part of the shale capping the second sequence. The distance to the marine shoreline is conjectural as the topography during deposition of the upper Mannville was probably quite subdued (Glaister, 1959); and, therefore, a small amount of submergence would flood a large area. At a low gradient of 0.05 m/km (0.25 ft/mi), a *relative* rise in sea level of 3 m (10 ft) would result in a transgression of 60 km (37 mi). Furthermore, it is well known that marine influences can extend far inland (40 km, 25 mi) in estuarine channels (Dorjes and Howard, 1975). The thin sandstone (518.0–521.5 m, 1,699 to 1,711 ft) overlying the shale was deposited in this brackish water, possibly as a crevasse splay.

The shale/sandstone unit (518 to 523 m, 1,699 to 1,716 ft) illustrates how lithology affects the type of data that is available to the geologist. Both rock types were deposited in brackish water but few trace fossils are evident in the shale; however, they are abundant in the sandstone. Conversely, the shale contains some marine microplankton, but their recovery from the sandstone would probably be poor. Microplankton have a low preservation potential in high-energy, coarse-grained settings. Trace fossils are interfacial phenomena and are best developed where a range of sediment textures is available. Therefore, shales are more amenable to palynology, whereas ichnology is better suited to sequences with an interbedding of different grain sizes, even where conglomerates are included.

Conclusions.—The thick sandstone (523 to 558 m, 1,716

TABLE 1.—DISTRIBUTION OF MARINE MICROPLANKTON IN THE UPPER MANNVILLE.

x 1-5 specimens present

F Flood of specimens

○ Illustrated in Figure 8

		Well 15-12-55-13W4	542.0 m (1778 ft)	522.8 m (1715 ft)	521.5 m (1711 ft)	Well 10D-29-47-1W4	567.2 m (1861 ft)	566.0 m (1857 ft)	540.7 m (1774 ft)	539.3 m (1769.5 ft)	538.0 m (1765 ft)	Well 6B-15-53-2W4	456.2 m (1496.8 ft)	449.9 m (1476 ft)	442.1 m (1450.5 ft)	Well 14-14-55-15W4	563.8 m (1850 ft)	561.7 m (1843 ft)	555.7 m (1823 ft)	553.4 m (1815.8 ft)	544.0 m (1785 ft)	Well 2-30-43-2W4	605.9 m (1987.9 ft)	605.0 m (1984.9 ft)	600.5 m (1970.1 ft)
DINOFLAGELLATES	*Cleistosphaeridium multispinosum*			x							(x)													x	x
	Cleistosphaeridium sp. A							(x)																	
	Cleistosphaeridium sp. B								(x)																
	Dapsilidinium sp.								(x)																
	Laciniadinium sp.																x	x							
	Muderongia sp. A																						(F)	x	
	Muderongia sp.B						x	x		(x)	(x)		x				F				x		x		x
	Odontochitina operculata																							(x)	
	Spinidinium sp.A										(x)			x										x	x
	Spinidinium sp.B							(x)			x														
	Spiniferites ramosus																								(x)
ACRITARCHS	*Baltisphaeridium* sp. cf. *B. longispinosum*				x			x											x	x					
	Micrhystridium stellatum							x	(x)		x														(x)
	Micrhystridium sp.		x		(x)														x	x				x	
	Veryhachium europaeum forma 1								(x)																
	Veryhachium europaeum forma 2																		x						
	Veryhachium formosum forma 2		x		(x)																				
	Veryhachium irregulare														(x)										
	Veryhachium lairdi							x	(x)		x									x					
	Veryhachium reductum			x	x			x	x	x			(x)						x	x					
	Veryhachium reductum forma *breve*																								(x)
	Veryhachium rhomboidium																		x					(x)	

to 1,831 ft) is composed of at least two vertically stacked fluvial deposits. It is a minimum of two because the tops of sequences can be removed during 'stacking' and amalgamated partial sequences are very difficult to distinguish in core (Fig. 10). Thus, the crossbedded sandstone facies may each represent more than one sequence. The fluctuations in grain size and in the size of the sedimentary structures may represent partial sequences or alternatively may simply represent variations in discharge in a single channel. Intraclasts and/or lag deposits may or may not be present at the base of a sequence; however, they may also be present at several levels within a single sequence (Bridge and Diemer, 1983). It is the recognition of the upper part of a fluvial sequence that is important in delineating stacked deposits (Fig. 10D) and this is the evidence upon which the interpretation of two (minimum) sequences is based. The permeability data also suggest two sequences, as values in the main channel facies are generally greater than 2 darcies, whereas those in the channel top facies are less (Fig. 5).

Oil saturation data are an independent line of evidence that corroborates the sedimentological interpretation of an amalgamated sandstone. The water saturation from 539 to 542 m (1,768 to 1,778 ft) is problematic if the sandstone is interpreted as a single genetic unit, because the low oil saturation is unrelated to matrix content. The water zone occurs, however, at the base of the upper sequence, which indicates that the sequences are separate reservoir units. There is probably only partial overlap of the sequences, and the shales at the top of the lower deposit isolate the reservoirs, at least with respect to heavy oil.

Testing of the interpretation of amalgamated sandstones in this area (Hairy Hill) is difficult because of the scarcity of core material. Two additional cores (8-17-55-11W4; 11-35-55-13W4) have been examined, however, and each contains evidence that the thickness of the sandstone is due to amalgamation.

Both the gamma ray and spontaneous potential logs show the break between the two sequences in well 15-12-55-13W4

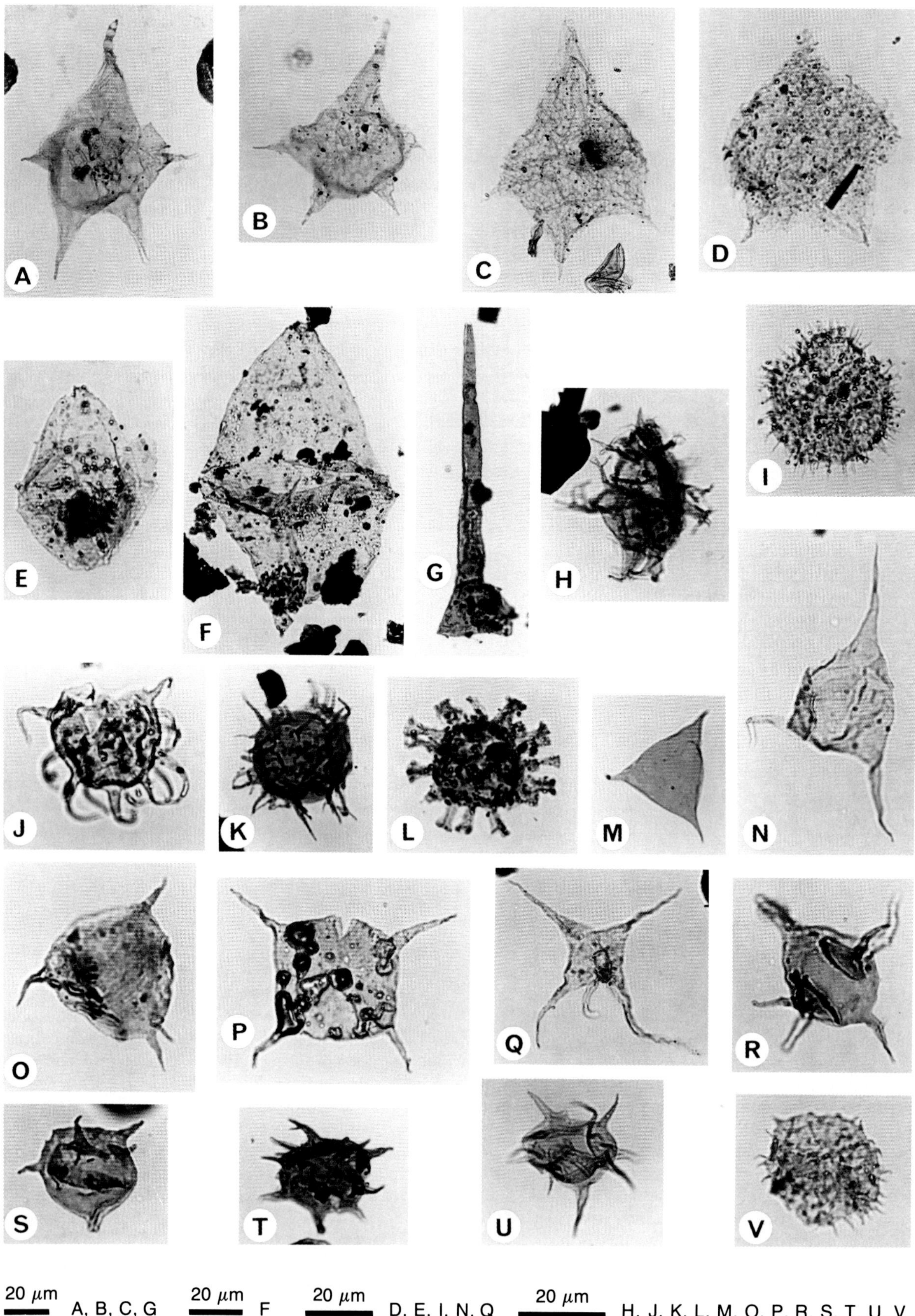
A
B
C
D
E
F
G
H
I
J
K
L
M
N
O
P
Q
R
S
T
U
V
20 μm A, B, C, G
20 μm F
20 μm D, E, I, N, Q
20 μm H, J, K, L, M, O, P, R, S, T, U, V

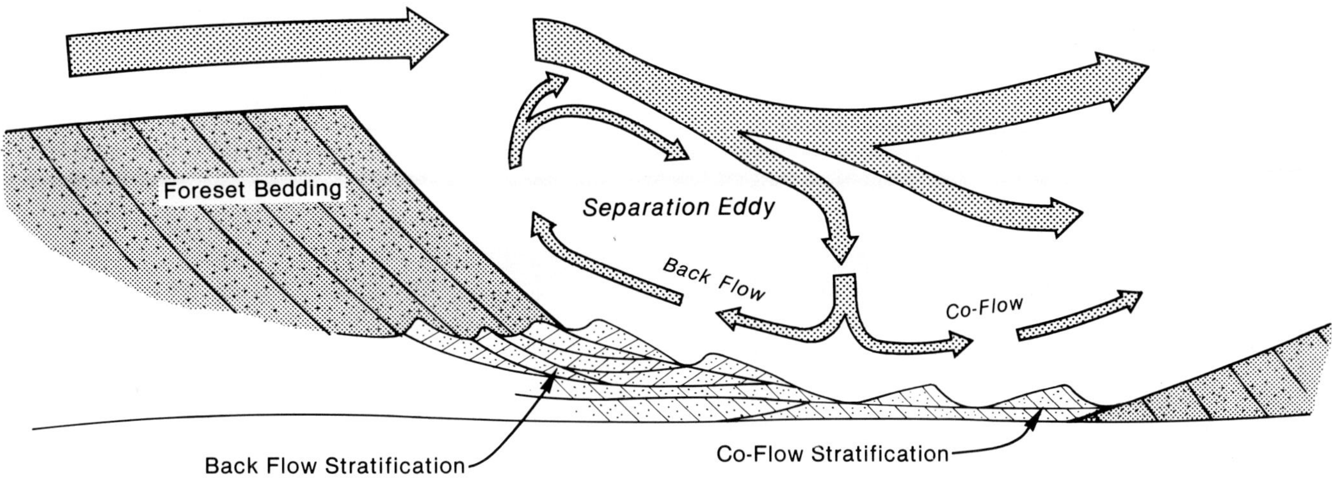

FIG. 9.—Schematic illustration of a large bedform advancing over back flow and co-flow ripples (modified after Reineck and Singh, 1980).

(Fig. 3). Many of the thick sandstones in other wells display similar breaks in their log profiles, suggesting the possibility of amalgamation. In addition, thick sandstones with smooth, unbroken log profiles frequently break up into two or more units in nearby wells (e.g., thick sandstone in well 10-26-54-10W4 breaking up in well 10-33-54-10W4).

A cross section through the Hairy Hill gas field (Figs. 11, 12) corroborates the interpretation of stacked deposits. The sandstones have fluvial-type log profiles and because of their proximity to well 15-12-55-13W4, they are assumed to be fluvial. The channel sandstones fluctuate substantially in thickness and their bases occur at various levels

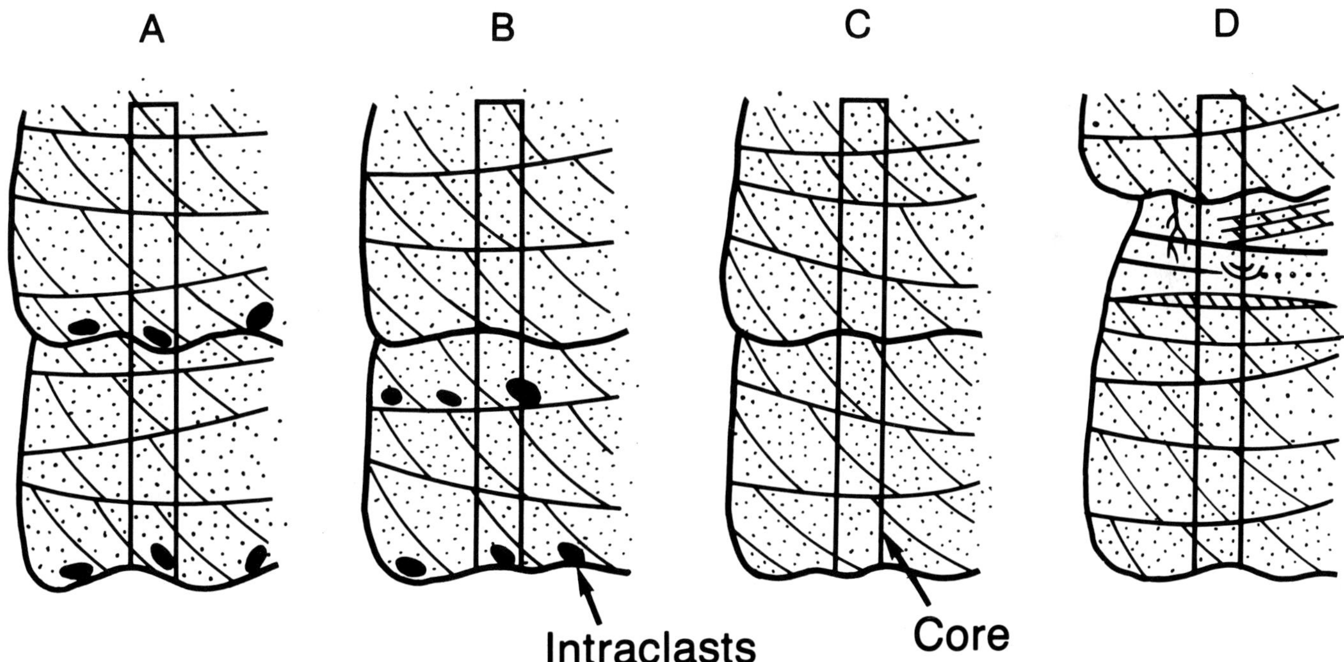

FIG. 10.—Schematic representation of stacked channel deposits and the inherent difficulty of recognizing these in core. Intraclasts could be used to identify bases of sequences in example A, but this would result in misinterpretation in example B. Absence of intraclasts makes example C appear as one sequence. Two fluvial sequences are readily apparent when the upper part of the first sequence is preserved (example D).

FIG. 8.—Photomicrographs of upper Mannville dinoflagellates and acritarchs from wells 2-30-43-2W4, 10D-29-47-1W4, 6B-15-53-2W4 and 15-12-55-13W4. (A) *Muderongia* sp. A, (B) *Muderongia* sp. A, (C) *Muderongia* sp. B, (D) *Muderongia* sp. B, (E) *Spinidinium* sp. A, (F) *Spinidinium* sp. B, (G) Apical horn of *Odontochitina operculata,* (H) *Spiniferites ramosus,* (I) *Cleistosphaeridium multispinosum,* (J) *Cleistosphaeridium* sp. A, (K) *Cleistosphaeridium* sp. B, (L) *Dapsilidinium* sp., (M) *Veryhachium reductum* forma *breve,* (N) *Veryhachium reductum,* (O) *Veryhachium europaeum* forma 1, (P) *Veryhachium lairdi,* (Q) *Veryhachium rhomboidium,* (R) *Veryhachium formosum* forma 2, (S) *Veryhachium irregulare,* (T) *Micrhystridium stellatum,* (U) *Micrhystridium stellatum,* (V) *Micrhystridium* sp.

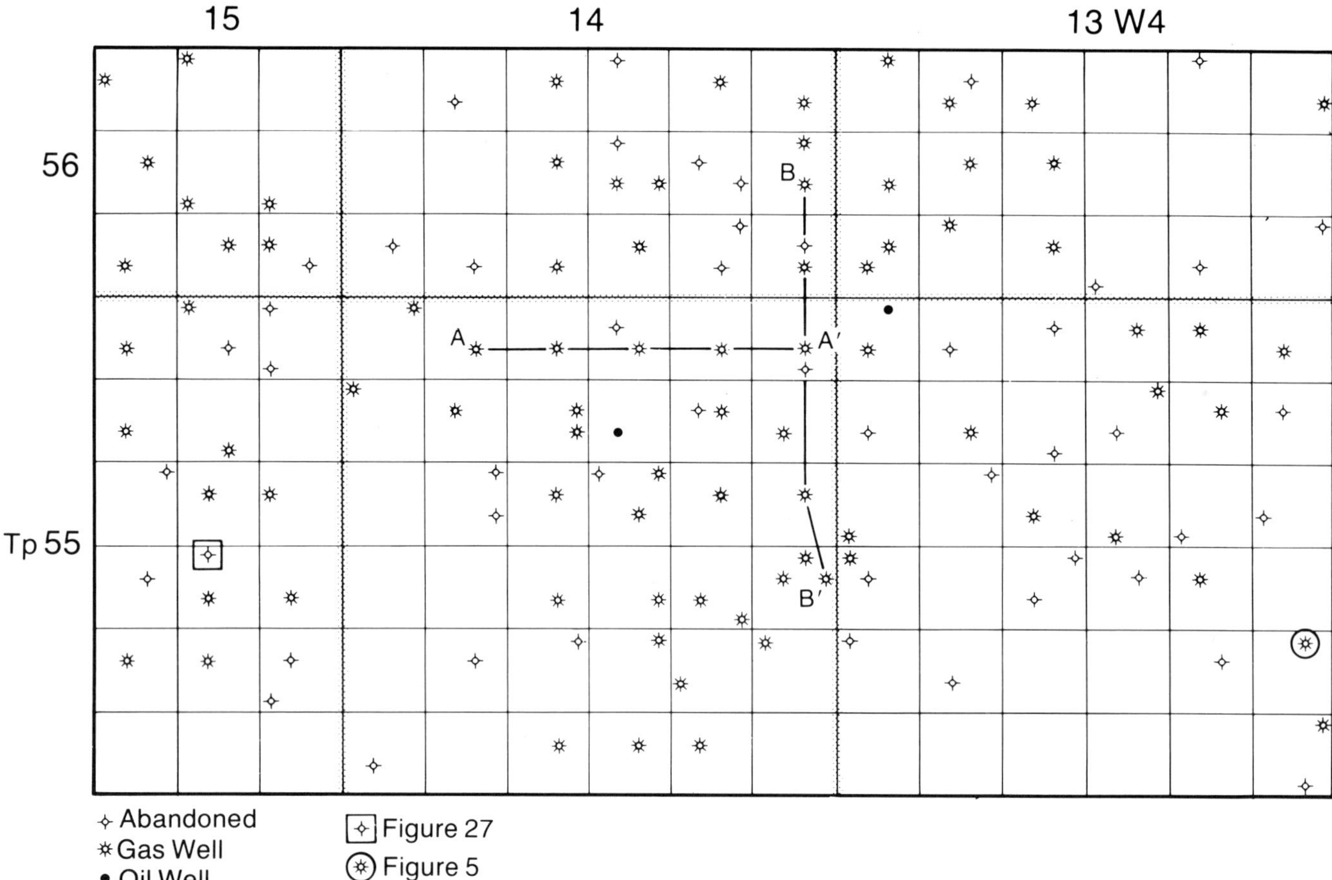

FIG. 11.—Location of cross sections in Figures 12 and 26, Hairy Hill area. Wells 15-12-55-13W4 (Fig. 5) and 14-14-55-15W4 (Fig. 27) are also highlighted.

in the upper Mannville. There appear to be two channel sandstones separated by a finer grained, coarsening-upward sequence (possibly a crevasse splay) in well 7-34-55-14W4. The irregular distribution of the sandstones supports the interpretation of multiple paleochannels. Because of the three-dimensional variability in overlap of the fluvial deposits, occasional breaks in lateral continuity between sandstones in adjacent wells is to be expected. Gas pressure data show that this indeed occurs. Gas in well 7-33-55-14W4 is in a different reservoir than the gas in wells 7-34 and 7-35-55-14W4 (Fig. 13). Putnam and Oliver (1980, Fig. 4), however, and Putnam (1982a, 1982b, 1983) interpreted that the channel sandstones in each of these wells were deposited by a single, vertically aggrading stream. Geophysical logs and reservoir data strongly suggest that this is not a single sandstone and thus is not the product of one channel.

Well 10D-29-47-1W4

A thick sandstone (541 to 565 m, 1,776 to 1,855 ft) has been cored (Fig. 14) in the upper Mannville in this well. A detailed examination of the core, however, reveals that it is very different from the thick sandstone in well 15-12-55-13W4. The shale facies above and below the sandstone are also described and interpreted.

Black and gray shale facies (565 to 574 m, 1,855 to 1,884 ft).—

The sandstone is underlain by a thick (9 m, 29 ft) laminated shale (Fig. 15) that is black for all but the uppermost meter, which is medium gray. The black shale has a higher organic carbon content (≃4%) than the gray shale (≃0.7%). Both the black and gray shales contain scattered siderite beds and nodules, usually 1 to 3 cm (0.4 to 1.2 in.) thick. The basal and mid-portions of the black shale are devoid of marine microplankton or trace fossils, but the upper part is weakly burrowed and contains one species of dinoflagellate (Table 1). Trace fossils are not evident in the medium gray shale, but acritarchs and dinoflagellates are relatively common (Fig. 8).

Interpretation.—The black and gray shale facies represents sedimentation in a low-energy environment that received abundant organic detritus. The setting may have been a lake for the lower and middle parts of the shale; however, the environment had a weak marine influence during deposition of the upper part of the black shale. The increase in numbers and diversity of acritarchs and dinoflagellates in the medium gray shale indicates that the setting was a marine (brackish) bay immediately prior to deposition of the sandstone.

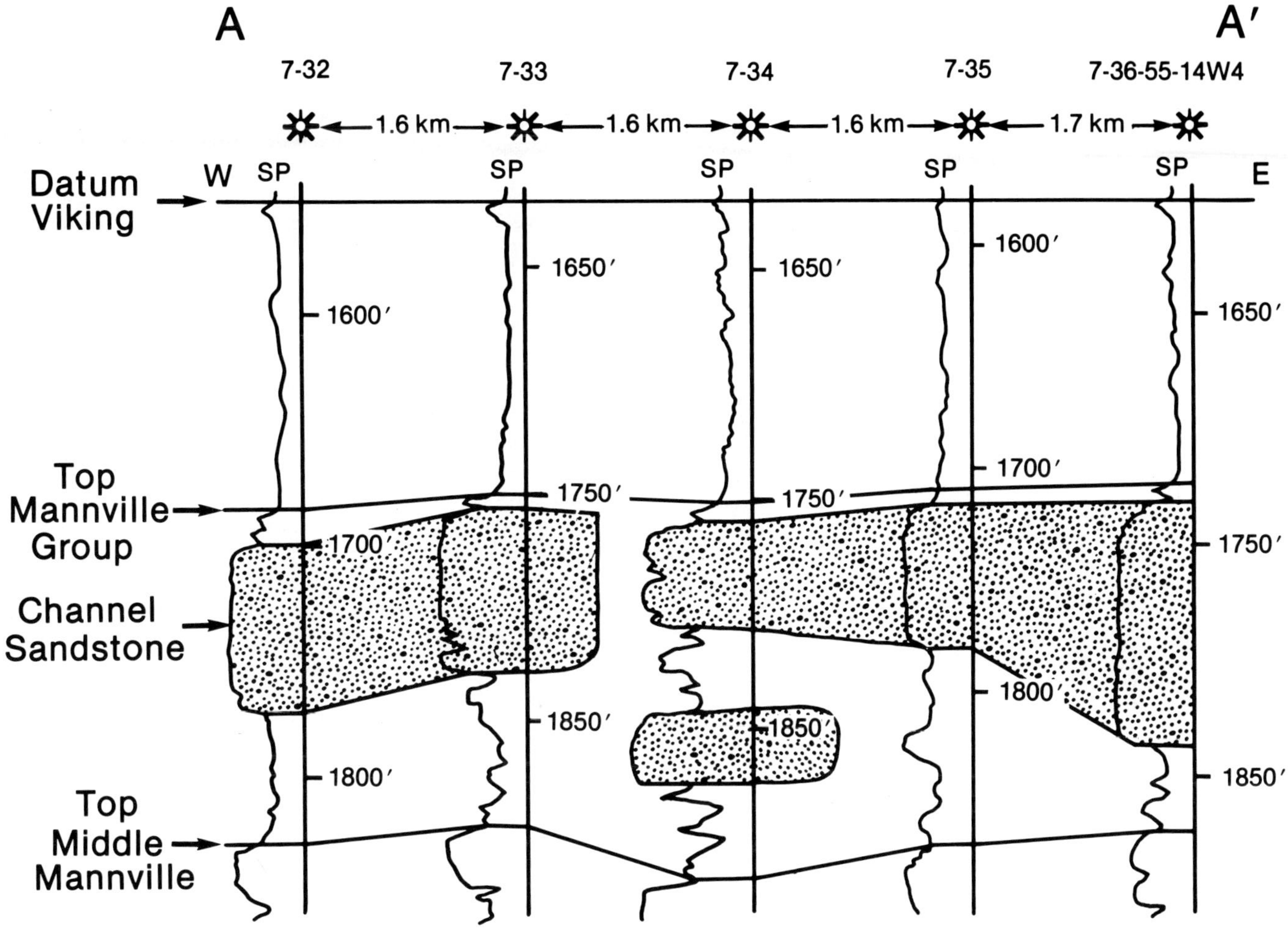

FIG. 12.—Stratigraphic cross section oriented west-east, Hairy Hill area, Alberta. Channel deposits, some of which are probably amalgamated, occur at various levels in the upper Mannville. The break between wells 7-33 and 7-34-55-14W4 is based on reservoir pressure data (Fig. 13).

Massive sandstone facies (559 to 565 m, 1,834 to 1,855 ft).—

The sandstone in this interval is very fine-grained, predominantly massive, contains 5 to 15% matrix and in the basal 20 cm (8 in.) has siltstone clasts oriented at various angles (Fig. 15). The contact with the underlying shale is poorly preserved but it is presumed to be sharp. In the upper part of the facies, the grain size increases slightly, but remains very fine-grained, some ripple cross-stratification is apparent and a set of crossbeds, 15 cm (6 in.) thick, caps the sequence. Kaolinite (most abundant), illite (abundant), montmorillonite and chlorite are all present in the clay fraction. The sandstone is stained with heavy oil except for the basal 1.2 m (4 ft) (Fig. 15).

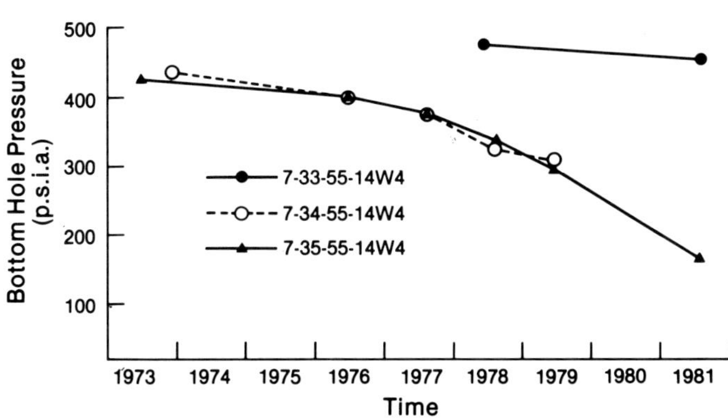

FIG. 13.—Reservoir pressure data, Hairy Hill gas field.

Interpretation.—There is no evidence of current activity during the deposition of the sandstone in the lower part of this facies, and the very fine grain size and irregular orientation of the siltstone clasts together suggest that the depositional process was suspension fallout. Tractive processes are evident in the upper part, and thus it appears that current strength increased toward the top of the interval, accompanied by a slight increase in grain size. Environmental interpretations that are compatible with this are crevasse splays or thin distributary mouth bar deposits, both of which are shallowing-upward sequences. The sharp contact with the underlying shale is characteristic of the more proximal parts of splay deposits (Coleman and Prior, 1982). In distributary mouth bars, a sharp basal contact is more prevalent in the central part of the bar (Horne, 1979).

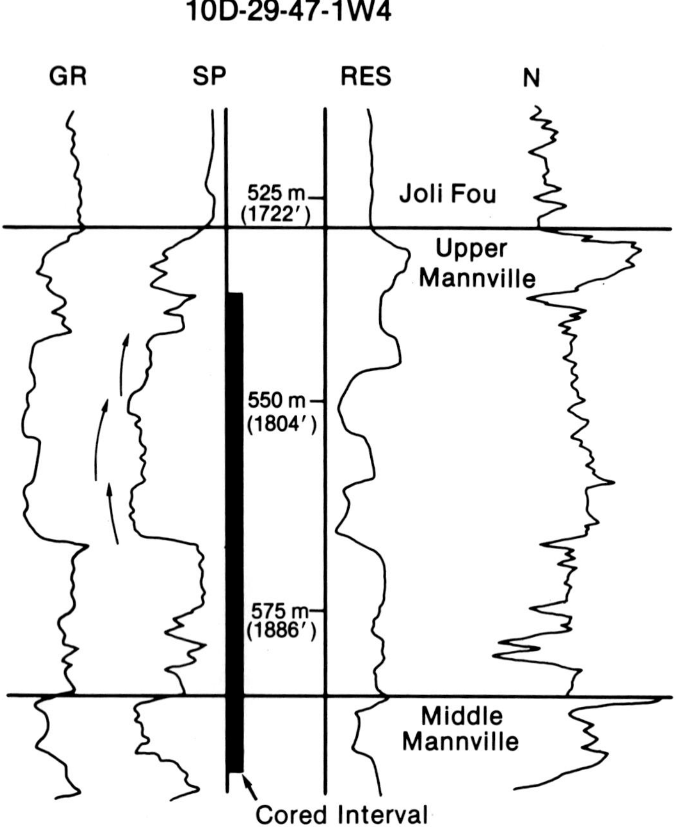

FIG. 14.—Geophysical logs and cored interval for 10D-29-47-1W4.

Crossbedded sandstone facies (550 to 559 m, 1,803 to 1,834 ft).—

At a depth of 559 m (1,834 ft), there is an abrupt coarsening in grain size to fine-grained, a significant increase in matrix content and the oil stain disappears. The permeability of the sandstone is greatly reduced (Fig. 15) because of the abundant (25 to 44%) kaolinite matrix. Thick (as much as 70 cm, 28 in.) sets of high-angle crossbeds are common, as are intervals which appear more massive, and some of the crossbedding is outlined by laminae composed largely of white kaolinite 'grains' and/or brown siderite grains. Foreset geometry is tangential to concave and reactivation surfaces are not apparent. Carbonaceous debris is occasionally concentrated at the base of the crossbeds. Abnormally high values (44%) for matrix content from grain size analyses are partly due to the breakup of 'clay' grains during sieving. The sandstone in this facies is carbonaceous throughout, fines upward and loses its high matrix content at the top of the interval, as shown by the permeability data. Kaolinite, by far the most abundant, and illite are the only clay minerals present in greater than trace amounts (Fig. 15). The contact with the overlying unit is gradational.

Interpretation.—The clay mineralogy, oil saturation, sedimentary structures and textures indicate that the facies in this interval is distinctly different from the one below. The sharp basal contact, fining-upward grain size and the presence of thick sets of high-angle crossbeds in the lower to middle portions suggest that this is a channel deposit. An abundant clay matrix, white kaolinite grains and brown siderite grains also occur at or near the bases of channel deposits in other wells (6D-32-52-7W4; 8-17-55-11W4), but the high matrix content usually does not persist throughout the sandstone. The absence of reactivation surfaces in the crossbeds and the vertical sequence of facies in this well suggest that the channel was a distributary.

Bioturbated sandstone facies (541 to 550 m, 1,776 to 1,803 ft).—

The sandstone in this interval is very fine-grained, carbonaceous, contains a few thin interbedded shales and occasionally fines to a coarse-grained siltstone. The core is in poor condition but where sedimentary structures are visible, the sandstone is horizontally bedded to ripple-laminated. Clay matrix is less abundant than in the underlying facies, the proportion of kaolinite is lower and montmorillonite and chlorite are occasionally present. There is a light oil stain from 546 to 548 m (1,790 to 1,798 ft) and above that the core is saturated with heavy oil. Trace fossils (*Skolithos*, Fig. 16; *Palaeophycus* and *Planolites*) are identifiable in several places and a bioturbate texture from 546 to 547 m (1,791 to 1,795 ft) indicates faunal reworking of the sediment.

Interpretation.—Sedimentary structures, textures, clay mineralogy and ichnology show that this represents a third facies, distinctly different from the other two. The *Skolithos* possess clay linings, which suggest there was a marine influence during deposition, since nonmarine producers of *Skolithos* (i.e., insects) do not line their burrows (Ratcliffe and Fagerstrom, 1980). The poor condition of the core in this interval inhibits detailed sedimentology so the sandstone is interpreted as a marine shoreline deposit, based on the trace fossil evidence. Possible depositional environments include a storm washover fan or a splay deposit in a lagoon or shallow bay.

Siltstone/shale facies (539 to 541 m, 1,768 to 1,776 ft).—

This facies consists of siltstones and shales that contain thin (1 to 3 cm, 0.4 to 1.2 in.) sandstone interbeds which are more numerous toward the base and top. Siderite nodules and beds as much as 5 cm (2 in.) thick are present in the shale. Contorted and faulted bedding is common in the sandy zones. Trace fossils (*Planolites, Skolithos* with a clay lining) indicate a marine influence during deposition. This is corroborated by marine acritarchs and dinoflagellates (Fig. 8) that have been recovered from two samples of the shale and from one sample of a shale slightly higher in the core (Table 1).

Interpretation.—The siltstone/shale facies represents sedimentation in a low-energy, brackish water environment such as a lagoon or shallow bay. The contorted bedding suggests that the core is located near the margin of such a setting, where the depositional surface was inclined.

Conclusions.—A wide range of data, including sedimentological, mineralogical, ichnological, petrophysical and oil saturation, indicates that there are three distinct facies in the sandstone (541 to 565 m, 1,776 to 1,855 ft). Thus, the thickness (24 m, 79 ft) of the sandstone is due to an amal-

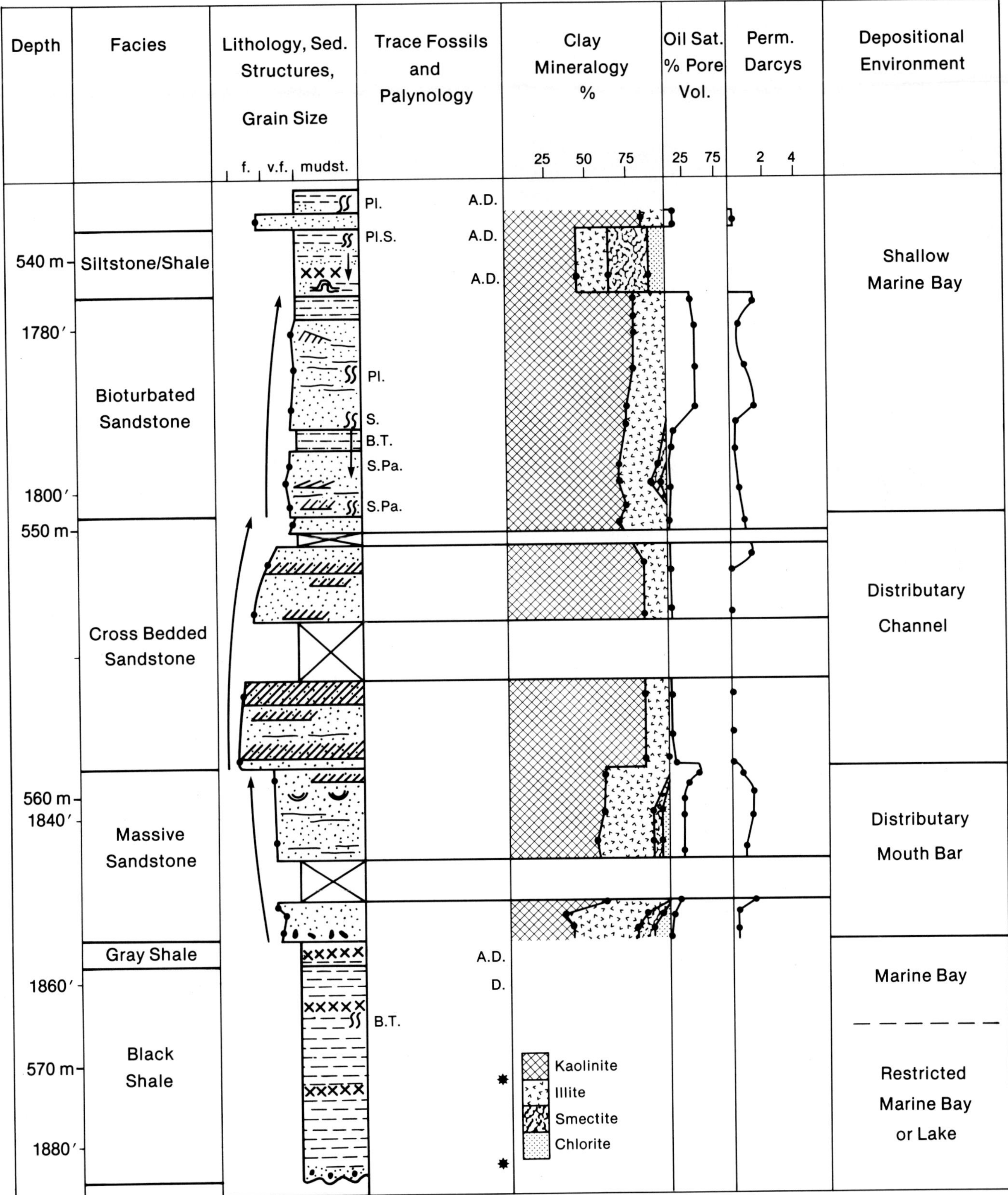

FIG. 15.—Core-derived data and sedimentologic description for well 10D-29-47-1W4. Legend in Figure 4.

FIG. 16.—Bioturbated sandstone facies. *Skolithos* displaying prominent wall linings in very fine-grained sandstone. 10D-29-47-1W4, depth approximately 550 m (1,803 ft).

gamation of three different facies that represent the following depositional environments: distributary mouth bar, channel, and shallow marine bay. Correlation to other wells shows that the geometry of each of the facies is different (Fig. 17). In particular, the shaly interval that occurs between the distributary mouth bar and shallow marine bay facies in wells 10D-28 and 7A-32-47-1W4 has been replaced by the channel facies in well 10D-29-47-1W4, resulting in the thick amalgamated sandstone.

The three facies are apparent on logs (Fig. 14), although without core, delineation of them would be tenuous. The resistivity of the shallow marine bay facies is anomalously high when compared with the residual heavy oil saturations derived from core analyses (45 to 50%). Some of the oil was probably lost during and after coring.

Marine shales occur immediately above and below the sandstone and indicate that it was deposited in a marine shoreline setting. This interpretation is supported by the ichnological data from the bioturbated sandstone facies.

A summary of the depositional events for this well, starting with the black shale facies at the base of the sandstone, is as follows. Fine-grained sediment and organic matter were deposited in a quiet water environment that was either freshwater or restricted marine. Microplankton recovered from the gray shale facies indicate, however, that marine conditions were well established prior to deposition of the overlying sandstone. The break from black to gray shale signals a change in detrital influx, which probably corresponds to the time that fluvial channels began debouching into the bay. Deposition of the coarse sediment of the distributary mouth bar and overlying channel filled this part of the bay. The thickness (≃16 m, 50 ft) of these two facies compares favorably to the thickness of similar facies in Carboniferous deltaic deposits in Kentucky (Horne and others, 1978, Fig. 6) and gives an approximate water depth for the bay. After abandonment of the channel, marine shoreline conditions were established during deposition of the bioturbated sandstone and siltstone/shale facies.

Well 6B-15-53-2W4

A thick sandstone (430.5 to 448 m, 1,412.5 to 1,471 ft) occurs at the top of the upper Mannville in well 6B-15-53-2W4 (Fig. 18). The sandstone has a 3-m (10 ft) shale break near its base but it has a rather blocky signature on the spontaneous potential log, similar to that of channel deposits. Examination of the core, however, reveals that the entire upper Mannville is composed of coarsening-upward sequences which vary in thickness from 4 to 17 m (12 to 55 ft). There are five of these sequences, four of which are shown in Figure 19. A typical coarsening-upward sequence is composed of three facies; mudstone, mudstone/sandstone, and bioturbated rippled sandstone. Each facies is described and interpreted below using data from the entire interval.

Mudstone facies.—

The lowest part of a coarsening-upward sequence is composed of interbedded siltstones and shales with a few thin, very fine-grained sandstones (Fig. 19). The mudstones (siltstones/shales) are generally massive in appearance to parallel-laminated. Siderite beds and nodules occur in the shales and some of the siderites contain diagenetic pyrite. Also present are pyrite nodules enclosed by a brown 'ring' of iron minerals. The mudstones are pervasively bioturbated except for one interval (441 to 443 m, 1,446 to 1,454 ft). The thin (<10 cm, 4 in.) interbedded sandstones are massive to parallel-laminated with moderate bioturbation. Slightly thicker (10 to 30 cm, 4 to 12 in.) sandstones, as well as some of the thinner sandstones and siltstones, are both wave- and current-rippled. Some of the current ripples contain thin flasers of shale and/or carbonaceous debris. Thicker sandstones tend to have less bioturbation than thinner ones. The thickest sandstone (457 to 458 m, 1,498 to 1,501 ft) varies from the norm in that it is composed of thin (several centimeters) graded beds (both inverse and normal grading) with no apparent ripple stratification. The graded beds are "event" sedimentation units that could have been deposited by one of several different processes.

Trace fossils present in the fine-grained rocks include *Palaeophycus, Planolites, Skolithos* (lined), *Rosselia, Chondrites, Teichichnus* and *Gyrolithes*. The degree of bioturbation and the assemblage of trace fossils indicate that the sediment was deposited in a marine setting, which is

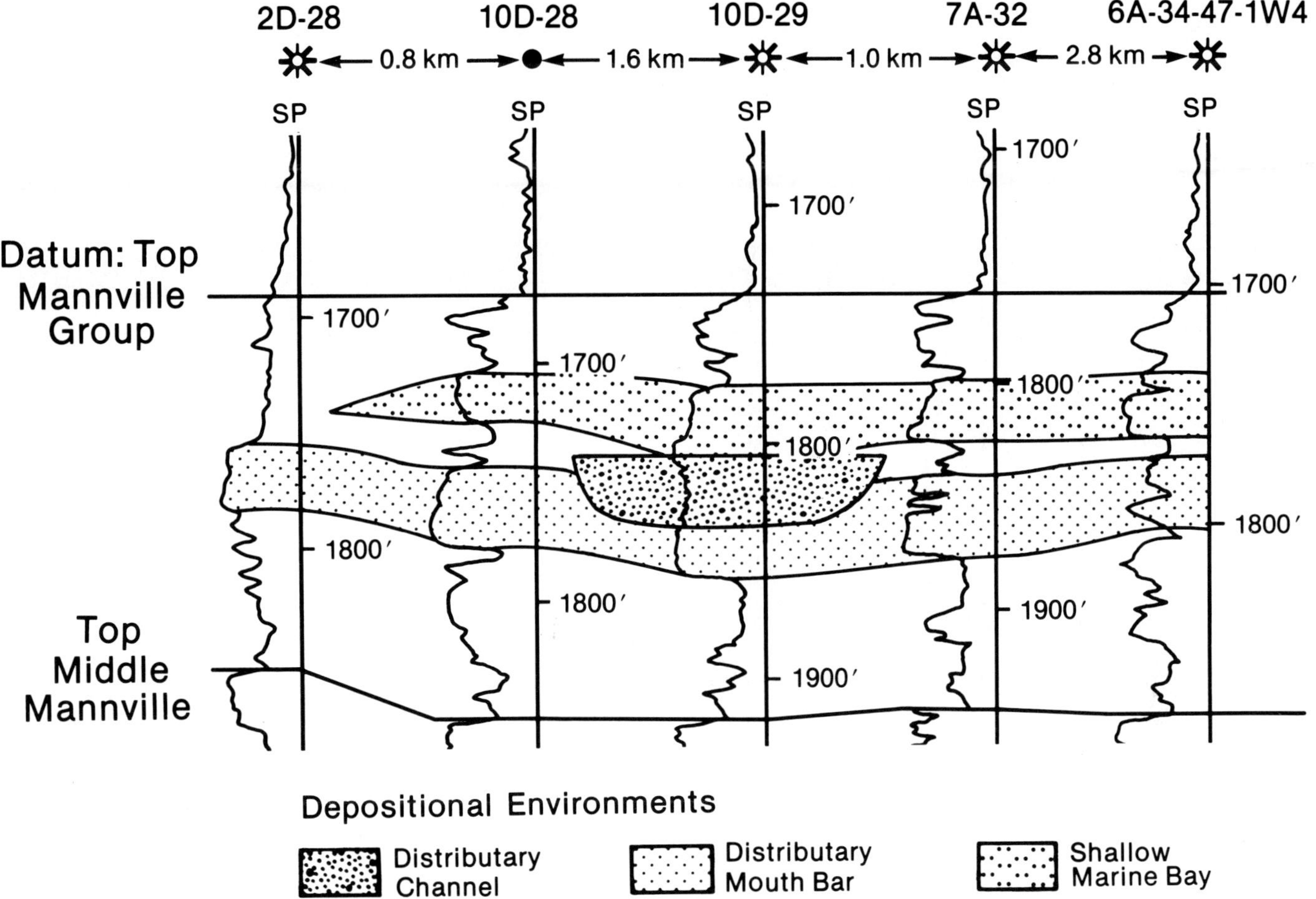

FIG. 17.—Stratigraphic cross section showing different geometries for the three sandstone facies in well 10D-29-47-1W4. Note how the shale in wells 10D-28 and 7A-32 has been cut out by the channel deposit to create one thick sandstone in well 10D-29.

corroborated by palynology. Acritarchs and dinoflagellates are present in two of the three shales that were sampled (Fig. 19), although they are quite sparse (Table 1). This is surprising as each of the coarsening-upward sequences is similar in its sedimentology and ichnology and it was expected that shales from each sequence would yield similar palynological results.

Interpretation.—The fine-grained nature of the rocks is indicative of low-energy conditions. The thin, rippled siltstones and sandstones represent occasional periods of weak wave and current activity. Infrequently, strong but episodic currents deposited coarse sediment as shown by the graded beds at 457 m (1,499 ft). The thin flasers in the ripple cross-stratification could be the result of pulsating currents (i.e., tidal), although the evidence for this is tenuous. Possible depositional environments include protected, quiet-water bays/lagoons or offshore areas, below the shoreface zone. The trace fossil assemblage is mixed as *Skolithos, Gyrolithes* and *Palaeophycus* are dwelling burrows, whereas the others are feeding burrows. A mixed assemblage indicates shallow, subtidal conditions or an intermediate position between the *Skolithos* and *Cruziana* ichnofacies of Seilacher (1967). The relatively low diversity of the assemblage suggests, however, that the environment was rather restricted. The low numbers and diversity of microplankton also indicate that the environment was very stringent (Table 1).

Thus, when the sedimentological, ichnological and palynological data are combined, the most appropriate depositional environment is a lagoon or shallow, protected bay.

Mudstone/sandstone facies.—

The middle part of a sequence is composed of interbedded mudstones (shales/siltstones) and sandstones. Bedding thickness of the mudstones is generally several centimeters but the sandstones thicken from several centimeters in the lower part to 10 to 20 cm (4 to 8 in.) toward the top of the facies. The amount of mudstone decreases as the sandstones thicken upward. The sandstones are generally very fine-grained, carbonaceous, and contain horizontal laminae, trough cross-stratification and a few wave ripples. Flaser bedding, formed by carbonaceous and clay drapes, is common. The sandstones in this interval can be cemented with carbonate as at 448 to 449 m (1,470 to 1,473 ft). At this level, the mudstone/sandstone facies is thin (1 m, 3 ft) and because the sandstones are cemented, they do not show up on the spontaneous potential log. Therefore, the overlying sandstone (443 to 449 m, 1,455 to 1,473 ft) has a sharp base on the spontaneous potential log, but the gamma-ray

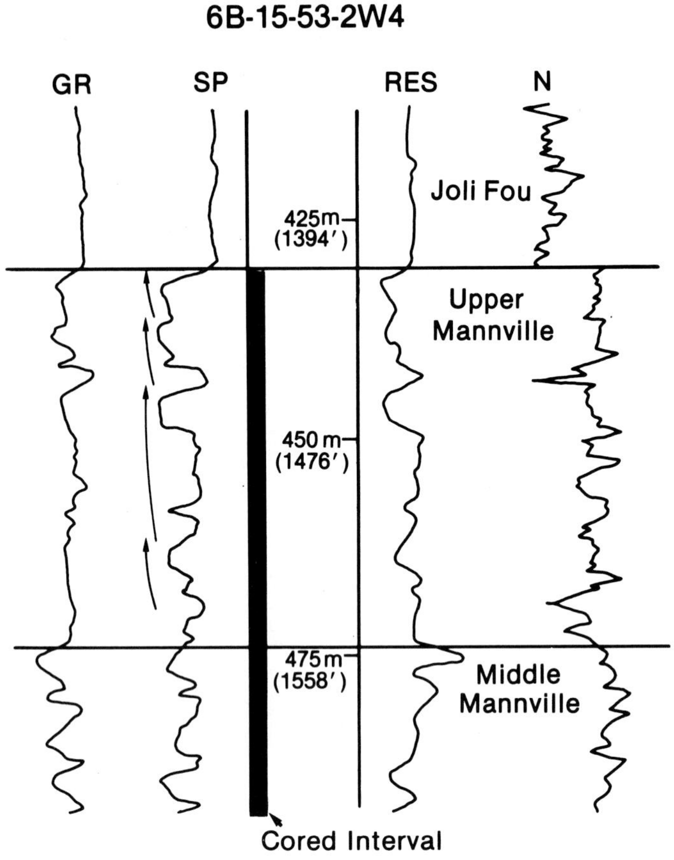

FIG. 18.—Geophysical logs and cored interval for 6B-15-53-2W4.

log picks up the mudstone/sandstone facies and shows more of a coarsening-upward profile (Fig. 18). The bedding of both the mudstones and sandstones is disrupted by bioturbation and in places it is obliterated (Fig. 20). The shales contain abundant shrinkage cracks, infilled with sandstone or siltstone. In plan view the shrinkage cracks are irregular in orientation and spindle-shaped, rather than polygonal in outline. In cross section the cracks are subvertical to vertical and are generally crenulated (Fig. 21). Diagenetic pyrite occurs in mudstones and sandstones, whereas siderite is present only in the mudstones.

Planolites and *Skolithos* are the most common trace fossils, but *Palaeophycus* and *Teichichnus* are also present. Some of the sandstones contain vertical escape structures which show that sand deposition was very rapid (event sedimentation). The trace fossils indicate that the sediment was deposited in a marine setting, but samples of this interval from two sequences are devoid of marine microplankton (Fig. 19).

Interpretation.—The interbedded mudstones and sandstones represent a time when increasingly coarse sediment was being supplied to the shallow bay or lagoon. The sand was transported into the bay by traction currents, as evidenced by the ripple cross-stratification.

The shrinkage cracks are attributed to syneresis, rather than desiccation, because of the irregular form and absence of any signs of subaerial exposure. Syneresis cracks may form because of large fluctuations in salinity (Burst, 1965). Plummer and Gostin (1981) stated that syneresis cracks can form at the sediment-water interface or substratally, with load casts frequently associated with the latter. Load casts are not present in this sequence, so it is concluded that the syneresis cracks formed at or near the sediment-water interface. Moreover, the crenulated form indicates that the cracks formed before the sediment had undergone much compaction.

Large influxes of fresh water are required to cause fluctuations in salinity in a marine bay. As the syneresis cracks are best developed in the interbedded sand/mudstone interval, it appears that the influx of fresh water was associated with the deposition of the sand.

Bioturbated rippled sandstone facies.—

The upper parts of the sequences are composed of very fine-grained, carbonaceous sandstone with a few thin interbedded shales near the base. The lower to middle parts of the sandstones contain abundant horizontal to low-angle laminae and trough cross-laminae. There are a few thin beds that grade from very fine-grained sandstone to black, carbonaceous debris (e.g., 446.2 m, 1,470 ft), indicating "event" sedimentation. The upper parts of the sandstones are more massive with scattered horizontal laminae and ripple cross-stratification. Low-angle scour surfaces are common throughout. Diagenetic pyrite is occasionally present in the sandstones and at 433 m (1,420 ft) pyrite is enclosed by a brown ring of iron minerals.

The thickest coarsening-upward sequence (Fig. 19) and the lowermost sequence (469 to 473 m, 1,538 to 1,553 ft), which is not depicted in Figure 19, are capped by thin coals which are *in situ*, as indicated by extensive rooted zones in the underlying rocks. There is also a thin (5 cm, 2 in.) black carbonaceous shale that is coal-like at the top of another sequence (435.4 to 443.3 m, 1,428.5 to 1,454.5 ft). The coal at 443.5 m (1,455 ft) contains pyrite as identified by X-ray diffraction.

Trace fossils are more abundant in the lower parts of the sandstones and root structures are common in the upper portions. *Palaeophycus* and *Planolites* are the most common trace fossils, but *Teichichnus, Rosselia, Monocraterion* and escape structures (Fig. 22) are also present. *Monocraterion, Rosselia* and *Teichichnus* suggest that the sand was deposited in a marine environment, which is consistent with the data from the underlying fine-grained rocks. The escape structures indicate rapid emplacement of some of the sand beds followed by periods of low deposition or nondeposition. This corroborates the sedimentological interpretation for the graded beds.

Interpretation.—The sandstones are interpreted as crevasse splay deposits because of the abundance of carbonaceous detritus, the episodic nature of the sedimentation and their thicknesses (1.5 to 5 m, 5 to 16 ft). Trace fossils show that the lower parts of the splays were deposited in marine water. The rooted nature of the upper parts of the sandstones indicates a shoaling upward, whereas the coals that cap two sequences represent complete infill of the bay and the establishment of freshwater conditions. The sandstones

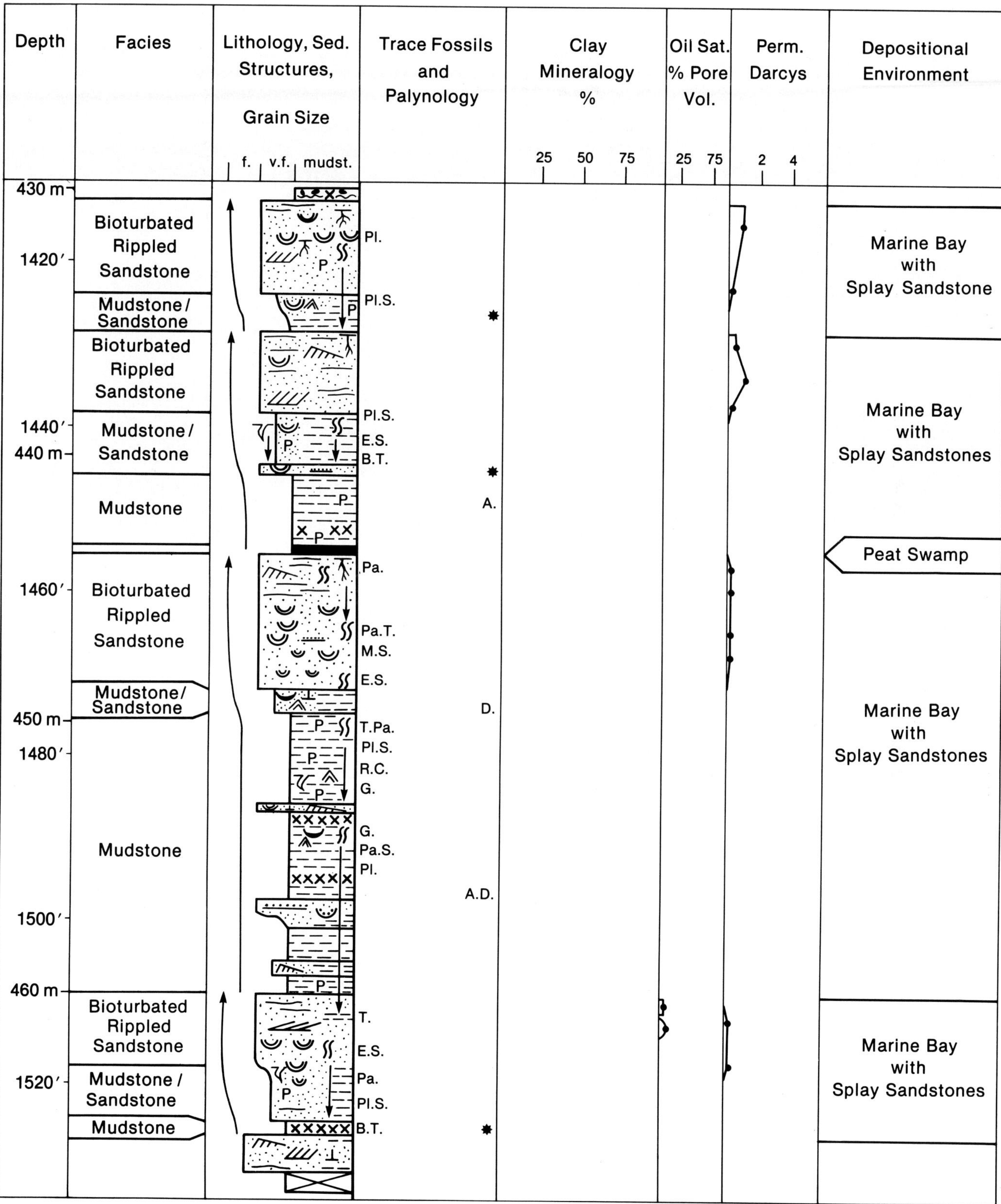

FIG. 19.—Core-derived data and sedimentologic description for well 6B-15-53-2W4. Legend in Figure 4. P in lithology column denotes pyrite.

FIG. 20.—Mudstone/sandstone facies. Complete disruption of bedding by bioturbation. 6B-15-53-2W4, depth approximately 440 m (1,444 ft).

FIG. 21.—Mudstone/sandstone facies. Syneresis cracks in both cross section and bedding plane views. 6B-15-53-2W4, depth approximately 439 m (1,440 ft).

are the upper parts of overall coarsening-upward, and shoaling-upward, sequences.

Conclusions.—The coarsening-upward sequences, which grade from mudstones in the lower part to sandstones at the top, were deposited in protected (low-energy) marine bays or lagoons. A typical series of events for one of the bays is as follows. Initially, detrital influx into the bay was fine-grained and at low rates, as indicated by the moderate to high degree of bioturbation of the mudstones. After a crevasse developed in the levee of a nearby channel, coarser sediment and plant debris were suddenly supplied to the bay. The overall sedimentation rate was greatly increased but was also highly variable, with flood events in the channel corresponding to "event" sedimentation units on the splay. As the bay shallowed, the splay sands provided a platform for plant growth. Final filling of the bay caused abandonment of the crevasse splay and in two instances the establishment of a freshwater peat swamp. The absence of detrital influx and compaction of the sediments resulted in subsidence and transgression of the sea. This terminated freshwater plant growth and established a low-energy, shallow marine bay and the series of events was repeated. This cyclical process resulted in the stacking of five coarsening-upward sequences. A corollary of this interpretation is that the thicknesses of the bay-fill sequences (4 to 17 m, 12 to 55 ft) give an approximate water depth for the bays (contemporaneous subsidence and compaction are two sources of error).

Only two of the five sequences are capped by coals, so the peat swamps may have been localized or the peats may have been eroded. The rooted nature of the upper parts of the sandstones, however, indicates that the shallowing of the bays did occur and the very carbonaceous shale at the top of one sequence probably represents a high ash peat. The horizontal lamination in the upper parts of the sandstones could be wave-formed or could represent upper flow regime conditions established during flood events.

The crevasse splay deposits form a thick sandstone at the top of the interval (430.5 to 448 m, 1,412.5 to 1,471 ft) because the mudstones at the bases of the uppermost two

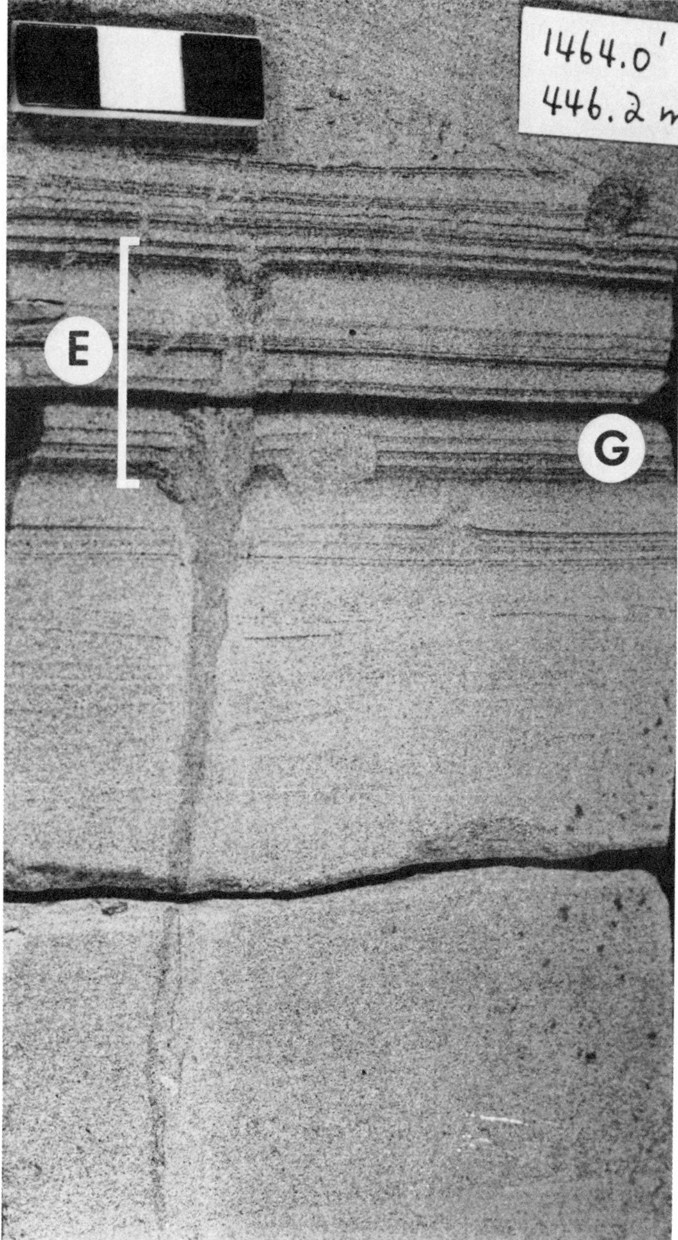

FIG. 22.—Bioturbated rippled sandstone facies. *Monocraterion* with escape structures (E). Note graded laminae (G). 6B-15-53-2W4, depth approximately 446 m (1,464 ft).

sequences are thin. These two sequences each represent a more proximal splay position.

The interpretation of crevasse splays is supported by the syneresis cracks in the interbedded mudstone/sandstone interval. Low discharge of fresh water from the channel onto the splay would keep the bay brackish, and high discharge during times of flood would quickly reduce the salinity of the water in the bay. Post-flood conditions would allow the salinity to increase again until the next major flood. The large fluctuations in salinity result in the formation of syneresis cracks.

The crevasse splay deposits in this well are very similar to lower delta plain crevasse splays of coal-bearing rocks in eastern Kentucky (Horne, 1979) and to crevasse splay sequences (Fig. 23) of the modern Mississippi delta (Saxena, 1976; Coleman, 1976). In particular, the cyclical sequence of events (marine bay-splay-peat-marine bay, etc.) described by Saxena (1976) for the interdistributary bays is a striking analog to that interpreted for the formation of the stacked crevasse splay deposits of this well.

DISCUSSION

Data from the three wells discussed in this paper show that thick sandstones in the upper Mannville can be formed by an amalgamation of channel facies, by an amalgamation of channel and marine shoreline facies and by the stacking of the proximal portions of crevasse splays (Figs. 24, 25). Considering the distance between the three wells (Fig. 1), it is almost certain that some of the thick sandstones in the intervening wells have formed in other ways. Therefore, to link sandstones of a given minimum thickness and ascribe one origin to them is unrealistic. Depositional environments can vary considerably over a large area and within one environment, there can be several possible ways to form thick sandstones.

Recognition of the stacked channel deposits in the Hairy Hill area is very important for optimum reservoir depletion. Interpretation of the sandstones as one genetic unit would result in a recovery scheme designed for a single reservoir when, in fact, there are two separate gas pools (Fig. 13). The development of an accurate depositional model is necessary to predict reservoir geometry successfully.

The detection of brackish water sedimentation in the channels in the western part of the study area signalled that the streams were flowing into a marine body of water. The recognition of marine shoreline facies near the Saskatchewan border indicated that the sea lay to the east. Thus, within our study area, depositional conditions changed from mainly continental in the west to marine shoreline in the east. The eastward paleoflow direction indicates a western provenance for the sediment, which is consistent with the lithofeldspathic mineralogy of many upper Mannville sandstones (Vigrass, 1977; Putnam, 1982b).

REGIONAL CONSIDERATIONS

The east-west cross section in the Hairy Hill area (Fig. 12) shows an abundance of channel deposits in the upper Mannville. A north-south cross section, with a common well (7-36-55-14W4), shows that the channel sandstones occur in a specific belt and are flanked by fine-grained deposits (Figs. 11, 26).

Data from well 14-14-55-15W4 (Fig. 11), which penetrates the fine-grained rocks flanking the channel sandstones, corroborates the interpretation that the upper Mannville is more continental to the west. The upper Mannville in this well is 36 m (118 ft) thick (Fig. 27) and was sampled in 3-m (10 ft) intervals by one of the authors (Singh, 1964). The only marine indicators (several species of acritarchs) were found over an 11-m thick (35 ft) zone (Singh, 1964, Table 4; 548.9 to 559.6 m, 1,801 to 1,836 ft). A cursory examination of the core by the present authors revealed that

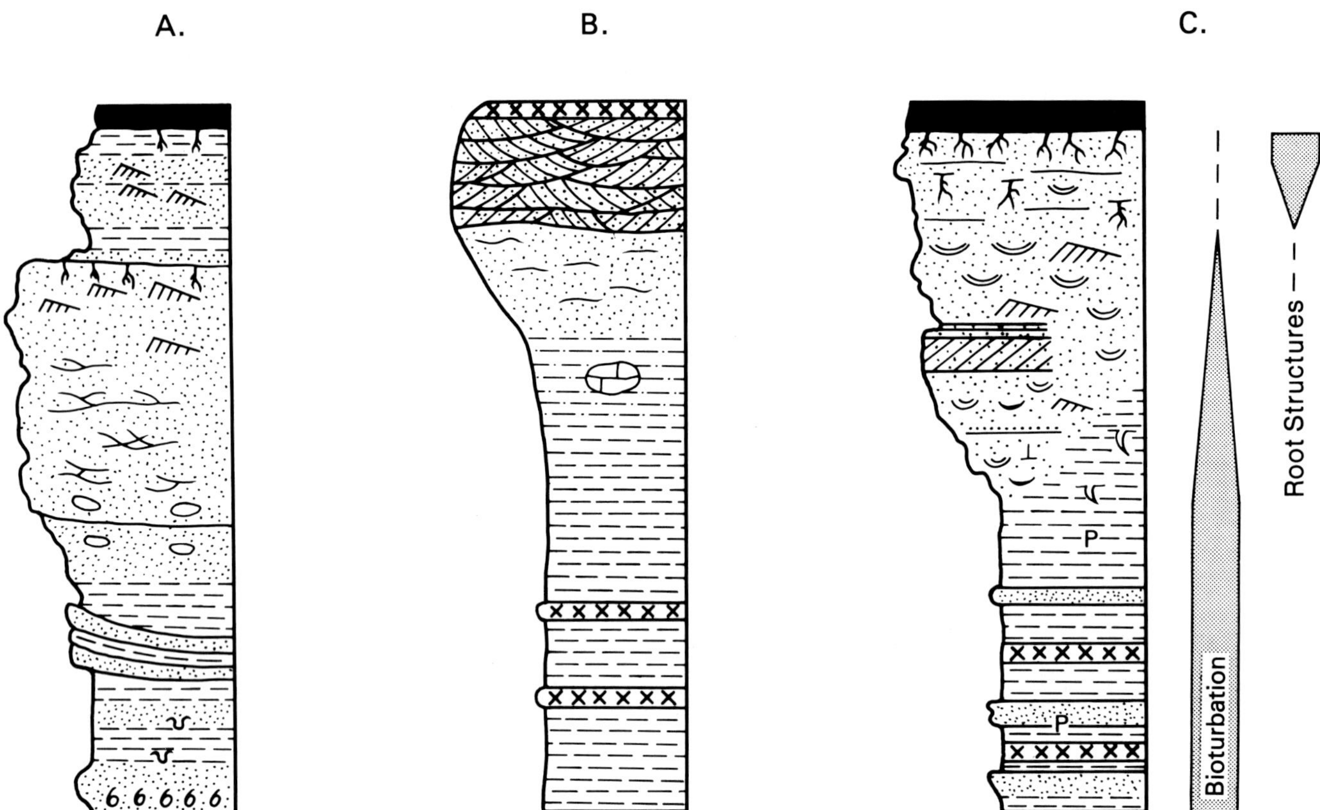

FIG. 23.—Comparison of bay-fill deposits. (A) Mississippi delta type (modified after Coleman and Prior, 1982). (B) Carboniferous of Kentucky (modified after Baganz and others, 1975). (C) Upper Mannville. Bioturbation, carbonaceous detritus, siderite beds or nodules and root structures (upper parts) are common to all three. The presence of pyrite is noted in examples A and C.

the upper Mannville has a strongly continental appearance. It is composed largely of carbonaceous shale (frequently rooted) and contains four coal beds and three reddish, mottled (soil?) horizons. Bioturbation is not apparent except in several thin sandy beds that contain *Planolites* and in one zone, *Palaeophycus*. Because the previous sampling interval was large (Singh, 1964), selected beds were resampled for palynology. Two samples (555.7 m, 1,823 ft; 553.4 m, 1,815.8 ft) confirmed the existence of a zone containing exclusively acritarchs (Table 1). Below this an interval (561.7 to 563.8 m, 1,843 to 1,850 ft) containing a moderate number of dinoflagellates and a flood of *Muderongia* sp B was discovered. A sample from a reddish, mottled bed (559.8 m, 1,836.5 ft) located between these two marine zones was barren of microplankton, pollen and spores (Fig. 27). This strengthens the interpretation of the reddish beds as soil horizons. In addition, one species of dinoflagellate was recovered from a bioturbated bed (544 m, 1,785 ft) near the top of the core.

The marine palynological indicators are not as well developed in this well as they are in one of the wells to the east (10D-29-47-1W4). The other data (coal beds, soil? horizons, lack of bioturbation) also strongly suggest that this area was more continental. The presence of the dinoflagellates and acritarchs, however, indicates that the marine influence periodically extended at least this far to the west. This is consistent with the marine influence that was detected in the abandonment phases of nearby channels (15-12-55-13W4) and reinforces the concept that several lines of evidence should be used to reconstruct paleoenvironments.

The interpretation of marine shoreline sedimentation near the Alberta-Saskatchewan border is strengthened by data from well 2-30-43-2W4. The upper Mannville in this well is predominantly sandstone, but samples from the interbedded shales yielded some of the best specimens of acritarchs and dinoflagellates (Fig. 8). The lowest marine shale (605.9 m, 1,987.9 ft) is 21.9 m (72 ft) below the top of the Mannville (Fig. 28) and overlies the what appears to be a channel sandstone, based on its log profile. The shale contains a flood of *Muderongia* sp A and a few *Muderongia* sp B, indicating a low salinity, marginal marine environment. There is a distinct increase in the diversity of the marine microplankton in samples taken at 605 m (1,984.9 ft) and 600.5 m (1,970.1 ft), in which a few specimens of dinoflagellates normally found in an open marine environment (*Spiniferites ramosus* and *Odontochitina operculata*) are present. These are accompanied by other dinoflagellates and acritarchs (Table 1). This assemblage is diagnostic of a brackish bay or estuarine environment.

On the Alberta side of Lloydminster, Orr and others (1977) interpreted the upper Mannville as a deltaic sequence with mixed marine/nonmarine affinities. Of particular interest are two cores that they described. The first core (Orr and others, 1977, Fig. 7; well 11A-4-49-1W4) contained several progradational sequences consisting of burrowed mud-

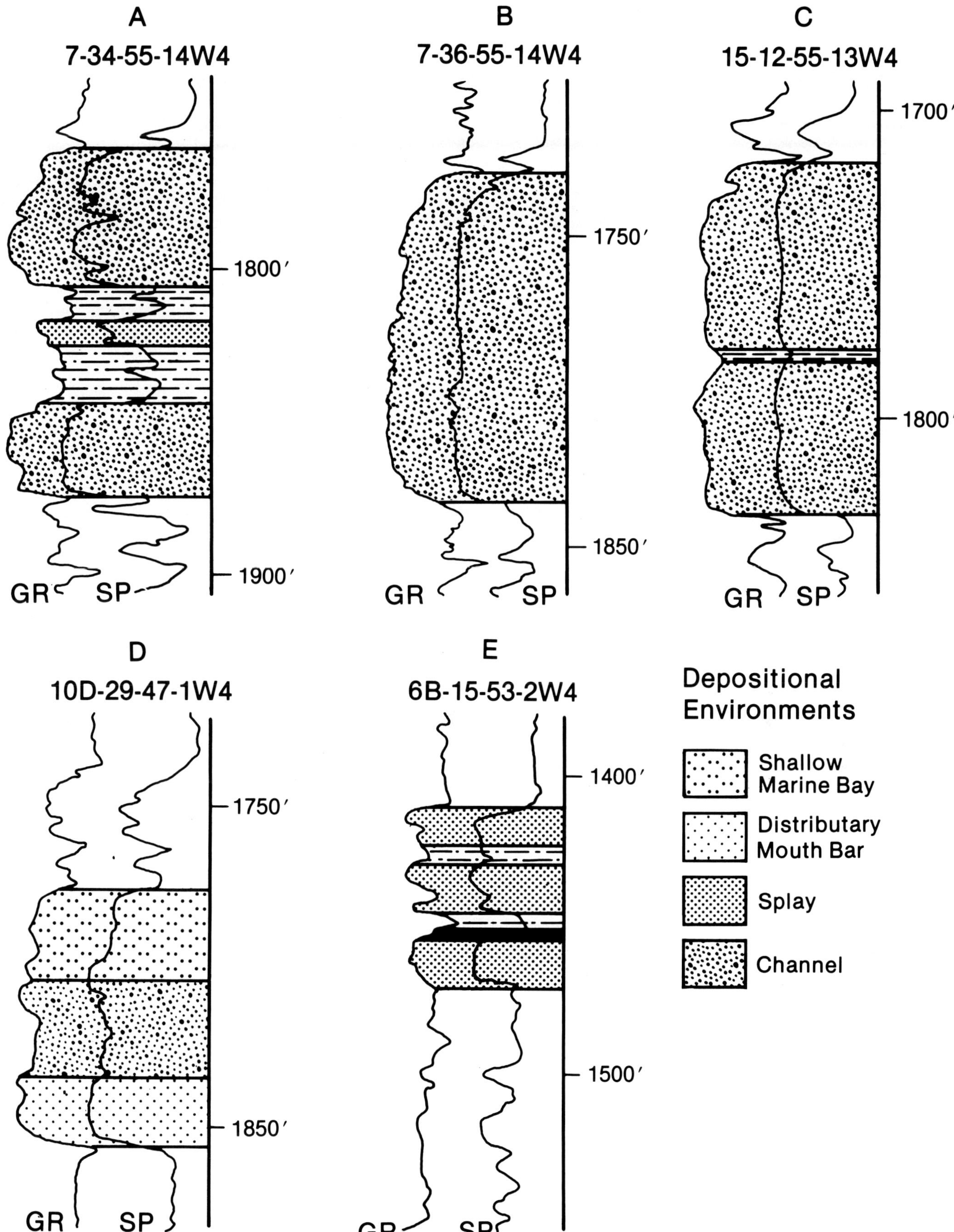

FIG. 24.—Typical depositional environments of thick upper Mannville sandstones in study area (Fig. 1).

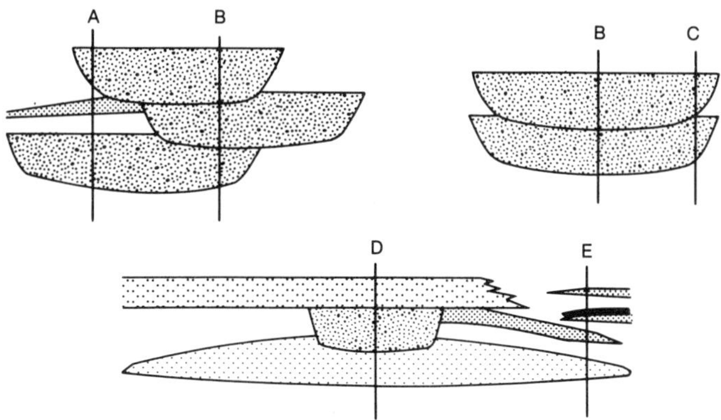

FIG. 25.—Schematic cross sections showing the genesis of thick sandstones. Letters correspond to those in Figure 24.

stones grading into rippled and parallel-laminated sandstones, occasionally capped by coals. Two fluvial deposits were also described. The coarsening-upward sequences are very similar to those found in well 6B-15-53-2W4 and their interpretation (Orr and others, 1977) of a shallow-water, restricted marine environment is similar to ours. Orr and others (1977) stated that the second core consisted of stacked point bar deposits that overlie a marine lagoonal facies. Thus, stacked fluvial deposits and marine shoreline sedimentation have both been previously interpreted for the upper Mannville.

Several authors have described brackish to marine conditions in the upper Mannville on the Saskatchewan side of Lloydminster, which fit with the data from this paper. Van Hulten (1984) interpreted sandstones in the Pikes Peak field 40 km (25 mi) east of Lloydminster as tidal/estuarine in origin, based on microfauna and the degree of bioturbation. Lorsong (1980) stated that the upper Mannville in the Celtic area, approximately 20 km (12 mi) north of Pikes Peak, was deposited in a nearshore marine setting. Fuglem (1970, Fig. 8) recovered marine foraminifera from the upper Mannville in a well 16 km (10 mi) east of Lloydminster and concluded that restricted marine fauna are locally present. Gross (1980) conducted a regional study of the Mannville based on log character, so marine versus nonmarine environments were not dealt with in detail; however, he stated that in the McLaren interval, blanket-type sandstones are more common in Saskatchewan than in Alberta and this type of sandstone is characteristic of marine deposits (Coleman, 1976).

Brackish water indicators in our study area suggested that a marine embayment lay to the east. The marine shoreline facies described by others in Saskatchewan corroborate this interpretation. Therefore, recognition of brackish water sediments can signal the change from a fluvial channel/floodplain system to a complex marine shoreline setting; the concomitant change in the geometry of the sand bodies is of paramount importance in both exploration and production geology. As well, detection of a marine influence in channel deposits can have important ramifications as to the type of fill. The Ogeechee River along the Georgia coast

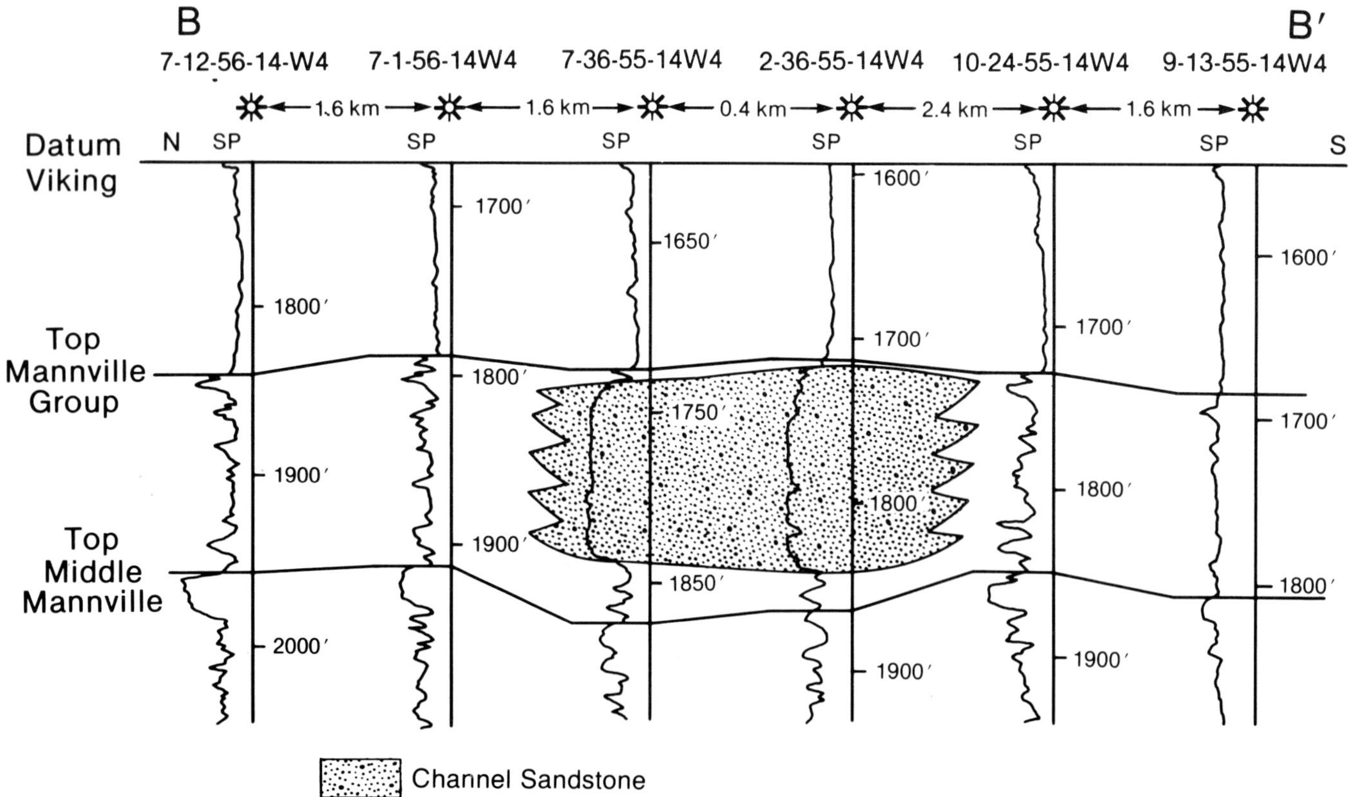

FIG. 26.—Stratigraphic cross section oriented north-south, Hairy Hill area, Alberta. Channel sandstones are flanked by fine-grained rocks.

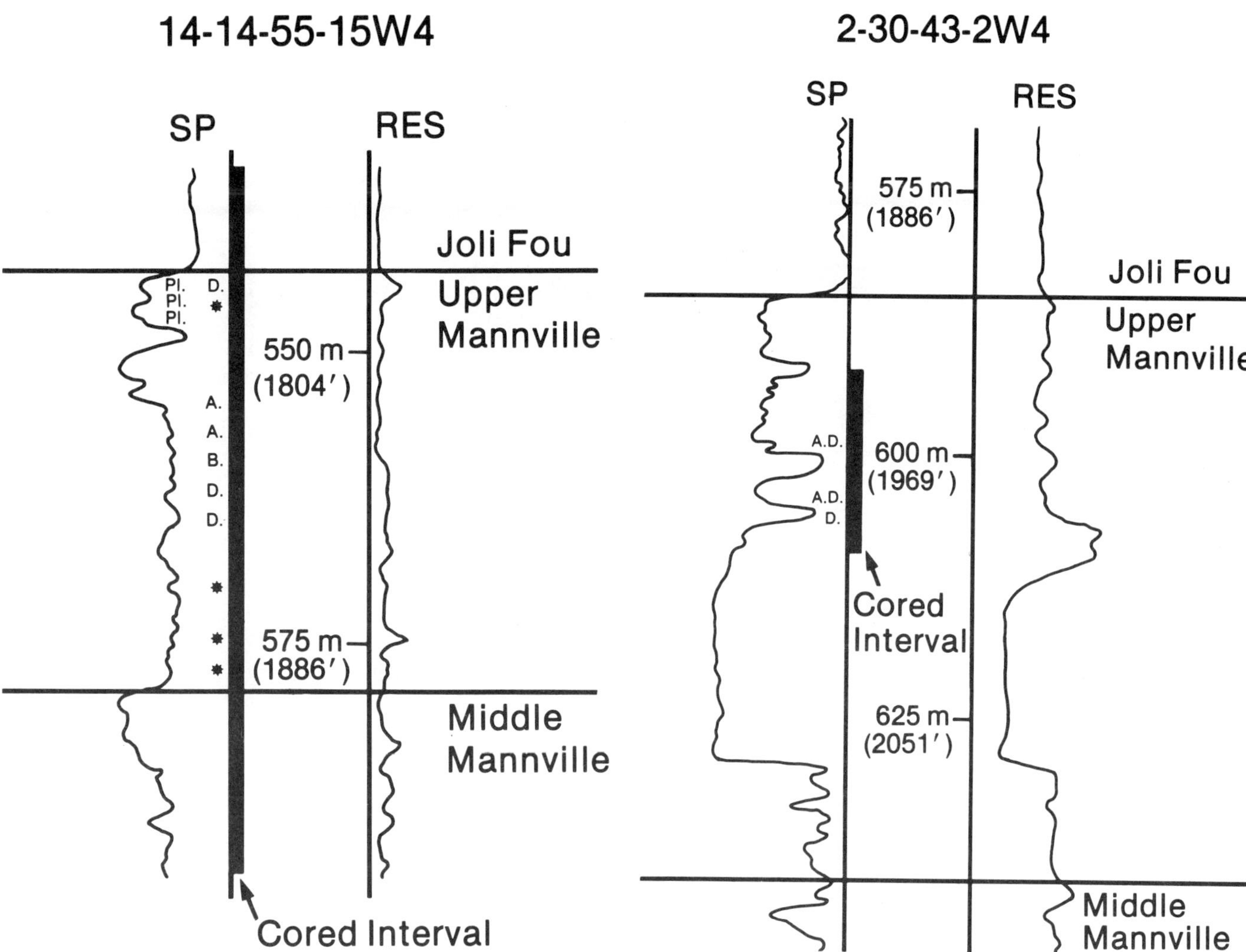

FIG. 27.—Geophysical logs, cored interval, palynological and ichnological data for well 14-14-55-15W4. Symbols as in Figure 4.

FIG. 28.—Geophysical logs, cored interval and palynological data for well 2-30-43-2W4. Symbols as in Figure 4.

has a tidal component and displays a downstream trend of sand (fluvial) to mud (upper limit of saline influence) to sand (maximum tidal strength; Dorjes and Howard, 1975). Ray Rahmani (pers. commun.) has identified this type of channel fill in the Upper Cretaceous Horseshoe Canyon Formation near Drumheller, Alberta. Thus, the recognition of brackish water deposits has predictive value for both local and regional geological interpretations.

SEDIMENTOLOGICAL MODEL

The lack of closely spaced cores hinders the development of a detailed sedimentological model for the upper Mannville; however, based on our data, a fluvially dominated deltaic environment is proposed. The overall setting of fluvial channels emptying into a marine embayment is delta-like and crevasse splay and distributary mouth bar sandstones are key components of the delta model, as are the interdistributary fine-grained mudstones.

The channel sandstones in the western part of the study area have a limited marine influence either within the sandstones or in the contiguous fine-grained rocks. Thus, they would represent fluvial channels to upper delta plain distributaries. The rivers tended to flow along certain paths, resulting in the stacking of fluvial deposits. There are several possible reasons for this. Horne and others (1978) described stacked channel deposits in an upper delta plain setting in which the channels were localized by faulting. Another possibility is that highs or lows on the underlying Wainwright ridge, a north-south-trending topographic high developed on the sub-Cretaceous unconformity, affected the east-west-trending channels. Facies from the other two wells (10D-29-47-1W4; 6B-15-53-2W4) fit into a lower delta plain setting, as the marine influence is more pronounced (Fig. 29). Thus, the delta was prograding in an easterly direction.

One question that typically arises with a deltaic interpretation is, "Where are the prodelta shales?" A 9-m-thick (29 ft) shale, part of which may be lacustrine, occurs beneath the sandstone in well 10D-29-47-1W4 and thicker shales should occur in a seaward direction, which is probably to the northeast of the study area. One factor which influences the thickness of deltaic sequences, and hence the thickness

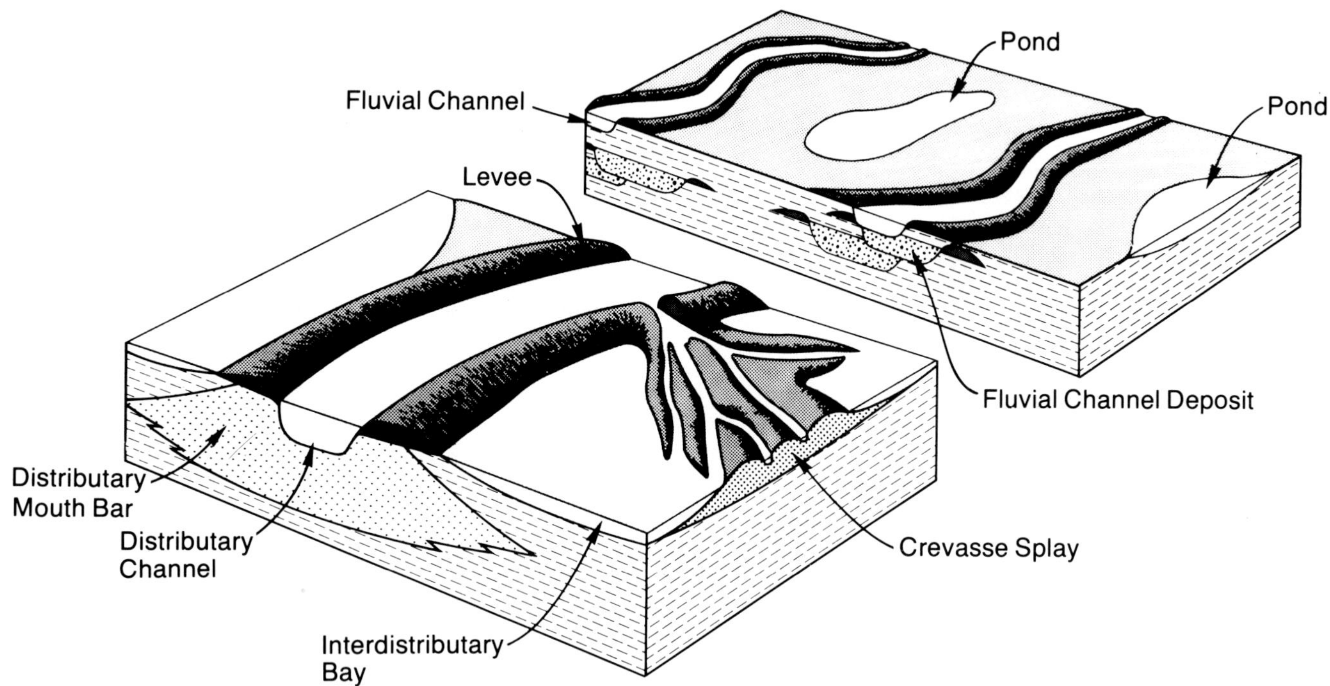

FIG. 29.—Schematic sedimentological model for the upper Mannville in the study area.

of the prodelta shales, is the depth of water into which the delta was building. Another significant factor is rate of subsidence. In the Lloydminster area, the regressive sequences lower in the section (e.g., Sparky, Rex, Cummings) are not very thick (generally less than 20 m, 66 ft) so it is inferred that the water depth was never very great, and the rate of subsidence was probably low in this area throughout the Mannville. The shallow-water depth would preclude the accumulation of a significant thickness of offshore fines, and the only thick (≥10 m, 33 ft) shale with a well developed open marine fauna in the Mannville (Clearwater Formation) occurs well to the north in the Athabasca oil sands deposit (north of Township 75). Baganz and others (1975) stated that the shallow-water depths of the inland sea prevented the development of thick prodelta shales in the Carboniferous sequences in eastern Kentucky.

Although a fluvially dominated delta best fits the data which we have presented, other sedimentological models may be more applicable elsewhere as conditions change along any given shoreline (e.g., a delta flanked by barrier island complexes). This is especially true as the size of the channel deposits suggests that any delta or deltas were relatively small.

COMPARISON OF MIDDLE AND UPPER MANNVILLE

In the Lloydminster area, shales that underlie the formations in the middle Mannville typically have relatively well developed marine affinities (Vigrass, 1977; Putnam, 1982a) so that the coarsening-upward regressive sequences (e.g., Lloydminster) are recognized as marine. The shales were probably deposited in more of an offshore position rather than in shallow, nearshore bays. Therefore, the transgressions of the Boreal Sea (Jeletzky, 1971), which occurred in middle Mannville time, extended farther inland, beyond the study area. In contrast, the limit of the upper Mannville transgression occurs within, or not far beyond, our study area, so that only brackish marine indicators are present. This is illustrated by well 14-14-55-15W4, where the upper Mannville contains few marine microplankton, but in the strata equivalent to the middle Mannville, dinoflagellates and acritarchs are relatively abundant (Singh, 1964, Table 4). To the east, core from wells such as 15-22-55-1W4 contains abundant microplankton, indicating the marine nature of parts of the middle Mannville.

The distinct change from the middle Mannville to the upper Mannville has been noted by Vigrass (1977) and others. We believe that one of the reasons for this is the relatively limited extent of the transgression during upper Mannville time.

There appears to be a relationship between the geology of the upper Mannville and isopach maps of the entire Mannville interval. The marine influence in the upper Mannville increases toward the east, where the Mannville isopach thickens (≥200 m, 656 ft) in Saskatchewan (Ranger, 1984; Orr and others, 1977). The isopach 'thick' coincides approximately with the salt solution edge of the underlying Devonian evaporites (Orr and others, 1977). Thus, solution of the Devonian evaporites may have controlled the upper Mannville marine embayment either directly by penecontemporaneous salt solution and/or indirectly by greater compaction of the thicker Mannville section in the area of salt removal.

RECOGNITION OF BRACKISH WATER DEPOSITS

Subtidal, brackish water environments are found, for example, in shallow bays, lagoons, estuaries and on delta platforms. Such marginal marine environments are subject to large fluctuations in physical, chemical and biological parameters (Dorjes and Howard, 1975), which create a physiologically stressful habitat for numerous organism groups (Rhoads, 1975). This results in a diminished faunal content, making the recognition of these facies more difficult. A high rate of sedimentation can further dilute the marine signal and be accompanied by a strong input of continental indicators. As well, brackish water deposits may comprise only a small proportion of a sequence dominated by continental facies. In contrast, sediment deposited in an offshore position is usually easier to recognize as marine because of the stability of the environment. Abundant and diverse types of microfossils, as well as macrofossils, are generally present in these rocks.

Brackish water deposits are important becaue they distinguish marine from nonmarine shoreline sequences where there are no associated offshore deposits. Examples of this can be seen in the Upper Cretaceous Blackhawk Formation in east central Utah (Fig. 30), which contains multiple transgression/regression cycles. Toward the limit of a transgression, the fine-grained offshore component pinches out, resulting in a partial regressive sequence. Mudstones present in this part of the section tend to be brackish. Another important observation is that this facies occurs at or near the limits of the transgressions, as is the case in the upper Mannville. Thus, brackish water deposits can be very important even if their volumetric proportions are low. Data from the upper Mannville indicate that an integrated approach combining ichnological, palynological and sedimentological criteria constitutes a powerful method for the recognition of marine shoreline zones in the rock record.

Sedimentary Structures

Whereas various sedimentary structures have been attributed to specific depositional processes (e.g., flaser bedding formed by tidal currents), most sedimentary structures formed by currents or waves can be found in both marine and nonmarine environments. The sedimentary structure which we feel is most useful in recognizing brackish conditions is syneresis cracks. Syneresis cracks have been described and interpreted in this paper, but because of their importance, a more complete review is given here.

Syneresis cracks can apparently form either subaqueously at the sediment-water interface or substratally (Plummer and Gostin, 1981). Substratal syneresis cracks form by a loading—dewatering—cracking process and are accompanied by load structures. This type of syneresis crack is probably not indicative of marine-influenced clay deposition. According to Plummer and Gostin (1981), syneresis cracks that form at the sediment-water interface can be the result of the flocculation of non-swelling clays that crack upon early compaction. This is the result of sedimentation under marine conditions. In muds with as little as 2% swelling clay, syneresis cracks can form at the sediment-water

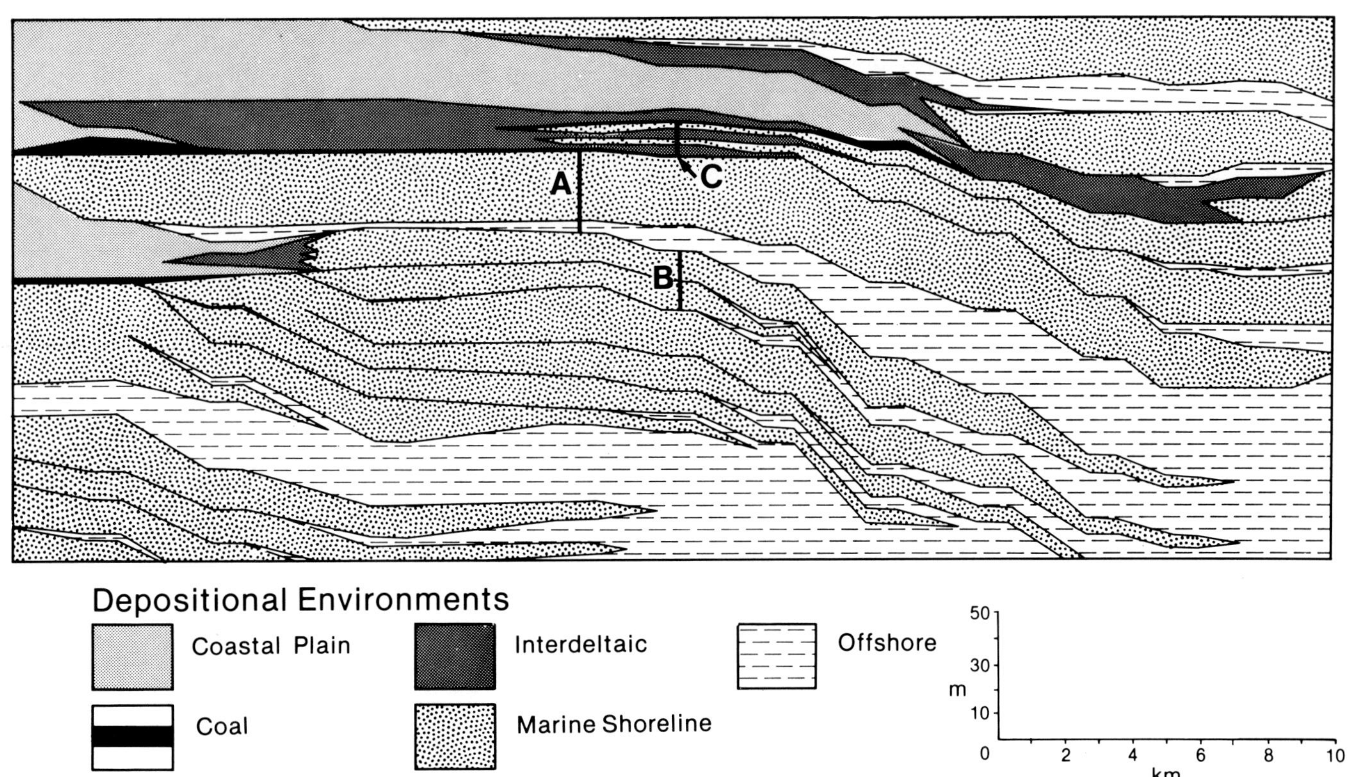

FIG. 30.—Depositional systems and facies, Upper Cretaceous Blackhawk Formation, Utah (modified after Balsley and Parker, 1983). Interdeltaic facies is largely composed of brackish water deposits. (A) A complete regressive sequence. (B) Two partial regressive sequences. (C) Two partial regressive sequences with associated brackish water deposits.

interface when there are large fluctuations in salinity (Burst, 1965; Plummer and Gostin, 1981). The changes in salinity result in contraction of the lattices of the swelling clay minerals, thereby causing the cracks. We interpret the shrinkage cracks in well 6B-15-53-2W4 to have formed in this way, and this type of structure is an excellent indicator of marine shoreline deposition.

Donovan and Foster (1972) interpreted shrinkage cracks, which were very similar to those from well 6B-15-53-2W4, as syneresis cracks formed by fluctuations in salinity. They concluded, however, that the depositional setting was a lake, which seems somewhat incompatible with their mechanism of crack formation. It is possible that the environment was a restricted marine bay rather than a lake. Shrinkage cracks have been described from freshwater sequences (e.g., Picard, 1966), but these have a different morphology and were probably not formed by a syneresis mechanism.

We feel that true syneresis cracks which are not associated with load structures can provide strong evidence that the depositional environment was subject to large fluctuations in salinity. This type of environment would typically be found in a marginal marine setting, as opposed to lakes, freshwater channels or in an offshore marine position.

Diagenetic Minerals

Horne and others (1978) summarized data which indicate that Carboniferous coals in the eastern United States contain greater amounts of iron sulfides where they are overlain by marine to brackish water deposits than where the coals are overlain by freshwater deposits. The iron sulfides are early diagenetic minerals that form when the overlying sediment is being deposited. Thus, the quality of coal (sulfur content) is dependent on the type of deposit that overlies the coal.

Horne and others (1978) attributed the formation of pyrite to sulfur-reducing organisms which are restricted to marine or brackish water. This explanation is not really complete, however. Pyrite is stable under low Eh (reducing) and low pS^{2-} (high sulfide) conditions (Berner, 1971). In a marine setting where there is abundant organic matter, aerobic bacteria utilize all of the free oxygen (O_2) in metabolizing the organic matter. Further decomposition of the organic matter is then accomplished by anaerobic bacteria that reduce, in succession, NO_3^-, NO_2^-, MnO_2 and finally SO_4^{2-} (Stumm and Morgan, 1970). Thus, under strongly reducing conditions, sulfate (SO_4^{2-}) is reduced to sulfide (H_2S) in the following manner (Berner, 1980):

$$2\,CH_2O + SO_4^{2-} \rightarrow H_2S + 2HCO_3^-.$$

The sulfide reacts with any iron present ultimately to form pyrite. The limiting factors in this process are the amount and type of iron compounds, the concentration of organic compounds and the availability of dissolved sulfate. Marine sediments usually contain some reactive iron compounds and dissolved sulfate, so the most important limiting factor is the amount of decomposable organic matter (Berner, 1971). Therefore, in a marine setting with sufficient organic matter, reducing conditions should result in the formation of diagenetic pyrite.

Siderite is stable under low Eh (reducing) and high pS^{2-} (low sulfide) conditions. Berner (1971) concluded that these conditions are rare in marine sediments because marine water contains abundant dissolved sulfate and thus, strongly reducing conditions are accompanied by the anaerobic conversion of sulfate to sulfide. For this reason Berner (1980) has suggested that siderite is a useful indicator of freshwater conditions. Siderite can form in a marine setting in either of two ways, however. If there is sufficient organic matter to cause reducing conditions but insufficient organic matter to result in sulfate reduction (post-oxic environment; Berner, 1981), siderite, and not pyrite, will form. The other scenario is where there is abundant organic matter and all of the sulfate is reduced and used (pyrite and other iron sulfides have formed), but there is an excess of reduced iron so siderite forms. This can be accompanied by the production of methane (Berner, 1981). In summary, reducing conditions in marine settings should produce diagenetic pyrite, whereas siderite may or may not form.

In nonmarine settings the limiting factor in pyrite formation is the availability of dissolved sulfate. Sulfate can be 100 (Berner and Raiswell, 1984) to 230 (Stumm and Morgan, 1970) times more abundant in marine water than in fresh water. Thus, reducing conditions in freshwater environments typically result in the formation of diagenetic siderite because there is very little sulfate to be reduced to sulfide. Freshwater lakes with high concentrations of sulfate are rare and usually result from the local weathering of gypsum (Berner and others, 1979).

As previously stated, diagenetic pyrite occurs throughout the core from well 6B-15-53-2W4 and it is frequently associated with siderite. There is a significant amount of organic matter in the coarsening-upward sequences, so reducing conditions probably developed at a shallow depth of burial. Sulfate was reduced to sulfide, resulting in the formation of pyrite, and after the sulfide was consumed, siderite formed. The tendency for pyrite and siderite to occur in finer grained beds indicates that reducing conditions were probably easier to maintain in the less permeable beds.

The intimate association of pyrite and siderite in the core is consistent with a brackish water depositional environment. Brackish water contains more dissolved sulfate than typical fresh water, which allows pyrite to form but not as much as normal marine water so that siderite forms when the reduced sulfate has been depleted.

The thick black shale that occurs below the sandstone in well 10D-29-47-1W4 contains siderite but no visible pyrite. The bulk of the shale is interpreted herein as a restricted marine bay or lacustrine deposit, based on ichnological and palynological data. The presence of siderite indicates that reducing conditions did occur, probably at a shallow depth of burial, so a marine influence should have resulted in the growth of diagenetic pyrite. Thus, the absence of pyrite is consistent with the interpretation of freshwater, or very low salinity, conditions.

Throughout this discussion of pyrite and siderite, it has been assumed that these minerals were forming during early diagenesis, or at very shallow depths of burial. If pyrite and/or siderite form during later diagenesis, their presence or absence has no significance as to the depositional setting. Berner (1981) stated, however, that the succession of early diagenetic environments (oxic, post-oxic, sulfidic and methanic) are achieved within the first few meters of burial. We have no data to prove conclusively that the siderite and pyrite in the cores from wells 6B-15-53-2W4 and 10D-29-47-1W4 have been formed at shallow depths of burial. The presence of siderite grains in some of the sandstones in both wells suggests that siderite was forming nearby as an early diagenetic mineral, which was subsequently eroded and deposited as detrital grains. Thin sections show that the siderite is not an alteration product. Differential compaction around siderite or pyrite nodules is indicative of early diagenetic formation, but this was not discernible in our core.

Cassagrande and others (1977), Martens and Goldhaber (1978) and Postma (1982) studied modern freshwater to brackish water systems and all reported a correlation between diagenetic pyrite and a marine influence during deposition. Postma (1982) also cautioned, however, that siderite could form in brackish or marine environments after the reduced sulfate has been depleted, and as a result is not a foolproof environmental indicator.

Berner and others (1979) proposed that the ratio of FeS_2/FeS can be used as a paleosalinity indicator because the abundant sulfate in marine environments favors the formation of pyrite (FeS_2) rather than monosulfide minerals. Berner and Raiswell (1984) observed that sedimentary rocks deposited in fresh water tend to have a much higher organic carbon-to-pyrite sulfur (C/S) ratio (>10) than those deposited in marine environments (0.5 to 5). Thus, the ratio could be used to distinguish freshwater from marine rocks. The lower C/S ratio in marine deposits is again due to the relative abundance of sulfate in salt water. There is obviously merit in both of these techniques, but we feel that the less sophisticated alternative of noting the presence or absence of iron sulfide minerals can be meaningful to the petroleum geologist.

In summary, diagenetic pyrite and siderite may be unrelated to the depositional environment of the host rock as they can be late diagenetic features. The presence of pyrite, however, especially if it is common, can indicate a marine influence during the deposition of the host, or surrounding (e.g., pyrite in coal) rock. The co-existence of pyrite and siderite suggests the possibility of brackish water deposition. If siderite occurs by itself, freshwater deposition should be considered; however, caution must be taken that microcrystalline pyrite is not present. As with syneresis cracks, pyrite and siderite are not unequivocal, but they offer another line of evidence that can be used when interpreting depositional environments.

Ichnology

The strong imprint of bioturbation on the brackish water deposits of the upper Mannville is not an isolated occurrence. The shallow interdistributary bays of the modern Mississippi delta support a brackish water fauna that includes small oyster reefs, but the characteristic feature of the bay-fill sediments is bioturbation (Coleman, 1976). Coleman (1976) described an ancient bay-fill sequence from Carboniferous rocks in Kentucky and noted that the lower part of the sequence consists of highly burrowed shales and silty shales, whereas body fossils are 'scattered'. Brackish water faunas occur in the interdeltaic facies of the Blackhawk Formation (Upper Cretaceous) but bioturbation is also prominent, varying from weak to moderate to locally intense (Balsley and Parker, 1983).

Thus, it appears that trace fossils are more abundant than body fossils in at least some clastic brackish water deposits. The abundance of trace fossils can be explained as follows. Sediments of estuaries and other brackish water environments are effective at dampening salinity fluctuations. Sanders and others (1965) found that while the salinity of the bottom water in an estuary ranged from 2.3 to 29.3‰ during a tidal cycle at one station, the salinity in the sediment below 5 cm (2 in.) of depth was uniform and steady at 20.5‰ (Fig. 31). Therefore, the deep infaunal habitat serves as a refugium (Rhoads, 1975), buffering the organism against rapid and extreme salinity variations. As a result, infaunal organisms are more abundant than epifaunal organisms in low salinity waters (Sanders and others, 1965). This is consistent with Alexander and others (1935), who showed that burrowing organisms are able to penetrate farther inland in an estuary with epifaunel organisms which are subject to the fluctuating salinities of the overlying water column (Fig. 32). It must be stressed that this 'abundant' bioturbation is relative because brackish waters are generally reduced in species numbers with respect to both fresh water and fully saline water (Dorjes and Howard, 1975). Therefore, sediment deposited in fully saline settings is

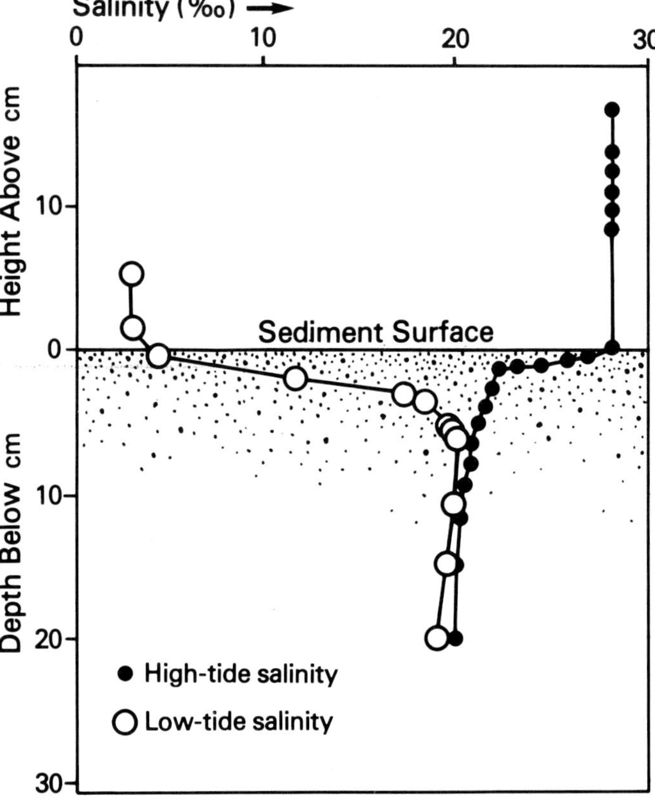

FIG. 31.—Salinity fluctuations in water and sediment at high and low tides, Pocasset River estuary, Massachusetts (modified after Sanders and others, 1965).

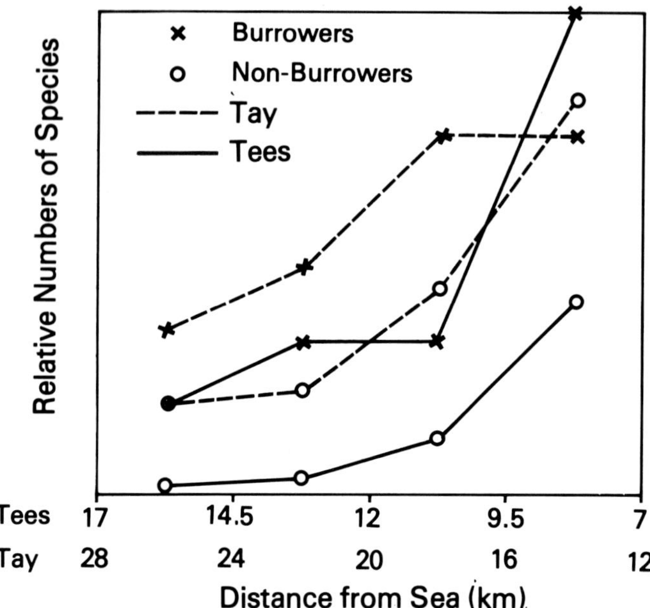

FIG. 32.—Relative density of burrowing and nonburrowing organisms in the Tees and Tay estuaries, England (modified after Alexander and others, 1935).

generally more bioturbated than brackish water sediment. The species minimum occurs at approximately 5 to 7‰ (salinity), with other major breaks at 18‰ and 30‰. Thus, the degree of bioturbation would obviously depend upon the degree of brackishness, as well as other factors (e.g., range of salinity fluctuations, rate of sedimentation, texture of sediment).

Although there are many freshwater organisms, the fauna is dominated almost entirely by two groups, branchiopods (freshwater crustaceans) and insects (Croghan, 1983). Biogenic structures created by these groups generally have a low preservation potential (Frey and others, 1984). Therefore, freshwater deposits tend to have a limited trace fossil assemblage.

The lack of body fossils in brackish deposits is the result of the fauna having a low preservation potential. There is a high proportion of infauna in this setting and infauna tend to be soft-bodied. As well, with decreasing salinity the reduction of species in groups forming a calcareous skeleton is greater than in those lacking such a skeleton (Remane and Schlieper, 1971). The number of shelly benthic invertebrates can be further reduced by local oxygen deficiencies and the production of H_2S resulting from salinity stratification of the overlying water column (Remane and Schlieper, 1971). These conditions are characteristic of many brackish bays or silled fjords with high concentrations of organic matter and limited water exchange rates. Rhoads and Morse (1970) indicated that most shelled invertebrates cannot tolerate dissolved oxygen concentrations of less than 1.0 ml/l; however, soft-bodied organisms which do not require oxygen for shell construction can maintain themselves at appreciably lower concentrations (i.e., 0.1 ml/l). Therefore, soft-bodied organisms may be the only fauna able to survive in very restricted brackish water environments.

The second aspect of the problem, which is not unique to brackish water deposits, is that even where a shelly fauna is relatively well developed, the preservation potential of the shells is low compared to the preservation potential of trace fossils. Post-mortem shell dissolution by acidic pore water can be responsible for the complete removal of calcareous shelled organisms in many marginal marine environments (Oomkens, 1974; Goldring and others, 1978; Flessa and Brown, 1983). Investigations in Long Island Sound by Aller (1982) and Great Bay Estuary by Hines and others (1982) indicated that carbonate dissolution is accentuated in bioturbated areas, where alkalinity buildup during sulphate reduction is prevented by both burrow irrigation and the oxidation of solid-phase sulphides during particle reworking. As well, diagenetic factors which have an adverse effect on shelly fauna tend to enhance biogenetic structures (Seilacher, 1964). Thus, although shelly benthic faunas may be important in some instances (Hudson, 1963; Fursich, 1981), biogenic structures (which record the activities of infaunal organisms) may represent one of the only preserved clues to the original biotic component of many brackish water environments.

In subsurface studies, the use of body fossils is further inhibited by the limited sample that the geologist obtains (core). Even if body fossils occur in a particular facies, they may not be present in a core, especially if the fossils have a patchy distribution (e.g., oyster reefs). In contrast, ichnology is amenable to core studies because of the size and density of most trace fossils. Even in facies where ichnofossils are sparse, there can be beds or zones of intense bioturbation that will be present in cored material (c.f., Balsley and Parker, 1983).

Although there exists an unfortunate paucity of studies dealing specifically with biogenic structures in modern brackish water environments, work by Howard and Frey (1973, 1975) on the estuaries of the Georgia coast indicate that (1) the diversity and abundance of biogenic structures increases seaward; (2) distinct biogenic structures and bioturbate textures are also more diverse and abundant along the margins of the estuaries than in the deeper channels; (3) distinct burrows and dwelling tubes characterize most muds, whereas sandy muds or muddy sands may exhibit both distinct burrows and various types of bioturbate textures; and (4) coarser sediments tend to lack biogenic sedimentary structures. In addition, they also recognized that the entire suite of estuarine biogenic structures generally consists of both vertical and horizontal burrows or burrow systems and thus does not fall conveniently into any of Seilacher's (1967) universal ichnofacies. Instead, the assemblage tends to be composed of a mixture of structures typical of both the *Skolithos* and the *Cruziana* ichnofacies. This assemblage is characterized by burrows that, if preserved, would consist of *Skolithos, Monocraterion, Thalassinoides, Ophiomorpha, Chondrites, Planolites* and *Palaeophycus* (Howard and Frey, 1973, 1975). Burrows typical of freshwater deposits (i.e., *Scoyenia* and insert perturbations) were not observed. Likewise, although sediments and physical sedimentary structures in tidal stream bars are very similar to those in terrestrial bars (Frey and Howard, 1969), the biogenic structures in the two are very different (Pryor, 1967).

The assemblage identified by Howard and Frey (1973, 1975) is very similar to that described for the upper Mannville (Fig. 33). The general characteristics of this assemblage are more important than the individual trace fossils and these are (a) low diversity; (b) forms typically found in marine environments (brackish water fauna can be considered more of an impoverished marine assemblage than a true mixture of freshwater and marine elements); (c) structures constructed by organisms which employ a non-specialized feeding strategy; and (d) vertical and horizontal ichnofossils which are common to both the *Skolithos* and *Cruziana* ichnofacies.

Although brackish marginal marine environments are widespread in the modern setting, few have been documented from the rock record. Notable exceptions include parts of the following units: the Middle Jurrassic Great Estuarine Group (Hudson, 1963, 1980) of Great Britain; the Upper Jurassic of Portugal (Fursich and Schmidt-Kittler, 1980; Fursich, 1981); the Lower Cretaceous Wealden Group of Great Britain (Allen, 1975; Stewart, 1981, 1983); the McMurray Formation of Alberta (Pemberton and others, 1982); the Upper Cretaceous Blackhawk Formation of Utah (Balsley and Parker, 1983); and the Eocene Bagshot Beds of southern England (Goldring and others, 1978). These units all contain a suite of biogenic structures which conform to the general characteristics stated above and which, in most instances, have been used as evidence of at least some saline influence.

In summary, we emphasize that trace fossils are similar to the other criteria discussed in this section in that they should not be solely relied upon for interpreting a marine influence during deposition. An integrated approach utilizing all available physical, biological and chemical data is preferred (Fursich, 1981).

Palynology

The upper Mannville subgroup contains a restricted assemblage of marine microplankton (Fig. 8) in certain shaly intervals, which indicates a brackish, marginal marine environment of deposition. Those dinoflagellates whose diversity and abundance increase significantly in sediments deposited under open marine conditions are not present in this unit. The sediments deposited under very low saline conditions exclusively contain ceratioid dinoflagellates tentatively assigned to *Muderongia* sp. A and *Muderongia* sp. B in the present study. It appears that these ceratioid dinoflagellates are especially adapted to very low saline conditions occurring in upper delta plain settings. Floods of similar *Muderongia* spp. have been reported from brackish intervals in the Wealden beds of England by Batten and Eaton (1980) and Batten (1982). Additional species of dinoflagellate genera *Spinidineum, Cleistosphaeridium, Dapsilidinium* and *Laciniadinium* appear in strata deposited under slightly more saline, estuarine or lagoonal conditions. These species were presumably able to tolerate subnormal

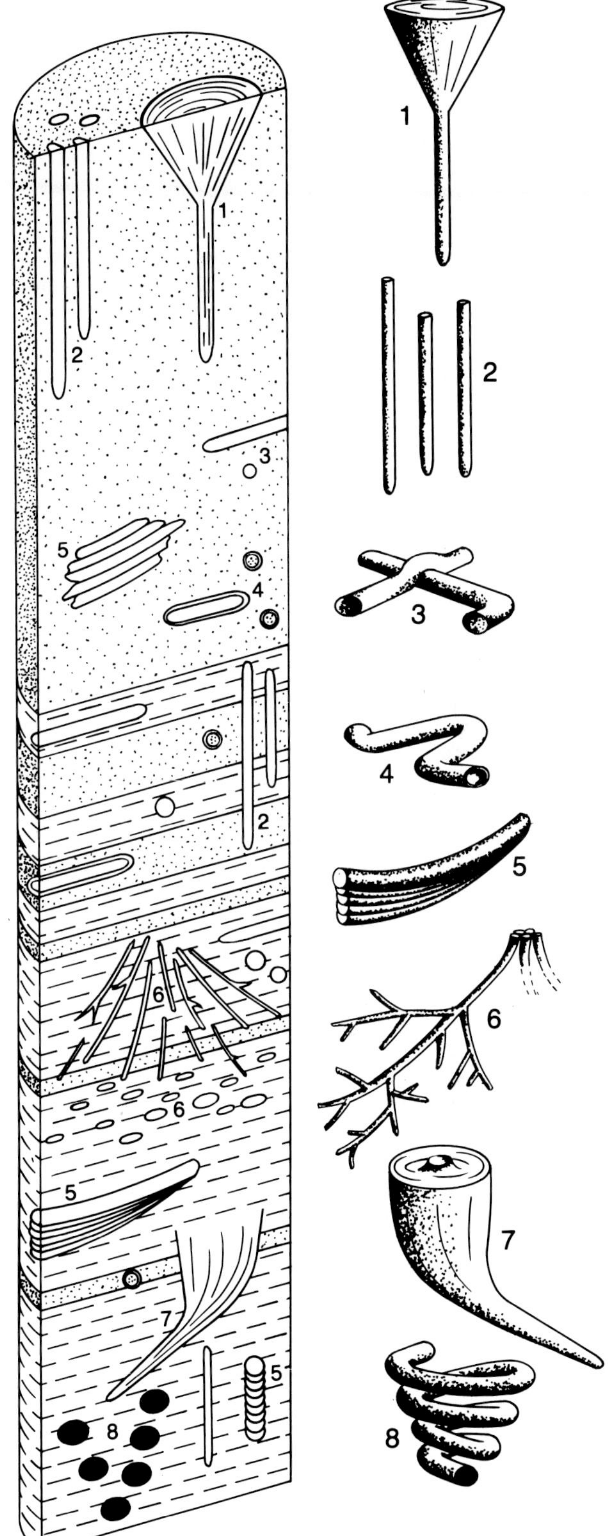

FIG. 33.—Schematic diagram of trace fossils found in brackish water deposits of the upper Mannville. Assemblage contains elements of both the *Skolithos* and *Cruziana* ichnofacies.

saline conditions. The presence of the open marine dinoflagellates *Spiniferites ramosus* and *Odontochitina operculata* in well 2-30-43-2W4 along the Alberta-Saskatchewan border suggests an occasional intermingling of predominantly estuarine brackish water with open marine waters of normal salinity to the east. Thus, variations in the diversity of the dinoflagellates reflect fluctuating salinities during the deposition of upper Mannville strata. A comparable, restricted assemblage of dinoflagellates with a similar distribution pattern has been described by Batten (1982) from the brackish intervals in the Wealden beds of England. In conditions of reduced salinity, dinoflagellates show a reduction in species diversity. Under very low saline conditions, the assemblage is commonly dominated by a single species. These findings are compatible with those of Hulburt (1963), who recorded a reduction in the species diversity of modern phytoplankton in the estuaries and bays of New England as compared to those found in the open-sea and deep-ocean habitats north of Bermuda.

A restricted assemblage of acritarch genera *Veryhachium* and *Micrhystridium* is also present in the upper Mannville subgroup. In the westernmost well, 14-14-55-15W4, the microplankton assemblage is composed exclusively of acritarchs at certain levels (Table 1). This may confirm Burger's (1980) observation that some acritarchs proliferate under brackish conditions adverse to dinoflagellate growth. The restricted assemblage of acritarchs also indicates a marginal marine environment. According to Wall (1965), assemblages dominated by a few acritarch species indicate an inshore environment, whereas species-rich, heterogeneous assemblages denote an offshore habitat. Acritarchs consist of unrelated organisms from diverse habitats, however. Therefore, in order to draw precise paleoecological conclusions, it is necessary to study the environmental behavior of individual taxa.

In well 6B-15-53-2W4, marine strata at certain levels have a high density of trace fossils and syneresis cracks, indicating wide fluctuations in salinity, but are devoid of marine microplankton. It is possible that few microplankton are capable of surviving in the inhospitable environment created by sudden and large fluctuations in salinity caused by flood waters or tidal currents. Their preservation in the bioturbated strata is also likely to be adversely affected by the presence of large populations of benthic deposit feeders. Moreover, the dinoflagellates and acritarchs are usually not preserved in the sandy intervals, and the presence of biogenic structures in such strata is often the only means available for detecting a marine influence.

Summary

We have proposed criteria from several geological disciplines for the recognition of brackish water deposits. Although the Lower Cretaceous Wealden Group of southern England is largely continental (Allen, 1975; 1981), Stewart (1983) found intervals that could have been deposited in brackish water. These intervals bear a strong resemblance to parts of the upper Mannville as they contain syneresis cracks and a restricted assemblage of trace fossils (*Ophiomorpha*) and microplankton (*Muderongia*). Thus, the features of the Wealden support the criteria which we have outlined in this paper.

CONCLUSIONS

(1) Thick sandstones in the upper Mannville have formed in several ways, including: stacked fluvial deposits, stacked crevasse splays and amalgamated channel and marine shoreline facies.
(2) The upper Mannville changes from a predominantly continental nature in the west to a marginal marine setting in the eastern part of the study area.
(3) The facies described in the paper are consistent with a fluvially dominated delta, although other depositional models may be more suitable elsewhere.
(4) Brackish water deposits are very important because they signal the change from a continental to a marine setting and tend to occur at or near the limits of transgressions.
(5) An integrated approach, utilizing data from several geological disciplines, is most useful in the delineation of brackish water facies. Criteria which aid in the recognition of these facies are syneresis cracks, pyrite and siderite, and a limited assemblage of certain trace fossils and microplankton.

ACKNOWLEDGMENTS

The authors thank Ray Rahmani, Rod Tillman, Barry Hatt, Bob Weimer, David James, Mike Ranger and Bob Frey for critically reading, and suggesting many improvements to, an earlier version of the paper. Shaun O'Connell provided the data for well 2-30-43-2W4, Joan Checholik typed the manuscript, Dan Magee drafted the figures, and Bruno Untergasser helped with identification of the microplankton.

REFERENCES

ALEXANDER, W. B., SOUTHGATE, A., AND BASSINDALE, R., 1935, Survey of the River Tees. Part II. The estuary—chemical and biological: Department of Scientific Industrial Research, Water Pollution Research Technical Paper Number 5, 171 p.

ALLEN, J. R. L., 1970, Physical Processes of Sedimentation: George Allen and Unwin, London, 248 p.

——— 1980, Sand waves: a model of origin and internal structure: Sedimentary Geology, v. 26, p. 281–328.

ALLEN, P., 1975, The Wealden of the Weald: a new model: Proceedings of the Geologists' Association, v. 86, p. 389–438.

——— 1981, Pursuit of Wealden models: Journal of the Geological Society of London, v. 138, p. 375–405.

ALLER, R. C., 1982, Carbonate dissolution in nearshore terrigenous muds: the role of physical and biological reworking: Journal of Geology, v. 90, p. 79–95.

BAGANZ, B. P., HORNE, J. C., AND FERM, J. C., 1975, Carboniferous and recent Mississippi lower delta plains: a comparison: Gulf Coast Association of Geological Societies Transactions, v. 25, p. 183–191.

BALSLEY, J. K., AND PARKER, L. R., 1983, Cretaceous wave-dominated delta, barrier island, and submarine fan depositional systems: Book Cliffs, east central Utah: American Association of Petroleum Geologists Field Guide, 285 p.

BATTEN, D. J., 1982, Palynofacies and salinity in the Purbeck and Wealden of southern England, *in* Banner, F. T., and Lord, A. R., eds., Aspects of Micropalaeontology: George Allen and Unwin, London, p. 278–295.

——— AND EATON, G. L., 1980, Dinoflagellates and salinity variations in the Wealden (Lower Cretaceous) of southern England (abs.): 5th International Palynological Conference, Cambridge, Abstracts, p. 32.

BERNER, R. A., 1971, Principles of Chemical Sedimentology: McGraw-Hill, New York, 240 p.

——— 1980, Early Diagenesis, a Theoretical Approach: Princeton University Press, Princeton, 241 p.

——— 1981, A new geochemical classification of sedimentary environments: Journal of Sedimentary Petrology, v. 51, p. 359–365.

——— AND RAISWELL, R., 1984, C/S method for distinguishing fresh-water from marine sedimentary rocks: Geology, v. 12, p. 365–368.

——— BALDWIN, T., AND HOLDREN, G. R., JR., 1979, Authigenic iron sulfides as paleosalinity indicators: Journal of Sedimentary Petrology, v. 49, p. 1345–1350.

BRIDGE, J. S., AND DIEMER, J. A., 1983, Quantitative interpretation of an evolving ancient river system: Sedimentology, v. 30, p. 599–623.

BURGER, D., 1980, Palynological studies in the Lower Cretaceous of the Surat Basin, Australia: Bureau of Mineral Resources, Geology and Geophysics, Australia, Bulletin 189, 106 p.

BURST, J. F., 1965, Subaqueously formed shrinkage cracks in clay: Journal of Sedimentary Petrology, v. 35, p. 348–353.

CALDWELL, W. G. E., NORTH, B. R., STELCK, C. R., AND WALL, J. H., 1978, A foraminiferal zonal scheme for the Cretaceous system in the interior plains if Canada, *in* Stelck, C. R., and Chatterton, B. D. E., eds., Western and Arctic Canadian Biostratigraphy: Geological Association of Canada Special Paper 18, p. 495–575.

CASAGRANDE, D. J., SIEFERT, K., BERSCHINSKI, C., AND SUTTON, N., 1977, Sulfur in peat-forming systems of the Okeefenokee Swamp and Florida Everglades: origins of sulfur in coal: Geochimica et Cosmochimica Acta, v. 41, p. 161–167.

CHRISTOPHER, J. E., 1984, The Lower Cretaceous Mannville Group, northern Williston Basin region, Canada, *in* Stott, D. F., and Glass, D. J., eds., The Mesozoic of Middle North America: Canadian Society of Petroleum Geologists Memoir 9, p. 109–126.

COLEMAN, J. M., 1976, Deltas: Processes of Deposition and Models for Exploration: Continuing Education Publication Company, Champaign, Illinois, 102 p.

——— AND PRIOR, D. B., 1982 Deltaic environments of deposition, *in* Scholle, P. A., and Spearing, D., eds., Sandstone Depositional Environments: American Association of Petroleum Geologists Memoir 31, p. 139–178.

COLLINSON, J. D., AND THOMPSON, D. B., 1982, Sedimentary Structures: George Allen and Unwin, London, 208 p.

CROGHAN, P. C., 1983, Osmotic regulation and the evolution of brackish—and fresh-water faunas: Journal of the Geological Society, v. 140, p. 39–46.

DONOVAN, R. N., AND FOSTER, R. J., 1972, Subaqueous shrinkage cracks from the Caithness Flagstone Series (Middle Devonian) of northeast Scotland: Journal of Sedimentary Petrology, v. 42, p. 309–317.

DORJES, J., AND HOWARD, J. D., 1975, Estuaries of the Georgia coast, U.S.A.: sedimentology and biology. IV. Fluvial-marine transition indicators in an estuarine environment, Ogeechee River-Ossabaw Sound: Senckenbergiana Maritima, v. 7, p. 137–179.

FLESSA, K. W., AND BROWN, T. J., 1983, Selective solution of macroinvertebrate calcareous hard parts: a laboratory study: Lethaia, v. 16, p. 193–205.

FREY, R. W., AND HOWARD, J. D., 1969, A profile of biogenic sedimentary structures in a Holocene barrier island—salt marsh complex, Georgia: Gulf Coast Association of Geological Societies Transactions, v. 19, p. 427–444.

——— PEMBERTON, S. G., AND FAGERSTROM, J. A., 1984, Morphological, ethological, and environmental significance of the ichnogenera *Scoyenia* and *Ancorichnus*: Journal of Paleontology, v. 58, p. 511–528.

FUGLEM, M. D., 1970, Use of core in evaluation of productive sands, Lloydminster area, *in* Mesozoic core seminar: Saskatchewan Geological Society and Department of Mineral Resources, 12 p.

FURSICH, F. T., 1981, Salinity-controlled benthic associations from the Upper Jurassic of Portugal: Lethaia, v. 14, p. 203–223.

——— AND SCHMIDT-KITTLER, N., 1980, Biofacies analysis of Upper Jurassic marginally marine environments of Portugal. I. The carbonate dominated facies at Cabo Espichel, Estremadura: Geologisches Rundschau, v. 69, p. 945–981.

GLAISTER, R. P., 1959, Lower Cretaceous of southern Alberta and adjoining areas: American Association of Petroleum Geologists Bulletin, v. 43, p. 590–640.

GOLDRING, R., BOSENCE, D. W. J., AND BLAKE, T., 1978, Estuarine sedimentation in the Eocene of southern England: Sedimentology, v. 25, p. 861–876.

GROSS, A. A., 1980, Mannville channels in east-central Alberta, *in* Beck, L. S., Christopher, J. E., and Kent, D. M., eds., Lloydminster and Beyond: Geology of Mannville Hydrocarbon Reservoirs: Saskatchewan Geological Society Special Publication Number 5, p. 33–63.

HARMS, J. C., SOUTHARD, J. B., AND WALKER, R. G., 1982, Structures and sequences in clastic rocks: Society of Economic Paleontologists and Mineralogists Short Course 9, Lecture Notes, 249 p.

HINES, M. E., OREM, W. H., LYONS, W. B., AND JONES, G. E., 1982 Microbial activity and bioturbation-induced oscillations in pore water chemistry of estuarine sediments in spring: Nature, v. 299, p. 433–435.

HORNE, J. C., 1979, The Louisa-Pikeville area, *in* Ferm, J. C., and Horne, J. C., eds., Carboniferous Depositional Environments in the Appalachian Region: Department of Geology, University of South Carolina, Columbia, 760 p.

——— FERM, J. C., CARUCCIO, F. T., AND BAGANZ, B. P., 1978, Depositional models in coal exploration and mine planning in Appalachian region: American Association of Petroleum Geologists Bulletin, v. 62, p. 2379–2411.

HOWARD, J. D., AND FREY, R. W., 1973, Characteristic physical and biogenic sedimentary structures in Georgia estuaries: American Association of Petroleum Geologists Bulletin, v. 57, p. 1169–1184.

——— AND, ———, 1975, Estuaries of the Georgia coast, U.S.A.: sedimentology and biology. II. Regional animal-sediment characteristics of Georgia estuaries: Senckenbergiana Maritima, v. 7, p. 33–103.

HUDSON, J. D., 1963, the recognition of salinity-controlled mollusc assemblages in the Great Estuarine Series (Middle Jurassic) of the Inner Hebrides: Palaeontology, v. 6, p. 318–326.

——— 1980, Aspects of brackish-water facies and faunas from the Jurassic of north-west Scotland: Proceedings of the Geological Association, v. 91, p. 99–105.

HULBURT, E. M., 1963, The diversity of phytoplanktonic populations in oceanic, coastal, and estuarine regions: Journal of Marine Research, v. 21, p. 81–93.

JARDINE, D., 1974, Cretaceous oil sands of western Canada, *in* Hills, L. V., ed., Oil Sands Fuel of the Future: Canadian Society of Petroleum Geologists Memoir 3, p. 50–67.

JELETZKY, J. A., 1971, Marine Cretaceous biotic provinces and paleogeography of western and arctic Canada: illustrated by a detailed study of ammonites: Geological Survey of Canada Paper 70-22, 92 p.

LENTIN, J. K., AND PUTNAM, P. E., 1981, Depositional environments of the Mannville Group—new light on an old problem (abs.): Geological Association of Canada Program with Abstracts, p. A-35.

LORSONG, J. A., 1980, Geometry of nearshore sand bodies in the upper Mannville Group, Celtic field, Saskatchewan, *in* Beck, L. S., Christopher, J. E., and Kent, D. M., eds., Lloydminster and Beyond: Geology of Mannville Hydrocarbon Reservoirs: Saskatchewan Geological Society Special Publication Number 5, p. 236–266.

MARTENS, C. S., AND GOLDHABER, M.B., 1978, Early diagenesis in transitional sedimentary environments of the White Oak River estuary, North Carolina: Limnology and Oceanography, v. 23, p. 428–441.

OOMKENS, E., 1974, Lithofacies relations in the late Quaternary Niger delta complex: Sedimentology, v. 21, p. 195–222.

ORR, R. D., JOHNSTON, J. R., AND MANKO, E. M., 1977, Lower Cretaceous geology and heavy-oil potential of the Lloydminster area: Bulletin of Canadian Petroleum Geology, v. 25, p. 1187–1221.

PEMBERTON, S. G., FLACH, P. D., AND MOSSOP, G. D., 1982, Trace fossils from the Athabasca oil sands, Alberta, Canada: Science, v. 217, p. 825–827.

PETTIJOHN, F. J., POTTER, P. E., AND SIEVER, R., 1973, Sand and Sandstone: Springer-Verlag, New York, 618 p.

PICARD, M. D., 1966, Oriented, linear-shrinkage cracks in Green River Formation (Eocene), Raven Ridge area, Uinta basin, Utah: Journal of Sedimentary Petrology, v. 36, p. 1050–1057.

PLUMMER, P. S., AND GOSTIN, V. A., 1981, Shrinkage cracks: desiccation or synaeresis?: Journal of Sedimentary Petrology, v. 51, p. 1147–1156.

POSTMA, D., 1982, Pyrite and siderite formation in brackish and fresh-water swamp sediments: American Journal of Science, v. 282, p. 1151–1183.

PRYOR, W. A., 1967, Biogenic directional features on several recent point-bars: Sedimentary Geology, v. 1, p. 235–245.

PUTNAM, P. E., 1980, Fluvial deposition within the upper Mannville of west-central Saskatchewan: stratigraphic implications, *in* Beck, L. S., Christopher, J. E., and Kent, D. M., eds., Lloydminster and Beyond: Geology of Mannville Hydrocarbon Reservoirs: Saskatchewan Geological Society Special Publication Number 5, p. 197–216.

——— 1982a, Aspects of the petroleum geology of the Lloydminster heavy oil fields, Alberta and Saskatchewan: Bulletin of Canadian Petroleum Geology, v. 30, p. 81–111.

——— 1982b, Fluvial channel and sandstones within upper Mannville (Albian) of Lloydminster area, Canada—geometry, petrography and paleogeographic implications: American Association of Petroleum Geologists Bulletin, v. 66, p. 436–459.

——— 1983, Fluvial deposits and hydrocarbon accumulations: examples from the Lloydminster area, Canada, *in* Collinson, J. D., and Lewin, J., eds., Modern and Ancient Fluvial Systems: International Association of Sedimentologists Special Publication Number 6, p. 517–532.

——— AND OLIVER, T. A., 1980, Stratigraphic traps in channel sandstones in the upper Mannville (Albian) of east-central Alberta: Bulletin of Canadian Petroleum Geology, v. 28, p. 489–508.

RANGER, M. J., 1984, The paleotopography of the pre-Cretaceous erosional surface in the western Canada basin (abs.), *in* Stott, D. F., and Glass, D. J., eds., The Mesozoic of Middle North America: Canadian Society of Petroleum Geologists Memoir 9, p. 570.

RATCLIFFE, B. C., AND FAGERSTROM, J. A., 1980, Invertebrate lebensspuren of Holocene floodplains: their morphology, origin and paleoecological significance: Journal of Paleontology, v. 54, p. 614–630.

REINECK, H. E., AND SINGH, I. B., 1980, Depositional Sedimentary Environments: Springer-Verlag, New York, 549 p.

REMANE, A., AND SCHLIEPER, C., 1971, Biology of Brackish Water: John Wiley and Sons, New York, 372 p.

RHOADS, D. C., 1975, The paleoecological and environmental significance of trace fossils, *in* Frey, R. W., ed., The Study of Trace Fossils: Springer-Verlag, New York, p. 147–160.

——— AND MORSE, J., 1970, Evolutionary and ecologic significance of oxygen deficient marine basins: Lethaia, v. 4, p. 413–428.

SANDERS, H. L., MANGELSDORF, P. C. JR., AND HAMPSON, G. R., 1965, Salinity and faunal distribution in the Pocasset River, Massachusetts: Limnology and Oceanography, v. 10, p. 216–229.

SAXENA, R. S., 1976, Modern Mississippi Delta-Depositional Environments and Processes: New Orleans Geological Society Guidebook, 125 p.

SEILACHER, A., 1964, Biogenic sedimentary structures, *in* Imbrie, J., and Newell, N. D., eds., Approaches to Paleoecology: John Wiley and Sons, New York, p. 296–316.

———, 1967, Bathymetry of trace fossils: Marine Geology, v. 5, p. 413–428.

SINGH, C., 1964, Microflora of the Lower Cretaceous Mannville Group, east-central Alberta: Alberta Research Council Bulletin 15, 239 p.

SMITH, D. G., AND PUTNAM, P. E., 1980, Anastomosed river deposits: modern and ancient examples in Alberta, Canada: Canadian Journal of Earth Sciences, v. 17, p. 1396–1406.

STELCK, C. R., 1958, Stratigraphic position of the Viking sand: Alberta Society of Petroleum Geology Journal, v. 6, p. 2–7.

———, 1975, The upper Albian *Miliammina manitobensis* zone in northeastern British Columbia, *in* Caldwell, W. G. E., ed., The Cretaceous System in the Western Interior of North America: Geological Association of Canada Special Paper 13, p. 251–275.

——— AND KRAMERS, J. W., 1980, *Freboldiceras* from the Grand Rapids Formation of north-central Alberta: Bulletin of Canadian Petroleum Geology, v. 28, p. 509–521.

STEWART, D. J., 1981, A field guide to the Wealden Group of the Hastings area and the Isle of Wight, *in* Elliott, T., ed., Field Guides to Modern and Ancient Fluvial Systems in Britain and Spain: International Fluvial Conference, Keele University, United Kingdom, p. 3.1–3.32.

———, 1983, Possible suspended-load channel deposits from the Wealden Group (Lower Cretaceous) of southern England, *in* Collinson, J. D., and Lewin, J., eds., Modern and Ancient Fluvial Systems: International Association of Sedimentologists Special Publication Number 6, p. 369–384.

STUMM, W., AND MORGAN, J. J., 1970, Aquatic chemistry: an introduction emphasizing chemical equilibria in natural waters: John Wiley and Sons, New York, 583 p.

VAN HULTEN, F. F. N., 1984, Petroleum geology of Pikes Peak heavy oil field, Waseca Formation, Lower Cretaceous, Saskatchewan, *in* Stott, D. F., and Glass, D. J., eds., The Mesozoic of Middle North America: Canadian Society of Petroleum Geologists Memoir 9, p. 441–454.

VIGRASS, L. W., 1977, Trapping of oil at intra-Mannville (Lower Cretaceous) disconformity in Lloydminster area, Alberta and Saskatchewan: American Association of Petroleum Geologists Bulletin, v. 61, p. 1010–1028.

VUKE, S. M., 1984, Depositional environments of the early Cretaceous western interior seaway in southwestern Montana and the northern United States, *in* Stott, D. F., and, Glass, D. J., eds., The Mesozoic of Middle North America: Canadian Society of Petroleum Geologists Memoir 9, p. 127–144.

WALL, D., 1965, Microplankton, pollen, and spores from the Lower Jurassic in Britain: Micropaleontology, v. 11, p. 151–190.

WIGHTMAN, D. M., TILLEY, B. J., AND LAST, W. M., 1981, Stratigraphic traps in channel sandstones in the upper Mannville (Albian) of east-central Alberta: discussion: Bulletin of Canadian Petroleum Geology, v. 29, p. 622–625.

WILLIAMS, G. D., 1963, The Mannville Group (Lower Cretaceous) of Alberta: Bulletin of Canadian Petroleum Geology, v. 11, p. 350–368.

SEDIMENTOLOGY AND SUBSURFACE GEOLOGY OF DELTAIC FACIES, ADMIRE 650′ SANDSTONE, EL DORADO FIELD, KANSAS

R. W. TILLMAN[1] AND D. W. JORDAN[2]

Cities Service Oil and Gas Corporation Research Tulsa, Oklahoma

ABSTRACT: El Dorado field, which is located in southeastern Kansas, was discovered in 1915 by Cities Service Company. The Admire 650′ sandstone reservoir (Permian Wolfcampian) is present at 650 ft (197 m), ranges in thickness from 11 to 23 ft (3 to 7 m) and has produced 36.5 million barrels of oil through primary and secondary recovery methods. A secondary waterflood was carried out in the field during the 1950s. A tertiary oil recovery project covering 51 acres (21 ha) was initiated in the Admire 650′ sandstone in El Dorado field in 1974. Deltaic facies are encountered at El Dorado field; the various deltaic facies are quite variable in reservoir quality. Documentation of the geometries and locations of reservoir and non-reservoir facies allows prediction of the distribution of flow units within the field.

Three types of sandstone reservoirs are recognized in the Admire 650′ sandstone. The major producing reservoir facies is the *Distributary Channel Sandstone Facies,* which includes *High- and Low-Energy Subfacies.*

The Admire 650′ sandstone was initially deposited from splay channels that prograded into a muddy interdistributary bay. Following deposition of some of the splays, distributary channels prograded into the enhanced oil recovery pilot project area (El Dorado Micellar-Polymer Demonstration Project). Some parts of the project area, however, contain substantially thicker distributary channel sandstones than others. In the southeastern corner no channel sandstones were deposited; the area remained an interdistributary bay environment. The final phase of deposition was one in which transgressive marine deposits spread over the whole area depositing limestones and shales.

The *Distributary Channel Sandstone Facies* is most prevalent along the western margin of the northern (Chesney) and southern (Hegberg) leases, as well as across the center of the project area. Flow directions obtained from oriented cores indicate northward and northeastward paleocurrent flow within a distributary channel system which bifurcates within the lease areas.

Reservoir heterogeneities depicted on cross sections, fence diagrams, and isopach maps were recognized in cores, logs, and whole-core analysis. Abundant high quality geologic data were available for this project and were used to formulate detailed reservoir descriptions. The reservoir descriptions were utilized to some degree by engineers to better model the field.

One engineering parameter which strongly supports the presence of geologic heterogeneity in El Dorado was pressure-transient analysis. Pressure-transient ratios, as much as 14 in areas of recognizable elongate sandstone body deposition, indicate many areas of strongly preferred transmissibility.

Distributary Channel Sandstones average 436 md permeability and 28% porosity, whereas splay channel sandstones average 567 md and 27% porosity. Splay sandstones are thinner and discontinuous and, hence, do not contribute as much to overall production.

The variation in permeability and porosity of the Admire 650′ sandstone is affected by diagenesis. Permeability and porosity are reduced by clay laminae, deformation of ductile rock fragments, mica and micaceous laminae, quartz, the secondary leaching of feldspar, and calcite cement.

INTRODUCTION

A detailed sedimentologic analysis of complex sandstones such as those found in the deltaic deposits of El Dorado can contribute to the economic success of any reservoir management program. Variation in reservoir quality in sandstone reservoirs is due, in large part, to depositional and diagenetic processes. Where high-energy river-dominated distributaries occur, high porosities and permeabilities will be encountered. Where sluggish streams deposit overbank deposits or fill abandoned channels, very poor porosities and permeabilities are expected. This study documents the distribution, size, boundaries and reservoir quality of numerous genetic units referred to as facies.

The initial geologic analysis (Phase I) was limited to seven cores. As more core data and oriented cores became available, Phase II was initiated. Phase III included a wide range of data and produced the final geologic model.

A variety of sedimentologic, subsurface geologic and engineering techniques were used to model the Admire 650′ sandstone. Detailed core descriptions, oriented cores, transmissibility tests, fence diagrams, and geologic cross sections and maps form the basis of the Admire 650′ model. Using this information, it is possible within the project area to position and program injection and production wells to yield optimal results, and to simulate in a reproducible way the expected performance of the reservoir.

El Dorado field, located in south-central Kansas (Figs. 1, 2) was discovered by Wichita National Gas Company in 1915 (later Cities Service Company) (Fig. 3). It was a major producing field during World War I and greatly aided the United States war effort at that time (Kentworth, 1977). Production from the multi-reservoir field peaked at a daily average of 79,720 barrels in 1918 (Reeves, 1929).

The field originally comprised 6,200 acres (2,509 ha) (Fath, 1921) and contains 11 reservoirs, ranging in age from Ordovician to Permian (Reeves, 1929). The Admire 650′ sandstone (Admire Group, Wolfcampian Series, Lower Permian) lies 650 ft (197 m) below the ground surface and is the only reservoir involved in the enhanced oil recovery pilot project discussed herein. A field-wide isopach map of the Admire 650′ sandstone, constructed with data available in 1974, shows the location of the pilot area (Fig. 4). Primary production methods yielded only 23 million barrels from the Admire 650′ sandstone, less than one-fifth of the 108 million barrels of oil-in-place in the Admire (Jewett and Abernathy, 1945). Secondary recovery by waterflood and airdrive produced another 25.7 million barrels from the Admire 650′. Secondary recovery wells were drilled on a 6.4-acre (2.6 ha) spacing, and the flood continued for 13 years (R. Miller, pers. commun.). It was predicted that half

[1]Present address: Consulting Sedimentologist, 4555 S. Harvard Ave., Tulsa, Oklahoma 74135

[2]Present address: H. N. Fisk Laboratory of Sedimentology, Department of Geology, University of Cincinnati, Cincinnati, Ohio 452210

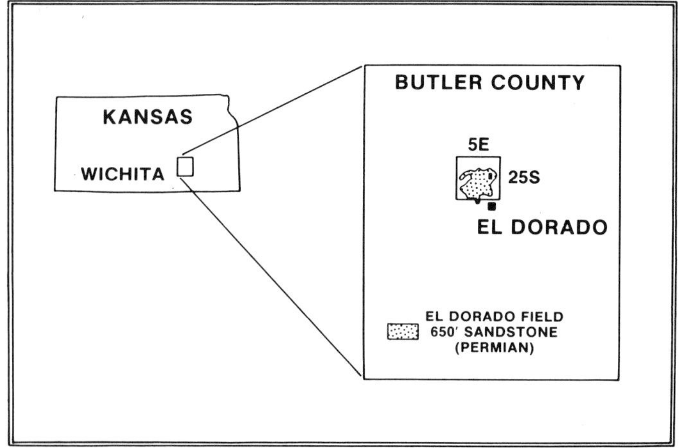

FIG. 1.—Area of study in southeastern Kansas and extent of Admire 650′ sandstone (stipled). Pilot project area indicated by black rectangle (upper right).

of the remaining 71.5 million barrels could be recovered by micellar-polymer flooding (Fig. 5).

The El Dorado Micellar-Polymer Demonstration Project was a cooperative venture between Cities Service Oil Company and the United States Energy Research and Development Administration (ERDA) under ERDA Contract Number E(34-1)-0003. Eight annual reports on the project were published by the U.S. Department of Energy (DOE): Rosenwald (1978, 1979), Rosenwald and others (1976, 1978), Miller and Rosenwald (1980), Van Horn (1981, 1982, 1983). The area defined in this study comprises 51 acres (21 ha) and is divided into a northern (Chesney) and a southern (Hegberg) lease in Sections 21 and 28, T25S R5E, Butler County, Kansas. During three phases of geologic study, 61 wells were drilled within the pilot project area on a very dense spacing of approximately 1.27 acres (0.5 ha) per well. The well pattern for the tertiary pilot project was designed so that there were only 264 ft (80 m) between production and injection wells. This spacing is significantly denser than the 6.4 acres (2.6 ha) per well for the secondary recovery phase. The increased density in the pilot area allowed more rapid monitoring of arrival of injection fluids at producing wells.

Geologic studies of the reservoir sandstones were carried out in three phases during the course of the tertiary recovery pilot project. Phase I was completed in 1975. *Preliminary* (Phase I) log cross sections suggested that the log patterns throughout the pilot project area were similar (Fig. 6) and that a certain degree of reservoir homogeneity should be expected throughout the pilot area. During Phase II, which was completed in 1978, seven cores, including three oriented cores, were studied by geologists at the Cities Service Research Laboratory. Studies of the cores indicated that the reservoir sandstones were not homogeneous and that they were deposited in a deltaic setting which included deposits formed in delta distributary channels, swamps, and bays. At that time, on the basis of core study, the Admire 650′ sandstone was divided into three genetic units (Fig. 7). The lower unit (Unit 1) is basically a claystone or shale. Unit 2, the major reservoir sandstone, is divided into upper and lower parts, the upper part being a sandstone containing mainly trough crossbedding; it also contains ripples and climbing ripples. The lower part of Unit 2 is composed of sandstone interbedded with shale.

The uppermost unit recognized in the cores at that time is Unit 3, which consists of several lithologies. It locally contains lignite at the base, biomicrite, fossiliferous limestone, claystones, shales and very minor amounts of siltstone. This threefold classification was refined in subsequent studies involving numerous additional cores. In addition to the work done at Cities Service, Bascle (1976) briefly discussed sedimentary facies and diagenesis and constructed cross sections which generally defined the reservoir geometry.

By the time Phase III of the study was completed in 1981, a total of 25 cores were used to map boundaries of facies and predict directions of preferred transmissibilities and flow of fluids. Reservoir quality and distribution were determined by utilizing detailed descriptions of lithology, facies types, and facies geometry. In nearly all wells, facies are identified on gamma-ray and resistivity logs so that 33 additional noncored wells could be utilized in the construction of cross sections, isopach maps, and fence diagrams. Tertiary recovery modeling was aided to some degree by utilizing geologic core data to outline the geometry and effectiveness of flow of the zones to be flooded. Detailed core descriptions for five wells (MP-112, MP-118, MP-205, MP-211, MP-213), which include typical descriptions of all facies, are in the Appendix. Also included in the Appendix are photos of the cores and logs and data on permeability and porosity in some of the wells.

X-ray diffraction, scanning electron microscope, and thin section data were used to determine diagenetic and petrologic controls on permeability and porosity. Whole-core analysis provided detailed information on diagenetic and facies control of residual oil saturation. The distribution of unproduced oil was important in designing the micellar-polymer flood, so that fluids would sweep the areas which still contained abundant oil. The most porous and permeable portions of the reservoir were probably less likely to produce large quantities of additional oil because they were more effectively swept during primary and secondary production. Since the micellar-polymer flooding in El Dorado was a pilot test project, it was expected that potential performance over a large area could be predicted on the basis of the detailed geologic conclusions summarized in this paper.

FACIES DESCRIPTIONS

Sediments are deposited by processes which vary both in type and intensity from location to location within the framework of a delta. A study of the sedimentary structures in cores was used to determine the processes of deposition, which in turn allowed recognition of environments of deposition. Nine depositional facies are recognized in the Admire 650′ sandstone.

The model suggested for the Admire 650′ sandstone is one in which a deltaic distributary channel system prograded over a broad area of shallow marine deposits. Splay

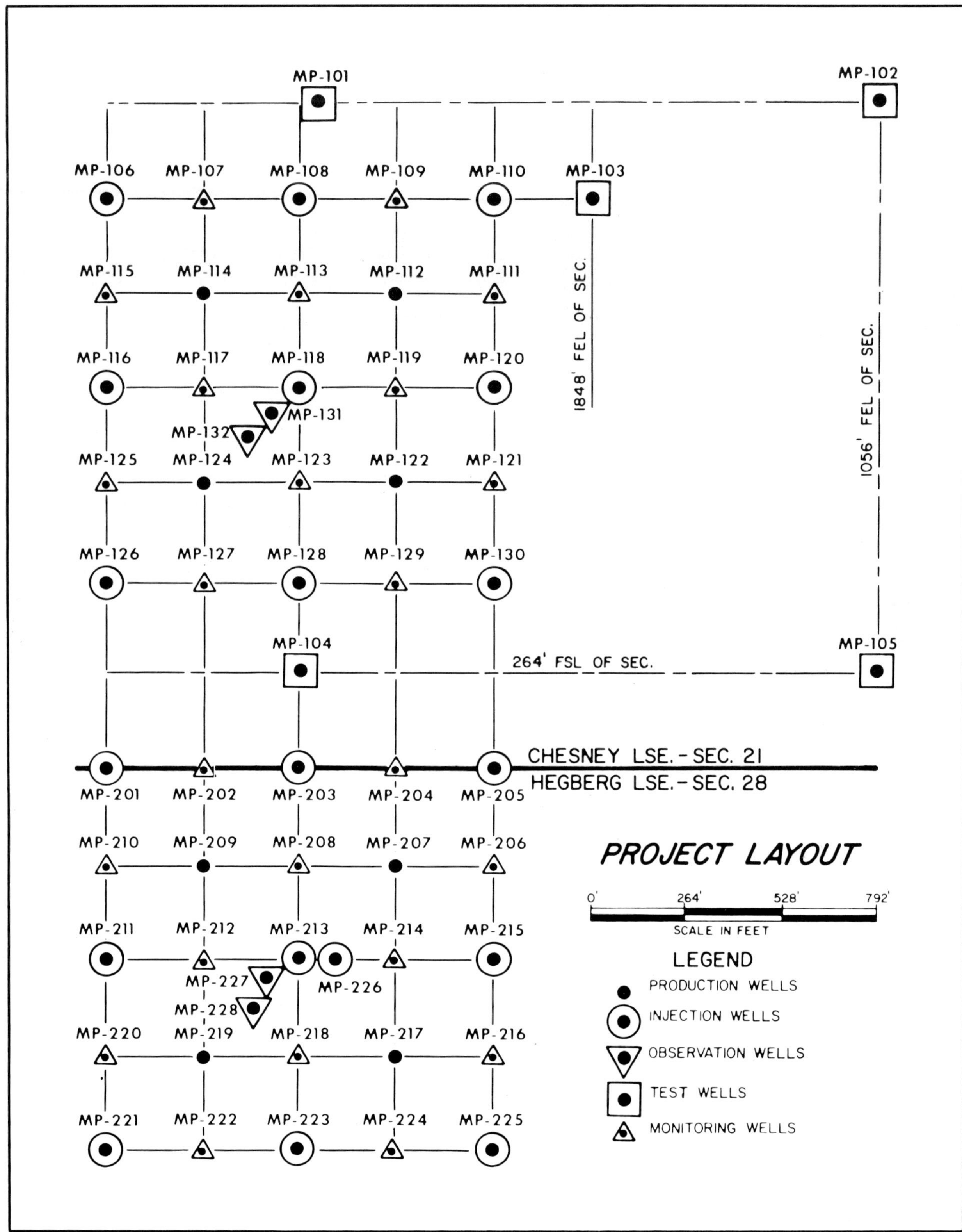

FIG. 2.—Project layout map, Admire 650′ sandstone, Butler County, Kansas (area shown by black rectangle in Fig. 1). Pilot area comprises 51 acres (21 ha). The northern portion of the project is on the Chesney lease and the southern portion of the project is on the Hegberg lease. Refer to this map for well numbers and locations.

FIG. 3.—El Dorado field, Kansas in the 1920s. The field was discovered by Wichita Natural Gas, Stapleton No. 1, October 5, 1915, at a depth of 670 ft.

channels locally breached channel levees and deposited sediment in interdistributary bays. During the final stage of delta-building, subsidence occurred and marine environments transgressed the deltaic deposits. The association of facies in a delta is given in Figure 8.

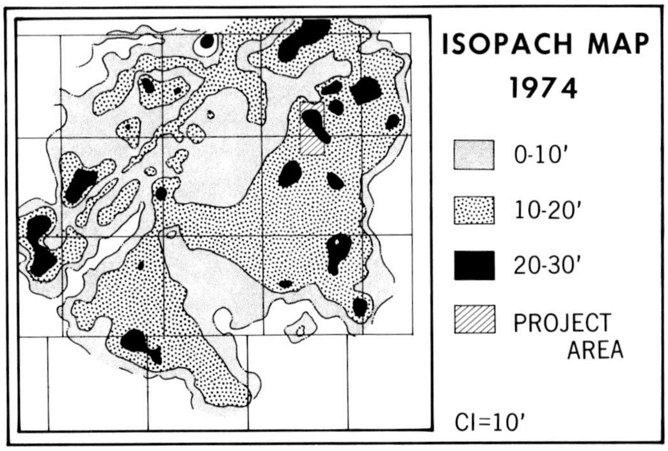

FIG. 4.—Isopach map of Admire 650′ sandstone in El Dorado field. Compiled in 1974 on the basis of subsurface "electric" and drillers logs. Project area marked by diagonally ruled area.

Figure 9 is a typical stratigraphic summary section which summarizes the details of cores studied in the Phase III geologic study; descriptions and environmental interpretations

EL DORADO FIELD 650′ SAND

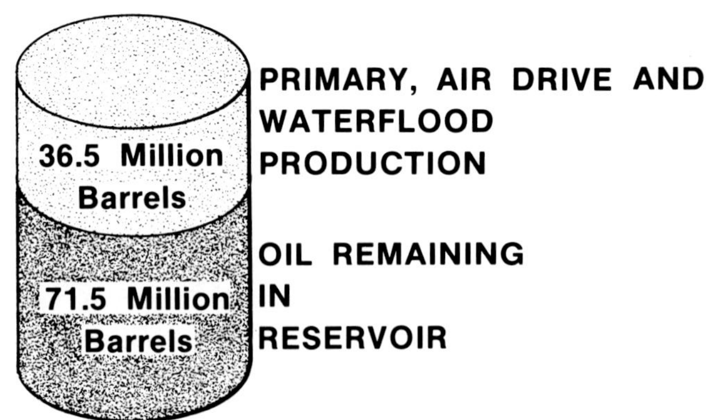

FIG. 5.—Comparison of oil remaining in the Admire 650′ sandstone reservoir with oil already produced. Original oil-in-place was 108 million barrels. It is expected that half of the remaining oil can be recovered by tertiary recovery methods.

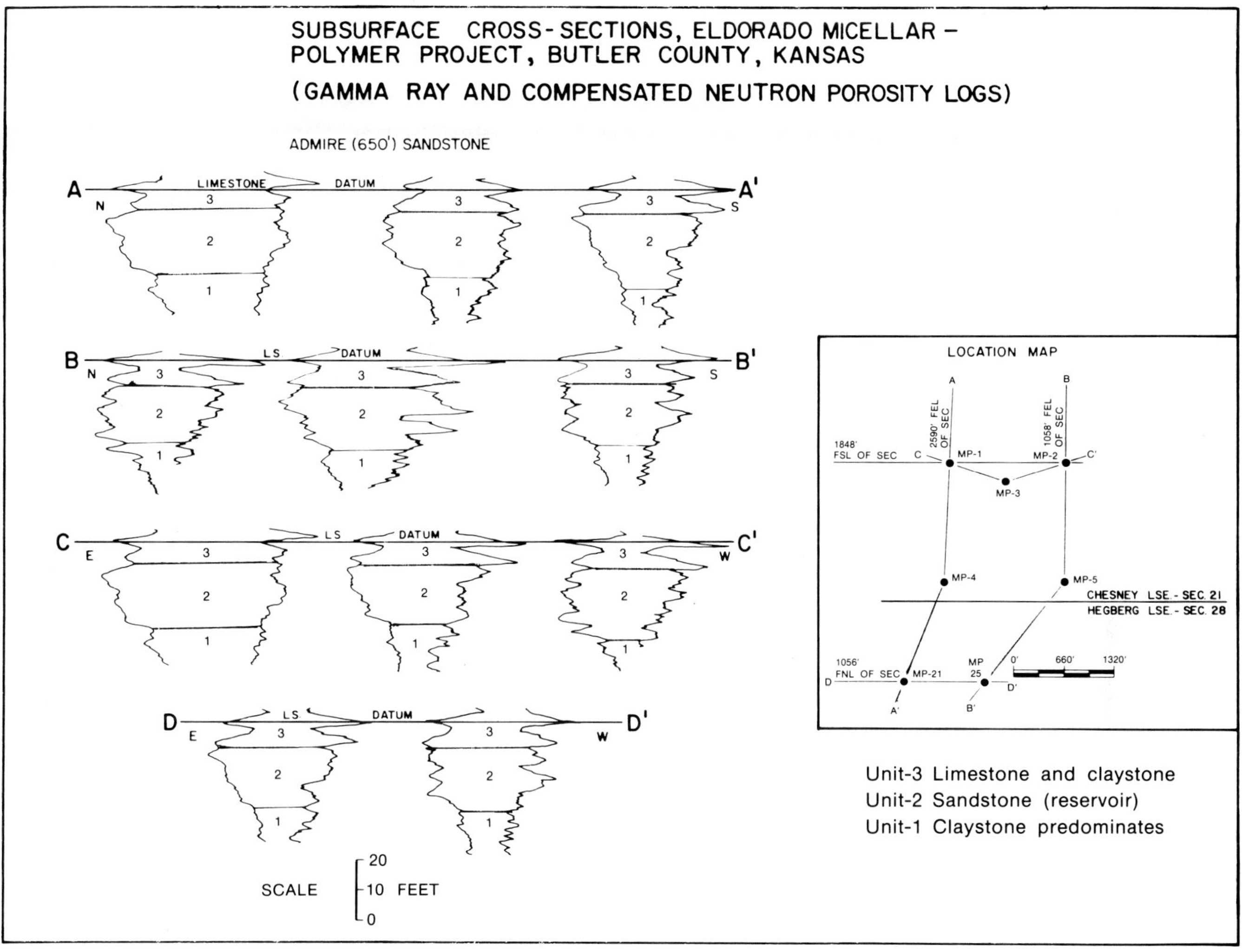

FIG. 6.—Preliminary subsurface cross sections, Admire 650′ sandstone, El Dorado micellar-polymer project (Phase I). Units 1, 2, and 3 are numbered in order of deposition.

are included for each unit. Descriptions for four additional cored wells are included in the Appendix.

Interdistributary Bay Silty Shale Facies

The *Interdistributary Bay Silty Shale Facies* underlies the Admire 650′ sandstone throughout the project area. The total thickness of this unit was not cored in any of the wells; on gamma-ray logs the facies is indicated to be about 20 ft (6 m) thick. This facies occurs at the base of several of the wells described in detail in the Appendix.

The *Interdistributary Bay Silty Shale Facies* can be described using three descriptors, lithology, and physical and biogenic sedimentary structures, each of which is described on the basis of the percentage of the volume of the unit that it occupies. This facies is composed predominantly of laminated silty shale (75%) and includes thin (1/4–1/2 in. thick,

ADMIRE (650') SANDSTONE

El Dorado Field, Butler County, Kansas

GENERAL LITHOLOGIC VERTICAL SEQUENCE COLUMN

Environment	Unit	Lithology / structures
Marine Bay	UNIT 3	Biomicrite; Claystone & minor Siltstone laminations; Biomicrite
Marsh		Lignite
Distributary Channels (MINOR INTERTIDAL)	UNIT 2 upper	ADMIRE (650') SANDSTONE; Fine-medium Sandstone; Troughs, ripples, climbing ripples; Root mottled fine sandstone - top
Splay Channels & Intertidal	lower	Fine-medium Sandstone with interlaminated clay/siltstone
Inter-Distributary Bay	UNIT 1	Claystone with interlaminated lenticular silt/sandstone

METERS 2 1 — 6 3 FEET

FIG. 7.—General lithologic vertical sequence column for Admire 650′ sandstone as determined during Phase I geologic study. Description of the lithologies and sedimentary structures are listed on the right. Interpretations of environments of deposition are given on the left.

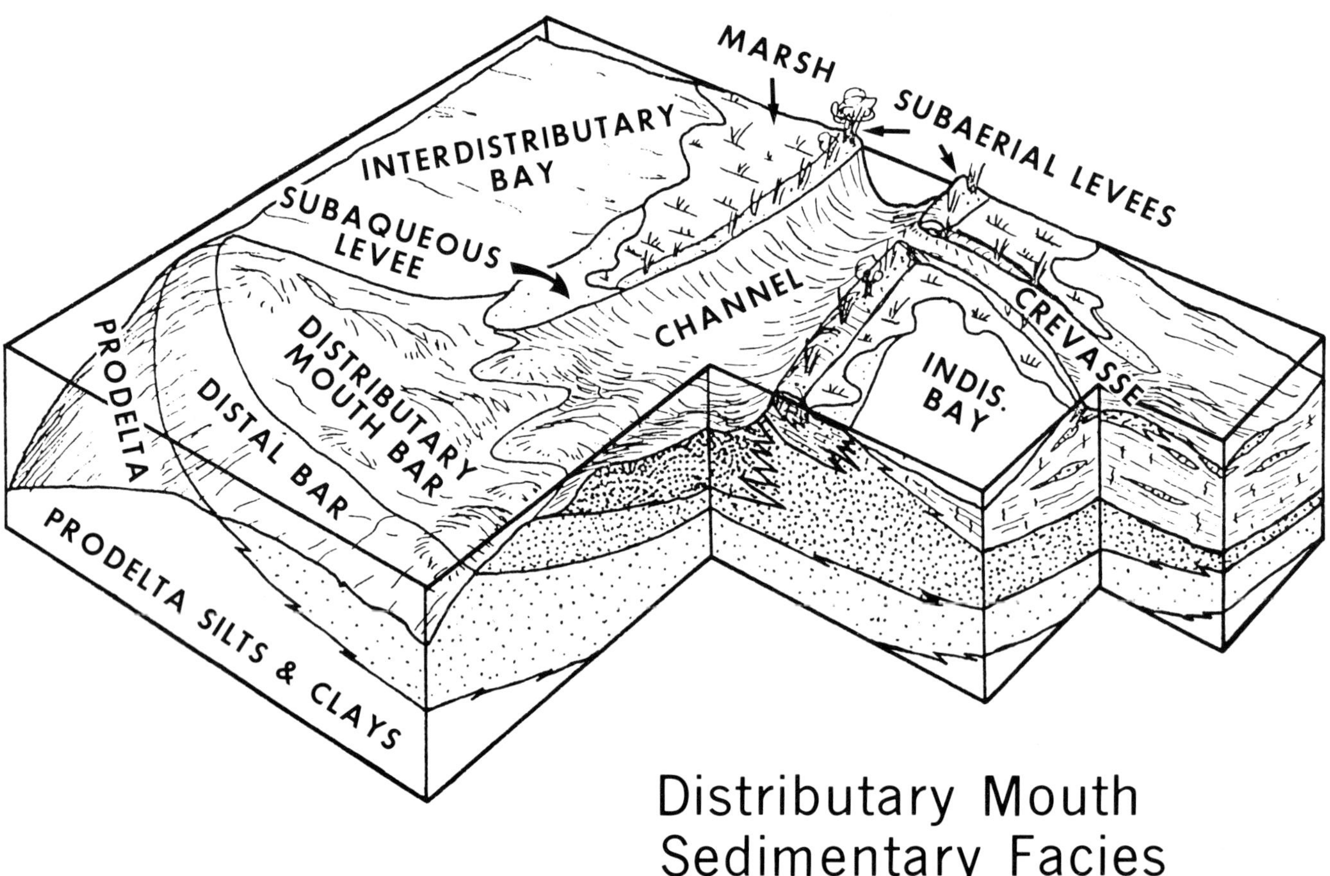

FIG. 8.—Depositional facies typical of deltaic environments. Note the break in channel levee which allowed a crevasse splay sand to build outward into an interdistributary bay. These environments (facies types) are present in the Admire 650′ sandstone (modified from Lankford, pers. commun., 1969).

0.63 to 1.3 cm) siltstone ripples (flasers). Horizontally bedded shale and siltstone (10%) and deformed bedding (5%) are locally present. Burrows, usually of low diversity, are horizontal, silt-filled, and 1/8–1/4 in. (0.32–0.63 cm) in diameter. Permeability and porosity are very low (<0.1 md and 15%).

Ripple bedding (typically asymmetrical ripples) in this facies is the result of deposition by unidirectional currents. Symmetrical ripples are rare. Woody fragments and other carbonaceous debris locally occur on the bedding planes of shales and indicate a contribution of terrestrial sediments. Ripped up siltstone clasts and wispy siltstone laminae occur in the shale (Fig. 10). Deformed bedding may result from earthquake-generated sliding of soft sediments, deformation during storms, or from postdepositional dewatering and concurrent soft-sediment deformation. Subhorizontal beds are present in some cores above and below the deformed beds (Fig. 10B). Deformation of this type took place following deposition of the underlying nondeformed beds and prior to the deposition of the overlying nondeformed beds. This type of deformation probably occurs most commonly in the prodelta area of modern deltas; however, it also may occur in interdistributary bays. Soft-sediment deformation on a variety of scales is observed in many modern Mississippi delta-front cores (Coleman, 1976).

Interdistributary bays in modern deltas are typically elongate to circular in plan view, depending upon their position relative to distributary channels (Coleman, 1976). They receive fine-grained material from distributary channels that leave their banks during floods, and coarser grained material from splays that prograde into the bay. The presence of relatively abundant burrows suggests that the bays were open to normal marine waters. The bay waters were brackish, however, resulting in a low diversity of the trace fossils. The low diversity probably reflects more stressful conditions which are more common in brackish water bays than in open marine water bodies. In cores of the Admire 650′ sandstone, an upward decrease in the amount of burrowing in the interdistributary bay facies may coincide with the progressive increase of channel- and splay-sand influx.

Interbedded Interdistributary Bay, Splay Channel, and Thin Beach Sandstone Facies

This facies overlies and is interbedded with the *Interdistributary Bay Silty Shale Facies*. Aspects of this facies are also contained in the overlying *Distributary Channel Sandstone Facies*. Detailed descriptions of this facies are given in Figure 9 and in well MP-118 in the Appendix.

The *Interbedded Interdistributary Bay, Splay Channel,*

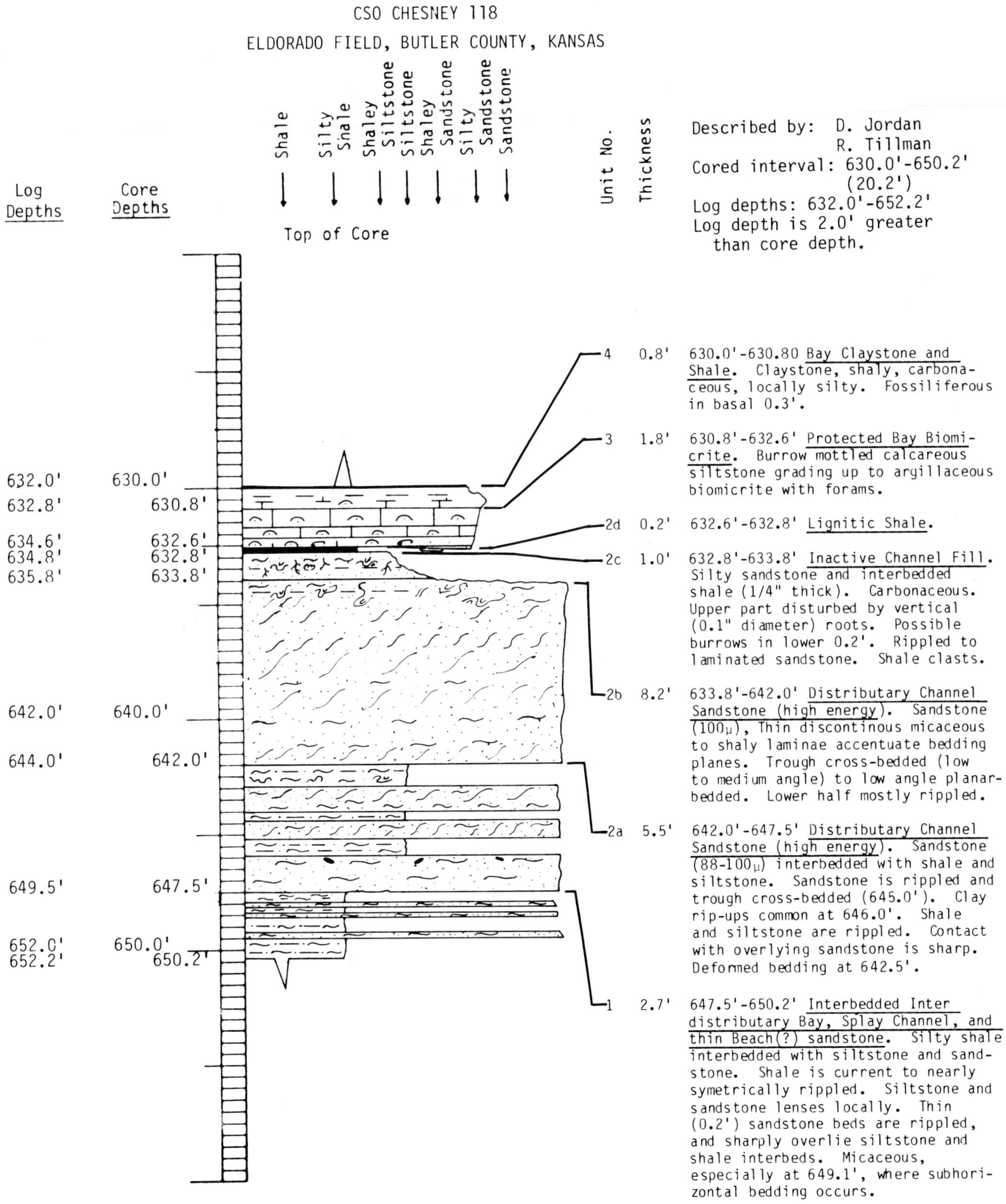

CSO CHESNEY 118

ELDORADO FIELD, BUTLER COUNTY, KANSAS

Shale | Silty Shale | Shaley Siltstone | Siltstone | Shaley Sandstone | Silty Sandstone | Sandstone

Unit No. | Thickness

Described by: D. Jordan
R. Tillman
Cored interval: 630.0'-650.2' (20.2')
Log depths: 632.0'-652.2'
Log depth is 2.0' greater than core depth.

Log Depths | Core Depths

Top of Core

Log Depths	Core Depths
632.0'	630.0'
632.8'	630.8'
634.6'	632.6'
634.8'	632.8'
635.8'	633.8'
642.0'	640.0'
644.0'	642.0'
649.5'	647.5'
652.0'	650.0'
652.2'	650.2'

4 0.8' 630.0'-630.80 Bay Claystone and Shale. Claystone, shaly, carbonaceous, locally silty. Fossiliferous in basal 0.3'.

3 1.8' 630.8'-632.6' Protected Bay Biomicrite. Burrow mottled calcareous siltstone grading up to argillaceous biomicrite with forams.

2d 0.2' 632.6'-632.8' Lignitic Shale.

2c 1.0' 632.8'-633.8' Inactive Channel Fill. Silty sandstone and interbedded shale (1/4" thick). Carbonaceous. Upper part disturbed by vertical (0.1" diameter) roots. Possible burrows in lower 0.2'. Rippled to laminated sandstone. Shale clasts.

2b 8.2' 633.8'-642.0' Distributary Channel Sandstone (high energy). Sandstone (100μ), Thin discontinous micaceous to shaly laminae accentuate bedding planes. Trough cross-bedded (low to medium angle) to low angle planar-bedded. Lower half mostly rippled.

2a 5.5' 642.0'-647.5' Distributary Channel Sandstone (high energy). Sandstone (88-100μ) interbedded with shale and siltstone. Sandstone is rippled and trough cross-bedded (645.0'). Clay rip-ups common at 646.0'. Shale and siltstone are rippled. Contact with overlying sandstone is sharp. Deformed bedding at 642.5'.

1 2.7' 647.5'-650.2' Interbedded Inter distributary Bay, Splay Channel, and thin Beach(?) sandstone. Silty shale interbedded with siltstone and sandstone. Shale is current to nearly symetrically rippled. Siltstone and sandstone lenses locally. Thin (0.2') sandstone beds are rippled, and sharply overlie siltstone and shale interbeds. Micaceous, especially at 649.1', where subhorizontal bedding occurs.

FIG. 9.—Typical geologic section observed in well MP-118 which contains the Admire 650′ sandstone reservoir. Interpreted environments of deposition of individual units are underlined. Included are marine, brackish, and nonmarine environments of deposition. Lithology, sedimentary and biogenic structures characteristic of the specific environment are given for each sedimentary facies (or genetic unit). Similarly numbered units on adjacent wells may or may not be equivalent. The location of this well is shown in Figure 2.

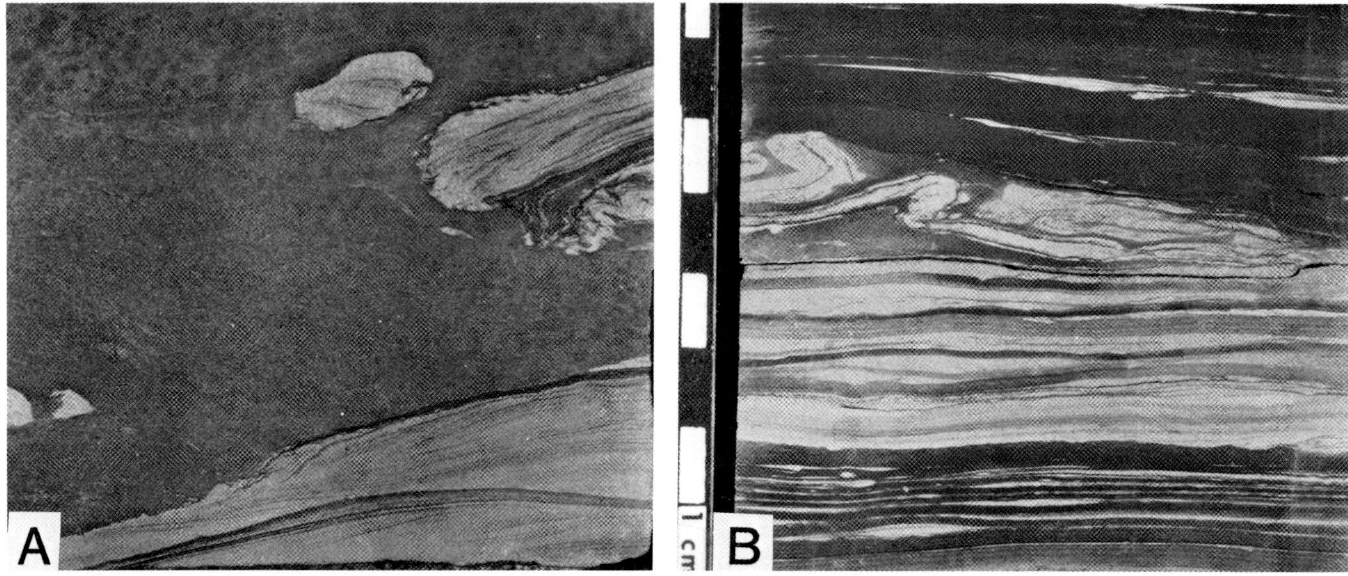

FIG. 10.—*Interdistributary Bay Silty Shale Facies.* (A) Silty shale containing siltstone rip-up clasts. Note rippled laminae are preserved in clasts. High-energy currents (possibly a storm) may have deposited the silty shale and siltstone rip-up clasts. MP-122, Unit 1, 651.0 ft (198.4 m). (B) Interbedded silty shale (dark) and siltstone (light). Siltstone and shale are rippled and slightly burrowed. Note soft-sediment deformation and truncation at the top of the siltstone bed. Earthquakes or storms could have disturbed the layer, then resumption of quiet-energy conditions allowed deposition of clays. MP-122, Unit 1, 654.2 ft (504.2 m).

and Thin Beach Sandstone Facies is composed of thin sandstones containing varying amounts of interbedded shale and siltstone (Fig. 11). The thickness of the facies locally exceeds 20 ft (6 m), but averages 8 ft (2.4 m). The facies is thickest in the southeastern part of the project area. The primary sedimentary structures in splay sandstones are moderate- to low-angle trough crossbedding and ripple bedding. Asymmetrical ripples are abundant and suggest unidirectional depositional currents. Interbedded silty shales contain abundant rippled siltstones in the form of lenticular bedding. Sandstone beds are commonly 2 to 3 ft (60 to 90 cm) thick and shales are 1 to 2 ft (30 to 60 cm) thick.

Thin (0.1–0.5 ft, 3 to 15 cm) interbeds of plane-laminated to low-angle crossbedded sandstone are locally common and are interpreted as beach deposits (Fig. 11D) that were formed by intermittent reworking by waves and currents at the edges of splay sands. Coleman (1976) also found, in the Mississippi River delta, that small beaches are located near natural levees where coarser grain-size material is available for reworking. These beach deposits are usually thin and discontinuous and are composed of sets of subhorizontal plane laminations truncating one another at low angles (well MP-118, Fig. 9, and in Appendix).

Splay sandstones are characterized by high sandstone content, abundant ripple-bedding, crossbedding and rip-up clasts. These are features similar to those in distributary channels; however, they differ in size, thickness (areal), geometry, location, and amount and type of intercalated sediments. Splays are typically lobate in shape, oriented at a high angle to flow of the main channel(s), and intercalated with interdistributary bay muds. Excellent discussions of splay channel deposition are given by Reineck and Singh (1975) and Coleman (1976).

The varying thicknesses of splay sandstones are related to their proximity to a main channel (Fig. 11C). The sandstones are thickest near the distributary channel and thin toward the flood basin (bay) in a wedge-like fashion. Many of the cored sandstones in the interbedded facies show this thinning phenomenon. Thin interbeds of sandstone and shale change upsection to thicker sets of sandstone and minor amounts of shale as the main part of the crevasse splays, with time, "approached" the core site. A vertical stacking of splays yields a localized sandstone deposit that is interbedded with fine-grained siltstones or shales. Sandstones at any one location may represent parts of several splays. Hence, it is very difficult to correlate individual splay sandstone beds between wells. Instead, "packets" of splays are correlated in cross sections.

Porosity and permeability values are higher in the sandstones of this facies than in the overlying channel facies. Some sandstones in this facies have a porosity of 29% and permeability of 1,570 md. These high values are in thin beach sandstones from which most interstitial matrix material has been winnowed. Other high porosity and permeability values are related to the process of splay deposition that selectively sorts sands deposited by flows in the upper flow regime. These upper flow regime deposits are characterized by horizontal laminations and may have been deposited adjacent to the source channel. Average porosities and permeabilities for splay sandstones, based on whole-core analysis, are 28% and 628 md. Highly porous and permeable sandstones in this facies are very local and discontinuous.

Distributary Channel Sandstone Facies

The *Distributary Channel Sandstone Facies* constitutes the major reservoir of the Admire 650′ sandstone. *Low-*

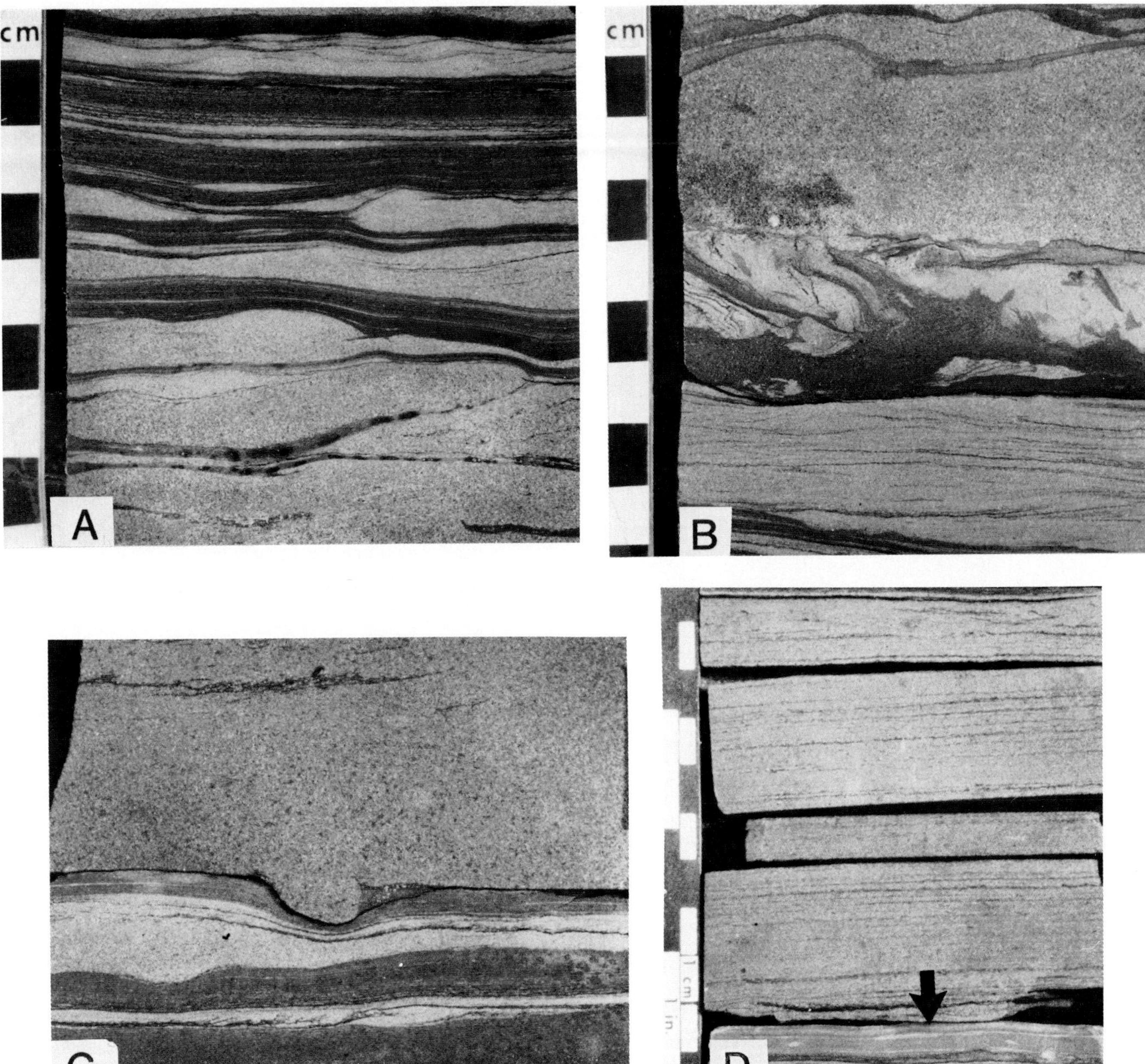

FIG. 11.—*Interbedded Interdistributary Bay, Splay Channel, and Thin Beach Sandstone Facies.* (A) Thinly interbedded rippled sandstone and shale. Note lenticular nature of sandstone in lower half. Sandstones are interpreted to have been deposited at the end of a splay channel which built into interdistributary bay muds. MP-225, Unit 2, 675.8 ft (815.6 m). (B) Soft-sediment deformed "wispy" siltstone in shale sandwiched between a lower rippled siltstone and an upper rippled sandstone. Thin deformed zone is the result of a periodic event, possibly a storm or an earthquake that temporarily resuspended sediment in the interdistributary bay environment. Sandstone may be the result of pulses of splay channel sandstone building into the bay. MP-225, Unit 2, 677.0 ft (816.0 m). (C) Thin rippled sandstone, burrowed at its base, overlying a rippled silty shale. Sandstone probably deposited in a distal part of a splay channel which built into an interdistributary bay. Burrows, in general, indicate marine or brackish water conditions. MP-225, Unit 2, 674.0 ft (205.4 m). (D) Subhorizontally laminated sandstone overlying a burrowed silty shale (below arrow). Sandstone formed by reworking a splay channel into a brackish marine bay beach deposit at the fringe of a splay channel. MP-211, Unit 1, 675.9 ft (206.0 m).

and High-Energy Channel Subfacies are designated within the *Distributary Channel Sandstone Facies*. Sedimentary features present in these subfacies suggest periodically decelerating flow rates. Both subfacies are predominantly sandstone; however, the *Low-Energy Subfacies* contains greater than 10% shale and more rippled bedding than the *High-Energy Channel Subfacies*. The distribution of the *Low-Energy Channel Subfacies* along the margins of the distributary channels suggests that the margins were areas of slower flow relative to the main part of the channel. Along the margins, more fine-grained material was deposited. Some overbank sediments may be included within the *Low-En-*

ergy Channel Subfacies. These features are common in ancient delta-sandstone reservoirs (Sneider and others 1978) and are also recognized in modern deltas (Woolen, 1976). Descriptions of typical *High-Energy Distributary Channel Subfacies* are given in wells MP-112, MP-118, MP-205 and MP-211 in the Appendix. The *High-Energy Channel Subfacies* contains significantly more trough crossbedding and only locally interbedded siltstone and shale. Distinction between the two subfacies may not always be possible or desirable. The *High-Energy Channel Subfacies* averages 8 ft

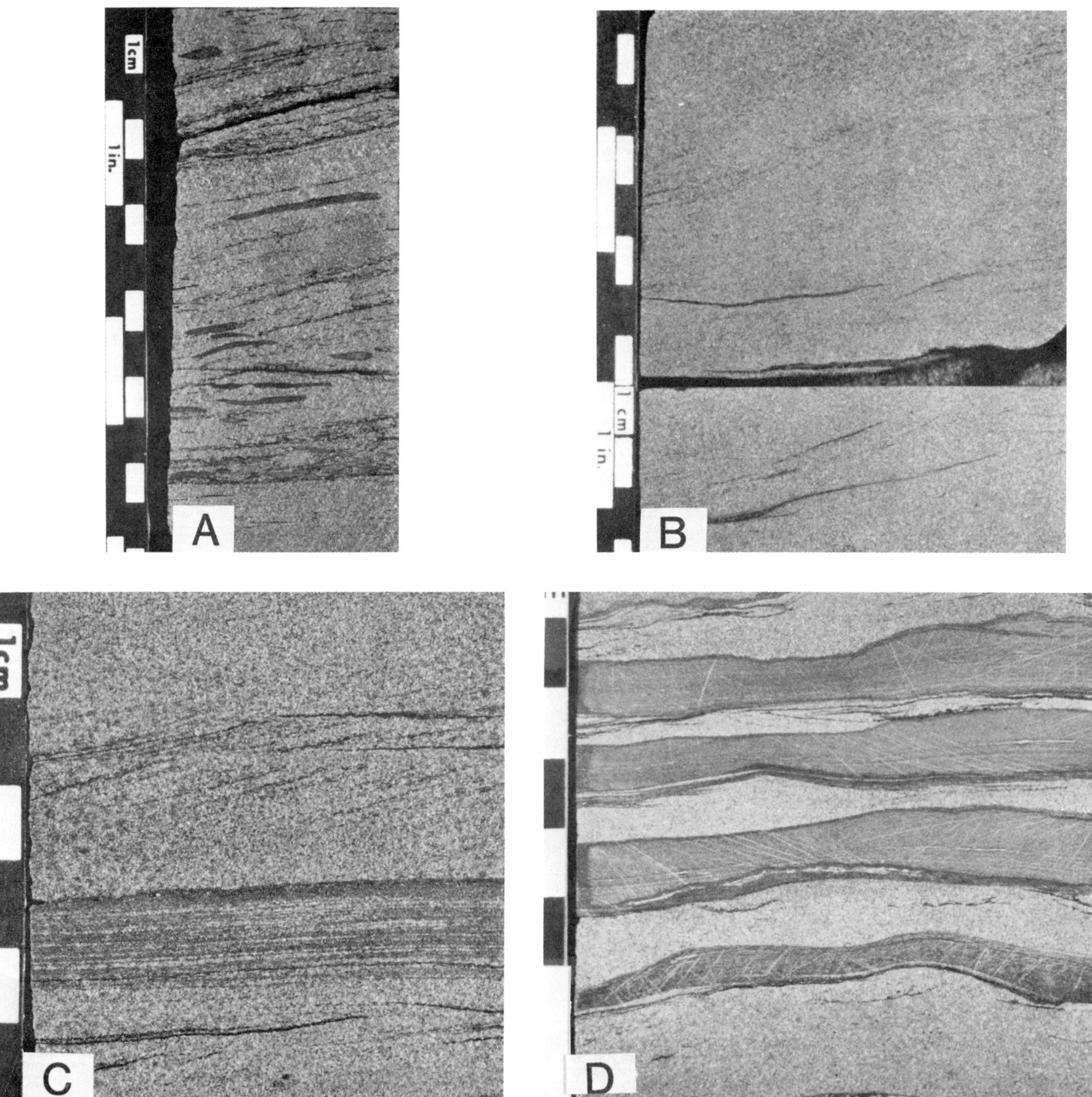

FIG. 12.—*Distributary Channel Sandstone Facies*. (A) Trough crossbedded sandstone with micaceous shale laminae. Elongate, rounded rip-up clasts lie along trough laminae. Clasts were ripped up by a high to moderately high-energy current. MP-119, *Low-Energy Distributary Channel Sandstone Facies*, Unit 3a, 636.4 ft (140.0 m). (B) Moderately high-angle trough crossbedded sandstone overlying a rippled sandstone and thin shale. MP-126, *High-Energy Distributary Channel Sandstone Facies*, Unit 1b, 663.8 ft (202.3 m). (C) Rippled to horizontally laminated sandstone. Micaceous shale laminae outline ripples. Horizontally interlaminated sandstone and shale overlies a rippled sandstone. Note tangential bases of laminae in upper rippled zone against a "reactivation surface," which is evidence of erosion and migration of another ripple set across the surface. MP-118, *High-Energy Distributary Channel Sandstone Facies*, Unit 2b, 641.4 ft (195.5 m). (D) Cyclically interbedded thin, rippled sandstone and shale. Thin sandstone interbeds deposited by pulses of sand that were periodically deposited in quiet water with clay (overbank/levee?). MP-221, *Low-Energy Distributary Channel Sandstone Facies*, Unit 2a, 674.6 ft (205.6 m).

(2.4 m) in thickness and the *Low-Energy Subfacies* averages 5.5 ft (1.7 m) in thickness. The distributary channel is not present in the southeastern part of the project area. Instead, the southeastern area contains only interbedded splay and bay sediments. Detailed descriptions of wells having typical *Low-Energy Distributary Channel Subfacies* are given in wells MP-112, MP-118 and MP-205 in the Appendix.

The distributary channel sandstones contain low- and high-angle trough crossbedding, ripples, and subhorizontal bedding (Fig. 12). Where siltstone and shale are locally ripped up and redeposited along foreset laminae (Fig. 12A), higher rates of flow are suggested. Asymmetric ripples formed by unidirectional currents are common (Fig. 12C). Some ripples occur superimposed on trough crossbeds (ripples-on-troughs). Climbing ripples are also locally present. The *Low-Energy Distributary Channel Subfacies* locally contains ripples and interbedded shale (Fig. 12D), which may have been deposited by waves in marginal marine environments. A bifurcating flow pattern within the channels was determined by using *oriented cores*. Details of the oriented core study are in a following section.

Dish structures are found at the top of some channel sandstones (Figs. 13A, B) and were formed during compaction by dewatering. Other types of deformed bedding (Fig. 14B) were formed by slumping of channel walls or levees. Coleman and Gagliano (1965) referred to this deformation as "intraformational recumbent folds" in the modern Mississippi delta.

The porosity of the *Distributary Channel Sandstone Facies* averages 22.1% based on petrographic analysis and 28% based on whole core analysis. Permeabilities average 489 md. Typical porosities and permeabilities are given for wells MP-116, MP-205, MP-213 and MP-225 in the Appendix. The lower permeability values of the distributary sandstones, when compared with splay and beach sandstones, may be attributed to poor sorting and fluctuating flow velocities. Splay sandstones are better sorted near the channel and become silty near the distal areas. The better sorted sandstones near the source/feeder channel usually have better reservoir parameters than sandstones in the distal part of splays.

Inactive Channel-Fill Facies

A channel-filling phase consisting of deposition of fine-grained sand, silt and clay occurs during late stages as the channels are filled and abandoned. The phase of deposition is represented in the Admire 650′ sandstone by the *Inactive Channel-Fill Facies* (Fig. 14). It is typically less than 2 ft (0.6 m) thick and in this study is distinguished from the underlying *Distributary Channel Sandstone Facies* only when it exceeds 1 ft (0.3 m) in thickness.

Rippled and deformed bedding are the principle sedimentary structures in this facies. Many of the ripples were probably formed in shallow ponds at the top of the abandoned channel. Deformation results from dewatering of sediments, animal burrowing, and root mottling. Rip-up clasts are derived from bank slumping. Burrows within this facies are nearly vertical and are filled with silt (Fig. 14D). Well developed root casts truncate laminae, bifurcate or thin downward and contain thin carbonaceous rims. Carbonaceous debris is quite common in this facies and attests to the abundance of plant material in surrounding environments. Detailed descriptions of this facies are given in the Appendix in wells MP-118 and MP-213.

The *Inactive Channel-Fill Facies* is poorly sorted due to intermixing of silt and sand-size grains, organic matter, and clay. Slump-derived sediments contribute to local heterogenity. Permeability and porosity average 139 md and 24%, respectively, based on whole core analyses.

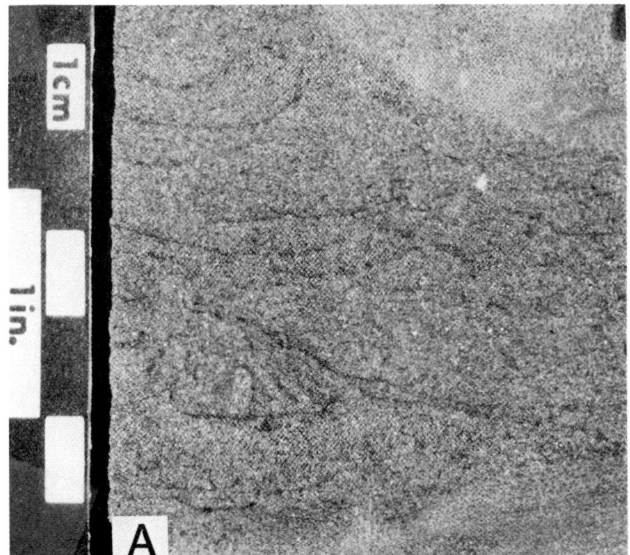

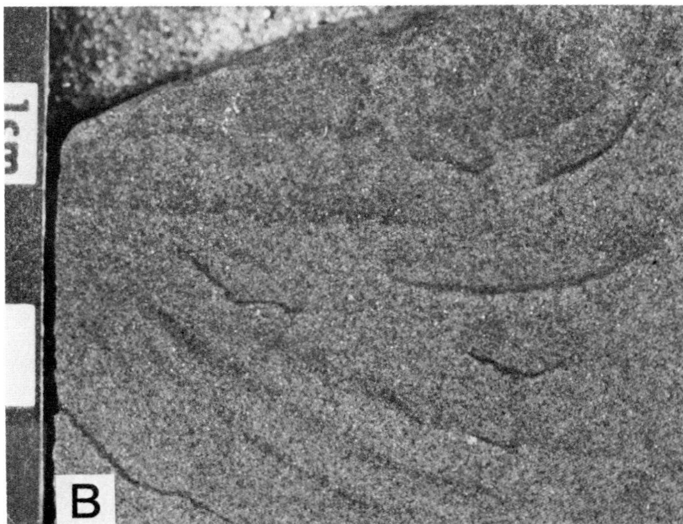

FIG. 13.—Crescent-shaped "dish structures" resulting from dewatering of sandstones. Dishes outlined by darker (finer?) material along curved planes which presumably formed by expulsion of water during loading of overlying bed. These features also are common to turbidites. *High-Energy Distributary Channel Sandstone Facies*, MP-118, Unit 2b, 634.3–635.0 ft (193.3–193.6 m).

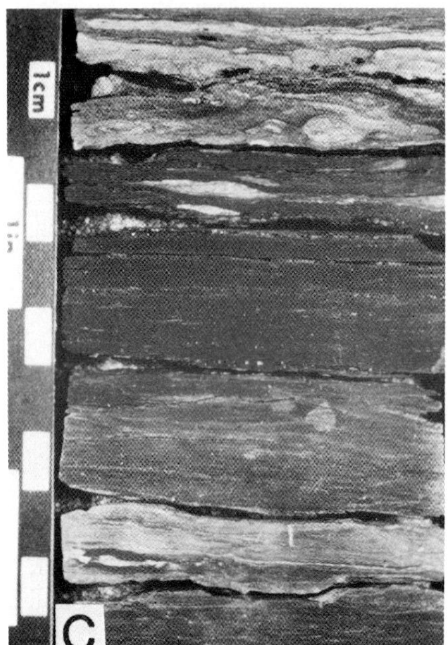

FIG. 14.—*Inactive Channel-Fill Facies*, associated sediments and *Lignitic Shale Facies*. (A) Sandstone containing angular, locally-derived, silty shale rip-up clasts overlies rippled sandstone. Rip-up clasts interpreted to have been either derived from the adjacent distributary channel walls (levees), which collapsed and added to the channel fill, or else they were ripped up from fine-grained mid-channel-fill sediment deposited during slack river flow. MP-221, Unit 2c, 664.0 ft (202.4 m). (B) Uppermost part of the *Distributary Channel Sandstone Facies* which directly underlies shales (mud plug) of the *Inactive Channel-Fill Facies*. Deformed bedding (recumbent folding) possibly due to dewatering of sands, differential compaction, or channel bank failure and slumping. MP-119, Unit 3b, 623.9 ft (190.2 m). (C) Black carbonaceous shale of the *Lignitic Shale Facies* sharply overlying a burrowed silty shale of the *Inactive Channel-Fill Facies*. Leaf impressions, typical of luxuriant plant growth in an abandoned channel system, are common along the lignitic shale bedding planes. Uppermost 0.1 ft (3 cm) is burrowed to root-mottled and rippled, shaly siltstone. MP-118, Unit 2d, 632.6 ft (192.8 m). (D) Subvertical roots cutting sandstone and shale laminae (upper half) and silt-filled circular burrows (lower half). Interbedded to interlaminated silty sandstone and shale of the *Inactive Channel-Fill Facies*. MP-118, Unit 2c, 633.0 ft (192.9 m).

Lignitic Shale Facies

Commonly overlying the *Inactive Channel-Fill Facies* is a thin *Lignitic Shale Facies*. This shale is typically 0.1 to 0.5 ft (3 to 15 cm) thick, gray to black in color, carbonaceous, and exhibits a gradational base and a sharp top (Figs. 14C, D). It is easily distinguished on gamma-ray and resistivity logs, so it can be used as a correlation marker across the area.

Carbonaceous debris in this facies includes macerated, woody fragments and other organic matter. The base of the lignitic shale is silty, locally rippled, and root-mottled. In some wells (i.e., MP-213) thin lignitic shale beds grade upward into an *Interdistributary Bay Shale*. The *Lignitic Shale Facies* forms during the final stages of channel fill and subsequent development of marsh or swamp environments. A detailed description of this facies is given in well MP-118 in the Appendix.

Protected Bay Biomicrite Facies

Unconformably overlying the *Lignitic Shale Facies* is either a *Protected Bay Biomicrite Facies* or *Lagoonal Shale Facies*. This facies consists of burrow-mottled calcareous siltstone containing coal chips and is commonly overlain by an argillaceous biomicrite or wackestone (Fig. 15) containing both abraded and unabraded fragments of gastropods, bivalves, ostracods, foraminifers (including fusulinids), echinoderms, and trilobites. Many of these fossils contain geopetal structures (Fig. 16). The base of the facies is locally sharp. *Teichichnus*-like burrows may project into the underlying lignitic shale and may be filled with either calcareous siltstone or sandstone.

The thickness of the *Protected Bay Biomicrite Facies* varies from 1.6 to 3.1 ft (0.5 to 1.0 m). This variation may be in response to slight changes in the depositional topography of the underlying facies. The log signature of the biomicrite facies (lower part of unit 3, Fig. 6) is consistent and is easily correlated throughout the project area. The biomicrite may locally (at well MP-205) overlie a *Lagoonal Shale Facies*.

The abundance of unabraded marine fossils, the presence of an organic-rich argillaceous matrix, and the position of this facies stratigraphically above a marsh or channel fill suggests that the biomicrite was deposited in a shallow-water, low-energy environment, possibly in a protected bay. Detailed descriptions of this facies are given for wells MP-112, MP-118 and MP-205 in the Appendix.

Lagoonal Shale Facies

The *Lagoonal Shale Facies* is local, overlying the *Lignitic Shale Facies* and underlying the *Protected Bay Biomicrite Facies*. It is carbonaceous (in contrast to all the other shale facies except the *Lignitic Shale Facies*). It also is calcareous, is burrowed, and contains thin sandstone stringers. A detailed description of this facies is given in well MP-205 in the Appendix.

Protected Bay Claystone and Shale Facies

The *Protected Bay Claystone and Shale Facies* consists of interbedded gray shaly claystone, silty shale, and shale. The claystone appears massive, is gray, organic-rich, burrowed, and possibly root-mottled (Fig. 17). It commonly splinters in core sections. Plant debris, pyrite, and siderite nodules are common. Silt content increases upward (especially in well MP-112, Unit 6). The lower half of the facies is predominantly shale and is slightly calcareous. Abraded fossil-hash zones (0.1 ft, 3 cm thick) are locally interbedded near the base. The fossil fragments are similar to the underlying biomicrite facies. Ripple- and plane-laminations are common in these zones.

This facies varies in thickness from 2.2–4.2 ft (0.7–1.3 m). It was deposited in a protected bay where a toxic substrate was generally unfavorable for carbonate-secreting organisms. Storm-deposited fossil-hash zones contain particles which were locally derived from the biomicrite facies.

Open Marine Limestone Facies

The uppermost unit in the cored sequences (in well MP-211) is the *Open Marine Limestone Facies*. It consists of abraded fragments of bivalves and echinoderms within an argillaceous matrix (Fig. 18). This facies is commonly burrow-mottled. Some burrows extend into the underlying shale.

The *Open Marine Limestone Facies* is recognized on gamma-ray and resistivity logs; it is the datum for the cross sections and fence diagrams. Thickness of this facies in cores averages 3.0 ft (0.9 m), although its total thickness was not

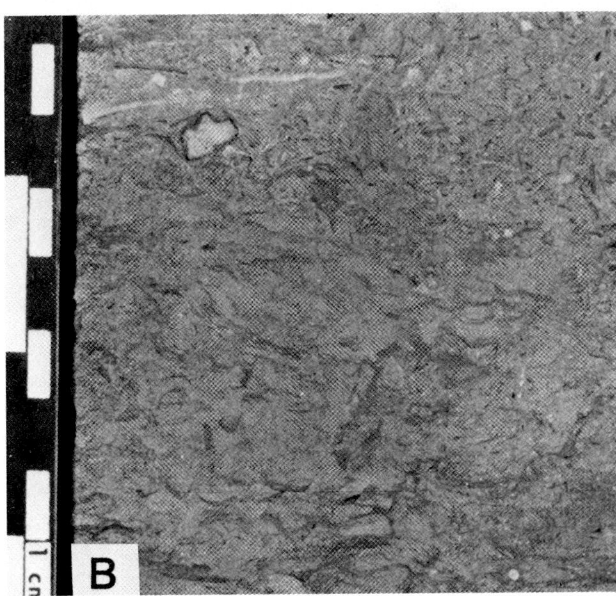

FIG. 15.—*Protected Bay Biomicrite Facies*. (A) Argillaceous biomicrite containing gastropods, bivalves, echinoderms, and trilobites (lower half of photo) overlain by *Protected Bay Claystone and Shale Facies* (upper half of photograph). Note decrease in shale content upward. MP-112, Unit 4, 613.6 ft (187.0 m). (B) Burrow-mottled calcareous sandstone; locally lies at the base of the *Protected Bay Bio-micrite Facies*. Some burrows are infilled with calcareous sandstone and sculptured into the underlying *Lignitic Shale Facies*. MP-112, Unit 4, 614.7 ft (187.4 m).

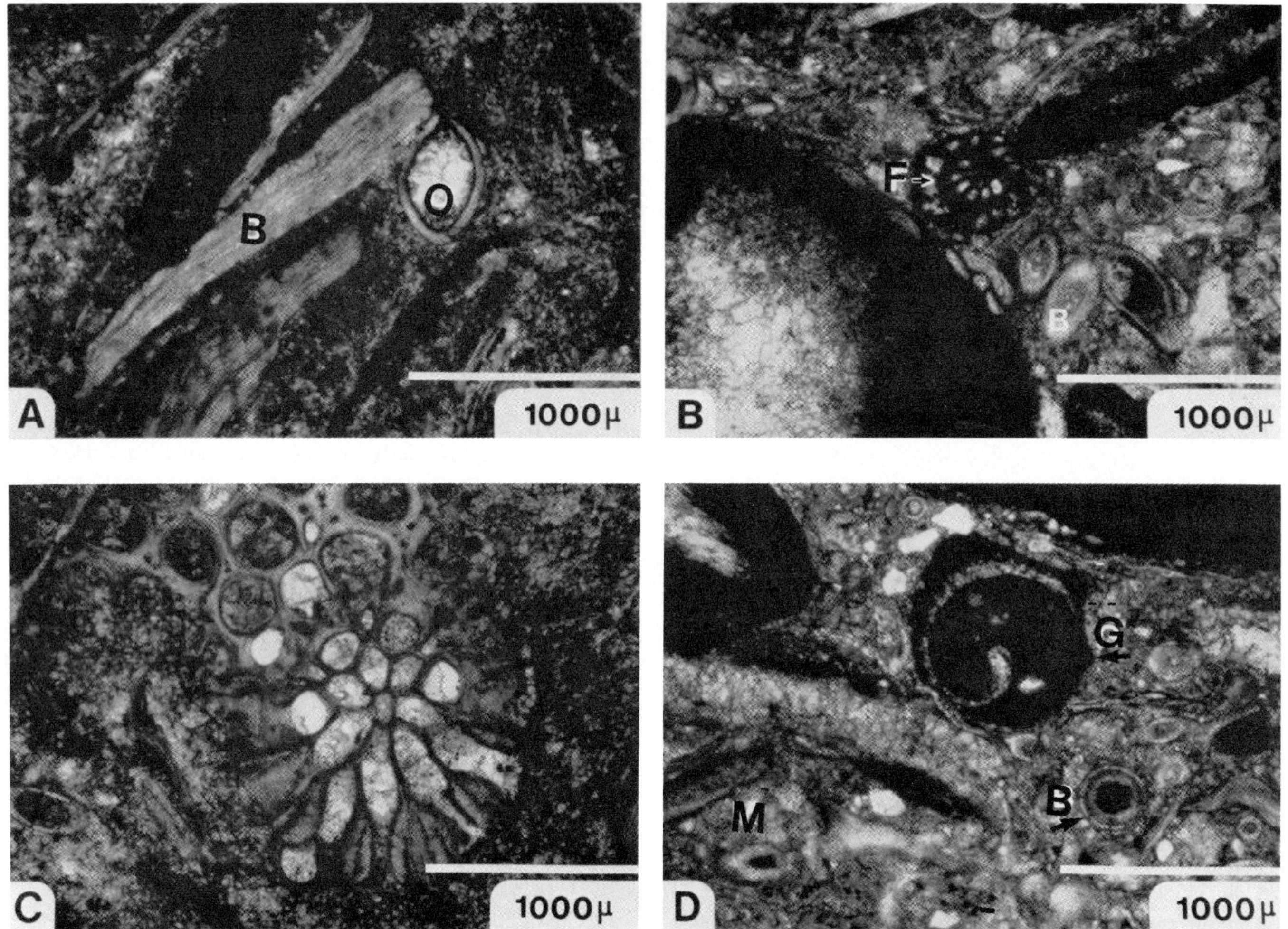

FIG. 16.—*Protected Bay Biomicrite Facies*. Thin section photomicrographs, plane polarized light. (A) Brachiopod (B) and ostracod (O) debris in a muddy (micrite) matrix. Ostracod displays geopetal structure where the original void in the shell was not filled by mud. Instead, coarse cement (sparry calcite) later filled the void. Lack of rounding and abrasion on grains suggests that these remains are "in place" in a low-energy environment. Admire 650′ sandstone. MP-213, 665.1 ft (202.7 m). (B) Brachiopod spine (B), fusulinid foraminifera (F), and mollusca (lower left) in a micritic matrix. Dark colored rims surrounding the mollusca fragments may represent destruction of shell wall by boring organisms. Permeability is 0.0 md. MP-213, 665.5 ft (202.8 m). (C) Large bryozoan, with zooecia filled with sparry calcite cement, in an organic-rich micrite matrix. Permeability is 0.12 md. MP-213, 665.1 ft (202.7 m). (D) Brachiopod spine (B), gastropod (G) and other molluscs (M) in a micrite or microspar matrix. Note mud filling of gastropod, and surrounding coarser calcite cement which has been partially destroyed by borings (dark rim). Permeability is 0.42 md. MP-213, 667.0 ft (203.3 m).

cored. This facies was deposited in an open marine environment where wave action was responsible for abrading shells of marine organisms. A detailed description of this facies is given in well MP-211 in the Appendix.

ORIENTED CORE ANALYSIS

When data for orienting cores are obtained during the coring process, or subsequently as described by Bleakly and others (1985a, b), flow directions may be determined for the currents which deposited the sandstones. Three oriented cores (MP-112, MP-126, and MP-211, Fig. 2) were included in the seven which were taken during Phase II of the study.

Twelve trough crossbed transport directions were measured and oriented in the Admire sandstone in core MP-211, which is located on the western border of the southern (Hegberg) block (Fig. 2). A major component of transport is to the north; a second major component is to the northeast (Fig. 19). *If* it were appropriate to average these two flow directions, the direction of transport would be almost directly north.

Transport directions of 11 trough crossbeds were measured in the Admire in the oriented core from well MP-126 located on the west side of the northern (Chesney) block and north of MP-211. Results at this location indicate a very strong easterly component and a minor northerly component (Fig. 20). This flow pattern is significantly different from that observed in well MP-211.

Ten paleocurrent readings were taken on troughs in the Admire in well MP-112, which is located in the eastern part of the northern (Chesney) block. A strong easterly and a minor westerly component are observed (Fig. 21).

FIG. 17.—*Protected Bay Claystone and Shale Facies*. Dark colored shaly claystone interbedded with lighter colored calcareous silty shale. Root traces cut laminae. MP-221, Unit 4, 657.0 ft (200.3 m).

The patterns of transport directions recognized in the three oriented cores are significantly different; however, as will be discussed in the next section, the areal distribution and relationship to geologic facies of these cores help put the variety of patterns into perspective. Figure 22 summarizes the Phase II oriented core transport direction results.

DEPOSITIONAL MODEL

Modern Analog

The Admire 650′ sandstone in the El Dorado pilot project area was deposited in a deltaic system similar to parts of the modern Mississippi River delta. The West Bay of the Mississippi delta, studied by Coleman (1976), is slightly larger than the area of deposition of the Admire 650′ sandstone, but it can be applied as an analog. It included a major distributary from which numerous crevasse splay subdeltas were formed. Several scales of subdeltas were formed by the crevassing process. The interdistributary bay which adjoined the delta was gradually filled by the crevasses.

A series of maps ranging as far back as 1845 (Fig. 23A) serve to illustrate the delta building and abandonment process. In 1875, a major flood increased the volume of water carried into the distributary channel adjacent to West Bay (Fig. 23B); consequently, floodwaters broke through the levee into the bay, creating a new system of subdelta distributary and splay channels. At the point of the breakthrough, the levee may have been low and, maybe, that year there was a rainy spring in the central United States. In the period from 1845 to 1875 the former interdistributary bay was filled in and evolved into a delta containing numerous distributary channels and distributary mouth bars.

By 1922, the distributary mouth bars submerged and many of the channels became inactive (Fig. 23C). The Mississippi delta is sinking, and as it sinks, low-lying areas become inundated as the interdistributary bay encroaches and

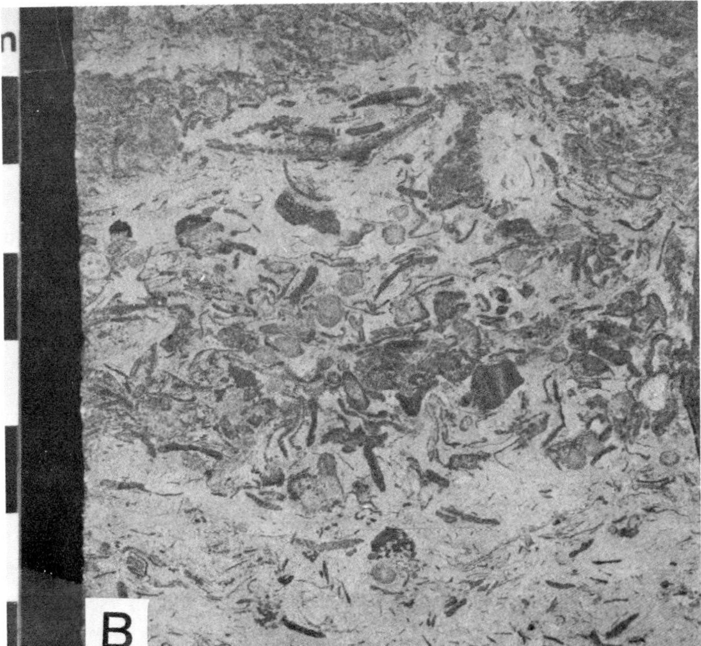

FIG. 18.—*Open Marine (Bay) Limestone Facies*. (A) Argillaceous and burrow-mottled biomicrite containing echinoderm and bivalve debris. MP-211, Unit 6, 648.5 ft (197.7 m). (B) Argillaceous limestone containing fusulinids, molluscs, and ostracods. This facies may represent a slightly more restricted marine environment (bay) than that in (A). MP-221, Unit 5, 655.2 ft (199.7 m).

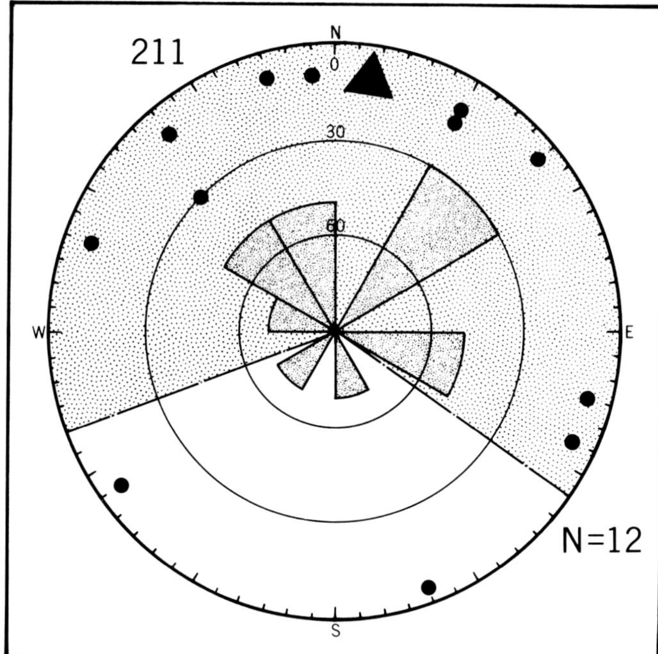

FIG. 19.—Transport directions in distributary and splay channels in oriented core MP-211, which is located along the west edge of southern (Hegberg) lease. Twelve measurements (black spots) were taken from 657.0–676.0 ft (200.3–206.1 m). Dip angle for individual measurements increases from 0° at edge through 30° and 60° to 90° dip at center. See text for discussion and interpretations; see Figure 2 for well location.

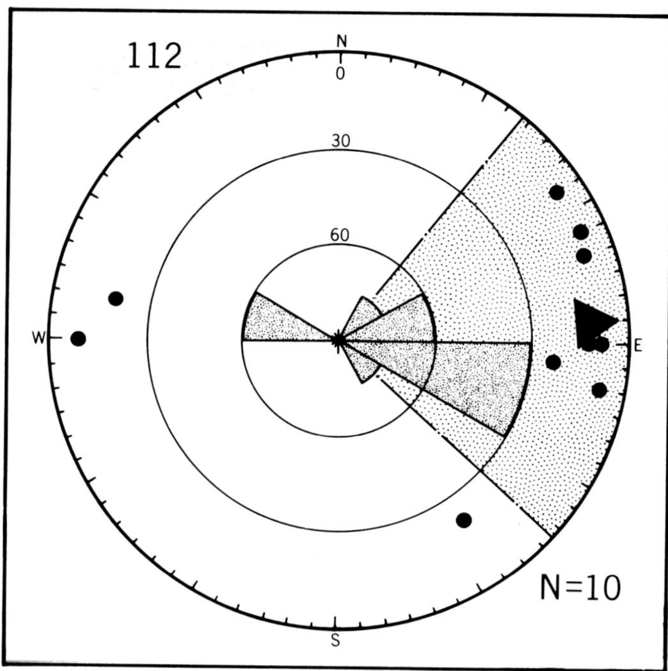

FIG. 21.—Transport direction in splay channels in oriented core MP-112. Ten directional measurements were made on trough crossbeds from 613.0-633.0 ft (186.8–192.9 m). Well located in northeast portion of northern (Chesney) lease; see Figure 2 for well location.

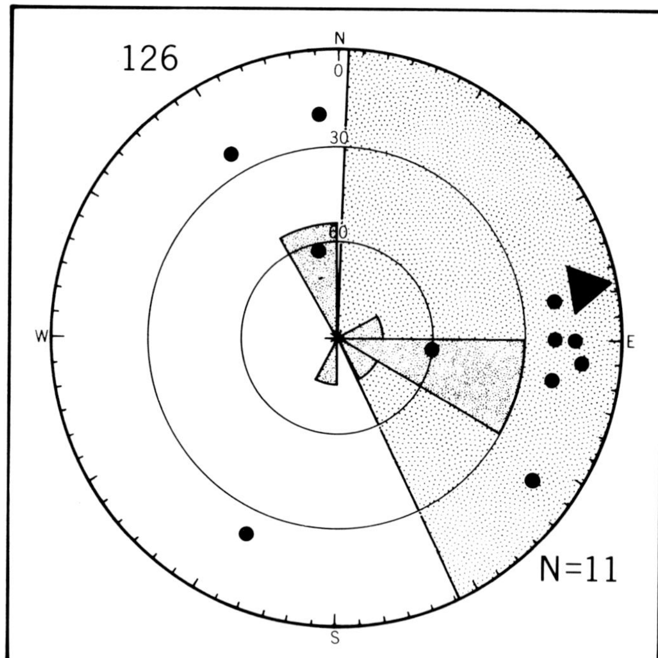

FIG. 20.—Transport directions in oriented core MP-126. Eleven measurements on trough crossbeds were taken from 655.5–669.0 ft (199.8–203.9 m). North and east transport directions are indicated. Well located in southwest corner of northern (Chesney) lease; see Figure 2 for location.

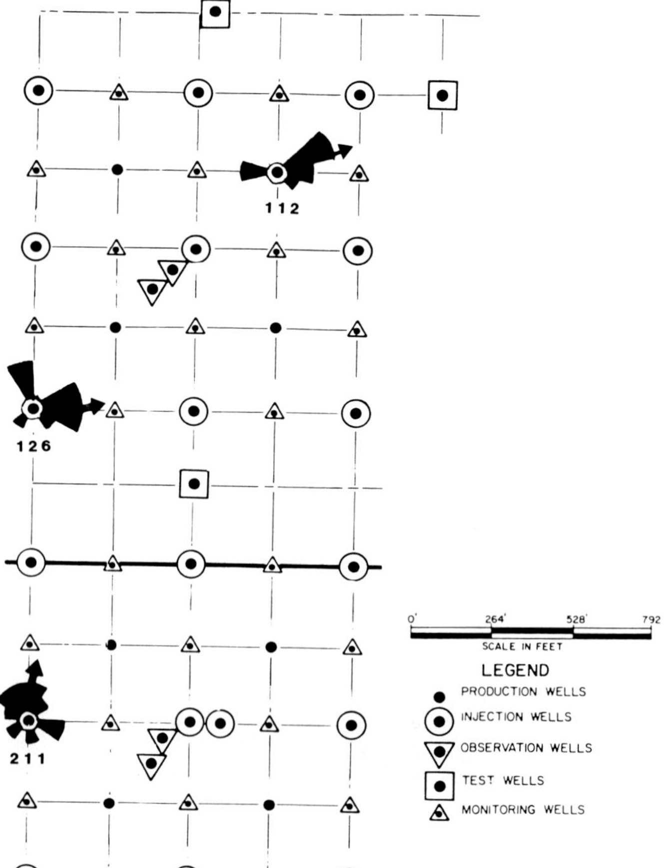

FIG. 22.—Compilation of transport directions in wells MP-112, MP-126 and MP-211. Note the variety of patterns east and west, north and east, northeast and north, respectively. See text for interpretation.

LA TOURETTE 1845

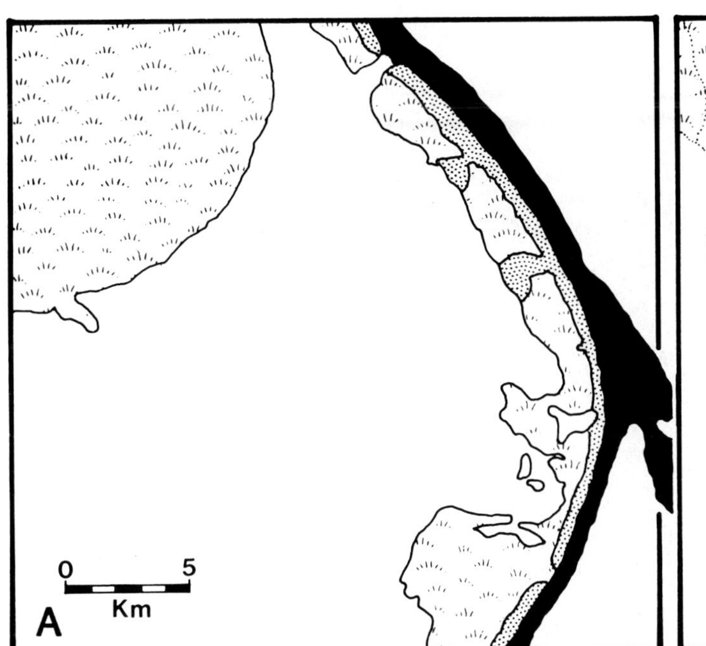

HOWELL 1875

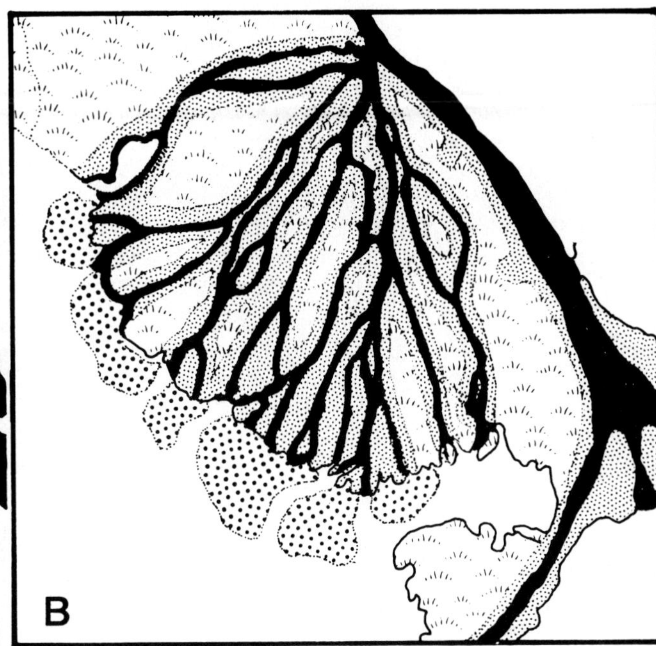

USC&GS 1922

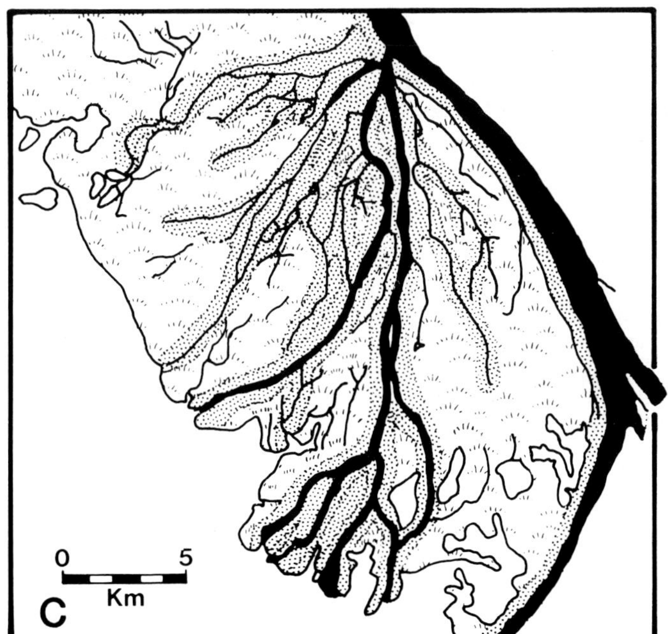

AERIAL PHOTO 1958

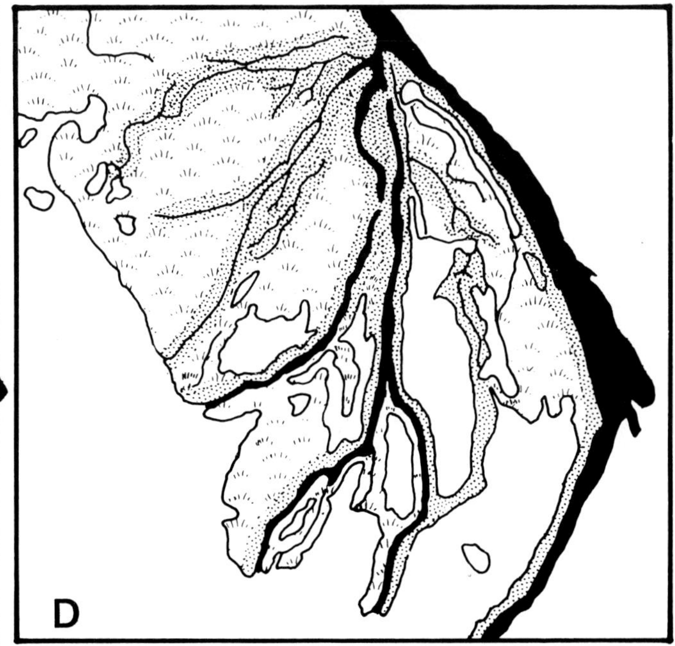

FIG. 23.—West Bay, Mississippi River delta. (A) West Bay, interdistributary bay. Map made by La Tourette in 1845 showing break in levee (near top) which occurred on the southwest bank of South Pass in 1839. Water depths in West Bay at that time were 7–10 m. Deposition in West Bay prior to levee break was limited to sediments formed by overbank levee deposition and interdistributary bay processes. (B) West Bay subdelta, 1875. Subdelta has prograded south, nearly filling West Bay. South Pass distributary channel, smaller distributaries and crevasse channel fills. Levee deposits line the channels. (C) West Bay subdelta, 1922. The subdelta is in an early stage of abandonment. Active channels (black), inactive channels and levee deposits (dense stipple) are shown. At this stage, few lakes and tidal creeks had formed. Luxuriant plant growth was attempting to keep pace with the subsiding basin. (D) Subsiding subdelta, West Bay 1958, Mississippi River delta. Subsidence is dominant. Lakes and bays were enlarging and almost all flow in channels was choked off. (After Coleman, 1976, Fig. 29.)

submerges the deposits near the fringe of the nearly inactive delta.

The distributaries, which were initiated by crevasses prior to 1875, extended southward, building longer and longer channels. As the distributary channels lengthened, the more sluggish they became, and the more difficulties they had in capturing sediment from the main distributary. The distributaries clogged with fine-grained material until they no longer were active and no longer were acquiring sand.

By 1858 the interdistributary bay encroached over a large portion of the delta; significant portions of the deltaic deposits were covered by brackish bays and lakes (Fig. 23D). Most of the marsh areas and most of the minor distributary channels were submerged. Only the larger sand-filled channels were still emergent. By 1977 all the deposits recorded in Figure 23 were submerged. In 122 years, the cycle was completed and the area returned to the 1845 configuration of distributary bays.

Admire 650′ Sandstone Model

The Admire 650′ sandstone was deposited in a similar manner, probably within a similar time framework. During Phase III of the geologic study, all the data from the earlier phases were combined with new data to yield a very detailed geologic synthesis of the pilot area. The earliest sediments deposited in the Admire 650′ sandstone pilot area consisted of interdistributary bay muds. During the second depositional stage, distributary channels prograded into the area, depositing significant amounts of sand. Subsequently, sand and silt flows cut through the distributary channel margins forming splays that partially filled some of the interdistributary bays. Eventually, distributary channels migrated over the splay-and-bay deposits (except in the southeastern part of the project area where there is no distributary channel sandstone). A major distributary channel flowed northward along the western margin of the area and branched to the east near the north edge of the southern (Hegberg) lease. Other areas received thinner channel margin sediments. These relationships are summarized in Figure 24.

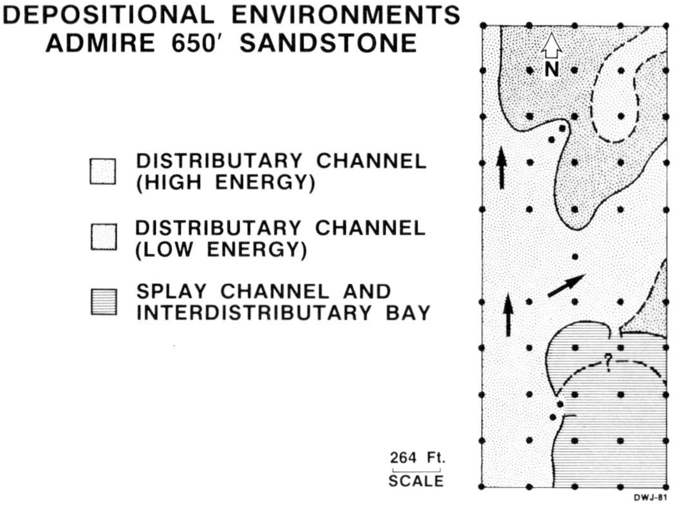

With time the larger channels lost competence and capacity, and parts of their upper portions were filled with fine-grained material deposited as levees and in swamps. Subsequent submergence of these sediments allowed deposition of interdistributary bay muds. These were, in turn, transgressed by marine bay sediments. The interdistributary bay shales and siltstones above and below the sandstone reservoirs provided seals for entrapment of hydrocarbons. Based upon the log characteristics of overlying sediments, it is surmised that similar sequences probably also occur stratigraphically higher in the section.

When oriented core interpretations are superimposed over the interpretations of environments of deposition (Fig. 25) a pattern of channel flow is evident. At well MP-211, paleocurrent directions of trough crossbeds coincide with a northeast- and a north-trending channel. In well MP-126, the predominantly easterly transport conforms with the northeast-trending channel, and the northerly component conforms with the north-flowing channel.

In well MP-112, in the northeast part of the northern

DEPOSITIONAL ENVIRONMENTS
ADMIRE 650′ SANDSTONE

DISTRIBUTARY CHANNEL (HIGH ENERGY)
DISTRIBUTARY CHANNEL (LOW ENERGY)
SPLAY CHANNEL AND INTERDISTRIBUTARY BAY

N
264 Ft.
SCALE
DWJ-81

FIG. 24.—Depositional model showing areal distribution of facies which occurs most commonly through the reservoir interval of the Admire 650′ sandstone in the El Dorado micellar-polymer pilot project. Model includes time of maximum channel development. Flow directions in channels are indicated by arrows. Well numbers are indicated in Figure 2.

(Chesney) lease, bidirectional flow is observed. The opposing directions of flow may be the result of flow in splay channels combined with tidal currents. The ripples, or small troughs, formed during flow in one direction are opposed to the bedforms formed when the flow was in the opposite direction. This particular core is in an area of predominantly splay channels.

Secondary Waterflood Results

Injection rates were monitored during the secondary waterflooding phase in the 650′ sandstone at El Dorado field. Figure 26 demonstrates how higher values of water injection are associated with the areas of good channel development throughout the northern (Chesney) lease and in the northwest portion of the southern (Hegberg) lease block. Poor response in the southeastern quarter of the project area is related to the discontinuous nature of the splay channel sandstones, which are the only reservoirs in the area. Lateral variations similar to those observed during the secondary flooding should be expected in micellar polymer flooding in areas outside the pilot area.

RESERVOIR HETEROGENITY

Sandstone oil reservoirs are rarely homogeneous. Facies changes and varying amounts of fine-grained material combine to produce vertical and lateral discontinuities. It is readily apparent from the preceding discussion that the Admire 650′ sandstone is a heterogeneous reservoir. Channel sandstones pinch out and change character laterally. The *High-Energy Distributary Channel Subfacies* contains the highest quality sandstone reservoirs and their thickness may be mapped (Fig. 27). The lower quality sandstones in the *Low-Energy Distributary Channel Subfacies* have a more limited distribution (Fig. 28). Channel margin sandstones of the *Low-Energy Distributary Channel Subfacies* and late-stage channel-

DEPOSITIONAL ENVIRONMENTS
ADMIRE 650′ SANDSTONE

DISTRIBUTARY CHANNEL (HIGH ENERGY)

DISTRIBUTARY CHANNEL (LOW ENERGY)

SPLAY CHANNEL AND INTERDISTRIBUTARY BAY

N

264 Ft.
SCALE

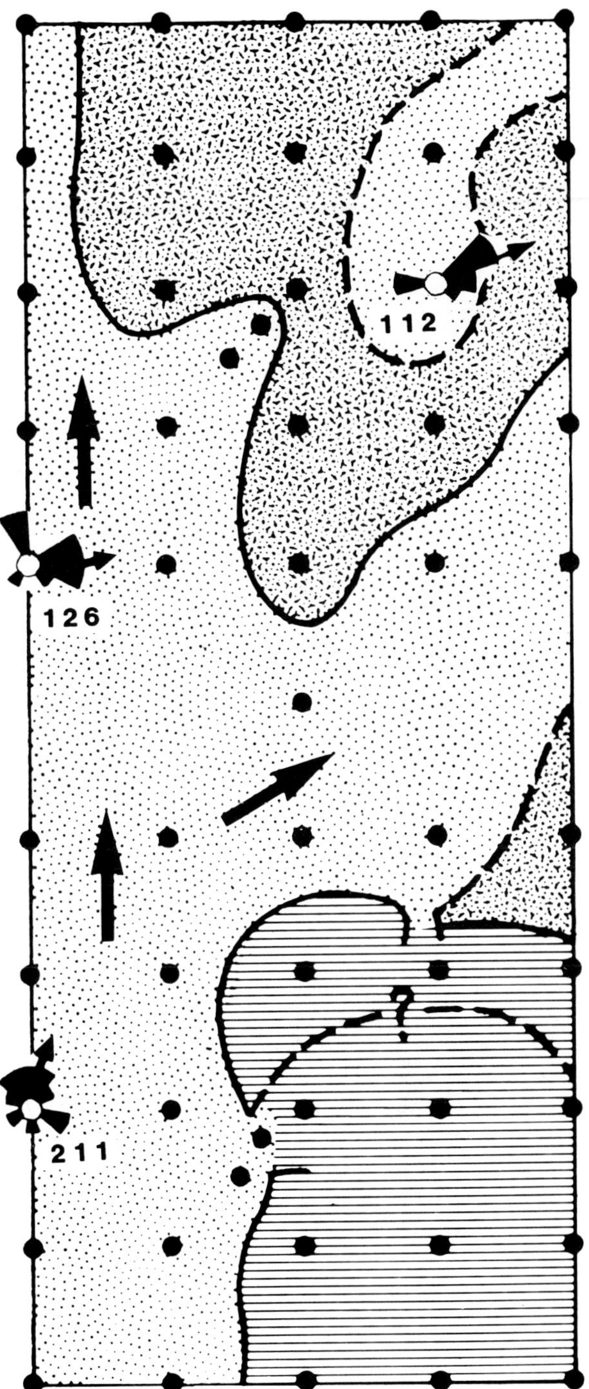

FIG. 25.—Oriented core flow directions superimposed on summary map of Admire 650′ sandstone environments of deposition (Fig. 24). For details of orientation data, see Figures 19–21.

fill mudstones contain thicker intervals of finer grained sediments than do the *High-Energy Channel Subfacies*. Splay channels are relatively discontinuous and highly variable in thickness and extent. Since the splays seek out low areas, however, they will eventually cover a wide area with thin, relatively discontinuous sandstones (Fig. 29). The variations in thickness and the distribution of the facies over the pilot area are shown as a fence diagram in Figure 30.

Figures 31A and 31B demonstrate the variability in reservoir sandstone thickness over a short lateral distance. Section A-A′ indicates that a uniformly thick channel sandstone lies along the western margin of both the southern and northern leases. The channel is well developed at wells MP-125 and MP-126 and no channel margin (*Low-Energy Channel Subfacies*) is developed. Quite different channel characteristics are shown in cross section B-B′ (Fig. 31), 528 ft (161 m) to the east of A-A′. In the northern (Chesney) lease area, much of the channel sandstone in B-B′ is

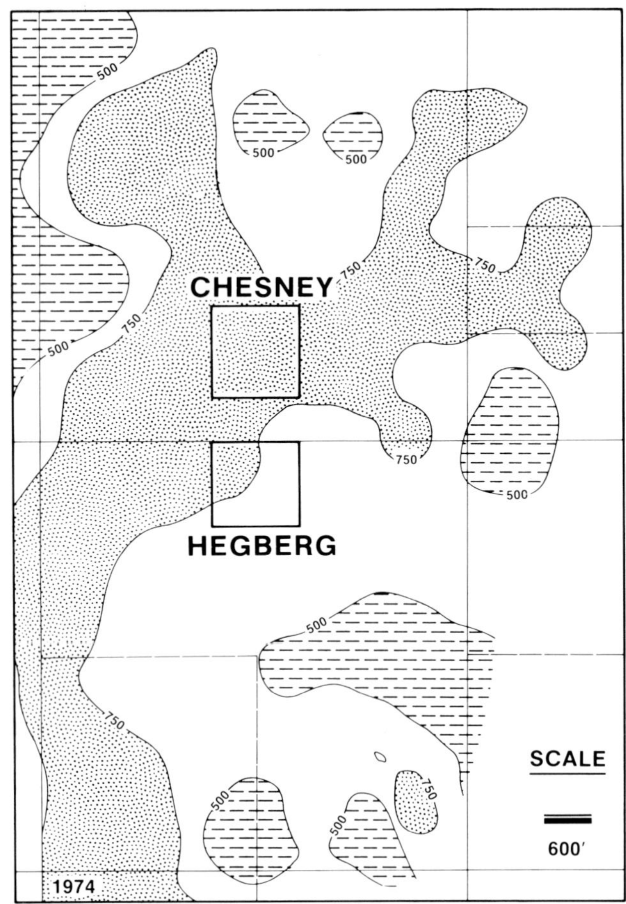

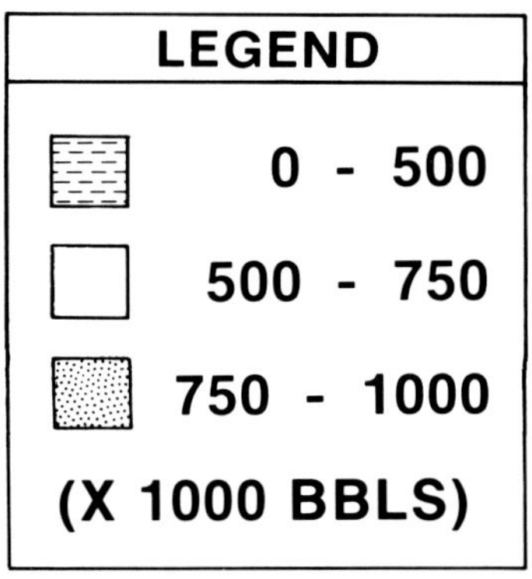

FIG. 26.—Map of cumulative water injectivity. Values obtained during secondary waterflood of El Dorado field. High values (750,000 to 1 million bbls) occur where channels are well developed. Low values are associated with areas of poor or no channel development.

a lower quality reservoir associated, in part, with finer grained channel margin facies. In the southern (Hegberg) lease, the *Low-Energy Channel Subfacies* are almost absent (Fig. 30).

Cross sections C-C′ and D-D′ in Figure 32 also show lateral facies changes. In C-C′, the *High-Energy Channel Subfacies* developed at well MP-116 grades into *Lower-Energy Channel Subfacies* toward wells MP-117 and MP-118 and may not be continuous with the channel sandstone in MP-119 and MP-120 due to a possible "shoal area" near MP-122 (see also fence diagram, Fig. 30).

Cross section D-D′ documents the lateral change from thick, high-energy channel sandstone in the west-central part of the project area to predominantly shale with thin interbedded splay sandstones in wells MP-217 and MP-225. D-D′ may be subdivided into three areas of deposition: (1) channel development at wells MP-201 and MP-209, (2) interdistributary bay and thin splays at MP-213, and (3) thicker splay deposition at wells MP-217 and MP-225.

The vertical sequence of facies at injection well MP-213 (Section D-D′, Fig. 32) is anomalous in that neither *High-* or *Low-Energy Channel Subfacies* sandstones are present in this well, which was cored and extensively studied (see detailed description in the Appendix). All the lateral communication between MP-213 and adjacent wells must be through the locally discontinuous sandstones of the *Interbedded Bay, Splay, and Thin Beach Facies*. Figure 30 shows the abrupt thinning of the distributary channel between wells MP-227 and MP-213, which are only 90 ft (27 m) apart. Other wells in which channel facies are absent are those in the southeast corner of the southern (Hegberg) lease (MP-215, MP-223, MP-224, MP-225, Fig. 30). These relatively abrupt local variations in the pilot area may be considered to be typical of the field as a whole.

The difference in the distribution of the *High- and Low-Energy Channel Subfacies* observed between the northern (Chesney) and southern (Hegberg) leases lead to different sequences of facies. In the southern lease, the *Low-Energy Channel Subfacies* is observed in only two wells (MP-205, MP-221). Hence, the vertical sequence of reservoir facies over most of the southern lease is *Interbedded Bay, Splay and Beach Facies* overlain by a *High-Energy Distributary Channel Subfacies* (Fig. 30).

In the northern (Chesney) lease a three-layer sequence of reservoir types is common except along the south and west sides of the lease. The three-layer sequence from bottom to top is *Interbedded Facies, Low-Energy Distributary Channel Subfacies,* and *High-Energy Distributary Channel Subfacies*. On the west side, in the area where the channel facies are thickest, high-energy channel deposits lie directly on interdistributary bay shales (Fig. 32). In portions of the southern area of the northern lease the *Low-Energy Channel Subfacies,* sandstones are commonly absent, giving a

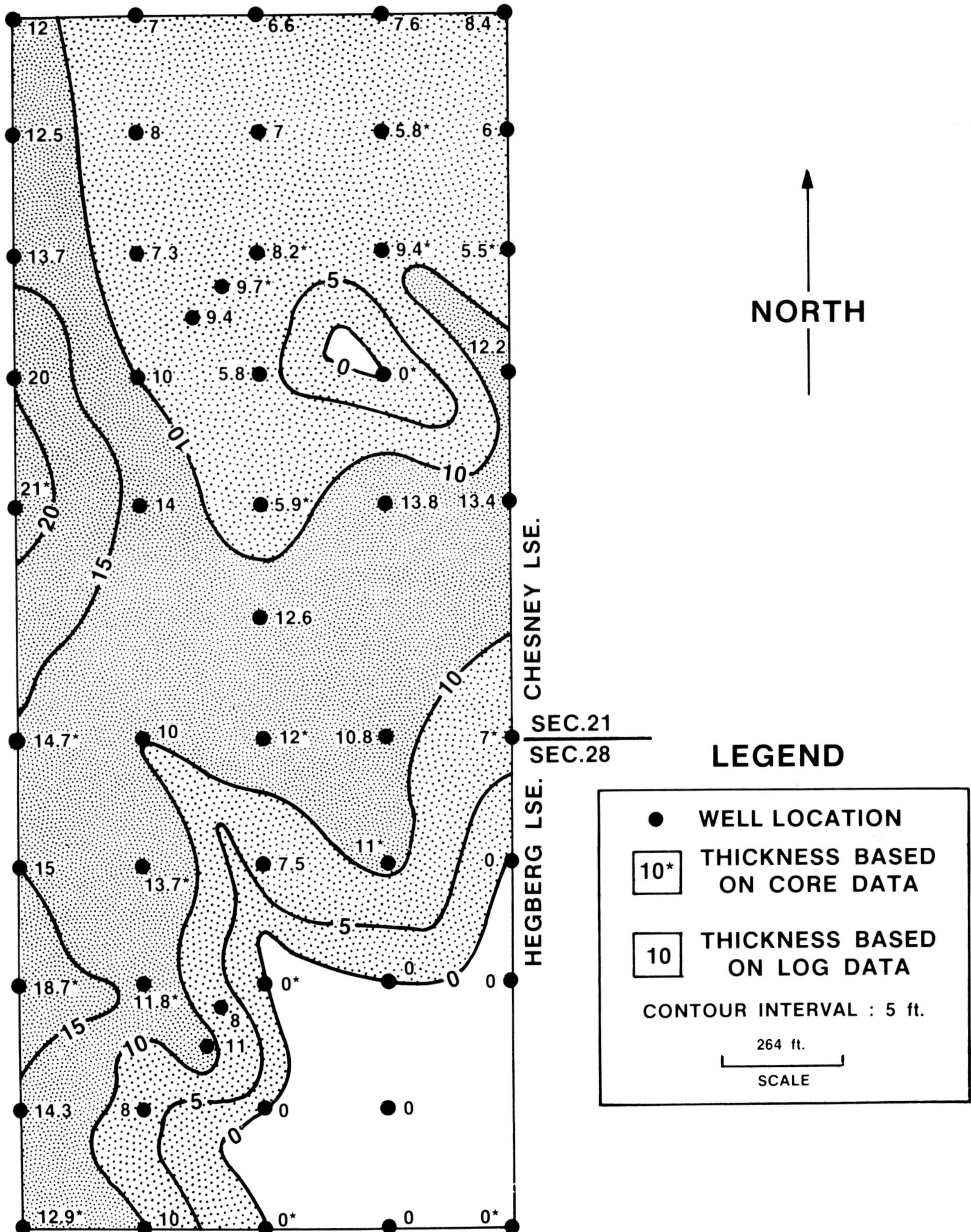

FIG. 27.—Isopach map of *High-Energy Distributary Channel Sandstone Subfacies* which forms the highest quality reservoir in the Admire 650′ sandstone. Note sandstone is thickest along the western margin and across the center of the project area. Thickness determined primarily from cores; however, logs were used as indicated. Thickness of sandstone greater than 10 ft (3 m) is shown by heavier stipling. See Figure 2 for individual well designations.

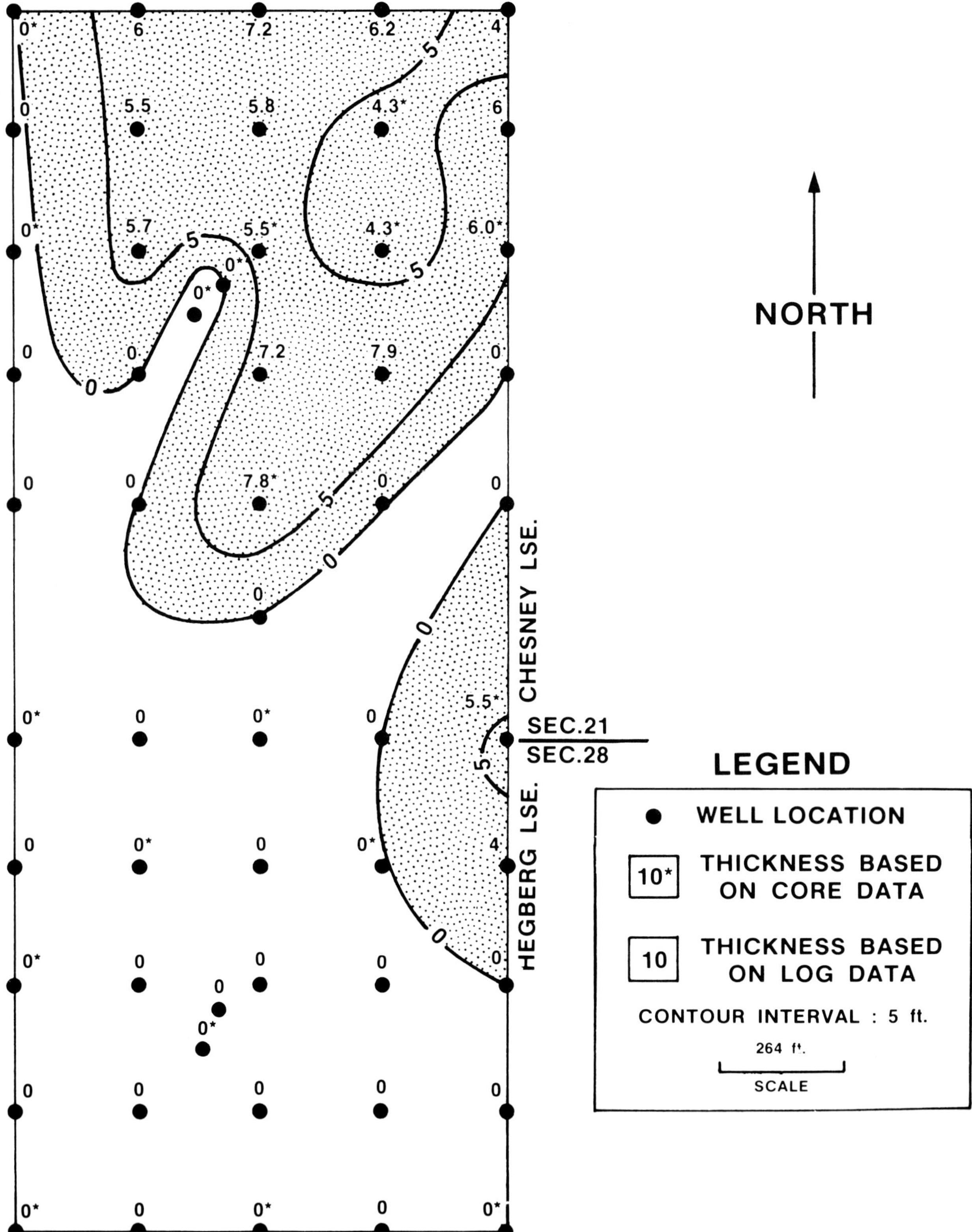

FIG. 28.—Isopach map of *Low-Energy Distributary Channel Sandstone Subfacies*. Facies is marginal to *High-Energy Disributary Channel Sandstone Subfacies* (Fig. 27) in the northern lease. There is a general lack of this subfacies in the southern lease. See Figure 2 for individual well designations.

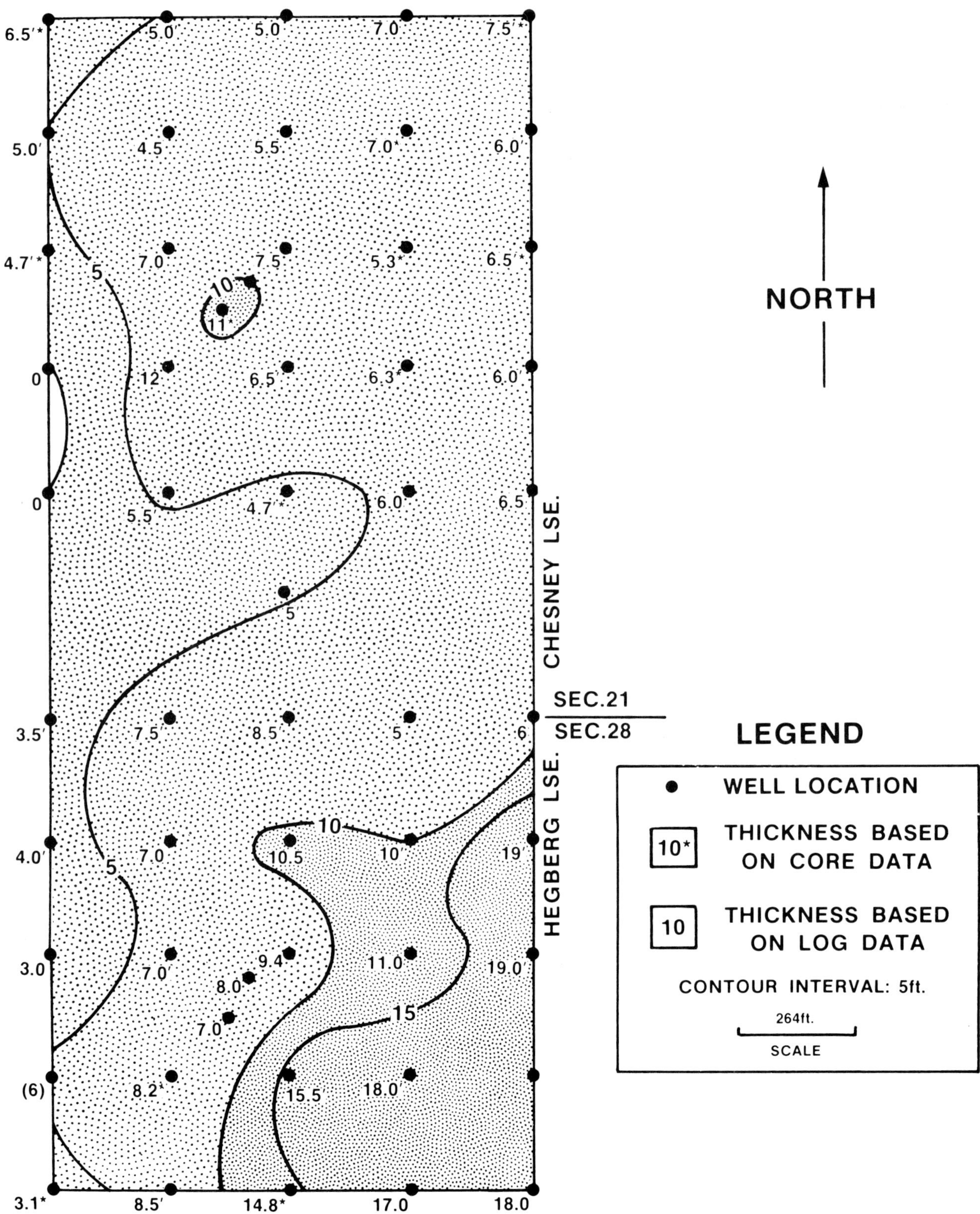

FIG. 29.—Isopach map of the *Interbedded Interdistributary Bay, Splay, and Beach Facies*. This facies most commonly lies directly above the interdistributary bay shales. The southeast corner of the lease contains significant thicknesses of this facies. The western margin of the map area contains the thinnest sections of this facies. Thickness of facies greater than 10 ft (3 m) is shown by heavier stipling.

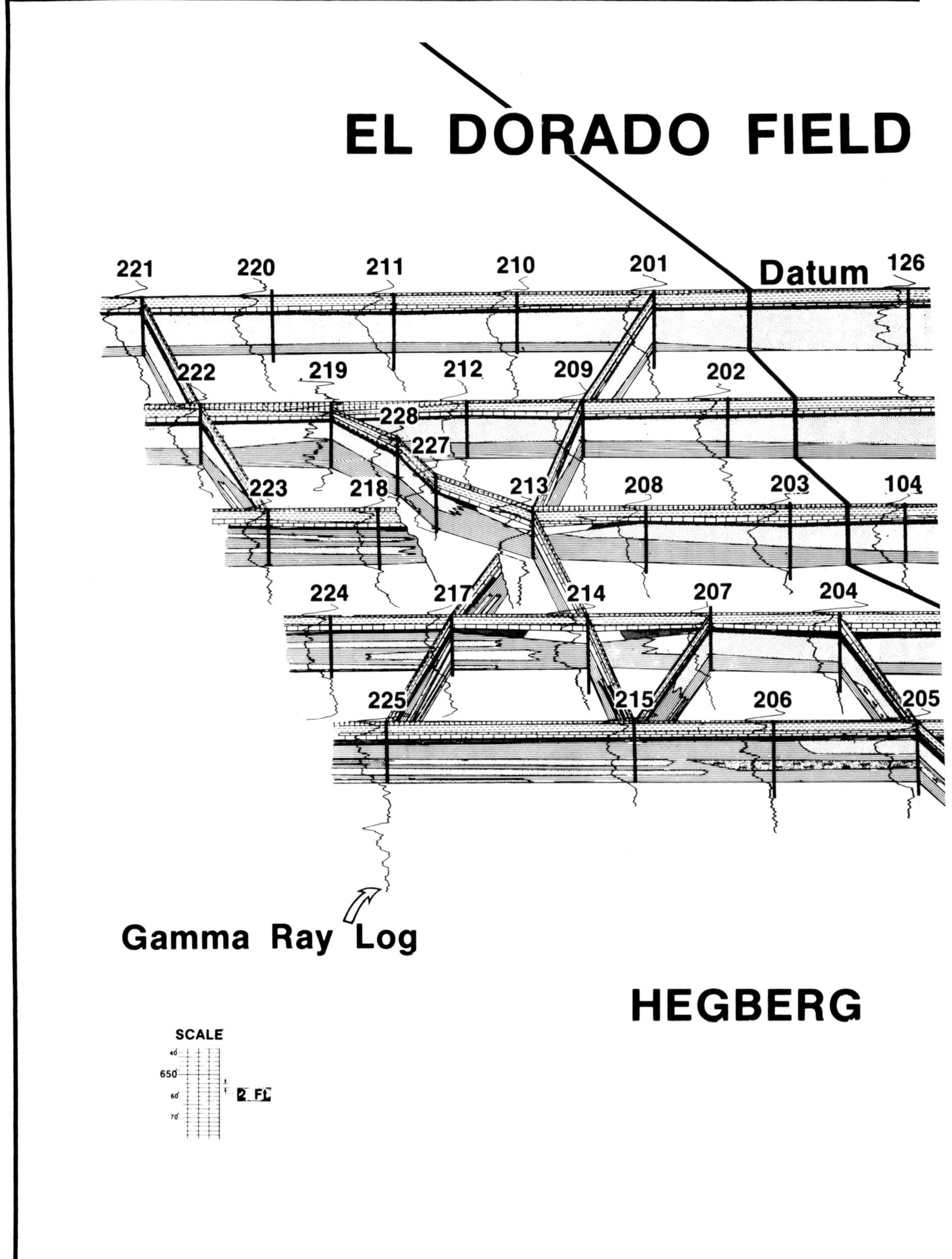

FIG. 30.—Fence diagram constructed from log and core data in patterns are indicated in legend in Figure 31.

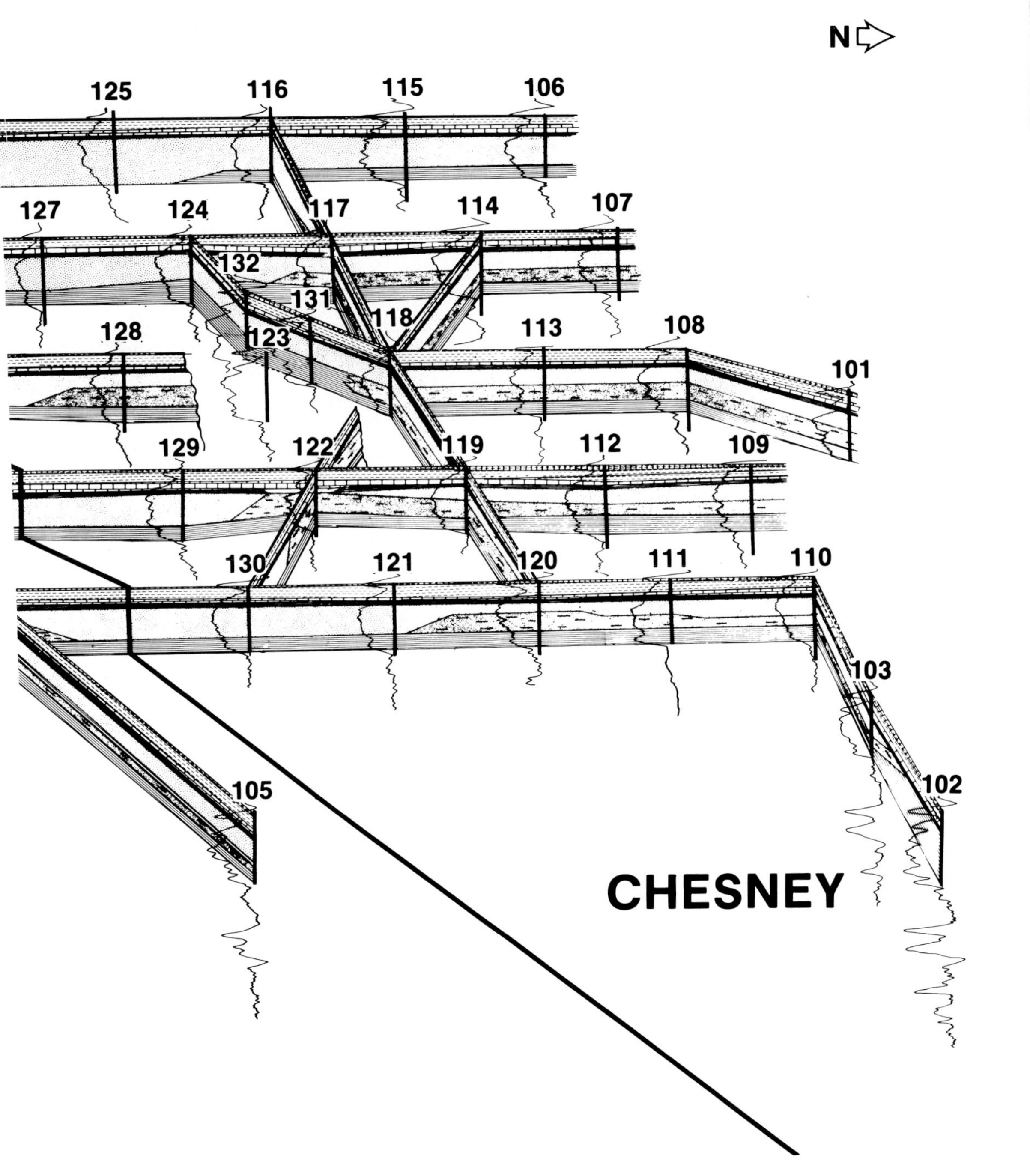

El Dorado field, Kansas. Gamma-ray logs and facies types are indicated. Facies

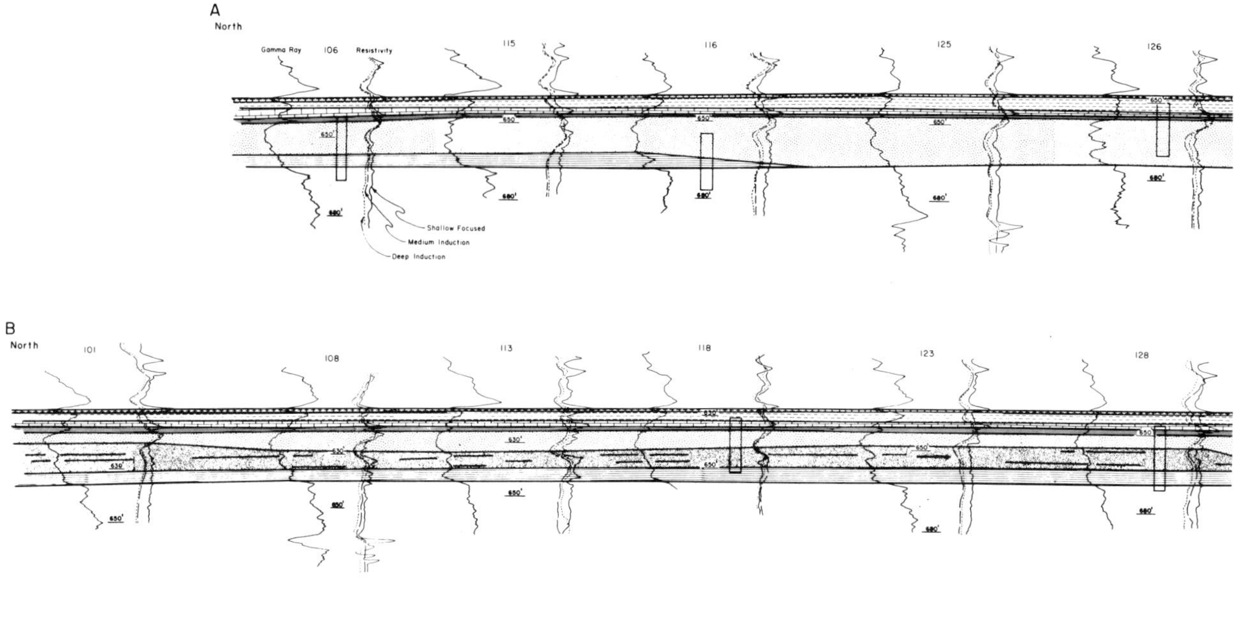

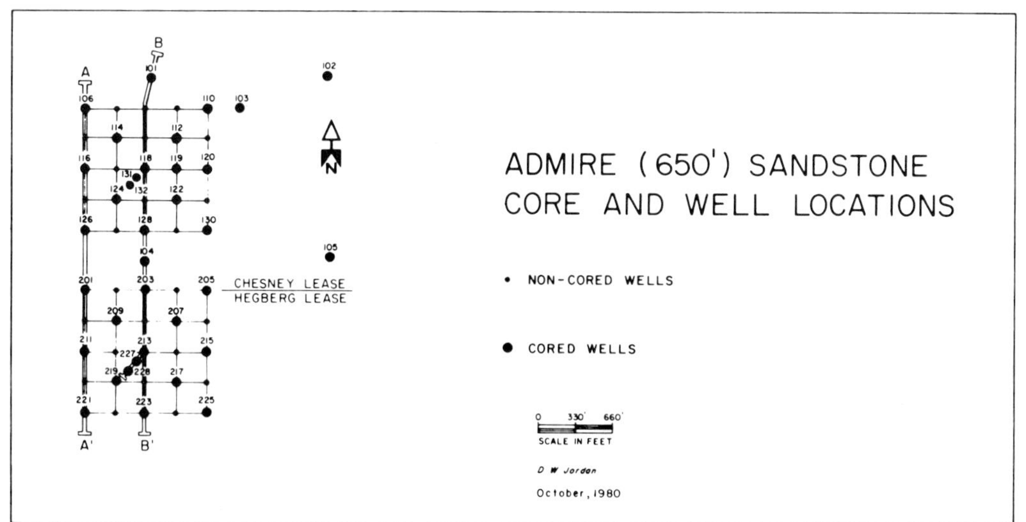

FIG. 31.—North-south facies cross sections. A-A′ lies along the western

sequence similar to that in the southern (Hegberg) lease (Fig. 30).

Reservoir quality is a function of the thickness and distribution of the distributary and splay sandstones. The shales interbedded in the *Low-Energy Channel Facies* apparently only slightly affect reservoir quality of the sandstones, but the shale intervals within the sandstones must be subtracted from the gross thickness to arrive at a net sandstone thickness. Permeabilities are marginally higher in the *High-Energy* compared to the *Low-Energy Distributary Channel Subfacies* (Fig. 33). Figures 34 and 35, and data for individual wells (Appendix) indicate that porosity, and residual oil saturations are similar in the sandstones in the *Low- and High-Energy Distributary Channel Subfacies*. In contrast, splay and associated sandstones of the *Interbedded Facies* are very permeable but lack continuity and, thus, may provide only locally acceptable reservoir quality units.

Correlation of individual beds of the splay sandstone deposits on cross sections and fence diagrams is difficult because of the discontinuous nature of splay deposits. Shales are likely to be more continuous than sandstones in the *Interbedded Interdistributary Bay, Splay Channel, and Thin Beach Sandstone Facies*. Shales in this facies may completely enclose splay sandstone packets and seriously affect fluid flow. On the other hand, interbedded shales within the *Distributary Channel Sandstone Facies* (both *High- and Low-Energy Subfacies*) are less likely to be continuous and should inhibit fluid flow to a lesser extent. Sneider and oth-

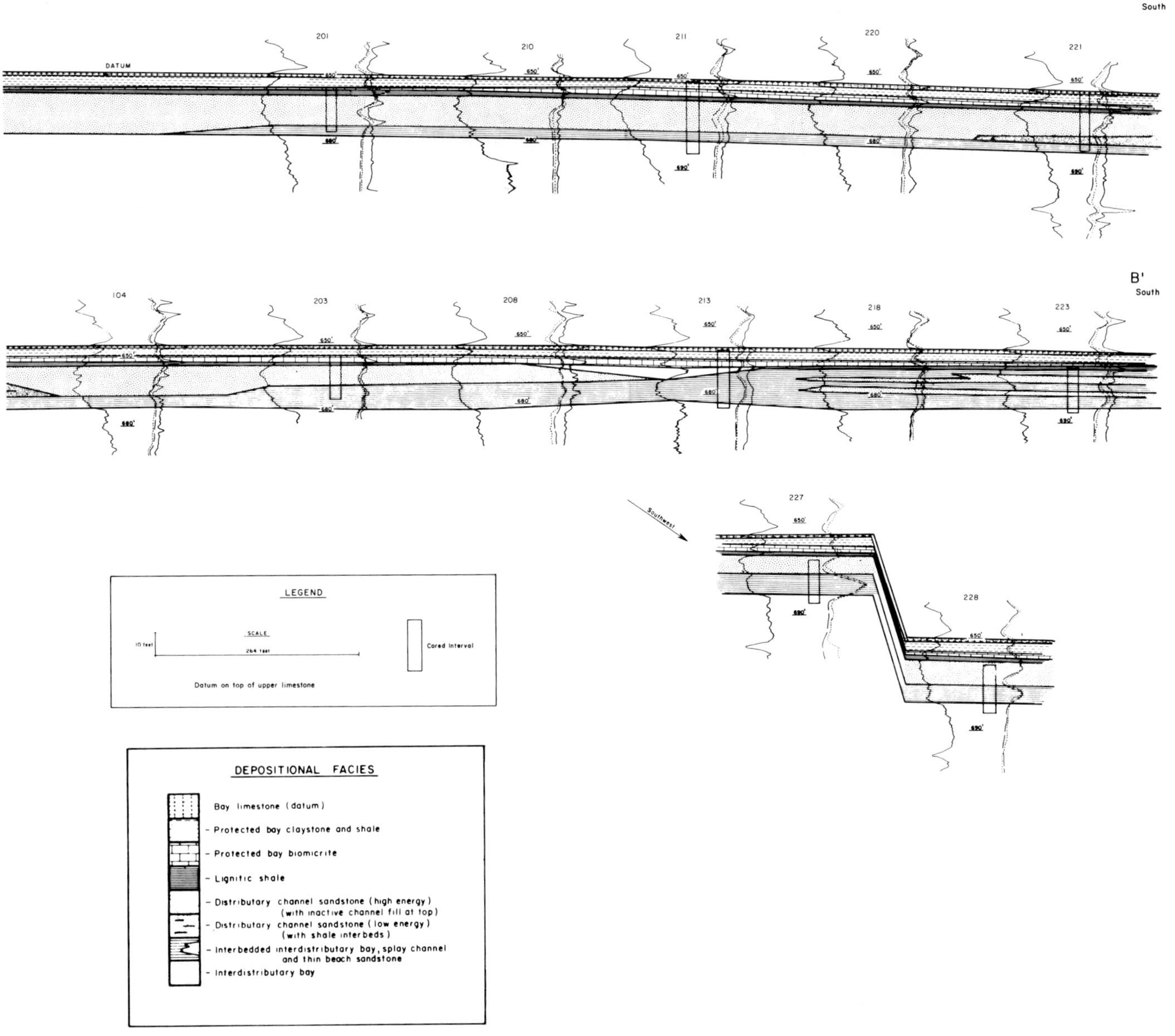

boundary of both leases and B-B′ lies two well locations to the east.

ers (1978) found that thin clay/silt beds in high-energy distributary channels are generally less than 1 ft (30 cm) thick and as much as a few tens of feet in maximum areal extent.

Pilot Study Location

To test two distinctly different types of micellar fluids, one oil-soluble slug and one high-water-content slug, two blocks of equal size were selected for study. The operator planned to inject into the Admire 650′ sandstone micellar fluids costing between $9.00 and $14.00 per barrel. Based on geologic findings in Phase I, it was recommended that the comparison of the two processes be carried out on the blocks laterally adjacent in a north-south rather than an east-west configuration. The north-south configuration (Fig. 4) was quite acceptable since, in addition to the interpreted mediocre quality reservoir in the eastern block, it was also found to be in the gas cap for the Admire 650′ sandstone.

Transmissibility

Determining the relative transmissibility of fluids within a reservoir is important in tertiary flooding of reservoirs. The major objective is to produce the most oil possible in any given area, but some areas will yield more oil by tertiary flooding than others. Equal allocations of production batteries and other surface equipment over the whole area in deltaic reservoirs of the type found in the Admire 650′ sandstone will not maximize economics. Instead, fluid-handling capabilities should be maximized in directions of best

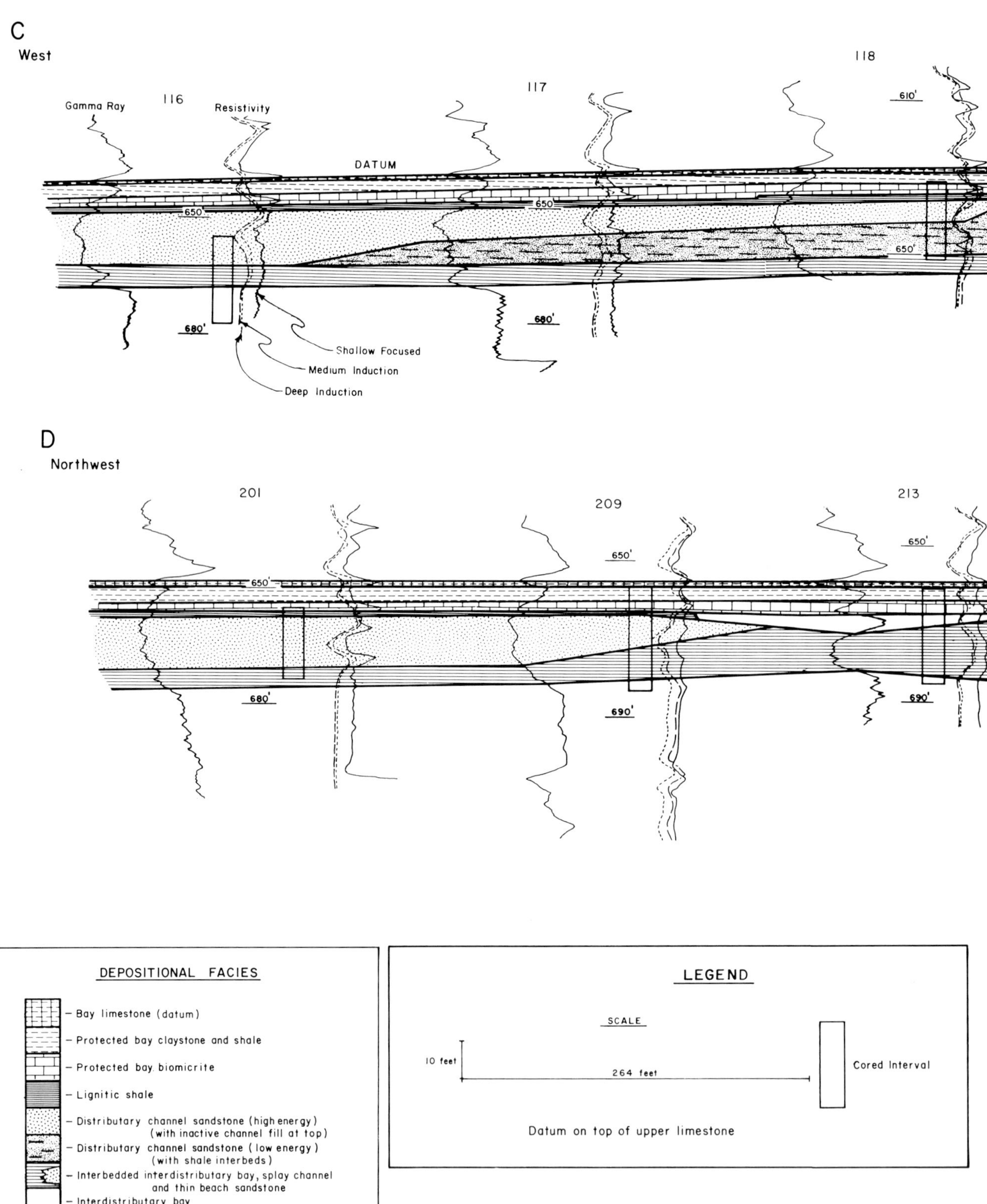

FIG. 32.—West-east (C-C′) and northwest-southeast (D-D′) cross

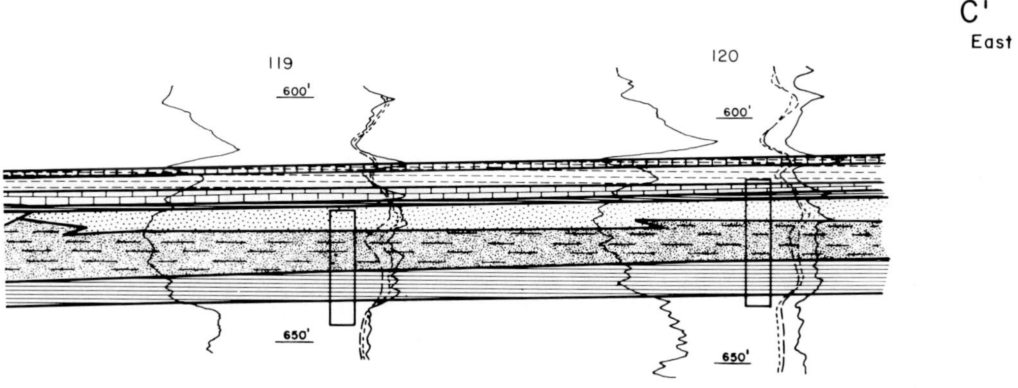

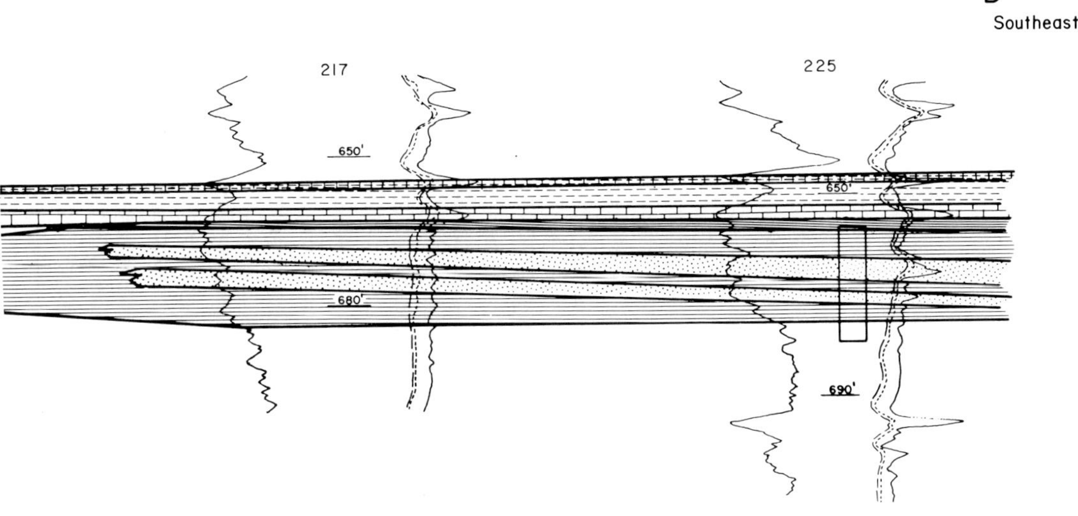

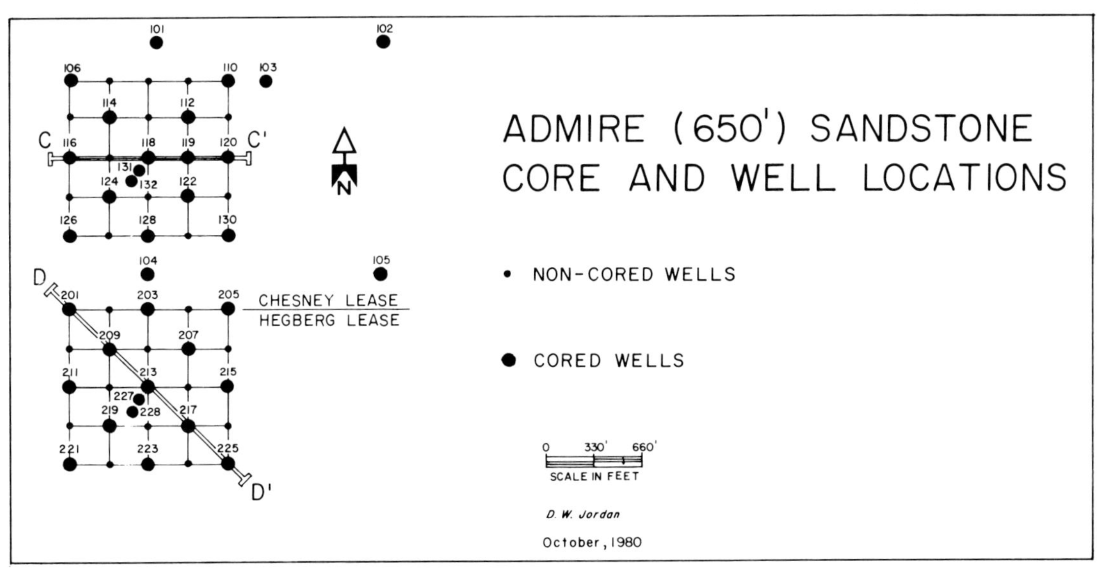

sections. Locations indicated on map. See text for discussion.

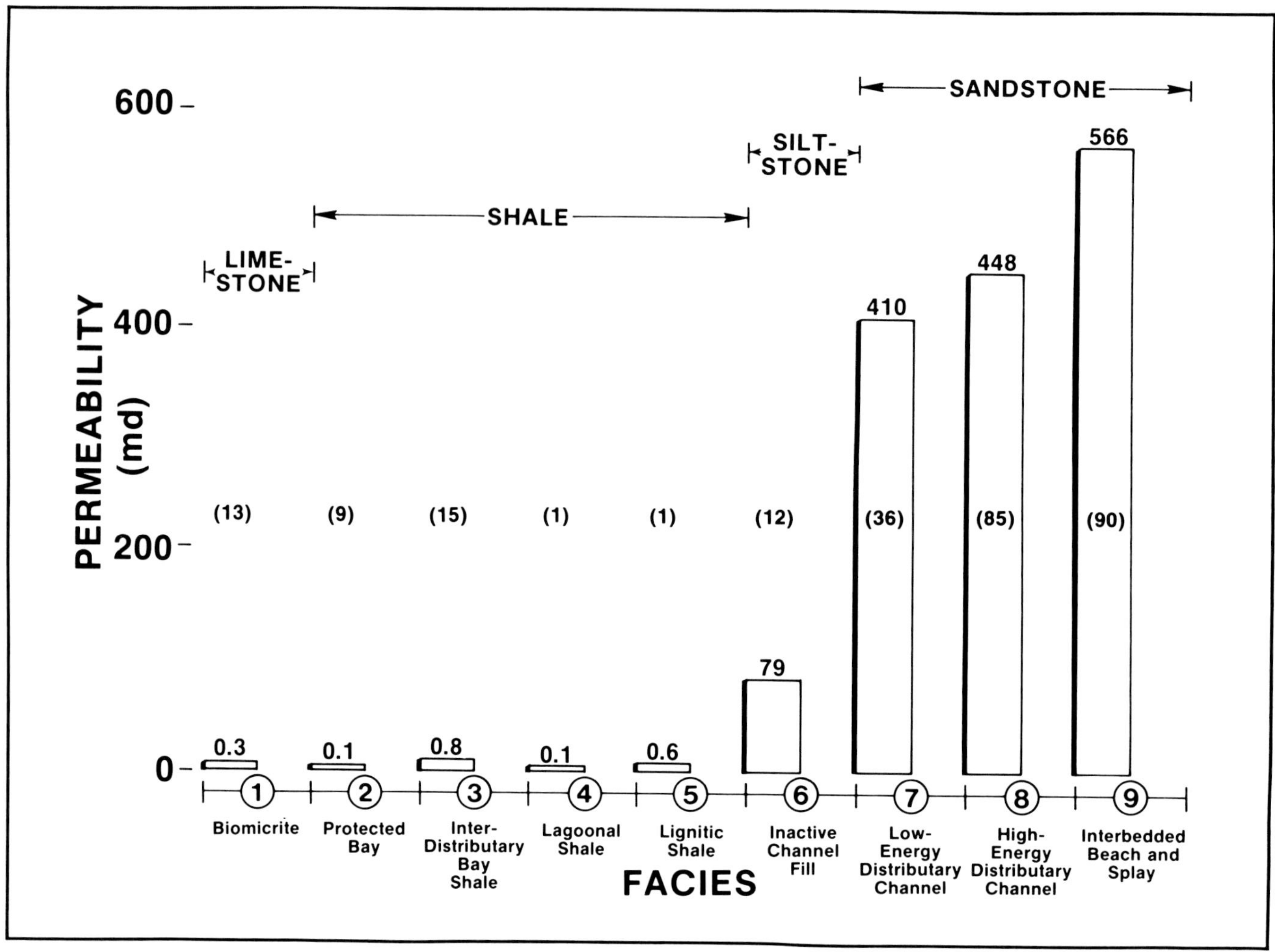

FIG. 33.—Mean permeability of the facies rcognized in cores. Lithology is indicated in the upper portion of the figure. Numbers in parenthesis are numbers of values used. Values for this figure and Figures 34 and 35 are based on data from 14 wells: MP-106, MP-110, MP-116, MP-120, MP-122, MP-128, MP-201, MP-203, MP-205, MP-207, MP-213, MP-221, MP-223 and MP-225. See Figure 2 for locations of these wells.

transmissibilities and capacities and trimmed where there will be predictably low hydrocarbon-flow rates. Consideration should also be given, however, to determining where the oil may have been almost completely flushed in high-flow areas during primary and secondary waterflood operations.

Two types of data were used in the project area to determine direction and magnitude of transmissibilities, namely oriented core data and measured rate of pressure changes (interference testing). Several of the cores were oriented so that true geographic directions could be determined; pairs of plugs were drilled at right angles to each other at 1-ft (30 cm) intervals.

In wells MP-112, MP-126, and MP-211 (Fig. 2) pairs of plugs were taken with east-west (perpendicular to channel flow) and north-south (flow-parallel) orientations. Foot-by-foot plots in all three wells (i.e., well MP-112, Fig. 36) show no real difference in permeability for the two directions. Results from these oriented cores do not suggest a directional preference for transmissibility.

Swift and Brown (1976) made a more detailed study of transmissibility, using five-spot well patterns (Figs. 37 and 38). Using interference testing, they measured the rate of pressure change at four wells surrounding an injection well. Fluid was introduced at the injection well and the rise in fluid level in the adjoining "production" wells was carefully timed and measured. The higher the rate of given fluid rise in the "production" well(s), the better the transmissibility of any given well relative to the others.

Transmissibility measurements made in a *homogeneous* reservoir would have nearly equal values among all the wells; however, the readings between wells in both leases varied considerably (Figs. 37 and 38). In some areas, transmissibilities varied less than in others. In the northern (Chesney) lease, which includes a large proportion of sandstone-filled distributary channels, several areas have similar transmissibilities.

If the transmissibility values for the northern (Chesney) lease are superimposed over the geology (Fig. 39), many of the variations in transmissibility can be explained. On this map, constructed from limited cores available in 1978 (Phase II), locations for distributary channels, interdistrib-

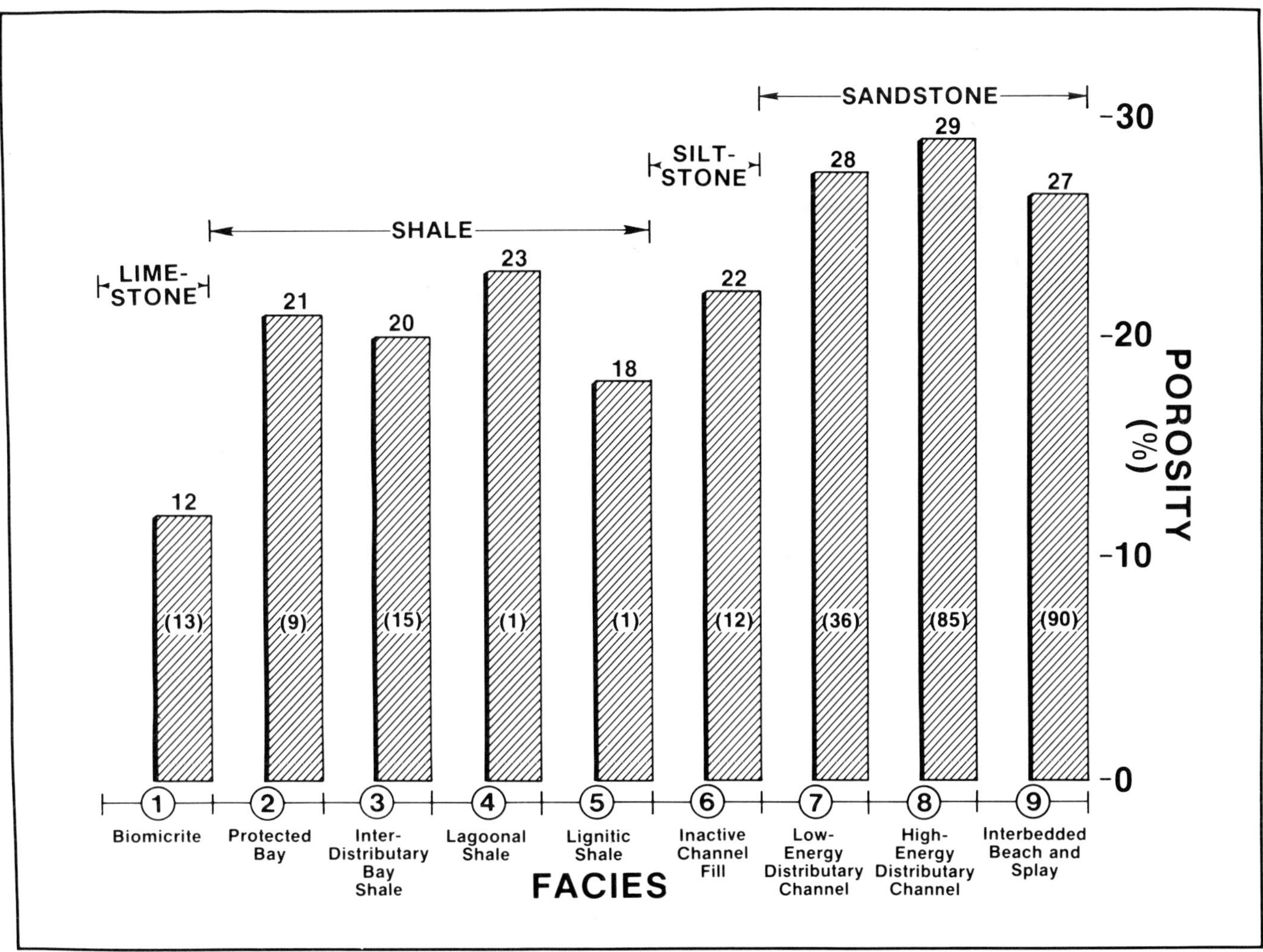

FIG. 34.—Mean porosity of facies recognized in cores. Lithology for facies is indicated at top of figure.

utary bay deposits and splay channels are shown. All of the high-transmissibility values are in areas containing relatively thick distributary and splay channels. Where interdistributary bay mudstone deposits are dominant, there is more shale and most of the low-transmissibility values occur. On the west side of both leases, within the channel, there is some variation, but the north-south component is consistently high. In the area composed predominantly of splay channels, transmissibilities are lower. In the east-central portion of the lease, away from the major channels, wide variations in transmissibility are observed (Fig. 39). In the area east and north of the distributary channels, there is a very strong east-west transmissibility, whereas a very poor transmissibility occurs north-south. At that five-spot location, one cannot inject in the middle and expect to produce equal amounts of fluids at the four surrounding locations.

In the southern (Hegberg) lease there is a wide variation in transmissibility values. When the geology is superimposed on the transmissibilities (Fig. 40), many of the variations may be explained. The distributary channel along the west side trends to the north-northeast and contains most of the high or moderate values. Most of the lowest transmissibility values fall within areas which are predominantly interdistributary bay deposits. Substantially more micellar fluids and oil will be moved along maximum-transmissibility directions, whereas much less fluid will be produced in areas of low transmissibility.

PETROPHYSICS

In order to understand better the producing characteristics of individual reservoir facies in the field, average values for permeability, porosity, and fluid and oil saturation were calculated for each facies using data from 14 cores which are representative of both leases (Figs. 33, 34, and 35).

The rock types associated with the reservoir at El Dorado are sandstone, siltstone, shale, and limestone. Only the sandstones are believed to yield significant hydrocarbons. Permeability values are similar for limestone and shale, averaging 0.4 md (Fig.33). The average permeability of the siltstones is 79 md. The sandstone facies (Fig. 33) can be subdivided on the basis of permeability variations. The *Interbedded Bay, Splay and Beach Sandstone Facies* has sig-

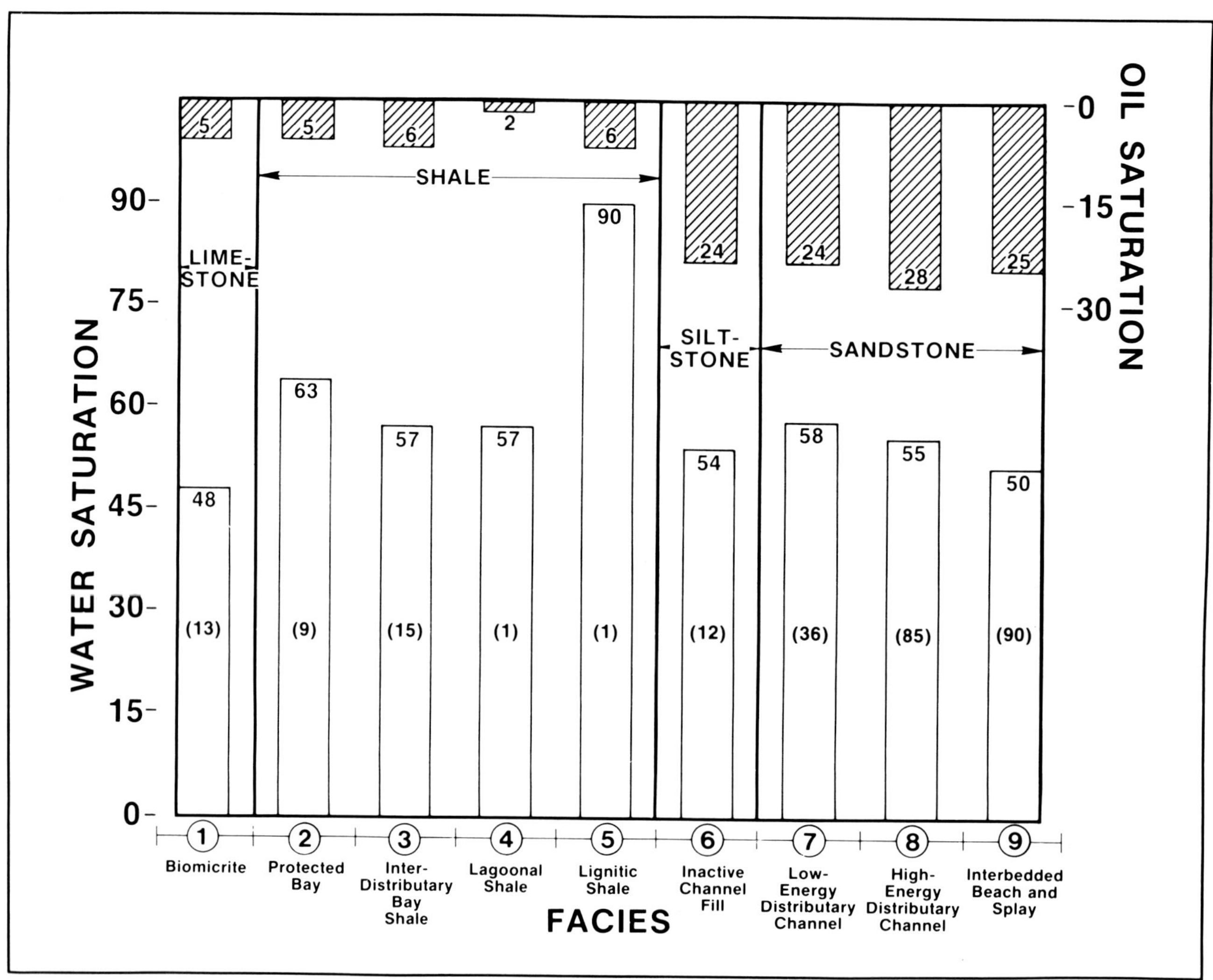

ASSOCIATIONS	FACIES
I	1. Bay Biomicrite
II	2. Bay Shale 3. Interdistributary Bay 4. Lagoon (shale) 5. Lignitic Shale
III	6. Inactive Channel Fill Siltstone
IV	7. Distributary Channel Sandstone (Low Energy) 8. Distributary Channel Sandstone (High Energy)
V	9. Interbedded Channel, Splay, and Beach Sandstone.

Fig. 35.—Mean values of water and oil saturation for each facies. Waterflooding of the reservoirs in the Admire 650′ sandstone has had a strong effect on the present distribution and amount of hydrocarbons in the reservoir units. Note that more oil remains in the *Low-Energy Distributary Channel Subfacies* than in the *High-Energy Subfacies*. This apparent anomaly results from better sweeping of the higher quality reservoir during secondary waterflooding.

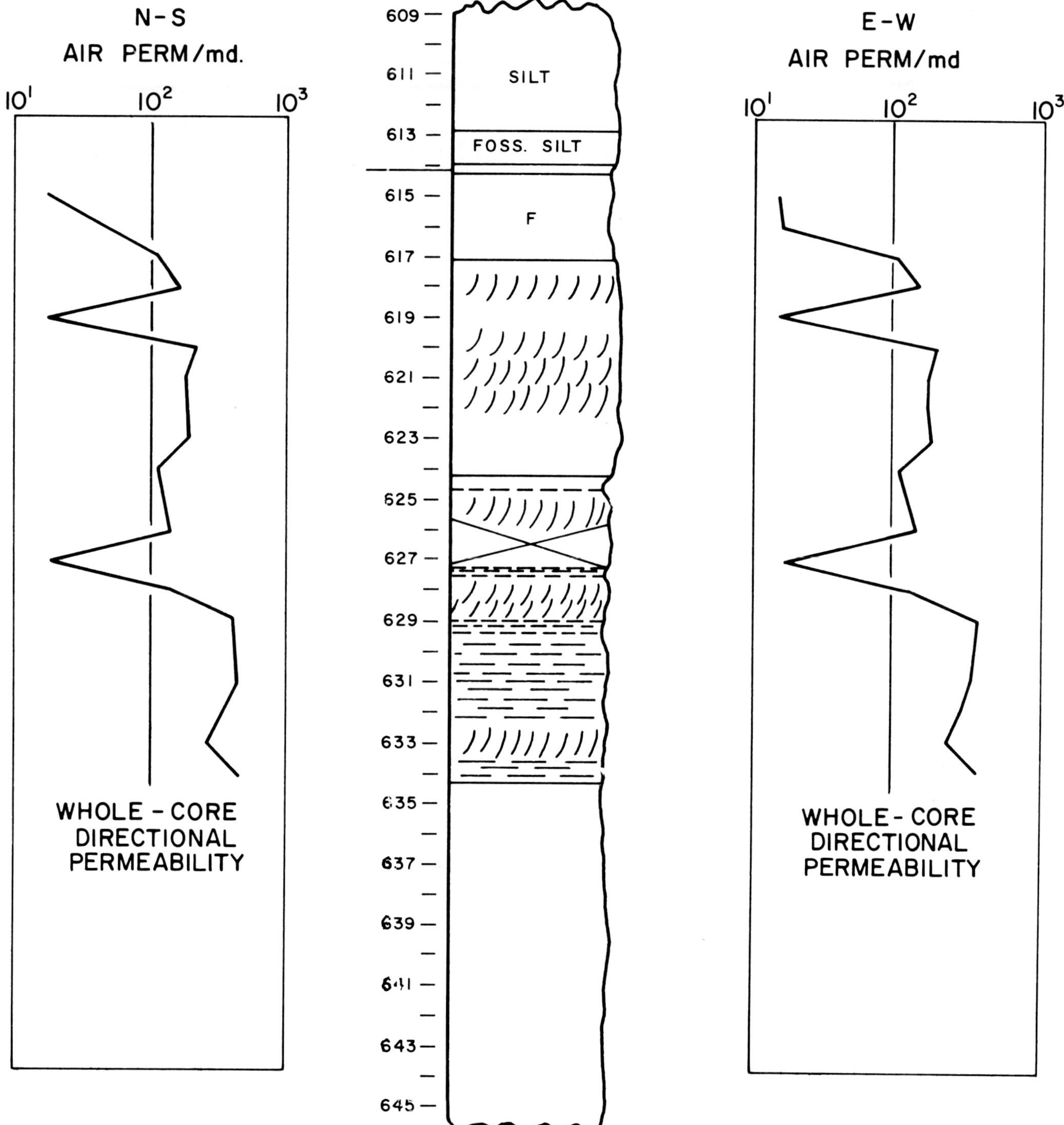

FIG. 36.—Directional permeability plot for oriented core MP-112. North-south and east-west permeabilities are plotted by depth. Permeability distribution from Flesch (pers. commun.).

nificantly higher permeabilities (566 md) than the *High- and Low-Energy Distributary Channel Subfacies* (410,448 md).

The range in porosity among the various lithologies, shale, siltstone, and sandstone, is surprisingly small (18–29%, Fig. 34). The *High-Energy Distributary Channel Subfacies* averages 28% porosity, however, and consistently have higher porosity values than any other of the nine facies types recognized in this study.

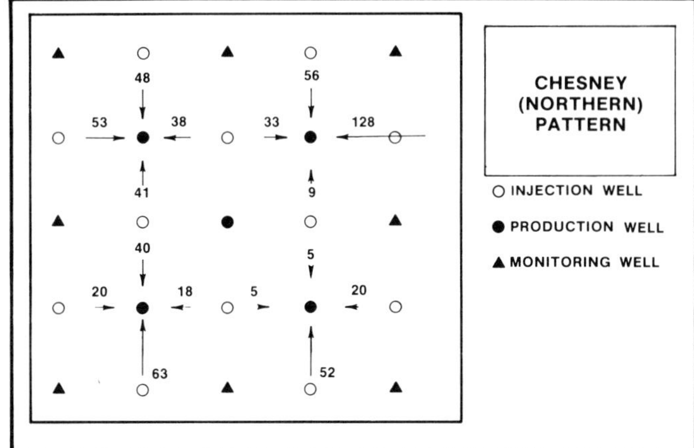

FIG. 37.—"Five-spot" well patterns for northern (Chesney) lease. Numerical transmissibilities between wells are shown. Higher values indicate better potential fluid flow. Strong inhomogeneities are suggested by wide range and erratic distribution of values. (After Swift and Brown, 1976.)

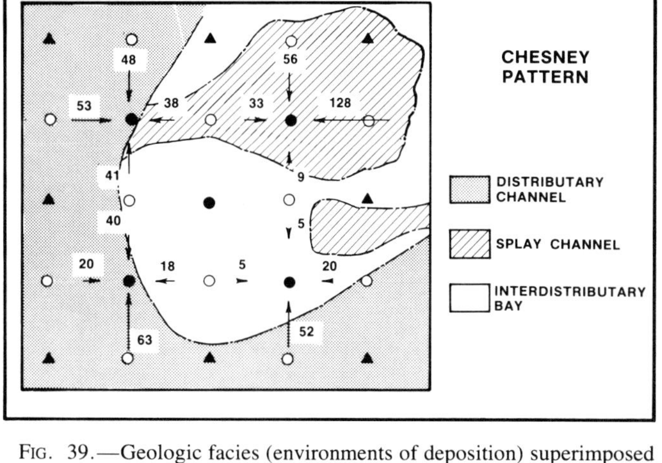

FIG. 39.—Geologic facies (environments of deposition) superimposed over transmissibilities (Fig. 37). Geologic interpretations are based on Phase I and II data only. (Compare with Figs. 25 and 27–29.) High values fall within distributary channel and splay channel facies. Low values are mostly limited to *Interdistributary Bay Facies*.

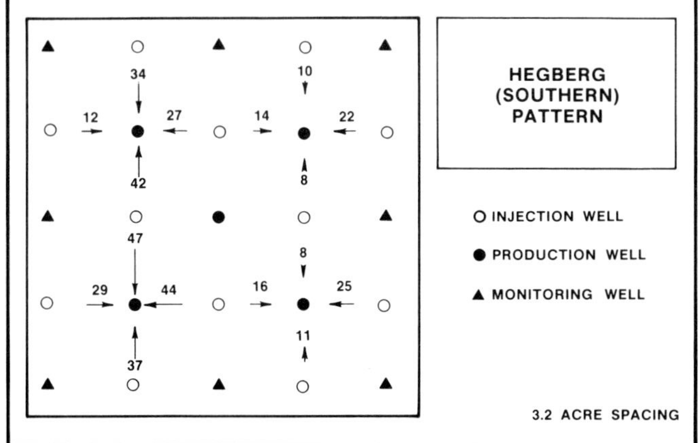

FIG. 38.—"Five-Spot" well patterns for southern (Hegberg) lease. Numerical transmissibilities between wells are shown. Higher values indicate better potential fluid flow. Strong lateral inhomogeneities are suggested by wide range and erratic distribution of values. (After Swift and Brown, 1976.)

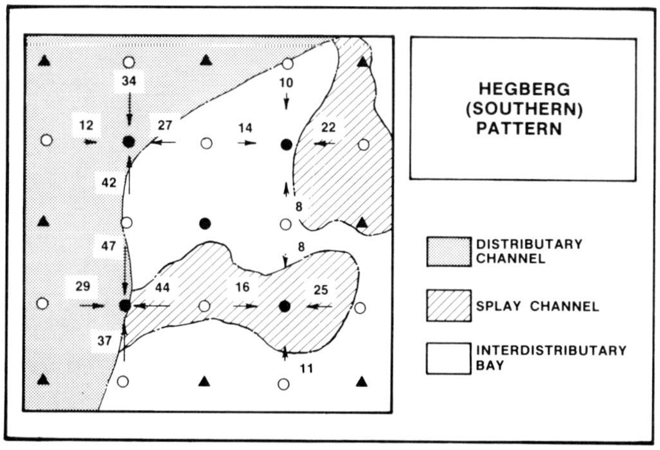

FIG. 40.—Geologic facies (environments of deposition) superimposed over transmissibilities (Fig. 38). Geologic interpretations are based on Phase I and II data only. (Compare with Figs. 25 and 27–29.) Higher values indicate better transmissibility of fluids. Strong lateral inhomogeneities are suggested by wide range and erratic distribution of values. High values fall within *Distributary Channel* and *Splay Channel Facies*. Low values are mostly limited to *Interdistributary Bay Facies*.

When porosity and permeability are cross-plotted, the number of separable reservoir groups is decreased to five (I through V, Fig. 41). The separation of data sets is based primarily on variations in permeability. Only limestones and shales are effectively differentiated on the basis of variations in porosity.

Within the channel sandstones, the slightly lower permeability values typical of the *Low-Energy Channel Subfacies* are the result of deposition of more clayey materials in these facies than in the *High-Energy Channel Subfacies*. Clay occurs in three places: within the pores, as thin laminae or lamina sets and as beds (thicker than 1/2 in., 3 cm). Deposition within the pores affects both vertical and horizontal flow. Deposition of laminae (1/32–1/2 in., 0.08–3 cm thick) or as beds (thicker than 1/2 in., 3 cm) affects vertical flow more than horizontal flow of fluids within the reservoir. Where a series of clayey beds is deposited in succession, significant barriers to both lateral and vertical flow may be observed.

Water saturations vary from 48–63% among the facies, except within lignitic shale which has a 90% water saturation (Fig. 35). Oil saturations are low in the shales and limestones (2–6%) and nearly identical in the siltstone and sandstone facies (24–28%, Fig. 35). These numbers may not be unusual because the cores from which the values were calculated have been partially flushed of oil in the more permeable intervals during secondary waterflood. The effects of the small variation in mean values of oil saturation, among the two channel facies and the interbedded sandstone facies, are difficult to evaluate.

The likelihood of recovering significant additional hydrocarbons in the siltstone (*Inactive Channel Fill*) is un-

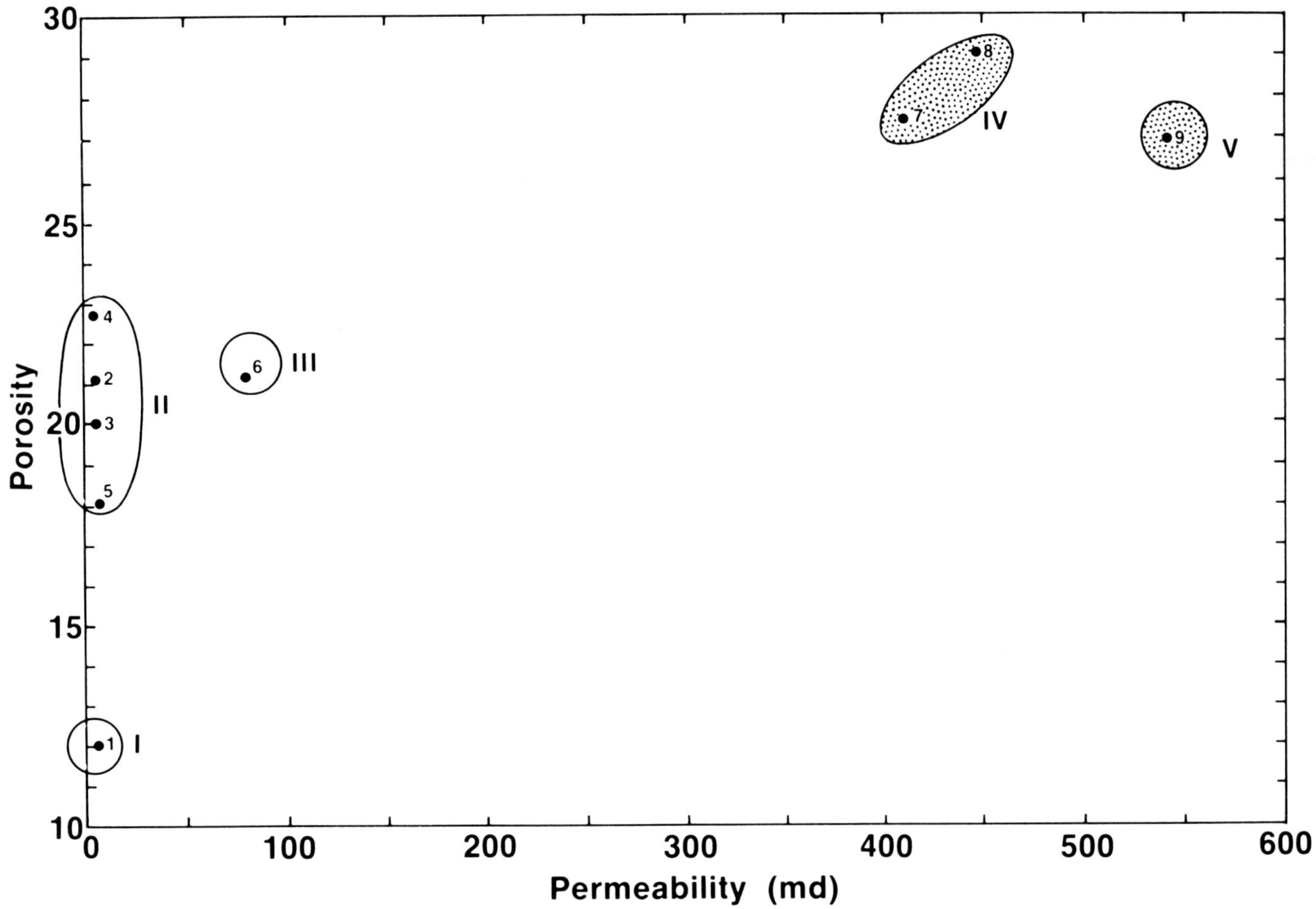

ASSOCIATIONS	FACIES
I	1. Bay Biomicrite
II	2. Bay Shale 3. Interdistributary Bay 4. Lagoon (shale) 5. Lignitic Shale
III	6. Inactive Channel Fill Siltstone
IV	7. Distributary Channel Sandstone (Low Energy) 8. Distributary Channel Sandstone (High Energy)
V	9. Interbedded Channel, Splay, and Beach Sandstone.

FIG. 41.—Associations of facies in the Admire 650′ sandstone based on porosity *vs.* permeability. Individual associations of facies are indicated by Roman numerals. The facies in each association are also listed. Reservoir facies (7, 8) form the major reservoir association (IV). Except for associations I and II, separation among the groups is based primarily on permeability variations.

likely due to low permeability (78 md). *Inactive Channel Fill* is conspicuous at the top of the reservoir unit in well MP-213 (Figs. 30, 32) and has created a number of producing problems, as explained elsewhere.

Water saturation (Fig. 35) varies inversely with the permeability of the reservoir facies (Fig. 33). The effect on water saturation by secondary waterflooding is unknown.

Based on the geologic model developed for the project area, continuity among three or more wells which contain sandstone reservoir facies should be greatest in the *High-Energy Channel Subfacies*, moderate in the *Low-Energy Channel Subfacies*, and moderate to poor in the *Interbedded Sandstone Facies*. These communication estimates may be improved or varied due to the close spacing of wells in the pilot project. For moderately wide spacing (40 acres, 16 ha), these estimates should be accurate.

PETROLOGY AND DIAGENESIS

Introduction

Information on mineral content and diagenesis of the Admire 650′ sandstone reservoir was obtained from analysis of 39 thin sections. Scanning electron microscopy was utilized to further determine diagenetic fabrics. Clay minerals were identified by X-ray powder diffraction analyses (W. Almon, pers. commun.).

The average distributary channel sandstone in the Admire 650′ is a sublitharenite (as defined by Folk, 1974) and contains 80% quartz, 15% rock fragments and 5% feldspar (Table 1). The sandstones are fine-grained (mean size 0.14 mm), well sorted, and subangular to subrounded (Fig. 42). Most grains are in point or long contact, suggesting that only low to moderate compaction resulted and that the Admire 650′ sandstone may never have been buried very deeply.

Granitic and low-grade metamorphic rock fragments were probably derived from the Nemaha Ridge, a structural element near El Dorado field. The fragments include quartz-mica schist and granite. Ductile fragments consist of micaceous schist and sedimentary clay clasts. Plagioclase and orthoclase grains each compromise approximately 5% of the *Distributary Channel Sandstone Facies*. Muscovite and biotite mica grains typically form laminae associated with organic matter.

TABLE 1.—AVERAGE COMPOSITIONS, POROSITIES AND PERMEABILITIES OF SANDSTONES IN ADMIRE 650′ SANDSTONE RESERVOIR

Facies	Inactive Channel Fill	Distributary Channel Sandstone	Interbedded Bay, Splay, and Beach Sandstones
Quartz (monocrystalline)	54.7	59.0	57.3
Nonductile quartz-mica schist	2.7	2.2	4.3
Nonductile quartz-mica granite	7.9	17.1	19.9
Plagioclase feldspar	2.2	2.8	1.1
Potassium feldspar	6.6	2.8	1.1
Quartz-feldspar granite	—	0.2	1.0
Ductile mica schist	9.4	6.5	5.6
Clay clasts	5.1	0.7	1.0
Micas	11.4	7.0	5.8
Quartz/feldspar/lithic fragments	67/9/24	80/5/15	83/5/12
Clay matrix	14.9	2.4	1.6
Calcite and Siderite	0.4	3.9	5.5
Quartz overgrowths	0.3	0.2	1.4
Inter- and intragranular porosity *	17.3	23.0	22.3
Measured porosity +	24.5	28.0	26.3
Measured permeability (core)	139 md	489 md	628 md

+Note: Compositions from thin section analysis. Porosity and permeability from whole-core analysis.

Diagenesis

Diagenesis plays an important part in controlling the permeability and porosity of the Admire 650′ reservoir. Porosity and permeability in the Admire 650′ sandstone are reduced by (1) clay laminations, (2) clay minerals, (3) deformation of ductile rock fragments, (4) calcite cement, (5) mica and micaceous laminae, and (6) quartz overgrowths. Secondary porosity is enhanced by dissolution of plagioclase feldspar and dissolution of calcium carbonate cement.

Permeability is controlled largely by the diagenetic overprint of pore-filling cements and clays. A fairly good correlation (r = 0.78) exists between the types and amounts of porosity reducers ("pore volume") and permeability (Fig. 43). An example of diagenetically reduced permeability and porosity is shown in Figure 44.

Six types of materials which degrade porosity and permeability are recognized in the Admire 650′ sandstone. Three of these result from the primary distribution of shale, rock fragments and mica. The others, clay matrix, calcite cement and quartz cement, formed during diagenesis.

FIG. 42.—Sandstone exhibiting good porosity and permeability. Minor amounts of mica (M) and ductile rock fragment deformation (upper right). ϕ = 32.1%, K = 1150 md. SEM photomicrograph. Admire 650′ sandstone. *Distributary Channel Sandstone Facies*. MP-228, 672.0 ft (204.8 m).

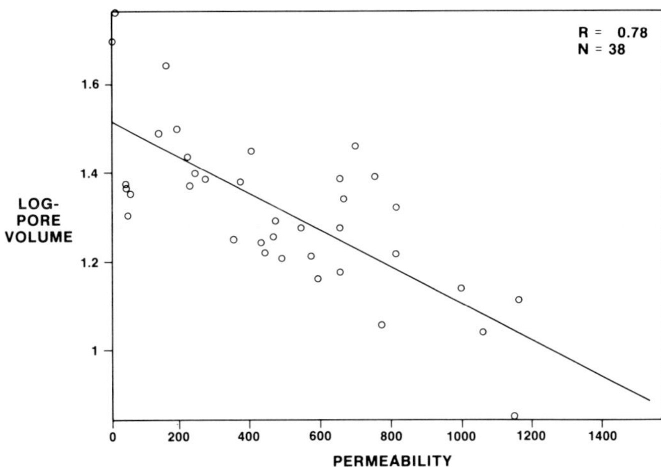

FIG. 43.—Plot of "log-pore volume" fill *versus* permeability. "Log-pore volume" is the amount of pore space filled by clays, ductile rock fragments, carbonate cement, micas, and quartz overgrowths. A fairly good correlation exists between the amount of these porosity reducers and the permeability in the reservoir sandstones.

Clay laminations.—

Deposition of clayey material as laminae and beds seriously affects porosity and permeability. Interconnection of pores is limited vertically due to clay laminae. Bedded claystone within the channel sandstone reservoir interval only locally forms barriers to overall horizontal and vertical flow.

Ductile rock fragments.—

Deformation of ductile rock fragments has a role in reducing permeability and porosity in the Admire 650′ sandstone. Deformation of rock fragments occurred during early burial. Locally, rock fragments containing appreciable amounts of mica (which act as zones of weakness) break up and partially occlude pore space. In most cases, malleable clay clasts, mica and schist rock fragments inhibit porosity and permeability by filling pore space between stable framework grains.

Mica and micaceous laminae.—

In addition to weakening the structure, fragments of biotite and muscovite reduce porosity and permeability by act-

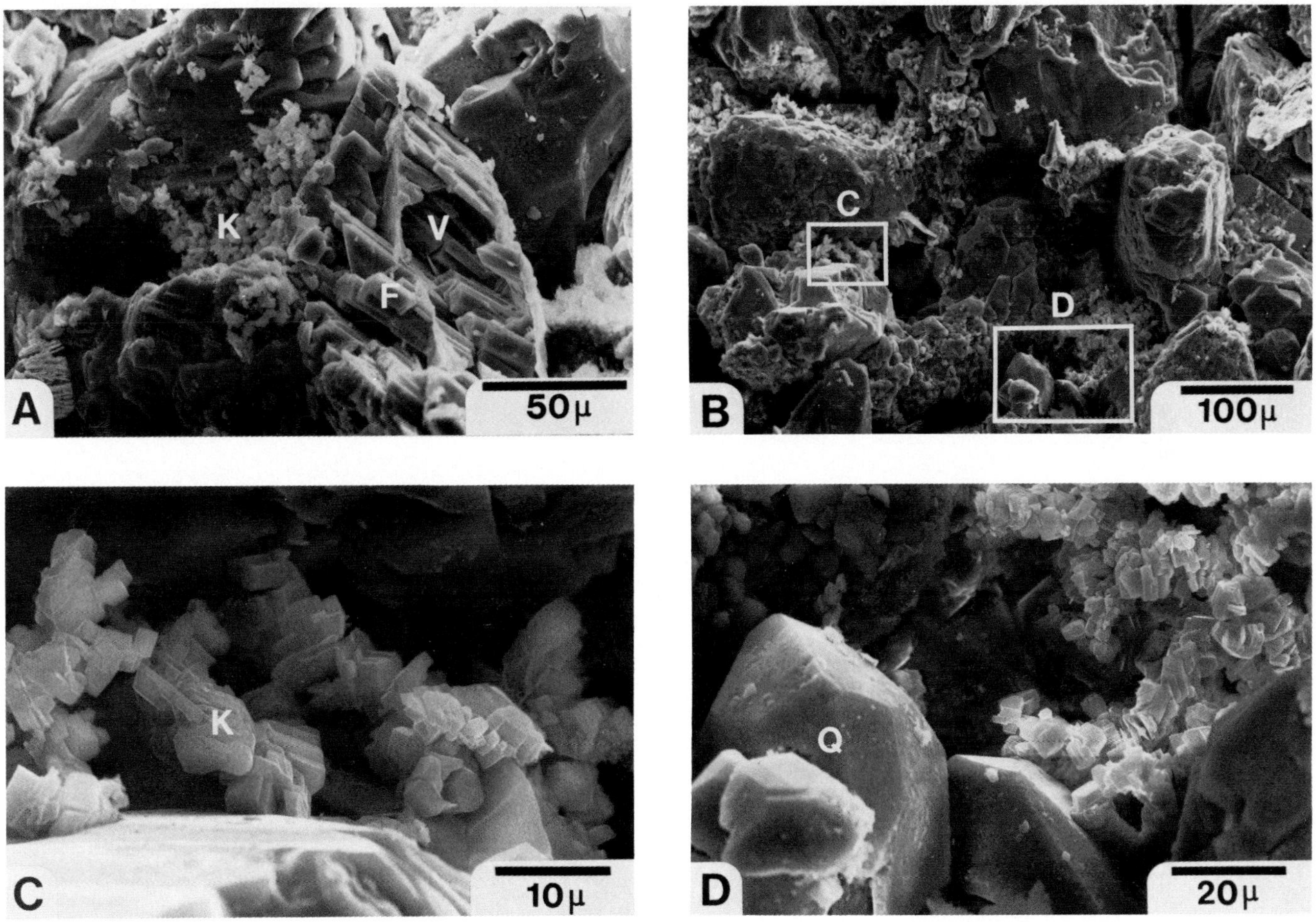

FIG. 44.—SEM photomicrographs of authigenic minerals. (A) Secondary porosity resulting from dissolution of feldspar. Large dissolution void (v) occurs in plagioclase feldspar (F) that has undergone partial dissolution along cleavage planes (conduits for fluids?). Kaolinite booklets (K) are commonly found associated with leached feldspars. *Distributary Channel Sandstone Facies*. MP-110, 613.0 ft (186.8 m). (B) Authigenic kaolinite partially fills intergranular pore space (outline in area C) and authigenic quartz (outlined in area D). K = 430 md. (C) Enlarged view of area outlined by rectangle C in (B). Booklets of kaolinite (K) contribute to reduction of intergranular porosity. (D) Enlarged view of area outlined by rectangle D in (B). Euhedral quartz overgrowths (Q) formed prior to formation of kaolinite. *Distributary Channel Sandstone Facies*. MP-110, 613.0 ft (186.8 m).

ing as barriers (bridges) across pore throats (Figure 45). This phenomenon may account for two samples having similar porosities but very dissimilar permeabilities (e.g., pore space is not filled, but fluid flow is blocked by mica flakes lying across possible fluid conduits). Vertical permeability is locally reduced by mica and organic-rich laminations, which serve to define sedimentary structures, including trough crossbedding, ripple-bedding, and parallel laminations.

Clay minerals.—

The dominant clay mineral in the Admire sandstone is kaolinite. Authigenic, well crystallized (hexagonal), loosely attached vermiform kaolinite booklets bridge pore throats (Fig. 45B) and clearly postdate quartz overgrowth formation. Kaolinite is intimately associated with dissolution of feldspar. Keller (1977) was among the first to document the leaching of feldspar followed by pervasive replacement and crystallization of kaolinite. Whitaker (1978) reported formation of authigenic kaolinite from feldspar in the Brent Sand in the North sea.

Kaolinite comprises 6% of the total rock in one sample, MP-228, 666.7 ft (203 m), although it typically makes up less than 2% of the total rock in those samples which were analyzed. Kaolinite can be dispersed by turbulence that occurs at high production or injection rates and by changes in formation fluid properties. Once dispersed, the clay migrates and can lodge in pore throats, impairing mobility.

Illite is identified by X-ray diffraction (1–4% of the total rock, averaging 1.2%), but it was rarely observed in thin section and scanning electron microscopy. Perhaps the presence of this clay reflects occurrence in argillaceous and micaceous schist rock fragments rather than as a pore-filling authigenic clay. Chlorite (an acid-sensitive, high-capillarity clay mineral) and smectite (a swelling clay) are minor constituents and should not inhibit permeability.

Calcite cement.—

Calcite cement occludes some pore space and may be restricted to upper parts of the *Distributary Channel Sandstone Facies*. Possible sources for calcium may have been through albitization of plagioclase feldspar, the transformation of smectite to illite, calcite dissolution from shales, micrite pressure solution and, or more likely, dissolution of calcite fossils.

More than one stage of calcium carbonate cementation in the Admire 650′ sandstone is suggested by thin section analyses. In some samples, calcite cementation postdates formation of quartz overgrowths because calcite cement embays quartz grains.

In some wells (MP-203, MP-205, MP-223) siderite (iron carbonate) is present in the reservoir sandstones. Siderite is a problem because an iron hydroxide gel may form where acidization (HCl) of the reservoir may be required. The siderite zone is most prevalent near the top of the *Distributary Channel Sandstone Facies* and is typically less than 1 ft (0.3 m) thick. This zone is probably not laterally continuous and should not be a major permeability barrier.

Quartz overgrowths.—

Of minor importance in reducing permeability and porosity is the presence of authigenic quartz forming euhedral overgrowths on some quartz grains. Quartz overgrowths rarely occlude the entire pore space (Fig. 46). In thin sections, quartz overgrowths account for the angularity of quartz grains when, in fact, the host grain is well rounded. Quartz growths predate authigenic kaolinite formation (Fig. 46) and calcite cementation. The source of the silica for quartz overgrowths may have been from the dissolution and reprecipitation of quartz at grain-to-grain contacts early in the diagenetic history of the sandstone.

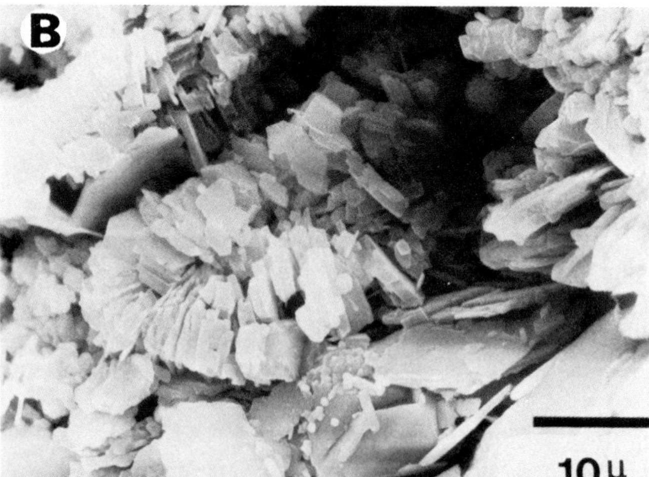

FIG. 45.—SEM photomicrographs showing porosity reducers in the Admire 650′ sandstone. (A) Mechanical deformation (bending) and pore bridging of mica or rock fragments (M). Note also light clay(?) coating on quartz grain (Q). Admire 650′ sandstone. *Distributary Channel Sandstone Facies*. MP-228, 672.0 ft (204.8 m). (B) Booklets of vermiform kaolinite from rectangular area of (A). Well crystallized clay indicates authigenic origin. *Distributary Channel Sandstone Facies*. MP-228, 672.0 ft (204.8 m).

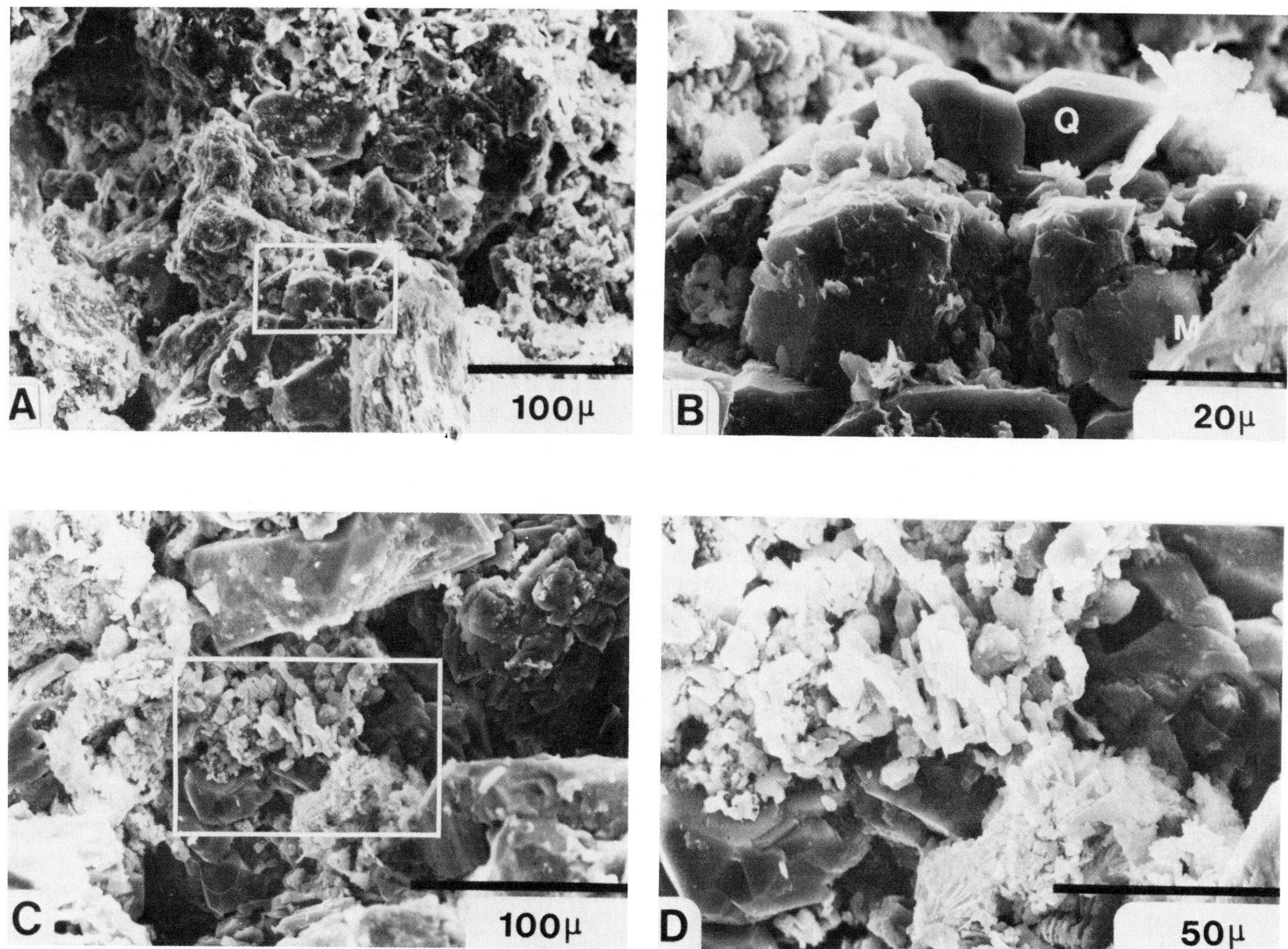

FIG. 46.—SEM photomicrographs of quartz overgrowths and dissolution of plagioclase. (A) Pore space partially filled with quartz overgrowths and clay. Sharp euhedral crystal faces are characteristic of quartz overgrowths (rectangular outline). Coatings on quartz grains (clay minerals and mica flakes) reduce permeability and porosity. *Distributary Channel Sandstone Facies*. MP-205, 646.0 ft (196.9 m). (B) Enlarged view of rectangle outlined in (A) showing authigenic quartz overgrowths (Q) and mica flakes (M). Authigenic clay (possibly chlorite) appears to predate authigenic quartz. *Distributary Channel Sandstone Facies*. MP-205, 646.0 ft (196.9 m). (C) Pore space containing secondary (intragranular porosity resulting from partial dissolution of plagioclase feldspar. Quartz overgrowths (in lower left portion of rectangle) and clay minerals (kaolinite, in lower right) partially destroy intergranular porosity. *Distributary Channel Sandstone Facies*. MP-228, 671.9 ft (204.8 m). (D) Enlarged view of rectangle outlined in (C) showing dissolution of plagioclase, upper part of photograph quartz overgrowths (lower left), and kaolinite (lower right). *Distributary Channel Sandstone Facies*. MP-228, 671.9 ft (204.8 m)

Secondary Porosity

Secondary porosity in the Admire 650′ sandstone is formed by dissolution of plagioclase feldspar and dissolution of calcium carbonate cement and is of minor importance. Most of the Admire 650′ sandstone production is from primary intergranular porosity.

Leaching of plagioclase feldspar.—

Secondary intragranular porosity occurs in many plagioclase grains through partial ("honeycomb texture") or complete dissolution (Figs. 46C, D). Apparently, cleavage planes acted as conduits for dissolving interstitial fluids. The release of ions from the feldspar probably contributed to the formation of authigenic kaolinite.

Calcite cementation typically predates dissolution of feldspar, but locally postdates the dissolution within feldspar cleavage planes. Calcium-rich interstitial water may have leached feldspar and precipitated calcium carbonate cement.

Leaching of calcium carbonate cement.—

Calcium carbonate has locally undergone partial dissolution and created the "intra-cement" porosity described by Schmidt and McDonald (1979b). The majority of calcite cement has not undergone dissolution, however.

Partial leaching of the calcite cement is contemporaneous and postdates feldspar dissolution. This release of calcium ions may have contributed to the precipitation of calcite cement at other sites and, together with silica ions released during feldspar dissolution, precipitation of kaolinite.

Paragenetic Sequence

Schmidt and McDonald (1979a) presented four stages of burial diagenesis and porosity development. With progressive burial they recognized a (an) (1) immature stage (mechanical compaction of clean uncompacted sands), (2) semimature stage (chemical compaction or dissolution of sand grains at points of contact, and reprecipitation as cements), (3) mature stage (absence of primary porosity and formation of secondary porosity—substage A, and destruction of secondary porosity by chemical compaction—substage B), and a (4) supermature stage (little or no primary or secondary porosity).

The paragenetic sequence for the Admire 650′ sandstone (Fig. 47) can, in part, be related to the stages listed above. A discussion of the paragenetic sequence in the Admire 650′ sandstone follows.

Burial and compaction allowed mica and micaceous laminae to restrict permeability by blocking interconnected pore throats. During the immature stage of diagenesis, primary porosity is partially reduced by compaction of mica, schist and argillaceous rock fragments. Chemical reduction of primary porosity during the semimature stage followed, with partial dissolution of quartz along grain contacts and reprecipitation of the silica quartz overgrowths. Calcite cement may have formed through dissolution of calcareous fossils and later reprecipitation.

At this stage, organic matter and surrounding shale presumably underwent thermomaturation. Carbon dioxide from released carbonic acid may have been responsible for the dissolution of plagioclase feldspar. Both of these effects probably created secondary porosity.

During a later, semimature stage, kaolinite and other clays precipitated from ions released during dissolution of feldspar. Calcium carbonate cement was precipitated in some pore spaces. Liquid hydrocarbons, generated in intercalated shales, may have migrated into the Admire 650′ reservoir at this stage, inhibiting diagenesis in the sandstones.

Burial diagenesis probably did not extend into the mature stage; most primary porosity is intact. It is concluded that the maximum depth of burial for the Admire 650′ sandstone probably did not exceed 6,000 ft (1.8 km), based on interpretations of similar relationships presented by Schmidt and McDonald (1979a).

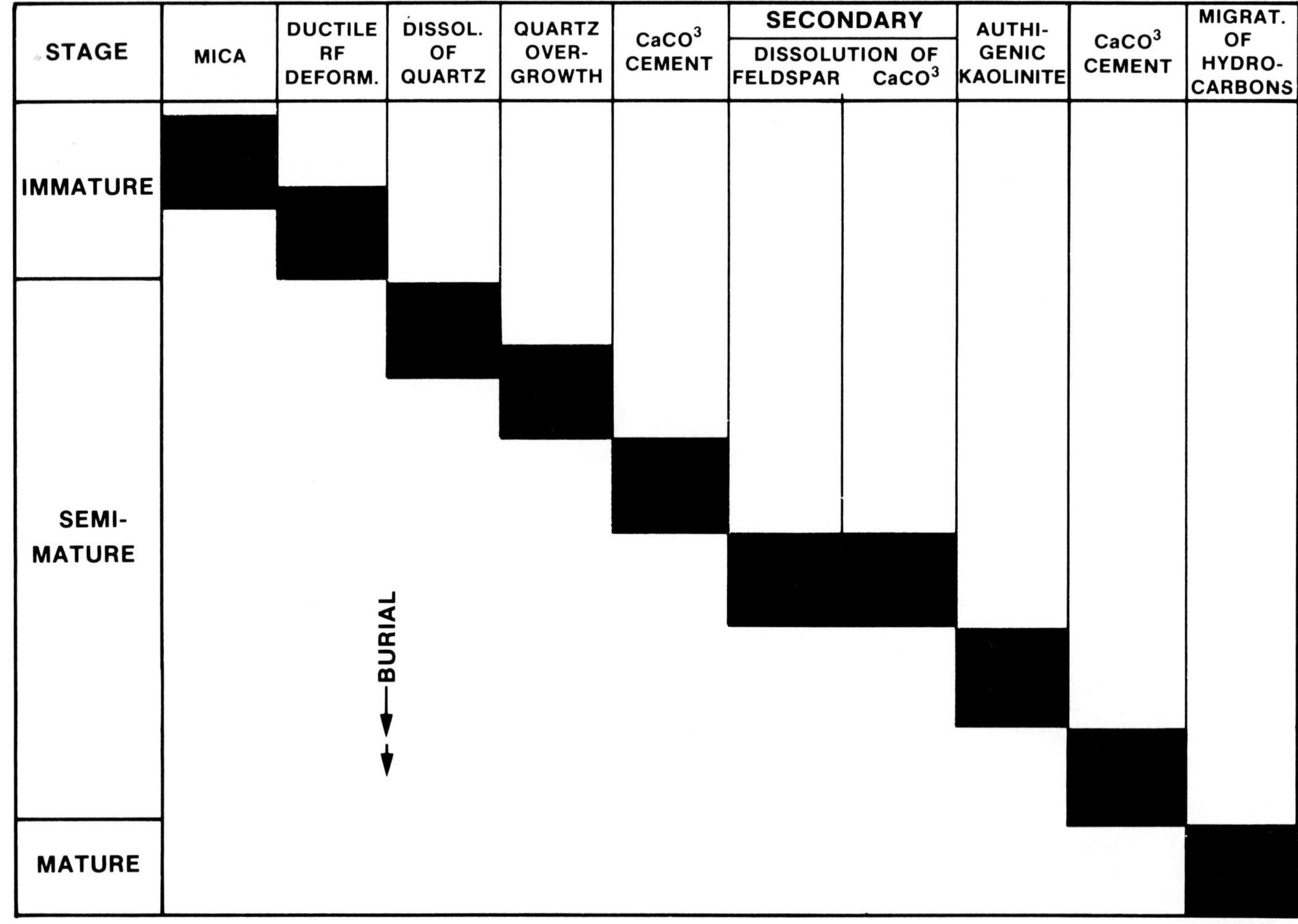

FIG. 47.—Paragenesis of the Admire 650′ sandstone, El Dorado field. Mica and micaceous laminae initially restricted permeability. Ductile rock fragment deformation further restricted porosity and permeability. Quartz was later dissolved and reprecipitated as overgrowths. Calcite cement was precipitated, then dissolved, along with plagioclase feldspar. Authigenic kaolinite then formed. Finally, a secondary stage of calcite cementation was followed by hydrocarbon emplacement.

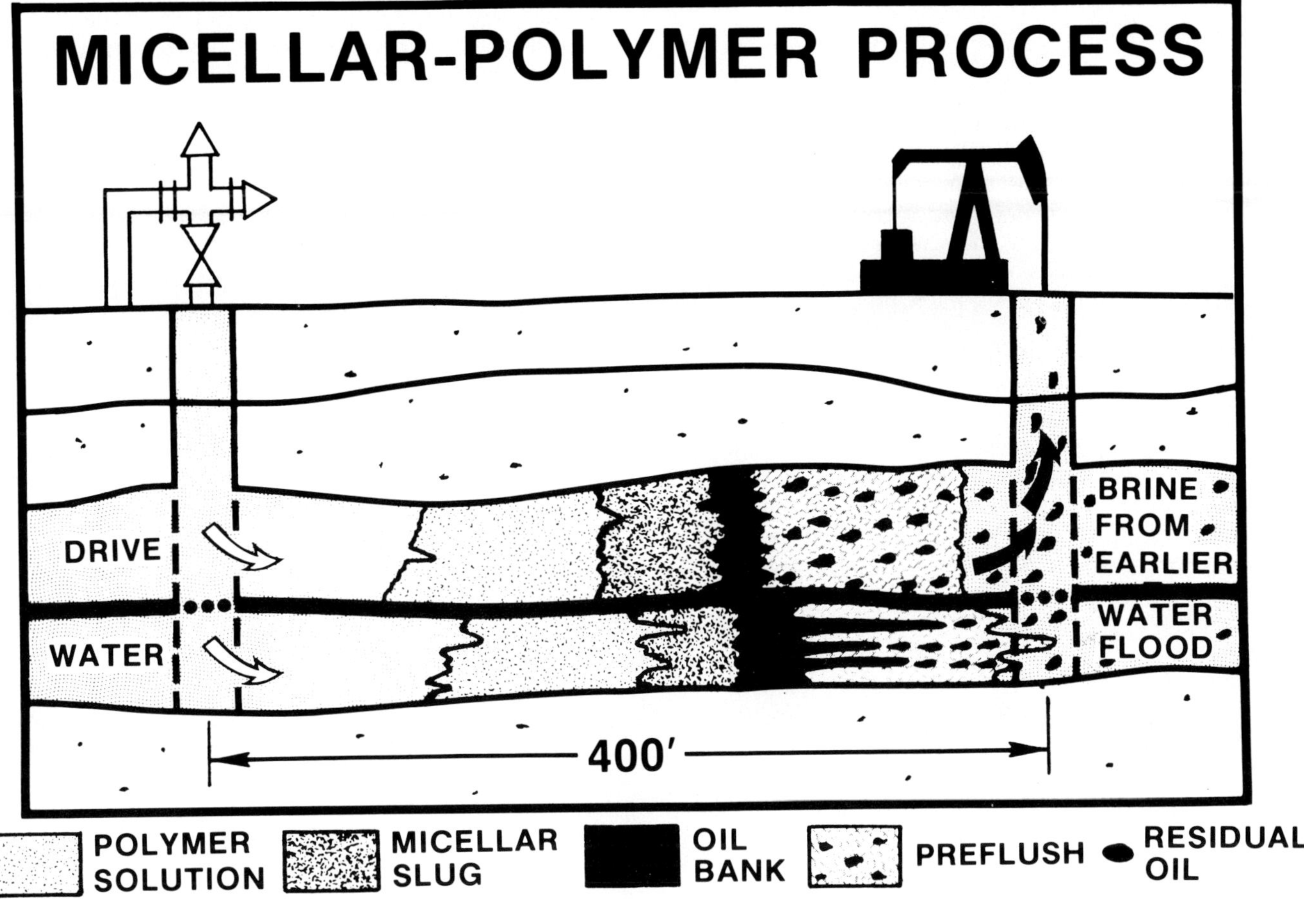

FIG. 48.—Diagram showing idealized fluid movement between micellar-polymer injection and production wells. The fluids move at different rates in the two layers separated by the heavy black line.

FLOOD OPERATION

During the tertiary micellar-polymer waterflood, five-spot patterns were used for injection and production in both leases. Four producing wells, at the centers of the five-spots, are located in each lease. Each producer is expected to produce fluids from all of the four surrounding injectors.

Tertiary recovery injection of fluids was begun in both the Chesney and Hegberg leases on November 18, 1975 (R. J. Miller, pers. commun.). Pre-flush fluids were injected at that time (Fig. 48). Injection of micellar fluids was begun in the southern (Hegberg) pattern on March 23, 1977, and in the northern (Chesney) pattern on November 16, 1977. The first oil to be produced from the bank, ahead of the tertiary flood, was produced in late 1978.

SUMMARY AND CONCLUSIONS

The Admire 650′ sandstone in the El Dorado field is a heterogeneous deltaic deposit in which distributary channels form the main sandstone reservoirs. Crevasse splays also make a minor contribution to the production. Nine depositional facies are recognized in cores and on logs. They generally form an idealized vertical (stratigraphic) sequence from bottom to top, which includes: *Interdistributary Bay Silty Shale; Interbedded Interdistributary Bay, Splay Channel, and Thin Beach Sandstone; Distributary Channel Sandstone (High- and Low-Energy Subfacies); Inactive Channel Fill; Lignitic Shale; Protected Bay Biomicrite; Lagoonal Shale; Protected Bay Claystone and Shale;* and *Open Marine Limestone*.

Permeable sandstones in the *Distributary-Channel Sandstone Facies* are distributed over most of the project area but are absent in the southeast quarter of the more southerly (Hegberg) lease. The thickest sandstones in the pilot area are present along the western margin and in the east-central portion and are coincident with the thickest accumulation of *High-Energy Distributary Channel Sandstone Subfacies*. Splay deposits, which occur primarily in the *Interbedded Facies*, have high permeabilities but lack continuity. Reservoir inhomogeneities, depicted on cross sections, fence diagrams, and isopach maps, show considerable lateral facies variability.

Primary porosity and permeability of Admire 650′ sandstones are reduced by authigenic clays, ductile rock fragments, calcium carbonate cement, mica, and quartz overgrowths. Minor secondary porosity resulted from leaching of plagioclase feldspar and calcium carbonate cement.

A wide variety of types of geological data was available in this enhanced oil recovery project; such data should be fully utilized to yield an adequate reservoir description prior to implementation of chemical or other types of secondary or tertiary recovery techniques.

ACKNOWLEDGMENTS

Initial geologic studies on the El Dorado field were initiated and, for the most part, supervised in later stages by the senior author. Gary Flesch carried out a geologic analysis of several cores in 1974 and 1975. Barbara Bascle completed a M.S. thesis, at the University of South Carolina, on the field in 1976. Oriented core analysis was carried out with the help of Gary Flesch, James Rennison and the junior author. Discussions and editorial assistance were provided by a number of Cities Service geologists and engineers, especially James Miller and James Ebanks. Drafting and other support services were supported by Cities Service. Some geologic analysis was carried out in the later stages of this study utilizing funds from ERDA Contract No. E(34-1)-0003. The final draft of the paper was reviewed by James Ebanks, Roger Slatt, Mario Caputo, Greg Hinterlong and James Beckett; however, any errors are the responsibility of the authors. Typing of the manuscript was by Janice Brewer and Maggie Draughon.

REFERENCES

BASCLE, B., 1976, Geological study of the Admire 650′ sandstone, micellar-polymer flood project, El Dorado field, Butler County, Kansas: Unpublished M.S. Thesis, University of South Carolina, 187 p.

BLEAKLY, D. C., VAN ALSTINE, D. R., AND PACKER, D. R., 1985a, Core orientation I, controlling errors minimizes risk and cost in core orientation: Oil and Gas Journal, December 2, p. 103–109.

———,———, AND ———, 1985b, How to evaluate orientation data, quality control, core orientation: conclusion: Oil and Gas Journal, December 9, p. 46–54.

COLEMAN, J. M., 1976, Deltas: processes of deposition, models for exploration: Continuing Education Publishing Company, Inc., Champaign, Illinois, 102 p.

———, AND GAGLIANO, S. M., 1965, Sedimentary structures: Mississippi River deltaic plain, *in* Middleton, G. V., ed., Primary Sedimentary Structures and their Hydrodynamic Interpretation: Society of Economic Paleontologists and Mineralogists Special Publication 12, p. 133–148.

FATH, A. E. 1921, Geology of El Dorado oil and gas field, Butler County, Kansas: State Geological Survey of Kansas Bulletin 7, 102 p.

FOLK, R. L., 1974, Petrology of Sedimentary Rocks: Hemphill Publishing Company, Austin, Texas, 182 p.

JEWETT, J. M., AND ABERNATHY, G. E., 1945, Oil and gas in eastern Kansas: State Geological Survey of Kansas Bulletin 57, p. 69–77.

KELLER, W. D., 1977, Scanning electron micrographs of kaolins collected from diverse environments of origin—IV. Georgia kaolin and kaolinizing source rocks: Clays and Clay Minerals, v. 25, p. 311–345.

KENTWORTH, L. P., 1977, Oil Hill, the town that Cities Service built: Towanda Publishing Company, Towanda, Kansas, 116 p.

MILLER, R. J., AND ROSENWALD, G. W., eds., 1980, El Dorado Micellar-Polymer Demonstration Project, Fifth Annual Report, September 1978-August 1979: U.S. Department of Energy, DOE/ET/13070-53, 204 p.

REEVES, J. R., 1929, El Dorado field, Butler County, Kansas, *in* Structure of Typical American Oil Fields, v. 2: A Symposium of the American Association of Petroleum Geologists, p. 160–167.

REINECK, H. E., AND SINGH, I. B., 1975, Depositional sedimentary environments: Springer-Verlag, New York, 439 p.

ROSENWALD, G. W., ed., 1978, El Dorado Micellar-Polymer Demonstration Project, Fourth Annual Report, September 1977-August 1978: U.S. Department of Energy, DOE/ET-78-C-03-1800, 234 p.

———, ed., 1979, El Dorado Micellar Polymer Demonstration Project, Fifth Annual Report, September 1978–August 1979: U.S. Department of Energy, DOE/ET/13070-53, 227 p.

———, MILLER, R. J., AND VAIROGS, J., eds., 1976, El Dorado Micellar-Polymer Demonstration Project, Second Annual Report, July 1975–May 1976: U.S. Department of Energy, DOE/E(34-1)-0003, 155 p.

———,———, AND ———, eds., 1978, El Dorado Micellar-Polymer Demonstration Project, Third Annual Report, June 1976–August 1977: U.S. Department of Energy, DOE/ET/786-C-02-4100, 191 p.

SCHMIDT, V., AND MCDONALD, D. A., 1979a, The role of secondary porosity in the course of sandstone diagenesis, *in* Scholle, P. A., and Schluger, P. R., eds., Aspects of Diagenesis: Society of Economic Paleontologists and Mineralogists Special Publication 26, p. 175–207.

———, AND ———, 1979b, Texture and recognition of secondary porosity in sandstones, *in* Scholle, P. A., and Schluger, P. R., eds., Aspects of Diagenesis: Society of Economic Palentologists and Mineralogists Special Publication 26, p. 209–225.

SNEIDER, R. M., TINKER, C. N., AND MECKEL, L. D., 1978, Deltaic environment reservoir types and their characteristics: Journal of Petroleum Technology, v. 20, p. 1538–1546.

SWIFT, S. C., AND BROWN, L. P., 1976, Interference testing for reservoir definition, the state of the art: Society of Petroleum Engineers, SPE Paper 5809, Improved Oil Recovery Symposium of SPE of AIME, Tulsa, Oklahoma, p. 165–175.

VAN HORN, L. E., ed., 1981, El Dorado Micellar-Polymer Demonstration Project, Sixth Annual Report, September 1979–August 1980: U.S. Department of Energy, DOE/ET/13070-63, 375 p.

———, 1982, El Dorado Micellar-Polymer Demonstration Project, Seventh Annual Report, September 1980–August 1981: U.S. Department of Energy, DOE/ET/13070-69, 569 p.

———, 1983, El Dorado Micellar-Polymer Demonstration Project Final Project Report and Eighth Annual Report, September 1981–November 1982: U.S. Department of Energy, DOE/ET/13070-92, 232 p.

WHITAKER, J. H. MCD., 1978, Diagenesis of the Brent Sand Formation: a scanning electron microscope study, *in* Whalley, W. B., ed., Scanning Electron Microscopy in the Study of Sediments (A Symposium): GeoAbstracts, Norwich, England, p. 363–380.

WOOLEN, I. D., 1976, Three-dimensional facies distribution of the Holocene Santee Delta, *in* Hayes, M. O., and Kana, T. W., eds., Terrigenous Clastic Depositional Environments: Coastal Research Division, Department of Geology, University of South Carolina, Technical Report No. 11, p. II-52 to II-63.

FIG. A1.—Geologic (sedimentologic) description of well MP-112, Sec. 21 T25S R5E, Kansas. Location of well shown in Figure 2.

CSO CHESNEY 112
ELDORADO FIELD, BUTLER COUNTY, KANSAS

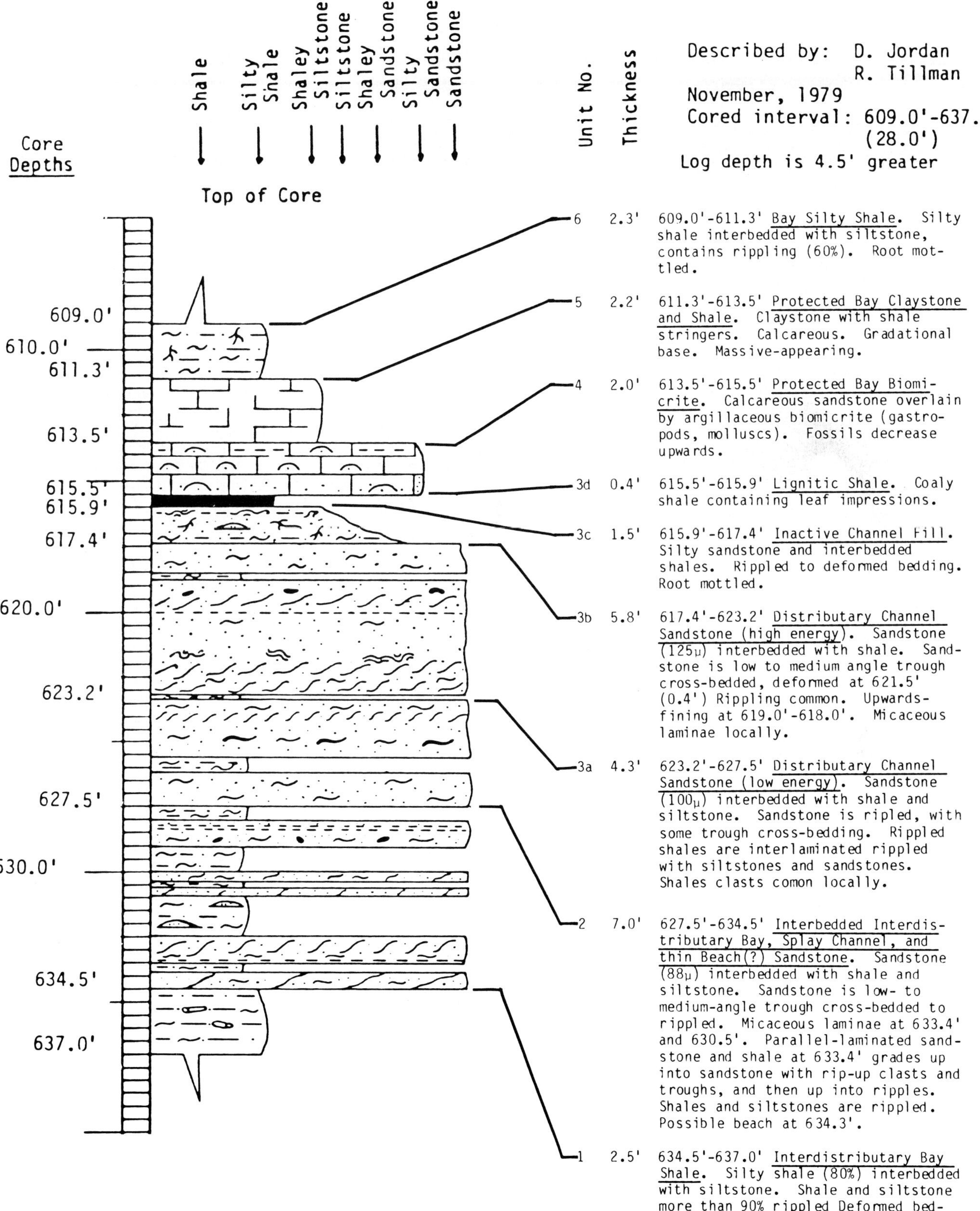

FIG. A2.—Core photos of well MP-112, Sec. 21 T25S R5E. Units are numbered in the order in which they were deposited, from bottom to top. Detailed descriptions of units are included in Figure A1.

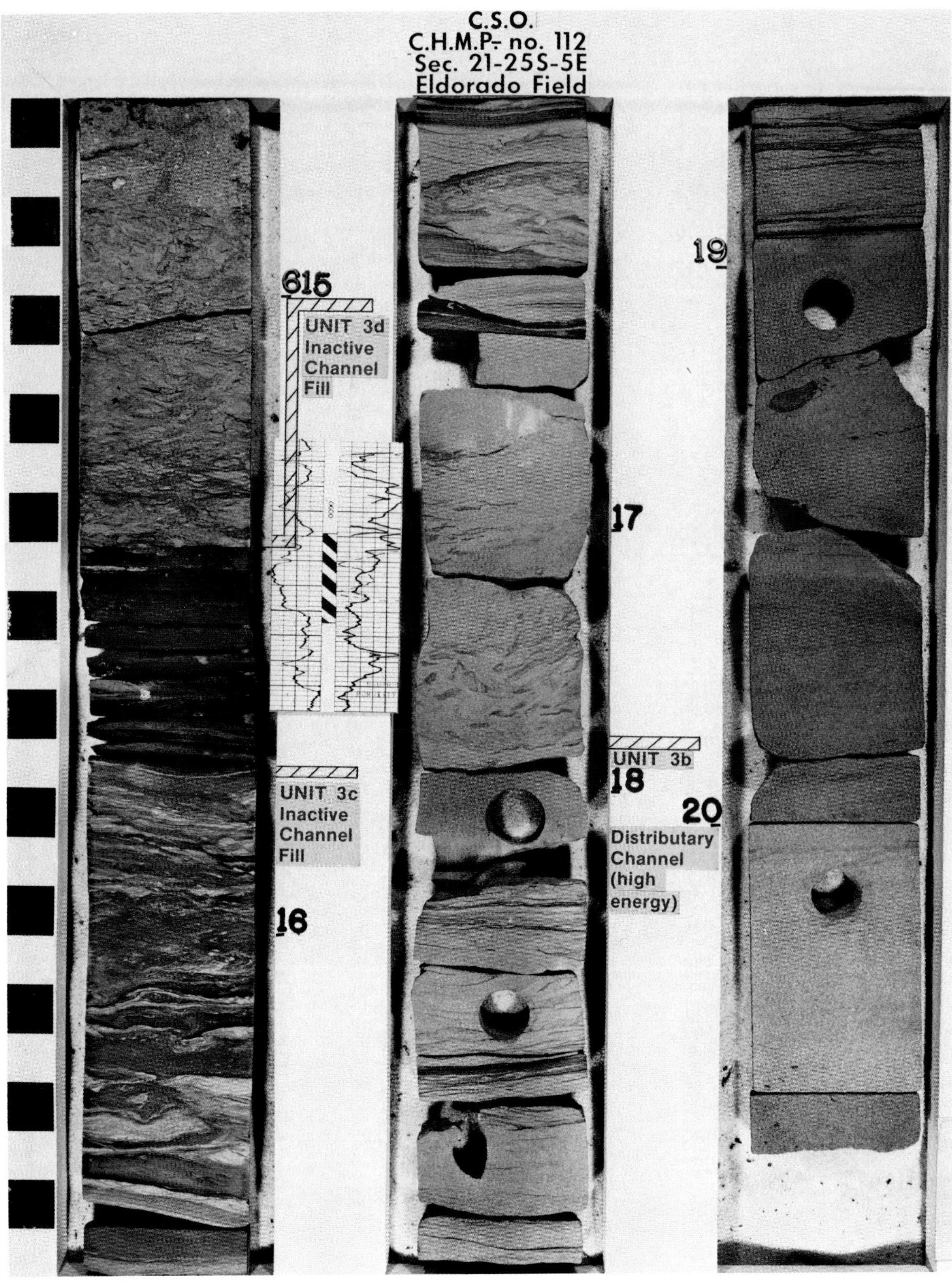
C.S.O.
C.H.M.P.- no. 112
Sec. 21-25S-5E
Eldorado Field
615
UNIT 3d
Inactive
Channel
Fill
UNIT 3c
Inactive
Channel
Fill
16
17
UNIT 3b
18
Distributary
Channel
(high
energy)
19
20

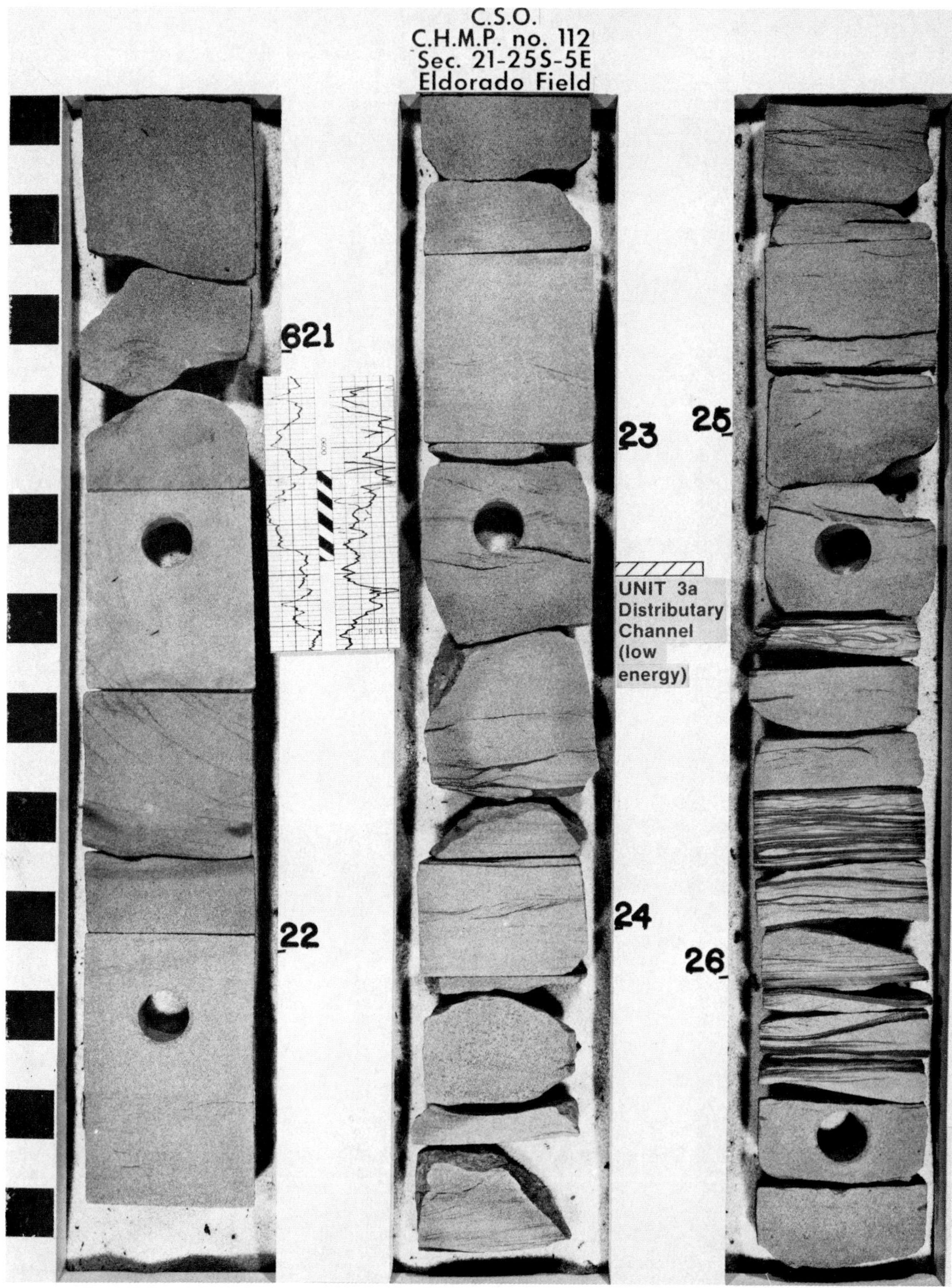
C.S.O.
C.H.M.P. no. 112
Sec. 21-25S-5E
Eldorado Field
621
22
23
24
25
26
UNIT 3a
Distributary
Channel
(low
energy)

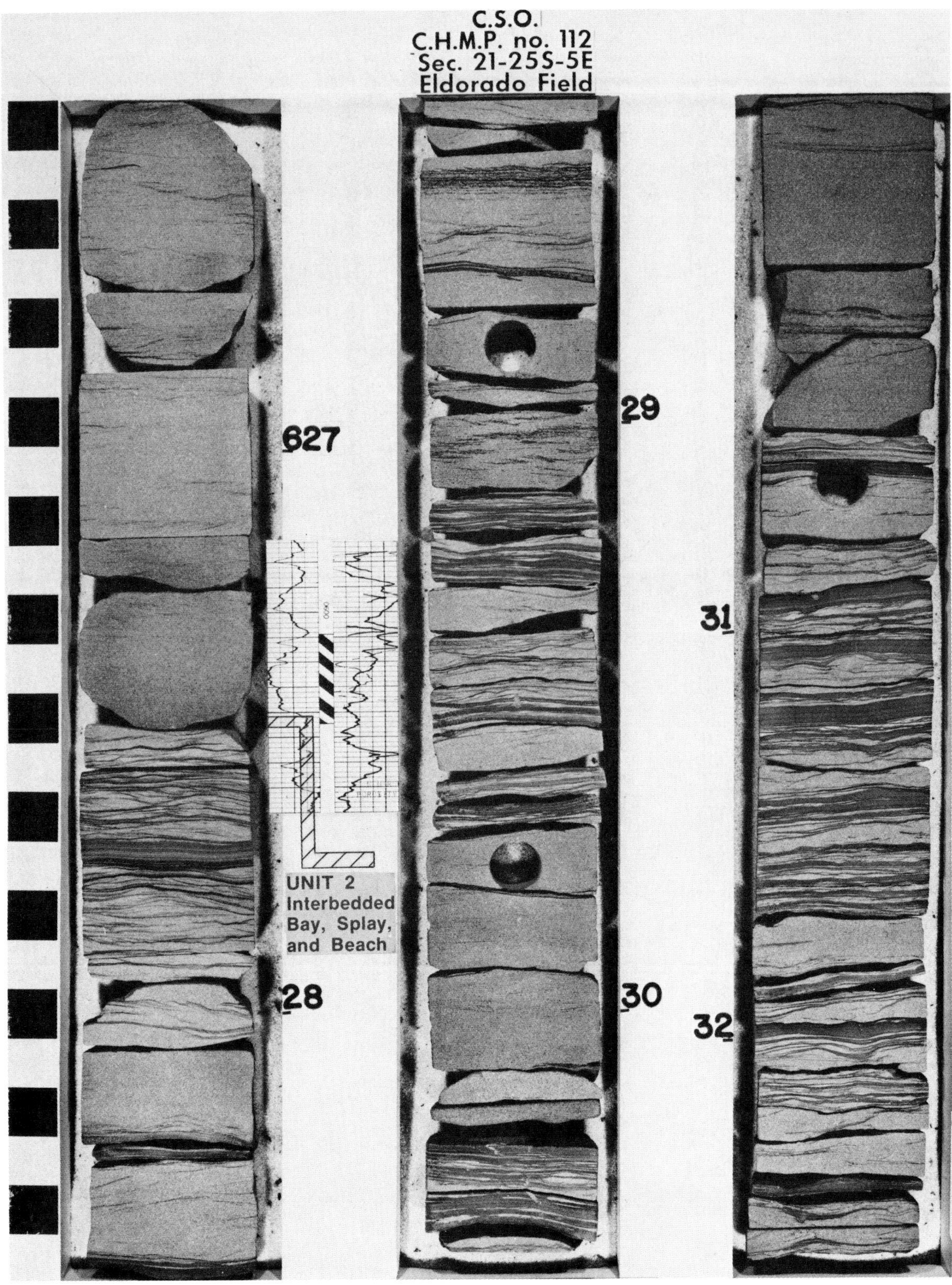
C.S.O.
C.H.M.P. no. 112
Sec. 21-25S-5E
Eldorado Field
627
28
29
30
31
32
UNIT 2
Interbedded
Bay, Splay,
and Beach

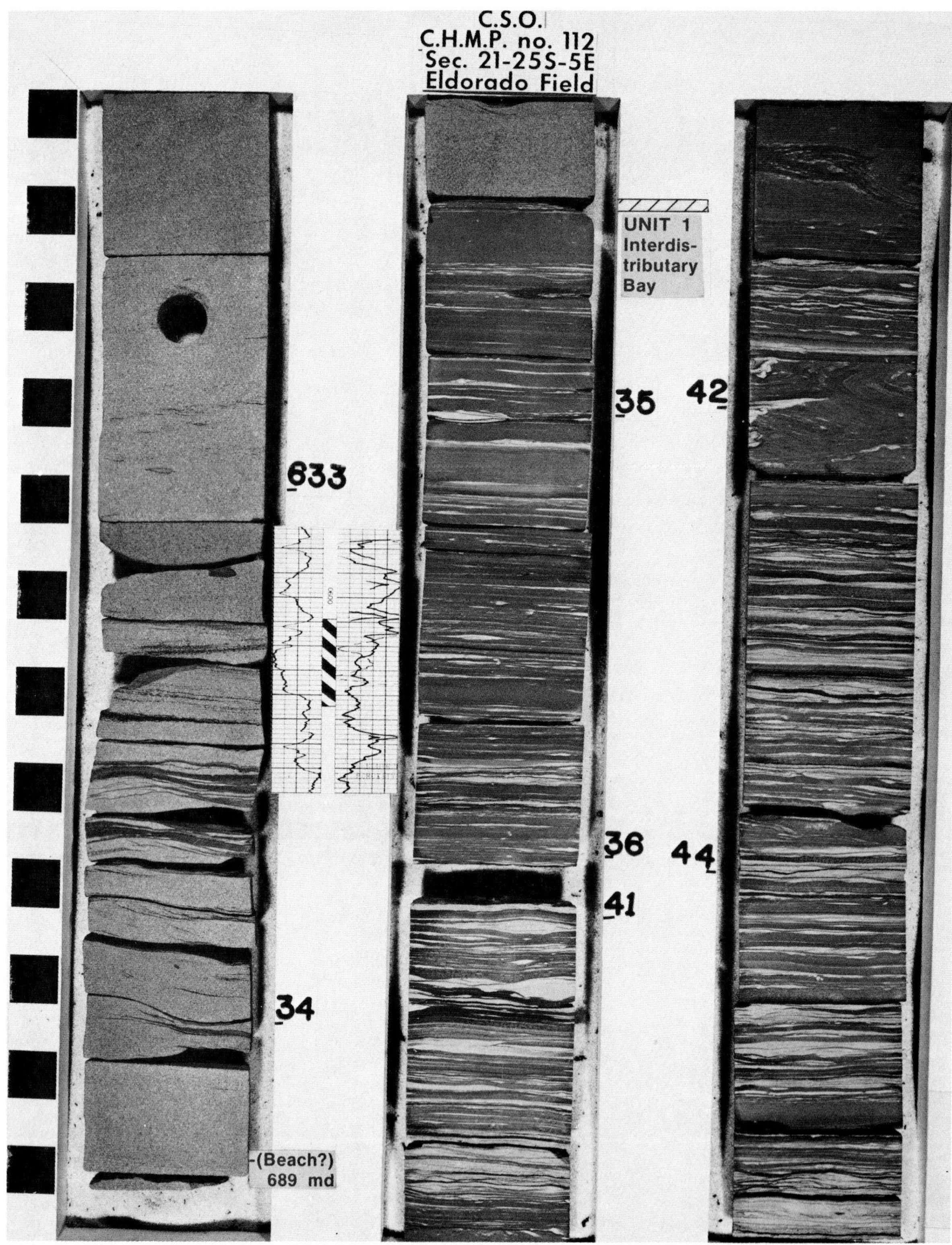
C.S.O.
C.H.M.P. no. 112
Sec. 21-25S-5E
Eldorado Field
UNIT 1
Interdis-
tributary
Bay
633
34
35
36
41
42
44
(Beach?)
689 md

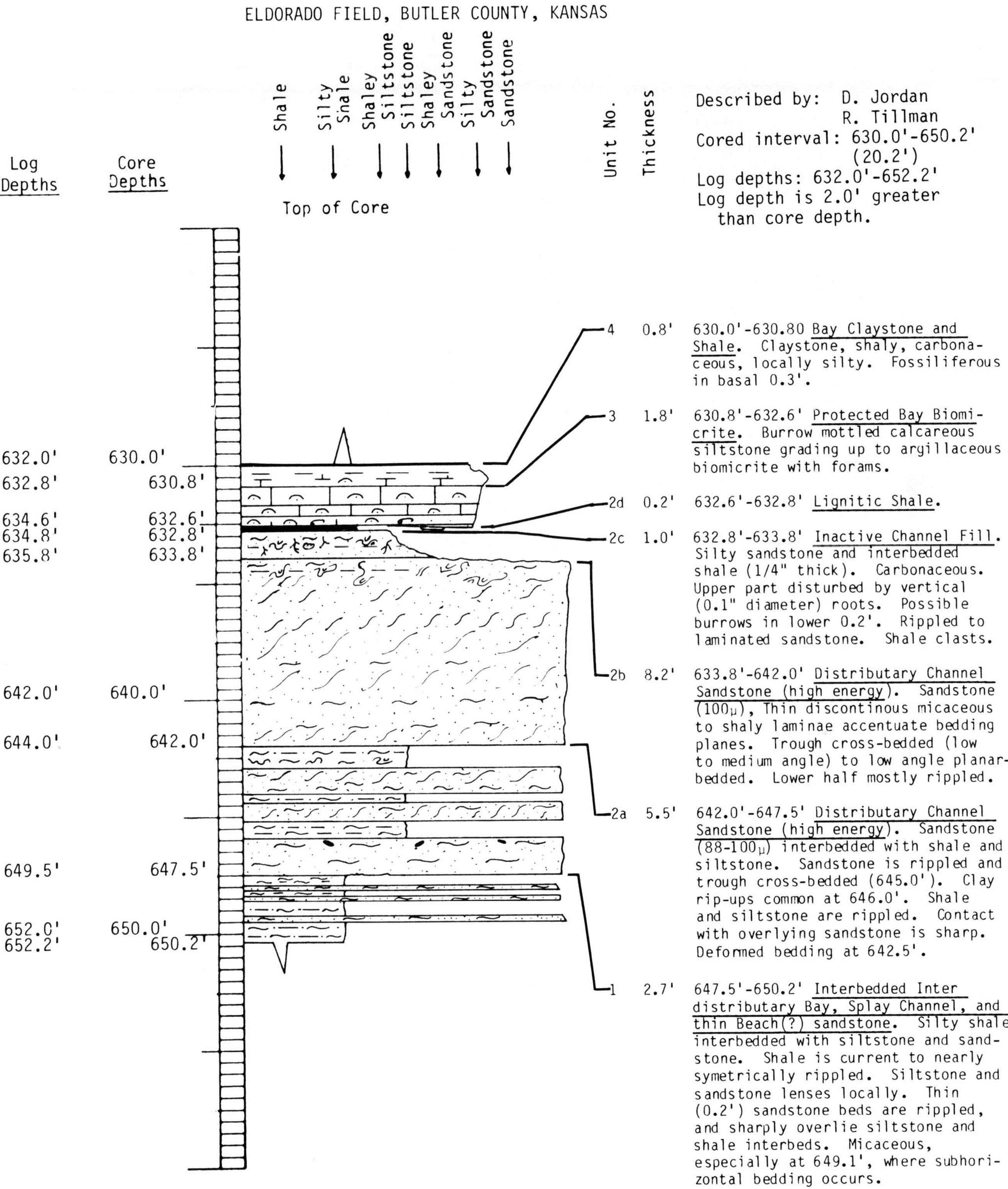

FIG. A3.—Geologic (sedimentologic) description of well MP-118, SE Sec. 21 T25S R5E. Location of well shown in Figure 2.

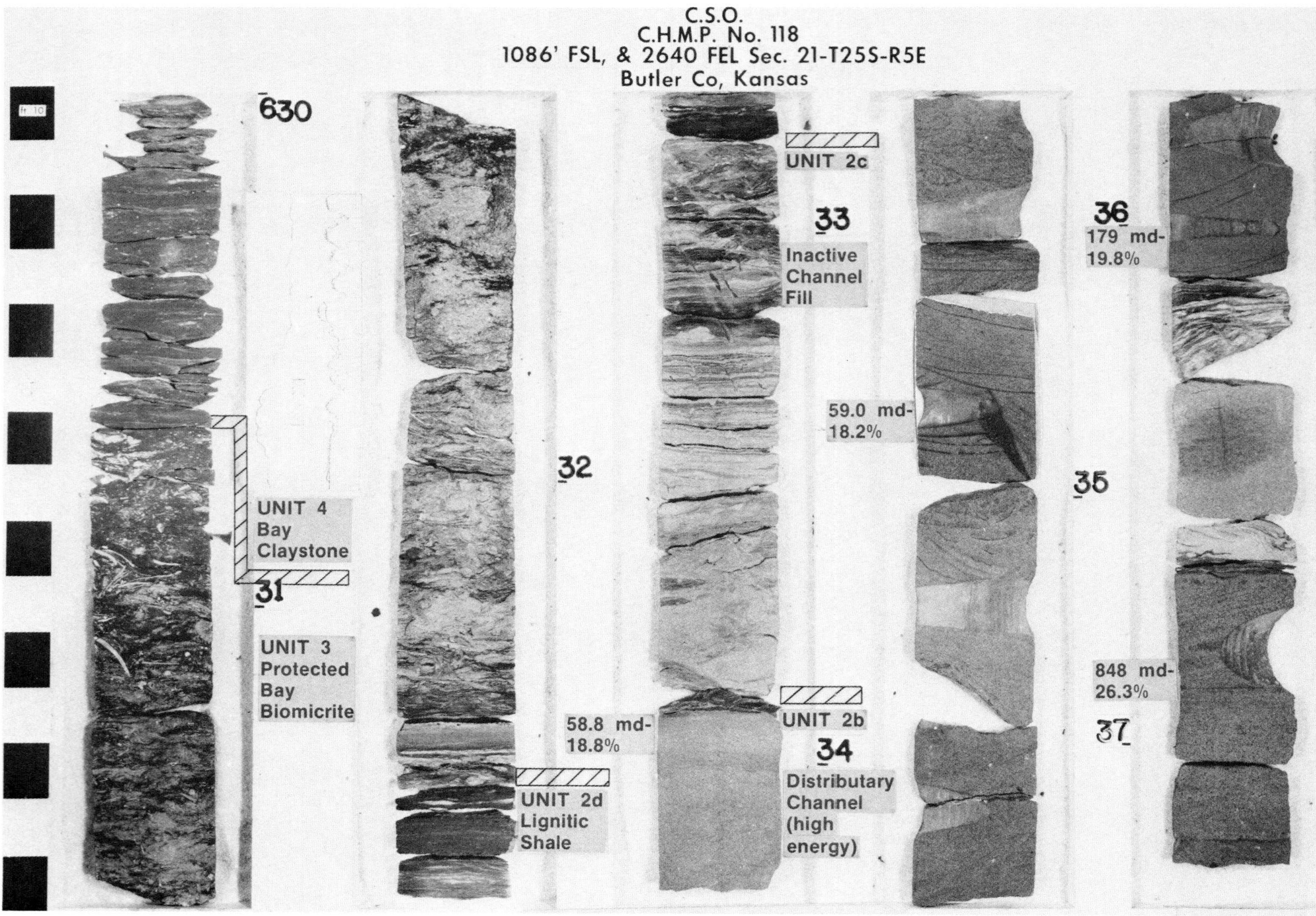

Fig. A4.—Core photos of well MP-118, Sec. 21 T25S R5E. Units are numbered in the order in which they were deposited, from bottom to top. Units are not correlatable by number in adjacent wells. Detailed descriptions of units are included in Figure A3. Porosity and permeability values are given on a foot-by-foot basis for individual Units 1 and 2 in Table A1.

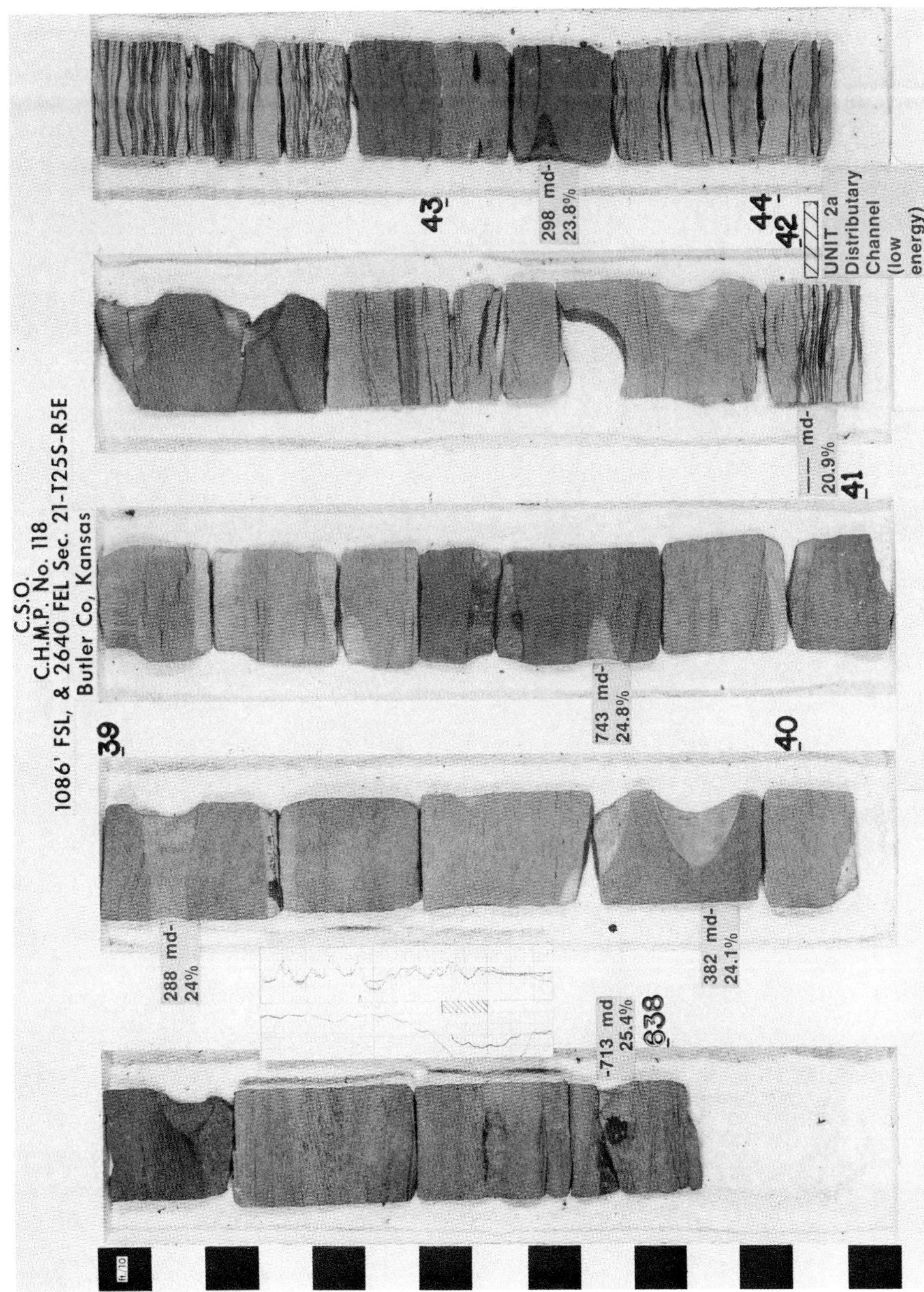
C.S.O.
C.H.M.P. No. 118
1086' FSL, & 2640 FEL Sec. 21-T25S-R5E
Butler Co, Kansas
39
40
41
42
43
44
288 md-
24%
382 md-
24.1%
743 md-
24.8%
--- md-
20.9%
298 md-
23.8%
-713 md
25.4%
638
UNIT 2a
Distributary
Channel
(low
energy)
ft/10

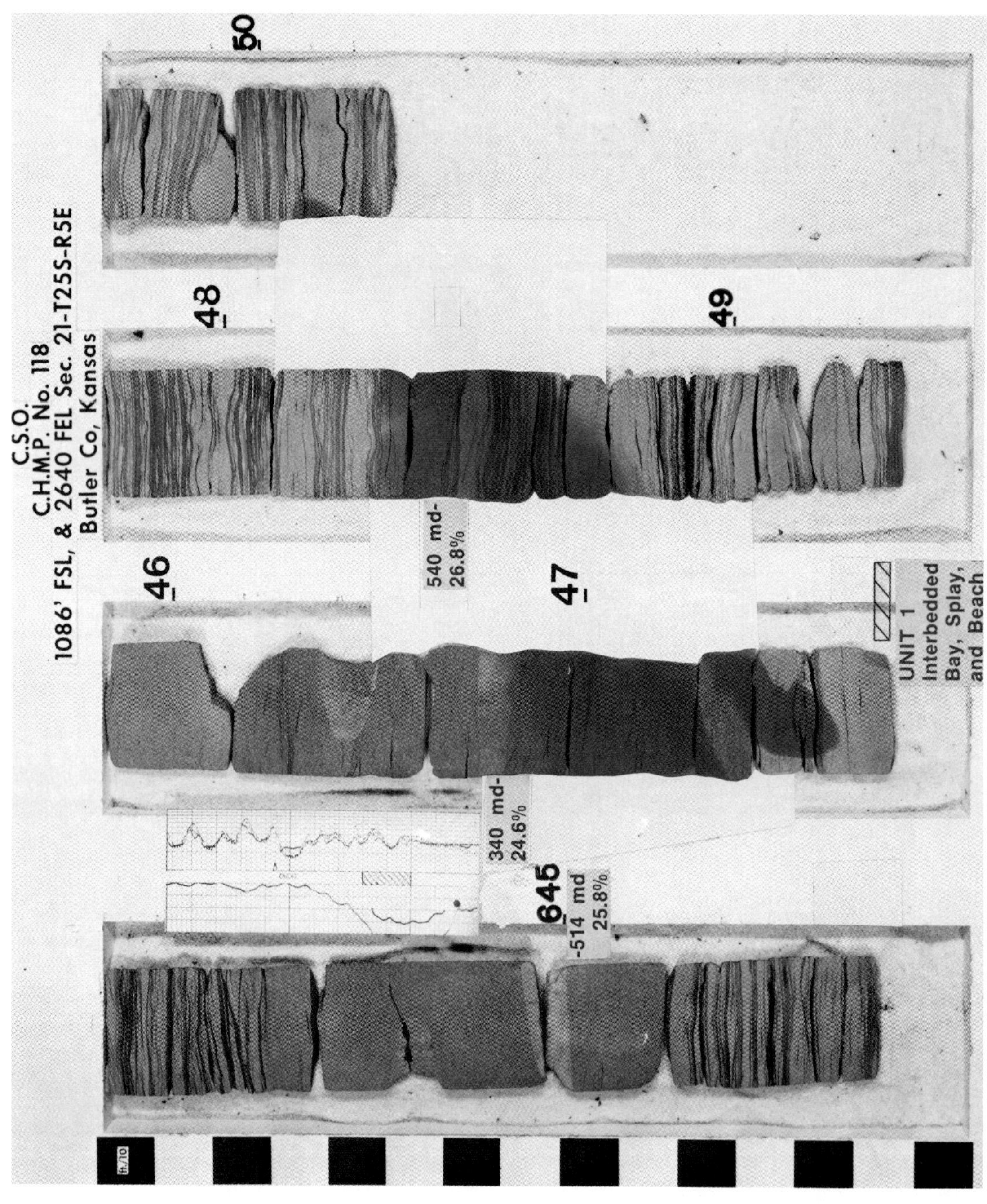
C.S.O.
C.H.M.P. No. 118
1086' FSL, & 2640 FEL Sec. 21-T25S-R5E
Butler Co, Kansas
46
47
48
49
50
645
540 md-
26.8%
340 md-
24.6%
-514 md
25.8%
UNIT 1
Interbedded
Bay, Splay,
and Beach
ft./10

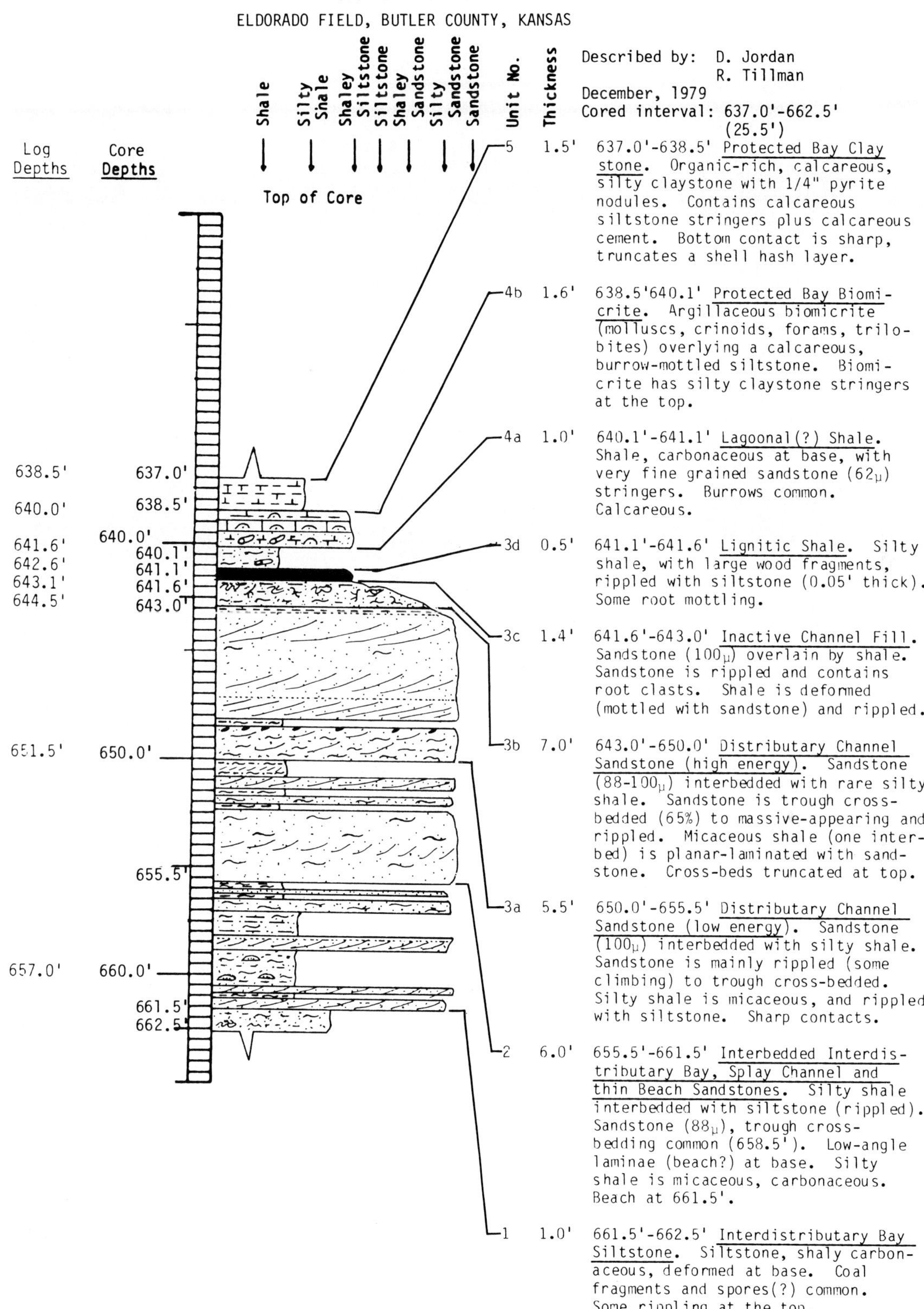

FIG. A5.—Geologic (sedimentologic) description of well MP-205. NE Sec. 28 T25S R5E. Location of well shown in Figure 2.

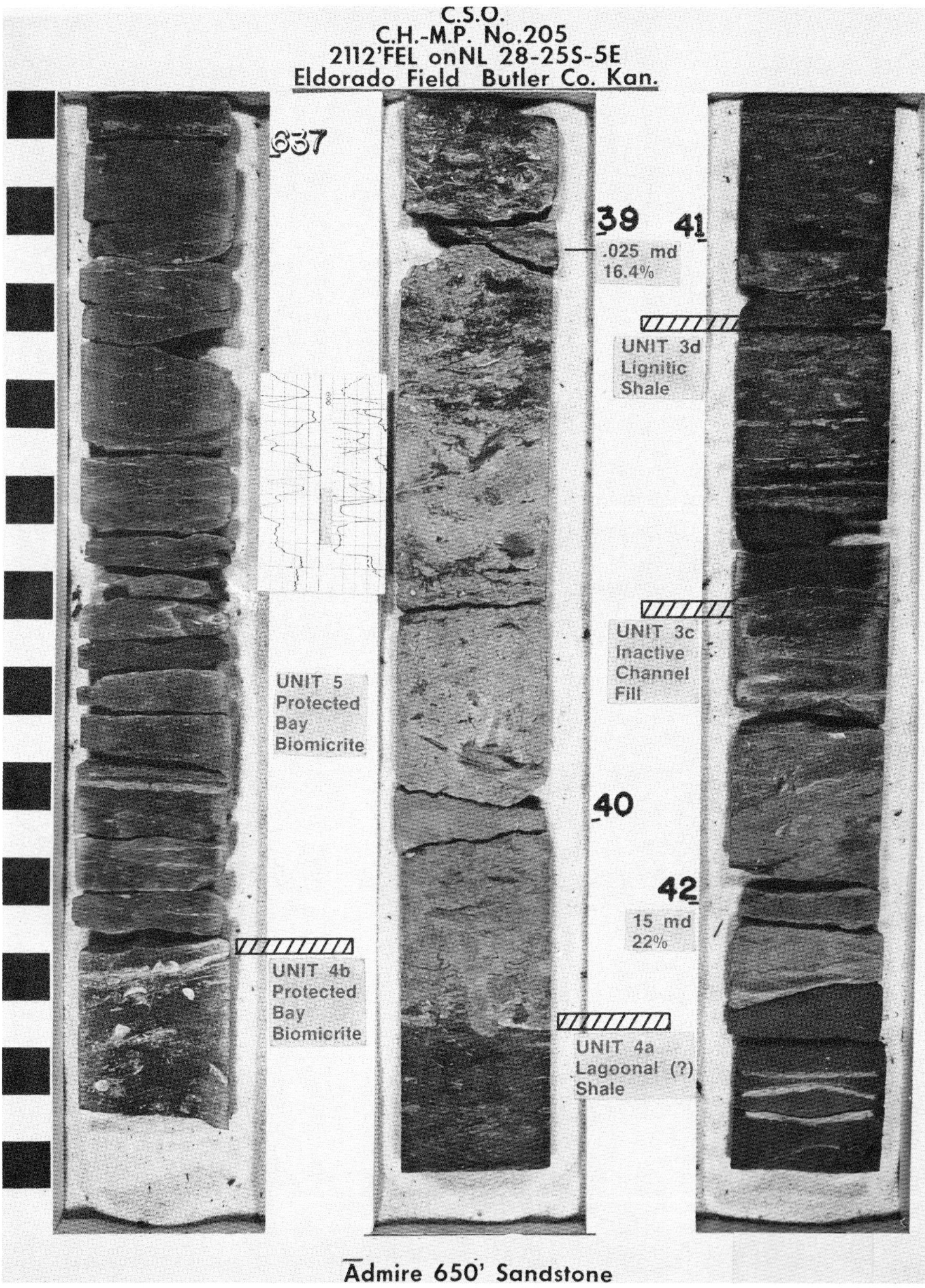

FIG. A6.—Core photos well MP-205, NE Sec. 28 T25S R5E. Units are numbered in the order in which they were deposited, from bottom to top. Units are not correlatable by number to adjacent wells. Detailed descriptions of units are included in Figure A5. Porosity and permeability values on a foot-by-foot basis and for individual units are given in Table A2.

Admire 650' Sandstone

Admire 650' Sandstone

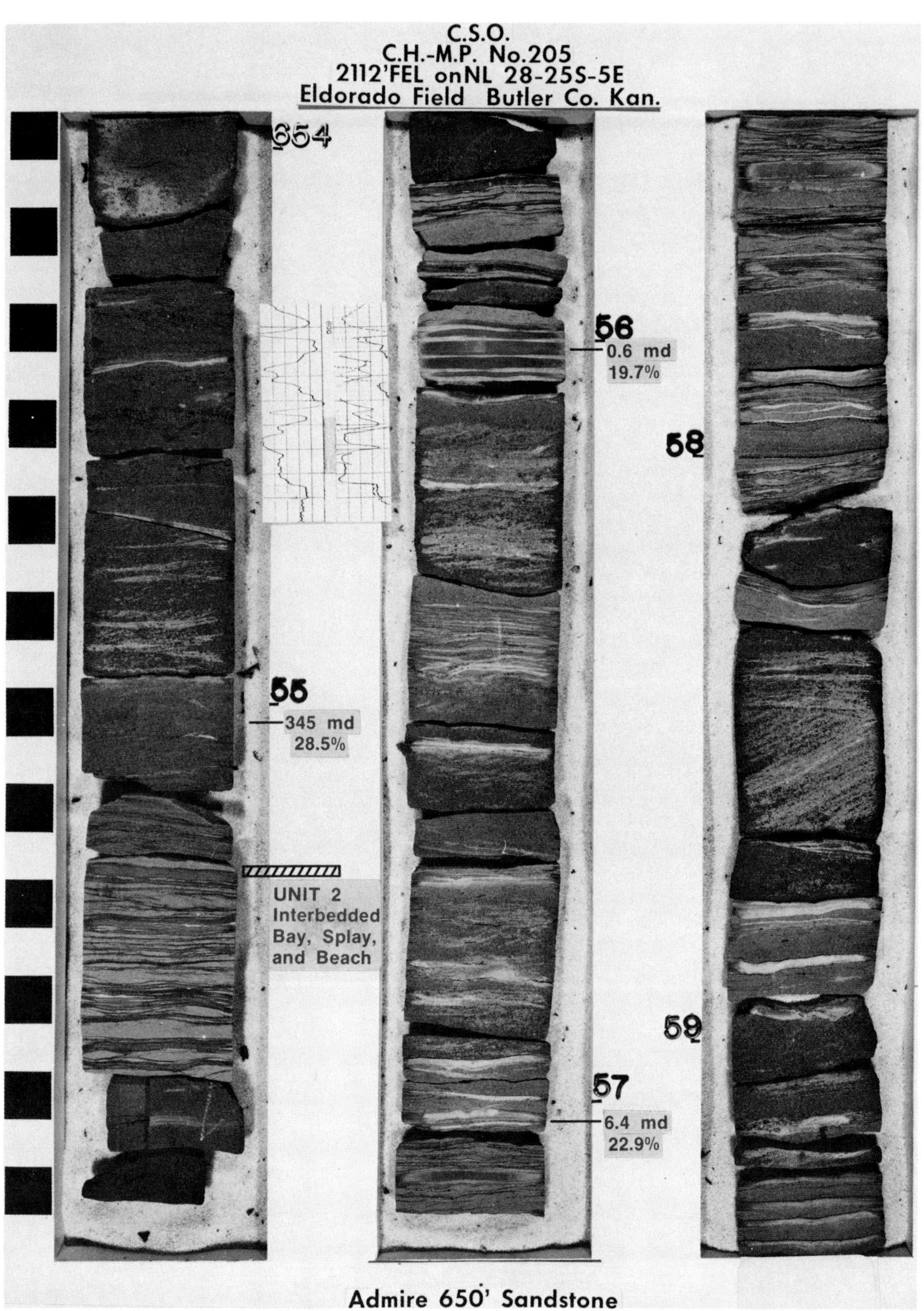
C.S.O.
C.H.-M.P. No.205
2112'FEL onNL 28-25S-5E
Eldorado Field Butler Co. Kan.
654
55
345 md
28.5%
UNIT 2
Interbedded
Bay, Splay,
and Beach
56
0.6 md
19.7%
57
6.4 md
22.9%
58
59
Admire 650' Sandstone

C.S.O.
C.H.-M.P. No.205
2112'FEL onNL 28-25S-5E
Eldorado Field Butler Co. Kan.
660
570 md
29.5%
61
-(Beach?)
650 md
28.7%
UNIT 1
Inter-
distributary
Bay
62
Admire 650' Sandstone

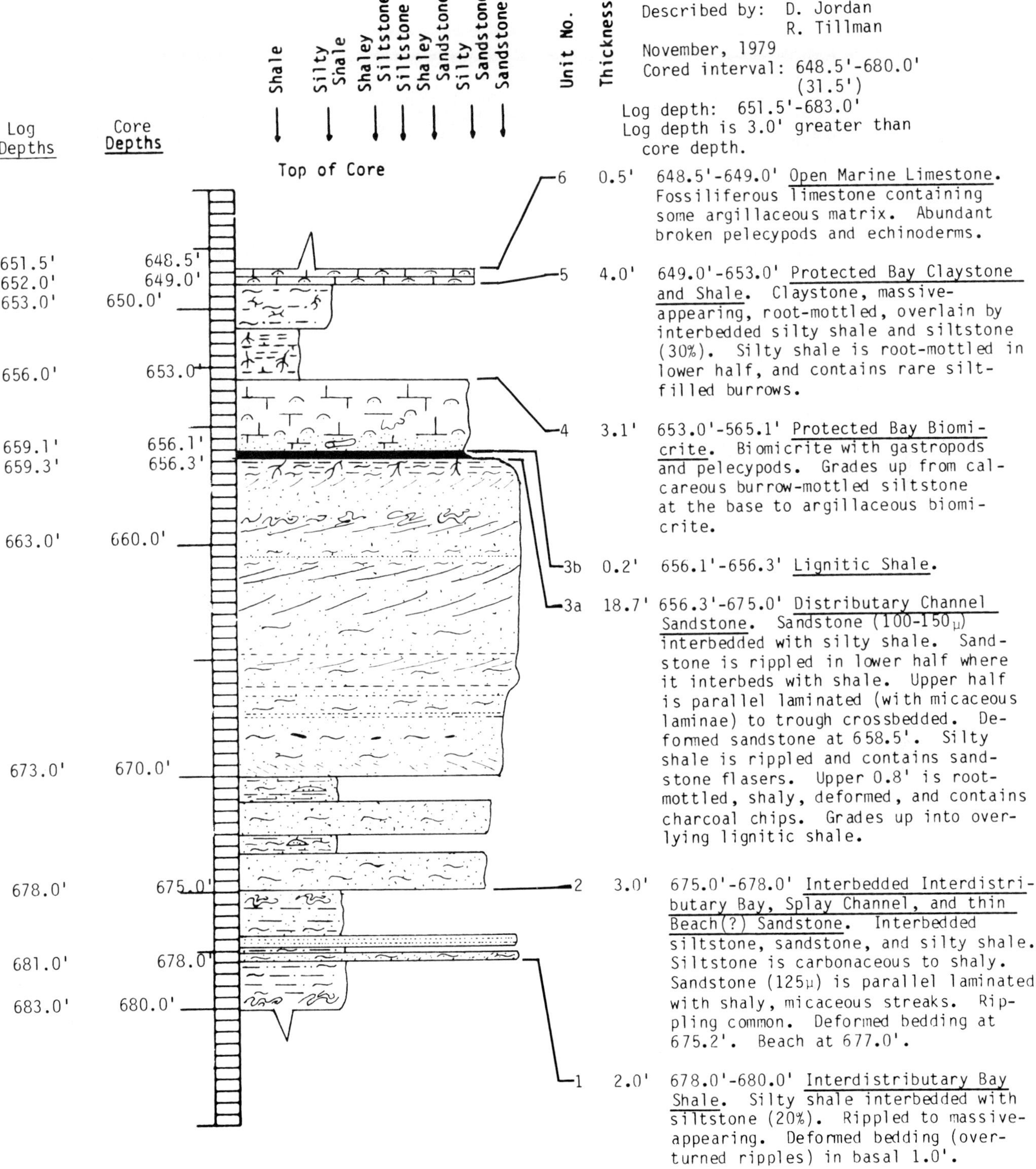

FIG. A7.—Geologic (sedimentologic) description of well MP-211, Sec. 28 T25S R5E. Location of well shown in Figure 2.

FIG. A8.—Core photos of well MP-211, Sec. 28 T25S R5E. Units are numbered in the order in which they were deposited, from bottom to top. Units are not correlatable by number in adjacent wells. Detailed description of units are given in Figure A7.

C.S.O.
C.H.M.P. no. 211
Sec. 21-25S-5E
Eldorado Field
UNIT 3a
Distributary
Channel
654
55
56
UNIT 3b
Lignitic
Shale
57
58
59
60

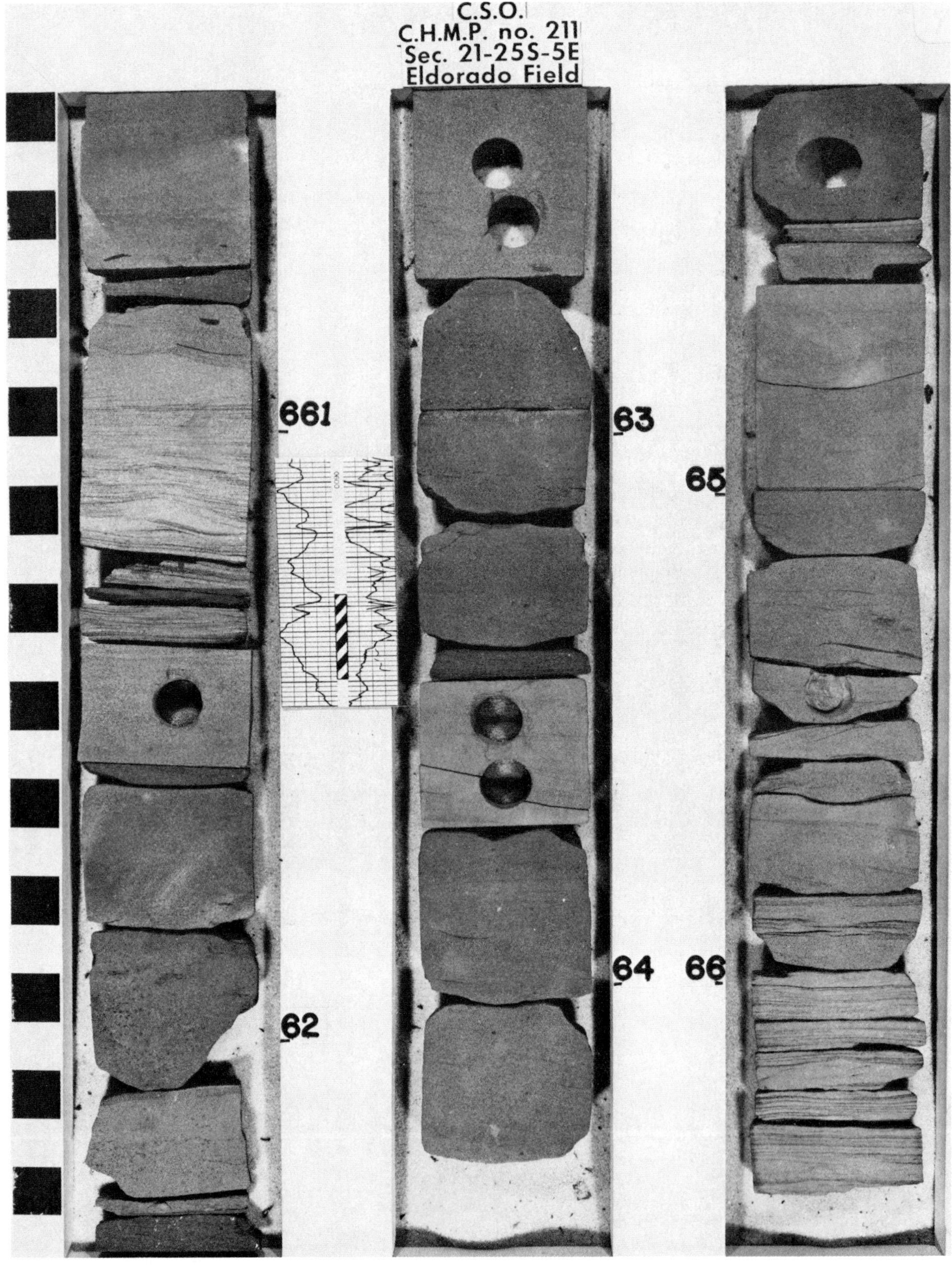
C.S.O.
C.H.M.P. no. 211
Sec. 21-25S-5E
Eldorado Field
661
62
63
64
65
66

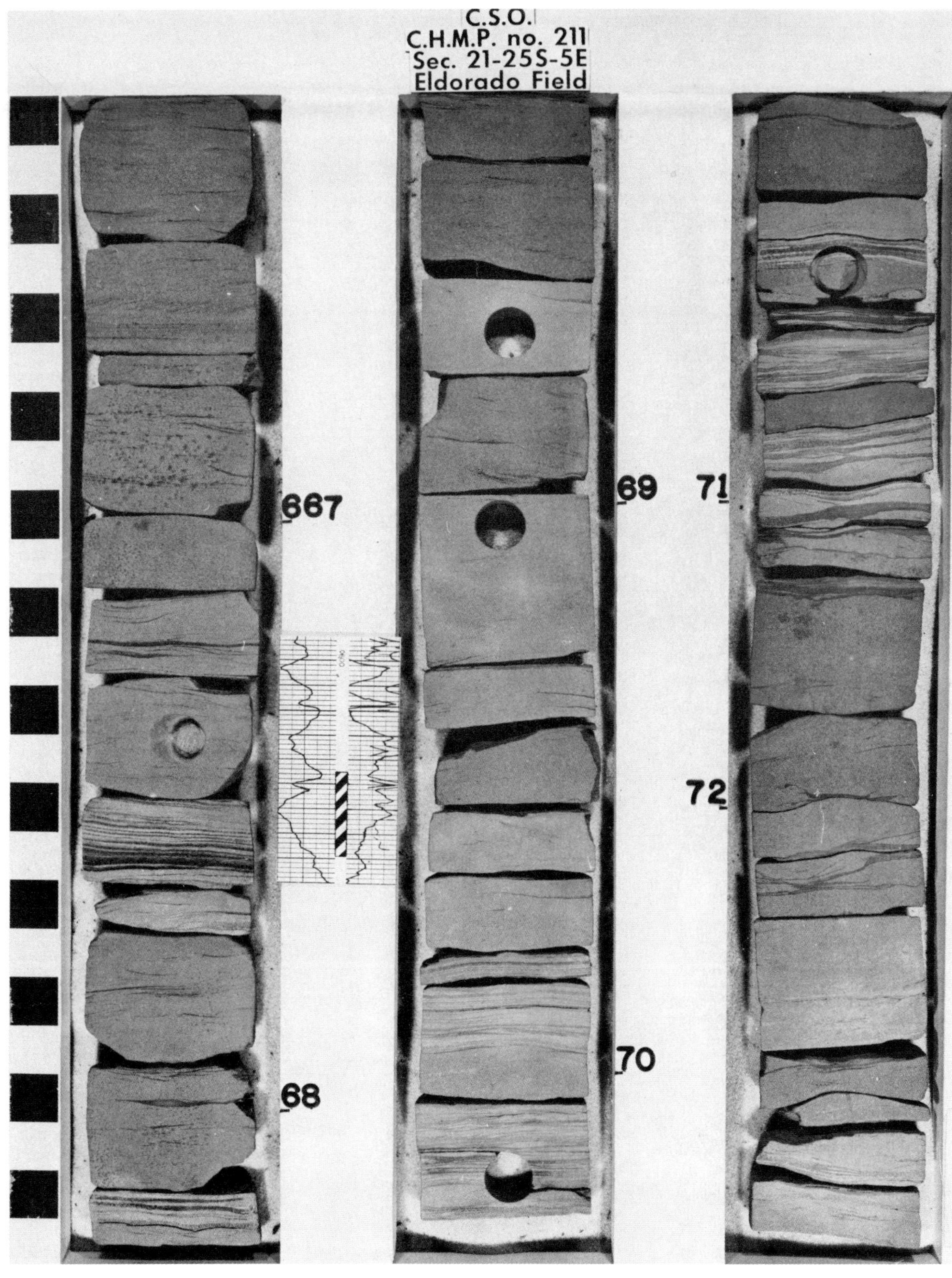
C.S.O.
C.H.M.P. no. 211
Sec. 21-25S-5E
Eldorado Field
667
68
69
70
71
72

C.S.O.
C.H.M.P. no. 211
Sec. 21-25S-5E
Eldorado Field
673
74
75
UNIT 2
Interbedded
Bay, Splay,
and Beach
76
77
-(Beach?)
78
UNIT 1
Inter-
distributary
Bay
79
80

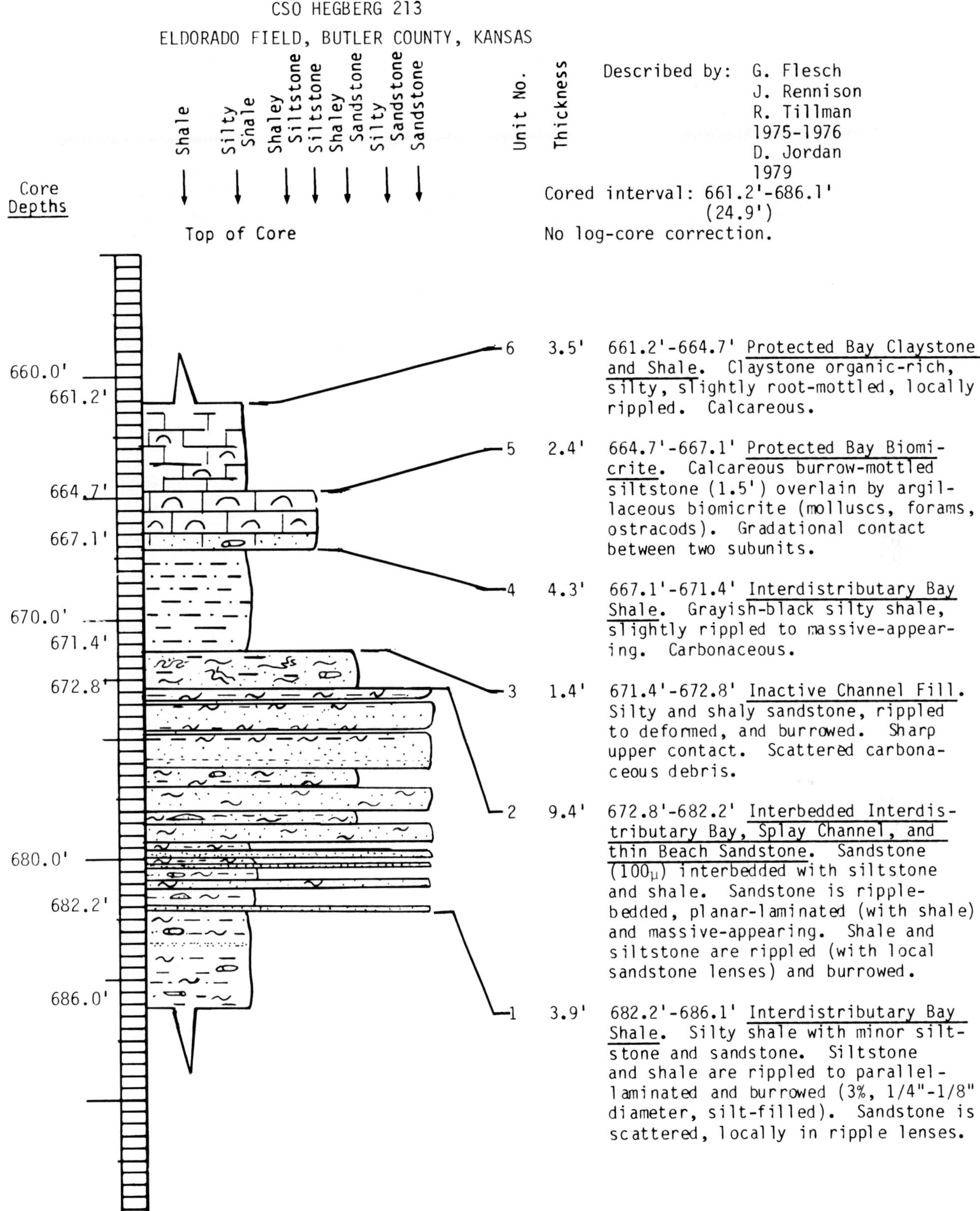

FIG. A9.—Geologic (sedimentologic) description of well MP-213, Sec. 28 T25S R5E. This well is anomalous because it contains no active channel-fill sandstones. Location of well shown in Figure 2.

FIG. A10.—Core photos of well MP-213, Sec. 28 T25S R5E. Units are numbered in the order in which they were deposited, from bottom to top. Units are not correlatable by number to adjacent wells. Detailed descriptions of units are included in Figure A9. Porosity and permeability values on a foot-by-foot basis and for indivifual units are given in Table A3.

C.S.O.
C.H.M.P.-213
Eldorado Field
Butler Co, Kan.
668
69
70
71
72
UNIT 3
Inactive
Channel
Fill
UNIT 2
Interbedded
Bay, Splay,
and Beach

C.S.O.
C.H.M.P.-213
Eldorado Field
Butler Co, Kan.
673
230 md
25.4%
74
270 md
28.8%
75
750 md
27.8%
350 md
27%
76
10.4 md
22.2%
820 md
28.9%
77
1410 md
29.3%
78
- md
27.9%
630 md
26.2%
380 md
26.9%
79

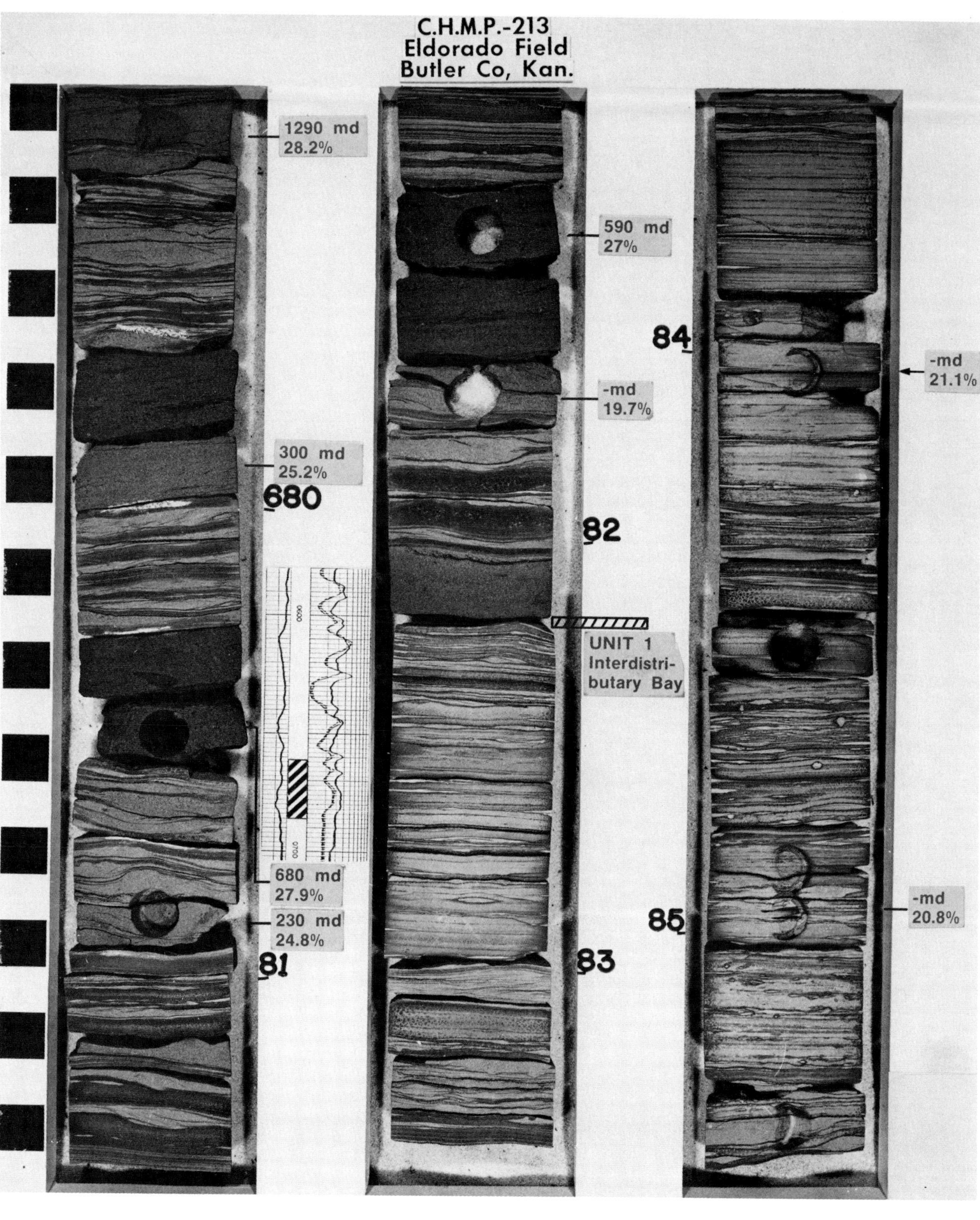
C.H.M.P.-213
Eldorado Field
Butler Co, Kan.
1290 md
28.2%
300 md
25.2%
680
680 md
27.9%
230 md
24.8%
81
590 md
27%
-md
19.7%
82
UNIT 1
Interdistri-
butary Bay
83
84
-md
21.1%
85
-md
20.8%

Table A1.—ROUTINE CORE ANALYSIS RESULTS VERSUS DEPTH FOR WELL MP-118

	Depth (ft)	Permeability to air, MD	Porosity, %	Grain Density, GM/CC	Bulk Density, GM/CC
UNIT 2-B	DISTRIBUTARY CHANNEL SANDSTONE FACIES (HIGH ENERGY)				
	634.0	58.8	18.8	2.55	2.07
	635.0	59	18.2	2.57	2.10
	636.0	179	19.8	2.54	2.03
	637.0	848	26.3	2.59	1.91
	638.0	713	25.4	2.55	1.90
	639.0	288	24.0	2.56	1.94
	640.0	382	24.1	2.58	1.96
	641.0	743	24.8	2.54	1.91
	642.0	Broken	20.9	2.56	2.03
MEANS		304	22.5	2.56	1.98
UNIT 2-A	DISTRIBUTARY CHANNEL SANDSTONE FACIES (LOW ENERGY)				
	643.0	298	23.8	2.58	1.96
	644.0	Broken	23.1	2.59	1.99
	645.0	514	25.8	2.61	1.94
	646.0	448	25.6	2.58	1.92
	647.0	340	24.6	2.59	1.95
	648.0	540	26.8	2.59	1.90
	649.0	338	26.0	2.61	1.95
MEANS		413	25.1	2.59	1.94
UNIT 1	INTERBEDDED INTERDISTRIBUTARY BAY, SPLAY CHANNEL, AND THIN BEACH SANDSTONE FACIES				
	650.0	Broken	23.3	2.60	1.99

Table A2.—ROUTINE CORE ANALYSIS RESULTS VS DEPTH FOR WELL MP-205

	Depth (ft)	Permeability to air, MD	Porosity %	Fluid Saturation Oil	Fluid Saturation Water
UNIT 2	INTERBEDDED INTERDISTRIBUTARY BAY, SPLAY CHANNEL, AND THIN BEACH SANDSTONE FACIES				
	656.0		19.67	27.1	72.9
	657.0		22.87	24.9	75.1
	658.0		23.57	43.7	45.0
	659.0		30.18	17.0	48.3
	660.0	570.0	29.49	29.1	49.1
	661.0		22.99	41.3	57.2
MEANS			24.80	30.5	57.9
UNIT 3A	DISTRIBUTARY CHANNEL SANDSTONE FACIES (LOW ENERGY)				
	651.0	463.0	29.39	20.2	57.3
	652.0		24.15	24.7	75.3
	653.0	348.0	29.29	21.4	58.0
	654.0	550.0	29.70	14.5	69.3
	655.0	345.0	28.57	19.4	49.5
MEANS		426.5	28.22	20.0	61.9
UNIT 3B	DISTRIBUTARY CHANNEL SANDSTONE FACIES (HIGH ENERGY)				
	644.0	213.0	26.87	23.1	49.8
	645.0	321.0	26.87	24.5	45.1
	646.0	136.0	25.67	19.6	73.1
	647.0	385.0	25.11	23.6	75.0
	648.0	302.0	27.90	16.8	65.6
	649.0	660.0	28.97	10.0	28.1
	650.0	20.0	16.55	33.5	66.5
MEANS		291.0	25.42	21.6	57.6
UNIT 3C	INACTIVE CHANNEL FILL FACIES				
	642.0	15.0	21.96	27.6	48.3
	643.0	Broken	16.51	56.3	43.7
MEANS			19.23	41.9	46.0
UNIT 4A	LAGOONAL (?) SHALE FACIES				
	641.0	0.12	22.73	2.1	56.9
UNIT 4B	PROTECTED BAY BIOMICRITE FACIES				
	639.0	0.02	16.42	0.3	46.2
	640.0	0.02	6.78	6.9	80.9
MEANS		0.02	11.6	3.6	63.6
UNIT 5	PROTECTED BAY CLAYSTONE FACIES				
	637.0	Broken	22.65	0.2	66.2
	638.0	Broken	16.10	0.8	85.0
MEANS			19.37	0.5	75.6

TABLE A3.—ROUTINE CORE ANALYSIS RESULTS VS DEPTH FOR WELL MP-213

		Depth (ft)	Permeability to air, MD	Porosity %	Fluid Saturation Oil	Fluid Saturation Water
UNIT 6		PROTECTED BAY CLAYSTONE AND SHALE FACIES				
		661.0	.000	22.40	8.0	57.7
		661.5	.040	22.56	7.1	71.6
		662.0	.030	23.19	12.2	62.8
		662.5	.140	23.39	9.1	66.6
		663.0	.020	22.87	8.9	54.0
		663.5	.010	18.42	8.3	58.9
		664.0	.015	23.67	4.4	55.0
		664.5	.290	20.86	4.7	56.9
	MEANS		.078	22.14	7.8	60.8
UNIT 5		PROTECTED BAY BIOMICRITE FACIES				
		665.0	.150	19.94	4.1	22.7
		665.5	.000	9.51	10.4	39.9
		666.0	.010	8.39	6.3	20.0
		666.5	.460	16.79	4.7	57.9
		667.0	.420	20.55	4.9	63.9
	MEANS		.28	15.03	6.1	40.9
UNIT 4		INTERDISTRIBUTARY BAY SHALE FACIES				
		667.0	Broken	25.98	.5	50.0
		668.0	.085	24.46	8.5	50.8
		668.5	Broken	26.69	5.7	42.0
		669.0	Broken	25.76	9.0	44.3
		669.5	Broken	25.70	4.8	51.5
		670.0	.010	24.68	4.1	57.0
		670.5	.390	25.33	3.3	51.8
		671.0	Broken	24.84	3.1	55.9
	MEANS		2.06	23.42	5.6	54.3
UNIT 3		INACTIVE CHANNEL FILL FACIES				
		671.5	340.000	27.83	23.1	38.8
		672.0	Broken	23.55	12.8	68.4
		672.5	18.0	24.82	6.5	58.1
	MEANS		179.0	25.40	14.1	55.1
UNIT 2		INTERBEDDED INTERDISTRIBUTARY BAY, SPLAY CHANNEL, AND THIN BEACH SANDSTONE FACIES				
		673.0	230.0	25.40	26.4	45.6
		673.5	57.0	25.94	25.0	39.7
		674.0	270.0	28.78	28.5	39.8
		674.5	460.0	27.85	36.4	34.9
		675.0	750.0	27.78	28.2	33.3
		675.5	350.0	26.97	20.4	33.2
		676.0	10.4	22.19	6.2	63.9
		676.5	820.0	28.87	36.6	30.4
		677.0	1410.0	29.30	34.4	31.0
		677.5	102.0	25.04	17.3	46.5
		678.0	Broken	27.89	18.4	41.6
		678.5	630.0	26.24	36.5	20.8
		679.0	380.0	26.86	19.3	38.7
		679.5	1290.0	28.22	31.0	17.5
		680.0	300.0	25.15	27.4	41.4
		680.5	680.0	27.93	37.3	21.9
		681.0	230.0	24.75	30.0	37.6
		681.5	590.0	27.09	33.8	25.8
		682.0	Broken	19.65	4.0	47.2
	MEANS		503.5	26.40	26.2	36.4
UNIT 1		INTERDISTRIBUTARY BAY SILTY SHALE FACIES				
		682.5	Broken	23.12	4.5	70.9
		683.0	Broken	18.55	5.3	68.1
		683.5	0.0	20.06	6.9	59.6
		684.0	Broken	21.08	11.4	56.9
		684.5	9.8	20.92	6.1	41.0
		685.0	Broken	20.78	—	—
	MEANS		5.9	20.75	6.8	59.3

SEDIMENTOLOGIC RESERVOIR STUDY OF A STEAM-DRIVE PROJECT IN DELTAIC RIVER SANDS, EAST TIA JUANA FIELD, VENEZUELA

C. KRUIT
EPGP MARAVEN, S.A.
Caracas, Venezuela[1]

ABSTRACT: A hexagon steam-flood pattern in the Middle Miocene Lower Lagunillas Sand Member was described geologically and petrophysically in anticipation of a fieldwide "M-6" steam drive in East Tia Juana field, Venezuela. Data were obtained in a seven-well hexagon centered on well 3054 and having an average spacing of 231 m (758 ft). Rubber-sleeve cores from five wells and sidewall samples from additional wells were examined. The sequence of productive sands is primarily fluvial deltaic and is overlain by a shallow marine interval. Lateral correlations emphasizing bases of channels and inundation planes were given preference over correlating log shapes. Log shapes were used in mapping local subfacies within the channels. Channel sand trends were found to be well defined on maps of Ø.So, where Ø.So values exceeded 0.3. Regional cross sections showing various classes of Ø.So were also used to illustrate the vertical and lateral sequences of active fill, abandoned fill, and interdistributary splay sands.

The productive sands recognized in the pilot area were designated separately as D-1, D-2 and D-3. Since no permeability boundaries were interpreted to exist between these sands, vertical communication was predicted to be very good.

Several wells contained only one fluvial sand subfacies. Additional oil produced from individual wells, as a result of steam injection in the D-1 sand, can be directly related to the type of channel subfacies encountered. In a well which was located primarily in the active channel fill, production of additional oil was 200,000 extra barrels; abandoned channel-fill wells produced from 0 to 50,000 extra barrels, and a well interpreted to produce primarily from a natural levee or flood basin produced only 20,000 extra barrels. These variations in production were predicted by construction of detailed geologic maps and cross sections.

INTRODUCTION

This paper considers methods and initial results of a detailed sedimentologic reservoir study of the Lower Lagunillas Formation in the East Tia Juana oil field located on the Bolivar Coast of Venezuela (Fig. 1). The study was initiated by the author in June 1978 in support of steam-drive operations (referred to as the "M-6 Project"), and the results presented here are an updated version of a paper that was presented at the 1979 United Nations Institute for Training and Research Conference in Edmonton, Canada (Kruit, 1979). The main producing interval is composed of a complex of river channel sands and associated coastal deposits which have a gross thickness of approximately 130 ft (40 m). The project area measures approximately 2.7 × 2.7 km (1.7 × 1.7 mi) and includes 151 wells with an average spacing of 231 m (758 ft). The structure in the project area is monoclinal with a dip of about 3.5° and with little or no faulting.

Most of the oil produced in the M-6 area originates from the middle Miocene Lower Lagunillas Sand Member (Fig. 2). The reservoir was subdivided by Soto (1977) into D-1, D-2 and D-3 sands. These sands are overlain by a poorly productive sequence of transgressive sand and shale lenses (C-sands), which grade upward into a shallow marine interval of sands, silts and shales of wide lateral distribution. The reservoir sands occur between 1,250 ft (380 m) and 2,200 ft (667 m) below the surface. The first well was drilled in 1935, but the main development took place in the early 1950s. All the Lower Lagunillas Formation reservoir sands appear to be saturated with heavy oil, and oil/water contacts have been observed only outside the project area. The oil gravity is approximately 12° API.

Primary production started in 1948 and was followed by a period of alternating steam injection ("steam soak") from 1967 until 1977. Continuous steam injection ("steam drive") started in three pilot wells in 1975 and 1976, whereas 17 additional injector wells became operative in 1978. The production wells occupy a hexagonal pattern around the injector wells (Herrera, 1977; Van Der Knaap, 1980).

A number of techniques were used in the study, including (a) correlation of a transgressive shallow-marine interval (overlying the coastal complex) to obtain a datum marker; (b) regional correlation of gross sedimentologic features of the productive interval, using log cross sections to show depositional cycles and certain lithofacies units (channel sands, offlap sands, etc.); (c) development of maps and depositional models for each cycle; (d) construction of regional sections of Ø.So; and (e) preparation of three detailed sand/shale correlation sections for each injection hexagon using 1:200-scale logs. Evidence for irregular advance of steam and fluids was then compared with the hexagon correlation model.

OBJECTIVES OF STUDY

A geologic study for a reservoir management project should provide reservoir engineers with a detailed evaluation of subsurface conditions for the efficient planning, developing and monitoring of the operations. Hence, the following objectives of study were formulated jointly with the reservoir engineers. The objectives were (a) to determine whether the productive sands are subdivided vertically by continuous-horizon impermeable layers and whether obstructions in horizontal flow should be taken into account; (b) to produce net sand (h) × porosity (Ø) × oil-saturation maps (So), (h.Ø.So); (c) to produce sections of the entire productive sand interval and of its possible members; and (d) to produce detailed correlation sections of all steam-injection hexagons.

Presented in this paper is the sedimentologic approach used to achieve these objectives; particular attention was paid to a single steam-drive hexagon (injector well LSE-3054 and surrounding production wells). This is one of three early pilot hexagons having a record of 2 years of steam-

[1]Present address: Nassau Odyckstratt 18, 2596 AH Den Haag, The Netherlands

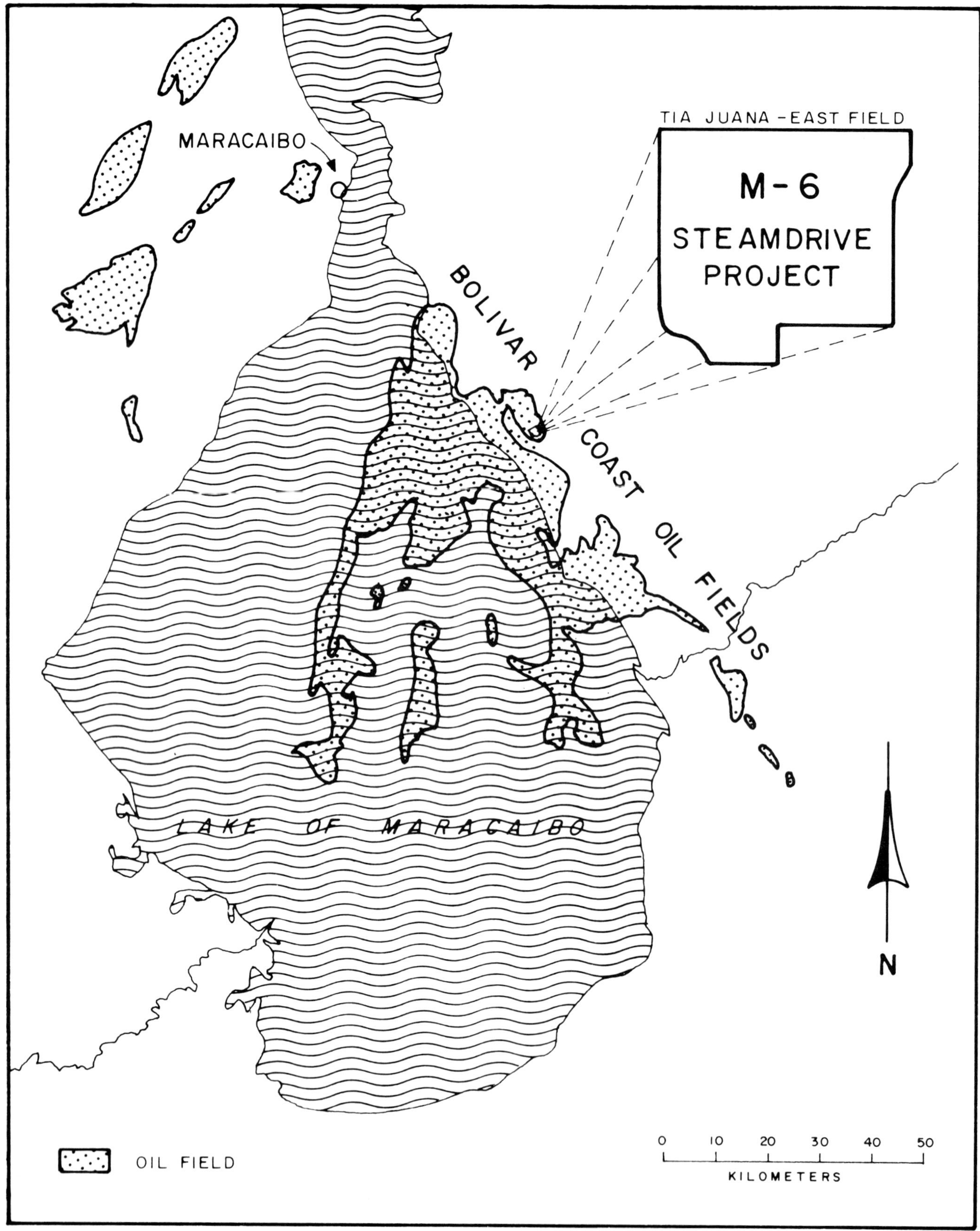

FIG. 1.—Location of west Venezuelan oil fields and the M-6 project area.

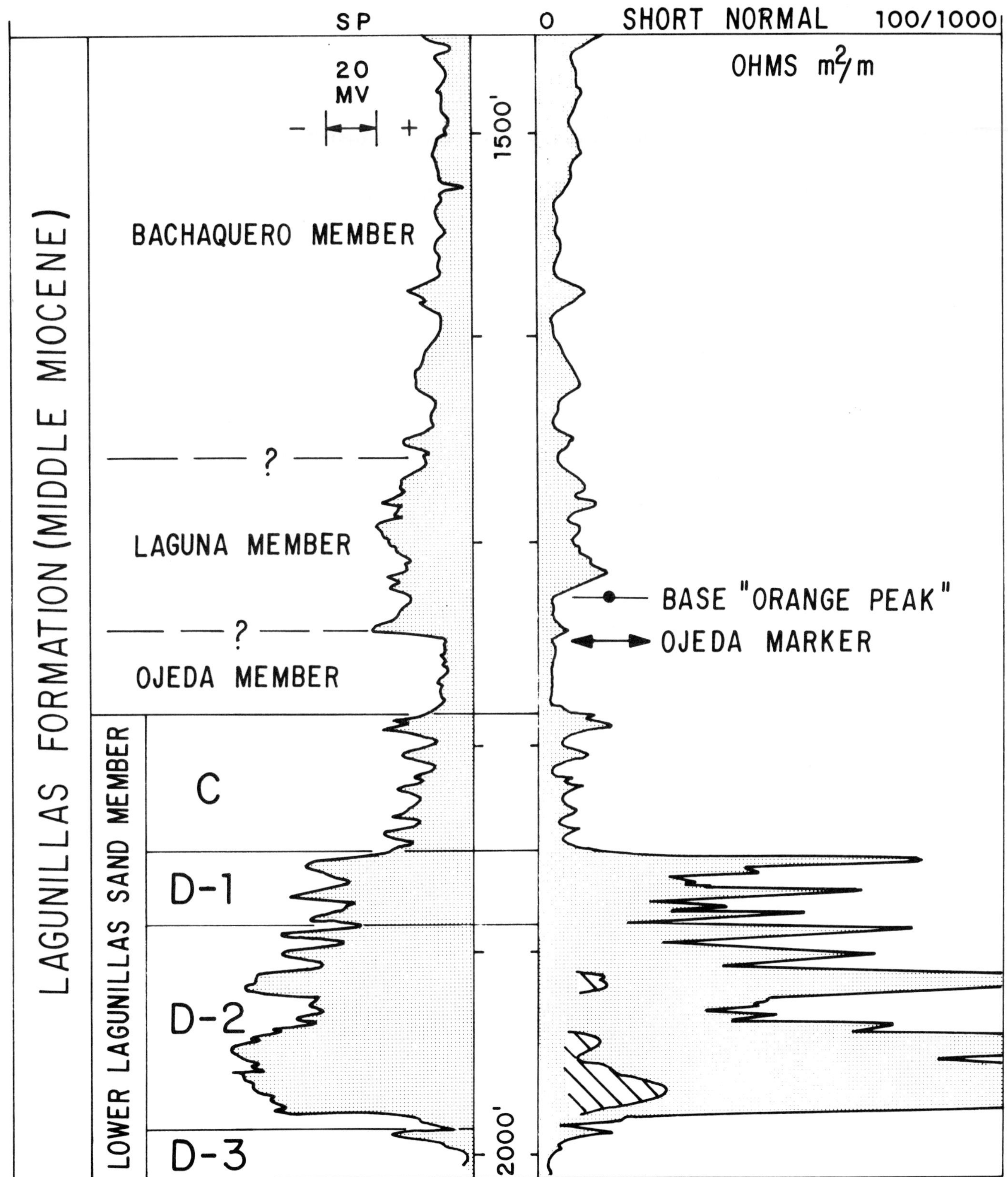

FIG. 2.—Type log of the Middle Miocene Lagunillas Formation in the M-6 steamdrive area.

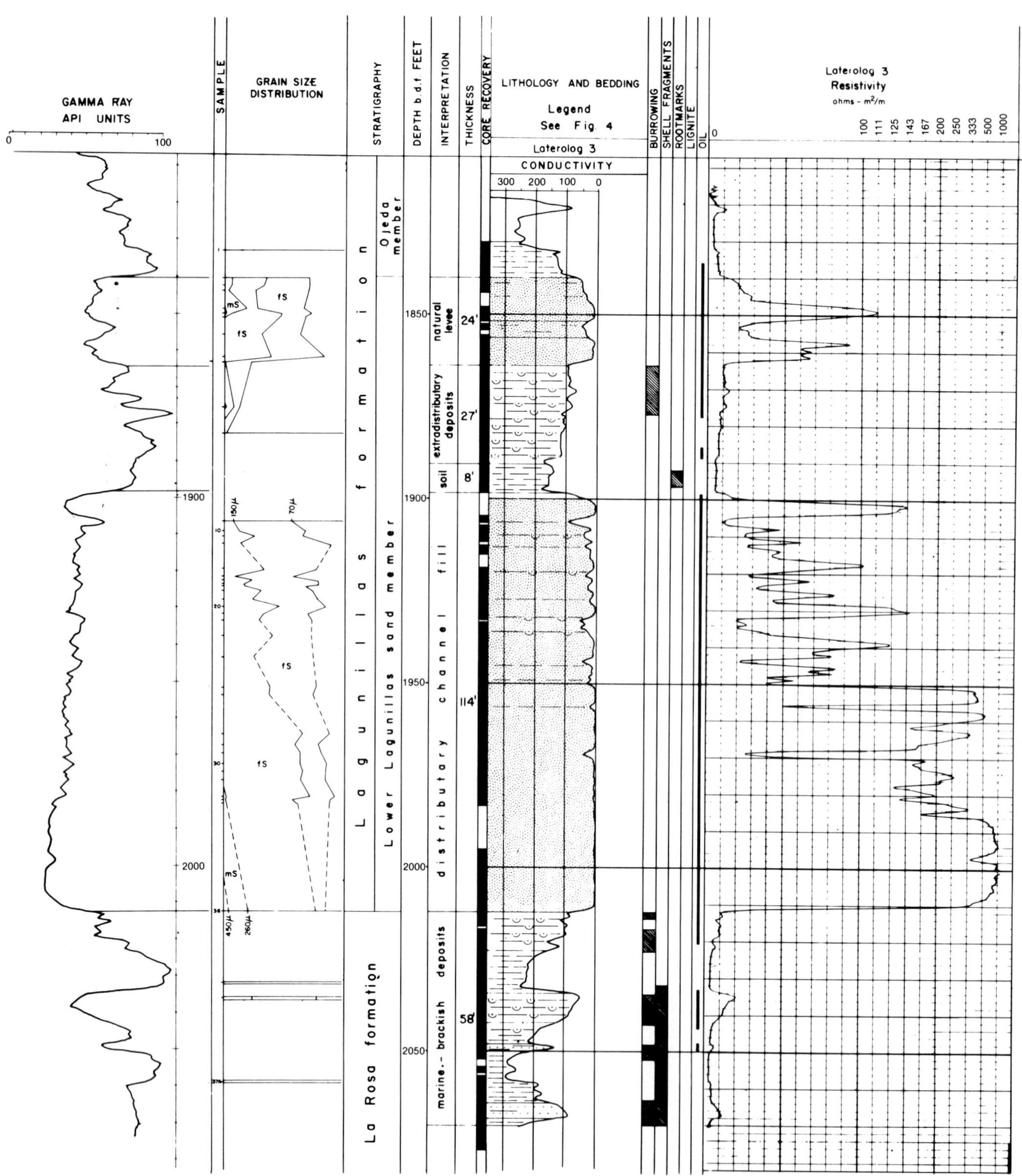

FIG. 3.—Well LSE-3054—core analysis and comparison with log data. See Figure 4 for legend. (After F. R. Van Veen, pers. commun., 1967.)

drive history uncontaminated by the larger scale steam-drive operations that started in the surrounding area 2 years later. (For details of history see Figure 17).

SEDIMENTOLOGIC APPROACH

Available Subsurface Information

Well logs.—

Good quality 1/200- and 1/1,000-scale electric logs (SP and 16″ and 64″ normal logs) were available for the majority of wells in the study area. In 10 of the wells laterologs had been run instead of normal electric logs. Density, gamma-ray and micrologs were available only sporadically. Where necessary, 1/200-scale logs with reasonable resolution quality were obtained by processing digitized 1/1,000-scale logs.

Petrophysical well log interpretations.—

Computer-processed log interpretations were available for all wells. Detailed subdivisions of productive and nonproductive zones with corresponding net-sand counts and Ø.So values were also available (Suarez, pers. commun., 1976).

Core data.—

Information from rubber-sleeve cores covering the productive interval in five wells included core photographs, lithologic descriptions, environmental interpretations, analytical data on grain-size distributions, and porosities and permeabilities. An example of sedimentologic information obtained from cores taken in well LSE-3054 is illustrated in Figures 3 and 4.

Sidewall sample data.—

In order to amplify the environmental information obtained from core studies, a series of sidewall samples at 2-ft (0.6 m) intervals was shot in three wells. The three wells were ones which had to be redrilled for reasons of production technology. A visual grain-size comparator study of these samples provided sequences of vertical grain-size variation, which were of value in determining environments of deposition (Fig. 5).

Previous geologic studies.—

Subdivision of the reservoir complex was made in the C, D-1, D-2 and D-3 sand members on the basis of 1/1,000-scale log correlations (Suarez, pers. commun., 1976; Soto, 1977).

Correlation of Datum Horizon

The fluvial-deltaic reservoir sands are overlain by a transgressive sequence of sand and shale lenses (thickness

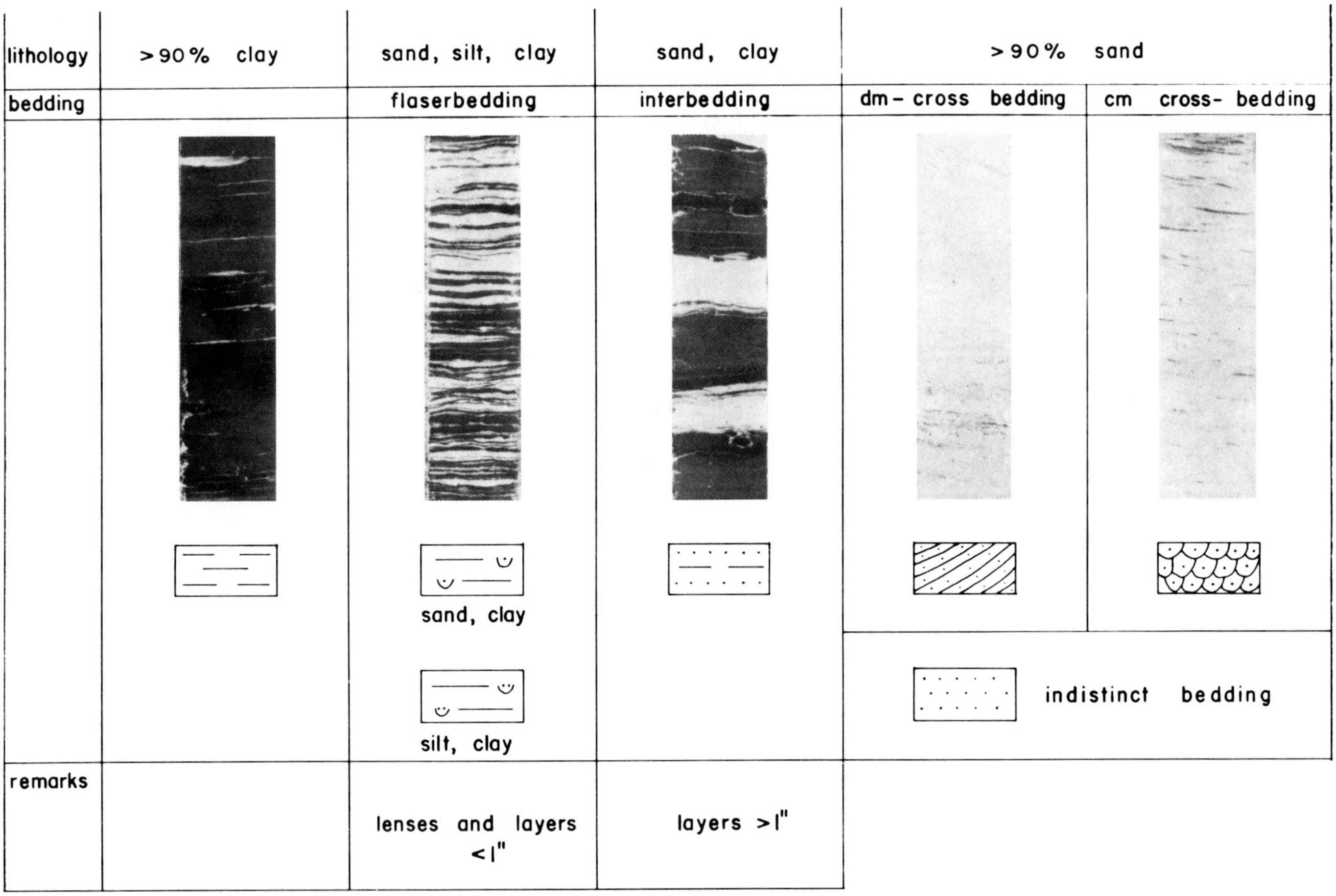

lithology	>90% clay	sand, silt, clay	sand, clay	>90% sand	
bedding		flaserbedding	interbedding	dm-cross bedding	cm cross-bedding
		sand, clay silt, clay		indistinct bedding	
remarks		lenses and layers <1"	layers >1"		

FIG. 4.—Photos of typical sedimentary structures which are keyed to units in Figure 3. (After F. R. Van Veen, pers. commun.)

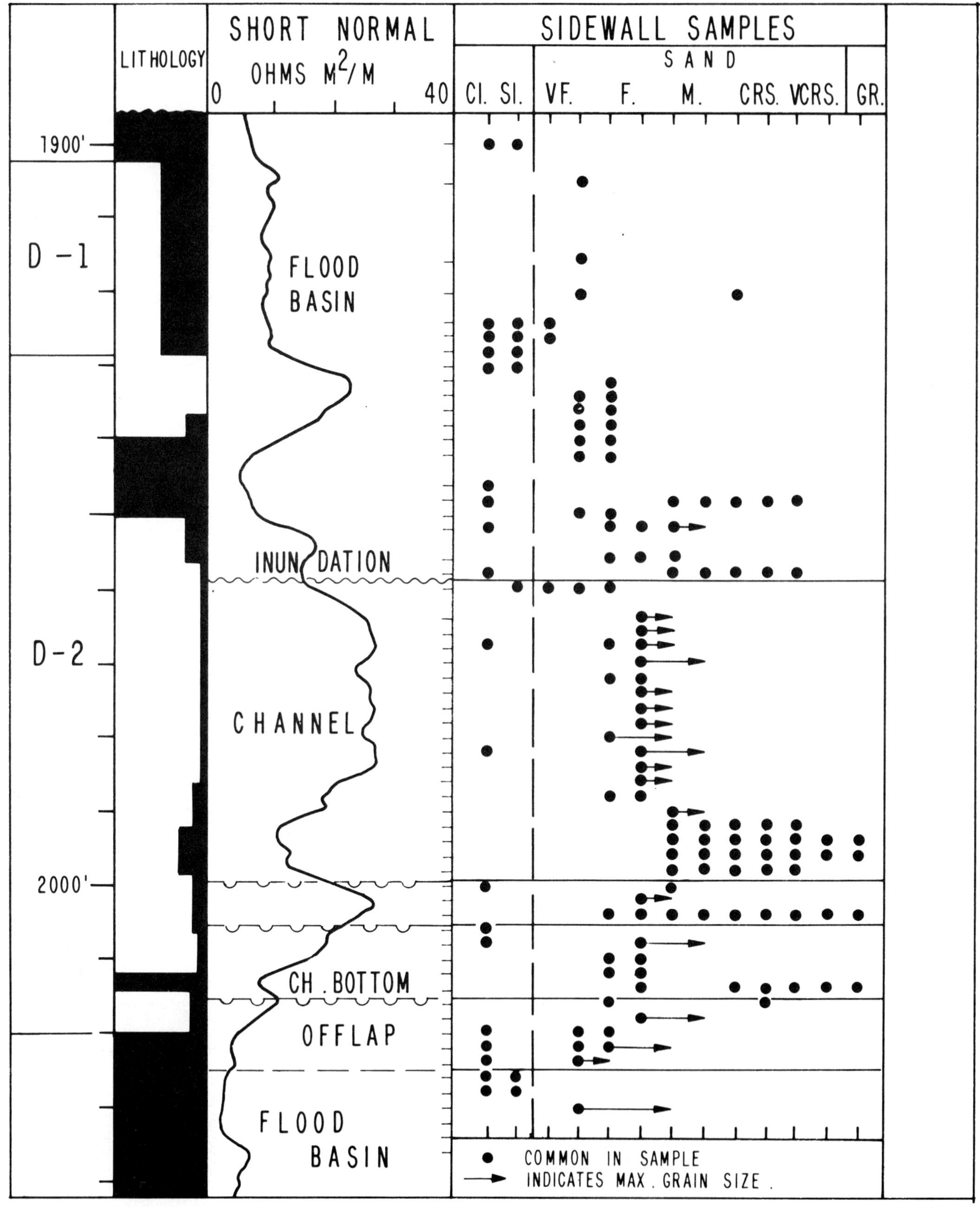
LITHOLOGY
SHORT NORMAL
OHMS M²/M
0
40
SIDEWALL SAMPLES
SAND
Cl. Sl.
VF.
F.
M.
CRS. VCRS.
GR.
1900'
D -1
FLOOD BASIN
INUNDATION
D-2
CHANNEL
2000'
CH. BOTTOM
OFFLAP
FLOOD BASIN
COMMON IN SAMPLE
INDICATES MAX. GRAIN SIZE.

around 120 ft or 36 m, including the C sands) which grades upward into a shallow-marine interval of interbedded sands, silts and shales of wide lateral extent. Detailed log correlations of the lowermost 80 ft (24 m) of the shallow-marine sequence were obtained without difficulty. The base of the lowermost sand peak of this easily correlatable interval was selected as the datum horizon (base "orange peak," Fig. 2) for correlation of the sections of the underlying reservoir sands.

A structure map of the datum horizon shows a southwest-dipping monocline with only minor faulting or flexuring (Soto, 1977). This structure map was used to establish, by interpolation, the datum positions in 23 wells that had been cased prior to logging.

Correlation of the Productive Interval

Cross sections oriented in three different directions were prepared for the entire project area. Since detailed correlations within a complex of largely fluviatile sand deposits may give misleading results, in this study more value was attached to horizontal (lateral) correlations than to log similarity. This was possible because of the availability of a reliable datum horizon not far above the sands and the absence of tectonic complications. Initial correlations were made by tracing recognizable sedimentologic features such as channel bottoms and inundation planes ("*Correlation of main sedimentological outline*"). A general classification of depositional conditions, based on the evidence of log shapes (Figs. 6, 7), was supplemented, where possible, with core data and sidewall sample information.

A second phase of correlation evaluated the continuity within the general sedimentologic pattern ("*infill correlation*") of all sand and shale units that were established during the petrophysical evaluation. (Detailed correlations within the fluvial-deltaic reservoir are shown in Figs. 12 and 13.) By purposely abstaining from "forcing" certain sand or shale correlations over the entire area, a realistic picture of the lateral and vertical (reservoir) sand contacts is obtained. A three-dimensional picture is also obtained which shows, to a limited extent, depositional cyclicity and the orientation of channel sand trends. Several log-shape patterns were used in mapping and identifying lithofacies units, such as "offlapping sands" and "clay-filled abandoned channels."

Preparation of Maps

Net-sand maps.—

Computer-generated sand distribution maps of the major sand intervals were produced using a petrophysical data bank.

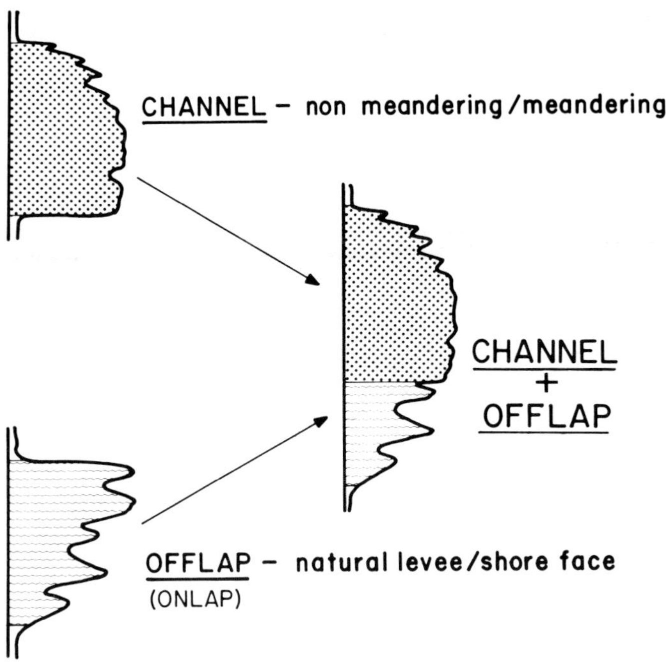

FIG. 6.—Hypothetical log shapes for coastal and channel reservoir sands.

The net-sand distribution of the D-1 interval is shown in Figure 8. The distribution pattern appears rather erratic because the map incorporates sands of diverse lithofacies units, *h.1Ø.So maps* ("oil column maps"). Maps showing the net thickness of the oil column, in combination with porosity and oil saturation information, allow an evaluation of horizontal oil distribution (Fig. 9). In a fluviatile area, the maximum-value trends shown on such maps should correspond closely to channel or meanderbelt trends. This type of map is more definitive than the net-sand map of the same interval (Fig. 8), and at least two major river-sand courses are clearly defined.

Ø.So sand isopach map.—

A crude relation exists between Ø.So and grain-size distribution, partly because coarse sands have higher oil saturations (So values) than fine sands due to the difference in connate water. In contrast, porosities (Ø) should be rather uniform in unconsolidated sands at reservoir depth. Differentiation can be made between (coarse-grained) channel sands and (finer grained) offlap sands on the basis of petrophysical information. Comparison of numerous computer-printed isopach maps of sands with different Ø.So minimum values indicates that channel sand trends show best on isopach maps

FIG. 5.—Environmental analysis from sidewall samples, taken with 2-ft (0.6 m) spacing (well LSE-1853-A). The grain size of sidewall samples was determined with a grain-size comparator. For each sample, grain-size classes of common presence only are recorded. The recognition of depositional environments depends on vertical grain-size sequences and on the general knowledge of the local sequential development obtained from the study of a few rubber-sleeve cores taken in this area. The lithology column is based only on petrophysical log analysis (black = silt/shale; white = reservoir sand). Illustrated is one of the limitations of the use of petrophysical log shapes in environmental reconstruction. On the basis of log shapes only, a channel bottom would have been postulated at 1,985 ft (605 m), whereas the interval between 1,985 and 2,020 ft (605–616 m) would have been considered to represent two cycles of flood basin and offlap deposition. Sidewall cores indicate that the relatively low resistivity of the interval between 1,990 and 2,000 ft (606–610 m) appears to be related to the rather poor sorting of medium to very coarse sands, which leads to a higher water saturation and consequently to a lower resistivity in comparison to the much better sorted fine-grained sands between 1,965 and 1,990 ft (606–616 m).

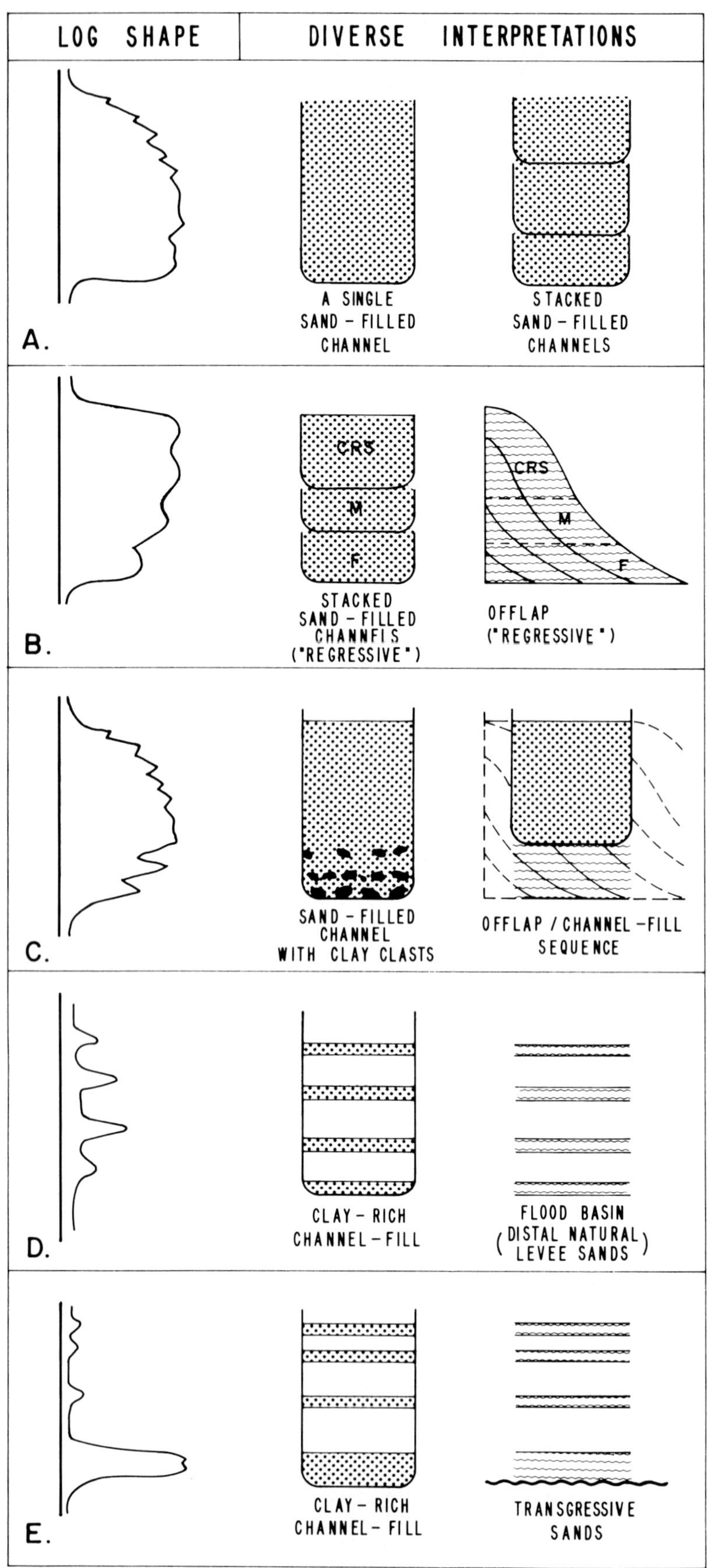

FIG. 7.—Sequences which may be recognized using log-shape interpretation and which may allow determination of potential environments of deposition.

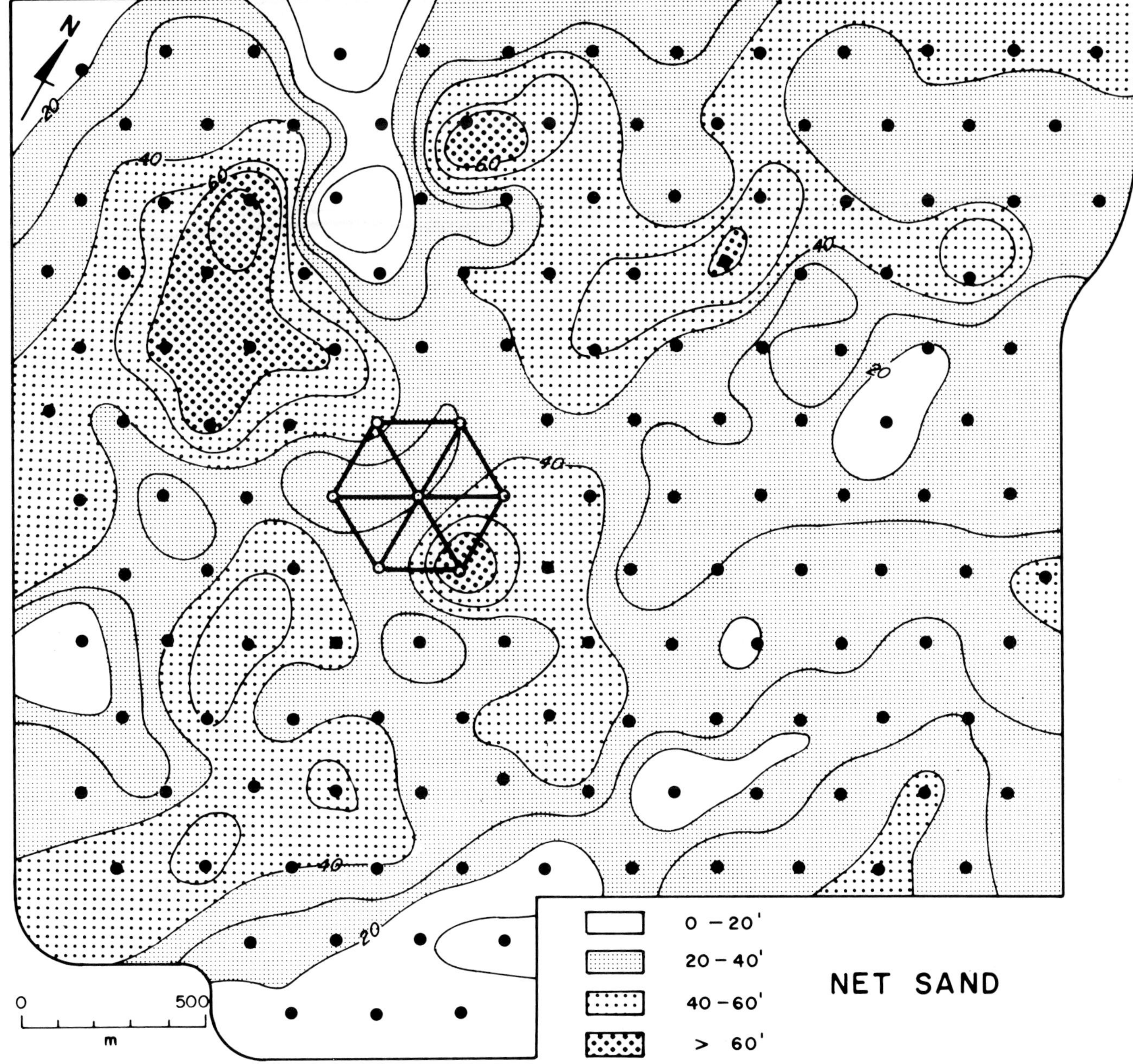

FIG. 8.—Net-sand thickness of the D-1 interval. The position of the steamdrive hexagon LSE-3054 is shown.

of sands with Ø.So values greater than 0.3 (Fig. 10).

At least two major trends with high Ø.So values are well expressed in the D-1 sand (Fig. 10); thickness trends indicate an approximate NNE/SSW orientation of the main river channels. It should be noted that values of 0.3 or higher do not occur in the southern and northwestern parts of the map.

REGIONAL DEPOSITIONAL MODELS

After the individual depositional cycles were identified, the next step in the sedimentologic approach was to prepare a regional depositional model for each cycle. The feasibility of developing such models depends largely on the accuracy with which depositional environments and cycles can be delineated. The D-1 cycle is reasonably well defined over a large part of the project area (Fig. 11), and within the cycle numerous facies are recognized. The major reservoir is a continuous channel trending northeast-southwest. A seemingly discontinuous minor channel sand trend also is observed and is interpreted as a meanderbelt deposit. A linear trend of clay deposition flanking the main channel sand trend is tentatively interpreted to be a clay-filled channel course ("clay plug"). Intervals with resistivities that in-

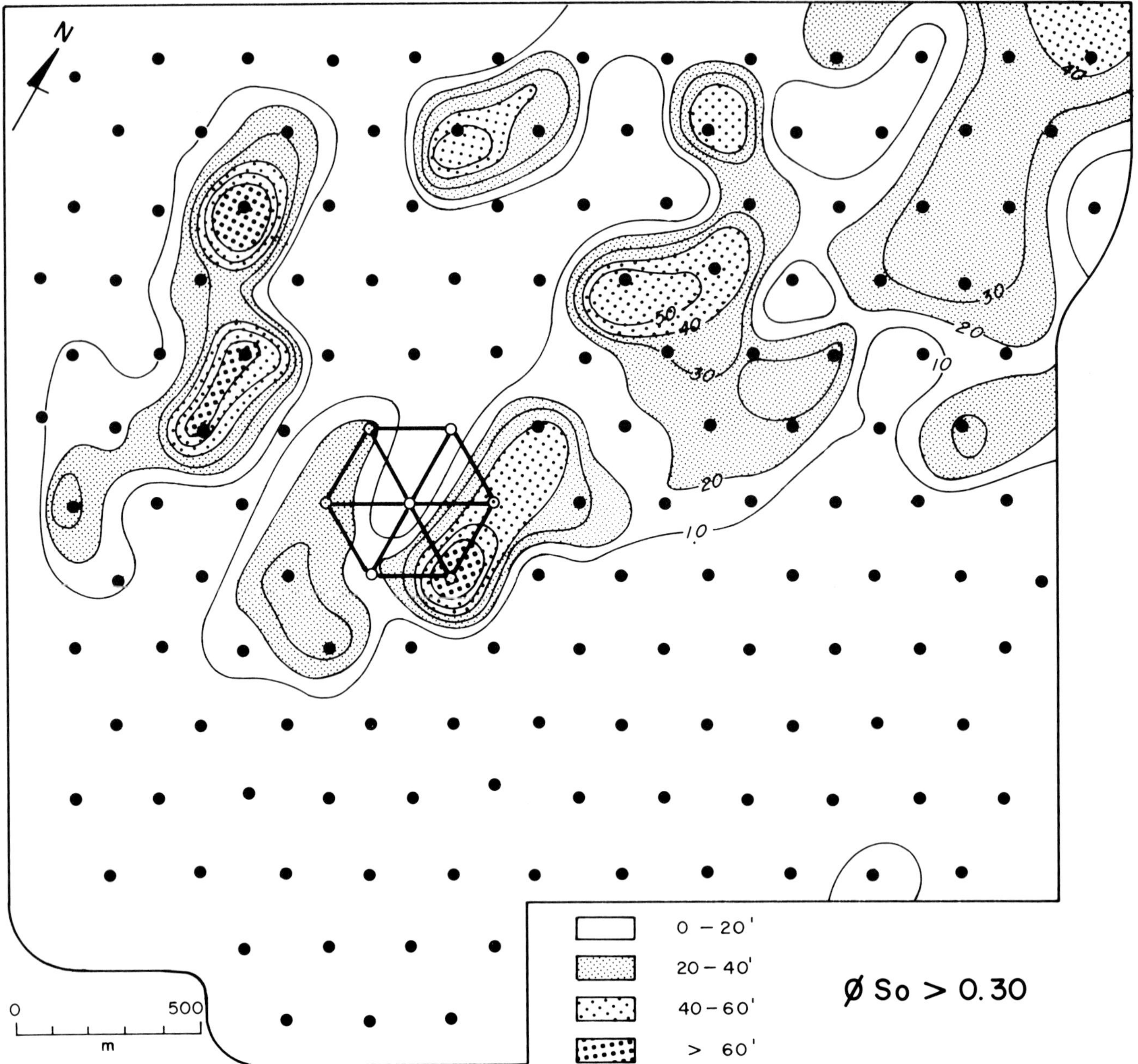

FIG. 9.—Oil column isopach map (h.Ø.So), net-sand thickness × porosity × oil saturation of the D-1 interval.

crease upward are interpreted as probable natural levee and/or shallow bay deposits. These environmental interpretations were based primarily on evidence from full-diameter cores and were supported by a series of closely spaced side-wall samples.

Information from cores and sidewall samples is usually available in only a very limited number of wells (less than 5% of the wells in the M-6 project area), however, which means that in the majority of the wells the environmental evidence has to be derived from well log interpretation (log shapes and Ø.So values). Within the fluviatile coastal domain, two basic types of log patterns mainly occur for sand deposits. These log patterns may occur individually or in combination (Figs. 6, 7). A log pattern which exhibits a maximum resistivity at or near the base of the sequence and yields a characteristic "bell-shaped log" is considered to be typical of a broad spectrum of sands deposited in channels, such as nonmeandering river channels, point bars, tidal channels, etc. A second type of log shape occurs where the maximum resistivity is at or near the top of the sequence and yields a characteristic "funnel-shaped log." This log pattern is definitive of several sand types, including lateral accretion of interdistributary sands such as natural levees and various types of shoreline sands.

Interpretations obtained from these basic log patterns are subject to a number of limitations, as shown in Figure 7.

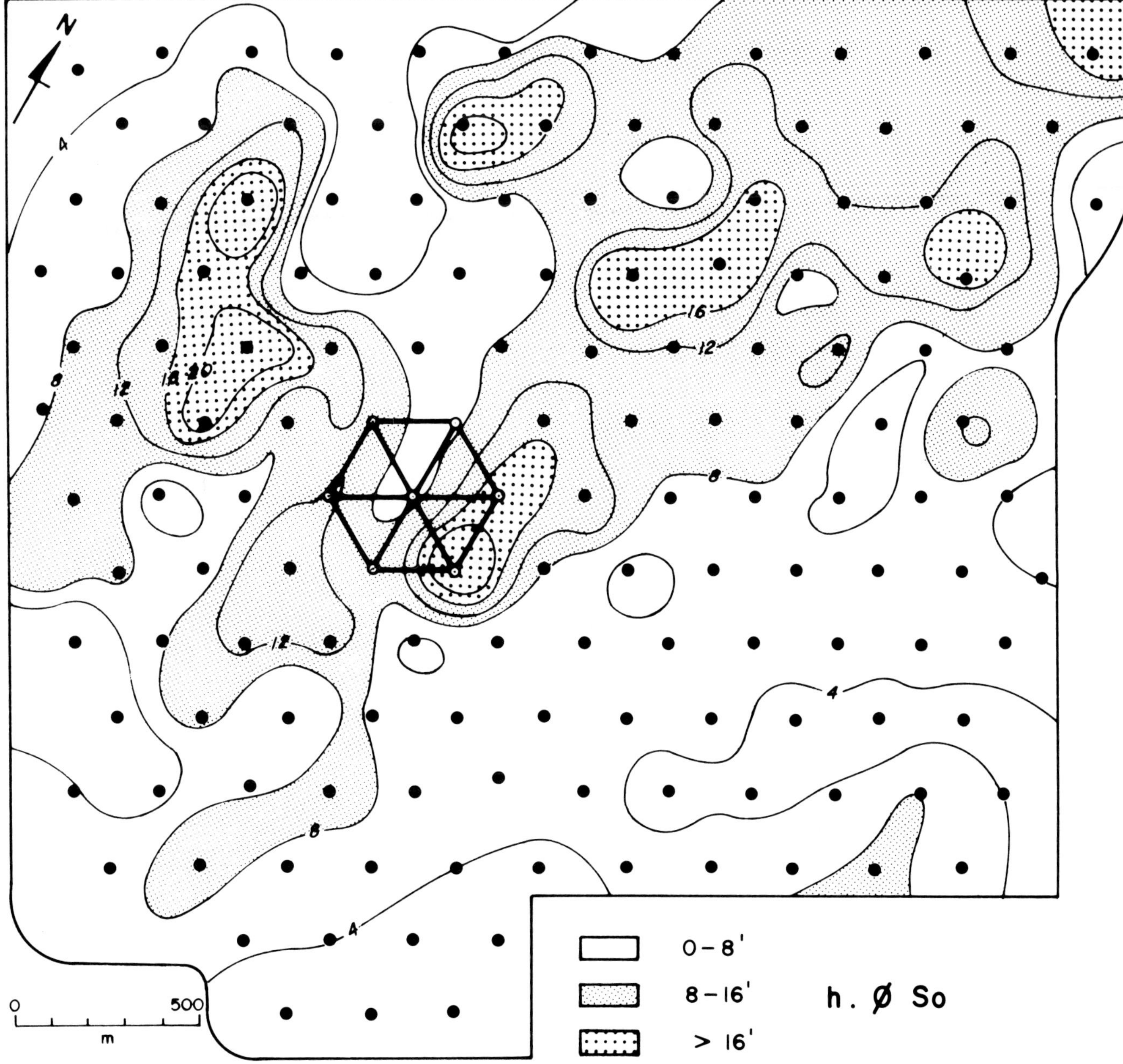

FIG. 10.—Thickness of sands with Ø.So values exceeding 0.3 in th D-1 interval.

Separation of individual genetic channel sand units within an amplified multistory channel sand body is commonly very difficult (Fig. 7A). Hence, any thick channel sand interval observed on well logs should possibly be considered to be a sequence of several genetic stacked channels, each having a different geometry and trend. Certain channel sands may show a resistivity sequence which increases upward (Fig. 7B). These would have the same appearance as offlap sequences of coastal barrier deposits or channel-mouth bars. Log shapes of channel sand sequences may be altered drastically by the presence of clay clasts and may simulate onlap/offlap deposition (Fig. 7C). In most instances there is little or no distinction between the log shapes of clay-rich channel fills, natural levee deposits, and flood basin deposits (Fig. 7D). The log shapes of the bottommost sands in clay-filled channels may be identical to those resulting from "transgressive" sands covering an inundation plane (Fig. 7E).

Although the problems in interpretation of log shape are recognized, in a number of cases, environmental identification does appear to be possible if cores are used to calibrate the logs and the log shapes are then used to interpret adjacent wells and wells on a more regional scale.

An environmental analysis of this type requires intensive cross-checking among individual wells, cross sections and maps. In this way, the regional sections produced in the

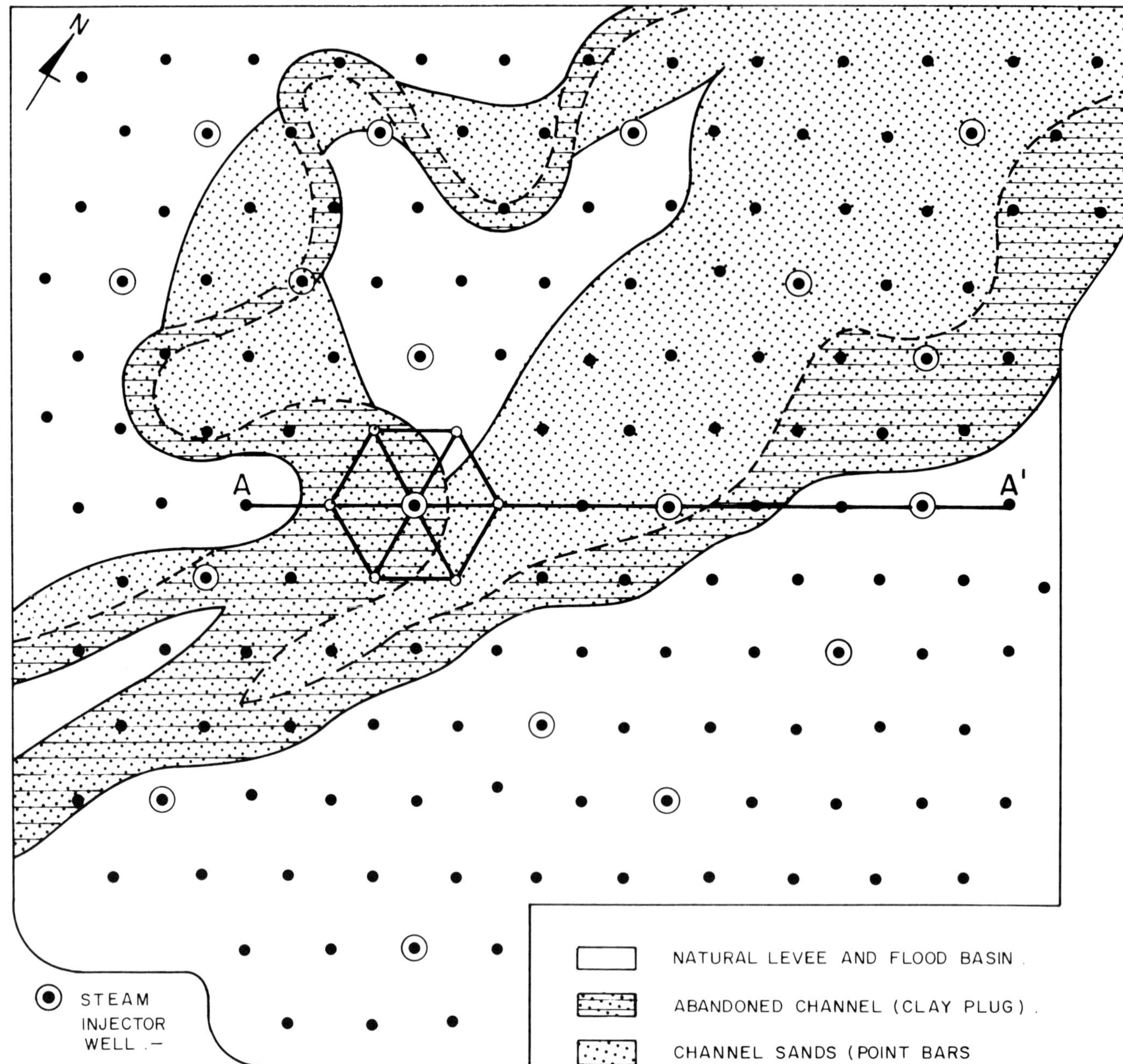

FIG. 11.—Depositional model of the D-1 interval. Section A-A′ is location of data shown in Figures 12 and 13.

initial phase of the project are gradually refined and remolded into *environmental correlation sections* of the type shown in Figure 12.

Regional Sections of Ø.So

The *environmental correlation sections* (Fig. 12) provide a good view of sand distribution and of possible sand continuity; however, they provide only limited information on grain size and permeability distributions. In order to incorporate grain size and permeability into the data base, environmental sections have been remodeled into sections showing the distribution of Ø.So classes (Fig. 13). These sections provide the best approximation to grain size and permeability variations that can be derived from the subsurface information available. Also, the occurrence of cut-and-fill types of channel sands and nonerosional interdistributary splay sands can be inferred from detailed correlations among the sections.

Hexagon Correlation Sections

In order to predict and monitor steam and fluid movements, three radial correlation sections were prepared for

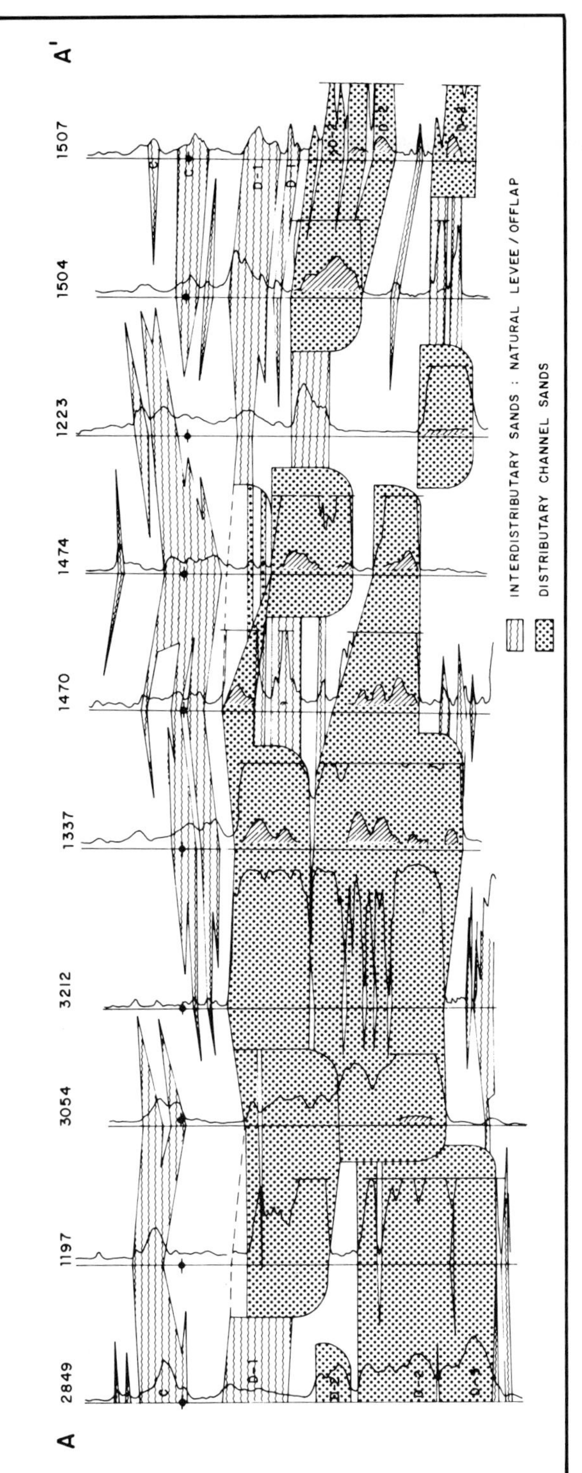

FIG. 12.—Environmental analysis of reservoir sands; compare with Figure 13. Location of west to east cross section is shown on Figure 11.

each steam-injection hexagon. The sections were made with 1/200-scale electric logs, which allowed detailed plots of sand/shale distributions and of vertical Ø.So variations.

For each hexagon, three sections were oriented in different directions in order to show not only sand/shale continuities from the injector well to the six production wells, but also a clear view of three-dimensional reservoir development within the hexagon (Figs. 14–16).

Hexagon correlations should, of course, correspond to concepts developed using gross regional correlations, but they should at the same time show the continuity or discontinuity for each sand/shale layer observed. Inferred continuities should be based on direct sedimentological evidence; however, reliance on what one may call "sedimentologic inspiration" can often not be avoided because of the usual complexities in the distribution of fluviatile sediments. To prepare the suite of detailed sections for the hexagon is very time consuming; thus, to reduce the time required to prepare individual sections, a computer program was used which utilized digitized logs and a petrophysical data bank.

ANISOTROPIC PRODUCTION BEHAVIOR OF STEAM-INJECTION HEXAGON LSE-3054

Steam was continuously injected into well LSE-3054. Flowmeter surveys showed that 20–33% of the steam was injected into the D-1 sands (see dotted steam-intake curves in Figs. 14–16). It is likely, however, that much of the remainder of the steam will also reach the D-1 sand, because core samples indicate that there is virtually no permeability boundary between the two intervals, apart from a few apparently noncontinuous thin clay lenses and laminae (Fig. 3).

Production results in this hexagon are shown in Figure 17 for wells with varying production capabilities. One example well had an exceptionally good reaction to steam drive (LSE-1347), a second well had an average reaction (LSE-1858), and the third well had a rather slow reaction (LSE-1408). The hexagon correlation sections suggest that there is a reasonable reservoir interconnection between the steam injector well and these three production wells. No good comparison can be made with the results of the other three production wells of the same hexagon, because they were stimulated by temporary steam injection ("steam soak") during the same time period. For this reason the results of these wells are not included in Figure 17 and Table 1. Table 1 indicates a direct relationship between the amount of additional oil produced as a result of the steam flood and the quality and depositional characterstics of the D-1 sand in the producing wells. Those wells located outside the continuous channel produced diminished amounts of, or no, oil. The well in one of the abandoned channels produced only 25% as much as the well within the continuous (active) channel. A second abandoned channel well had no additional production. The well which produced only from the natural levee or flood basin had only 10% additional production.

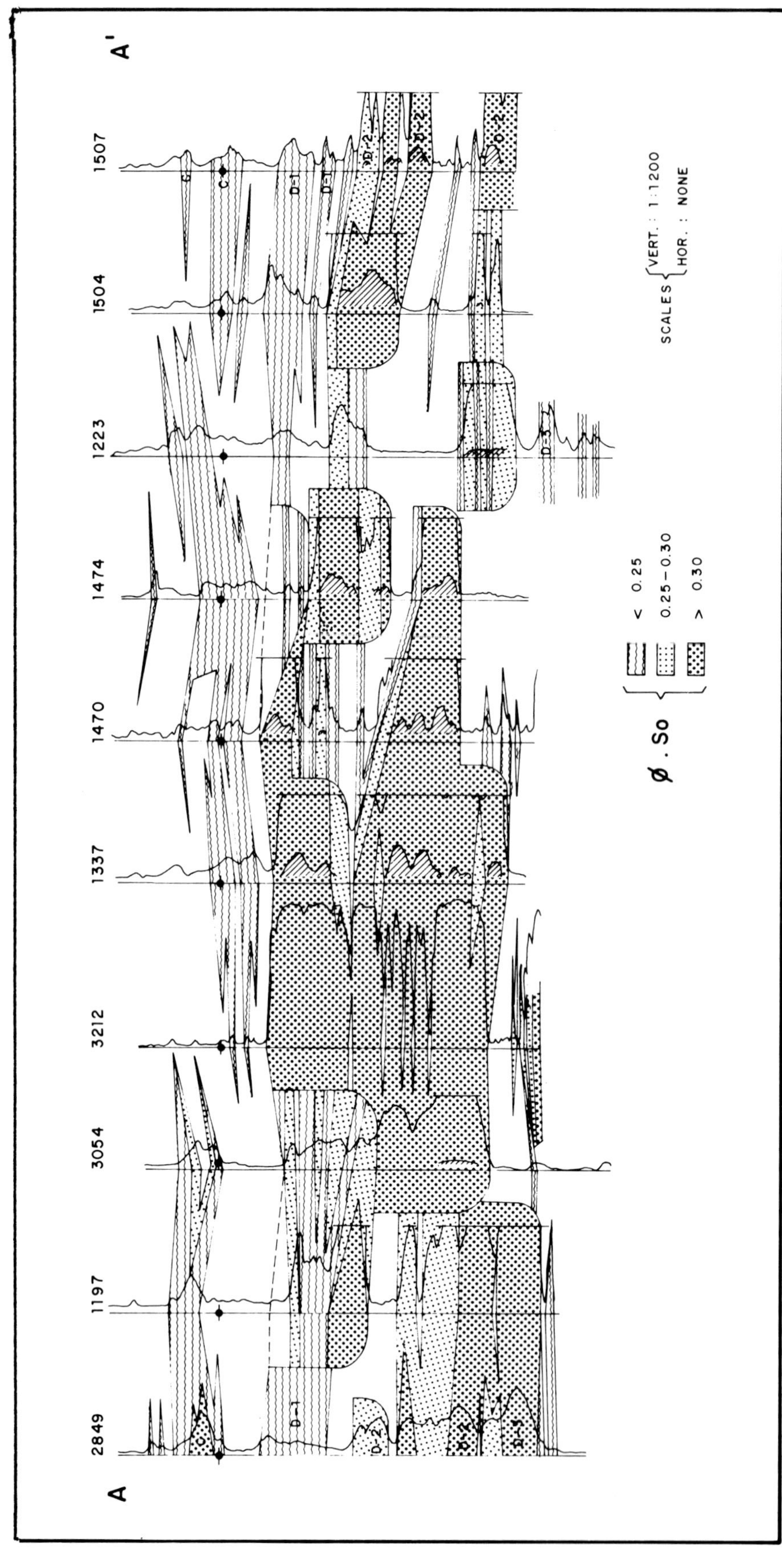

FIG. 13.—Lateral correlation of Ø.So. Location of west-to-east cross section is shown in Figure 11.

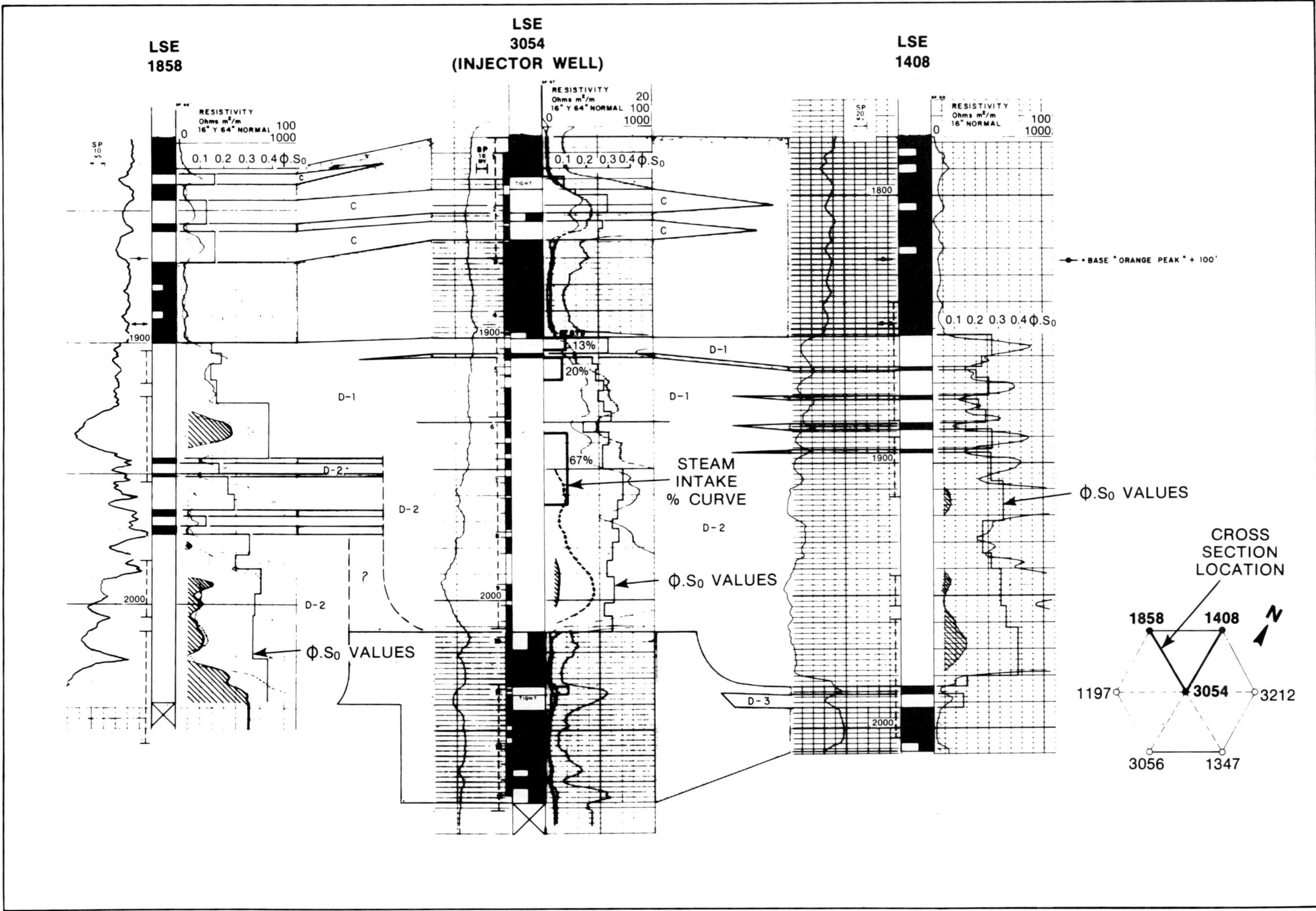

FIG. 14.—Cross section through injection hexagon in which LSE-3054 is an injection well. Stepped curve to the right of central column indicates percentage of steam entering. Vertical scale 1:500. No horizontal scale.

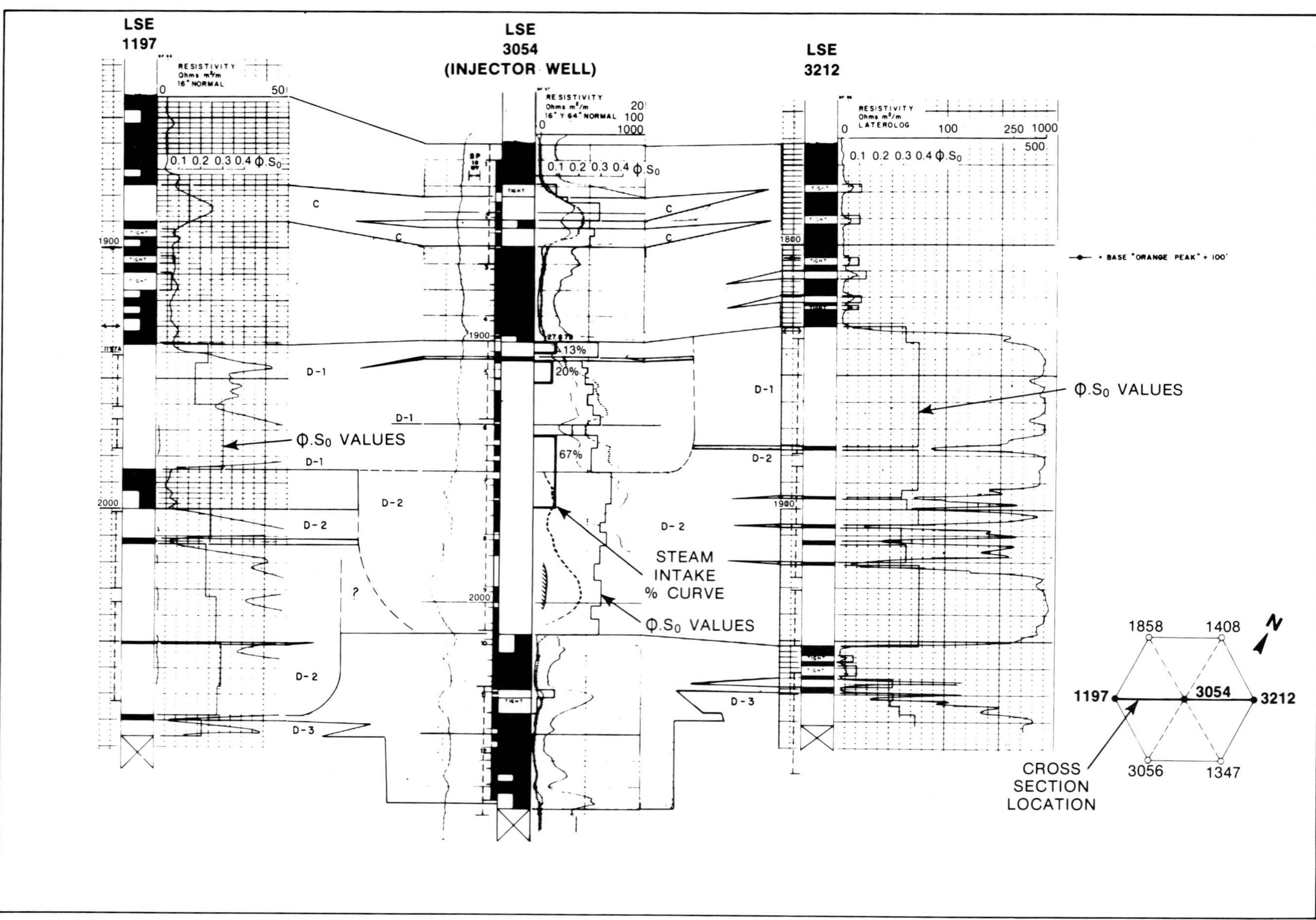

FIG. 15.—Southwest-northeast cross section through an injection hexagon in which well LSE-3054 is injection well. Vertical scale 1:600. No horizontal scale.

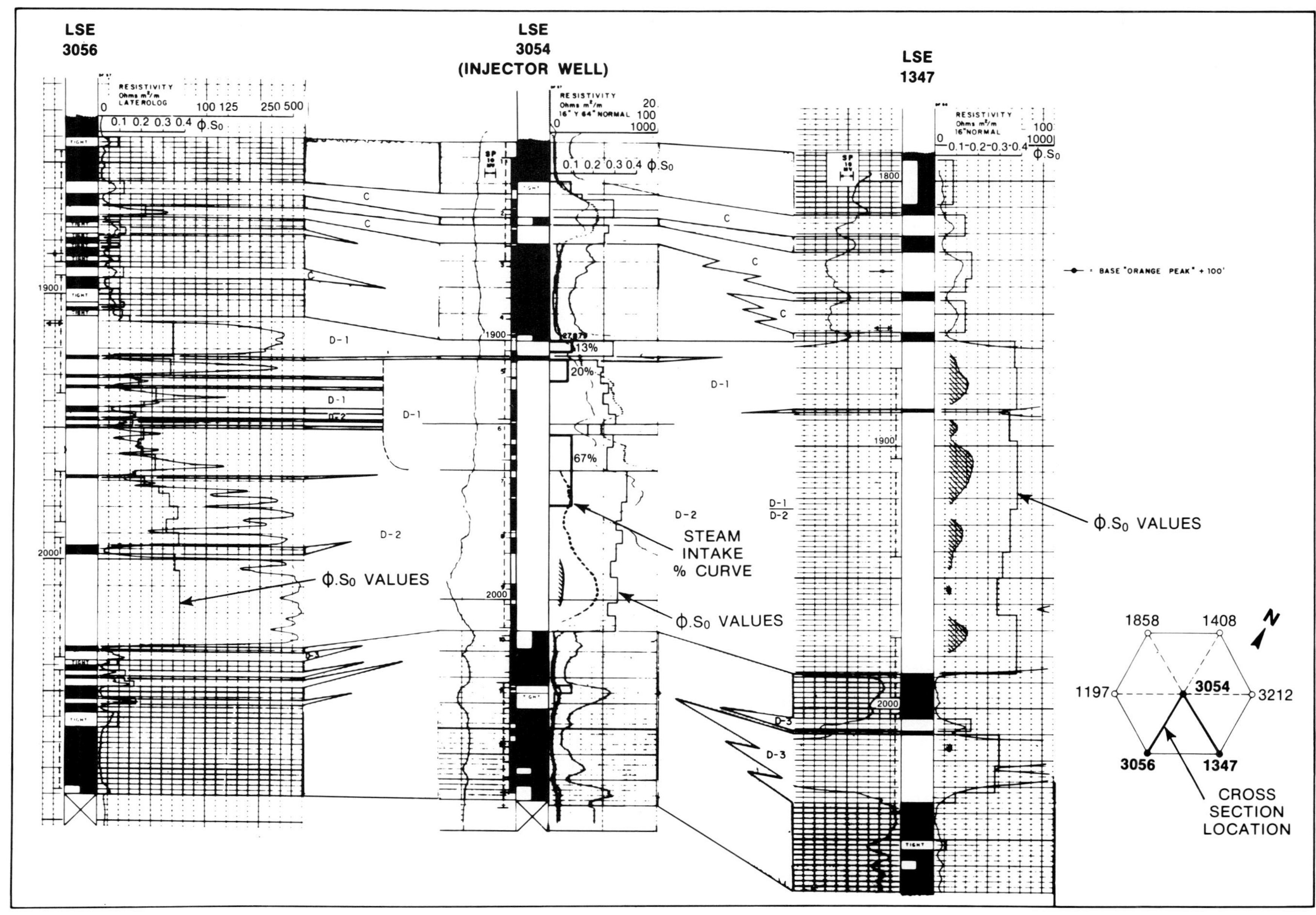

FIG. 16.—Cross section through injection hexagon in which well LSE-3054 is an injection well. Vertical scale 1:600. No horizontal scale.

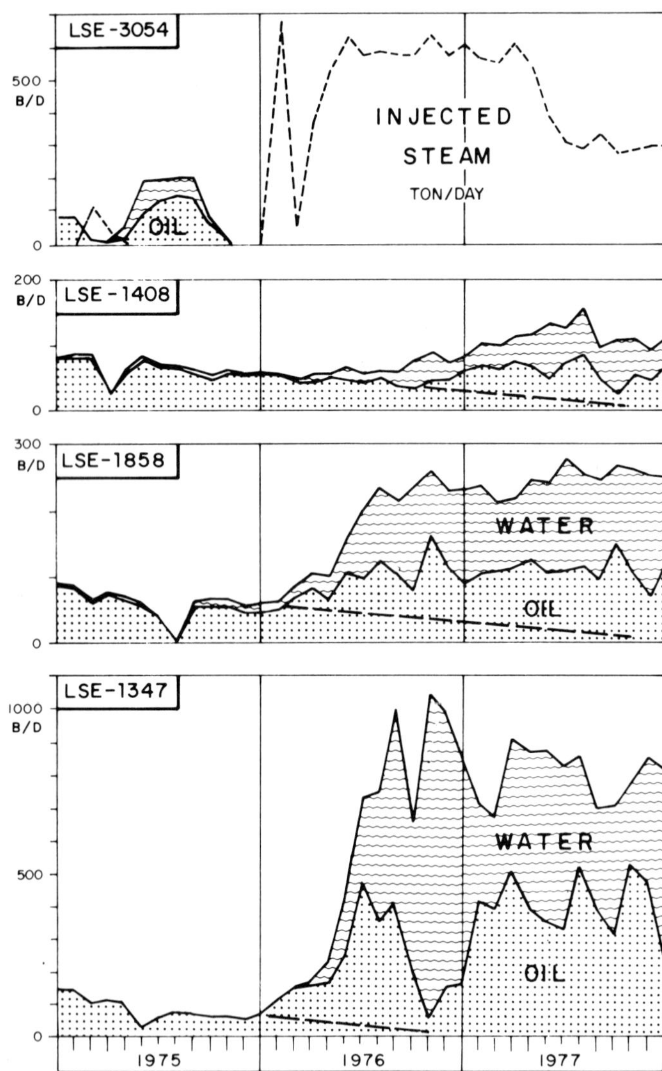

FIG. 17.—Performance of steam-injection hexagon LSE-3054 1 year prior to steam drive (1975) and during 2 years of steaming (1976–1977). The distance between the steam injector well and the oil production wells is about 231 m. Four types of results are illustrated: LSE-3054, a steam-injector well, was originally an oil producer but was converted into a continuous steam injector ("steam drive") in January 1976. Steam had also been injected in February 1975 in order to stimulate the oil production during the rest of that year ("steam soak"). LSE-1408, an oil production well, had a rather slow reaction to continuous steam injection in LSE-3054. The amount of extra oil produced by steam drive can be visualized by taking into account the predicted decline in oil production rate over the period 1976–1977 (dashed line is extrapolation of decline). LSE-1858, an oil production well, is an example of average reaction to continuous steam injection in LSE-3054. LSE-1347, an oil production well, is an example of exceptionally strong reaction to continuous steam injection in LSE-3054.

TABLE 1.—CHARACTERISTICS OF D-1 SAND IN STEAM-INJECTION HEXAGON LSE-3054

Well	Net sand (ft)	Oil column h.Ø.So (ft)	Sand with Ø.So >0.3 (ft)	Depositional environment	Extra oil produced 1976/1977 (bbls)
LSE-3054	30	7.48	0	Abandoned channel	0 (steam injector)
LSE-1408	30.5	8.04	0	Natural levee or flood basin	20,000
LSE-1858	42.5	10.05	20.5	Abandoned channel	50,000
LSE-1347	62.5	20.97	62.5	Channel	200,000

CONCLUSION

The major contribution of this sedimentologic subsurface study for steam-drive operations was the preparation of very detailed injection hexagon correlation cross sections, which show well-to-well sand/shale distributions based on petrophysical interpretations that were initially calibrated from full-diameter core interpretations. These hexagon cross sections are used to show reservoir hetergeneity, to predict the variability of production results, and to depict reservoir-continuity distributions. Such results may also be utilized to predict production capabilities through computer simulation or by other means. By utilizing the techniques discussed herein, we have significantly improved our capability to predict steam and fluid movements. Further work of this type will improve planning and implementation of future steam-injection patterns.

ACKNOWLEDGMENTS

The author conveys his gratitude to Maraven S.A. and Petroleos de Venezuela S.A. for granting permission for the publication of this paper. I am also grateful to my professional colleagues and especially to Ir. Theo Verhaegh for the many discussions on aspects of reservoir engineering. Dr. Jack Gibson Smith, R. W. Tillman and Barbara H. Lidz read the manuscript and improved the style of writing.

REFERENCES

HERRERA, A. J., 1977, The M-6 Steam Drive Project, design and implementation: Journal of Canadian Petroleum Technology, v. 16, p. 62–71.

KRUIT, C., 1979, Sediment-geological study of a steam-drive project in deltaic river sands, *in* the Future of Heavy Crude and Tar Sands: McGraw Hill, Inc., New York, p. 321–326.

SOTO, C., 1977, Estudio geologico sedimentologico para un proyecto de recuperacion secondaria en la Costa Bolivar: V Congreso Geologico Venezolano, Caracas 1977, Memoria Tomo IV, p. 1493–1503.

VAN DER KNAAP, W., 1980, M-6 Steam Drive Project, preliminary results of a large scale field test: Society of Petroleum Engineers, SPE Paper 9452, Dallas, Texas, 8 p.

GEOLOGIC FACIES ANALYSIS FOR ENHANCED OIL RECOVERY, BARTLESVILLE SANDSTONE, GREENWOOD COUNTY, KANSAS

DOUGLAS W. JORDAN[1] AND RODERICK W. TILLMAN[2]
Cities Service Research, Tulsa, Oklahoma

ABSTRACT: Five depositional facies occur in five oriented cores obtained from the Bartlesville Sandstone (Cherokee Group, Middle Pennsylvanian) in the Madison Unit "B" enhanced recovery pilot project in Greenwood County, Kansas. By orienting the cores and studying the flow directions determined from high-angle crossbedding, a model for predicting the trend of the channel sandstone was constructed. Environments of deposition associated with the channels are (1) *Interdistributary Bay Siltstone, Shale, and Sandstone Facies;* (2) *Crevasse Splay Sandstone and Siltstone Facies;* (3) *Overbank Shale and Siltstone Facies;* (4) *Distributary Channel Sandstone Facies;* and (5) *Delta Plain Shale and Siltstone Facies.*

These facies may be recognized on gamma-ray logs as well as on permeability and grain-size plots. The *Distributary Channel Facies* is the major reservoir facies and overlies brackish to continental delta plain sediments. It is overlain by brackish and marine facies, including overbank and interdistributary bay shales and siltstones. The channel had low sinuousity, as is evident from the narrow trend of the sandstone and the blocky (rather than "bell-shape") nature of the gamma-ray log. An overall fining-upward sequence typical of fluvial-deltaic sediments is suggested by an upward decrease in the grain size, a decrease in the scale of sedimentary structures, and an increase in the amount of interstitial clay matrix.

Variations among the sandstone facies are also reflected by variations in permeability values. Controls on permeability that differ among the various facies are packing, the presence of ductile rock fragments, the amount and morphology of authigenic and detrital clay, and sorting. The permeability of the *Distributary Channel Facies* decreases from 60 to 100 md in trough crossbedded sandstones typical of the lower and middle part of the channel fill to 20 to 60 md in rippled sandstones, siltstones, and shales of the upper part of the channel. Low permeability values (0–20 md) are characteristic of facies composed primarily of siltstone and shale.

The *Distributary Channel Facies* is continuous across the project area and its thickness varies in a predictable manner. Numerous and detailed dip directions measured primarily in trough crossbeds in four of the five oriented cores that were used in the study indicate that the channel flowed in a west-southwesterly direction across the project area. This flow direction is parallel to the maximum elongation direction of the sand body. The *Overbank, Crevasse Splay,* and *Interdistributary Bay Facies* are not continuous across the whole project area. These facies filled the topographically low areas adjacent to the channel.

INTRODUCTION

Five oriented cores, 51 to 91 ft (15–28 m) in length, from the Bartlesville Sandstone in the Madison "B" Unit, Seeley-Wicke field in Greenwood County, Kansas, were submitted for geological analysis in support of an ongoing micellar-polymer injection pilot project. The primary objective of the pilot project was to determine the oil recovery efficiency of the micellar-polymer process in the Bartlesville Sandstone in the Madison area. The purpose of this study was to interpret the environments of deposition of the reservoir sandstone, establish permeability and porosity controls and trends, determine favorable reservoir facies for tertiary stimulation, and geologically predict the efficiency of tertiary recovery techniques.

A mini-test, utilizing four wells that are not included in this study, was initiated in 1977 by the Cities Service Oil and Gas Corporation in order to evaluate the feasibility of using the surfactant-polymer process to recover oil after waterflood and to determine some of the reservoir parameters that cannot be determined in the laboratory. The pilot study originally covered an area of 22.5 acres (9 ha; Fig. 1). The pilot area has since been expanded to cover 40 acres (16 ha) with the addition of 16 outside producing wells.

Seeley-Wicke field, located on the eastern flank of the Nemaha Ridge in southeastern Kansas, was discovered in 1928. Depth to the Bartlesville Sandstone is less than 2,000 ft (610 m), making these wells very economical to drill. Total estimated primary and secondary production from the Bartlesville Sandstone in this field through 1979 was 12.5 million barrels of oil, with 18.6 million barrels remaining in place. Eight major Bartlesville Sandstone fields exist in Greenwood County, and 40% of the original 183 million barrels of oil-in-place in these fields has also been recovered by primary and secondary methods.

The site for the pilot project was selected because only 40% of the original oil-in-place had been recovered, isopach maps showed favorable reservoir thickness, and the site was near the successful mini-test. The parameters calculated from this mini-test are shown in Table 1.

The Bartlesville Sandstone, or "shoestring sand" as it has been called by some, has been interpreted by Bass (1934) and others as being of marine origin (barrier island) based on, among other criteria, its plan view geometrical resemblance to the modern United States Atlantic coastline. The sedimentary structures, paleocurrent direction indicators, and facies relationships that are described in this study, however, support work by later authors in Oklahoma (Visher and others, 1971; Zeliff, 1976) and in Kansas (Hulse, 1979), who recognized the fluvial-deltaic nature of the unit. Hulse (1979) suggested that the Bartlesville trend of elongate sandstones corresponds closely to paleotopographic lows on the underlying pre-Pennsylvanian surface and that the sandstone is the result of perennial, low-gradient, meandering streams following the underlying lows.

Cores from the Cities Service Fromm A1-W, C7-W, D-4, G1-W, and G7-W wells (Fig. 1) were described, and the vertical sequence of environments of deposition in each core was determined (Siemers and Tillman, 1981). The cores were described in terms of lithology, sedimentary structures (trough crossbedding, rippling, planar laminations), type

[1]Present address: H. N. Fisk Laboratory of Sedimentology, Department of Geology, University of Cincinnati, Cincinnati, Ohio 45221

[2]Present address: Consulting Sedimentologist, 4555 S. Harvard, Tulsa, OK 74135

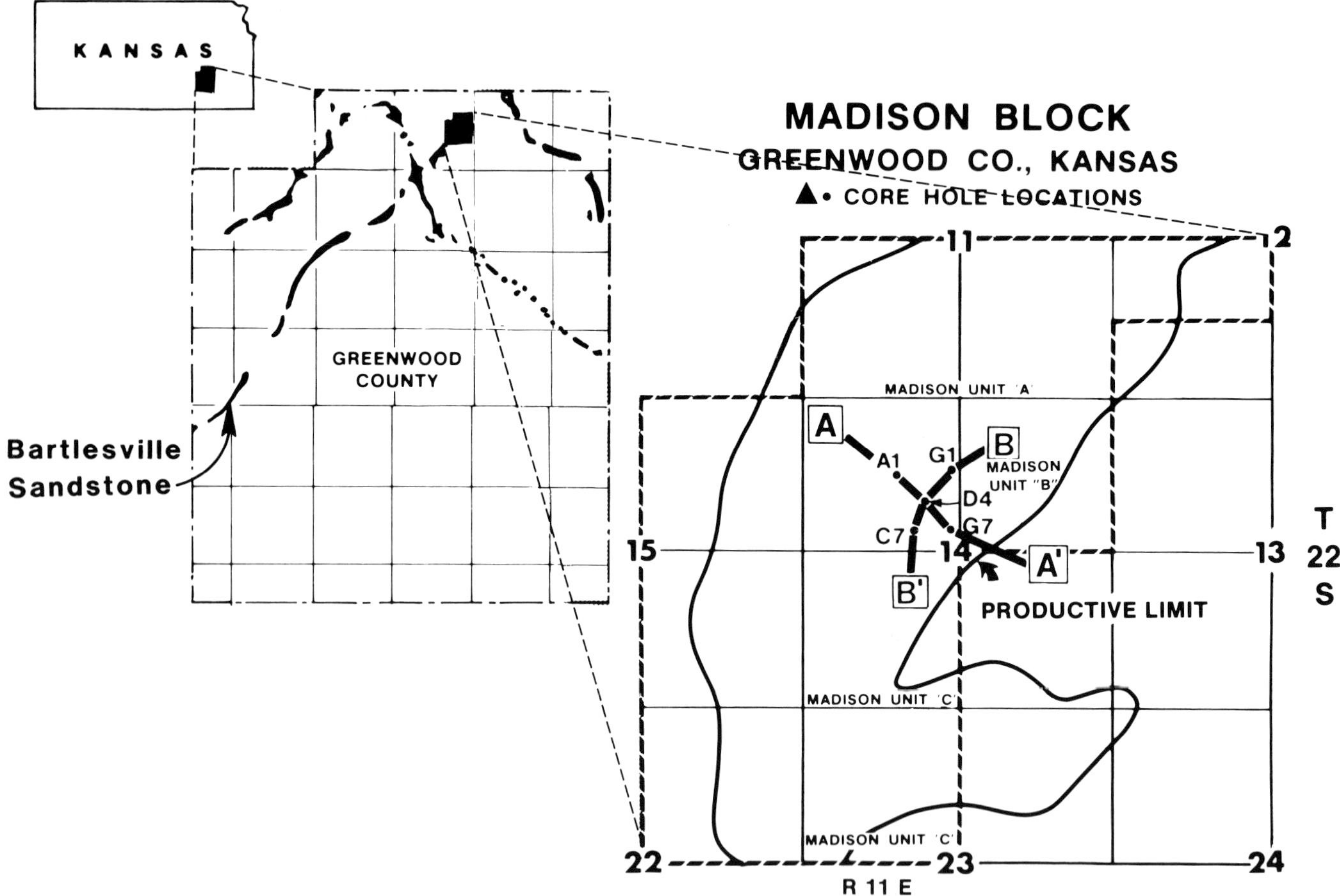

FIG. 1.—Location of five cored wells in the micellar-polymer pilot project study area, Seeley-Wicke field, Madison Unit "B", Greenwood County, Kansas. Boundaries of channel indicated as productive limit. Locations of subsurface cross sections A-A′ and B-B′ are shown.

and amount of burrows, types of cementation, nature of bedding contacts, and grain-size variations (Table 2). Permeability and porosity data were compared to the grain sizes, lithologies, sedimentary structures, log character, and the amount of fine-grained matrix in each facies in order to delineate the zones of best reservoir communication and potential transmissibility.

The cores were oriented using standard downhole orientation techniques. Once a concise log-to-core depth correlation was established for each well (depth differences between core and log depths of as much as 3.5 ft or 1.1 m were determined in the wells studied), the dip directions of crossbed laminae and bedding boundaries were measured. Transport directions of trough crossbed sets were tabulated for each core. Laminae comprising the crossbeds were commonly defined by minor changes in particle grain size as well as by concentrations of coaly debris. Over 100 measurements were obtained from *each* core and the results for each core were summarized on computer plots. The five computer-generated directional plots were placed on a map of the project area and the general transport directions of currents which deposited the crossbedded units were resolved. This oriented-core analysis is discussed in detail in a subsequent section of the paper.

TABLE 1.—PRODUCTIVE VALUES CALCULATED IN BARTLESVILLE SANDSTONE MINI-TEST

MEAN POROSITY: 19%	GROSS PAY: 50 ft (average)
MEAN PERMEABILITY (AIR): 42 md	DEPTH: 1900 ft
MEAN PERMEABILITY (WATER): 6 md	RESIDUAL OIL SATURATION AFTER WATERFLOOD: 34%
INITIAL RESERVOIR PRESSURE: 410 psig @ 1900 ft	

DEPOSITIONAL FACIES

Distributary channel sandstone facies.—

The *Distributary Channel Sandstone Facies* is the thickest sandstone unit of the Bartlesville Sandstone in the study area. It is the principal reservoir in the Bartlesville and, hence, is the prime unit of interest in the pilot study.

The thickness of this facies in the cores ranges from 62.9 ft (19.2 m) in the D-4 well (Fig. 2) to 28.8 ft (8.8 m) in the A1-W well (Fig. 3). The unit may be thicker in the G7-W well (the entire unit was not cored). An isopach map of the pilot area indicates that the thickness trends are aligned northeast-southwest (Fig. 4). The sandstone mean grain size ranges from lower very fine to upper fine-grained (62 to 220 μm, averaging 100 μm, or 0.1 mm). The sandstone is a litharenite containing a large amount of ductile micaceous rock fragments. Authigenic clays include illite and kaolinite, and other intergranular components consist of quartz overgrowths and carbonate cement. Interbedded siltstone comprises approximately 5% of this unit. The unit fines upward overall, and thinner fining-upward sandstone beds are commonly found throughout the unit (i.e., from 1,897.0–1,925.8 ft or 578–587 m in the A1-W well, Fig. 5).

This channel sequence was deposited after currents scoured into the underlying shale unit, and large mudstone and coal rip-up clasts derived from the underlying unit are found at its base (Fig. 6D). Rip-up clasts also occur higher in the channel sequence at the base of some of the thinner sandstone beds that have scoured surfaces. Hence, the unit may be interpreted as a multistoried sandstone. Overlying the basal zone of rip-up clasts or "channel lag" is either a sandstone that appears to be massive or a horizontally laminated sandstone and shale sequence (Fig. 6C). This unit is in turn overlain by 1- to 3-ft-thick sets of trough crossbedded sandstone (Fig. 6B) which contain carbonaceous and micaceous lamina sets (0.05–0.5 in. or 0.13–1.27 cm thick) having a narrow range of inferred flow directions. The sequence is capped by thinner sets of trough crossbedded sandstones and siltstones and by thin beds that appear to be massive. Rippled sandstones, siltstones, and silty shales lie at the top of the sequence (Fig. 6A).

The channel lag at the bottom of the channel sequence is interpreted to have been deposited by upper-flow regime currents. The overlying units reflect lower current energy and were deposited by traction and settling of fine-grained material from suspension.

The trough crossbeds commonly have coaly debris defining the laminae, and the ripple beds have silty shale, shale, and carbonaceous debris highlighting the laminations. Large wood fragments, as much as 8 cm long and 2.5 cm wide, are common in the sandstones. These fragments generally lie at the base of scour channels where one thinning-upward sequence rests directly on another thinning-upward sequence. In the D-4 well, a 0.15-ft-thick (4.6 cm) carbonaceous seam, composed of transported carbonaceous debris, was observed. In all cases where coaly laminae are present, they are discontinuous and do not form permeability barriers. Carbonate-cemented sandstones are found in only one core (D-4, near the base of a channel sandstone). This carbonate is scattered and patchy and provides no major permeability barriers. No burrows are observed in the channel facies.

The bases of channels are marked on gamma-ray logs by a sharp deflection to the left of the shale baseline (indicating lower radioactivity) and the logs have a blocky to serrated log profile (Figs. 2A, 3A). The stacked multistoried sandstones observed in the cores are recognized on log plots by the blocky nature of the gamma-ray log. The stacked nature of these sandstones is interpreted to be the result of velocity pulsations within the channel system. A classical "point bar" sequence typical of many channel sandstones is not present in the Bartlesville sandstones in Seely-Wicke field. The Bartlesville channels probably had low sinuosity, varying from straight to slightly meandering, and were controlled by the paleotopography of the underlying pre-Pennsylvanian depressions (Hulse, 1979). The *Distributary Channel Facies* was probably deposited in an outer deltaic plain environment.

Delta plain shale and siltstone facies.—

The *Delta Plain Facies* was cored in two of the five wells in this study (A1-W and D-4). It underlies the *Distributary Channel Facies,* consists primarily of siltstone and shale (Figs. 2B, 3B), and is readily recognized on gamma-ray logs. Both the siltstone and shale are typically gray or dark gray and contain large amounts (10–80%) of wood fragments and carbonaceous detritus. The shales often have slickenside surfaces (due to flowage during compaction). Brown sandy mudballs are commonly present; these are locally squeezed vertically into the surrounding shale (Fig. 7B). The *Delta Plain Facies* fines upward from silty shale and siltstone to shale. The upper contact is uneven and scoured into by the overlying unit.

Overbank shale and siltstone facies.—

The *Overbank Facies* is only observed in cores in the A1-W and D-4 wells. Generally, it is composed of approximately 75% siltstone and 25% silty shale with the principal sedimentary structure being asymmetrical (current) ripples (Fig. 7A). This facies is interpreted to be the result of sediment-charged floodwaters spilling over the levees of the channel. Reactivation surfaces at the top of the ripples are common. This facies is typically less than 2 ft (0.6 m) thick.

The *Overbank Facies* varies in character across the project area; at some locations, more siltstone, sandstone and burrowing are observed than at other locations. Variations in the distance from the channel and proximity to marine environments are the probable causes of this variation.

Crevasse splay sandstone and siltstone facies.—

The *Crevasse Splay Facies* contains several lithologies. Generally, it consists of siltstone, but sandstone, silty shale, and carbonaceous shale are also common. It has many of the sedimentary features associated with the main channel, including rip-up clasts, ripple and trough-crossbedding, and a fining-upward sequence (Fig. 7C). It is finer-grained than the *Distributary Channel Sandstone Facies,* however, and contains a greater percentage of ripples. Sandstones that appear massive are common. Coarsening-upward sequences, which presumably resulted from lateral progradation following channel levee breakthrough, are also present. The *Crevasse Splay Facies* commonly contains appreciable amounts of calcareous cement. Glauconite is also present. The thickness of the *Crevasse Splay Facies* ranges from 1.5 ft or 0.5 m (A1-W well, Fig. 3B) to greater than 7.2 ft or 2.2 m (D-4 well, Fig. 2B).

Interdistributary bay siltstone, shale, and sandstone facies.—

The *Interdistributary Bay Facies* consists mainly of siltstone, but shale and sandstone are also present. This facies resembles the *Overbank Facies* (with which it is laterally equivalent, based on detailed log correlations) but differs in that it contains abundant burrowing and less sandstone. Horizontal burrowing dominates in the shales, whereas vertical burrows prevail in sandstone interbeds and extend down into the underlying shale. An upward increase in siltstone content is associated with an upward increase in burrowing. The burrows are typically silt-filled, clay-lined, and 5-10 mm long.

The *Interdistributary Bay Facies* locally displays a coarsening-upward sequence from silty shale with siltstone lenses in the lower half of the unit to siltstone containing some low-angle troughs and abundant ripple bedding in the upper half. Reactivation surfaces are common, and there is some hint of bidirectional cross laminations which might be attributed to tidal currents (Fig. 7D). The thickness of the *Interdistributary Bay Facies* varies from 1.0 ft (0.3 m) in the G7-W and D-4 wells to 17 ft (5.2 m) in the C7-W well.

DEPOSITIONAL MODEL

The Bartlesville Sandstone in the study area is interpreted to have been deposited in a middle to outer delta-plain environment in a low-sinuosity distributary channel. The channel cut into underlying delta-plain mudstones and siltstones. Periodic flooding of the channel allowed fine-grained overbank sediments to be deposited adjacent to the channel. Numerous splay channels resulted from channel levee breakthrough, and these episodic units prograded into an interdistributary bay environment (Fig. 8).

Sedimentary structures and lateral facies variations, as observed in the vertical sequences, support a fluvial-deltaic model. Horizontally laminated sandstones, which include coaly shale and rip-up clasts near the base of fining-upward sandstones, suggest upper-flow regime deposition. The overlying trough and ripple bedding suggest that the deposits were formed by unidirectional traction currents. The elongate geometry of the trend observed in the pilot area supports this interpretation. Two cross sections across the project area, shown in Figures 9 and 10, demonstrate that the channel has a steep northwestern side (thinning from 62 ft or 18.9 m at D-4 to 29 ft or 8.8 m at A1-W) and a nearly flat base on the southeastern side. The channel thins, primarily by nondeposition or erosion at the top, from 62 ft (18.9 m) at D-4 to 56 ft (17.1 m) at G7-W.

The five cores utilized in this study were oriented using standard downhole orientation techniques. These techniques require special core barrels which have a scribing ring near the base of the core barrel. The scribing ring has lugs that are designed to cut three parallel grooves the whole length of the core (Fig. 11). The reference groove used in orienting the core is the "a-groove." The orientation of this groove is reported at 2- to 3-ft intervals. By use of an orientation jig (Fig. 12), the reference groove may be oriented at each data point. Prior to determining flow directions of trough crossbeds, a "north line" was traced the whole length

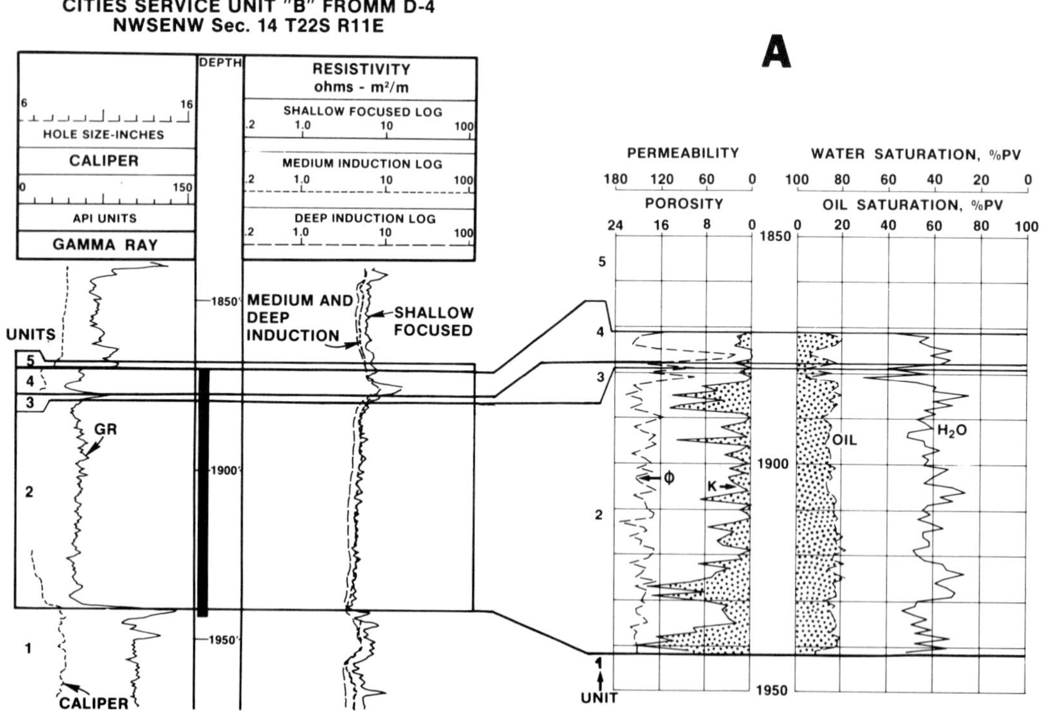

FIG. 2.—Cities Service Fromm D-4. (A) Depositional units correlated to the gamma-ray resistivity and petrophysical logs. Contacts of individual

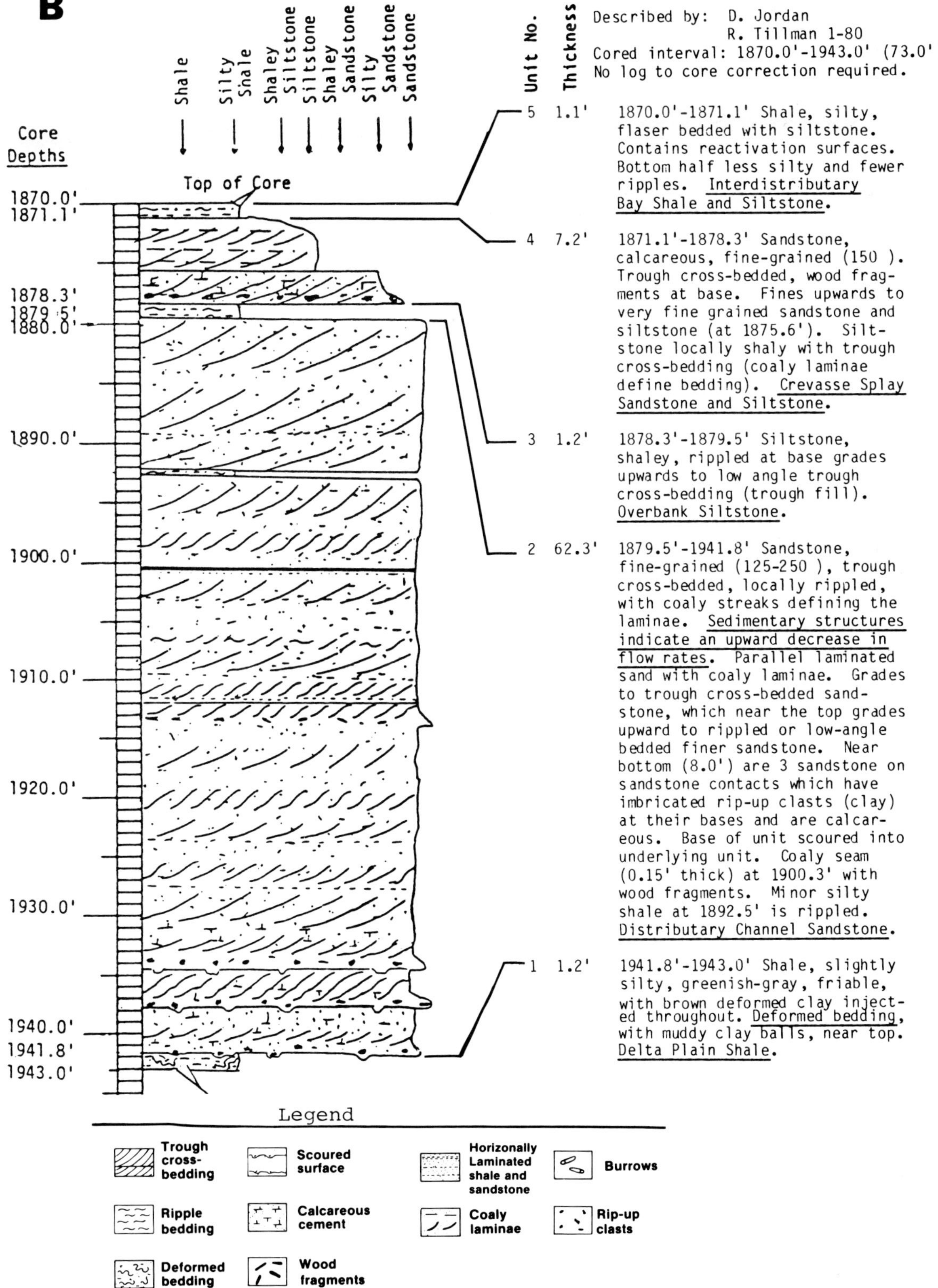

units are based on changes in sedimentary features and lithology in the cores. (B) Facies description of the cored interval (black bar in A).

of the cores and the basal boundary of individual crossbeds was traced on the cores with chalk. Crossbeds were particularly abundant in the distributary channel portions of the cores, and 100 or more crossbed dip directions were tabulated in *each* core.

A detailed directional sediment transport summary plot for well A1-W is shown in Figure 13. Summary data plots and locations for each cored well are shown in Figure 14. Paleocurrent directions were calculated from measurements obtained from the trough crossbeds. Measurement of transport directions based on the orientation of crossbedding were made in the laboratory. Generally, the patterns suggest a channel that flowed from east-northeast to west-southwest across the project area. Meandering within the channel may be responsible for some of the variability observed in the paleoflow directions. If we disregard the data from the Fromm D-4 well, however, the channels appear to have been deposited by relatively unidirectional currents with transport directions varying within a relatively small arc (77°). The standard deviation for flow directions for individual wells varies from 56° to 148°. Data included in one standard deviation includes 66% of the measurements. Dashed lines on the rose diagram plots (Figs. 13, 14) mark the boundaries of one standard deviation.

The Fromm D-4 core has a uniformly anomolous trend (Figs. 14, 15). The direction of trough crossbedding is almost 180° from the mean values for the four other cored wells. Personnel present at the well site and later study of the data strongly suggest that the deviation may be attributed to a systematic data collecting error at the well site. In any case, four of the five wells show flow parallel to the west-southwest trend of the sand body. This regional southwest trend of the sand body was also observed by Bass (1934), Bass and others (1937), Visher and others (1971), and Zeliff (1976).

PETROPHYSICAL TRENDS

The permeability, porosity, water and oil saturation values from the five cored wells were determined. The permeability data were compared to the geologic data (Figs. 2B, 3B) in order to investigate correlations with grain size,

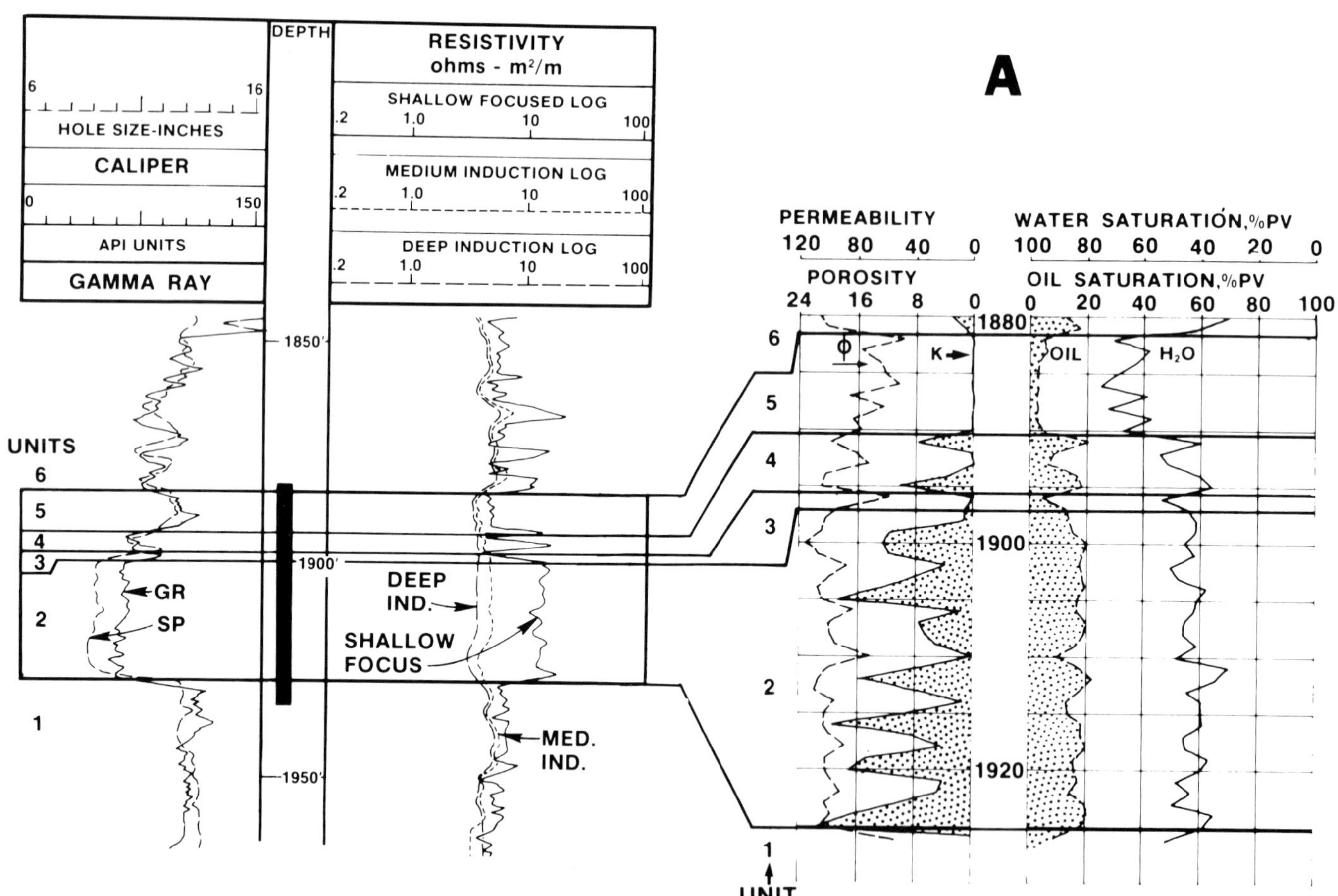

FIG. 3.—Cities Service Fromm A1-W. (A) Depositional units correlated to the gamma-ray resistivity and petrophysical logs. Contacts of individual (Unit 2) range from 5 to 105 md, and porosities range from 14 to 23%; average permeability is 46 md, average porosity is 20%. Residual oil centrations of carbonaceous laminae and bedding contacts containing impermeable or low-permeability streaks (i.e., rip-up clasts). (B) Facies

sedimentary structures, log character, and the type and amount of pore-filling constituents. Inflection points on the plots and on the gamma-ray logs denote very fine-grained sandstone or siltstone and correspond to low values of permeability. The *Interdistributary Bay Facies* is primarily siltstone and shale and has very low permeability values. The *Crevasse Splay Facies,* consisting of sandstone and siltstone, has variable permeabilities.

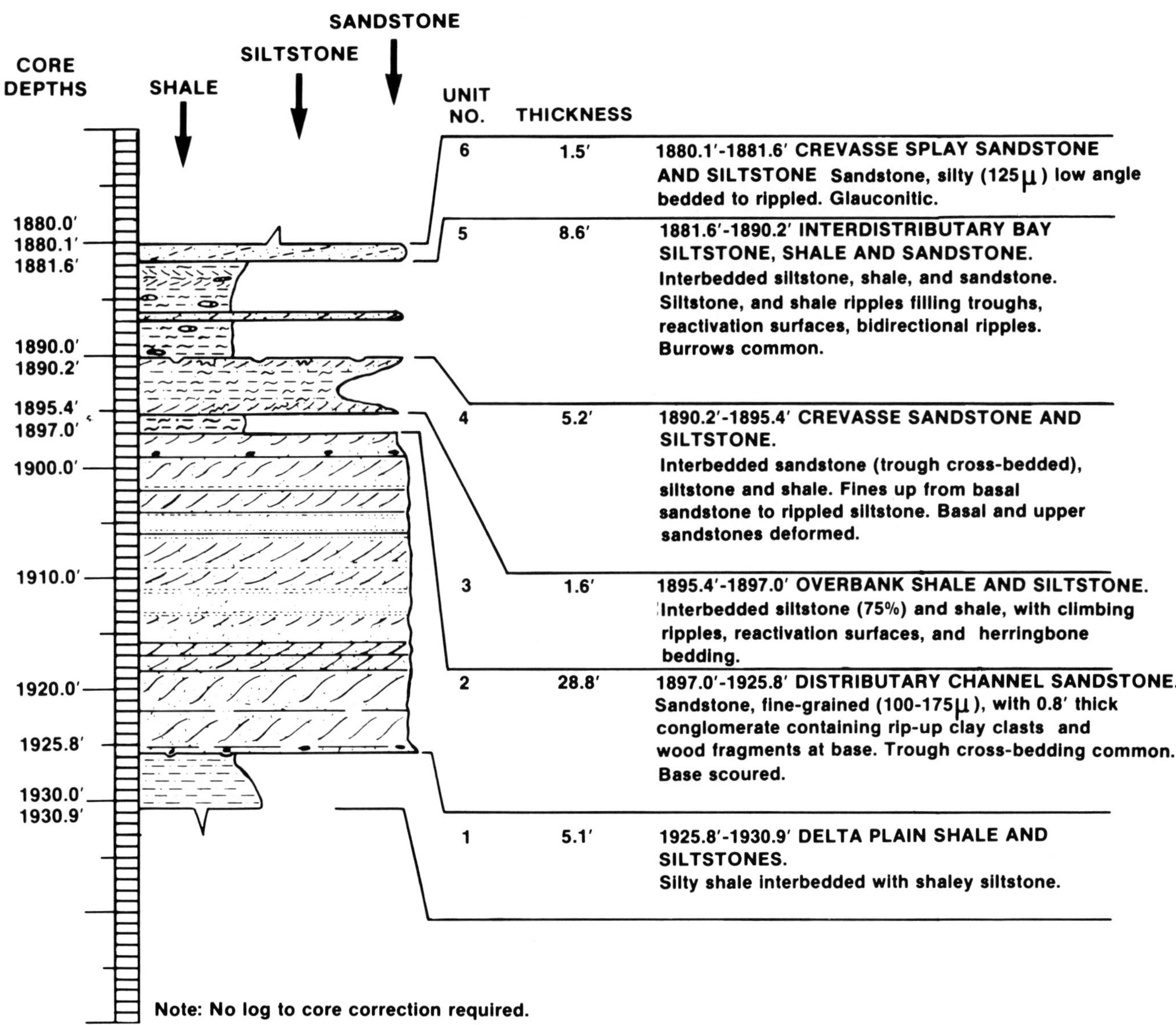

units are based on changes of sedimentary features and lithology observed in the core. Permeabilities in the *Distributary Channel Sandstone Facies* saturation values average 18% and water saturation values average 57%. Sharp decreases in permeability within Unit 2 are the result of high con- description of the cored interval (black bar in A).

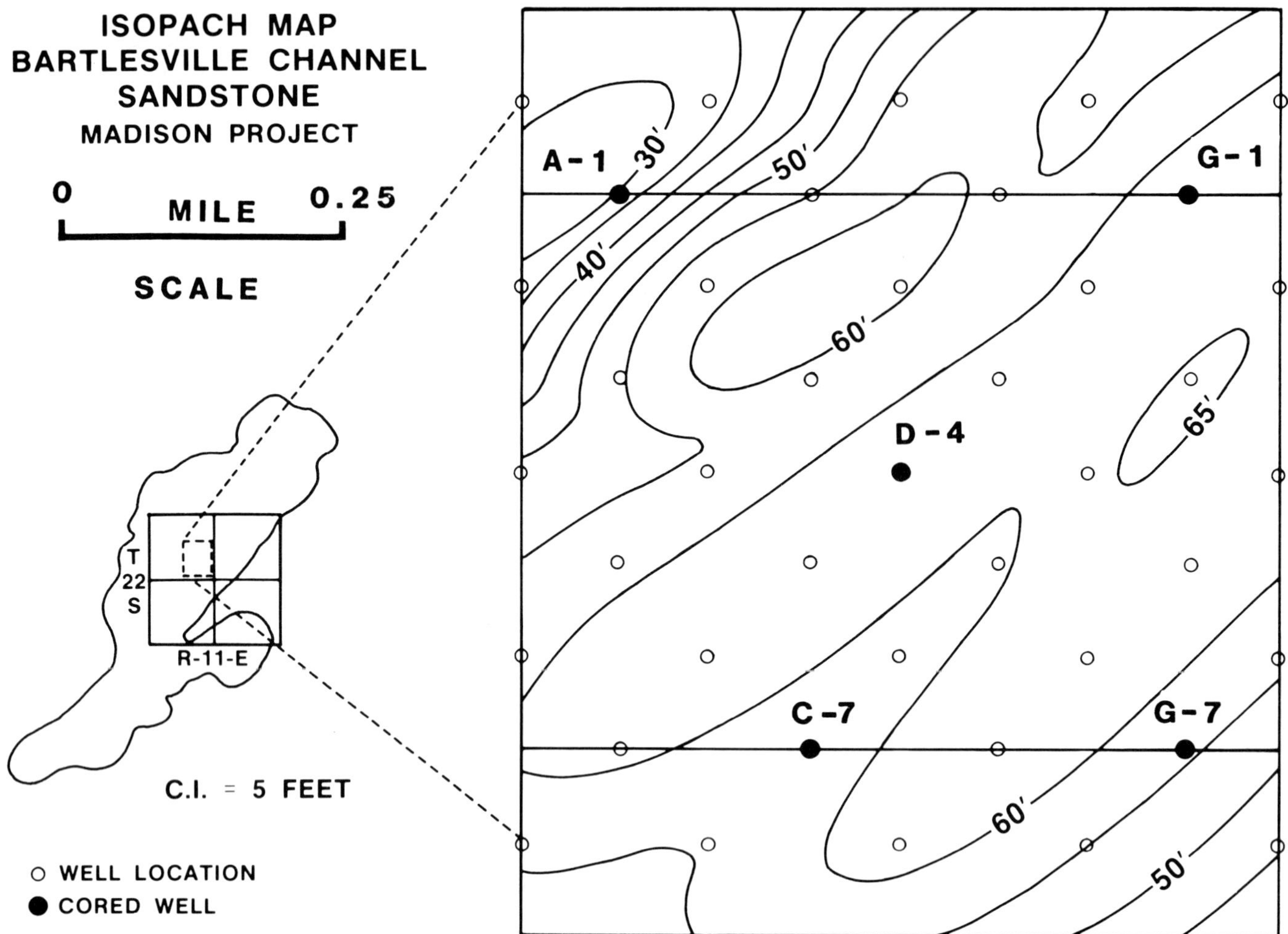

FIG. 4.—Isopach map of Bartlesville Sandstone in Madison project area (SE-NW Sec. 14 T22S R11E) of Seeley-Wicke field, Kansas.

Decreases in grain size and the scale of sedimentary structures (troughs to ripples) along with increases in radioactivity on the gamma-ray log and the amount of clay matrix are observed to parallel the decrease in permeability. Carbonate-cemented sandstones have essentially no permeability.

The scale and size trends of sedimentary structures (troughs and ripples) show an overall upward decrease within the channel facies. Horizontally laminated sandstones and carbonaceous debris are common in the basal portion of the *Distributary Channel Facies*. This is overlain by thick trough crossbed sets. These features indicate high current energy in the lower parts of the channel. Sandstones are better sorted and permeability values are highest (60–100 md) in the trough crossbedded portion of the channel. Permeability values decrease upward (20–60 md) as the amount of trough crossbedding decreases (decreasing current energy) and ripple-bedding (including clay drapes) increases. The productivity values for the test area are summarized in Table 1.

MINERALOGY AND DIAGENESIS

Thin section modal analyses were performed on 13 samples from the Fromm A1-W and D-4 wells by M. Wilson (pers. commun.). The samples were selected from the main reservoir facies (the distributary channel) in order to deter-

FIG. 5.—Sequence of full-diameter core from the Bartlesville Sandstone interval in the Fromm A1-W well.
1,880.1–1,881.6 ft 573.1–573.5 m *Crevasse Splay Facies* (1.5 ft +, 0.5 m)
1,881.6–1,890.2 ft 573.5–576.1 m *Interdistributary Bay Facies* (8.6 ft, 2.6 m)
1,890.2–1,895.4 ft 576.1–577.7 m *Crevasse Splay Facies* (5.2 ft, 1.6 m)
1,895.4–1,897.0 ft 577.7–578.2 m *Overbank Facies* (1.6 ft, 0.5 m)
1,897.0–1,925.8 ft 578.2–587.0 m *Distributary Channel Facies* (28.8 ft, 8.8 m)
1,925.8–1,930.9 ft 587.0–588.5 m *Delta Plain Facies* (5.1 ft, 1.6 m)

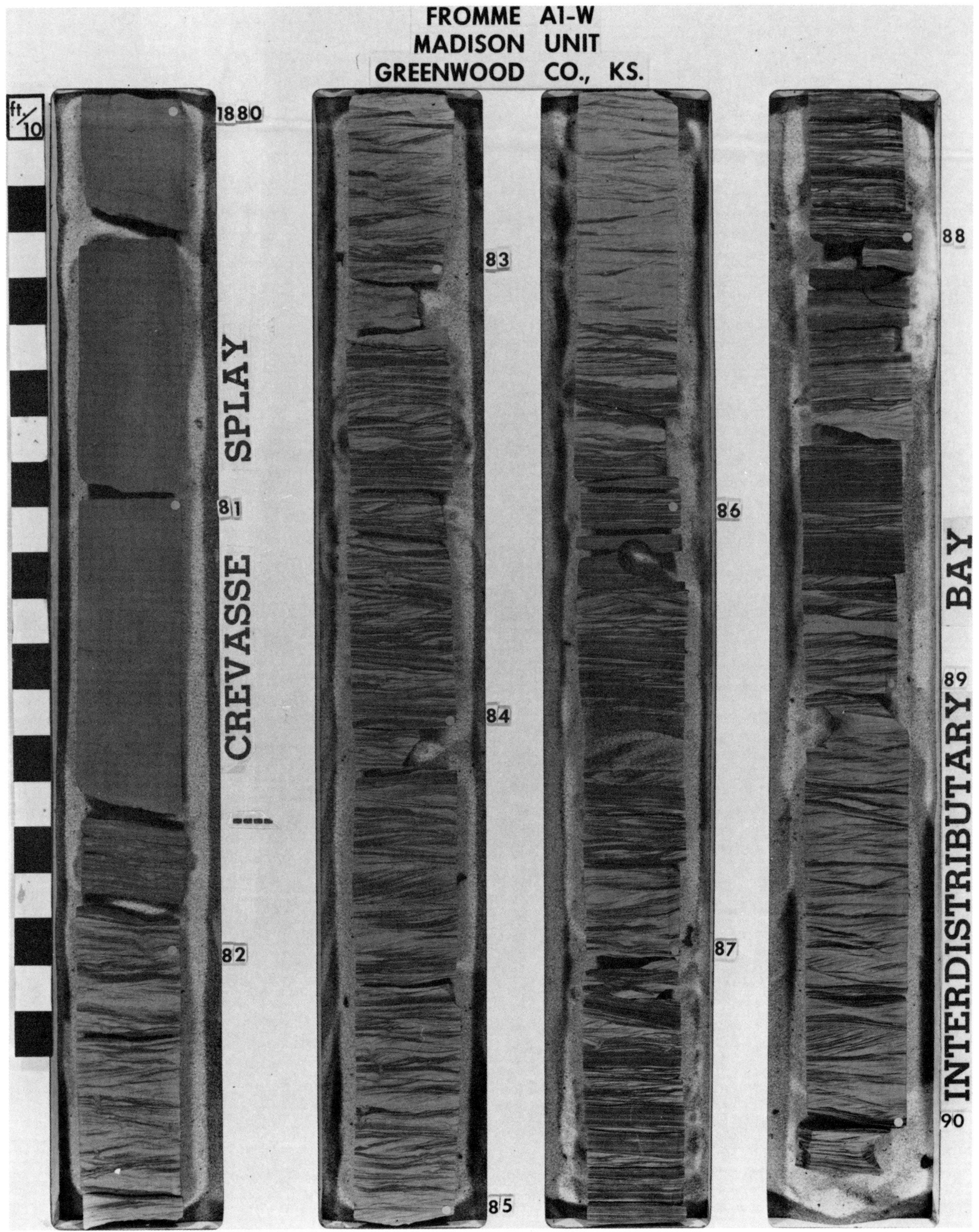
FROMME A1-W
MADISON UNIT
GREENWOOD CO., KS.
ft/10
1880
81
82
83
84
85
86
87
88
89
90
CREVASSE SPLAY
INTERDISTRIBUTARY BAY

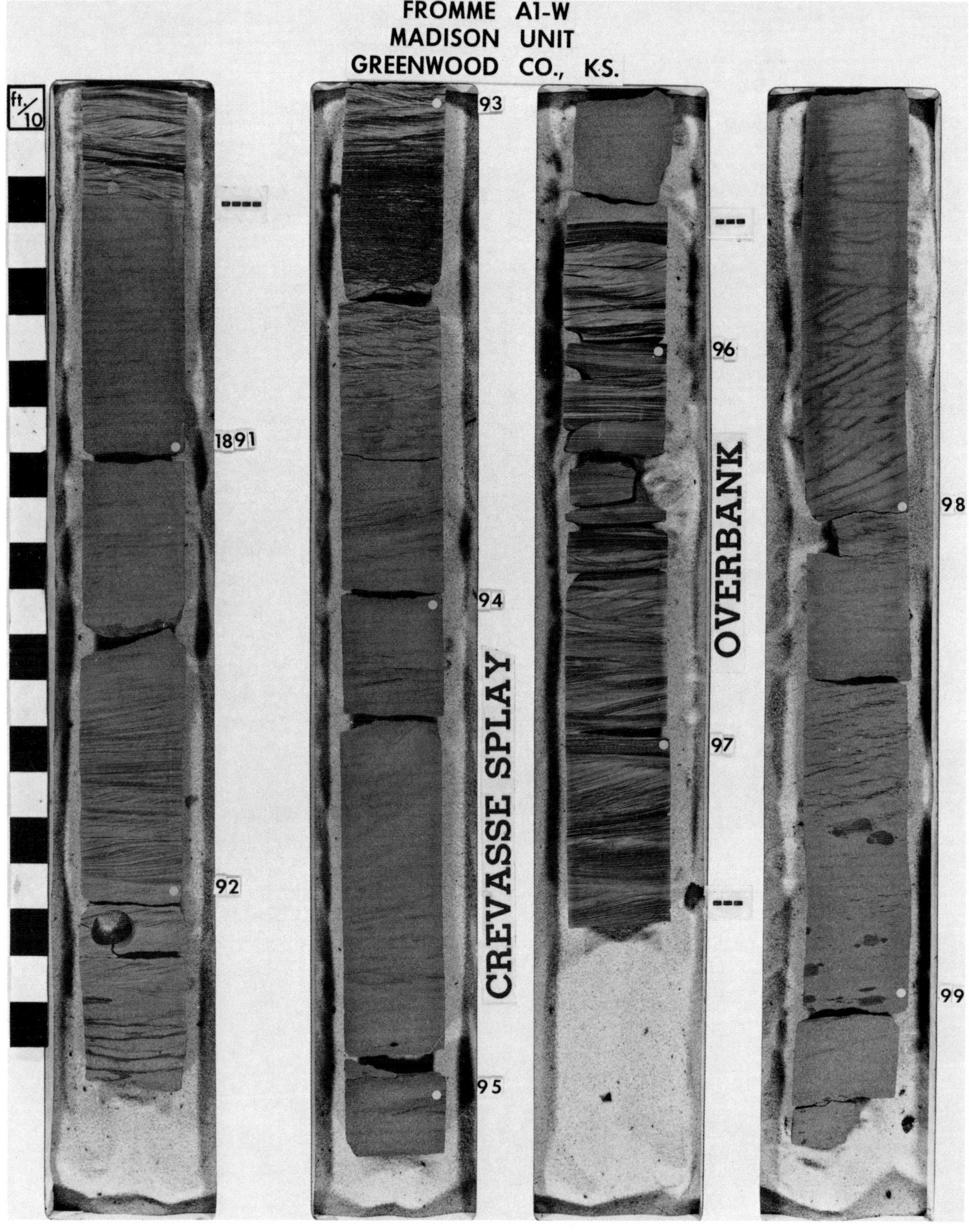
FROMME A1-W
MADISON UNIT
GREENWOOD CO., KS.
ft./10
1891
92
93
94
95
CREVASSE SPLAY
96
97
OVERBANK
98
99

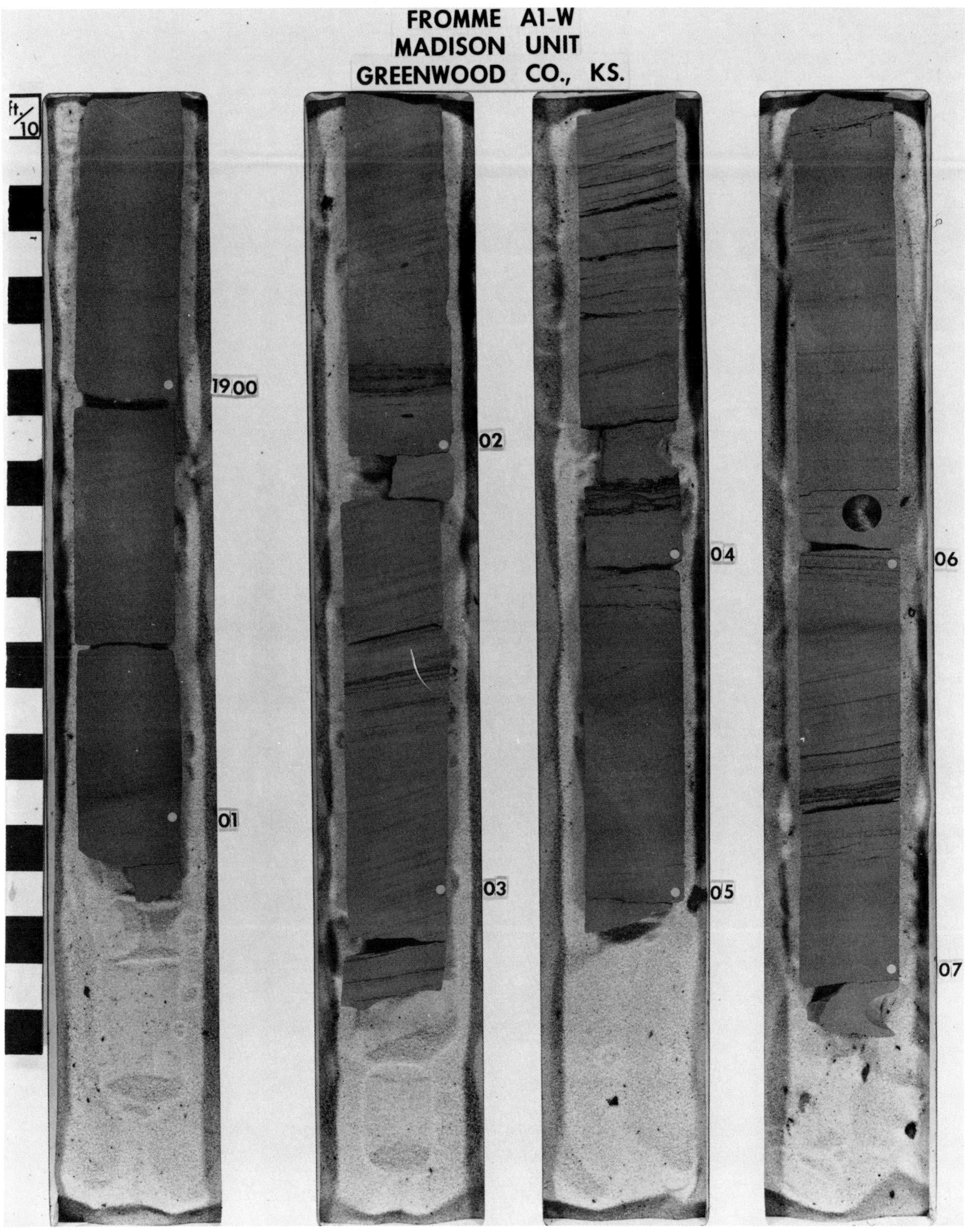
FROMME A1-W
MADISON UNIT
GREENWOOD CO., KS.
ft./10
1900
01
02
03
04
05
06
07

FROMME A1-W
MADISON UNIT
GREENWOOD CO., KS.
ft./10
1908
09
10
11
12
13
14
15
16
17

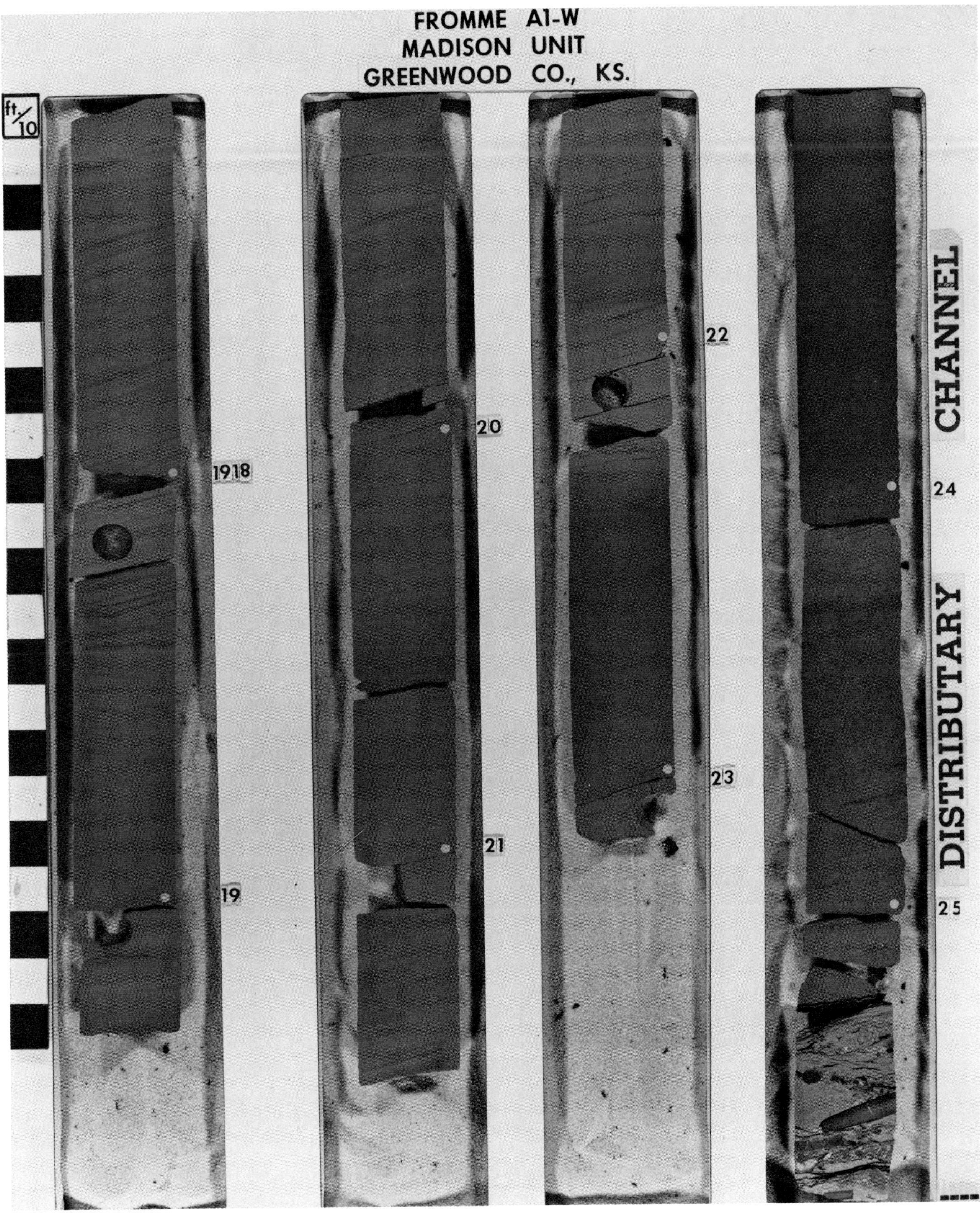
FROMME A1-W
MADISON UNIT
GREENWOOD CO., KS.
ft/10
1918
19
20
21
22
23
24
25
DISTRIBUTARY CHANNEL

FROMME A1-W
MADISON UNIT
GREENWOOD CO., KS.
ft./10
1926
27
28
29
30
DELTA PLAIN

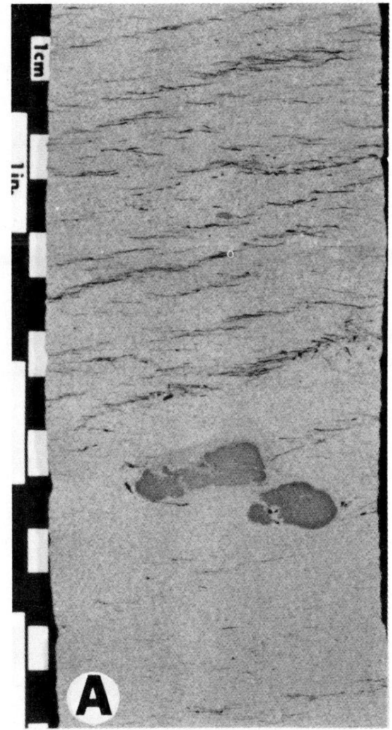

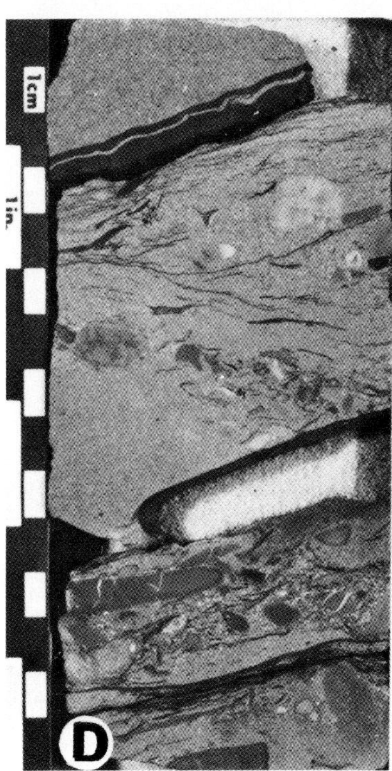

FIG. 6.—Closeup photographs of vertical sequence (D is base) of sedimentary features typically observed in cores of the *Distributary Channel Facies* in the Bartlesville Sandstone. (A) Climbing ripples in sandstone, in part, made distinct by concentrations of coaly shale. This type of ripple is typical of the uppermost portion of the distributary channel. The ripples "climb" toward the right in the photograph. Large rip-up mudstone clasts occur in the lower one-third of the photograph. Fromm A1-W. (B) Trough crossbedded sandstone typical of the lower to middle portion of the distributary channel. Two distinct crossbed sets occur in this photograph. Fromm A1-W, 1,906.0 ft (580.9 m). (C) Lower portion of the *Distributary Channel Sandstone-Facies*. These laminated sandstones and coaly siltstones are the result of high-energy currents in the channel. Fromm D-4, 1,925.8 ft (587.0 m). (D) Basal part of *Distributary Channel Sandstone Facies*. This basal lag consists of poorly sorted conglomeratic sandstone, shale rip-up clasts, and coaly debris deposited during periods of high velocity current flow. Fromm A1-W 1,925.3′.

mine mineralogical and diagenetic components in the sandstones. The sandstones are litharenites and feldspathic litharenites with framework grains consisting of quartz (68–76%), lithic fragments (14–28%), and feldspar (8–18%). Total pore-filling constituents average 14%, with low values of clay matrix (0.8–4.6% of the total rock; average 3%) and moderate amounts of authigenic components (7-17%; average 11%). Siderite cement ranges from 0.3–8%, calcite and dolomite from 0.4–8%, quartz overgrowths 0.4–7%, and kaolinite 1–3%. The majority of the pores are intergranular (6-26%) and permeability ranges from 15–83 md for the samples analyzed.

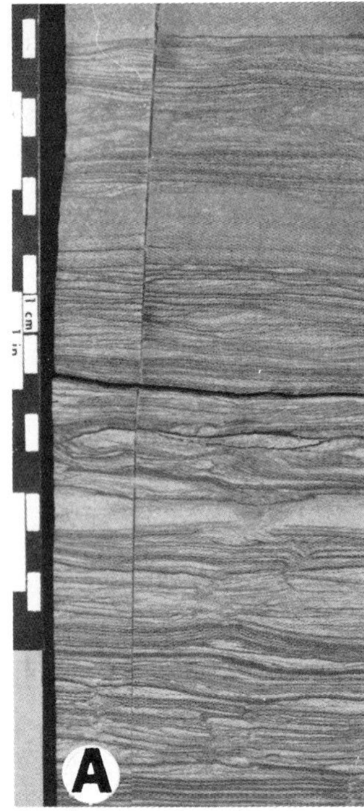

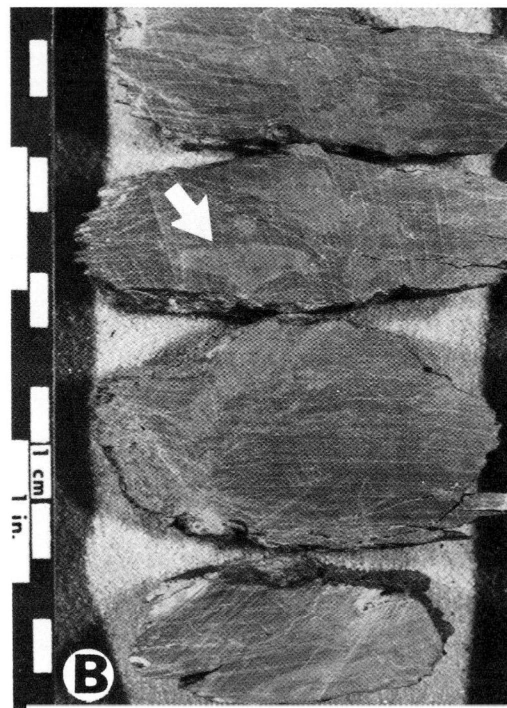

FIG. 7.—Photographs of sedimentary features typical of the *Low-Energy Deltaic Facies* in the Bartlesville Sandstone. (A) Photograph of rippled shale and siltstone in the core of *Overbank Shale and Siltstone Facies*. Note absence of burrowing. Thin laminated sandstones and coaly shales result from overbank flow during floods and are deposited adjacent to the distributary channel. Fromm C7-W, 1,901.2 ft (579.5 m). (B) *Delta Plain Shale and Siltstone Facies*. Note contorted mudstones (arrow) "squeezed" into surrounding shale. Fromm D-4, 1,941.3 ft (591.7 m). (C) Photograph of the *Crevasse Splay Sandstone and Siltstone Facies* in a core. Crevasse splays have characteristics similar to the main channel sand facies; however, in this case, an upward increase in the scale of sedimentary structures (from climbing ripples to trough crossbeds) is present and suggests upward-increasing current energy and proximity of the splay to the distributary channel. Fromm A1-W, 1,891.4 ft (576.5 m). (D) Photograph of the *Interdistributary Bay Siltstone, Shale, and Sandstone Facies*. Note presence of possible bidirectional ripple laminations (suggesting possible very shallow-water tidal influence). Very small burrows are indicated by arrow. Fromm C7-W, 1,861.4 ft (567.4 m).

Porosity has been reduced in the sandstone by quartz overgrowth cement, carbonate cement (primarily calcite, but minor dolomite and siderite are present), detrital clay matrix and authigenic clay, and ductile grain deformation. Low permeability values are associated with silty sandstone and sandstone having abundant carbonaceous laminations. Low permeability values are also characteristic of the sandy facies containing shale and siltstone (interdistributary bay, splays) which are, in part, a function of detrital clay matrix. The presence of authigenic filamentous illite and loosely

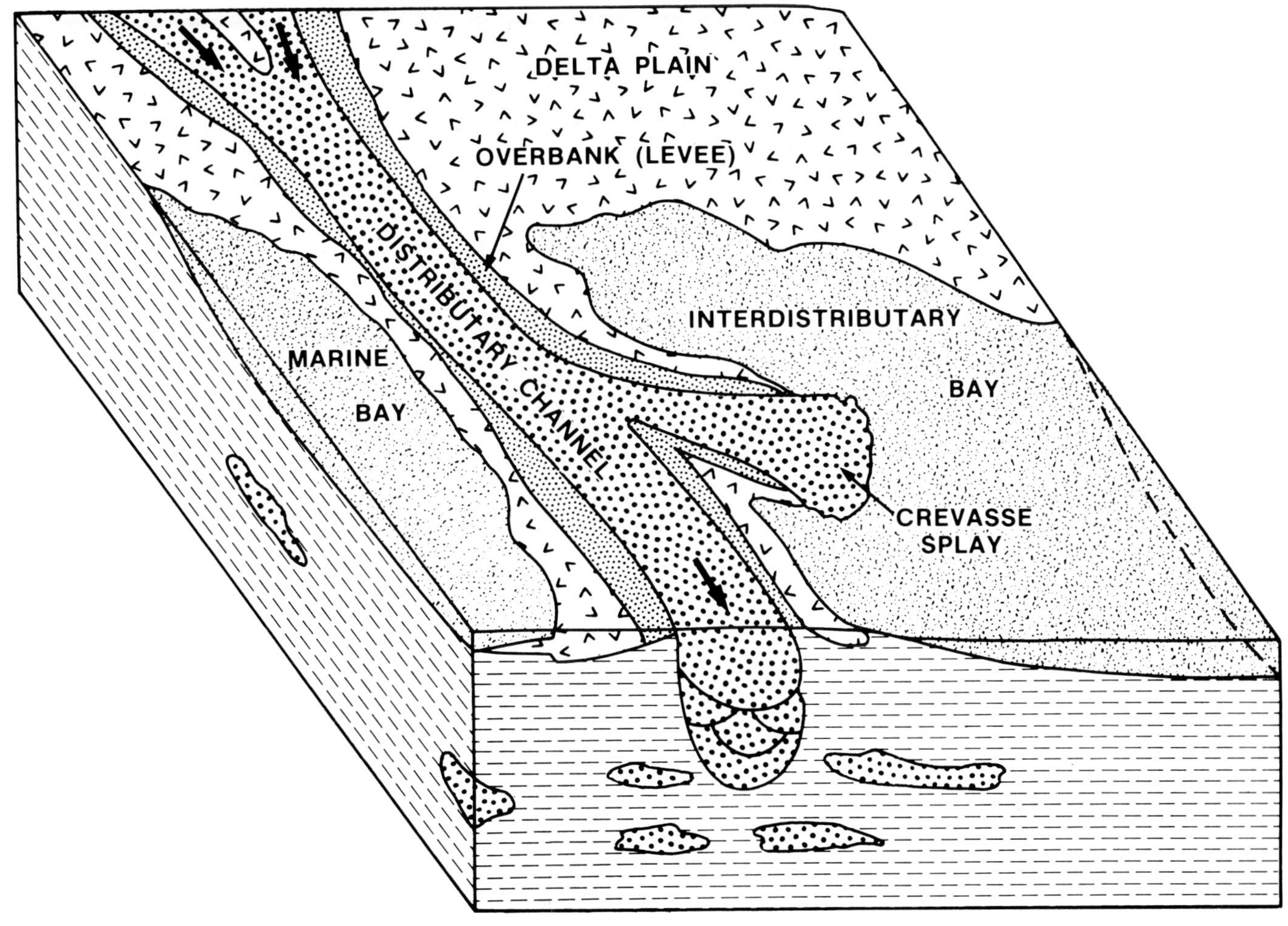

FIG. 8.—Depositional model of the Bartlesville Sandstone in the Seeley-Wicke field located in Greenwood County, Kansas. The producing interval in the field is restricted primarily to the distributary channel sandstone.

attached kaolinite could lead to reduced permeabilities and formation damage by migration of fines during micellar-polymer flooding.

In addition to petrographic studies, a series of fluid-sensitivity studies was carried out on selected Bartlesville Sandstone samples from the cores. The reservoir rock properties were determined to be of good quality prior to injection of tertiary flooding fluids. The injected preflush fluid used in the pilot study may have drastically altered the pore structure and nature of the pore-occluding minerals. Tests at the Cities Service Research Laboratory showed that during preflush, iron in the rock (as siderite, iron-rich dolomite, and iron-rich clay) dissolved and reprecipitated as a gelatinous iron hydroxide, which plugged pore throats and drastically reduced permeability. Potential problems with kaolinite, clay mobilization and bridging of pore throats was found to be controllable by the use of clay stabilizers. After the clay stabilizer was added, injectivity appeared to stabilize. It was also found that little iron hydroxide formed in the rock after the addition of the clay stabilizer. Hence, the use of the clay stabilizer may have prevented iron from reprecipitating (i.e., may have prevented the fluids from contacting the iron-bearing minerals).

SUMMARY AND CONCLUSIONS

Five sedimentary facies are recognized in five oriented cores from the Bartlesville Sandstone in the Madison "B"

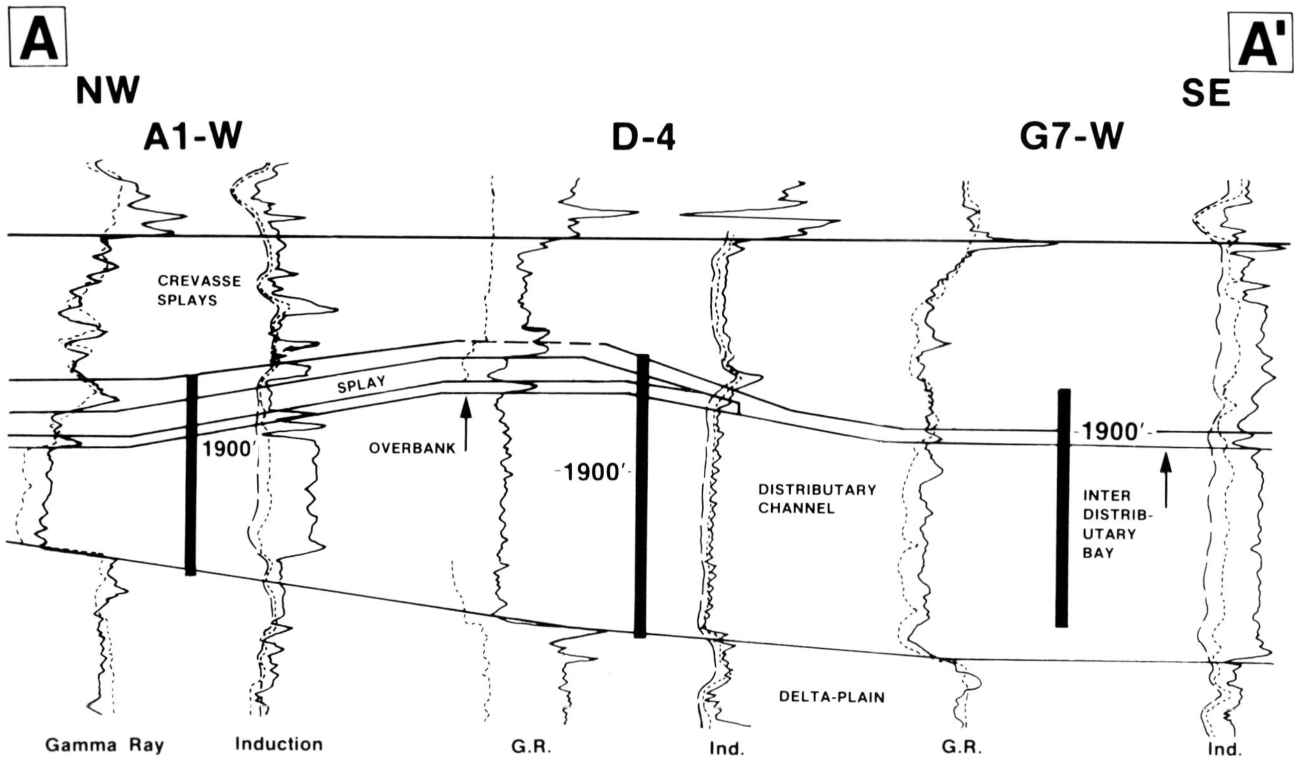

FIG. 9.—Cross section A-A′ is perpendicular to the channel (Fig. 1) in the Madison "B" study area of the Bartlesville Sandstone. Note that the channel facies thickens from northwest to southeast. The channel may be even thicker in the G7-W well than shown, but no core was obtained in the lower interval. Crevasse splays in the A1-W and D-4 wells are probably genetically related to the underlying channel.

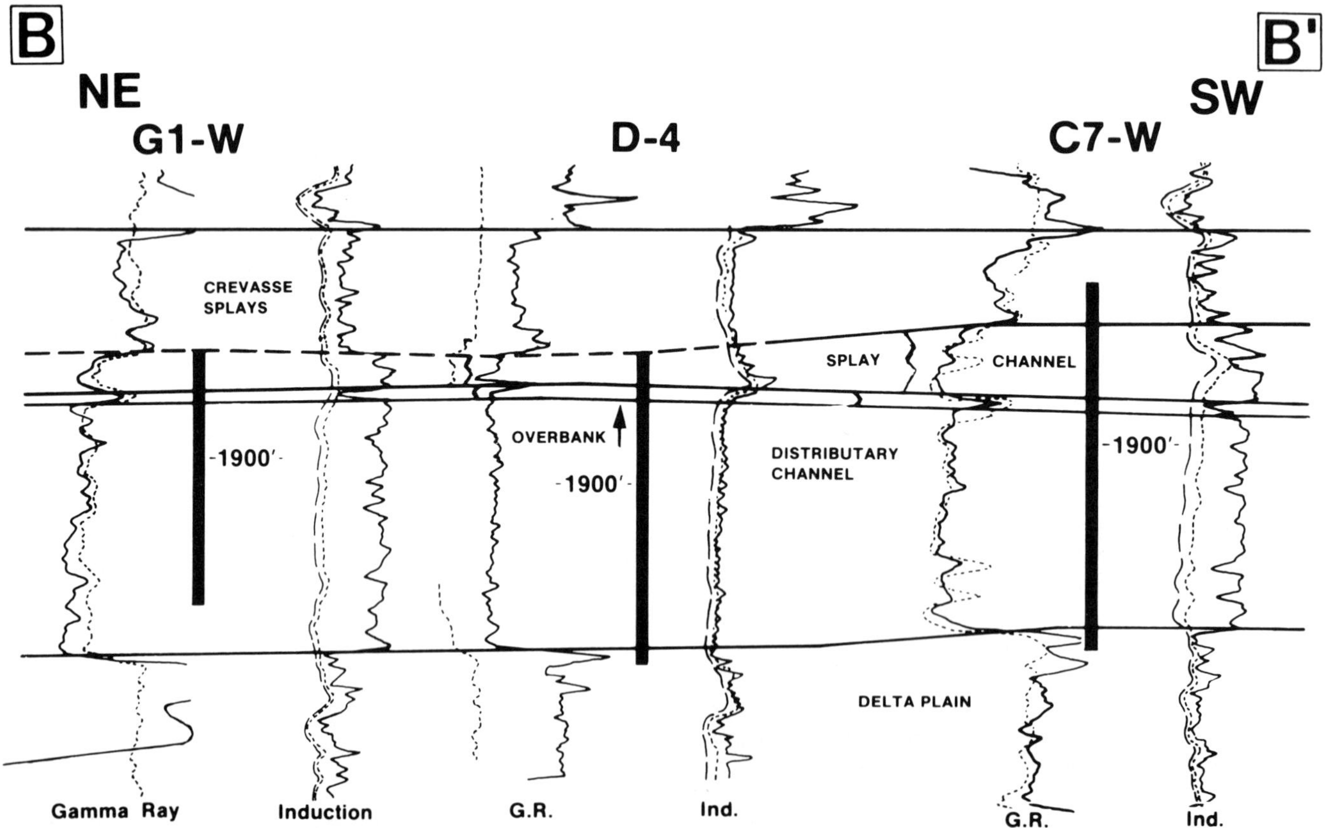

FIG. 10.—Cross section B-B′ which parallels the direction of sediment transport in the channel (Fig. 1). Note that the channel facies is nearly the same thickness throughout the length of the section and that the overlying facies are relatively discontinuous.

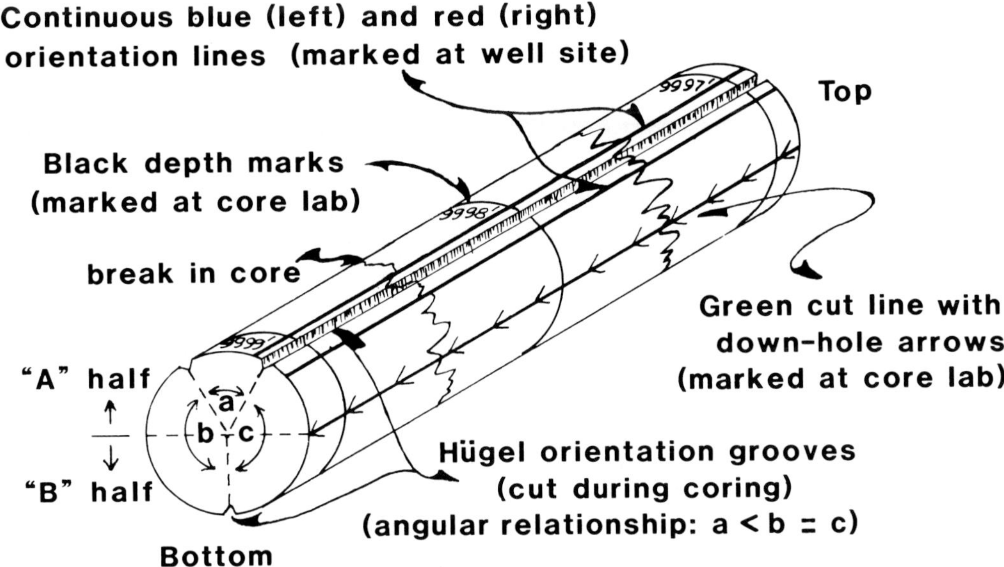

FIG. 11.—Core segment showing Hügel orientation grooves cut during coring. Note that two of the grooves are close together; the angles between grooves are $a < b = c$. The "*a*-groove" is the reference groove used to orient the core. Data supplied by the service company which oriented the core are compared to the reference groove at specified vertical intervals (usually 2 or 3 ft or 0.6 to 0.9 m), (modified from Siemers and Tillman, 1981).

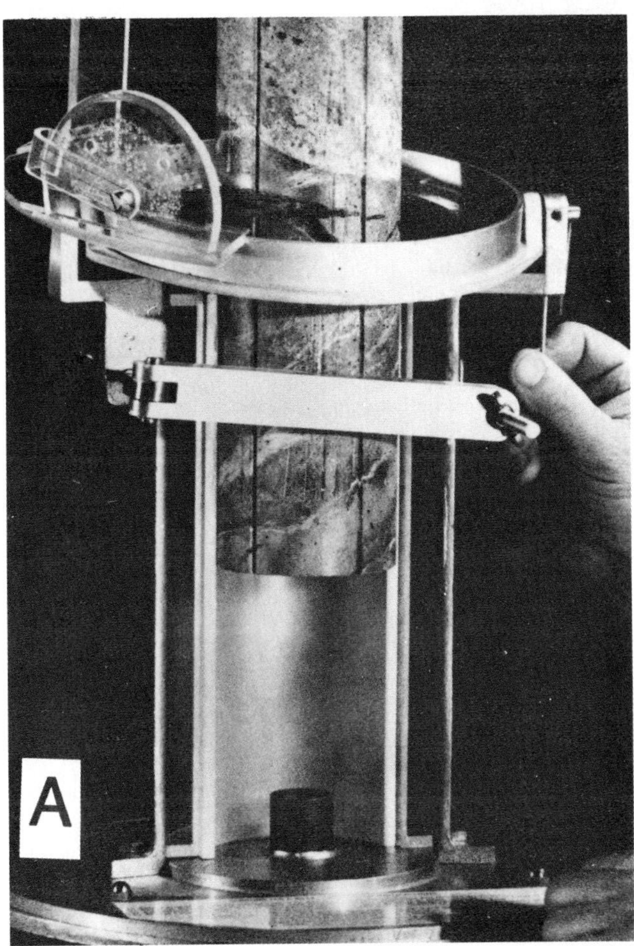

FIG. 12.—Core orientation devices. (A) Commercial downhole core orientation device. Note the paired Hügel grooves (black) on the core. A third groove, which is the one used to orient the core segment, is on the opposite side of the core. Dip and strike of crossbeds and fractures can be determined using this device. (B) Orientation device used to orient cores at Cities Service Research; similar to, but more portable than, the device shown in (A).

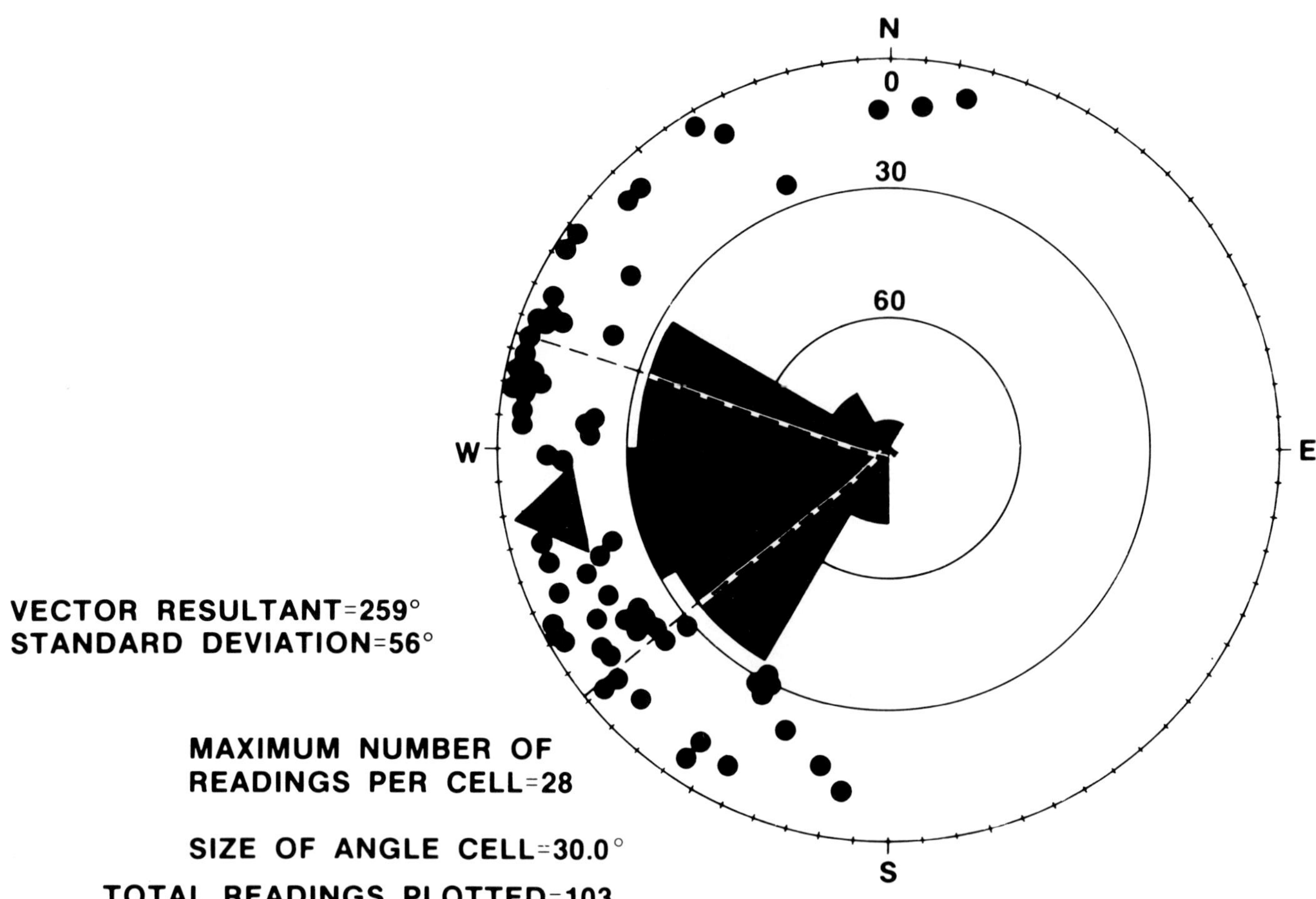

FIG. 13.—Typical detailed rose diagram of trough crossbedding directions for the *Distributary Channel Sandstone Facies* in oriented core in the Fromm A1-W well. The cored interval from which readings were taken is from 1,886.5 to 1,925.6 ft (575.0–587.0 m). Circles indicate 88 data points. Dip angle decreases from 90° at the center to zero at the edge of the diagram. The maximum trough crossbed dips observed were about 35°. The mean transport direction is 259° (arrow); the standard deviation is 56°. The area of one standard deviation is the area bordered by the dashed lines. Sixty-six percent of the data points is included in one standard deviation. The nearly horizontal aspect of bedding in the shales in the cores indicates that regional dip is low and no corrections for regional dip were made in calculating transport directions. The absence of any bidirectional transport direction is significant and probably eliminates the possibility of tidal influence during deposition.

Unit in Seely-Wicke field. The vertical sequence of facies observed from bottom to top is: (1) *Delta Plain Shale and Siltstone,* (2) *Distributary Channel Sandstone,* (3) *Overbank Shale and Siltstone* (4), *Interdistributary Bay Siltstone, Shale, and Sandstone,* and (5) *Crevasse Splay Sandstone and Siltstone*. The *Distributary Channel Facies* is the principal reservoir facies in the Bartlesville Sandstone and it is continuous across the pilot project area, although its thickness varies (from 28 to 62 ft or 8.5–18.9 m; Fig. 4). Facies overlying the channel, some of which contain sandstone, do not show such continuity.

The channel facies has several sequences of decreasing-upward mean grain size accompanied by similar cycles of decreasing permeability. The sequence of sedimentary structures reflects decreasing current energy upward (trough crossbeds changing to ripples). These charcteristics are typical of channelized fluvial-deltaic sequences. Dip directions of trough crossbeds indicate that the channel flowed to the west-southwest across the project area. Lower portions of the channel facies are expected to provide better responses to chemical flooding, because the detrital clay matrix is lower and the permeability is higher than in the upper portions of

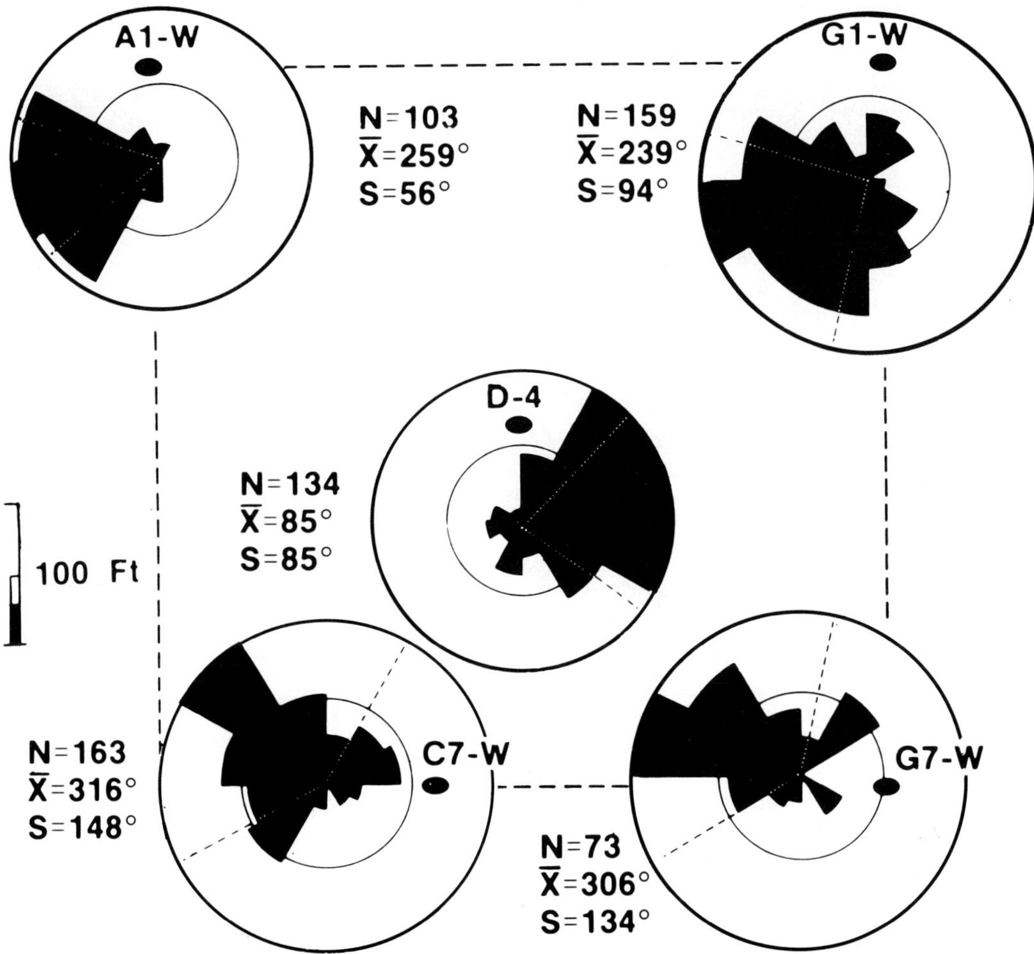

FIG. 14.—Diagram showing area distribution of the five oriented cores in the Madison "B" study area. Mean paleocurrent flow directions for each well are specified ($\bar{x}$). See caption for Figure 13 for explanation of diagrams.

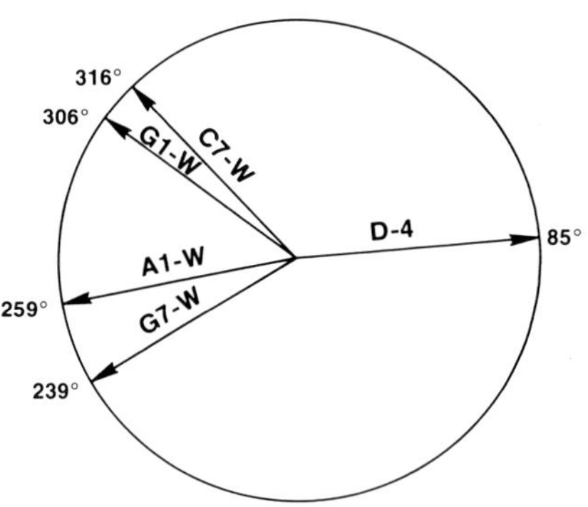

FIG. 15.—Mean paleocurrent transport directions calculated from trough crossbeds in the five oriented cores in the study area. Vector resultants for each core are plotted. Four of five cores suggest that the channel flowed to the west-southwest across the project area. The fifth core (D-4) probably had a systematic error in the data collection; see text for probable explanation.

the channel sandstone. Using a clay stabilizer to pretreat the clays in the reservoir prior to initiation of a miscible flood may prevent the occurrence of reduction in permeability by the migration of fluid-sensitive clays and formation of an iron-hydroxide gel.

ACKNOWLEDGMENTS

We express our thanks to a number of colleagues who assisted with this project. Jim Rine and Ken Helmold assisted in measuring paleocurrents on several oriented cores. Mike Wilson (consultant) carried out most of the petrographic analysis. Fruitful discussions with Ken Haugan and Curtis Burgess of the Cities Service Reservoir Management Staff in Tulsa aided in the description of the cores and the correlation of the well logs. Willis Waldorf prepared the cores for examination, and Pam Jenni and Fred Mason did the photography. The drafting department at the Cities Service Research Laboratory prepared the illustrations. Mary

TABLE 2.—DISTINGUISHING CHARACTERISTICS OF SEDIMENTARY FACIES IN THE BARTLESVILLE SANDSTONE, SEELEY-WICKE FIELD, MADISON UNIT "B"

CHARACTERISTICS \ FACIES		DELTA PLAIN	DISTRIBUTARY CHANNEL	OVERBANK	CREVASSE SPLAY	INTER-DISTRIBUTARY BAY/FLOOD PLAIN
FINING UPWARDS		R	A	N	C	R
COARSENING UPWARDS		N	R	N	C	C
RIP-UP CLASTS		R	C	N	C	R
GLAUCONITE		N	N	N	N	R
CALCAREOUS CEMENT		N	R	N	C	N
CLAY	GRAIN SIZE	C	R	C	C	C
SILT	GRAIN SIZE	C	R	A	C	A
VERY FINE SAND	GRAIN SIZE	R	C	R	C	R
FINE SAND	GRAIN SIZE	R	C	R	C	R
MEDIUM SAND	GRAIN SIZE	R	R	N	R	N
COARSE SAND	GRAIN SIZE	N	R	N	N	N
TROUGH CROSSBEDDING		N	A	N	C	R
RIPPLE BEDDING		N	C	A	C	A
"MASSIVE"-APPEARING		C	C	N	R	R
DEFORMED BEDDING		C	R	R	R	N
BURROWS		R	N	N	R	C

A-ABUNDANT (75-100%) C-COMMON (10-75%)
R-RARE (1-10%) N-NONE

Walker typed the manuscript. James Ebanks, John Hobson and Mary Walker reviewed the final manuscript.

REFERENCES

BASS, N. W., 1934, Origin of Bartlesville shoestring sands, Greenwood and Butler Counties, Kansas: American Association of Petroleum Geologists Bulletin, v. 18, p. 1313–1345.

———LEATHERNECK, C., DILLARD, W. R., AND KENNEDY, L. E., 1937, Origin and distribution of Bartlesville and Burbank shoestring oil sands in parts of Oklahoma and Kansas: American Association of Petroleum Geologists Bulletin, v. 21, p. 30–66.

HULSE, W. J., 1979, Depositional environment in the Bartlesville Sandstone in the Sallyards Field, Greenwood County, Kansas, *in* Hyne, N. J., ed., Pennsylvanian Sandstones of the Mid-Continent: Tulsa Geological Society, p. 327–336.

SIEMERS, C. T., AND TILLMAN, R. W., 1981, Recommendations for the proper handling of cores and sedimentological analysis of core sequences, *in* Siemers, C. T., Tillman, R. W., and Williamson, C. R., eds., Deep Water Clastic Sediments, a Core Workshop: Society of Economic Paleontologists and Mineralogists Core Workshop No. 2, p. 20–44.

VISHER, G. S., SANDRO, S. B., AND PHARES, R. S., 1971, Pennsylvanian delta patterns and petroleum occurrences in eastern Oklahoma: American Association of Petroleum Geologists Bulletin, v. 55, p. 1206–1230.

ZELIFF, C. W., 1976, Subsurface analysis, "Cherokee" Group (Pennsylvanian), northern Kingfisher County, Oklahoma: Shale Shaker, v. 27, p. 24–32.

PART III
AEOLIAN RESERVOIRS

COMPUTATION OF INITIAL WELL PRODUCTIVITIES IN AEOLIAN SANDSTONE ON THE BASIS OF A GEOLOGICAL MODEL, LEMAN GAS FIELD, U.K.

K. J. WEBER

Shell International Petroleum Maatschappÿ, The Hague, The Netherlands

ABSTRACT: Realistic well productivity calculations based on geologic models are an important aid in predicting field performance. For the Leman field, such models have been used to predict production potentials of untested wells and to judge the danger of water coning at an early stage (1973) of the field production history.

The Leman field is situated in the southern North Sea 48 km offshore from the English coast. The original reserves were about 10.5 Tcf when production started in 1968. The western half of the field is operated by Shell/Esso and the eastern unit by Amoco. At the time of the study, there were 10 platforms with 130 producing wells.

The Leman field reservoir rock is the Permian Rotliegendes Sandstone, which is 180–270 m thick and lies at a depth of about 2,000 m. The major and most productive part of the reservoir is composed of giant aeolian crossbed sets with an average thickness of 4.5 m. The orientation of the forset laminae is remarkably uniform. It is inferred that the laminae represent the lee slope of transverse dunes. The variation in permeability of the foreset laminae and the generally low permeability of the bottomset zones underlying these spoon-shaped crossbed sets cause a very heterogeneous permeability distribution. The heterogeneity is enhanced by variations in grain size and associated authigenic clay content and diagenesis.

No literature data were available on the length/width/thickness ratio of giant aeolian crossbed sets formed by transverse dunes. Outcrop studies in the De Chelly Canyon (Arizona) were thus carried out to gather information on the geometry of this type of crossbed set. The large horizontal extent of the crossbed sets (length approximately 200 × thickness), combined with the low permeability of the associated bottomsets, indicates that water coning will be minimal, an interpretation that has been corroborated by 10 years of production data. Initial well behavior is probably controlled by the properties of the thickest more permeable crossbed sets. Furthermore, some pairs of wells may be interconnected via continuous, fairly permeable beds, because the average well spacing is smaller than the average crossbed set length (900 m). Log correlations tend to confirm this conclusion.

INTRODUCTION

The primary aim of this paper is to outline the methods by which initial well productivity can be computed on the basis of log and core data from cored Leman wells. Second, it will be shown that because of the presence of a correlatable zonation in the Rotliegendes, the methods can also be applied for noncored wells on the basis of log data only.

Theoretical well productivities calculated on the basis of averaged log and core data rarely match tested productivities. Apart from complications caused by the drilling and completion of a well, the observed disparities are mostly due to reservoir inhomogeneities. Although it is usually possible to gain some insight into reservoir heterogeneity from pressure buildups, various other reservoir engineering methods and production logging, these methods do not necessarily lead to a detailed realistic reservoir model.

Furthermore, the above methods take time and can only be applied to completed wells. In this particular case, a realistic geological/petrophysical reservoir model could be developed because the petrophysical implications of the sedimentologic and diagenetic characteristics of the reservoir rock were understood. Also, a sufficient number of wells have been cored and a suitable core analysis program has been carried out.

Initial well productivities, computed on the basis of a geological/petrophysical model, core and log data, are useful for three purposes:

(1) If the computed initial productivities are roughly equal to the measured initial productivities, the geological/petrophysical model is probably basically correct. This model can then be used in reservoir engineering calculations of future well behavior. A realistic model is also an important aid in interpreting pressure buildup curves and pulse-testing results.

(2) The initial productivity of a given bed can be calculated prior to making the perforations. The perforations scheme for a well can thus be based on these computations. Comparison of computed and measured productivities can lead to the recognition of completion imperfections, such as plugging or poor cement bonds.

(3) Similarly, the total calculated initial productivity potential of a well can be used as an early indication of local reservoir quality. In the Leman field the reservoir quality changes rapidly, but several adjacent wells often have comparable production potentials, especially in east-west oriented sections. Early recognition of areas of high production potential can, therefore, influence the drilling pattern in a well cluster and thereby reduce the total number of wells to be drilled.

DESCRIPTION

In Leman field (Fig. 1), as in Groningen, the reservoir rock is Permian Rotliegendes Sandstone. The gas emanated from the coal in the underlying Carboniferous rock is trapped in the reservoir sandstone against the overlying Zechstein salt. The Rotliegendes Sandstone has a thickness of 180–270 m in the Shell/Esso concession area.

Much has already been written about the Rotliegendes Sandstone in the Leman area. Glennie (1970, 1972; Glennie and others, 1978) has written several papers on the sedimentation, whereas Nagtegaal (1973, 1979) has written about the grain-size distribution, sedimentary structures and the diagenesis of the Rotliegendes in the southern North Sea. A general geologic description of the Leman field was given by Van Veen (1975).

The Rotliegendes can be subdivided into three facies zones (Fig. 2). At the base, overlying the truncated Carboniferous, there is an interval of 30–60 m of wadi sediments which consists of low-porosity sandstones and some shales. About

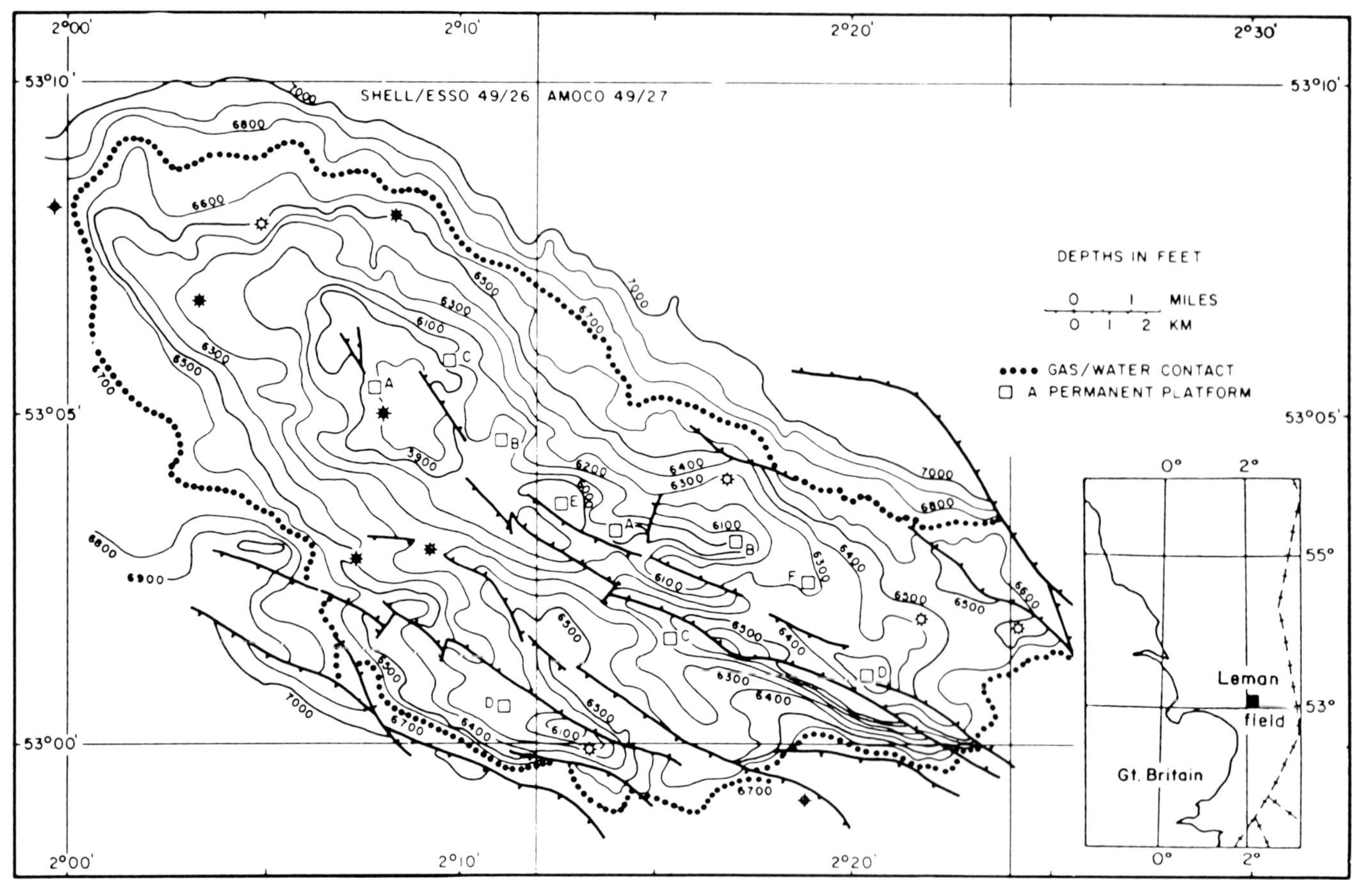

FIG. 1.—Interpretation of Leman gas field, showing contours on top of Rotliegendes in feet below sea level. (After van Veen, 1975.)

two-thirds of the reservoir (140–200 m) is composed of crossbedded aeolian sandstone, which locally has good reservoir-rock properties. The lateral facies distribution during the deposition of this interval is shown in Figure 3. This zone is truncated at the top by a zone of water-laid reworked aeolian sediments (6–30 m), which was formed during the transgression of the sea over the coastal desert at the close of the Rotliegendes period. This water-laid zone has a relatively low porosity and permeability and is of little importance with respect to the initial well productivity.

The west-northwest to east-southeast cross section shown in Figure 4 illustrates the subdivision of the Rotliegendes over a large part of the field. In this cross section, the faults have been omitted. (There is a large number of steep normal faults throughout the field; Fig. 1.) Most faults have throws of 100 ft (30.3 m) or less, but a few have a throw on the order of 200 to 300 ft (60.6 to 90.9 m).

Some of the faults running more or less along the axis of the field, from northwest to southeast, are thought to be possibly of a strike-slip nature. The larger faults have a considerable influence on the pressure distribution in the reservoir (Craig and others, 1977; Goldwater and others, 1978) and even small faults in well clusters can be partially sealing (Fig. 5). A complication is that the basal Zechstein salt effectively absorbs seismic energy, leaving the base of the upper Zechstein salt as the deepest reflection that can be mapped. Thus, only major faults observed on seismic records can be extrapolated downward to the Rotliegendes level.

In Figure 1, it can be seen that even in the structurally high areas, where the platforms are located, the wadi interval and the lower part of the aeolian zone are almost everywhere below the gas/water contact. Thus, this part of the formation is of little importance in the context of this paper.

SEDIMENTOLOGICAL ASPECTS

A detailed sedimentological study of the Rotliegendes in the Leman area is possible because of the large number of cored wells. In 10 wells in the Shell/Esso part of the field, an important part, or all, of the Rotliegendes was cored.

Beds with a sufficient permeability to contribute significantly to a well's initial productivity are largely restricted to the aeolian interval of the Rotliegendes. This interval is composed of large dune crossbed sets (Fig. 6). Each dune is truncated by its successor, and the resulting shape of the individual crossbed set is a flat oval disc with a spoon-shaped bottom. The average thickness of the crossbed sets is about 4.5 m. The dunes were of the transverse type, with sinuous crests oriented perpendicular to the wind direction (Glennie, 1972; Wilson, 1972; Fig. 7).

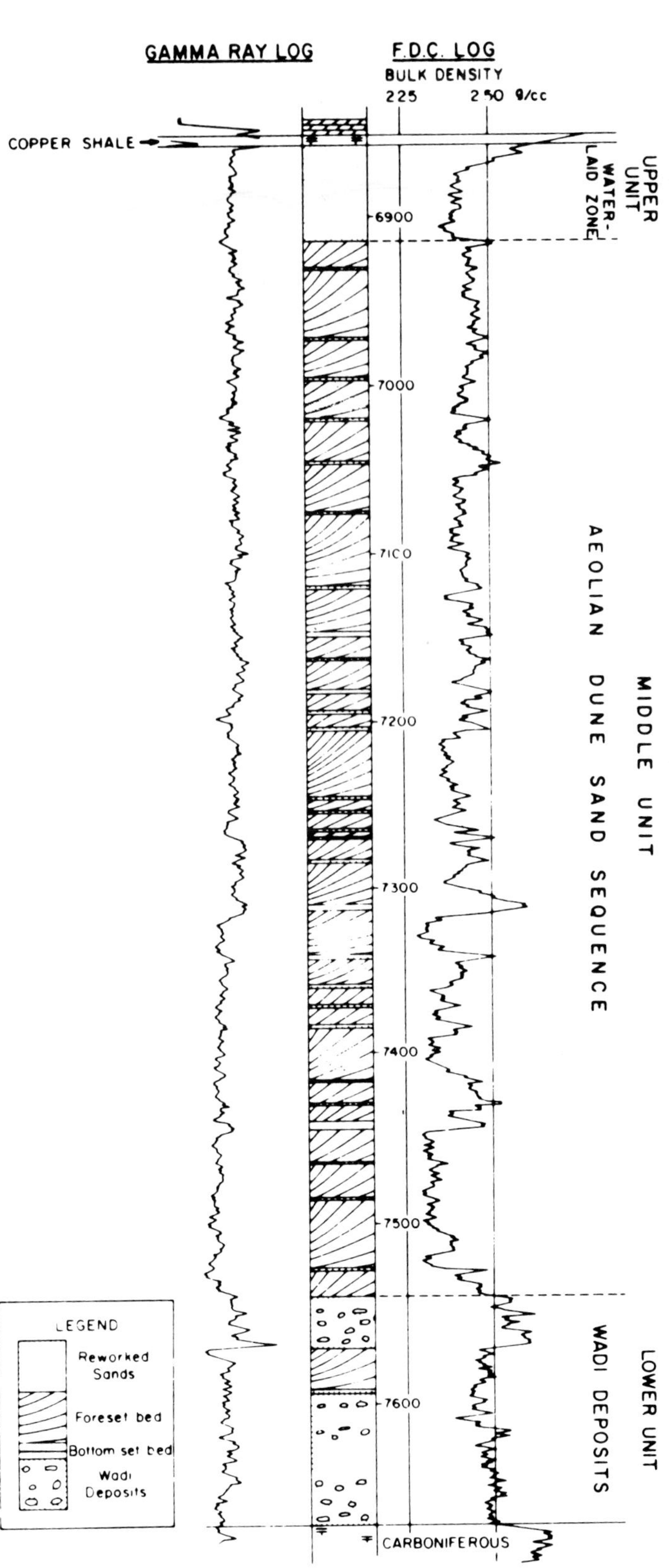

FIG. 2.—Type log of the Rotliegendes in the Leman gas field. (After van Veen, 1975.)

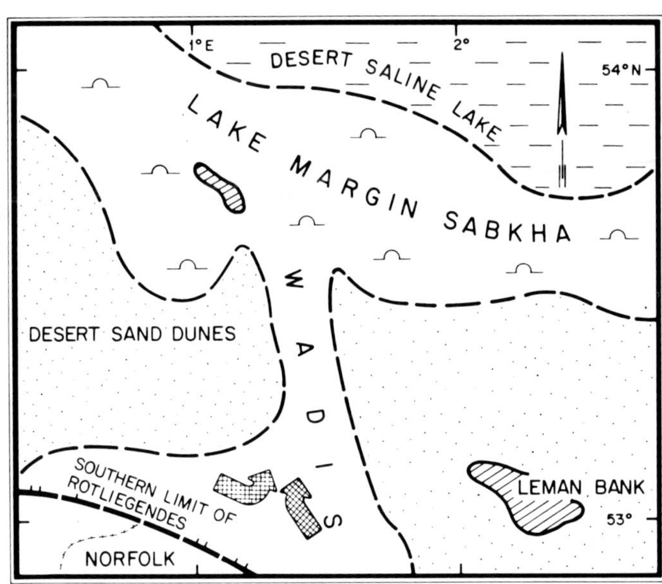

FIG. 3.—Facies distribution during deposition of desert sand dunes in the Leman field area. (After Glennie and others, 1978.)

The wind blew persistently from the east during deposition of the aeolian beds. The thick crossbed sets, with their steeply-dipping foreset laminae, give excellent sedimentary dips on dipmeter logs (Fig. 8).

An important problem with respect to the design of a geological reservoir model is the lack of information from the literature on the dimensions of ancient aeolian crossbed sets of this type. It is essential to know whether permeable crossbed sets extend over distances greater than the well spacing or if they are of restricted lateral continuity. Rotliegendes outcrops in England or Germany are not very suitable for studies of crossbed set dimensions because of their relatively small size. Hence, a study was made of very large outcrops of aeolian sandstone in the De Chelly Canyon in northern Arizona. The De Chelly Canyon deposits closely resemble the Rotliegendes dune interval.

The relationship between the facies, diagenesis and the reservoir quality in the Rotliegendes of the southern North Sea has been studied by Glennie and others (1978). The results stress the importance of the amount of interstitial matter on the porosity. This amount of interstitial matter differs for certain zones within the aeolian interval. Authigenic illite and chlorite are quite common. The pore-bridging nature of these clays has a very detrimental effect on permeability (Stalder, 1973; Fig. 9).

During aeolian transport, sand grains move over the windward dune slopes and slide down along the leeward dune slip face. Discrete coarser and finer sand streaks are formed. Bagnold (1941) has described the effect of bedload sorting due to gravitational sliding. The differences in grain size and sorting of the individual sand streaks can be quite large and, as a result, there can be a considerable difference in the permeability of adjacent sand laminae (Van Veen, 1975).

At the base of the leeward slope of a dune is deposited a mixture of sand grains rolling down the slope and fine

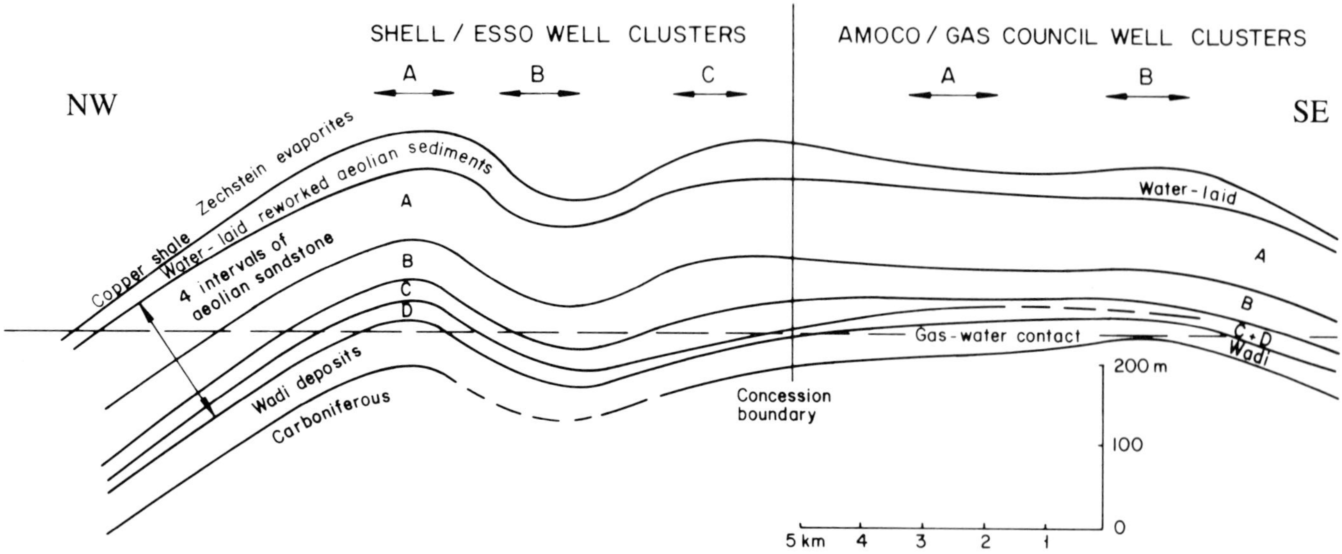

FIG. 4.—Cross section of Leman field showing zonation of the Rotliegendes reservoir.

particles suspended in the air and carried downward by turbulence behind the dune crest. This mixture is poorly sorted and forms a finely laminated, low-permeability layer beneath the zone of foreset laminae. This layer is called the bottomset bed. The foreset laminae merge tangentially with the laminae of the bottomset bed. In Leman field, the bottomset beds usually contain a higher percentage of carbonate cement than the foreset laminae zones. The poor sorting and the carbonate cement give the bottomset beds low porosity and high grain density, which make it possible to identify these beds on FDC (gamma-ray-resistivity) logs. Moreover, the relatively high clay content of the bottomset beds results in marked peaks on the gamma-ray logs (Fig. 8).

Besides the bottomset beds discussed above, there are also other dune-base sandstones which are deposited on essentially flat or gently sloping interdune areas (Nagtegaal, 1979). These sandstones are horizontally laminated with a marked bimodal grain-size distribution ranging from very coarse to fine-grained laminae. The bimodality is thought to result from selective deflation. For the geologic model, these various dune-base sandstones are grouped together.

Sometimes, intercalations of a few inches to a few feet of adhesion ripple deposits (Nagtegaal, 1973) are found in the aeolian interval. They occur as individual beds or as the lower parts of the bottomset beds. Except for some more prominent beds of this nature in wells 49/26-D405 and 49/26-2, these beds are not important in the aeolian zone and can be ignored in the model. In the neighboring Indefatigable field, adhesion ripple beds appear to be more common. The higher moisture content of the Rotliegendes interdune areas, suggested by abundant adhesion ripples, is probably due to a more seaward location and, presumably, lower topographic position.

The water-laid zone is the result of the Zechstein transgression. The log correlations show that the contact between the water-laid and aeolian zones is a subhorizontal erosional surface. The differences in thickness of the water-laid zone are caused by erosional relief at the top rather than at the base of this zone. The Copper Shale (or Kupferschiefer) conformably overlies the water-laid zone and thus was deposited on an undulating surface. In correlating the Rotliegendes, the Copper Shale is often drawn as a datum level, but this can easily lead to erroneous correlations.

The water-laid sands are composed of homogenized, partly reworked aeolian sands and have a much higher carbonate cement content than the dune crossbed sets. Slumping structures are commonly observed in the water-laid zone; however, log correlations indicate that, in general, the water-laid sands are deposited in extensive horizontal beds, which can usually be followed over a distance of several well spacings within a cluster.

Wadi sediments have been described in detail by Glennie (1972). For the calculation of initial well productivities these deposits are unimportant. They are usually situated near or underneath the gas/water contact and are therefore not perforated.

Aeolian crossbed set geometry.—

The literature on aeolian sandstones and recent dunes is mainly concerned with diagnostic characteristics of aeolian sediments, regional extent and stratigraphy of aeolian formations, descriptions of crossbedding types and measurements of foreset dip orientation. There is a distinct lack of numerical data on crossbed geometry and petrophysical properties.

In European outcrops of aeolian sands, it is difficult to determine crossbed set geometry because the outcrops are small in relation to the lateral extend of the crossbed sets. Large and well exposed aeolian outcrops are present, however, in the Colorado Plateau area of the United States. The Permian De Chelly Sandstone exposed in the De Chelly Canyon (Fig. 10) in northern Arizona has much in common with the Rotliegendes in the Leman field area. The type of crossbedding, the average crossbed-set thickness, the formation thickness and the striking uniformity of the foreset

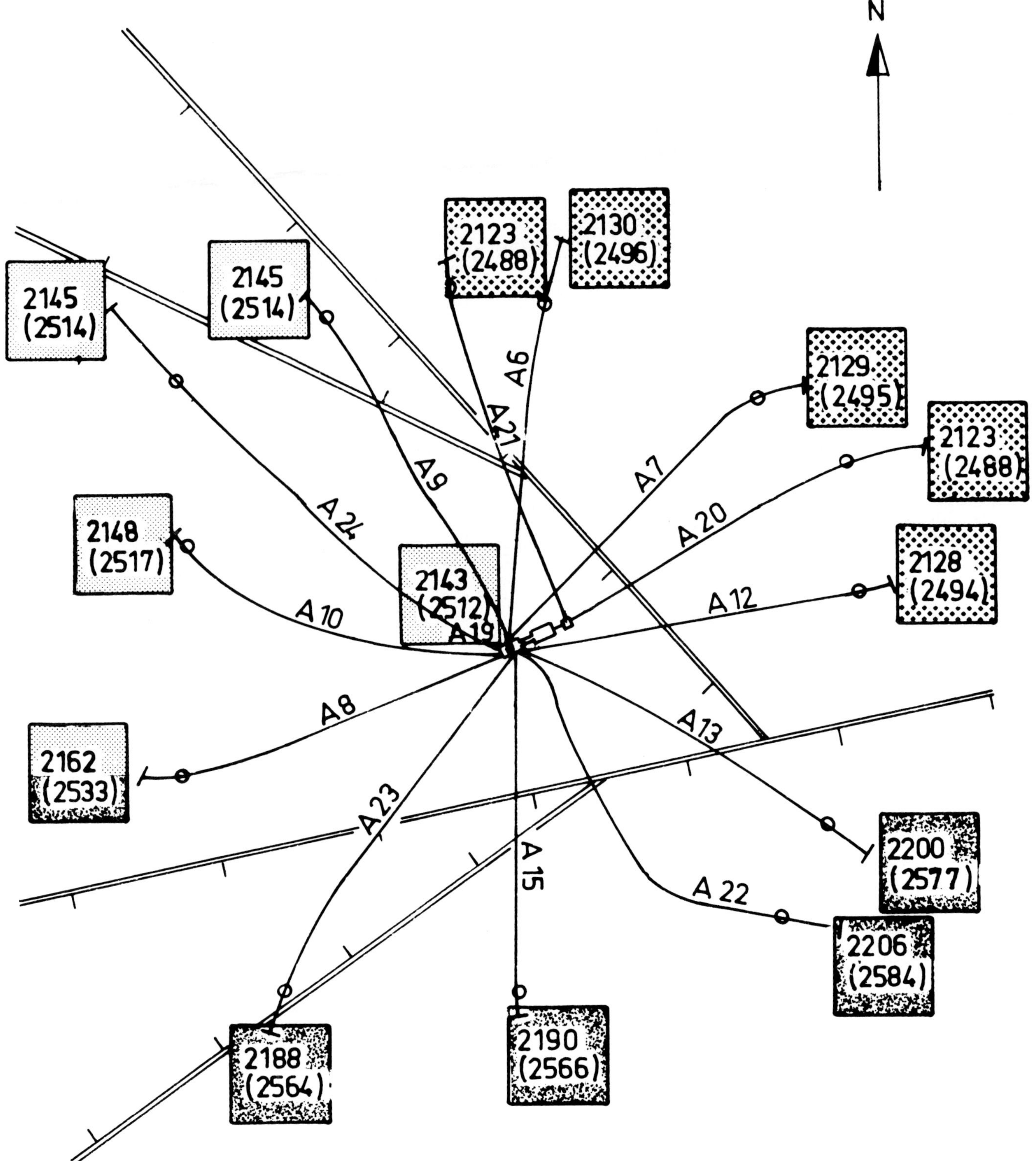

FIG. 5.—Closed-in tubing-head pressures and corresponding calculated static pressures at datum (6,500 ft TVSS)—"A" platform, September 1979. (All pressures psig.)

dip directions are all similar to the Rotliegendes characteristics (Stokes, 1961; Baars, 1961). Thus, the extensive and almost vertical canyon walls were studied to gain insight into the crossbed set geometry of fossil transverse dunes.

De Chelly Sandstone in De Chelly Canyon, Arizona.—

The Permian De Chelly Sandstone in northwestern Arizona is predominantly of aeolian origin. The source area

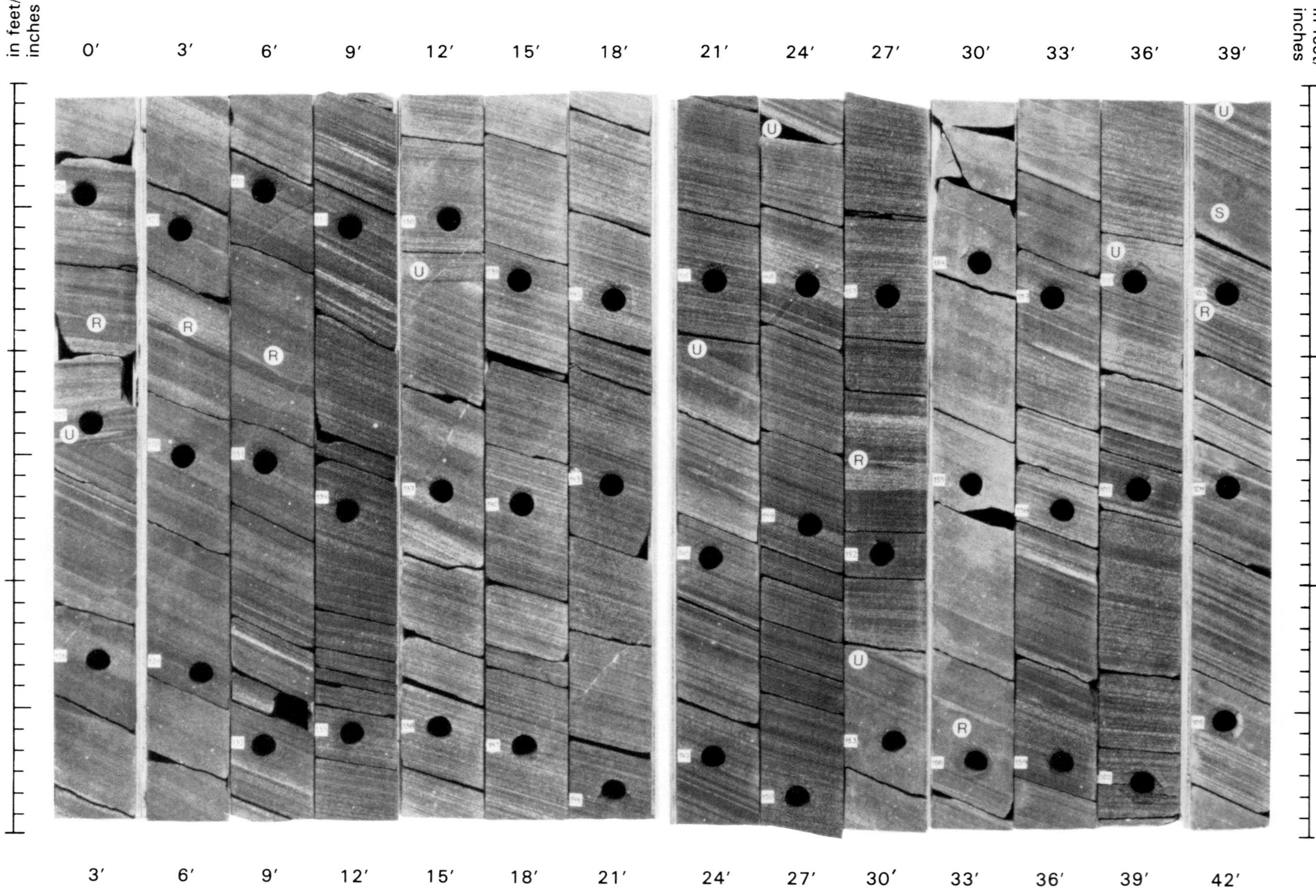

FIG. 6.—Typical example of dune crossbedding in the Rotliegendes of the southern North Sea, showing foreset laminae grading into bottomset (U) laminae. Locally, some weakly developed ripples can be observed (R). A small-scale intraformational slump is located at S.

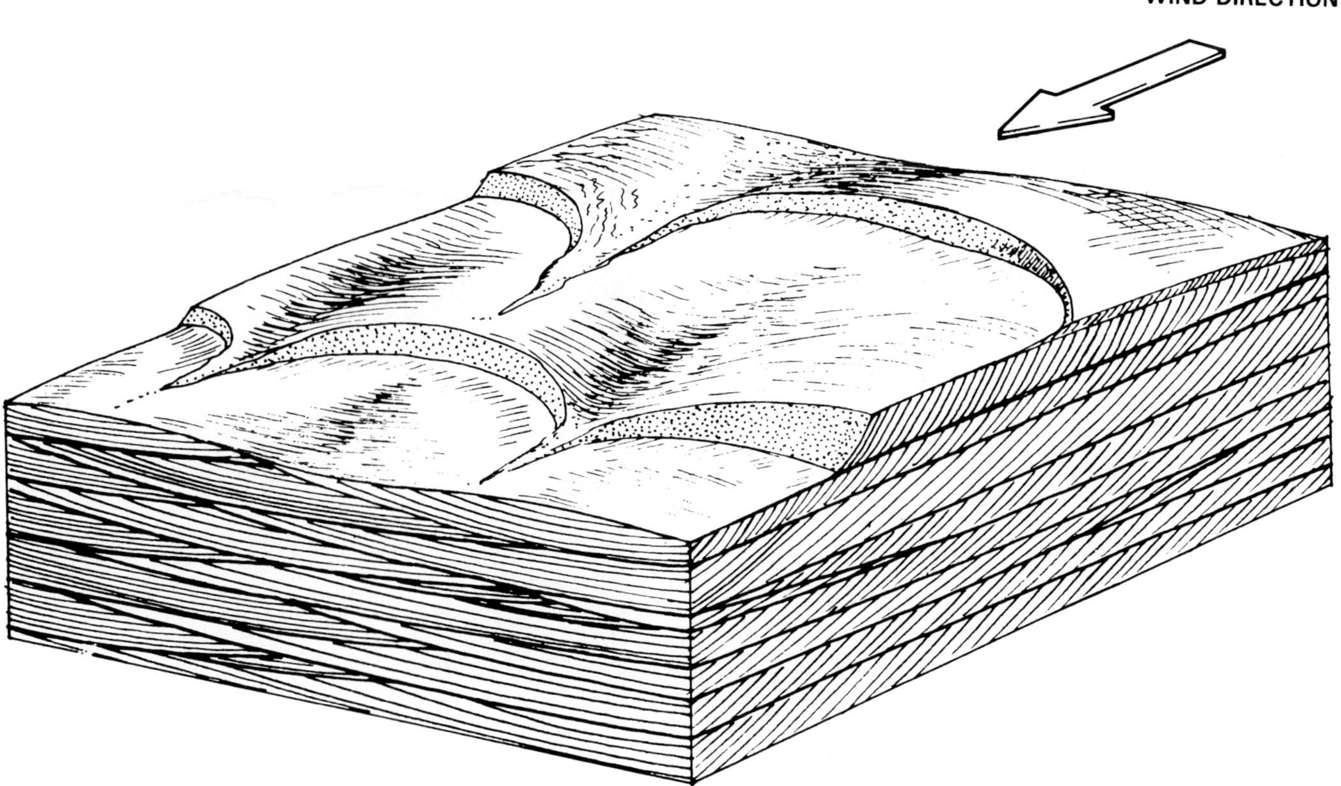

FIG. 7.—Model of transverse dune sedimentation.

was the Uncompahgre uplift to the northeast and the source material is thought to be the Cutler redbeds (Baars, 1961). To the south, the aeolian desert was bounded by a sea, which finally transgressed over the aeolian sands.

In the De Chelly Canyon, however, the De Chelly Sandstone is unconformably overlain by the Triassic Shinarump Member of the Chinle Formation, composed of fluviatile channel fills of coarse sandstone and conglomerate. The Shinarump is somewhat harder than the De Chelly Sandstone and forms the tops of the cliffs in the De Chelly Canyon. The De Chelly Sandstone is found over southeastern Utah, northeastern Arizona, and western and central New Mexico.

De Chelly is subdivided into a number of members of which the White House Member, with a thickness of as much as 570 ft (173 m), forms the almost vertical cliffs in the De Chelly Canyon. The White House is a fine- to medium-grained sandstone (average grain size 0.130 mm), often containing more than 10 percent feldspar. Coarse as well as very fine streaks occur frequently. The grains are usually well rounded. Successive laminae differ in the degree of cementation; the finer grained laminae are more strongly cemented than the coarser grained streaks. The cement is described as mainly ferric oxide and calcite (Baars, 1961), but it appears that no detailed study of the cementation has been made.

The White House Member is almost entirely composed of medium to large aeolian crossbed sets with a few associated smaller crossbed sets. Over a large area in the De Chelly Canyon, the average thickness of the crossbed sets is almost 20 ft (6.1 m), ranging from a few to over 100 ft (30.3 m). Most crossbed sets, however, are 10 to 35 ft (3 to 10.6 m) thick. The foreset dip direction is very consistent (Stokes, 1961; Baars, 1961), with an average of N210E.

Several suitable outcrops studied in the De Chelly Canyon area could be reached or observed from the roads and the lookout points of the De Chelly Canyon National Monument (Fig. 10). In one place, it is possible to descend into the canyon to the White House ruins area (Fig. 11). This area provided three good cliff sections for study, two parallel to the average foreset dip direction (outcrops 1 and 2, Fig. 12) and one perpendicular to this orientation (outcrop 5, Fig. 13). Another section parallel to the foreset dip was studied along a branch of the main canyon near the White Overlook (outcrop 3). An isolated rock massif, near the center of the canyon in the same area, provided a fourth section with this orientation (outcrop 4).

Good outcrops perpendicular to the sand transport direction are difficult to find because of the orientation of the main canyon system. The narrow Three Turkey Canyon is partly oriented in this direction. The data derived from this outcrop, however, must be regarded as tentative because of its irregular profile (outcrop 6).

FIELDWORK AND COMPUTATION OF CROSSBED SET GEOMETRY

To obtain numerical data on crossbed set geometry, detailed sketches and photographs were made of the six outcrops.

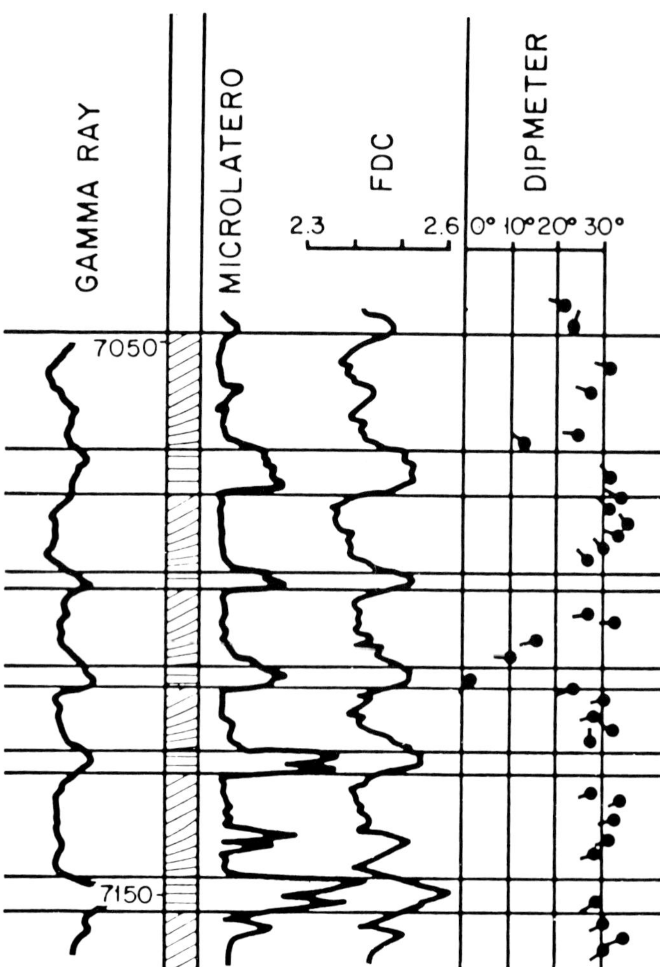

FIG. 8.—Dipmeter log showing unidirectional foreset dip on large dune crossbed sets in the Rotliegendes, Leman field.

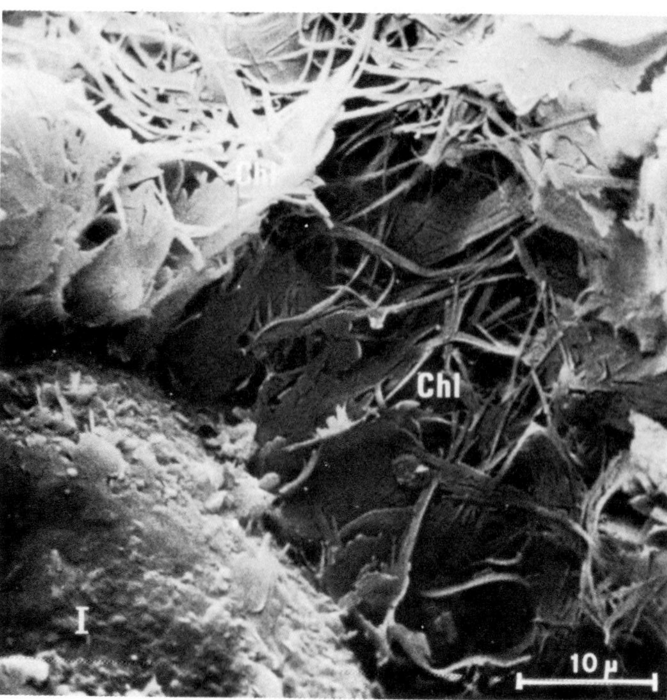

FIG. 9.—SEM photomicrograph of illite/chlorite cement, flaky to thread-bridge illite/chlorite (Chl), and illite coating (I).

In all of the studied outcrops in the De Chelly Canyon, relatively few crossbed sets can be observed over their entire width or length (Figs. 12, 13). Consequently, in order to estimate average dimensions of crossbed sets, an assumption has to be made.

In fluviatile festoon crossbedding the ratio between average height, width and length is quite constant (1:10:25). Because of the similarity between aeolian and fluviatile festoon crossbedding, it is assumed that these ratios will also be roughly constant for aeolian crossbed sets. On this assumption it even becomes possible to use outcrops where only a few crossbed sets are exposed over their entire width or length. This is clear when one considers a dip section composed of crossbed sets of equal size. If the outcrop width (O) is approximately equal to half the crossbed set length, a number of wedge-outs (E) will still be observed but not as a single complete set. When the number of superimposed crossbed sets (n) is sufficiently large, however, the crossbed length (L) can be estimated with the formula $L = 2n.O./E$ (Fig. 14).

In reality, the crossbed sets have unequal thickness and length, but using the above assumption, we can transform all the measured crossbed set dimensions to values for sets of equal size (Fig. 15). For a dip section, the measured length of a set (L_i) is multiplied by H_{av}/H_i, i.e., the average crossbed set height for the outcrop divided by the height of the relevant set. The number of wedge-outs for each set (E_i), which is, of course, 0, 1 or 2, is multiplied by H_i/H_{ay} to give each observed wedge-out equal weight. For strike sections, the maximum height of the crossbed sets was measured (H_{max}), but the procedure is the same as for the dip sections.

In cases where the outcrop orientations differ too much from those of true dip or strike sections to enable the observations to be used without corrections, the crossbed sets have been projected on a plane with the correct orientation and l_{proj} or W_{proj} are given. An example of the data recorded for one of the outcrops is given in Table 1.

The average ratios of crossbed set length over height and width over maximum height, can be calculated with the formula

$$\frac{L_{av}}{H_{av}} = \frac{2\sum^{i}\left(\frac{H_{av}}{H_i}\cdot L_i\right)}{H_{av}\sum^{i}\left(\frac{H}{H_{av}}\cdot E_i\right)} \tag{1}$$

and

$$\frac{W_{av}}{H_{max\ av}} = \frac{2\sum^{i}\left(\frac{H_{max\ av}}{H_{max\ i}}\cdot L_i\right)}{H_{max\ av}\sum^{i}\left(\frac{H_{max\ i}}{H_{max\ av}}\cdot E_i\right)}. \tag{2}$$

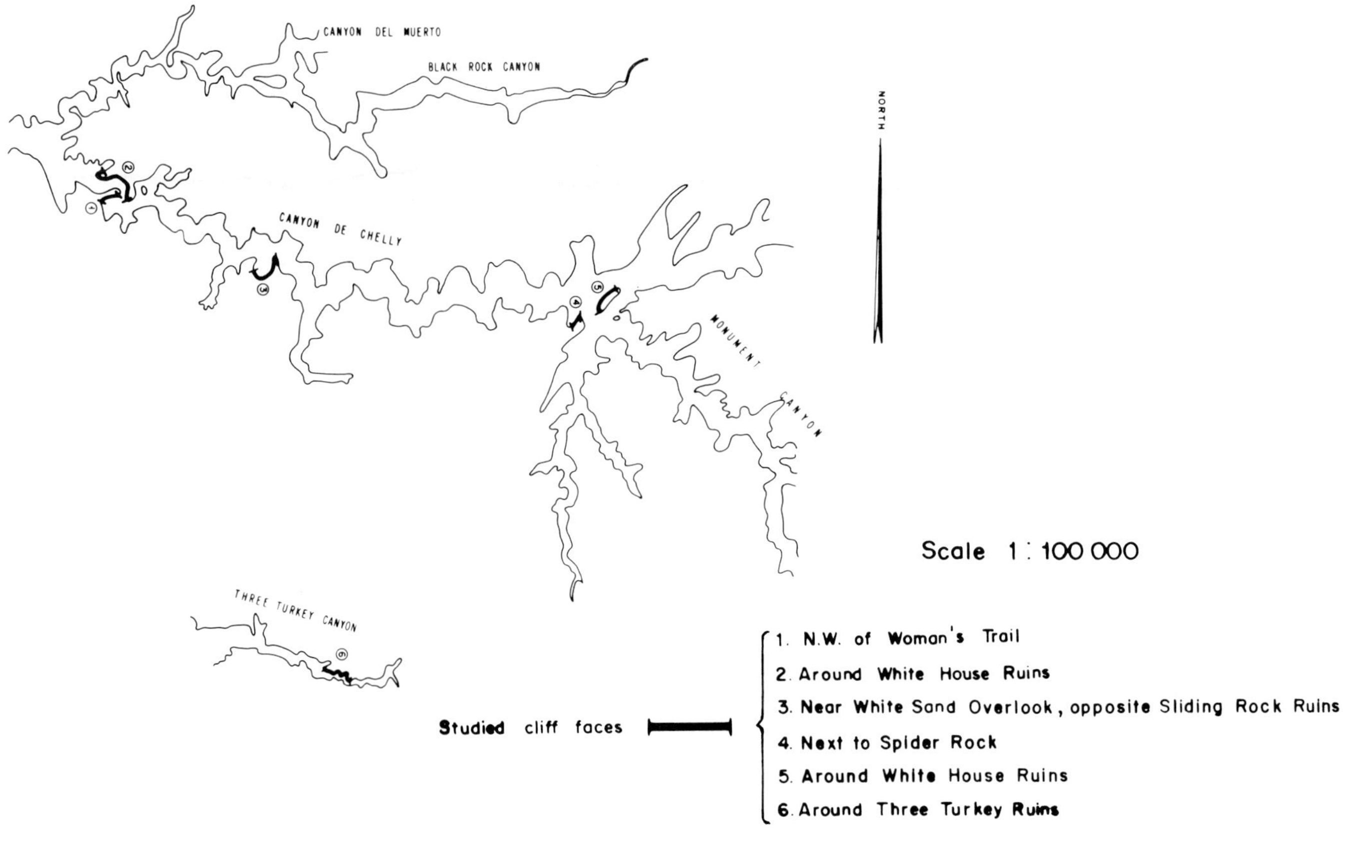

FIG. 10.—Sketch map of De Chelly Canyon area showing cliff faces studied.

The results of the calculations are given in Tables 2 and 3. The average crossbed set thickness is about 21 ft (6.4 m). From the figures, it was measured that the average maximum crossbed set thickness is roughly 1.5 × H_{av} or 32 ft (9.7 m). Formula 1 gives remarkably similar figures of L_{av}/H_{av} for all four cases, and it appears justified to conclude that this ratio must be around 200.

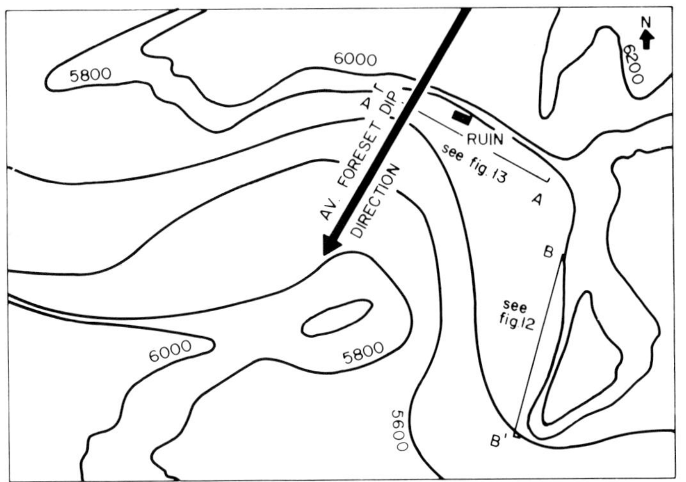

FIG. 11.—De Chelly Canyon, White House Ruin area. Contours are in feet above sea level.

Only two strike sections were studied in the canyon area, and of these the Three Turkey Canyon was difficult to measure. Thus, the two W_{av}/H ratios given cannot be regarded as providing sufficient information to decide on a characteristic average ratio. In relation to the L_{av}/H ratio, however, the obtained order of magnitude between 50 and 100 for W_{av}/H is quite probable. This would give an average ratio L/W of 2 to 4, which may be compared with a value of about 2.5 for fluviatile crossbed sets.

In the aeolian interval of the Rotliegendes in the Leman Bank field, the average thickness of the crossbed sets is also of the order of 20 ft (6.1 m). Assuming crossbed set geometries comparable with the above, the average length of these crossbed sets may well be about 4,000 ft (1,212 m). This suggests that at least part of the crossbed sets can be correlated between two cluster wells situated on a line parallel to the aeolian transport direction (east-west in the Leman Bank field). This is the case for an east-to-west row of wells in the A cluster (Fig. 16).

PETROPHYSICAL ASPECTS

In the four Shell/Esso cluster areas (Fig. 1) most of all of the Rotliegendes has been cored in wells 49/26-A6, -B240, -C370 and -D40. In addition, the topmost part of the reservoir was cored in well 49/26-1; in well B260 the lower half of the Rotliegendes was cored. The fully cored well 49/26-2 is fairly close to the A- and C-cluster areas

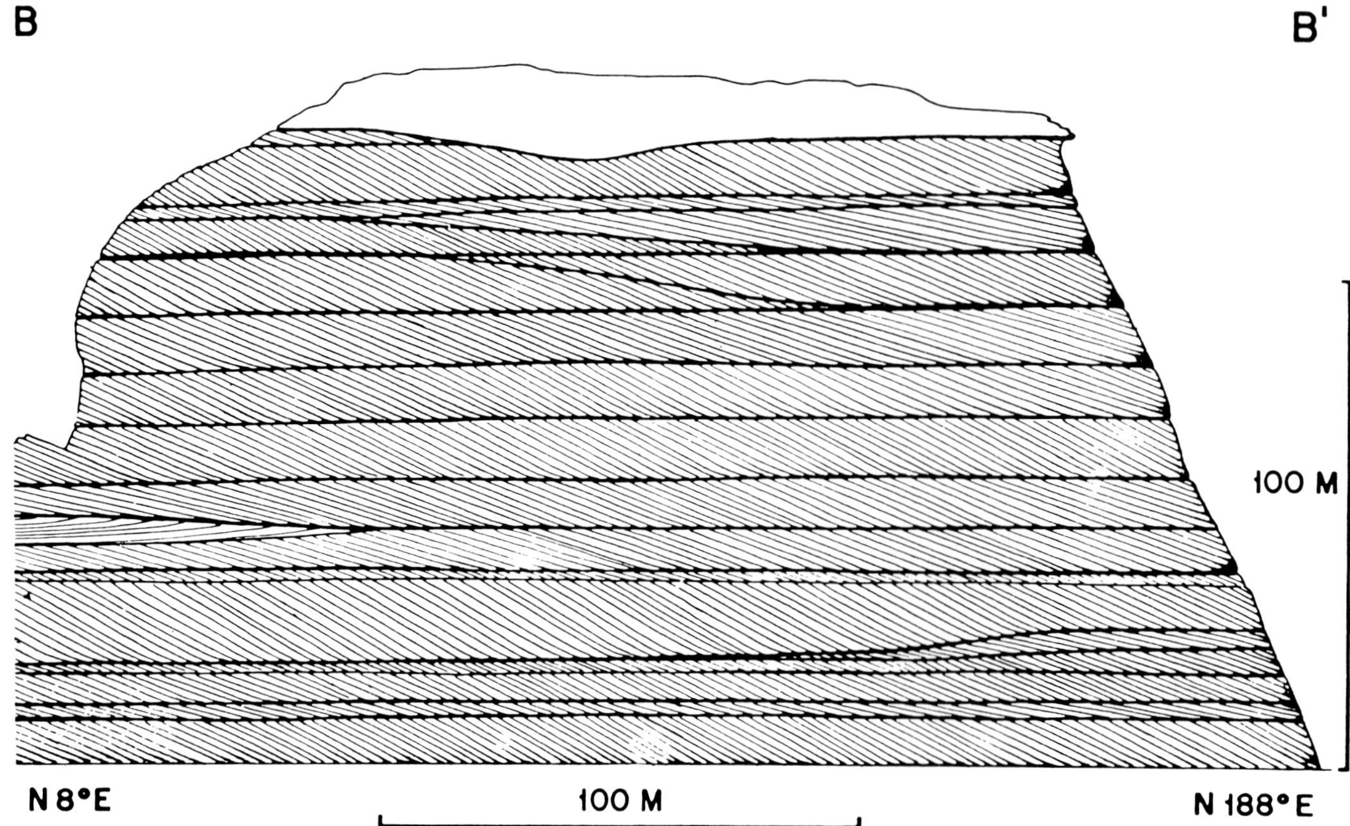

FIG. 12.—Cliff parallel to wind transport direction.

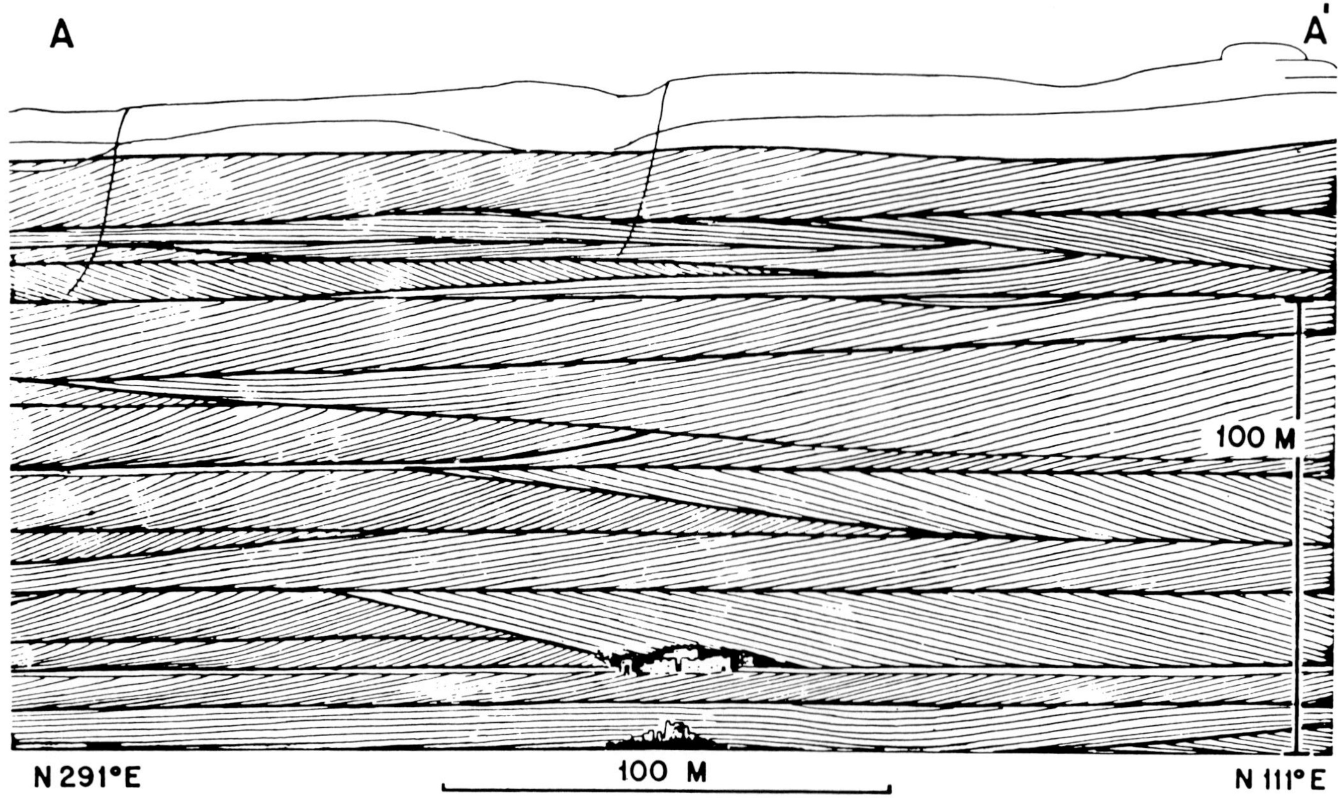

FIG. 13.—White House Ruin cliff. Oriented almost perpendicular to transport direction.

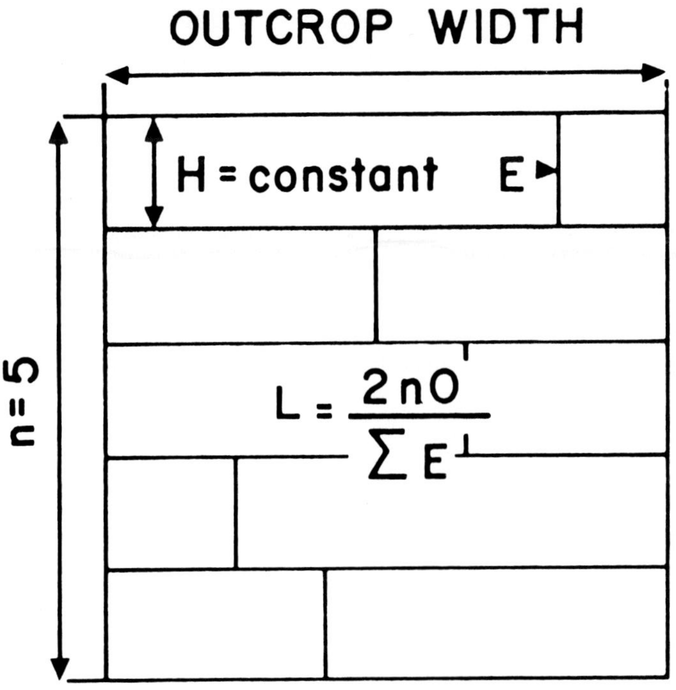

FIG. 14.—Formula for computing the length of incompletely exposed crossbed sets of equal size.

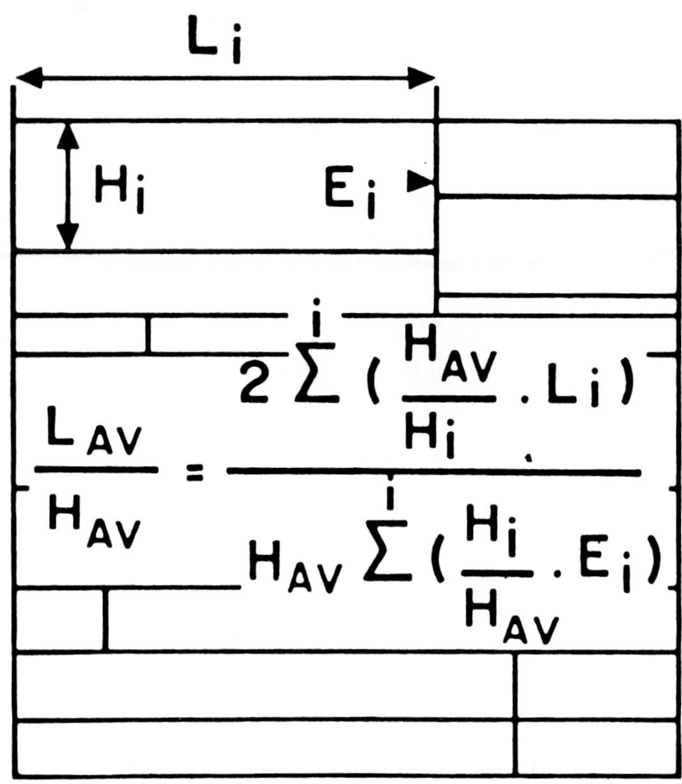

FIG. 15.—Formula for computing length/thickness ratio of incompletely exposed crossbed sets in cliff face.

TABLE 1.—CROSSBED SETS IN CLIFFS EAST OF WHITE SANDS OVERLOOK (OUTCROP)

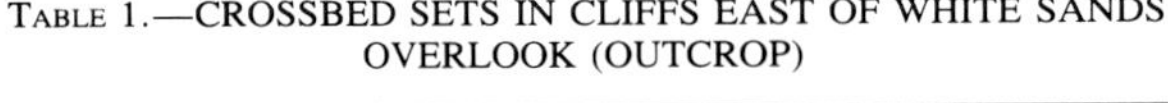

No. of crossbed-set sets	1) H ft	2) L_{proj} ft	$\frac{L_{proj}}{H}$	$\frac{H_{av}}{H_i} L_i$	4) $\frac{H_i}{H_{av}} E_i$
		3)			
1	14	1370)	98	1700	0.80
2	8	300)	37	650	0.46
3	17	1650)	97	1690	0.98
4	25	1700)	68	1180	1.44
5	17	1750)	103	1790	0.98
6	14	1900	136	2350	0.00
7	8	1900)	237	4130	0.46
8	9	1500)	166	2900	0.52
9	22	1030)	47	815	1.26
10	28	1600	57	995	0.00
11	10	(1350	135	2350	0.57
12	11	(670)	61	1060	1.26
13	11	360)	33	570	0.63
14	14	(1600	114	1980	0.80
15	31	1850)	60	1040	1.78
16	31	1900	61	1065	0.00
17	33	1900	57	1000	0.00
18	16	(950	59	1035	1.84
19	14	470)	33	580	0.80
20	8	360)	45	780	0.46
21	16	(2000	125	2170	0.92
22	16	800)	50	870	0.92
23	22	1700)	77	1340	1.26
24	23	1750	76	1320	0.00

1) H is height of crossbed set
2) L_{proj} is length of crossbed set as projected on plane N30-210E
3)) crossbed set wedges out on one side
( ) crossbed set wedges out on both sides
4) B is number of wedge-outs per crossbed set
Note: Orientation of northern part of cliff is roughly parallel to the transport direction. For measuring the crossbed-set geometry, the section has been projected on a plane oriented N30-210E.

and can also be used as an indicator for the development of the Rotliegendes in these areas. Additional cores are available from wells 49/26-3, 4, 5 and 14. All cores were sampled every foot (0.3 m) and care was taken that the plugs were oriented parallel to the laminae of the crossbedding. Thus, a considerable amount of data on permeability parallel to the laminae is available.

In well A6 an effort was made also to measure the vertical permeabilities. Plugs were drilled with their axes parallel to the cores axes at locations as close as possible to the horizontal samples. Since this is a deviated well, these permeabilities have been measured at a considerable angle to the actual vertical direction. In the hetergeneous rock, taking separate samples and horizontal permeability is not ideal. In the vertical well B260, cubic samples were therefore sawed from the core, which provided permeability measurements in three directions. The sample locations were

TABLE 2.—ESTIMATES OF THE RATIO OF CROSSBED-SET LENGTH OVER AVERAGE CROSSBED-SET THICKNESS FOR FOUR OUTCROPS IN THE DE CHELLY CANYON

Outcrop No.	H_{av}	$\sum^{i}\left(\frac{H_{av}}{H_i}\cdot L_i\right)$	$\sum^{i}\left(\frac{H_i}{H_{av}}\cdot E_i\right)$	$\frac{L_{av}}{H_{av}} = \frac{2\sum^{i}\left(\frac{H_{av}}{H_i}\cdot L_i\right)}{H_{av}\sum^{i}\left(\frac{H_i}{H_{av}}\cdot E_i\right)}$
1	22.3	11320	5.32	191
2	26.8	25835	9.67	199
3	17.4	35360	18.14	224
4	20.5	37485	14.89	246

TABLE 3.—ESTIMATES OF THE RATIO OF CROSSBED-SET WIDTH OVER MAXIMUM CROSSBED-SET THICKNESS FOR TWO OUTCROPS IN THE DE CHELLY CANYON

Outcrop No.	$H_{\max\,av}$ 1)	$\sum^{i}\left(\frac{H_{\max\,av}}{H_{\max\,i}}\cdot W_i\right)$	$\sum^{i}\left(\frac{H_{\max\,i}}{H_{\max\,av}}\cdot E_i\right)$	$\frac{W_{av}}{H_{\max\,av}}=\frac{2\sum^{i}\left(\frac{H_{\max\,av}}{H_{\max\,i}}\cdot W_i\right)}{H_{\max\,av}\sum^{i}\left(\frac{H_{\max\,i}}{H_{\max\,av}}\cdot E_i\right)}$	$\frac{W_{av}}{H_{av}}$
5	40.5	29200	21.18	68	102
6	33.7	12245	23.74	31	47

1) $H_{\max\,av}$ is about $1.5 \times H_{av}$

selected in such a way that a good description of the vertical permeability was obtained.

For Well A6, a special study was carried out to see how a detailed permeability analysis of the Rotliegendes could be made with a mini-permeameter. This study showed that adjacent foreset laminae of the dune crossbed sets can have strongly contrasting permeabilities (Fig. 17). The laminae appear to be quite homogenous and to have a fairly constant permeability. A certain degree of rhythmic succession of high- and low-permeability laminae was present everywhere. The best permeabilities were found in crossbed sets with thick laminae of 2 to 5 cm. The finely laminated crossbed sets are almost invariably of poor reservoir quality. The low vertical permeability is generally caused by the presence of thin (1 to 2 mm), dark, well cemented streaks of very fine grains. On the other hand, the most permeable thick laminae are hardly cemented at all, containing only a small amount of hematite cement and some authigenic clay (Nagtegaal, 1979).

When laboratory permeability measurements performed under atmospheric conditions are converted to *in situ* gas permeabilities, corrections for the Klinkenberg effect and the effect of rock decompression have to be made. Furthermore, multiplication by the relative permeability to gas

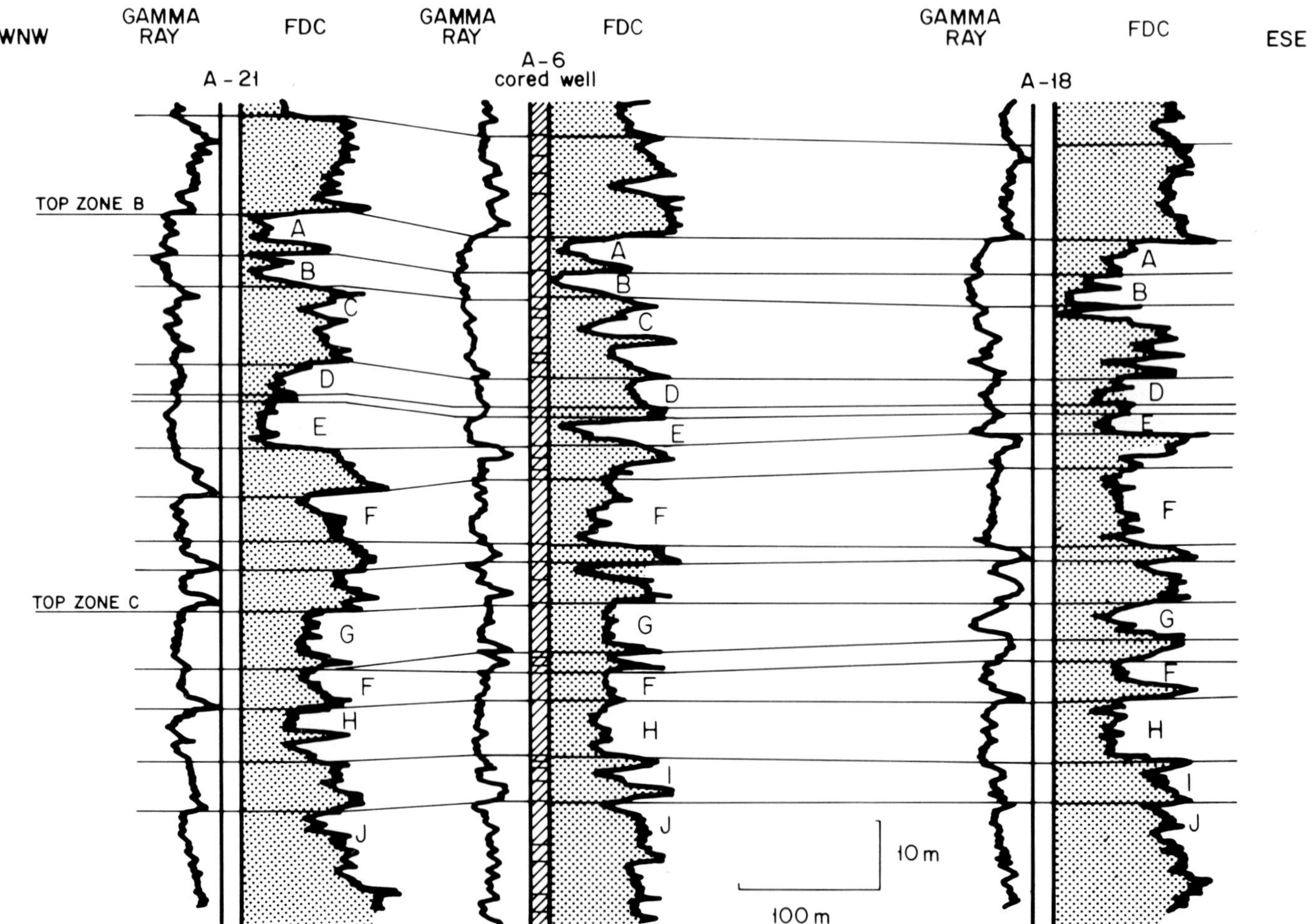

FIG. 16.—Section showing correlation of thick crossbed sets.

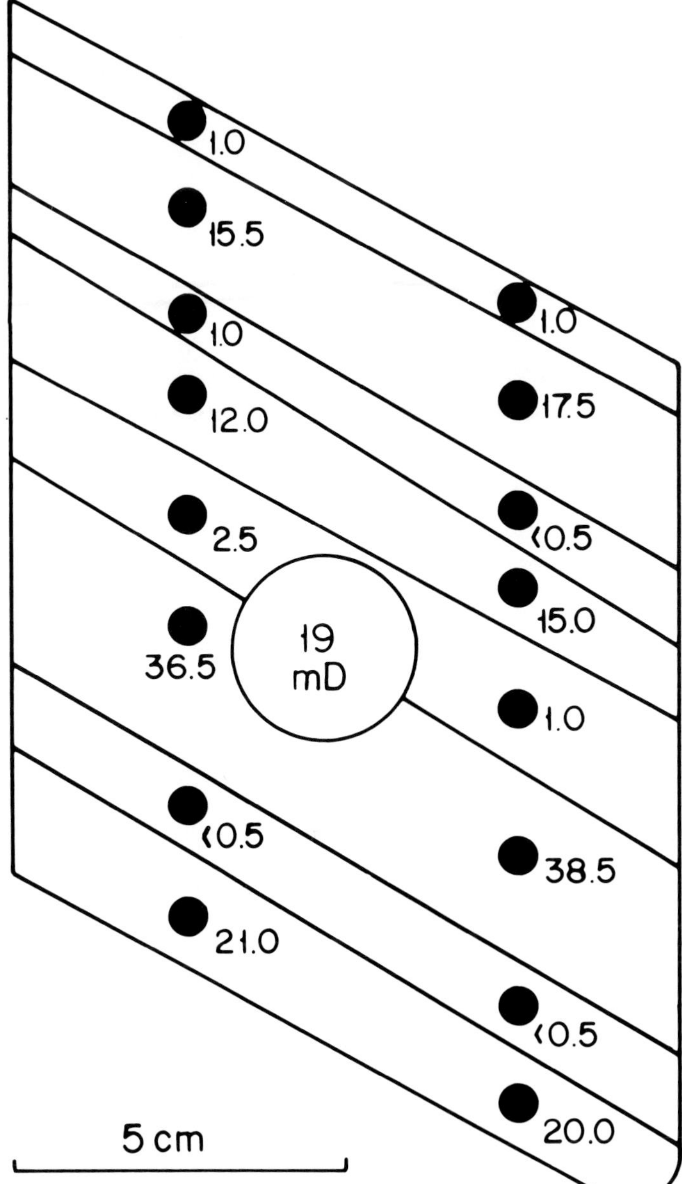

FIG. 17.—Aeolian core slab with a dense grid of mini-permeameter permeability readings. Aeolian crossbed set boundaries are indicated.

for the log-derived water saturation has to be done. Each of these corrections has been subject to a special study, including a series of measurements.

Because of the fairly high gas-flow velocities near the boreholes, inertia forces play an important role. There are several published studies on the coefficient of inertial resistance (or turbulence factor) as a function of porosity and permeability (Katz, 1959; Geertsma, 1974; Firoozabadi and Katz, 1979). The initial well productivities calculated with turbulence factors derived from the published relationships are systematically much too high. Studies of the diagenesis of the Leman Rotliegendes indicate fairly large amounts of illite and chlorite are present (Fig. 9). The fibrous growth of the clay minerals in the pores has a strong effect on the permeability (Stalder, 1973), and it was thought likely that the turbulence factors might also be unusually high. For this reason, a series of measurements was carried out on samples from well 49/26-A6. The results confirm the above idea and show that there is a good relationship between permeability and turbulence factor. The measurements were performed on dry samples, but work by Wong (1970) indicates that it is possible to read from the permeability/turbulence factor graph using the *in situ* gas permeabilities multiplied by the relative permeability to gas. The resulting graph of turbulence factor against permeability is given in Figure 18. In this case the permeability plotted is the effective inflow permeability of a well, i.e., after all necessary corrections have been carried out.

The calibration of the FDC (gamma-ray density) logs has been studied in some detail. Straightline correlations between porosity and bulk density appear to be applicable for reserve estimation. In the high porosity range (18–25%), however, calibration lines are clearly curved when logging in the gas reservoir, indicating the presence of some gas within the zone of investigation (Fig. 19). Below the gas/water contact, the high porosity zones have bulk densities on the logs which agree with the straight line calibrations.

Water saturation is calculated with the porosity derived from the FDC logs and the R_t derived from induction logs. The induction logs have a rather poor resolution, and it is not possible to obtain accurate R_t readings for thin crossbed sets and bottomset beds. For the thicker homogeneous crossbed sets the R_t values are probably substantially correct. A cementation factor (m) of 1.90 has been found by averaging the fairly large number of available measurements under simulated *in situ* stress from four wells. The saturation exponent (n) was not measured, but North Sea literature data (Fertl and others, 1972) indicate that 1.80 is a probable average value.

A figure of 0.0262 ohm m has been used for R_w. This value, together with the m- and n-exponents used, yields a relationship between porosity, R_t and S_w, as shown in Figure 20. Study of the gas-saturation distribution showed that there is a considerable effect of height above the gas/water contact on water saturation. Comparison of the water saturation of the rock at several elevations above the gas/water contact, plotted versus the rock permeability, indicates that even the more permeable rock has an appreciable transition zone.

ZONATION OF AEOLIAN INTERVAL

The aeolian zone, as a whole, does not show a porosity/permeability trend which could be used with any accuracy for predicting permeabilities using only FDC logs. The long transition zones preclude the possiblity of relating permeability to porosity and water saturation. Fortunately, however, it has been observed that when the average crossbed set porosities (excluding the bottomset beds) are plotted against the corresponding average permeabilities parallel to the foreset laminae, successive series of crossbed sets have the same porosity/permeability relationship trends.

Over the area of well clusters A, B and C, the aeolian interval can be subdivided into four zones with fairly well

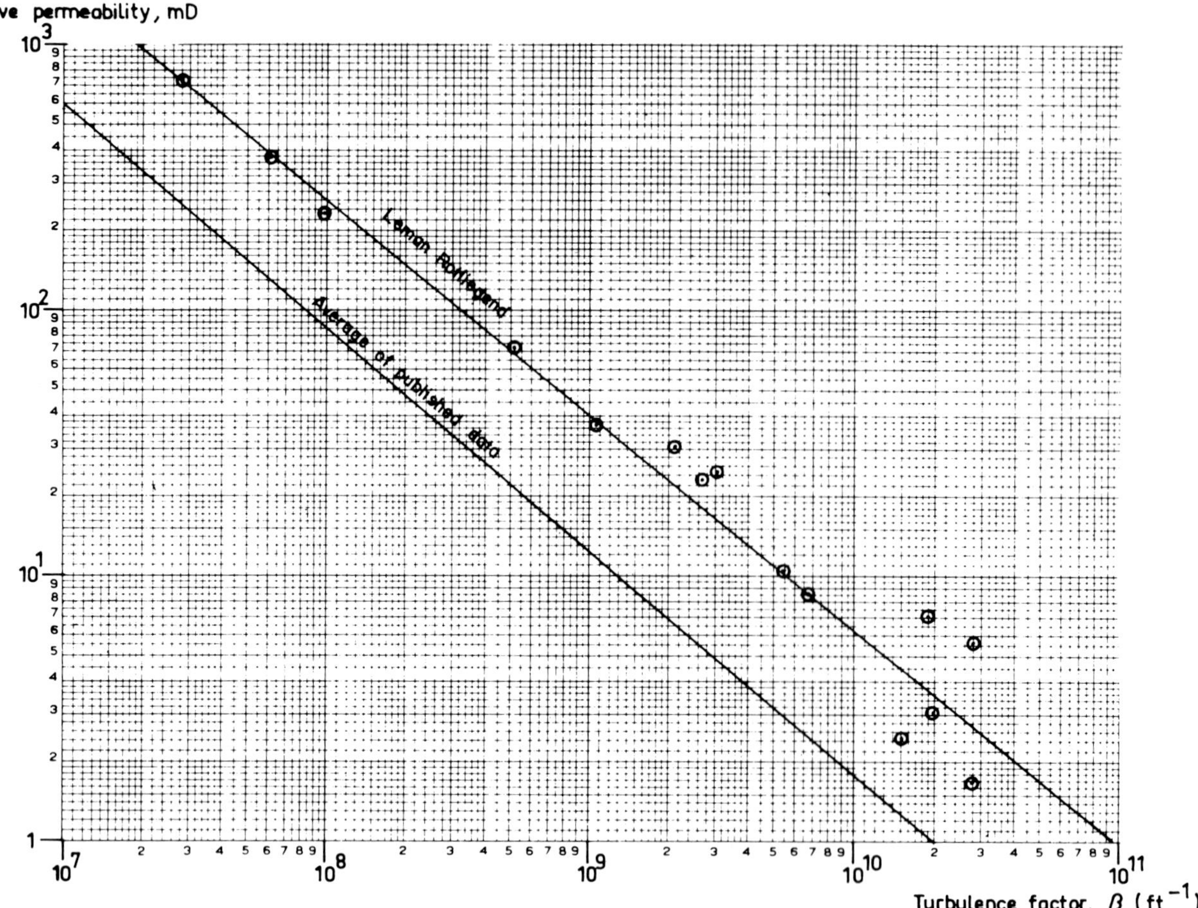

FIG. 18.—Relationship between effective permeability and turbulence factor.

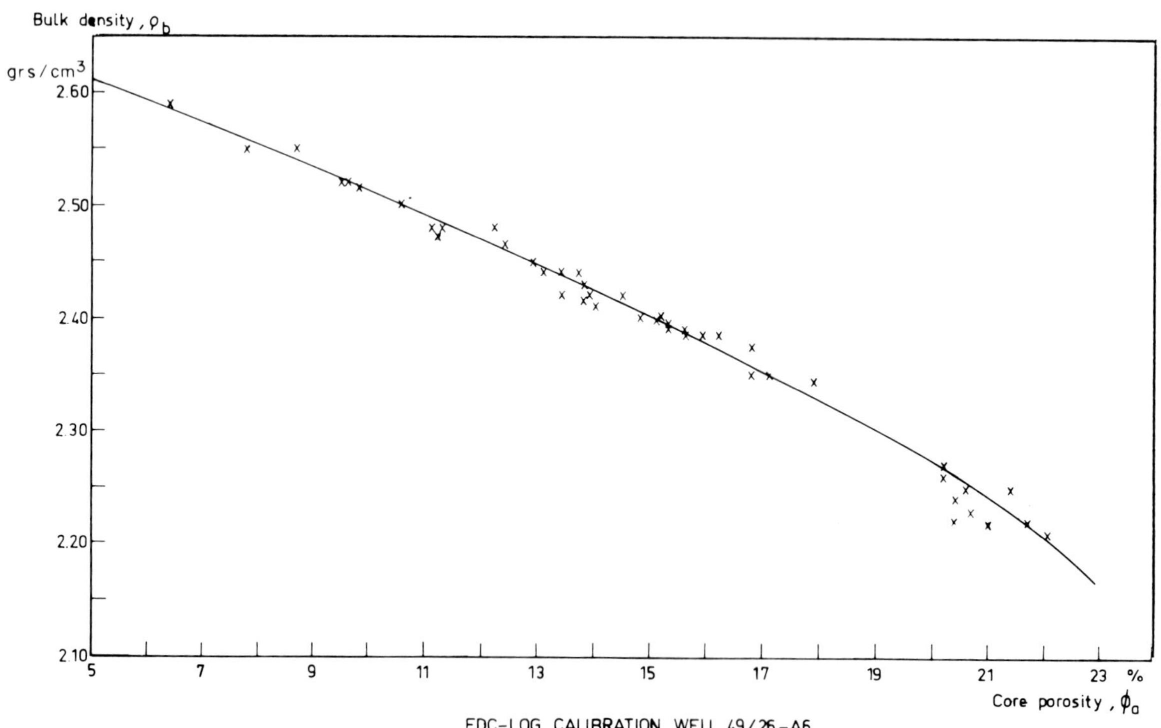

FIG. 19.—FDC log (gamma-ray density) calibration, well 49/26-A6.

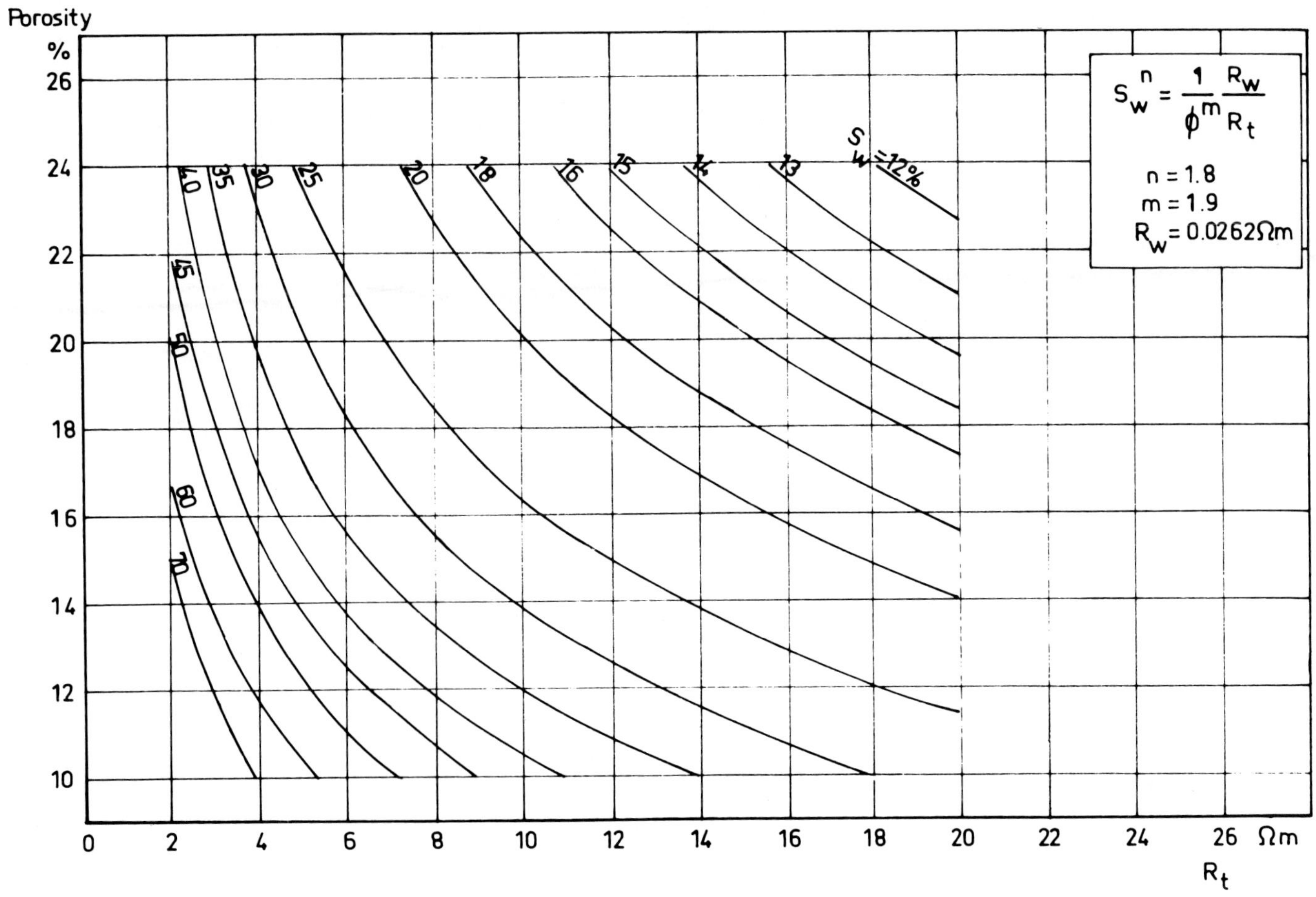

FIG. 20.—Relationship between true resistivity, porosity and water saturation in Leman reservoir.

defined porosity/permeability trends (Fig. 21). The zones are called A, B, C and D, from top to bottom. Zones A and C are of poor reservoir quality, whereas zones B and D are much more favorably developed. Zones A and B together form the bulk of the aeolian interval. In most wells

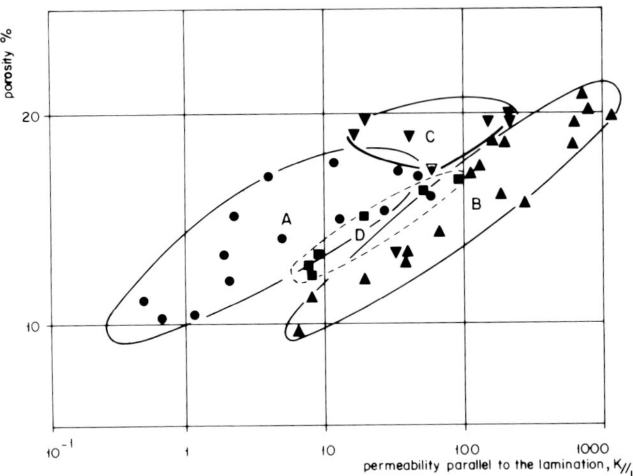

FIG. 21.—Porosity/permeability plot per crossbed set (bottomsets omitted). Zonation of Rotliegendes aeolian interval caused by differences in grain size distribution and diagenesis.

zones C and D are too close to or underneath the gas/water contact and are, thus, rarely perforated. The cause of the presence of the zonation is most probably a sedimentologic one, enhanced by diagenetic processes. Zone B has a lower gamma radiation than zone A, indicating a lower clay content. Also, zone B appears to be somewhat coarser grained on average. Even a small difference in clay content makes a large difference in permeability because of the druse-like overgrowth of chlorite and illite. The zones probably represent depositional cycles sedimented with somewhat different average wind velocities and climatic conditions.

The zonation of the aeolian interval is thought to extend over the entire field area (Fig. 4). There are, however, thickness variations as well as changes in the porosity/permeability relationship within zones. Parallel to the field's axis (NW-SE), i.e., parallel to the coastal trend and also roughly parallel to the paleowind direction, the changes in rock properties are more gradual than those in the NE-SW direction. An anomalous porosity/permeability relationship in faulted wells (49/26-3 and 29/26-14) and the lateral change in maximum depth of burial complicate the recognition of clear patterns in permeability development.

MODEL CONFIGURATION

From the foregoing it is possible to derive quite a realistic model configuration incorporating all the major properties of the Rotliegendes reservoir. In the first place, we have the subdivision into a water-laid and an aeolian interval. The water-laid interval can be considered to consist of horizontal beds of homogeneous sandstone without a horizontal permeabillity anisotropy. The vertical permeability will be low, however, because of the many low-permeability intervals. The aeolian interval consists of crossbed sets. The foreset laminae zones are the main producing intervals. They have a strong horizontal permeability anisotropy. The bottomset beds usually have a low horizontal and an extremely low vertical permeability (usually 0.1 md or less). Consequently, perforated crossbed sets are thought to produce initially as isolated sand bodies, with little cross flow from adjacent beds. The low vertical permeability of the bottomset beds also hampers water coning.

The lateral extent of the crossbed sets is a function of average thickness. Combined with the above characteristics, this leads to the model sketched in Figure 22. The effects of this type of formation heterogeneity on initial well behavior have been discussed by Kingston (1973). Initially, the more permeable crossbed sets have a high productivity, but this will soon drop to a level corresponding to an average effective Kh value in the drainage area of the well, reflecting the effect of the low vertical permeability.

Faults can have a considerable effect on the initial well productivity. The persistence of high productivity from a

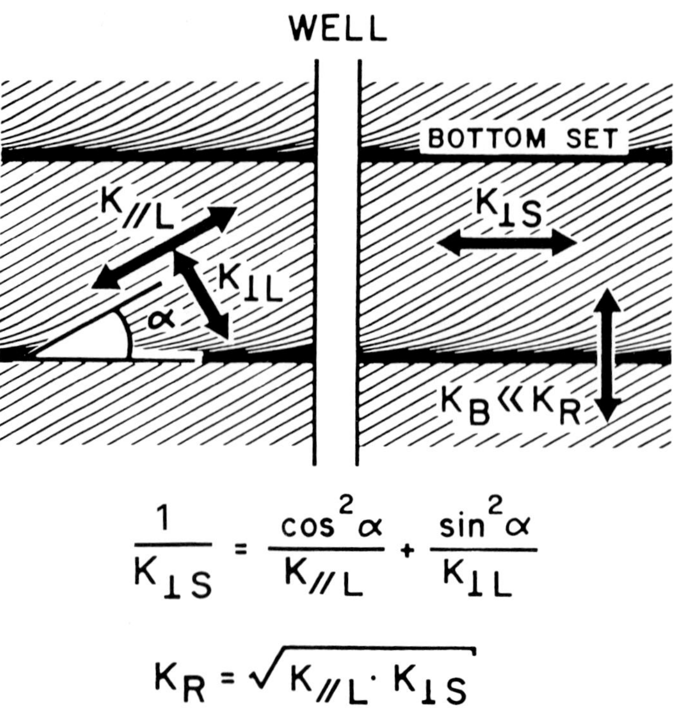

$$\frac{1}{K_{\perp S}} = \frac{\cos^2\alpha}{K_{//L}} + \frac{\sin^2\alpha}{K_{\perp L}}$$

$$K_R = \sqrt{K_{//L} \cdot K_{\perp S}}$$

FIG. 22.—Geologic well model for estimating radial inflow.

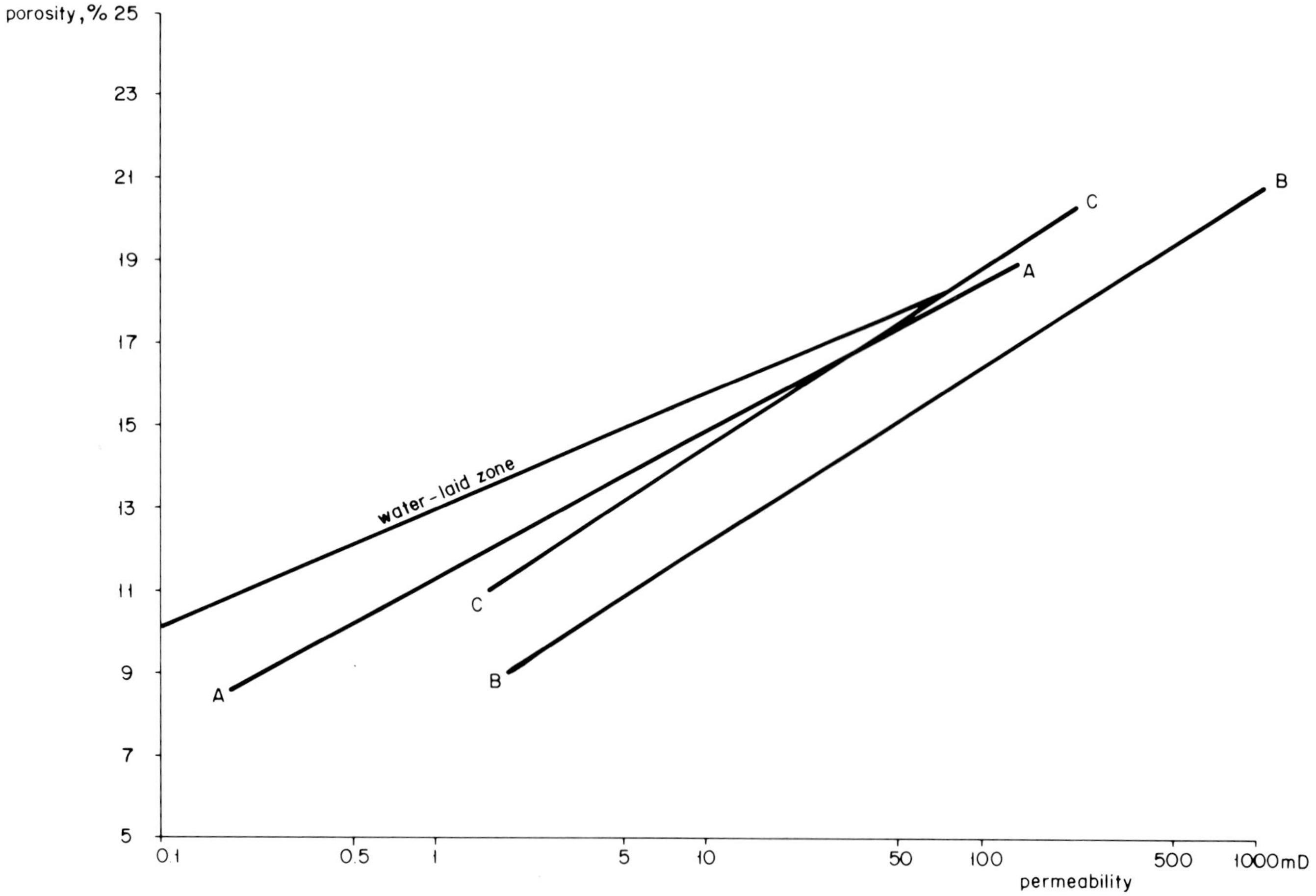

FIG. 23.—Relationship of porosity and permeability parallel to crossbed set laminae for zones A, B, and C and for the water-laid zones.

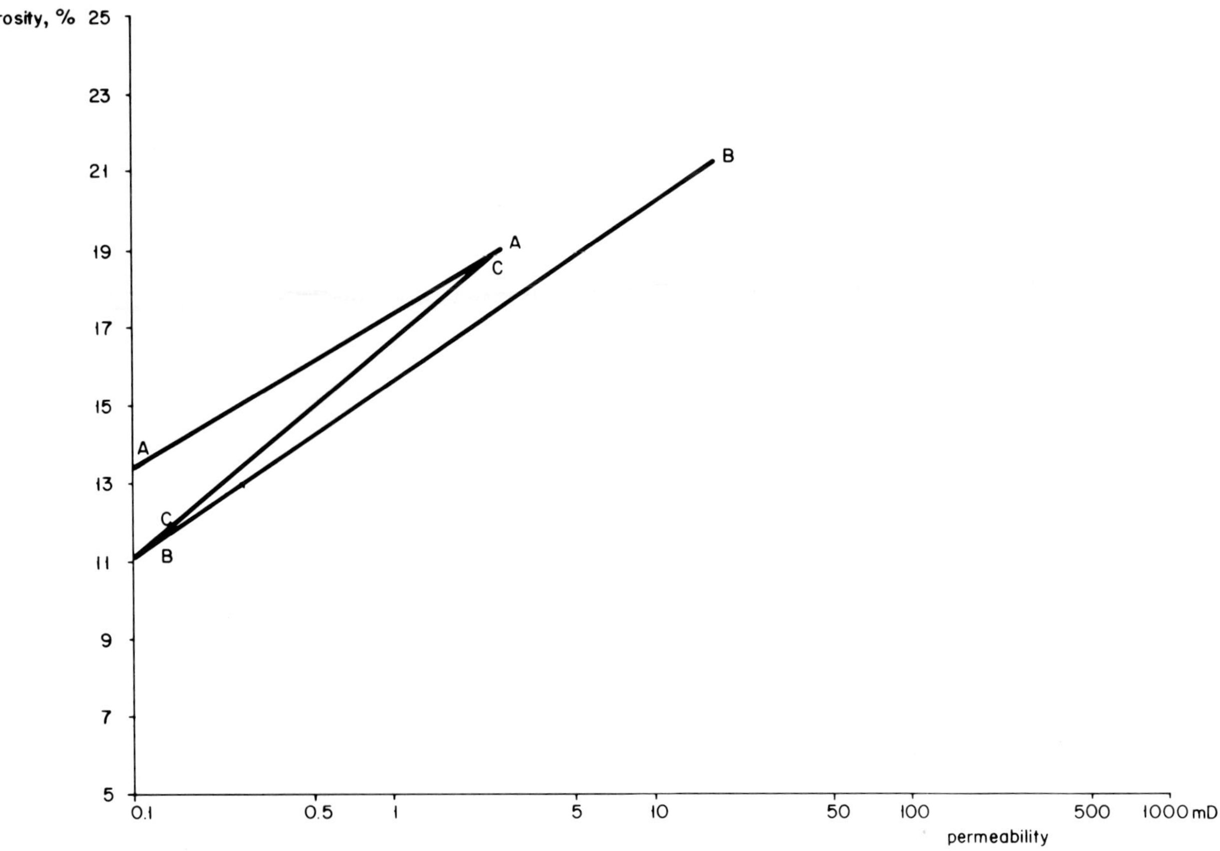

FIG. 24.—Relationship of porosity and harmonic mean of permeabilities measured perpendicular to crossbed set laminae for zones A, B, and C.

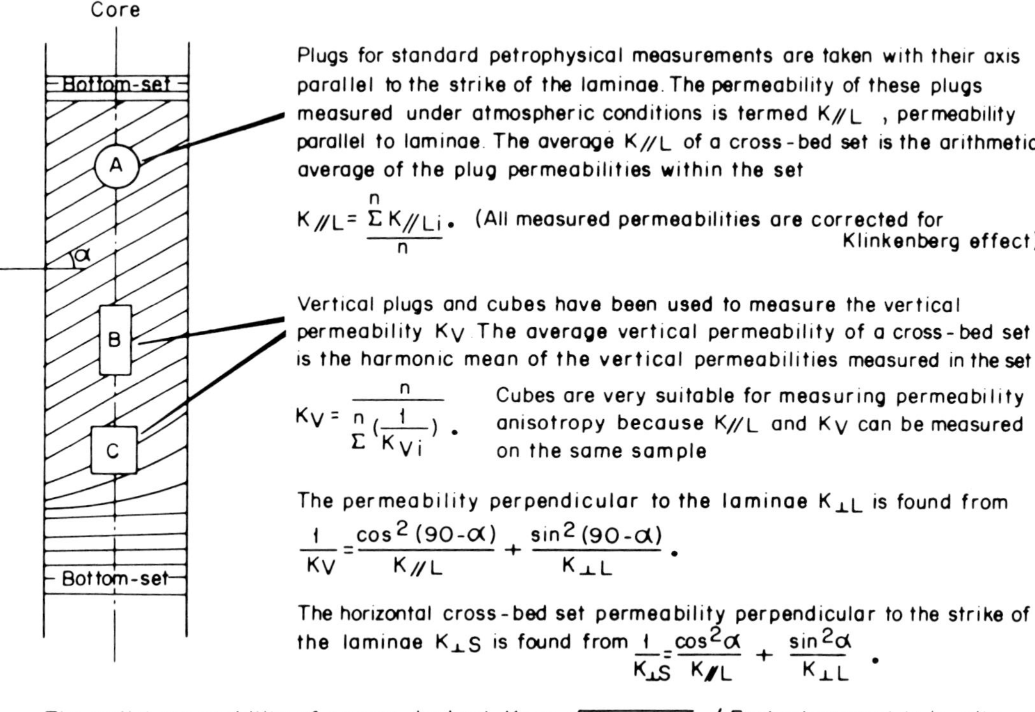

FIG. 25.—Calculation of effective inflow permeability of a crossbed set.

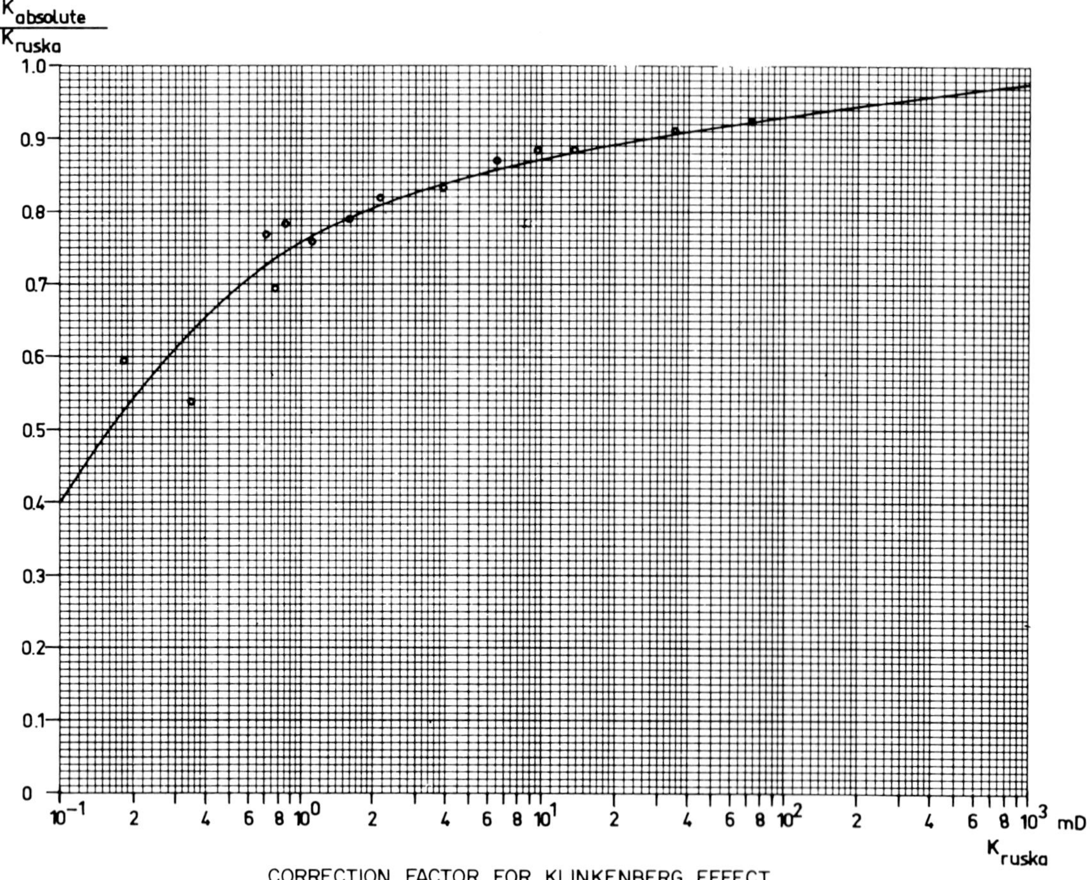

FIG. 26.—Correlation factor for Klinkenberg effect.

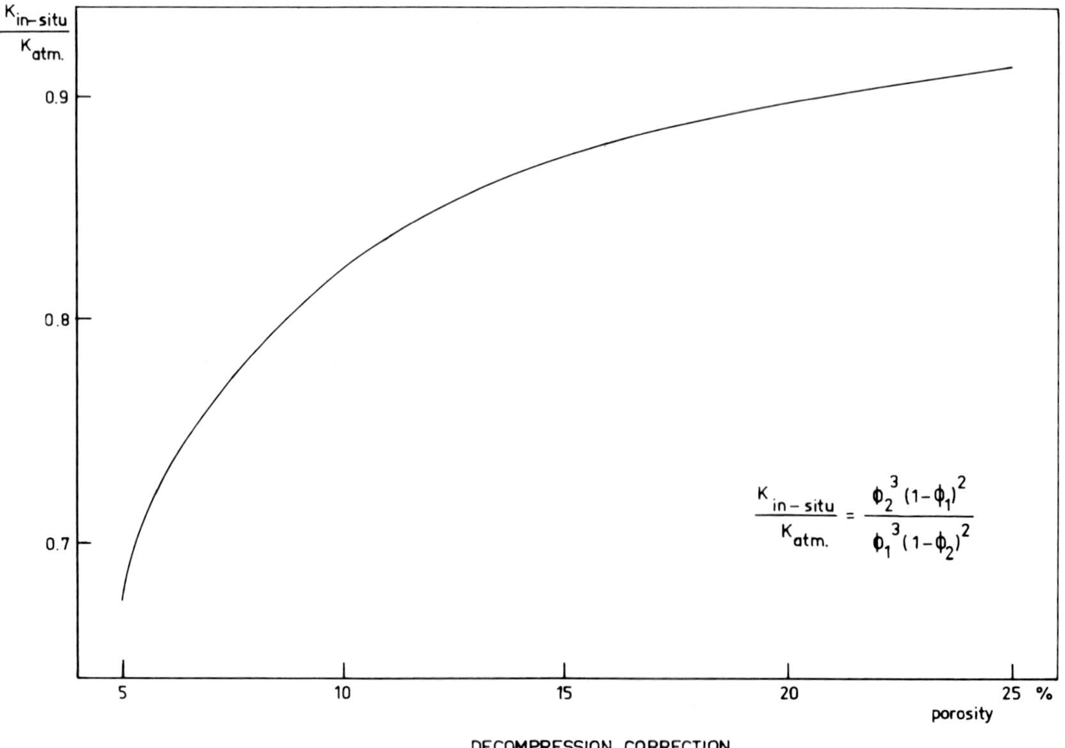

FIG. 27.—Presure correction curve.

highly permeable crossbed set is proportional to its lateral extent. In particular, when the fault throw is larger than the thickness of the favorably developed zone B, early decline in well productivity will be observed when a well close to this fault is tested. It has also been found that the faults subdivide the well cluster areas in the different pressure provinces with pressure drops over the fault zones of as much as about 80 psi (Fig. 5).

Figure 23 shows the average relationship between permeability parallel to the foreset laminae ($K//_L$) and porosity for the upper three aeolian zones. The relationship between horizontal permeability and porosity for the water-laid zone is also shown. Because of the homogeneous character of the water-laid sandstone and the horizontal bedding, there will hardly be any horizontal permeability anisotropy in this rock. In the aeolian rock, however, the difference in permeability of the successive foreset laminae causes a large horizontal anisotropy.

To calculate the influence of this anisotropy, additional oriented permeability measurements must be used. Permeabilities measured parallel to the core axis are available from wells A6 and B260. These permeabilities are called "vertical permeabilities" (K_v) in the core analysis reports. This nomenclature has been used, although, of course, the measurements are actually parallel to the borehole axis and A6 is a deviated well. This does not matter, however, as long as the angle between the laminae and the core axis for each crossbed set is measured.

The K_v data are grouped per crossbed set and, because of the fact that flow across instead of parallel to the laminae is now of concern, the harmonic mean is calculated for each crossbed set. Next, the averaged K_v data are converted into average permeabilities perpendicular to the laminae (K_{1L}) with Maasland's equation (1957). The resulting relationships between K_{1L} and porosity are shown in Figure 24.

In Figure 25 the calculation of the effective inflow permeability is given with all the necessary equations. All permeabilities used have to be corrected for the Klinkenberg effect (Fig. 26). Having obtained the regression lines representing the relationships between porosity and $K//_L$ and K_{1L}, respectively, for each zone, corresponding values of the horizontal permeability perpendicular to the strike of the laminae (K_{1L}) are calculated. The average horizontal inflow permeability of a given crossbed set is the square root of the product of $K//_L$ and K_{1L}, representing, respectively, the maximum and the minimum horizontal permeabilities.

At this stage corrections for the fact that the permeability figures used were measured under atmospheric stress conditions still has to be made. The rock compressibility of the Rotliegendes is not very large, but the correction of the

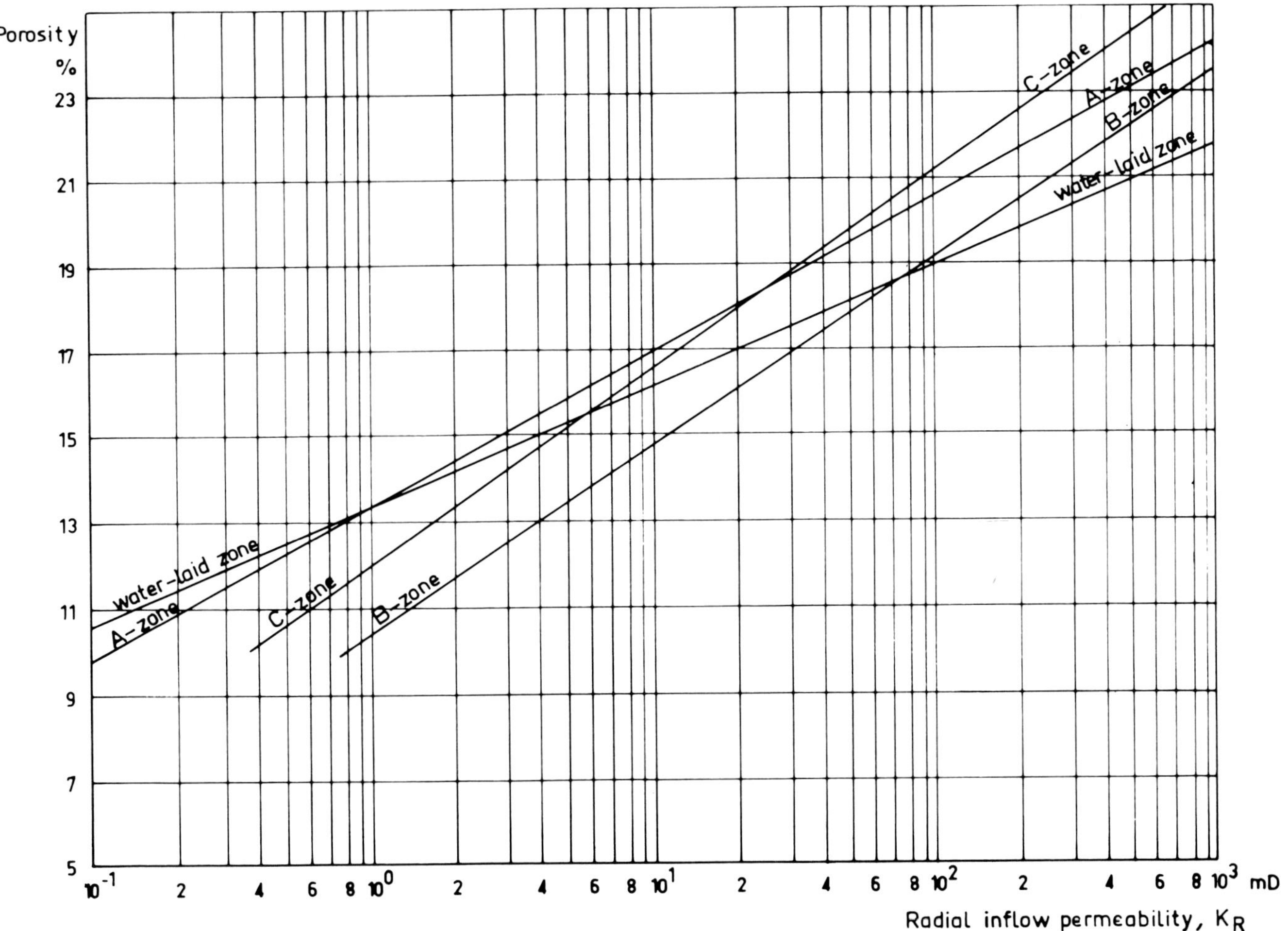

FIG. 28.—Radial inflow permeability for water-laid zones A, B, and C (all corrections applied, except the effect of water saturation).

permeability is still more than 10% for most of the rock. First, porosity is converted into *in situ* porosity, and then the corresponding permeability reduction is calculated with Kozeny's equation (Fig. 27).

The radial inflow permeabilities obtained for each zone are given in Figure 28. The *in situ* effective inflow permeabilities for a given crossbed set are calculated by reading the graphs for the average porosity of the crossbed set. Next, the radial inflow permeability is multiplied by the relative permeability at the water saturation derived from the logs (Fig. 29). The resulting effective inflow permeability (K_{eff}) is thought to be a good approximation of the actual inflow permeability in the proximity of the borehole, assuming horizontal gas flow in the crossbed sets.

CALCULATION OF INITIAL WELL PRODUCTIVITY

Kingston (1973) showed the results of calculations with a single-phase, two-dimensional, r-z, numerical simulator, based on the model described in the previous section. For a model consisting of a thin perforated layer of relatively high permeability in an otherwise homogeneous reservoir of 1 md vertical permeability, the measured kh value is that of the perforated interval only. Only toward the end of the buildup survey does the effect of the reservoir start to become appreciable. In this article it is assumed that during the initial production stage of a well, the perforated permeable beds behave as if they go through a succession of steady states.

For this reason the well known Forcheimer equation has been used to calculate the initial productivity of a well. In the calculation, interference effects due to the simultaneous production of other wells are not considered because they will generally be absent during the testing of a series of wells in a cluster. Furthermore, the skin factors cannot, of course, be taken into account in a prediction. In fact, one of the purposes of the calculation is to compare predicted productivity with actual productivity under influence of skin effects.

The following equation and properties have been used,

$$p_1^2 - p_2^2 = \frac{1424\ \mu z T Q \ln\left(\frac{r_1}{r_2}\right)}{Kh} + \frac{3.161 \times 10^{-12} \beta G Q^2 z T \left(\frac{1}{r_2} - \frac{1}{r_1}\right)}{h^2} \qquad (1)$$

where:

P_1 = CIBHP, psia (2,950 for the A and B clusters and 2,600 for the C cluster);

P_2 = FBHP, psia (calculations were made for values 500 and 1,000 psi lower than CIBHP);

r_1 = radius at which pressure equals CIBHP, ft (600 ft was used throughout);

r_2 = radius well, ft (for each well the corresponding well radius was used);

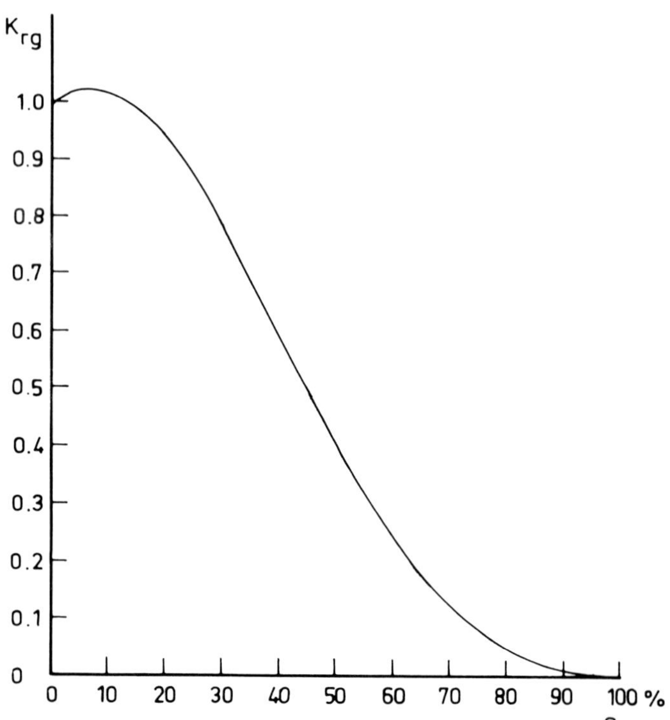

FIG. 29.—Relative gas permeability as a function of water saturation.

FIG. 30.—Gas production as a function of effective inflow permeability under semi-steady-state conditions.

TABLE 4.—COMPARISON OF MEASURED AND CALCULATED INITIAL PRODUCTIVITIES OF SHELL U.K. A- AND B-CLUSTER WELLS AND WELLS C370 AND D405

Well No.	Initial Production Rate, Mm scf/d — Measured at Drawdown		Calculated at Drawdown		Remarks
	500 psi	1000 psi	500 psi	1000 psi	
A 6	88	—	79	118	Cored well
A 7	88	—	123	175	Permeability of surrounding area somewhat lower
A 8	29	45	26	43	
A 9	(36)	(53)	100(26)	143(41)	Figures in brackets refer to partially perforated well
A 10	19	27	24	39	
A 12	30	47	26	41	Rock improves away from well bore*
A 13	((31))	((47))	30	47	Test unreliable. For the measured productivity the average is given of the two neighboring wells A12 & A22
A 15	13	22	41	62	62% of the calculated production comes from a single 23-ft-thick interval
A 18	((88))	—	89	138	Not tested. Should have produced as neighboring wells A6 and A7
A 19	46	65	100	151	Flowmeter survey indicates plugging of several intervals
A 20	86	—	63	96	Rock improves away from well bore*
A 21	((88))	—	68	101	No reliable test. Productivity should be similar to A6
A 22	±32	±46	59	92	Permeability decreases away from well bore*
A 23	33	±46	8	14	Faulted in Rotliegendes
A 24	65	—	84	128	Permeability decreases away from well bore*. Also, probably close to a fault
B210	14	26	15	24	Very permeable beds occur underneath the lowest perforated interval
B215	35	61	94(35)	139(54)	Flowmeter survey indicates that lowermost intervals are almost completely plugged. Figures in brackets refer to producing intervals
B220	11	17	10	16	
B235	19	28	11	15	Barefoot completion
B240	33	53	30	47	Cored well
B245	53	—	49	74	
B250	110	161	47	73	Water-laid zone may contain unusually permeable streaks
B255	14	25	11	17	
B260	48	70	44	67	Cored well
B265	±9	±15	7	11	
B270	29	49	24	38	
B275	113	163	78	122	Probably close to fault
B280	±26	±45	49	75	Half of the calculated production comes from a 27-ft interval
B285	44	60	49	75	
B290	62	90	70	104	
B295	35	46	23	36	Faulted in Rotliegendes, large negative skin
C370	10	18	12	20	Cored well
D405	25	±40	31	48	Cored well

* Deductions from pressure buildup tests
Double brackets indicate that no reliable test results are available and that a figure is used which is based on the test behavior of similar neighboring wells

μ = *in situ* gas viscosity, cP (0.02 cP was used, which may be slightly too high);
T = reservoir temperature, °R (585°R);
Q = flow rate, mscf/day;
K = effective inflow permeability for the gas, md;
h = formation thickness, ft (the along-hole thickness was used; a correction was applied later);
G = gas gravity (0.587);
β = turbulence factor, ft^{-1} (see Fig. 18).

The productivities per foot (0.3 m) of perforated interval were plotted versus the effective inflow permeability for drawdowns of 500 and 1,000 psi, respectively, for each of the borehole diameters encountered. These graphs are shown in Figure 30 for initial closed-in bottom-hole pressures of 2,950 psi (platforms A and B) and 2,600 psi (platform C), respectively. The next step is to read the productivity per foot of each perforated crossbed set (or interval of water-laid rock) from the graphs. These productivities per foot are multiplied by the corresponding bed thicknesses. Adding the individual contributions of each bed gives the total well productivity, provided the well is vertical.

Most of the wells are deviated, however, in which case the calculated productivities are based on bed thicknesses which are too large. Hence, a correction is based on the assumption that the main effect of deviating a hole is to increase the wall surface of the borehole. Thus, the Q obtained with equation 1 has to be multiplied by the ratio of true vertical to drilled thickness, and subsequently it also has to be multiplied by a factor related to the increase in borehole wall thickness of the slanted versus verticle borehole. A pseudo-well diameter (D_{pseudo}) is calculated which gives the hypothetical vertical hole the same wall surface as that of the deviated well (D_{true}) using the equation

$$D_{peudo} = D_{true} \frac{h_{drilled}}{h_{vertical}} . \quad (2)$$

It has been found that for each increase of 1 in. (2.54 cm) of D_{pseudo} over D_{true}, productivity becomes about 4 percent higher for both the 500 and the 1,000 psi drawdown.

CONCLUSIONS

The calculations of initial well productivity were carried out for all the A- and B-cluster wells tested. In addition, the calculation was performed for the cored wells (C370 and D405. The results are summarized in Table 4.

As a whole, the method appears to be reasonably accurate. For 500 psi drawdown the calculated initial productivities are probably within about 30% of the tested productivities for 23 out of 33 wells. Of the 10 wells with a poor fit, two are faulted (A23 and B295), one is close to a fault (B275), two are badly plugged (A19 and B215), and in two, the bulk of the calculated production comes from thin, very permeable intervals (A15 and B280). Furthermore, near well A32 the permeability changes rapidly; well B235 is a barefoot completion, whereas well B250 may have a few unusually permeable intervals in the water-laid zone.

It is clear that the calculation procedure leads to inaccurate figures when initial productivity comes from only a few thin, very permeable beds. Furthermore, it appears that for wells close to the boundary between a high- and a low-permeability area, the initial productivity is influenced by the lateral permeability change.

It can be concluded that the method presented in this report is sufficiently accurate for the purposes mentioned in the introduction. The field contains provinces with high and low productivity, which mainly reflect contrasts in permeability of zone B of the aeolian interval. Furthermore, the effect of faulting on productivity has been demonstrated, showing that in a faulted zone porosities may be relatively low, but the porosity/permeability is more favorable than in an unfaulted area.

ACKNOWLEDGMENTS

Much of the work involved in this study was carried out by F. C. Wonink. I also thank K. W. Glennie and P. J. C. Nagtegaal for their advice and cooperation. Permission of Shell UK Exploration and Production and of Shell Internationale Research Maatschappij B.V. to publish this paper is gratefully acknowledged.

REFERENCES

BAARS, D. L., 1961, Permian blanket sandstones of Colorado plateau, *in* Peterson, J. A., and Osmond, J. C, eds., Geometry of Sandstone Bodies: American Association of Petroleum Geologists, Tulsa, Oklahoma, p. 179–207.

BAGNOLD, R. A., 1941, The physics of blown sands and desert dunes: Methuen and Co., (reprinted 1954), London, 265 p.

CRAIG, F. F., WILCOX, P. J., BALLARD, J. R., AND NATION, W. R., 1977, Optimized recovery through continuing interdisciplinary cooperation: Journal of Petroleum Technology, v. 29, p. 755–760.

FERTL, W. H., CAVANAUGH, R. J., SHEPLER, J. C., AND SCOPE, B. G., 1972, Formation evaluation parameters of the Rotliegendes Sandstone, North Sea area: Society of Petroleum Engineers, SPE Paper 3747, 4 p.

FIROOZABADI, A., AND KATZ, D. L., 1979, An analysis of high velocity gas flow through porous media: Journal of Petroleum Technology, v. 29, p. 211–216.

GEERTSMA, J., 1974, Estimating the coefficient of inertial resistance in fluid flow through porous media: Society of Petroleum Engineers Journal, v. 14, p. 445–450.

GLENNIE, K. W., 1970, Desert sedimentary environments: Developments in Sedimentology 14, Elsevier, Amsterdam, 222 p.

———, 1972, Permian Rotliegendes of northwest Europe interpreted in light of modern desert sedimentation studies: American Association of Petroleum Geologists Bulletin, v. 56, p. 1048–1071.

———, MUDD, G. C., AND NAGTEGAAL. P. J. C., 1978, Depositional environment and diagenesis of Permian Rotliegendes Sandstones in Leman Bank and Sole Pit areas of the U.K., southern North Sea: Journal of the Geological Society, v. 135, part 1, p. 25–34.

GOLDWATER, M. H., COLLINS, P. A., AND TAYLOR, B. A., 1978, The use of GRASP, a finite element program, to model faulted gas reservoirs of the southern North Sea basin: European Offshore Petroleum Conference and Exhibition, Society of Petroleum Engineers, Ltd. London, United Kingdom, Paper 87, p. 219–232.

KATZ, D. L., CORNELL, D., KOBAYASHI, R., POETTMANN, F. H., VARY, J. A., ELENBAAS, J. R., and WEINANG, C. F., 1959, Handbook of Natural Gas Engineering: McGraw-Hill Book Company, Inc., New York, 802 p.

KINGSTON, P. E., 1973, Pressure evaluation methods—Leman Field: Society of Petroleum Engineers, SPE Paper 4306, 12 p.

MAASLAND, M., 1957, Drainage of agricultural lands: American Society of Agronomy Journal, v. 49, p. 216–285.

NAGTEGAAL, P. J. C., 1973, Adhesion-ripple and barchan-dune sands of the Recent Namib (SW Africa) and Permian Rotliegendes (NW Europe) deserts: Madoqua, series II, v. 2, nos. 63–68, p. 5–19.

———, 1979, Relationship of facies and reservoir quality in Rotliegendes desert sandstones, southern North Sea region: Journal of Petroleum Geology, v. 2, p. 145–158.

STALDER, P. J., 1973, Influence of crystallographic habit and aggregate structure of authigenic clay minerals on sandstone permeability: Geologie en Mijnbouw, v. 52, p. 217–220.

STOKES, W. L., 1961, Fluvial aeolian sandstone bodies in Colorado Plateau, *in* Peterson, J. A., and Osmond, J. C., eds., Geometry of Sandstone Bodies: American Association Petroleum Geologists, Tulsa, Oklahoma, p. 151–178.

VAN VEEN, F. R., 1975, Geology of the Leman Gas field, *in* Petroleum and the Continential Shelf of Northwest Europe: John Wiley and Sons, New York, v. 1, p. 223–231.

WILSON, I. G., 1972, Aeolian bedforms—their development and origins: Sedimentology, v. 19, p. 173–210.

WONG, S. W., 1970, Effect of liquid saturation on turbulence factors for gas-liquid systems: Journal of Canadian Petroleum Technology, v. 9, p. 274–278.

INDEX